THEORY OF MACHINES

AND MECHANISMS

Third Edition

John J. Uicker, Jr.

Professor of Mechanical Engineering
University of Wisconsin—Madison

Gordon R. Pennock

Associate Professor of Mechanical Engineering
Purdue University

Joseph E. Shigley

Late Professor Emeritus of Mechanical Engineering
The University of Michigan

New York ■ Oxford
OXFORD UNIVERSITY PRESS
2003

Oxford University Press

Oxford   New York
Auckland   Bangkok   Buenos Aires   Cape Town   Chennai
Dar es Salaam   Delhi   Hong Kong   Istanbul   Karachi   Kolkata
Kuala Lumpur   Madrid   Melbourne   Mexico City   Mumbai
Nairobi   São Paulo   Shanghai   Taipei   Tokyo   Toronto

Published by Oxford University Press, Inc.
198 Madison Avenue, New York, New York, 10016
http://www.oup-usa.org

Oxford is a registered trademark of Oxford University Press

ISBN 0–19–515598–X

Printing number: 9 8 7 6 5 4 3 2 1

Printed in the United States of America
on acid-free paper

This textbook is dedicated to the memory of the third author, the late **Joseph E. Shigley,** Professor Emeritus, Mechanical Engineering Department, University of Michigan, Ann Arbor, on whose previous writings much of this edition is based.

This work is also dedicated to the memory of my father, John J. Uicker, Emeritus Dean of Engineering, University of Detroit; to my mother, Elizabeth F. Uicker; and to my six children, Theresa A. Uicker, John J. Uicker III, Joseph M. Uicker, Dorothy J. Winger, Barbara A. Peterson, and Joan E. Uicker.

—John J. Uicker, Jr.

This work is also dedicated first and foremost to my wife, Mollie B., and my son, Callum R. Pennock. The work is also dedicated to my friend and mentor Dr. An (Andy) Tzu Yang and my colleagues in the School of Mechanical Engineering, Purdue University, West Lafayette, Indiana.

—Gordon R. Pennock

Contents

Preface

This book is intended to cover that field of engineering theory, analysis, design, and practice that is generally described as mechanisms and kinematics and dynamics of machines. While this text is written primarily for students of engineering, there is much material that can be of value to practicing engineers. After all, a good engineer knows that he or she must remain a student throughout their entire professional career.

The continued tremendous growth of knowledge, including the areas of kinematics and dynamics of machinery, over the past 50 years has resulted in great pressure on the engineering curricula of many schools for the substitution of "modern" subjects for those perceived as weaker or outdated. At some schools, depending on the faculty, this has meant that kinematics and dynamics of machines could only be made available as an elective topic for specialized study by a small number of students; at others it remained a required subject for all mechanical engineering students. At other schools, it was required to take on more design emphasis at the expense of depth in analysis. In all, the times have produced a need for a textbook that satisfies the requirements of new and changing course structures.

Much of the new knowledge developed over this period exists in a large variety of technical papers, each couched in its own singular language and nomenclature and each requiring additional background for its comprehension. The individual contributions being published might be used to strengthen the engineering courses if first the necessary foundation were provided and a common notation and nomenclature were established. These new developments could then be integrated into existing courses so as to provide a logical, modern, and comprehensive whole. To provide the background that will allow such an integration is the purpose of this book.

To develop a broad and basic comprehension, all the methods of analysis and development common to the literature of the field are employed. We have used graphical methods of analysis and synthesis extensively throughout the book because the authors are firmly of the opinion that graphical computation provides visual feedback that enhances the student's understanding of the basic nature of and interplay between the equations involved. Therefore, in this book, graphic methods are presented as one possible solution technique for vector equations defined by the fundamental laws of mechanics, rather than as mysterious graphical "tricks" to be learned by rote and applied blindly. In addition, although graphic techniques may be lacking in accuracy, they can be performed quickly and, even though inaccurate, sketches can often provide reasonable estimates of a solution or can be used to check the results of analytic or numeric solution techniques.

We also use conventional methods of vector analysis throughout the book, both in deriving and presenting the governing equations and in their solution. Raven's methods using complex algebra for the solution of two-dimensional vector equations are presented throughout the book because of their compactness, because they are employed so frequently in the literature, and also because they are so easy to program for computer evaluation. In the chapters dealing with three-dimensional kinematics and robotics, we briefly present an introduction to Denavit and Hartenberg's methods using transformation matrices.

With certain exceptions, we have endeavored to use U.S. Customary units and SI units in about equal proportions throughout the book.

One of the dilemmas that all writers on the subject of this book have faced is how to distinguish between the motions of two different points of the same moving body and the motions of coincident points of two different moving bodies. In other texts it has been customary to describe both of these as "relative motion"; but because they are two distinct situations and are described by different equations, this causes the student difficulty in distinguishing between them. We believe that we have greatly relieved this problem by the introduction of the terms *motion difference* and *apparent motion* and two different notations for the two cases. Thus, for example, the book uses the two terms, *velocity difference* and *apparent velocity,* instead of the term "relative velocity," which will not be found when speaking rigorously. This approach is introduced beginning with the concepts of position and displacement, used extensively in the chapter on velocity, and brought to fulfillment in the chapter on accelerations where the Coriolis component *always* arises in, and *only* in, the apparent acceleration equation.

Another feature, new with the third edition, is the presentation of kinematic coefficients, which are derivatives of various motion variables with respect to the input motion rather than with respect to time. The authors believe that these provide several new and important advantages, among which are the following: (1) They clarify for the student those parts of a motion problem which are kinematic (geometric) in their nature, and they clearly separate them from those that are dynamic or speed-dependent. (2) They help to integrate different types of mechanical systems and their analysis, such as gears, cams, and linkages, which might not otherwise seem similar.

Access to personal computers and programmable calculators is now commonplace and is of considerable importance to the material of this book. Yet engineering educators have told us very forcibly that they do not want computer programs included in the text. They prefer to write their own programs and they expect their students to do so too. Having programmed almost all the material in the book many times, we also understand that the book should not become obsolete with changes in computers or programming languages.

Part 1 of this book is an introduction that deals mostly with theory, with nomenclature, with notation, and with methods of analysis. Serving as an introduction, Chapter 1 also tells what a mechanism is, what a mechanism can do, how mechanisms can be classified, and some of their limitations. Chapters 2, 3, and 4 are concerned totally with analysis, specifically with kinematic analysis, because they cover position, velocity, and acceleration analyses, respectively.

Part 2 of the book goes on to show engineering applications involving the selection, the specification, the design, and the sizing of mechanisms to accomplish specific motion objectives. This part includes chapters on cam systems, gears, gear trains, synthesis of linkages, spatial mechanisms, and robotics.

Part 3 then adds the dynamics of machines. In a sense this is concerned with the consequences of the proposed mechanism design specifications. In other words, having designed a machine by selecting, specifying, and sizing the various components, what happens during the operation of the machine? What forces are produced? Are there any unexpected operating results? Will the proposed design be satisfactory in all respects? In addition, new dynamic devices are presented whose functions cannot be explained or understood without dynamic analysis. The third edition includes complete new chapters on the analysis and design of flywheels, governors, and gyroscopes.

As with all topics and all texts, the subject matter of this book also has limits. Probably the clearest boundary on the coverage in this text is that it is limited to the study of rigid-body mechanical systems. It does study multibody systems with connections or constraints between them. However, all elastic effects are assumed to come within the connections; the shapes of the individual bodies are assumed constant. This assumption is necessary to allow the separate study of kinematic effects from those of dynamics. Because each individual body is assumed rigid, it can have no strain; therefore the study of stress is also outside of the scope of this text. It is hoped, however, that courses using this text can provide background for the later study of stress, strength, fatigue life, modes of failure, lubrication, and other aspects important to the proper design of mechanical systems.

John J. Uicker, Jr.
Gordon R. Pennock

About the Authors

John J. Uicker, Jr. is Professor of Mechanical Engineering at the University of Wisconsin—Madison. His teaching and research specialties are in solid geometric modeling and the modeling of mechanical motion and their application to computer-aided design and manufacture; these include the kinematics, dynamics, and simulation of articulated rigid-body mechanical systems. He was the founder of the Computer-Aided Engineering Center and served as its director for its initial 10 years of operation.

He received his B.M.E. degree from the University of Detroit and obtained his M.S. and Ph.D. degrees in mechanical engineering from Northwestern University. Since joining the University of Wisconsin faculty in 1967, he has served on several national committees of ASME and SAE, and he is one of the founding members of the US Council for the Theory of Machines and Mechanisms and of IFToMM, the international federation. He served for several years as editor-in-chief of the *Mechanism and Machine Theory* journal of the federation. He is also a registered Mechanical Engineer in the State of Wisconsin and has served for many years as an active consultant to industry.

As an ASEE Resident Fellow he spent 1972–1973 at Ford Motor Company. He was also awarded a Fulbright–Hayes Senior Lectureship and became a Visiting Professor to Cranfield Institute of Technology in England in 1978–1979. He is the pioneering researcher on matrix methods of linkage analysis and was the first to derive the general dynamic equations of motion for rigid-body articulated mechanical systems. He has been awarded twice for outstanding teaching, three times for outstanding research publications, and twice for historically significant publications.

Gordon R. Pennock is Associate Professor of Mechanical Engineering at Purdue University, West Lafayette, Indiana. His teaching is primarily in the area of mechanisms and machine design. His research specialties are in theoretical kinematics, and the dynamics of mechanical motion. He has applied his research to robotics, rotary machinery, and biomechanics; including the kinematics, and dynamics of articulated rigid-body mechanical systems.

He received his B.Sc. degree (Hons.) from Heriot–Watt University, Edinburgh, Scotland, his M.Eng.Sc. from the University of New South Wales, Sydney, Australia, and his Ph.D. degree in mechanical engineering from the University of California, Davis. Since joining the Purdue University faculty in 1983, he has served on several national committees and international program committees. He is the Student Section Advisor of the American Society of Mechanical Engineers (ASME) at Purdue University, Region VI College Relations Chairman, Senior Representative on the Student Section Committee, and a member of the Board on Student Affairs. He is an Associate of the Internal Combustion Engine Division, ASME, and served as the Technical Committee Chairman of Mechanical Design, Internal Combustion Engine Division, from 1993 to 1997.

He is a Fellow of the American Society of Mechanical Engineers and a Fellow and a Chartered Engineer with the Institution of Mechanical Engineers (CEng, FIMechE), United Kingdom. He received the ASME Faculty Advisor of the Year Award, 1998, and was named the Outstanding Student Section Advisor, Region VI, 2001. The Central Indiana Section recognized him in 1999 by the establishment of the Gordon R. Pennock Outstanding Student Award to be presented annually to the Senior Student in recognition of academic achievement and outstanding service to the ASME student section at Purdue University. He received the ASME Dedicated Service Award, 2002, for dedicated voluntary service to the society marked by outstanding performance, demonstrated effective leadership, prolonged and committed service, devotion, enthusiasm, and faithfulness. He received the SAE Ralph R. Teetor Educational Award, 1986, and the Ferdinand Freudenstein Award at the Fourth National Applied Mechanisms and Robotics Conference, 1995. He has been at the forefront of many new developments in mechanical design, primarily in the areas of kinematics and dynamics. He has published some 80 technical papers and is a regular symposium speaker, workshop presenter, and conference session organizer and chairman.

Joseph E. Shigley   (deceased May 1994) was Professor Emeritus of Mechanical Engineering at the University of Michigan, Fellow in the American Society of Mechanical Engineers, received the Mechanisms Committee Award in 1974, the Worcester Reed Warner medal in 1977, and the Machine Design Award in 1985. He was author of eight books, including *Mechanical Engineering Design* (with Charles R. Mischke) and *Applied Mechanics of Materials.* He was Coeditor-in-Chief of the *Standard Handbook of Machine Design.* He first wrote *Kinematic Analysis of Mechanisms* in 1958 and then wrote *Dynamic Analysis of Machines* in 1961, and these were published in a single volume titled *Theory of Machines* in 1961; these have evolved over the years to become the current text, *Theory of Machines and Mechanisms,* now in its third edition.

He was awarded the B.S.M.E. and B.S.E.E. degrees of Purdue University and received his M.S. at the University of Michigan. After several years in industry, he devoted his career to teaching, writing, and service to his profession starting first at Clemson University and later at the University of Michigan. His textbooks have been widely used throughout the United States and internationally.

PART 1

Kinematics and Mechanisms

1 | The World of Mechanisms

1.1 INTRODUCTION

The theory of machines and mechanisms is an applied science that is used to understand the relationships between the geometry and motions of the parts of a machine or mechanism and the forces that produce these motions. The subject, and therefore this book, divides itself naturally into three parts. Part 1, which includes Chapters 1 through 4, is concerned with mechanisms and the kinematics of mechanisms, which is the analysis of their motions. Part 1 lays the groundwork for Part 2, comprising Chapters 5 through 13, in which we study the methods of designing mechanisms. Finally, in Part 3, which includes Chapters 14 through 23, we take up the study of kinetics, the time-varying forces in machines and the resulting dynamic phenomena that must be considered in their design.

The design of a modern machine is often very complex. In the design of a new engine, for example, the automotive engineer must deal with many interrelated questions. What is the relationship between the motion of the piston and the motion of the crankshaft? What will be the sliding velocities and the loads at the lubricated surfaces, and what lubricants are available for the purpose? How much heat will be generated, and how will the engine be cooled? What are the synchronization and control requirements, and how will they be met? What will be the cost to the consumer, both for initial purchase and for continued operation and maintenance? What materials and manufacturing methods will be used? What will be the fuel economy, noise, and exhaust emissions; will they meet legal requirements? Although all these and many other important questions must be answered before the design can be completed, obviously not all can be addressed in a book of this size. Just as people with diverse skills must be brought together to produce an adequate design, so too many branches of science must be brought to bear. This book brings together material that falls into the science of mechanics as it relates to the design of mechanisms and machines.

1.2 ANALYSIS AND SYNTHESIS

There are two completely different aspects of the study of mechanical systems, *design* and *analysis.* The concept embodied in the word "design" might be more properly termed *synthesis,* the process of contriving a scheme or a method of accomplishing a given purpose. Design is the process of prescribing the sizes, shapes, material compositions, and arrangements of parts so that the resulting machine will perform the prescribed task.

Although there are many phases in the design process which can be approached in a well-ordered, scientific manner, the overall process is by its very nature as much an art as a science. It calls for imagination, intuition, creativity, judgment, and experience. The role of science in the design process is merely to provide tools to be used by the designers as they practice their art.

It is in the process of evaluating the various interacting alternatives that designers find need for a large collection of mathematical and scientific tools. These tools, when applied properly, can provide more accurate and more reliable information for use in judging a design than one can achieve through intuition or estimation. Thus they can be of tremendous help in deciding among alternatives. However, scientific tools cannot make decisions for designers; they have every right to exert their imagination and creative abilities, even to the extent of overruling the mathematical predictions.

Probably the largest collection of scientific methods at the designer's disposal fall into the category called *analysis.* These are the techniques that allow the designer to critically examine an already existing or proposed design in order to judge its suitability for the task. Thus analysis, in itself, is not a creative science but one of evaluation and rating of things already conceived.

We should always bear in mind that although most of our effort may be spent on analysis, the real goal is synthesis, the design of a machine or system. Analysis is simply a tool. It is, however, a vital tool and will inevitably be used as one step in the design process.

1.3 THE SCIENCE OF MECHANICS

That branch of scientific analysis that deals with motions, time, and forces is called *mechanics* and is made up of two parts, statics and dynamics. *Statics* deals with the analysis of stationary systems—that is, those in which time is not a factor—and *dynamics* deals with systems that change with time.

As shown in Fig. 1.1, dynamics is also made up of two major disciplines, first recognized as separate entities by Euler in 1775:[1]

The investigation of the motion of a rigid body may be conveniently separated into two parts, the one geometrical, the other mechanical. In the first part, the transference of the body from a given position to any other position must be investigated without respect to the causes of the motion, and must be represented by analytical formulae, which will define the position of each point of the body. This investigation will therefore be referable solely to geometry, or rather to stereotomy.

It is clear that by the separation of this part of the question from the other, which belongs properly to Mechanics, the determination of the motion from dynamical principles will be made much easier than if the two parts were undertaken conjointly.

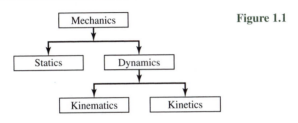

Figure 1.1

These two aspects of dynamics were later recognized as the distinct sciences of *kinematics* (from the Greek word *kinema,* meaning motion) and kinetics, and they deal with motion and the forces producing it, respectively.

The initial problem in the design of a mechanical system is therefore understanding its kinematics. *Kinematics* is the study of motion, quite apart from the forces which produce that motion. More particularly, kinematics is the study of position, displacement, rotation, speed, velocity, and acceleration. The study, say, of planetary or orbital motion is also a problem in kinematics, but in this book we shall concentrate our attention on kinematic problems that arise in the design of mechanical systems. Thus, the kinematics of machines and mechanisms is the focus of the next several chapters of this book. Statics and kinetics, however, are also vital parts of a complete design analysis, and they are covered as well in later chapters.

It should be carefully noted in the above quotation that Euler based his separation of dynamics into kinematics and kinetics on the assumption that they should deal with *rigid* bodies. It is this very important assumption that allows the two to be treated separately. For flexible bodies, the shapes of the bodies themselves, and therefore their motions, depend on the forces exerted on them. In this situation, the study of force and motion must take place simultaneously, thus significantly increasing the complexity of the analysis.

Fortunately, although all real machine parts are flexible to some degree, machines are usually designed from relatively rigid materials, keeping part deflections to a minimum. Therefore, it is common practice to assume that deflections are negligible and parts are rigid when analyzing a machine's kinematic performance, and then, after the dynamic analysis when loads are known, to design the parts so that this assumption is justified.

1.4 TERMINOLOGY, DEFINITIONS, AND ASSUMPTIONS

Reuleaux[2] defines a *machine*[3] as a "*combination of resistant bodies so arranged that by their means the mechanical forces of nature can be compelled to do work accompanied by certain determinate motions.*" He also defines a *mechanism* as an "*assemblage of resistant bodies, connected by movable joints, to form a closed kinematic chain with one link fixed and having the purpose of transforming motion.*"

Some light can be shed on these definitions by contrasting them with the term *structure.* A structure is also a combination of resistant (rigid) bodies connected by joints, but its purpose is not to do work or to transform motion. A structure (such as a truss) is intended to be rigid. It can perhaps be moved from place to place and is movable in this sense of the word; however, it has no *internal* mobility, no *relative motions* between its various members, whereas both machines and mechanisms do. Indeed, the whole purpose of a machine or mechanism is to utilize these relative internal motions in transmitting power or transforming motion.

A machine is an arrangement of parts for doing work, a device for applying power or changing its direction. It differs from a mechanism in its purpose. In a machine, terms such as force, torque, work, and power describe the predominant concepts. In a mechanism, though it may transmit power or force, the predominant idea in the mind of the designer is one of achieving a desired motion. There is a direct analogy between the terms structure, mechanism, and machine and the three branches of mechanics shown in Fig. 1.1. The term "structure" is to statics as the term "mechanism" is to kinematics as the term "machine" is to kinetics.

We shall use the word *link* to designate a machine part or a component of a mechanism. As discussed in the previous section, a link is assumed to be completely rigid. Machine components that do not fit this assumption of rigidity, such as springs, usually have no effect on the kinematics of a device but do play a role in supplying forces. Such members are not called links; they are usually ignored during kinematic analysis, and their force effects are introduced during dynamic analysis. Sometimes, as with a belt or chain, a machine member may possess one-way rigidity; such a member would be considered a link when in tension but not under compression.

The links of a mechanism must be connected together in some manner in order to transmit motion from the *driver,* or input link, to the *follower,* or output link. These connections, joints between the links, are called *kinematic pairs* (or just *pairs*), because each joint consists of a pair of mating surfaces, or two elements, with one mating surface or element being a part of each of the joined links. Thus we can also define a link as *the rigid connection between two or more elements of different kinematic pairs.*

Stated explicitly, the assumption of rigidity is that there can be no relative motion (change in distance) between two arbitrarily chosen points on the same link. In particular, the relative positions of pairing elements on any given link do not change. In other words, the purpose of a link is to hold constant spatial relationships between the elements of its pairs.

As a result of the assumption of rigidity, many of the intricate details of the actual part shapes are unimportant when studying the kinematics of a machine or mechanism. For this reason it is common practice to draw highly simplified schematic diagrams, which contain important features of the shape of each link, such as the relative locations of pair elements, but which completely subdue the real geometry of the manufactured parts. The slider-crank mechanism of the internal combustion engine, for example, can be simplified to the schematic diagram shown later in Fig. 1.3*b* for purposes of analysis. Such simplified schematics are a great help because they eliminate confusing factors that do not affect the analysis; such diagrams are used extensively throughout this text. However, these schematics also have the drawback of bearing little resemblance to physical hardware. As a result, they may give the impression that they represent only academic constructs rather than real machinery. We should always bear in mind that these simplified diagrams are intended to carry only the minimum necessary information so as not to confuse the issue with all the unimportant detail (for kinematic purposes) or complexity of the true machine parts.

When several links are movably connected together by joints, they are said to form a *kinematic chain.* Links containing only two pair element connections are called *binary* links; those having three are called *ternary* links, and so on. If every link in the chain is connected to at least two other links, the chain forms one or more closed loops and is called a *closed* kinematic chain; if not, the chain is referred to as *open*. When no distinction is made, the chain is assumed to be closed. If the chain consists entirely of binary links, it is *simple-closed; compound-closed* chains, however, include other than binary links and thus form more than a single closed loop.

Recalling Reuleaux' definition of a mechanism, we see that it is necessary to have a closed kinematic chain *with one link fixed*. When we say that one link is fixed, we mean that it is chosen as a frame of reference for all other links—that is, that the motions of all other points on the linkage will be measured with respect to this link, thought of as being fixed. This link in a practical machine usually takes the form of a stationary platform or base (or a housing rigidly attached to such a base) and is called the *frame* or *base link*. The question of whether this reference frame is truly stationary (in the sense of being an inertial reference frame) is immaterial in the study of kinematics but becomes important in the investigation of kinetics, where forces are considered. In either case, once a frame member is designated (and other conditions are met), the kinematic chain becomes a mechanism and as the driver is moved through various positions, called *phases,* all other links have well-defined motions with respect to the chosen frame of reference. We use the term *kinematic chain* to specify a particular arrangement of links and joints when it is not clear which link is to be treated as the frame. When the frame link is specified, the kinematic chain is called a *mechanism.*

In order for a mechanism to be useful, the motions between links cannot be completely arbitrary; they too must be constrained to produce the proper relative motions, those chosen by the designer for the particular task to be performed. These desired relative motions are obtained by a proper choice of the number of links and the kinds of joints used to connect them.

Thus we are led to the concept that, in addition to the distances between successive joints, the nature of the joints themselves and the relative motions that they permit are essential in determining the kinematics of a mechanism. For this reason it is important to look more closely at the nature of joints in general terms, and in particular at several of the more common types.

The controlling factor that determines the relative motions allowed by a given joint is the shapes of the mating surfaces or elements. Each type of joint has its own characteristic shapes for the elements, and each allows a given type of motion, which is determined by the possible ways in which these elemental surfaces can move with respect to each other. For example, the pin joint in Fig. 1.2a has cylindric elements and, assuming that the links cannot slide axially, these surfaces permit only relative rotational motion. Thus a pin joint allows the two connected links to experience relative rotation about the pin center. So, too, other joints each have their own characteristic element shapes and relative motions. These shapes restrict the totally arbitrary motion of two unconnected links to some prescribed type of relative motion and form the constraining conditions or constraints on the mechanism's motion.

It should be pointed out that the element shapes may often be subtly disguised and difficult to recognize. For example, a pin joint might include a needle bearing, so that two mating surfaces, as such, are not distinguishable. Nevertheless, if the motions of the individual rollers are not of interest, the motions allowed by the joints are equivalent and the pairs are of the same generic type. Thus the criterion for distinguishing different pair types is the relative motions they permit and not necessarily the shapes of the elements, though these may provide vital clues. The diameter of the pin used (or other dimensional data) is also of no more importance than the exact sizes and shapes of the connected links. As stated previously, the kinematic function of a link is to hold fixed geometric relationships between the pair elements. In a similar way, the only kinematic function of a joint or pair is to control the relative motion between the connected links. All other features are determined for other reasons and are unimportant in the study of kinematics.

When a kinematic problem is formulated, it is necessary to recognize the type of relative motion permitted in each of the pairs and to assign to it some variable parameter(s) for measuring or calculating the motion. There will be as many of these parameters as there are degrees of freedom of the joint in question, and they are referred to as the *pair variables.* Thus the pair variable of a pinned joint will be a single angle measured between reference lines fixed in the adjacent links, while a spheric pair will have three pair variables (all angles) to specify its three-dimensional rotation.

Kinematic pairs were divided by Reuleaux into *higher pairs* and *lower pairs,* with the latter category consisting of six prescribed types to be discussed next. He distinguished between the categories by noting that the lower pairs, such as the pin joint, have surface contact between the pair elements, while higher pairs, such as the connection between a cam and its follower, have line or point contact between the elemental surfaces. However, as noted in the case of a needle bearing, this criterion may be misleading. We should rather look for distinguishing features in the relative motion(s) that the joint allows.

The six lower pairs are illustrated in Fig. 1.2. Table 1.1 lists the names of the lower pairs and the symbols employed by Hartenberg and Denavit[4] for them, together with the number of degrees of freedom and the pair variables for each of the six:

The *turning pair* or *revolute* (Fig. 1.2a) permits only relative rotation and hence has one degree of freedom. This pair is often referred to as a pin joint.

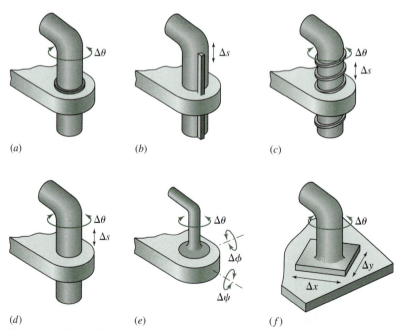

(a)   (b)   (c)

(d)   (e)   (f)

Figure 1.2 The six lower pairs: (a) revolute or pin; (b) prism; (c) helical; (d) cylindric; (e) spheric; and (f) planar.

TABLE 1.1 The Lower Pairs

Pair	Symbol	Pair Variable	Degrees of Freedom	Relative Motion
Revolute	R	$\Delta\theta$	1	Circular
Prism	P	Δs	1	Rectilinear
Screw	S	$\Delta\theta$ or Δs	1	Helical
Cylinder	C	$\Delta\theta$ and Δs	2	Cylindric
Sphere	G	$\Delta\theta, \Delta\phi, \Delta\psi$	3	Spheric
Flat	F	$\Delta x, \Delta y, \Delta\theta$	3	Planar

The *prismatic pair* (Fig. 1.2*b*) permits only a relative sliding motion and therefore is often called a sliding joint. It also has a single degree of freedom.

The *screw pair* or *helical pair* (Fig. 1.2*c*) has only one degree of freedom because the sliding and rotational motions are related by the helix angle of the thread. Thus the pair variable may be chosen as either Δs or $\Delta\theta$, but not both. Note that the screw pair reverts to a revolute if the helix angle is made zero and to a prismatic pair if the helix angle is made 90°.

The *cylindric pair* (Fig. 1.2*d*) permits both angular rotation and an independent sliding motion. Thus the cylindric pair has two degrees of freedom.

The *globular* or *spheric pair* (Fig. 1.2*e*) is a ball-and-socket joint. It has three degrees of freedom, a rotation about each of the coordinate axes.

The *flat pair* or *planar pair* (Fig. 1.2*f*) is seldom, if ever, found in mechanisms in its undisguised form. It has three degrees of freedom.

All other joint types are called higher pairs. Examples are mating gear teeth, a wheel rolling on a rail, a ball rolling on a flat surface, and a cam contacting its roller follower. Because there are an infinite number of higher pairs, a systematic accounting of them is not a realistic objective. We shall treat each as a separate situation as it arises.

Among the higher pairs there is a subcategory known as *wrapping pairs*. Examples are the connection between a belt and a pulley, between a chain and a sprocket, or between a rope and a drum. In each case, one of the links has one-way rigidity.

In the treatment of various joint types, whether lower or higher pairs, there is another important limiting assumption. Throughout the book, we assume that the actual joint, as manufactured, can be reasonably represented by a mathematical abstraction having perfect geometry. That is, when a real machine joint is assumed to be a spheric pair, for example, it is also assumed that there is no "play" or clearance between the joint elements and that any deviation in the spherical geometry of the elements is negligible. When a pin joint is treated as a revolute, it is assumed that no axial motion can take place; if it is necessary to study the small axial motions resulting from clearances between the real elements, the joint must be treated as cylindric, thus allowing for the axial motion.

The term "mechanism," as defined earlier, can refer to a wide variety of devices, including both higher and lower pairs. There is a more limited term, however, that refers to those mechanisms having only lower pairs; such a mechanism is called a *linkage*. A linkage, then, is connected only by the lower pairs of Fig. 1.2.

1.5 PLANAR, SPHERICAL, AND SPATIAL MECHANISMS

Mechanisms may be categorized in several different ways to emphasize their similarities and differences. One such grouping divides mechanisms into planar, spherical, and spatial categories. All three groups have many things in common; the criterion that distinguishes the groups, however, is to be found in the characteristics of the motions of the links.

A *planar mechanism* is one in which all particles describe plane curves in space and all these curves lie in parallel planes; that is, the loci of all points are plane curves parallel to a single common plane. This characteristic makes it possible to represent the locus of any chosen point of a planar mechanism in its true size and shape on a single drawing or figure. The motion transformation of any such mechanism is called *coplanar.* The plane four-bar linkage, the plate cam and follower, and the slider-crank mechanism are familiar examples of planar mechanisms. The vast majority of mechanisms in use today are planar.

Planar mechanisms utilizing only lower pairs are called *planar linkages;* they include only revolute and prismatic pairs. Although a planar pair might theoretically be included, this would impose no constraint and thus be equivalent to an opening in the kinematic chain. Planar motion also requires that all revolute axes be normal to the plane of motion and that all prismatic pair axes be parallel to the plane.

A *spherical mechanism* is one in which each link has some point that remains stationary as the linkage moves and in which the stationary points of all links lie at a common location; that is, the locus of each point is a curve contained in a spherical surface, and the spherical surfaces defined by several arbitrarily chosen points are all *concentric.* The motions of all particles can therefore be completely described by their radial projections, or "shadows," on the surface of a sphere with a properly chosen center. Hooke's universal joint is perhaps the most familiar example of a spherical mechanism.

Spherical linkages are constituted entirely of revolute pairs. A spheric pair would produce no additional constraints and would thus be equivalent to an opening in the chain, while all other lower pairs have nonspheric motion. In spheric linkages, the axes of all revolute pairs must intersect at a point.

Spatial mechanisms, on the other hand, include *no* restrictions on the relative motions of the particles. The motion transformation is not necessarily coplanar, nor must it be concentric. A spatial mechanism may have particles with loci of double curvature. Any linkage that contains a screw pair, for example, is a spatial mechanism, because the relative motion within a screw pair is helical.

Thus, the overwhelmingly large category of planar mechanisms and the category of spherical mechanisms are only special cases, or subsets, of the all-inclusive category spatial mechanisms. They occur as a consequence of special geometry in the particular orientations of their pair axes.

If planar and spherical mechanisms are only special cases of spatial mechanisms, why is it desirable to identify them separately? Because of the particular geometric conditions that identify these types, many simplifications are possible in their design and analysis. As pointed out earlier, it is possible to observe the motions of all particles of a planar mechanism in true size and shape from a single direction. In other words, all motions can be represented graphically in a single view. Thus, graphical techniques are well-suited to their solution. Because spatial mechanisms do not all have this fortunate geometry, visualization becomes more difficult and more powerful techniques must be developed for their analysis.

Because the vast majority of mechanisms in use today are planar, one might question the need for the more complicated mathematical techniques used for spatial mechanisms.

There are a number of reasons why more powerful methods are of value even though the simpler graphical techniques have been mastered:

1. They provide new, alternative methods that will solve the problems in a different way. Thus they provide a means of checking results. Certain problems, by their nature, may also be more amenable to one method than to another.
2. Methods that are analytic in nature are better suited to solution by calculator or digital computer than by graphic techniques.
3. Even though the majority of useful mechanisms are planar and well-suited to graphical solution, the few remaining must also be analyzed, and techniques should be known for analyzing them.
4. One reason that planar linkages are so common is that good methods of analysis for the more general spatial linkages have not been available until relatively recently. Without methods for their analysis, their design and use has not been common, even though they may be inherently better suited in certain applications.
5. We will discover that spatial linkages are much more common in practice than their formal description indicates.

Consider a four-bar linkage. It has four links connected by four pins whose axes are parallel. This "parallelism" is a mathematical hypothesis; it is not a reality. The axes as produced in a shop—in any shop, no matter how good—will be only approximately parallel. If they are far out of parallel, there will be binding in no uncertain terms, and the mechanism will move only because the "rigid" links flex and twist, producing loads in the bearings. If the axes are nearly parallel, the mechanism operates because of the looseness of the running fits of the bearings or flexibility of the links. A common way of compensating for small nonparallelism is to connect the links with self-aligning bearings, which are actually spherical joints allowing three-dimensional rotation. Such a "planar" linkage is thus a low-grade spatial linkage.

1.6 MOBILITY

One of the first concerns in either the design or the analysis of a mechanism is the number of degrees of freedom, also called the *mobility* of the device. The mobility* of a mechanism is the number of input parameters (usually pair variables) that must be controlled independently in order to bring the device into a particular position. Ignoring for the moment certain exceptions to be mentioned later, it is possible to determine the mobility of a mechanism directly from a count of the number of links and the number and types of joints that it includes.

To develop this relationship, consider that before they are connected together, each link of a planar mechanism has three degrees of freedom when moving relative to the fixed link. Not counting the fixed link, therefore, an n-link planar mechanism has $3(n-1)$ degrees of freedom before any of the joints are connected. Connecting a joint that has one degree of freedom, such as a revolute pair, has the effect of providing two constraints between the connected links. If a two-degree-of-freedom pair is connected, it provides one

---

*The German literature distinguishes between *movability* and *mobility*. Movability includes the six degrees of freedom of the device as a whole, as though the ground link were not fixed, and thus applies to a kinematic chain. Mobility neglects these and considers only the internal relative motions, thus applying to a mechanism. The English literature seldom recognizes this distinction, and the terms are used somewhat interchangeably.

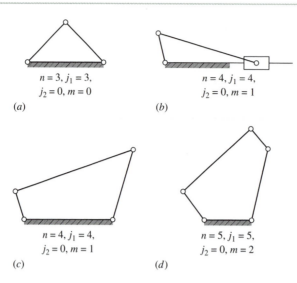

Figure 1.3 Applications of the Kutzbach mobility criterion.

$n = 3, j_1 = 3,$
$j_2 = 0, m = 0$

(a)

$n = 4, j_1 = 4,$
$j_2 = 0, m = 1$

(b)

$n = 4, j_1 = 4,$
$j_2 = 0, m = 1$

(c)

$n = 5, j_1 = 5,$
$j_2 = 0, m = 2$

(d)

constraint. When the constraints for all joints are subtracted from the total freedoms of the unconnected links, we find the resulting mobility of the connected mechanism. When we use j_1 to denote to number of single-degree-of-freedom pairs and j_2 for the number of two-degree-of-freedom pairs, the resulting mobility m of a planar n-link mechanism is given by

$$m = 3(n - 1) - 2j_1 - j_2 \tag{1.1}$$

Written in this form, Eq. (1.1) is called the *Kutzbach criterion* for the mobility of a planar mechanism. Its application is shown for several simple cases in Fig. 1.3.

If the Kutzbach criterion yields $m > 0$, the mechanism has m degrees of freedom. If $m = 1$, the mechanism can be driven by a single input motion. If $m = 2$, then two separate input motions are necessary to produce constrained motion for the mechanism; such a case is shown in Fig. 1.3d.

If the Kutzbach criterion yields $m = 0$, as in Fig. 1.3a, motion is impossible and the mechanism forms a structure. If the criterion gives $m = -1$ or less, then there are redundant constraints in the chain and it forms a statically indeterminate structure. Examples are shown in Fig. 1.4. Note in these examples that when three links are joined by a single pin, two joints must be counted; such a connection is treated as two separate but concentric pairs.

Figure 1.5 shows examples of Kutzbach's criterion applied to mechanisms with two-degree-of-freedom joints. Particular attention should be paid to the contact (pair) between the wheel and the fixed link in Fig. 1.5b. Here it is assumed that slipping is possible

Figure 1.4 Applications of the Kutzbach criterion to structures.

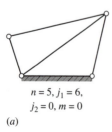

$n = 5, j_1 = 6,$
$j_2 = 0, m = 0$

(a)

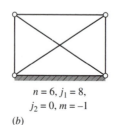

$n = 6, j_1 = 8,$
$j_2 = 0, m = -1$

(b)

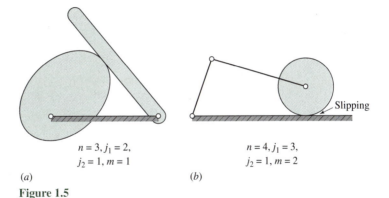

$n = 3, j_1 = 2,$
$j_2 = 1, m = 1$

(a)

$n = 4, j_1 = 3,$
$j_2 = 1, m = 2$

(b)

Figure 1.5

between the links. If this contact included gear teeth or if friction was high enough to prevent slipping, the joint would be counted as a one-degree-of-freedom pair, because only one relative motion would be possible between the links.

Sometimes the Kutzbach criterion gives an incorrect result. Notice that Fig. 1.6a represents a structure and that the criterion properly predicts $m = 0$. However, if link 5 is arranged as in Fig. 1.6b, the result is a double-parallelogram linkage with a mobility of 1 even though Eq. (1.1) indicates that it is a structure. The actual mobility of 1 results only if the parallelogram geometry is achieved. Because in the development of the Kutzbach criterion no consideration was given to the lengths of the links or other dimensional properties, it is not surprising that exceptions to the criterion are found for particular cases with equal link lengths, parallel links, or other special geometric features.

Even though the criterion has exceptions, it remains useful because it is so easily applied. To avoid exceptions, it would be necessary to include all the dimensional properties of the mechanism. The resulting criterion would be very complex and would be useless at the early stages of design when dimensions may not be known.

An earlier mobility criterion named after Grübler applies to mechanisms with only single-degree-of-freedom joints where the overall mobility of the mechanism is unity. Putting $j_2 = 0$ and $m = 1$ into Eq. (1.1), we find Grübler's criterion for planar mechanisms with constrained motion:

$$3n - 2j_1 - 4 = 0 \qquad (1.2)$$

From this we can see, for example, that a planar mechanism with a mobility of 1 and only single-degree-of-freedom joints cannot have an odd number of links. Also, we can find the simplest possible mechanism of this type; by assuming all binary links, we find

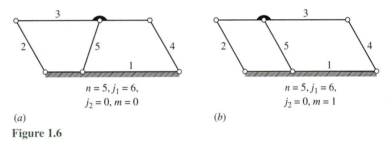

$n = 5, j_1 = 6,$
$j_2 = 0, m = 0$

(a)

$n = 5, j_1 = 6,$
$j_2 = 0, m = 1$

(b)

Figure 1.6

$n = j_1 = 4$. This shows why the four-bar linkage (Fig. 1.3c) and the slider-crank mechanism (Fig. 1.3b) are so common in application.

Both the Kutzbach criterion, Eq. (1.1), and the Grübler criterion, Eq. (1.2), were derived for the case of planar mechanisms. If similar criteria are developed for spatial mechanisms, we must recall that each unconnected link has six degrees of freedom; and each revolute pair, for example, provides five constraints. Similar arguments then lead to the three-dimensional form of the Kutzbach criterion,

$$m = 6(n - 1) - 5j_1 - 4j_2 - 3j_3 - 2j_4 - j_5 \tag{1.3}$$

and the Grübler criterion,

$$6n - 5j_1 - 7 = 0 \tag{1.4}$$

The simplest form of a spatial mechanism,* with all single-freedom pairs and a mobility of 1, is therefore $n = j_1 = 7$.

1.7 CLASSIFICATION OF MECHANISMS

An ideal system of classification of mechanisms would be one that allows the designer to enter the system with a set of specifications and leave with one or more mechanisms that satisfy those specifications. Though history[5] shows that many attempts have been made, none have been very successful in devising a completely satisfactory method. In view of the fact that the purpose of a mechanism is the transformation of motion, we shall follow Torfason's lead[6] and classify mechanisms according to the type of motion transformation. Altogether, Torfason displays 262 mechanisms, each of which is capable of variation in dimensions. His categories are as follows:

Snap-Action Mechanisms   The mechanisms of Fig. 1.7 are typical of snap-action mechanisms, but Torfason also includes spring clips and circuit breakers.

Linear Actuators   Linear actuators include:

1. Stationary screws with traveling nuts.
2. Stationary nuts with traveling screws.
3. Single- and double-acting hydraulic and pneumatic cylinders.

Fine Adjustments   Fine adjustments may be obtained with screws, including the differential screw of Fig. 1.8, worm gearing, wedges, levers and levers in series, and various motion-adjusting mechanisms.

Clamping Mechanisms   Typical clamping mechanisms are the C-clamp, the woodworker's screw clamp, cam- and lever-actuated clamps, vises, presses such as the toggle press of Fig. 1.7b, collets, and stamp mills.

Locational Devices   Torfason pictures 15 locational mechanisms. These are usually self-centering and locate either axially or angularly using springs and detents.

---

*Note that all planar mechanisms are exceptions to the spatial-mobility criteria. They have special geometric characteristics in that all revolute axes are parallel and perpendicular to the plane of motion and all prism axes lie in the plane of motion.

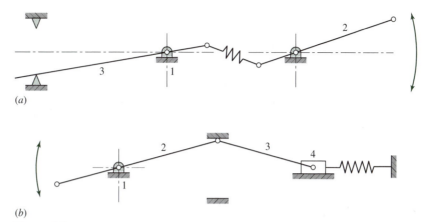

(a)

(b)

Figure 1.7 Typical snap-action, toggle, or flip-flop mechanisms used for switches, clamps, or fasteners. The elements of a mechanism are always numbered beginning with 1 for the base or frame, and 2 for the input or driving element. The mechanism of part (a) is bistable; that of (b) is a true toggle.

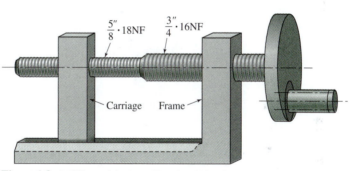

Figure 1.8 A differential screw. You should be able to determine the motion of the carriage resulting from one turn of the handle.

Ratchets and Escapements

There are many different forms of ratchets and escapements, some quite clever. They are used in locks, jacks, clockwork, and other applications requiring some form of intermittent motion. Figure 1.9 illustrates four typical applications.

The ratchet in Fig. 1.9a allows only one direction of rotation of wheel 2. Pawl 3 is held against the wheel by gravity or a spring. A similar arrangement is used for lifting jacks, which then employ a toothed rack for rectilinear motion.

Figure 1.9b is an escapement used for rotary adjustments.

Graham's escapement of Fig. 1.9c is used to regulate the movement of clockwork. Anchor 3 drives a pendulum whose oscillating motion is caused by the two clicks engaging the escapement wheel 2. One is a push click, the other a pull click. The lifting and engaging of each click caused by oscillation of the pendulum results in a wheel motion which, at the same time, presses each respective click and adds a gentle force to the motion of the pendulum.

The escapement of Fig. 1.9d has a control wheel 2 which may rotate continuously to allow wheel 3 to be driven (by another source) in either direction.

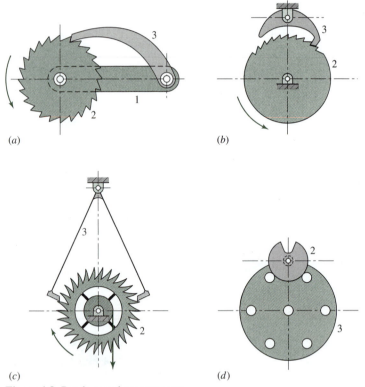

(a)

(b)

(c)

(d)

Figure 1.9 Ratchets and escapements.

Indexing Mechanisms   The indexer of Fig. 1.10a uses standard gear teeth; for light loads, pins can be used in wheel 2 with corresponding slots in wheel 3, but neither form should be used if the shaft inertias are large.

Figure 1.10b illustrates a Geneva-wheel indexer. Three or more slots (up to 16) may be used in driver 2, and wheel 3 can be geared to the output to be indexed. High speeds and large inertias may cause problems with this indexer.

The toothless ratchet 5 in Fig. 1.10c is driven by the oscillating crank 2 of variable throw. Note the similarity of this to the ratchet of Fig. 1.9a.

Torfason lists nine different indexing mechanisms, and many variations are possible.

Swinging or Rocking Mechanisms   The class of swinging or rocking mechanisms is often termed *oscillators;* in each case the output member rocks or swings through an angle that is generally less than 360°. However, the output shaft can be geared to a second shaft to produce larger angles of oscillation.

Figure 1.11a is a mechanism consisting of a rotating crank 2 and a coupler 3 containing a toothed rack that meshes with output gear 4 to produce the oscillating motion.

In Fig. 1.11b, crank 2 drives member 3, which slides on output link 4, producing a rocking motion. This mechanism is described as a *quick-return linkage* because crank 2 rotates through a larger angle on the forward stroke of link 4 than on the return stroke.

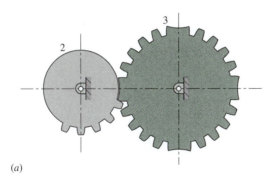

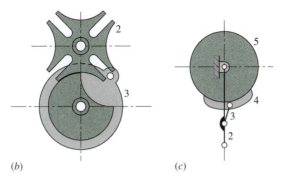

(a)

(b)   (c)

Figure 1.10 Several indexing mechanisms.

Figure 1.11c is a *four-bar linkage* called the *crank-and-rocker mechanism.* Crank 2 drives rocker 4 through coupler 3. Of course, link 1 is the frame. The characteristics of the rocking motion depend on the dimensions of the links and the placement of the frame points.

Figure 1.11d illustrates a *cam-and-follower mechanism,* in which the rotating *cam* 2 drives link 3, called the *follower,* in a rocking motion. There are an endless variety of cam-and-follower mechanisms, many of which will be discussed in Chapter 5. In each case the cams can be formed to produce rocking motions with nearly any set of desired characteristics.

Reciprocating Mechanisms   Repeating straight-line motion is commonly obtained using pneumatic and hydraulic cylinders, a stationary screw and traveling nut, rectilinear drives using reversible motors or reversing gears, as well as cam-and-follower mechanisms. A variety of typical linkages for obtaining reciprocating motion are shown in Figs. 1.12 and 1.13.

The *offset slider-crank mechanism* shown in Fig. 1.12a has velocity characteristics that differ from an on-center slider crank (not shown). If connecting rod 3 of an on-center slider-crank mechanism is large relative to the length of crank 2, then the resulting motion is nearly harmonic.

Link 4 of the *Scotch yoke mechanism* shown in Fig. 1.12b delivers exact harmonic motion.

The six-bar linkage shown in Fig. 1.12c is often called the *shaper mechanism,* after the name of the machine tool in which it is used. Note that it is derived from Fig. 1.12b by adding coupler 5 and slider 6. The slider stroke has a quick-return characteristic.

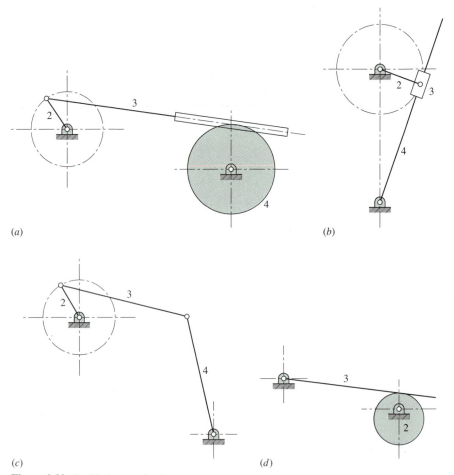

(a)

(b)

(c)

(d)

Figure 1.11 Oscillating mechanisms.

Figure 1.12*d* shows another version of the shaper mechanism, which is often termed the *Whitworth quick-return mechanism.* The linkage is shown in an upside-down configuration to show its similarity to Fig. 1.12*c*.

In many applications, mechanisms are used to perform repetitive operations such as pushing parts along an assembly line, clamping parts together while they are welded, or folding cardboard boxes in an automated packaging machine. In such applications it is often desirable to use a constant-speed motor; this will lead us to a discussion of Grashof's law in Section 1.9. In addition, however, we should also give some consideration to the power and timing requirements.

In these repetitive operations there is usually a part of the cycle when the mechanism is under load, called the advance or working stroke, and a part of the cycle, called the return stroke, when the mechanism is not working but simply returning so that it may repeat the operation. In the offset slider-crank mechanism of Fig. 1.12*a*, for example, work may be required to overcome the load F while the piston moves to the right from C_1 to C_2 but

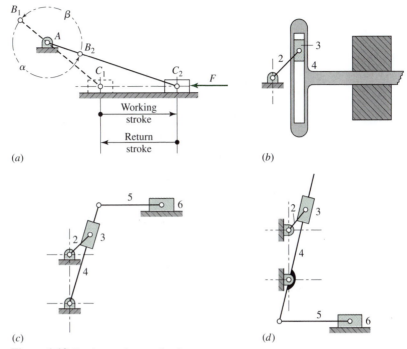

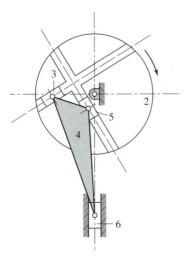

Figure 1.12 Reciprocating mechanisms.

not during its return to position C_1 because the load may have been removed. In such situations, in order to keep the power requirements of the motor to a minimum and to avoid wasting valuable time, it is desirable to design the mechanism so that the piston will move much faster through the return stroke than it does during the working stroke—that is, to use a higher fraction of the cycle time for doing work than for returning.

Figure 1.13 The Wanzer needle-bar mechanism. (As reported by A. R. Holowenko, *Dynamics of Machinery,* Wiley, New York, 1955, p. 54.)

A measure of the suitability of a mechanism from this viewpoint, called *advance-to return-time ratio,* is defined by the formula

$$Q = \frac{\text{time of advance stroke}}{\text{time of return stroke}} \qquad (a)$$

A mechanism for which the value of Q is high is more desirable for such repetitive operations than one in which Q is lower. Certainly, any such operations would use a mechanism for which Q is greater than unity. Because of this, mechanisms with Q greater than unity are called *quick-return mechanisms.*

Assuming that the driving motor operates at constant speed, it is easy to find the time ratio. As shown in Fig. 1.12a, the first thing is to determine the two crank positions AB_1 and AB_2, which mark the beginning and end of the working stroke. Next, noticing the direction of rotation of the crank, we can measure the crank angle α traveled through during the advance stroke and the remaining crank angle β of the return stroke. Then, if the period of the motor is τ, the time of the advance stroke is

$$\text{Time of advance stroke} = \frac{\alpha}{2\pi}\tau \qquad (b)$$

and that of the return stroke is

$$\text{Time of return stroke} = \frac{\beta}{2\pi}\tau \qquad (c)$$

Finally, combining Eqs. (a), (b), and (c), we get the following simple equation for time ratio:

$$Q = \frac{\alpha}{\beta} \qquad (1.5)$$

Notice that the time ratio of a quick-return mechanism does not depend on the amount of work being done or even on the speed of the driving motor. It is a kinematic property of the mechanism itself and can be found strictly from the geometry of the device.

We also notice, however, that there is a proper and an improper direction of rotation for such a device. If the motor were reversed in the example of Fig. 1.12a, the roles of α and β would also reverse and the time ratio would be less than 1. Thus the motor must rotate counterclockwise for this mechanism to have the quick-return property.

Many other mechanisms can be found with quick-return characteristics. Another example is the *Whitworth mechanism,* also called the *crank-shaper mechanism,* shown in Figs. 1.12c and 1.12d. Although the determination of the angles α and β is different for each mechanism, Eq. (1.5) applies to all.

Figure 1.14a shows a six-bar linkage derived from the crank-and-rocker linkage of Fig. 1.11c by expanding coupler 3 and adding coupler 5 and slider 6. Coupler point C should be located so as to produce the desired motion characteristic of slider 6.

A crank-driven toggle mechanism is shown in Fig. 1.14b. With this mechanism, a high mechanical advantage is obtained at one end of the stroke of slider 6. The synthesis of a

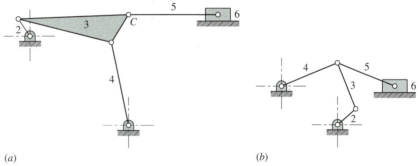

Figure 1.14 Additional six-bar reciprocating mechanisms.

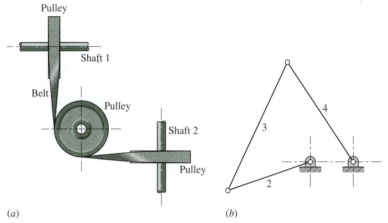

Figure 1.15 Two-shaft coupling mechanisms.

quick-return mechanism, as well as mechanisms with other properties, is discussed in some detail in Chapter 11.

Reversing Mechanisms   When a mechanism is desired which is capable of delivering output rotation in either direction, some form of reversing mechanism is required. Many such devices make use of a two-way clutch that connects the output shaft to either of two drive shafts turning in opposite directions. This method is used in both gear and belt drives and does not require that the drive be stopped to change direction. Gear-shift devices, as in automotive transmissions, are also in quite common use.

Couplings and Connectors   Couplings and connectors are used to transmit motion between coaxial, parallel, intersecting, and skewed shafts. Gears of one kind or another can be used for any of these situations. These will be discussed in Chapters 6 through 9.

Flat belts can be used to transmit motion between parallel shafts. They can also be used between intersecting or skewed shafts if guide pulleys, as shown in Fig. 1.15a, are used. When parallel shafts are involved, the belts can be open or crossed, depending on the direction of rotation desired.

Figure 1.15b shows the four-bar *drag-link mechanism* used to transmit rotary motion between parallel shafts. Here crank 2 is the driver and link 4 is the output. This is a very

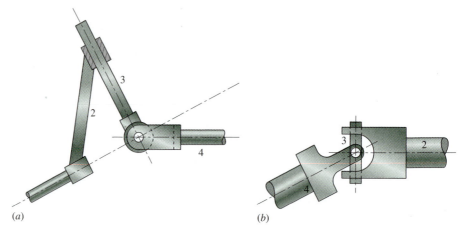

Figure 1.16 Coupling mechanisms for intersecting shafts.

interesting linkage; you should try to duplicate it using cardboard strips and thumbtacks for the joints in order to observe its motion. Can you devise a complete working model?

The *Reuleaux coupling* shown in Fig. 1.16*a* for intersecting shafts is recommended only for light loads.

Figure 1.16*b* shows *Hooke's joint* for intersecting shafts. It is customary to use two of these for parallel shafts.

Sliding Connectors   Sliding connectors are used when one slider (the input) is to drive another slider (the output). The usual problem is that the two sliders operate in the same plane but in different directions. The possible solutions are:

1. A rigid link pivoted at each end to a slider.
2. A belt or chain connecting the two sliders with the use of a guide pulley or sprocket.
3. Rack gear teeth cut on each slider and the connection completed using one or more gears.
4. Flexible cable connector.

Stop, Pause, and Hesitation Mechanisms   In an automotive engine a valve must open, remain open for a period of time, and then close. A conveyor line may need to halt for a period of time while an operation is being performed, and then continue its advance motion. Many similar requirements occur in the design of machines. Torfason classifies these as *stop and dwell, stop and return, stop and advance,* and so on. Such requirements can often be met using cam mechanisms (see Chapter 5), all indexing mechanisms including those of Fig. 1.10, ratchets, linkages at the limits of their motion, and gear-and-clutch mechanisms.

The six-bar linkage of Fig. 1.17 is a clever method of obtaining a rocking motion containing a dwell. This mechanism has a four-bar linkage consisting of frame 1, crank 2, coupler 3, and rocker 4 selected such that point C on the coupler generates the coupler curve shown by dashed lines. A portion of this curve fits very closely to a circular arc whose radius is the link length DC. Thus when point C traverses this portion of the coupler curve, link 6, the output rocker, will remain motionless. In the study of four-bar linkages, many similar coupler curves can be found.

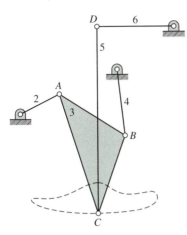

Figure 1.17 Six-bar stop-and-dwell mechanism.

Curve Generators   The connecting rod or coupler of a planar four-bar linkage may be imagined as an infinite plane extending in all directions but pin-connected to the input and output links. Then, during motion of the linkage, any point attached to the plane of the coupler generates some path with respect to the fixed link; this path is called a *coupler curve*. Two of these paths, namely, those generated by the pin connections of the coupler, are simple circles with centers at the two fixed pivots. However, other points can be found which trace much more complex curves.

One of the best sources of coupler curves for the four-bar linkage is the Hrones–Nelson atlas.[7] This book consists of a set of 11- by 17-in charts containing over 7000 coupler curves of crank-rocker linkages. Figure 1.18 is a reproduction of a typical page of this atlas. In each case, the length of the crank is unity and the lengths of other links vary from page to page to produce the different combinations. On each page a number of different coupler points are chosen and their coupler curves are shown. This atlas of coupler curves is invaluable to the designer who needs a linkage to generate a curve with specified characteristics.

The algebraic equation of a coupler curve is, in general, of sixth order; thus it is possible to find coupler curves with a wide variety of shapes and many interesting features. Some coupler curves have sections which are nearly straight line segments; others have circular arc sections; still others have one or more cusps or cross over themselves like a figure eight. Therefore it is often not necessary to use a mechanism with a large number of links to obtain a fairly complex motion.

Yet the complexity of the coupler-curve equation is also a hindrance; it means that hand calculation methods can become very cumbersome. Thus, over the years, many mechanisms have been designed by strictly intuitive procedures and proved with cardboard models, without the use of kinematic principles or procedures. Until quite recently, those techniques which did offer a rational approach have been graphical, avoiding tedious computations. Finally, with the availability of digital computers, and particularly with the growth of computer graphics, useful design methods are now emerging which can deal directly with the complicated calculations required without burdening the designer with the computational drudgery (see Section 12.9 for details on some of these design methods).

One of the more curious and interesting facts about the coupler-curve equation is that the same curve can always be generated by three different four-bar linkages. These are called *cognate linkages,* and the theory is developed in Section 11.11.

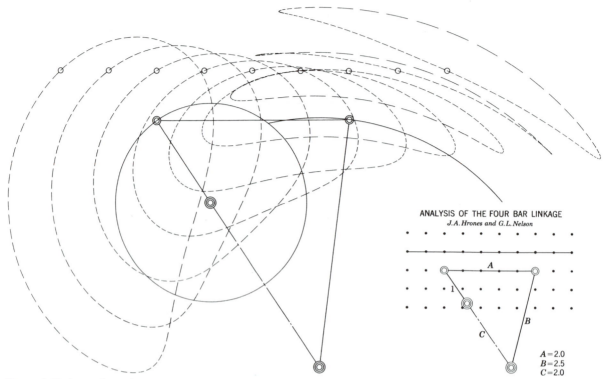

ANALYSIS OF THE FOUR BAR LINKAGE
J.A. Hrones and G.L. Nelson

$A = 2.0$
$B = 2.5$
$C = 2.0$

Figure 1.18 A set of coupler curves. (Reproduced by permission of the publishers, The Technology Press, M.I.T., Cambridge, MA, and John Wiley & Sons, Inc., New York, from J. A. Hrones and G. L. Nelson, *Analysis of the Four-Bar Linkage,* 1951.)

Straight-Line Generators   In the late seventeenth century, before the development of the milling machine, it was extremely difficult to machine straight, flat surfaces. For this reason, good prismatic pairs without backlash were not easy to make. During that era, much thought was given to the problem of attaining a straight-line motion as a part of the coupler curve of a linkage having only revolute connections. Probably the best-known result of this search is the straight-line mechanism developed by Watt for guiding the piston of early steam engines. Figure 1.19*a* shows Watt's linkage to be a four-bar linkage developing an approximate straight line as a part of its coupler curve. Although it does not generate an exact straight line, a good approximation is achieved over a considerable distance of travel.

Another four-bar linkage in which the tracing point P generates an approximate straight-line coupler-curve segment is Roberts' mechanism (Fig. 1.19*b*). The dashed lines in the figure indicate that the linkage is defined by forming three congruent isosceles triangles; thus $BC = AD/2$.

The tracing point P of the Chebychev linkage in Fig. 1.19*c* also generates an approximate straight line. The linkage is formed by creating a 3–4–5 triangle with link 4 in the vertical position as shown by the dashed lines; thus $DB' = 3$, $AD = 4$, and $AB' = 5$. Because $AB = DC$, we have $DC' = 5$ and the tracing point P' is the midpoint of link BC. Note that $DP'C$ also forms a 3–4–5 triangle and hence that P and P' are two points on a straight line parallel to AD.

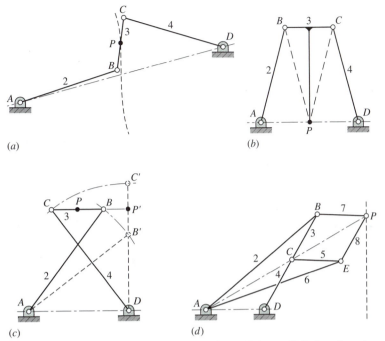

Figure 1.19 Straight-line mechanisms: (*a*) Watt's linkage; (*b*) Roberts' mechanism; (*c*) Chebychev linkage; and (*d*) Peaucillier inversor.

Yet another mechanism that generates a straight-line segment is the Peaucillier inversor shown in Fig. 1.19*d*. The conditions describing its geometry are that $BC = BP = EC = EP$ and $AB = AE$ such that, by symmetry, points A, C, and P always lie on a straight line passing through A. Under these conditions $AC \cdot AP = k$, a constant, and the curves generated by C and P are said to be *inverses* of each other. If we place the other fixed pivot D such that $AD = CD$, then point C must trace a circular arc and point P will follow an exact straight line. Another interesting property is that if AD is not equal to CD, point P can be made to trace a true circular arc of very large radius.

Figure 1.20 shows an exact straight-line mechanism, but note that it employs a slider.

The *pantagraph* of Fig. 1.21 is used to trace figures at a larger or smaller size. If, for example, point P traces a map, then a pen at Q will draw the same map at a smaller scale. The dimensions O_2A, AC, CB, BO_3 must conform to an equal-sided parallelogram.

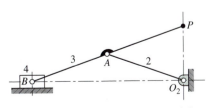

Figure 1.20 Scott–Russell exact straight-line mechanism; $AB = AP = O_2A$.

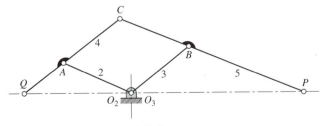

Figure 1.21 The pantagraph linkage.

Torfason also includes robots, speed-changing devices, computing mechanisms, function generators, loading mechanisms, and transportation devices in his classification. Many of these utilize arrangements of mechanisms already presented. Others will appear in some of the chapters to follow.

1.8 KINEMATIC INVERSION

In Section 1.4 we noted that every mechanism has a fixed link called the frame. Until a frame link has been chosen, a connected set of links is called a kinematic chain. When different links are chosen as the frame for a given kinematic chain, the *relative* motions between the various links are not altered, but their *absolute* motions (those measured with respect to the frame link) may be changed drastically. The process of choosing different links of a chain for the frame is known as *kinematic inversion.*

In an *n*-link kinematic chain, choosing each link in turn as the frame yields *n* distinct kinematic inversions of the chain, *n* different mechanisms. As an example, the four-link slider-crank chain of Fig. 1.22 has four different inversions.

Figure 1.22*a* shows the basic slider-crank mechanism, as found in most internal combustion engines today. Link 4, the piston, is driven by the expanding gases and forms the input; link 2, the crank, is the driven output. The frame is the cylinder block, link 1. By reversing the roles of the input and output, this same mechanism can be used as a compressor.

Figure 1.22*b* shows the same kinematic chain; however, it is now inverted and link 2 is stationary. Link 1, formerly the frame, now rotates about the revolute at *A*. This inversion of the slider-crank mechanism was used as the basis of the rotary engine found in early aircraft.

Another inversion of the same slider-crank chain is shown in Fig. 1.22*c*; it has link 3, formerly the connecting rod, as the frame link. This mechanism was used to drive the wheels of early steam locomotives, link 2 being a wheel.

The fourth and final inversion of the slider-crank chain has the piston, link 4, stationary. Although it is not found in engines, by rotating the figure 90° clockwise this mechanism

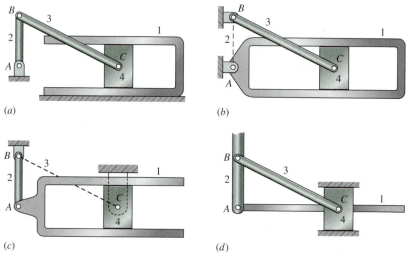

(a)   (b)

(c)   (d)

Figure 1.22 Four inversions of the slider-crank mechanism.

can be recognized as part of a garden water pump. It will be noted in the figure that the prismatic pair connecting links 1 and 4 is also inverted; that is, the "inside" and "outside" elements of the pair have been reversed.

1.9 GRASHOF'S LAW

A very important consideration when designing a mechanism to be driven by a motor, obviously, is to ensure that the input crank can make a complete revolution. Mechanisms in which no link makes a complete revolution would not be useful in such applications. For the four-bar linkage, there is a very simple test of whether this is the case.

Grashof's law states that *for a planar four-bar linkage, the sum of the shortest and longest link lengths cannot be greater than the sum of the remaining two link lengths if there is to be continuous relative rotation between two members.* This is illustrated in Fig. 1.23, where the longest link has length l, the shortest link has length s, and the other two links have lengths p and q. In this notation, Grashof's law states that one of the links, in particular the shortest link, will rotate continuously relative to the other three links if and only if

$$s + l \leq p + q \tag{1.6}$$

If this inequality is not satisfied, no link will make a complete revolution relative to another.

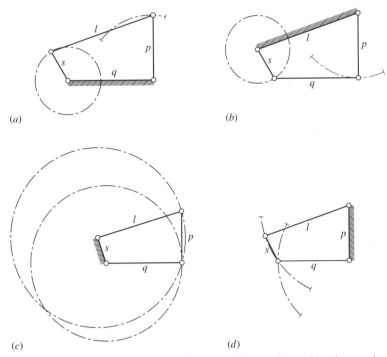

(a) (b)

(c) (d)

Figure 1.23 Four inversions of the Grashof chain: (a, b) crank-rocker mechanisms; (c) drag-link mechanism; and (d) double-rocker mechanism.

Attention is called to the fact that nothing in Grashof's law specifies the order in which the links are connected or which link of the four-bar chain is fixed. We are free, therefore, to fix any of the four links. When we do so, we create the four inversions of the four-bar linkage shown in Fig. 1.23. All of these fit Grashof's law, and in each the link s makes a complete revolution relative to the other links. The different inversions are distinguished by the location of the link s relative to the fixed link.

If the shortest link s is adjacent to the fixed link, as shown in Figs. 1.23a and 1.23b, we obtain what is called a *crank-rocker* linkage. Link s is, of course, the crank because it is able to rotate continuously; and link p, which can only oscillate between limits, is the rocker.

The *drag-link* mechanism, also called the *double-crank* linkage, is obtained by fixing the shortest link s as the frame. In this inversion, shown in Fig. 1.23c, both links adjacent to s can rotate continuously, and both are properly described as cranks; the shorter of the two is generally used as the input.

Although this is a very common mechanism, you will find it an interesting challenge to devise a practical working model that can operate through the full cycle.

By fixing the link opposite to s we obtain the fourth inversion, the *double-rocker* mechanism of Fig. 1.23d. Note that although link s is able to make a complete revolution, neither link adjacent to the frame can do so; both must oscillate between limits and are therefore rockers.

In each of these inversions, the shortest link s is adjacent to the longest link l. However, exactly the same types of linkage inversions will occur if the longest link l is opposite the shortest link s; you should demonstrate this to your own satisfaction.

Reuleaux approaches the problem somewhat differently but, of course, obtains the same results. In this approach, and using Fig. 1.23a, the links are named

s the crank	p the lever
l the coupler	q the frame

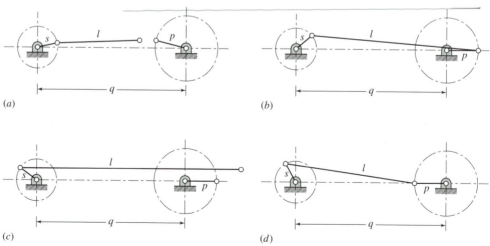

Figure 1.24 (*a*) Equation (1.7); $s + l + p < q$ and the links cannot be connected. (*b*) Equation (1.8); $s + l - p > q$ and s is incapable of rotation. (*c*) Equation (1.9); $s + q + p < 1$ and the links cannot be connected. (*d*) Equation (1.10); $s + q - p < l$ and s is incapable of rotation.

where l need not be the longest link. Then the following conditions apply:

$$s + l + p \geq q \tag{1.7}$$

$$s + l - p \leq q \tag{1.8}$$

$$s + q + p \geq l \tag{1.9}$$

$$s + q - p \leq l \tag{1.10}$$

These four conditions are illustrated in Fig. 1.24 by showing what happens if the conditions are not met.

1.10 MECHANICAL ADVANTAGE

Because of the widespread use of the four-bar linkage, a few remarks are in order here which will help to judge the quality of such a linkage for its intended application. Consider the four-bar linkage shown in Fig. 1.25. Since, according to Grashof's law, this particular linkage is of the crank-rocker variety, it is likely that link 2 is the driver and link 4 is the follower. Link 1 is the frame and link 3 is called the coupler because it couples the motions of the input and output cranks.

The *mechanical advantage* of a linkage is the ratio of the output torque exerted by the driven link to the necessary input torque required at the driver. In Section 3.17 we will prove that the mechanical advantage of the four-bar linkage is directly proportional to the sine of the angle γ between the coupler and the follower and inversely proportional to the sine of the angle β between the coupler and the driver. Of course, both these angles, and therefore the mechanical advantage, are continuously changing as the linkage moves.

When the sine of the angle β becomes zero, the mechanical advantage becomes infinite; thus, at such a position, only a small input torque is necessary to overcome a large output torque load. This is the case when the driver AB of Fig. 1.25 is directly in line with the coupler BC; it occurs when the crank is in position AB_1 and again when the crank is in

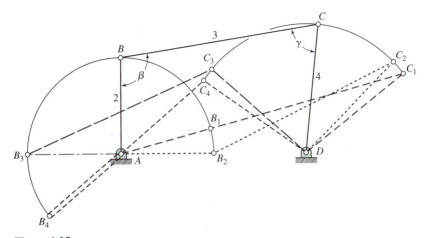

Figure 1.25

position AB_4. Note that these also define the extreme positions of travel of the rocker DC_1 and DC_4. When the four-bar linkage is in either of these positions, the mechanical advantage is infinite and the linkage is said to be in a *toggle* position.

The angle γ between the coupler and the follower is called the *transmission angle*. As this angle becomes small, the mechanical advantage decreases and even a small amount of friction will cause the mechanism to lock or jam. A common rule of thumb is that a four-bar linkage should not be used in the region where the transmission angle is less than, say, $45°$ or $50°$. The extreme values of the transmission angle occur when the crank AB lies along the line of the frame AD. In Fig. 1.25 the transmission angle is minimum when the crank is in position AB_2 and maximum when the crank has position AB_3. Because of the ease with which it can be visually inspected, the transmission angle has become a commonly accepted measure of the quality of the design of a four-bar linkage.

Note that the definitions of mechanical advantage, toggle, and transmission angle depend on the choice of the driver and driven links. If, in the same figure, link 4 is used as the driver and link 2 as the follower, the roles of β and γ are reversed. In this case the linkage has no toggle position, and its mechanical advantage becomes zero when link 2 is in position AB_1 or AB_4, because the transmission angle is then zero. These and other methods of rating the suitability of the four-bar or other linkages are discussed more thoroughly in Section 3.17.

NOTES

[1]. *Novi Comment. Acad. Petrop.,* vol. 20, 1775; also in *Theoria Motus Corporum,* 1790. The translation is by A. B. Willis, *Principles of Mechanism,* 2nd ed., 1870, p. viii.

[2]. Much of the material of this section is based on definitions originally set down by F. Reuleaux (1829–1905), a German kinematician whose work marked the beginning of a systematic treatment of kinematics. For additional reading see A. B. W. Kennedy, *Reuleaux' Kinematics of Machinery,* Macmillan, London, 1876; republished by Dover, New York, 1963.

[3]. There appears to be no agreement at all on the proper definition of a machine. In a footnote Reuleaux gives 17 definitions, and his translator gives 7 more and discusses the whole problem in detail.

[4]. Richard S. Hartenberg and Jacques Denavit, *Kinematic Synthesis of Linkages,* McGraw-Hill, New York, 1969, Chapter 2.

[5]. For an excellent short history of the kinematics of mechanisms, see Hartenberg and Denavit, *Kinematic Synthesis of Linkages,* Chapter 1.

[6]. See L. E. Torfason, "*A Thesaurus of Mechanisms,*" in J. E. Shigley and C. R. Mischke (Eds.), *Mechanical Designer's Notebooks, Volume 5, Mechanisms,* McGraw-Hill, New York, 1990, Chapter 1. Alternately, see L. E. Torfason, "*A Thesaurus of Mechanisms,*" in J. E. Shigley and C. R. Mischke (Eds.), *Standard Handbook of Machine Design,* McGraw-Hill, New York, 1986, Chapter 39.

[7]. J. A. Hrones and G. L. Nelson, *Analysis of the Four-Bar Linkage,* The Technology Press, M.I.T., Cambridge, MA, Wiley, New York, 1951.

PROBLEMS

1.1 Sketch at least six different examples of the use of a planar four-bar linkage in practice. They can be found in the workshop, in domestic appliances, on vehicles, on agricultural machines, and so on.

1.2 The link lengths of a planar four-bar linkage are 1, 3, 5, and 5 in. Assemble the links in all possible combinations and sketch the four inversions of each. Do these linkages satisfy Grashof's law? Describe each inversion by name—for example, a crank-rocker mechanism or a drag-link mechanism.

1.3 A crank-rocker linkage has a 100-mm frame, a 25-mm crank, a 90-mm coupler, and a 75-mm rocker. Draw the linkage and find the maximum and minimum values of the transmission angle. Locate both toggle positions and record the corresponding crank angles and transmission angles.

1.4 In the figure, point C is attached to the coupler; plot its complete path.

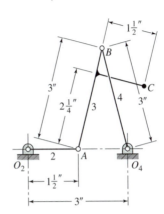

Figure P1.4

1.5 Find the mobility of each mechanism shown in the figure.

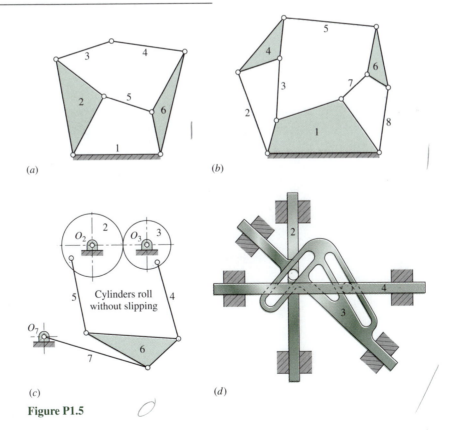

(a)

(b)

(c)

(d)

Figure P1.5

1.6 Use the mobility criterion to find a planar mechanism containing a moving quaternary link. How many distinct variations of this mechanism can you find?

1.7 Find the time ratio of the linkage of Problem 1.3.

1.8 Devise a practical working model of the drag-link mechanism.

1.9 Plot the complete coupler curve of the Roberts' mechanism of Fig. 1.19b. Use $AB = CD = AD = 2.5$ in and $BC = 1.25$ in.

1.10 If the crank of Fig. 1.8 is turned 10 revolutions clockwise, how far and in what direction will the carriage move?

1.11 Show how the mechanism of Fig. 1.12b can be used to generate a sine wave.

1.12 Devise a crank-and-rocker linkage, as in Fig. 1.11c, having a rocker angle of 60°. The rocker length is to be 0.50 m.

2 Position and Displacement

In analyzing motion, the first and most basic problem encountered is that of defining and dealing with the concepts of position and displacement. Because motion can be thought of as a time series of displacements between successive positions, it is important to understand exactly the meaning of the term *position;* rules or conventions must be established to make the definition precise.

Although many of the concepts in this chapter may appear intuitive and almost trivial, many subtleties are explained here which are required for an understanding of the next several chapters.

2.1 LOCUS OF A MOVING POINT

In speaking of the position of a particle or point, we are really answering the question: Where is the point or what is its location? We are speaking of something that exists in nature and are posing the question of how to express this (in words or symbols or numbers) in such a way that the meaning is clear. We soon discover that position cannot be defined on a truly absolute basis. We must define the position of a point in terms of some agreed-upon frame of reference, some reference coordinate system.

As shown in Fig. 2.1, once we have agreed upon the xyz coordinate system as the frame of reference, we can say that point P is located x units along the x axis, y units along the y axis, and z units along the z axis *from the origin O*. In this very statement we see that three vitally important parts of the definition depend on the existence of the reference coordinate system:

1. The *origin* of coordinates O provides an agreed-upon location from which to measure the location of point P.

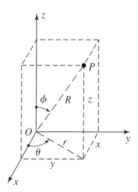

Figure 2.1 A right-handed three-dimensional coordinate system showing how point P is located algebraically.

2. The *coordinate axes* provide agreed-upon *directions* along which the measurements are to be made; they also provide known lines and planes for the definition and measurement of angles.
3. The unit distance along any of the axes provides a scale for quantifying distances.

These observations are not restricted to the Cartesian coordinates (x, y, z) of point P. All three properties of the coordinate system are also necessary in defining the cylindrical coordinates (r, θ, z), spherical coordinates (R, θ, ϕ), or any other coordinates of point P. The same properties would also be required if point P were restricted to remain in a single plane and a two-dimensional coordinate system were used. No matter how defined, the concept of the position of a point cannot be related without the definition of a reference coordinate system.

We note, in Fig. 2.1, that the direction cosines for locating point P are

$$\cos\alpha = \frac{x}{R}, \qquad \cos\beta = \frac{y}{R}, \qquad \cos\gamma = \frac{z}{R} \qquad (2.1)$$

where the angles α, β, and γ are, respectively, the angles measured form the positive coordinate axes to the line OP.

One means of expressing the motion of a point or particle is to define its components along the reference axes as functions of some parameter such as time:

$$x = x(t), \qquad y = y(t), \qquad z = z(t) \qquad (2.2)$$

If these relations are known, the position of P can be found for any time t. This is the general case for the motion of a particle and is illustrated in the following example.

EXAMPLE 2.1

Describe the motion of a particle P whose position changes with time according to the equations $x = a\cos 2\pi t$, $y = a\sin 2\pi t$, and $z = bt$.

SOLUTION

Substitution of values for *t* from 0 to 2 gives values as shown in Table 2.1. As shown in Fig. 2.2, the point moves with *helical motion* with radius *a* around the *z* axis and with a lead of *b*. Note that if $b = 0$, then $z(t) = 0$, the moving particle is confined to the xy plane, and the motion is a circle with its center at the origin.

TABLE 2.1 Example 2.1

t	x	y	z
0	a	0	0
1/4	0	a	b/4
1/2	−a	0	b/2
3/4	0	−a	3b/4
1	a	0	b
5/4	0	a	5b/4
3/2	−a	0	3b/2
7/4	0	−a	7b/4
2	a	0	2b

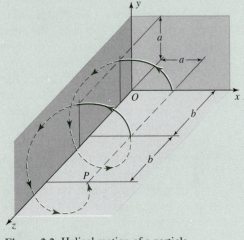

Figure 2.2 Helical motion of a particle.

We have used the words *particle* and *point* interchangeably. When we use the word *point*, we have in mind something that has no dimensions—that is, something with zero length, zero width, and zero thickness. When the word *particle* is used, we have in mind something whose dimensions are small and unimportant—that is, a tiny material body whose dimensions are negligible, a body small enough for its dimensions to have no effect on the analysis to be performed.

The successive positions of a moving point define a line or curve. This curve has no thickness because the point has no dimensions. However, the curve does have length because the point occupies different positions as time changes. This curve, representing the successive positions of the point, is called the *path* or *locus* of the moving point in the reference coordinate system.

If three coordinates are necessary to describe the path of a moving point, the point is said to have *spatial motion*. If the path can be described by only two coordinates—that is, if the coordinate axes can be chosen such that one coordinate is always zero or constant—the path is contained in a single plane and the point is said to have *planar motion*. Sometimes it happens that the path of a point can be described by a single coordinate. This means that two of its spatial position coordinates can be taken as zero or constant. In this case the point moves in a straight line and is said to have *rectilinear motion*.

In each of the three cases described, it is assumed that the coordinate system is chosen so as to obtain the least number of coordinates necessary to describe the motion of the point. Thus the description of rectilinear motion requires one coordinate, a point whose path is a *plane curve* requires two coordinates, and a point whose locus is a *space curve*, sometimes called a *skew curve*, requires three position coordinates.

2.2 POSITION OF A POINT

The physical process involved in observing the position of a point, as shown in Fig. 2.3, implies that the observer is actually keeping track of the relative location of two points, P and O, by looking at both, performing a mental comparison, and recognizing that point P has a certain location *with respect to point O*. In this determination two properties are noted, the distance from O to P (based on the unit distance or grid size of the reference coordinate system) and the *relative* angular orientation of the line OP in the coordinate system. These two properties, magnitude and direction, are precisely those required for a vector. Therefore we can also define the *position of a point* as the *vector from the origin of a specified reference coordinate system to the point*. We choose the symbol $\mathbf{R}_{PO}$ to denote the vector position of point P relative to point O, which is read *the position of P with respect to O*.

The reference system is, therefore, related in a very special way to what is seen by a specific observer. What is the relationship? What properties must this coordinate system have to ensure that position measurements made with respect to it are actually those of this observer? The key is that the coordinate system *must* be stationary with respect to this particular observer. Or, to phrase it in another way, the observer is always *stationary in* this reference system. This means that if the observer moves, the coordinate system moves too—through a rotation, a distance, or both. If there are objects or points fixed in this coordinate system, then these objects always appear stationary to the observer regardless of what movements the observer (and the reference system) may execute. Their positions with respect to the observer do not change, and hence their position vectors remain unchanged.

The actual location of the observer within the frame of reference has no meaning because the positions of observed points are always defined with respect to the origin of the coordinate system.

Often it is convenient to express the position vector in terms of its components along the coordinate axes:

$$\mathbf{R}_{PO} = R_{PO}^x \hat{\mathbf{i}} + R_{PO}^y \hat{\mathbf{j}} + R_{PO}^z \hat{\mathbf{k}} \qquad (2.3)$$

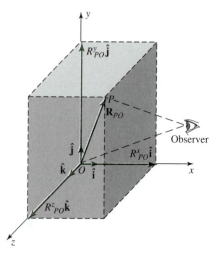

Figure 2.3 The position of a point defined by vectors.

where superscripts are used to denote the direction of each component. As in the remainder of this book, $\hat{\mathbf{i}}, \hat{\mathbf{j}}$, and $\hat{\mathbf{k}}$ are used to designate unit vectors in the x, y, and z axis directions, respectively. While vectors are denoted throughout the book by boldface symbols, the scalar magnitude of a vector is signified by the same symbol in italic, without boldface. The magnitude of the position vector, for example, is

$$R_{PO} = |\mathbf{R}_{PO}| = \sqrt{\mathbf{R}_{PO} \cdot \mathbf{R}_{PO}} = \sqrt{\left(R_{PO}^x\right)^2 + \left(R_{PO}^y\right)^2 + \left(R_{PO}^z\right)^2} \qquad (2.4)$$

The unit vector in the direction of $\mathbf{R}_{PO}$ is denoted by the same boldface symbol with a caret:

$$\hat{\mathbf{R}}_{PO} = \frac{\mathbf{R}_{PO}}{R_{PO}} \qquad (2.5)$$

2.3 POSITION DIFFERENCE BETWEEN TWO POINTS

We now investigate the relationship between the position vectors of two different points. The situation is shown in Fig. 2.4. In the last section, we found that an observer fixed in the xyz coordinate system would observe the positions of points P and Q by comparing each with the position of the origin. The positions of the two points are defined by the vectors $\mathbf{R}_{PO}$ and $\mathbf{R}_{QO}$.

Inspection of the figure shows that they are related by a third vector $\mathbf{R}_{PQ}$, the *position difference* between points P and Q. The figure shows the relationship to be

$$\mathbf{R}_{PQ} = \mathbf{R}_{PO} - \mathbf{R}_{QO} \qquad (2.6)$$

The physical interpretation is now slightly different from that of the position vector itself. The observer is no longer defining the position of P with respect to the origin but is now defining the position of P with respect to the position of Q. Put another way, the position of point P is being defined as if it were in another coordinate system $x'y'z'$ with origin at Q and axes directed parallel to those of the basic reference frame xyz, as shown in Fig. 2.4. Either point of view can be used for the interpretation, but you should understand both of them because we shall use both in future developments. Finally, it is worth remarking that having

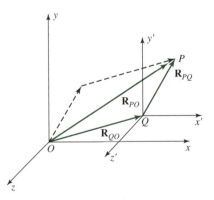

Figure 2.4 Definition of the position-difference vector $\mathbf{R}_{PQ}$.

the axes of $x'y'z'$ parallel to those of xyz is really a convenience and not a necessary condition. This concept will be used throughout the book, however, because it causes no loss of generality and simplifies the visualization when the coordinate systems are in motion.

Having now generalized our concept of relative position to include the position difference between any two points, we reflect again on the above discussion of the position vector itself. We notice that it is merely the special case where we agree to measure using the origin of coordinates as the second point. Thus, to be consistent in notation, we have denoted the position vector of a single point P by the dual subscripted symbol $\mathbf{R}_{PO}$. However, in the interest of brevity, we will henceforth agree that when the second subscript is not given explicitly, it is understood to be the origin of the observer's coordinate system,

$$\mathbf{R}_P = \mathbf{R}_{PO} \tag{2.7}$$

2.4 APPARENT POSITION OF A POINT

Up to now, in discussing the position vector our point of view has been entirely that of a single observer in a single coordinate system. However, it is often desirable to make observations in a secondary coordinate system, that is, as seen by a second observer in a different coordinate system, and then to convert this information into the basic coordinate system. Such a situation is illustrated in Fig. 2.5.

If two observers, one using the reference frame $x_1 y_1 z_1$ and the other using $x_2 y_2 z_2$, were both asked to give the location of a particle at P, they would report different results. The observer in coordinate system $x_1 y_1 z_1$ would observe the vector $\mathbf{R}_{PO_1}$, while the second observer, using the $x_2 y_2 z_2$ coordinate system, would report the position vector $\mathbf{R}_{PO_2}$. We note from Fig. 2.5 that these vectors are related by

$$\mathbf{R}_{PO_1} = \mathbf{R}_{O_2 O_1} + \mathbf{R}_{PO_2} \tag{2.8}$$

The difference in the positions of the two origins is not the only incompatibility between the two observations of the position of point P. Because the two coordinate systems are not aligned, the two observers would be using different reference lines for their measurements of direction; the first observer would report components measured along the $x_1 y_1 z_1$ axes while the second would measure in the $x_2 y_2 z_2$ directions.

A third and very important distinction between these two observations becomes clear when we consider that the two coordinate systems may be moving with respect to each

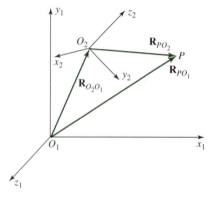

Figure 2.5 Definition of the apparent-position vector $\mathbf{R}_{PO_2}$ of point P.

other. Whereas point P may appear stationary with respect to one observer, it may be in motion with respect to the other; that is, the position vector $\mathbf{R}_{PO_1}$ may appear constant to observer 1 while $\mathbf{R}_{PO_2}$ appears to vary as seen by observer 2.

When any of these conditions exists, it will be convenient to add an additional subscript to our notation which will distinguish which observer is being considered. When we are considering the position of P as seen by the observer using coordinate system $x_1y_1z_1$, we will denote this by the symbol $\mathbf{R}_{PO_1/1}$ or, since O_1 is the origin for this observer,* by $\mathbf{R}_{P/1}$. The observations made by the second observer, done in coordinate system $x_2y_2z_2$, will be denoted as $\mathbf{R}_{PO_2/2}$ or $\mathbf{R}_{P/2}$. With this extension of the notation, Eq. (2.8) becomes

$$\mathbf{R}_{P/1} = \mathbf{R}_{O_2/1} + \mathbf{R}_{P/2} \tag{2.9}$$

We refer to $\mathbf{R}_{P/2}$ as *the apparent position* of point P to an *observer in coordinate system* 2, and we note that it is by no means equal to the apparent position vector $\mathbf{R}_{P/1}$ seen by observer 1.

We have now made note of certain intrinsic differences between $\mathbf{R}_{P/1}$ and $\mathbf{R}_{P/2}$ and found Eq. (2.9) to relate them. However, there is no reason why components of either vector must be taken along the natural axes of the observer's coordinate system. As with all vectors, components can be found along any convenient set of axes.

In applying the apparent-position equation (2.9), we must use a single consistent set of axes during the numerical evaluation. Though the observer in coordinate system 2 would find it most natural to measure the components of $\mathbf{R}_{P/2}$ along the $x_2y_2z_2$ axes, these components must be transformed into the equivalent components in the $x_1y_1z_1$ system before the addition is actually performed:

$$
\begin{aligned}
\mathbf{R}_{P/1} &= \mathbf{R}_{O_2/1} + \mathbf{R}_{P/2} \\
&= R_{O_2/1}^{x_1}\hat{\mathbf{i}}_1 + R_{O_2/1}^{y_1}\hat{\mathbf{j}}_1 + R_{O_2/1}^{z_1}\hat{\mathbf{k}}_1 + R_{P/2}^{x_1}\hat{\mathbf{i}}_1 + R_{P/2}^{y_1}\hat{\mathbf{j}}_1 + R_{P/2}^{z_1}\hat{\mathbf{k}}_1 \\
&= \left(R_{O_2/1}^{x_1} + R_{P/2}^{x_1}\right)\hat{\mathbf{i}}_1 + \left(R_{O_2/1}^{y_1} + R_{P/2}^{y_1}\right)\hat{\mathbf{j}}_1 + \left(R_{O_2/1}^{z_1} + R_{P/2}^{z_1}\right)\hat{\mathbf{k}}_1 \\
&= R_{P/1}^{x_1}\hat{\mathbf{i}}_1 + R_{P/1}^{y_1}\hat{\mathbf{j}}_1 + R_{P/1}^{z_1}\hat{\mathbf{k}}_1
\end{aligned}
$$

The addition can be performed equally well if all vector components are transformed into the $x_2y_2z_2$ system or, for that matter, into any other consistent set of directions. However, they cannot be added algebraically when they have been evaluated along inconsistent axes. The additional subscript in the apparent position vector, therefore, does not specify a set of directions to be used in the evaluation of components; it merely states the coordinate system in which the vector is defined, the coordinate in which the observer is stationary.

2.5 ABSOLUTE POSITION OF A POINT

We now turn to the meaning of absolute position. We saw in Section 2.2 that every position vector is defined relative to a second point, the origin of the observer's coordinate reference frame. It is a special case of the position-difference vector studied in Section 2.3, where the reference point is the origin of coordinates.

---

*Note that $\mathbf{R}_{PO_2/1}$ cannot be abbreviated as $\mathbf{R}_{P/1}$ because O_2 is not the origin used by observer 1.

In Section 2.4 we noted that it may be convenient in certain problems to consider the apparent positions of a single point as viewed by more than one observer using different coordinate systems. When a particular problem leads us to consider multiple coordinate systems, however, the application will lead us to single out one of the coordinate systems as primary or most basic. Most often this is the coordinate system in which the final result is to be expressed, and this coordinate system is usually considered stationary. It is referred to as the *absolute coordinate system*. The absolute position of a point is defined as its apparent position as seen by an observer in the *absolute coordinate system*.

Which coordinate system is designated as absolute (most basic) is an arbitrary decision and unimportant in the study of kinematics. Whether the absolute coordinate system is truly stationary is also a moot point because, as we have seen, all position (and motion) information is measured relative to something else; nothing is truly absolute in the strict sense. When analyzing the kinematics of an automobile suspension, for example, it may be convenient to choose an "absolute" coordinate system attached to the frame of the car and to study the motion of the suspension relative to this. It is then unimportant whether the car is moving or not; motions of the suspension relative to the frame would be defined as absolute.

It is common convention to number the absolute coordinate system 1 and to use other numbers for other moving coordinate systems. Because we adopt this convention throughout this book, absolute-position vectors are those apparent-position vectors viewed by an observer in coordinate system 1 and carry symbols of the form $\mathbf{R}_{P/1}$. In the interest of brevity and to reduce the complexity, we will agree that when the coordinate system number is not shown explicitly it is assumed to be 1; thus, $\mathbf{R}_{P/1}$ can be abbreviated as $\mathbf{R}_P$. Similarly, the apparent-position equation (2.9) can be written* as

$$\mathbf{R}_P = \mathbf{R}_{O_2} + \mathbf{R}_{P/2} \tag{2.10}$$

EXAMPLE 2.2

The path of a moving point is defined by the equation $y = 2x^2 - 28$. Find the position difference from point P to point Q if $R_P^x = 4$ and $R_Q^x = -3$.

SOLUTION

The y components of the two vectors can be written as

$$R_P^y = 2(4)^2 - 28 = 4 \quad \text{and} \quad R_Q^y = 2(-3)^2 - 28 = -10$$

Therefore, the two vectors can be written as

$$\mathbf{R}_P = 4\hat{\mathbf{i}} + 4\hat{\mathbf{j}} \quad \text{and} \quad \mathbf{R}_Q = -3\hat{\mathbf{i}} - 10\hat{\mathbf{j}}$$

The position difference from point P to point Q is

$$\mathbf{R}_{QP} = \mathbf{R}_Q - \mathbf{R}_P = -7\hat{\mathbf{i}} - 14\hat{\mathbf{j}} = 15.65\angle 243.4°$$

---

*Reviewing Sections 2.1 through 2.3 will show that the position difference vector $\mathbf{R}_{PQ}$ was treated entirely from the absolute coordinate system and is an abbreviation of the notation $\mathbf{R}_{PQ/1}$. We have no need to treat the completely general case $\mathbf{R}_{PQ/2}$, the apparent-position-difference vector.

2.6 THE LOOP-CLOSURE EQUATION

Our discussion of the position-difference and apparent-position vectors has been quite abstract so far, the intent being to develop a rigorous foundation for the analysis of motion in mechanical systems. Certainly, precision is not without merit, because it this rigor that permits science to predict a correct result in spite of the personal prejudices and emotions of the analyst. However, tedious developments are not interesting unless they lead to applications in problems of real life. Although there are yet many fundamental principles to be discovered, it might be well at this point to show the relationship between the relative-position vectors discussed above and some of the typical linkages met in real machines.

One of the most common and most useful of all mechanisms is the four-bar linkage. One example is the clamping device shown in Fig. 2.6. A brief study of the assembly drawing shows that, as the handle of the clamp is lifted, the clamping bar swings away from the clamping surface, thereby opening the clamp. As the handle is pressed, the clamping bar swings down and the clamp closes again. If we wish to design such a clamp accurately, however, things are not quite so simple. It may be desirable, for example, for the clamp to open at a given rate for a certain rate of lift of the handle. Such relationships are not obvious; they depend on the exact dimensions of the various parts and the relationships or

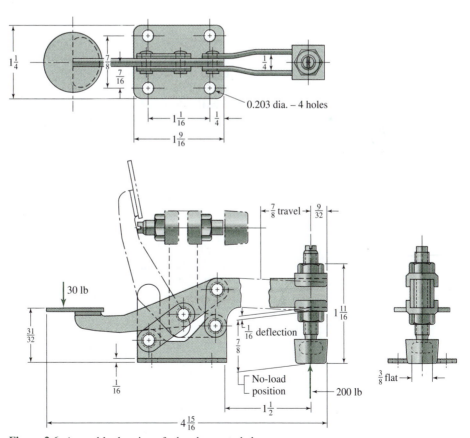

Figure 2.6 Assembly drawing of a hand-operated clamp.

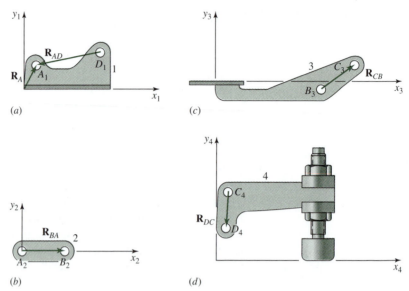

Figure 2.7 Detail drawings of the clamp of Fig. 2.6: (*a*) frame link; (*b*) connecting link; (*c*) handle; and (*d*) clamping bar.

interactions between the parts. To discover these relationships, a rigorous description of the essential features of the device is required. The position-difference and apparent-position vectors can be used to provide such a description.

Figure 2.7 shows the detail drawings of the individual links of the disassembled clamp. Although not shown, the detail drawings would be completely dimensioned, thus fixing once and for all the complete geometry of each link. The assumption that each is a rigid link ensures that the position of any point on any one of the links can be determined precisely relative to any other point on the same link by simply identifying the proper points and scaling the appropriate detail drawing.

The features that are lost in the detail drawings, however, are the interrelationships between the individual parts—that is, the constraints which ensure that each link will move relative to its neighbors in the prescribed fashion. These constraints are, of course, provided by the four pinned joints. Anticipating that they will be of importance in any description of the linkage, we label these pin centers A, B, C, and D and we identify the appropriate points on link 1 as A_1 and D_1, those on link 2 as A_2 and B_2, and so on. As shown in Fig. 2.7, we also pick a different coordinate system rigidly attached to each link.

Because it is necessary to relate the relative positions of the successive joint centers, we define the position difference vectors $\mathbf{R}_{AD}$ on link 1, $\mathbf{R}_{BA}$ on link 2, $\mathbf{R}_{CB}$ on link 3, and $\mathbf{R}_{DC}$ on link 4. We note again that each of these vectors appears constant to an observer fixed in the coordinate system of that particular link; the magnitudes of these vectors are obtainable from the constant dimensions of the links.

A vector equation can also be written to describe the constraints provided by each of the revolute (pinned) joints. Notice that no matter which position or which observer is chosen, the two points describing each pin center—for example, A_1 and A_2—remain coincident. Thus

$$\mathbf{R}_{A_2A_1} = \mathbf{R}_{B_3B_2} = \mathbf{R}_{C_4C_3} = \mathbf{R}_{D_1D_4} = 0 \tag{2.11}$$

Let us now develop vector equations for the absolute position of each of the pin centers. Because link 1 is the frame, absolute positions are those defined relative to an observer in coordinate system 1. Point A_1 is, of course, at the position described by $\mathbf{R}_A$. Next, we mathematically connect link 2 to link 1 by writing

$$\mathbf{R}_{A_2} = \mathbf{R}_{A_1} + \mathbf{R}^{\,0}_{A_2A_1} = \mathbf{R}_A \tag{a}$$

Transferring to the other end of link 2, we attach link 3:

$$\mathbf{R}_B = \mathbf{R}_A + \mathbf{R}_{BA} \tag{b}$$

Connecting joints C and D in the same manner, we obtain

$$\mathbf{R}_C = \mathbf{R}_B + \mathbf{R}_{CB} = \mathbf{R}_A + \mathbf{R}_{BA} + \mathbf{R}_{CB} \tag{c}$$

$$\mathbf{R}_D = \mathbf{R}_C + \mathbf{R}_{DC} = \mathbf{R}_A + \mathbf{R}_{BA} + \mathbf{R}_{CB} + \mathbf{R}_{DC} \tag{d}$$

Finally, we transfer back across link 1 to point A:

$$\mathbf{R}_A = \mathbf{R}_D + \mathbf{R}_{AD} = \mathbf{R}_A + \mathbf{R}_{BA} + \mathbf{R}_{CB} + \mathbf{R}_{DC} + \mathbf{R}_{AD} \tag{e}$$

and from this we obtain

$$\mathbf{R}_{BA} + \mathbf{R}_{CB} + \mathbf{R}_{DC} + \mathbf{R}_{AD} = 0 \tag{2.12}$$

This important equation is called the *loop-closure* equation for the clamp. As shown in Fig. 2.8, it expresses the fact that the mechanism forms a closed loop and therefore the polygon formed by the position-difference vectors through successive links and joints must remain closed as the mechanism moves. The constant lengths of these vectors ensure that the joint centers remain at constant distances apart, the requirement for rigid links. The relative rotations between successive vectors indicate the motions within the pinned joints,

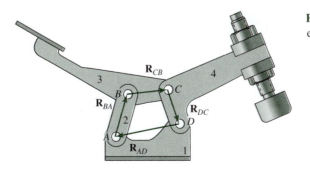

Figure 2.8 The loop-closure equation.

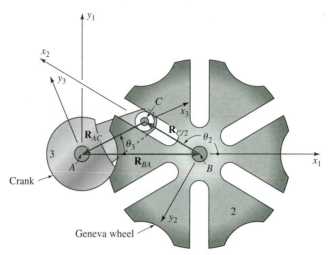

Figure 2.9 The Geneva mechanism or Maltese cross.

while the rotation of each individual position-difference vector shows the rotational motion of a particular link. Thus the loop-closure equation holds within it all the important constraints that determine how this particular clamp operates. It forms a mathematical description, or *model,* of the linkage, and many of the later developments in this book are based on this model as a starting point.

Of course, the form of the loop-closure equation depends on the type of linkage. This is illustrated by another example, the *Geneva* mechanism or *Maltese* cross, shown in Fig. 2.9. One early application of this mechanism was to prevent overwinding a watch. Today it finds wide use as an indexing device—for example, in a milling machine with an automatic tool changer.

Although the frame of the mechanism, link 1, is not shown in the figure, it is an important part of the mechanism because it holds the two shafts with two centers A and B a constant distance apart. Thus we define the vector $\mathbf{R}_{BA}$ to show this dimension. The left crank, link 3, is attached to a shaft, usually rotated at a constant speed, and carries a roller at C, running in the slot of the Geneva wheel. The vector $\mathbf{R}_{AC}$ has a constant magnitude equal to the crank length, the distance from the center of the roller C to the shaft center A. The rotation of this vector relative to link 1 will be used later to describe the angular speed of the crank. The x_2 axis is aligned along one slot of the wheel; thus the roller is constrained to ride along this slot. The vector $\mathbf{R}_{C/2}$ has the same rotation as the wheel, link 2. Also, its changing length $\Delta\mathbf{R}_{C/2}$ shows the relative sliding motion taking place between the roller on link 3 and the slot on link 2.

From the figure we see that the loop-closure equation for this mechanism is

$$\mathbf{R}_{BA} + \mathbf{R}_{C/2} + \mathbf{R}_{AC} = 0 \tag{2.13}$$

Notice that the term $\mathbf{R}_{C/2}$ is equivalent to $\mathbf{R}_{CB}$ because point B is the origin of coordinate system 2.

This form of the loop-closure equation is a valid mathematical model only as long as roller C remains in the slot along x_2. However, this condition does not hold throughout the entire cycle of motion. Once the roller leaves the slot, the motion is controlled by the two mating circular arcs on links 2 and 3. A new form of the loop-closure equation is required for this part of the cycle.

Mechanisms can, of course, be connected together forming a multiple-loop kinematic chain. In such a case more than one loop-closure equation is required to model the system completely. The procedures for obtaining the equations, however, are identical to those illustrated in the above examples.

2.7 GRAPHIC POSITION ANALYSIS

When the paths of the moving points in a mechanism lie in a single plane or in parallel planes, it is called a *planar* mechanism. Because a substantial portion of the investigations in this book deals with planar mechanisms, the development of special methods suited to such problems is justified. As we will see in the following section, the nature of the loop-closure equation often leads to the solution of simultaneous nonlinear equations when approached analytically and can become quite cumbersome. Yet, particularly for planar mechanisms, the solution is usually straightforward when approached graphically.

First let us briefly review the process of vector addition. Any two known vectors **A** and **B** can be added graphically as shown in Fig. 2.10*a*. After a scale is chosen, the vectors are drawn tip to tail in either order and their sum **C** is identified:

$$\mathbf{C} = \mathbf{A} + \mathbf{B} = \mathbf{B} + \mathbf{A} \tag{2.14}$$

Notice that the magnitudes and the directions of both vectors **A** and **B** are used in performing the addition and that both the magnitude and the direction of the sum **C** are found as a result.

The operation of graphical vector subtraction is illustrated in Fig. 2.10*b*, where the vectors are drawn tip to tip in solving the equation

$$\mathbf{A} = \mathbf{C} - \mathbf{B} \tag{2.15}$$

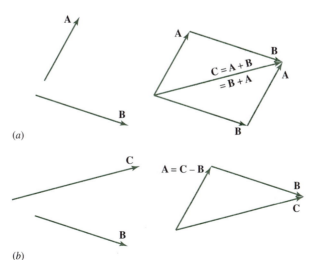

(*a*)

(*b*)

Figure 2.10 (*a*) Vector addition and (*b*) Vector subtraction.

These graphical vectors operations should be studied carefully and understood, because they are used extensively throughout the book.

A three-dimensional vector equation,

$$\mathbf{C} = \mathbf{D} + \mathbf{E} + \mathbf{B} \qquad (a)$$

can be divided into components along any convenient axes, leading to the three scalar equations

$$C^x = D^x + E^x + B^x, \qquad C^y = D^y + E^y + B^y, \qquad C^z = D^z + E^z + B^z \qquad (b)$$

Because they are components of the same vector equation, these three scalar equations must be consistent. If the three are also linearly independent, they can be solved simultaneously for three unknowns, which may be three magnitudes, three directions, or any combination of three magnitudes and directions. For some combinations, however, the problem is highly nonlinear and quite difficult to solve. Therefore, we shall delay consideration of the three-dimensional problem until Chapter 12 where it is needed.

A two-dimensional vector equation can be solved for two unknowns: two magnitudes, two directions, or one magnitude and one direction. Sometimes it is desirable to indicate the known ($\surd$) and unknown (o) quantities above each vector in an equation like this:

$$\begin{array}{cccc} \text{o}\surd & \surd\surd & \surd\surd & \text{o}\surd \\ \mathbf{C} & = \mathbf{D} + & \mathbf{E} + & \mathbf{B} \end{array} \qquad (c)$$

where the first symbol ($\surd$ or o) above each vector indicates its magnitude and the second indicates its direction. Another, equivalent form is

$$\begin{array}{cccc} \text{o}\surd & \surd\surd & \surd\surd & \text{o}\,\surd \\ C\hat{\mathbf{C}} & = D\hat{\mathbf{D}} + & E\hat{\mathbf{E}} + & B\hat{\mathbf{B}} \end{array} \qquad (d)$$

Either of these equations clearly identifies the unknowns and indicates whether a solution can be achieved. In Eq. (c) the vectors $\mathbf{D}$ and $\mathbf{E}$ are completely defined and can be replaced by their sum

$$\mathbf{A} = \mathbf{D} + \mathbf{E} \qquad (e)$$

giving

$$\mathbf{C} = \mathbf{A} + \mathbf{B} \qquad (2.16)$$

In like manner any plane vector equation, if it is solvable, can be reduced to a three-term equation with two unknowns.

Depending on the forms of the two unknowns, four distinct cases occur. These can be classified according to the unknowns; the four cases and the corresponding unknowns are presented in Table 2.2.

We will illustrate the solutions of these four cases graphically.

TABLE 2.2 Planar Vector Equation Unknowns

Case	Unknowns
1	Magnitude and direction of the same vector—for example, C, $\hat{\mathbf{C}}$
2	Magnitude of one vector and direction of another vector—for example, A, $\hat{\mathbf{B}}$
3	Magnitudes of two different vectors—for example, A, B
4	Directions of two different vectors—for example, $\hat{\mathbf{A}}$, $\hat{\mathbf{B}}$

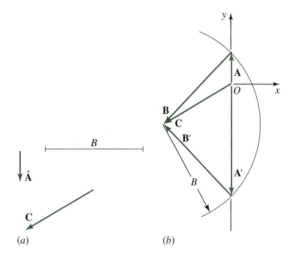

Figure 2.11 Graphical solution of case 2: (a) given $\mathbf{C}$, $\hat{\mathbf{A}}$, and B; (b) solution for A, $\hat{\mathbf{B}}$ and A', $\hat{\mathbf{B}}'$.

(a)   (b)

In case 1 the two unknowns are the magnitude and the direction of the same vector. This case can be solved by straightforward graphical addition or subtraction of the remaining vectors, which are completely defined. This case was illustrated in Fig. 2.10.

For case 2 a magnitude and a direction from different vectors—say, A and $\hat{\mathbf{B}}$—are to be found:

$$\overset{\sqrt{\sqrt{}} \quad \sqrt{o} \quad \sqrt{o}}{\mathbf{C} = \mathbf{A} + \mathbf{B}} \tag{2.17}$$

The solution, shown in Fig. 2.11, is obtained as follows:

1. Choose a coordinate system and scale factor and draw vector $\mathbf{C}$.
2. Construct a line through the origin of $\mathbf{C}$ parallel to $\hat{\mathbf{A}}$.
3. Adjust a compass to the scaled magnitude B and construct a circular arc with center at the terminus of $\mathbf{C}$.
4. The two intersections of the line and the arc define the two sets of solutions A, $\hat{\mathbf{B}}$ and A', $\hat{\mathbf{B}}'$.

For case 3, two magnitudes—say, A and B—are to be found:

$$\overset{\sqrt{\sqrt{}} \quad \sqrt{o} \quad \sqrt{o}}{\mathbf{C} = \mathbf{A} + \mathbf{B}} \tag{2.18}$$

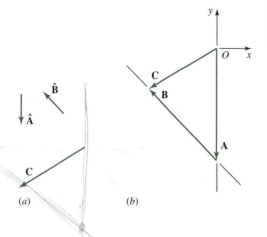

Figure 2.12 Graphical solution of case 3: (*a*) given **C**, **Â**, and **B̂**; (*b*) solution for *A* and *B*.

The solution, shown in Fig. 2.12, is obtained as follows:

1. Choose a coordinate system and scale factor and draw vector **C**.
2. Construct a line through the origin of **C** parallel to **Â**.
3. Construct another line through the terminus of **C** parallel to **B̂**.
4. The intersection of these two lines defines both magnitudes, *A* and *B*, which may be either positive or negative.

Note that case 3 has a unique solution unless the lines are collinear; if the lines are parallel but distinct, the magnitudes *A* and *B* are both infinite.

Finally, for case 4 the directions of two vectors, **Â** and **B̂**, are to be found:

$$\mathbf{C} = A + B \tag{2.19}$$

The steps in the solution are illustrated in Fig. 2.13.

1. Choose a coordinate system and scale factor and draw vector **C**.
2. Construct a circular arc of radius *A* centered at the origin of **C**.
3. Construct a circular arc of radius *B* centered at the terminus of **C**.
4. The two intersections of these arcs define the two sets of solutions **Â**, **B̂** and **Â′**, **B̂′**. Note that a real solution is possible only if **A** + **B** ≥ **C**.

The Slider-Crank Linkage   Let us now apply the procedures of this section to the solution of the loop-closure equation for the slider-crank linkage of Fig. 2.14*a*. Here link 2 is the crank, constrained to rotate about the fixed pivot *O*. Link 3 is the connecting rod and link 4 is the slider. In Fig. 2.14*b* we have replaced the link centerlines with vectors **R**$_A$ for the crank, **R**$_B$ for the frame, and **R**$_{BA}$ for the connecting rod. Then the loop-closure equation is

$$\mathbf{R}_B = \mathbf{R}_A + \mathbf{R}_{BA} \tag{f}$$

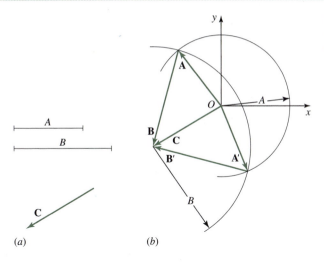

Figure 2.13 Graphical solution of case 4: (*a*) given **C**, *A*, and *B*; (*b*) solution for $\hat{\mathbf{A}}$, $\hat{\mathbf{B}}$ and $\hat{\mathbf{A}}'$, $\hat{\mathbf{B}}'$.

(*a*)   (*b*)

The problem of position analysis is to determine the values of all position variables (the positions of all points and joints) given the dimensions of each link and the value(s) of the independent variable(s), those chosen to represent the degree(s) of freedom of the mechanism.

In the slider-crank mechanism of Fig. 2.14 we choose to specify the position of slider 4. Then the unknowns are the angles θ_2 and θ_3 corresponding to the direction vectors

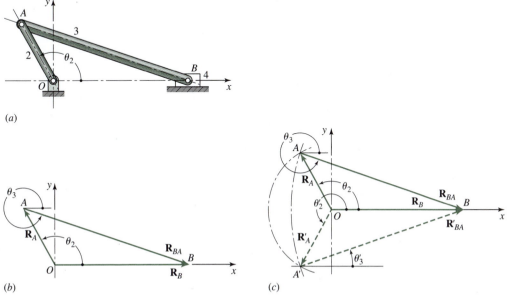

(*a*)

(*b*)   (*c*)

Figure 2.14 (*a*) Slider-crank mechanism with distance OB given and θ_2 and θ_3 the unknowns. (*b*) Vectors replace the link centerlines. (*c*) Position analysis using graphics.

$\hat{\mathbf{R}}_B$ and $\hat{\mathbf{R}}_{BA}$. After identifying the known dimensions of the links, we can write

$$\mathbf{R}_B = \mathbf{R}_A + \mathbf{R}_{BA} \tag{g}$$

We recognize this as case 4 of the loop-closure equation (see Table 2.1). The graphical solution procedure explained in Fig. 2.13 is now carried out in Fig. 2.14c. We note that two possible solutions are found, θ_2 and θ_3 and θ_2' and θ_3', and correspond to two different configurations of the linkage—that is, two positions of the links, both of which are consistent with the given position of the slider. Of course in a graphical solution we know in advance which of the two solutions is desired. In an analytical solution both sets of results are equally valid roots to the loop-closure equation, and it is necessary to choose between them according to the application.

The Four-Bar Linkage   The loop-closure equation for the four-bar linkage of Fig. 2.15a is

$$\mathbf{R}_A + \mathbf{R}_{BA} = \mathbf{R}_C + \mathbf{R}_{BC} \tag{h}$$

and the position of point P is given by the position-difference equation

$$\mathbf{R}_P = \mathbf{R}_A + \mathbf{R}_{PA} \tag{i}$$

Equation (i) contains three unknowns, but it can be reduced to two after Eq. (h) is solved by noticing the constant angular relationship between $\hat{\mathbf{R}}_{PA}$ and $\hat{\mathbf{R}}_{BA}$,

$$\theta_5 = \theta_3 + \alpha \tag{j}$$

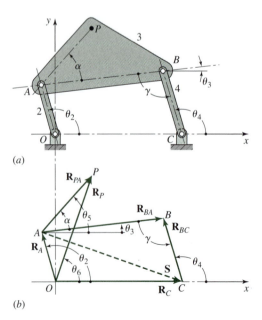

(a)

(b)

Figure 2.15 (a) Four-bar linkage showing coupler point P. The dimensions of the links together with the position of link 2 are given. The problem is that of defining the positions of the remaining links and of point P.
(b) Vector diagram showing the graphical solution for the open configuration of the linkage.

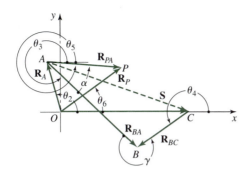

Figure 2.16 Vector diagram showing the graphical solution for the crossed configuration of the linkage of Fig. 2.15*a*.

The graphical solution for this problem is started by combining the two unknown terms of Eq. (h), thus locating the positions of points *A* and *C* as shown in Figs. 2.15*b* and 2.16.

$$\mathbf{S} = \mathbf{R}_C - \mathbf{R}_A = \mathbf{R}_{BA} - \mathbf{R}_{BC} \tag{k}$$

The solution procedure for case 4 (that is, two unknown directions) is then used to locate point *B*. Two possible solutions are found: the open configuration in Fig. 2.15*b* and the crossed configuration in Fig. 2.16.

Next we apply Eq. (*j*) to determine the two possible directions of $\hat{\mathbf{R}}_{PA}$. Equation (*i*) can then be solved by the procedures for case 1. Two solutions are finally achieved, as shown in Figs. 2.15*b* and 2.16, for the position of point *P*. Both are valid solutions to Eqs. (*h*) through (*j*), although the crossed configuration could not be obtained, in this instance, from the open configuration shown in Fig. 2.15*a* without first disassembling the mechanism.

From the slider-crank and four-bar linkage examples it is clear that graphical position analysis requires precisely the same constructions that would be chosen naturally in drafting a scale drawing of the mechanism at the position being considered. For this reason the procedure seems trivial and not truly worthy of the title "analysis." Yet this is highly misleading. As we shall see in the coming sections, the position analysis of a mechanism is a nonlinear algebraic problem when approached by analytical or computer methods. It is, in fact, the most difficult problem in kinematic analysis, and this is one primary reason that graphical solution techniques have retained their attraction for planar mechanism analysis.

2.8 ALGEBRAIC POSITION ANALYSIS

In this section we present the classical approach used in the position analysis of the slider-crank mechanism. Figure 2.17 shows the offset version that has been chosen for analysis. By making the offset equal to zero, the same equations can be used for the centered or symmetrical version. The notation in Fig. 2.17 shows that the connecting rod angle θ_3 has been selected in a manner that avoids use of negative angles in calculator and computer calculations.

The two problems that occur in the position analysis of the slider-crank mechanism are:

PROBLEM 1   Given the input angle θ_2, find the connecting rod angle θ_3 and the position x_B.

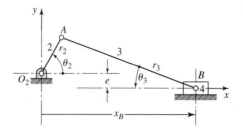

Figure 2.17 Notation used for the offset slider-crank mechanism.

PROBLEM 2   Given the position x_B, find the input angle θ_2 and the connecting rod angle θ_3.

Starting with problem 1, we define the position of point A with the equation set

$$x_A = r_2 \cos \theta_2 \quad \text{and} \quad y_A = r_2 \sin \theta_2 \tag{2.20}$$

Next, we note that

$$r_2 \sin \theta_2 = r_3 \sin \theta_3 - e \tag{a}$$

so that

$$\sin \theta_3 = \frac{1}{r_3}(e + r_2 \sin \theta_2) \tag{2.21}$$

From the geometry of Fig. 2.17 we see that

$$x_B = r_2 \cos \theta_2 + r_3 \cos \theta_3 \tag{b}$$

Then using the trigonometric identity

$$\cos \theta_3 = \pm(1 - \sin^2 \theta_3)^{1/2}$$

we have, from Eq. (2.21),

$$\cos \theta_3 = \pm \frac{1}{r_3} \left[r_3^2 - (e + r_2 \sin \theta_2)^2 \right]^{1/2} \tag{c}$$

We arbitrarily select the positive sign, which we see from Fig. (2.17) corresponds to the solution with the piston to the right of the crank pin. Then, substituting Eq. (c) into Eq. (b) gives

$$x_B = r_2 \cos \theta_2 + \left[r_3^2 - (e + r_2 \sin \theta_2)^2 \right]^{1/2} \tag{2.22}$$

Thus, with the angle θ_2 given, the unknowns θ_3 and x_B can be obtained by solving Eqs. (2.21) and (2.22).

Problem 2 requires that, given x_B, we solve Eq. (2.22) for the angle θ_2. This requires the use of a calculator or a computer together with a root-finding technique. Here we select

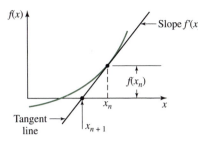

Figure 2.18 The Newton–Raphson method.

the well-known Newton–Raphson method.[1] This method can be explained by reference to Fig. 2.18. This figure is a graph of some function $f(x)$ versus x. Let x_n be a first approximation or a rough estimate of the root where $f(x) = 0$ which we wish to find. A tangent line drawn to the curve at $x = x_n$ intersects the x axis at x_{n+1}, which is a better approximation to the root. The slope of the tangent line is the derivative of the function at $x = x_n$ and is

$$f'(x_n) = \frac{f(x_n)}{x_n - x_{n+1}} \tag{d}$$

Solving for x_{n+1} gives

$$x_{n+1} = x_n - \frac{f(x_n)}{f'(x_n)} \tag{2.23}$$

Using a computer, for example, we can start a solution by entering a good estimate of x_n, solve for x_{n+1}, use this as the next estimate, and repeat this process as many times as are needed to get the result with satisfactory accuracy. This accuracy is evaluated by comparing $x_{n+1} - x_n$ with a small number ε on every run and stopping when $(x_{n+1} - x_n) < \varepsilon$.

We note that the root-finding programs built into some calculators utilize an approximation to obtain $f'(x_n)$. These sometimes will be of no value in solving Eq. (2.22), so we proceed as follows. In Eq. (2.22), replace the angle θ_2 with x, and let r_2, r_3, e, and x_B be given constants. Then

$$f(x) = r_2 \cos x + \left[r_3^2 - (e + r_2 \sin x)^2\right]^{1/2} - x_B \tag{e}$$

and

$$f'(x) = -r_2 \sin x - \frac{(e + r_2 \sin x)\cos x}{r_3^2 - (e + r_2 \sin x)^2} \tag{f}$$

These two equations can now be programmed with Eq. (2.23) to solve for the unknown value of the angle θ_2 when x_B is given. We note that the angle θ_2 will have two possible values. These may be found separately by using appropriate initial estimates.

An algebraic solution is possible if the eccentricity e is zero—that is, if the linkage is centered. For this case take Eqs. (a) and (b), square them, and add them together. Noting a trigonometric identity, the result is found to be

$$x_B^2 - 2x_B r_2 \cos\theta_2 + r_2^2 = r_3^2 \tag{g}$$

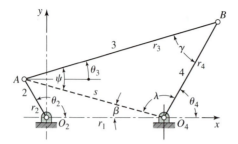

Figure 2.19 The crank-and-rocker four-bar linkage.

Solving for θ_2 gives

$$\theta_2 = \cos^{-1} \frac{x_B^2 + r_2^2 - r_3^2}{2x_B r_2} \tag{2.24}$$

The solution of problem 1 for the centered version is, of course, obtained directly from Eq. (2.22) by making $e = 0$.

The Crank-and-Rocker Mechanism   The four-bar linkage shown in Fig. 2.19 is called the *crank-and-rocker* mechanism. Thus link 2, which is the crank, can rotate in a full circle; but the rocker, link 4, can only oscillate. We shall generally follow the accepted practice of designating the frame or fixed link as link 1. Link 3 in Fig. 2.19 is called the *coupler* or *connecting rod*. With the four-bar linkage the position problem generally consists of finding the positions of the coupler and output link or rocker when the dimensions of all the members are given together with the crank position.

To obtain the analytical solution we designate s as the distance AO_4 in Fig. 2.19. The cosine law can then be written twice for each of the two triangles O_4O_2A and ABO_4. In terms of the angles and link lengths shown in the figure we then have

$$s = \left[r_1^2 + r_2^2 - 2r_1 r_2 \cos \theta_2 \right]^{1/2} \tag{2.25}$$

$$\beta = \cos^{-1} \frac{r_1^2 + s^2 - r_2^2}{2r_1 s} \tag{2.26}$$

$$\psi = \cos^{-1} \frac{r_3^2 + s^2 - r_4^2}{2r_3 s} \tag{2.27}$$

$$\lambda = \cos^{-1} \frac{r_4^2 + s^2 - r_3^2}{2r_4 s} \tag{2.28}$$

There will generally be two values of λ corresponding to each value of θ_2. If θ_2 is in the range $0 \leq \theta_2 \leq \pi$, the unknown directions are taken as

$$\theta_3 = \psi - \beta \tag{2.29}$$

$$\theta_4 = \pi - \lambda - \beta \tag{2.30}$$

However, if θ_2 is in the range $\pi \leq \theta_2 \leq 2\pi$, then

$$\theta_3 = \psi + \beta \tag{2.31}$$

$$\theta_4 = \pi - \lambda + \beta \tag{2.32}$$

Finally, it is worth noting that Eqs. (2.27) and (2.28) yield double values too, because they use arc cosines. These will always be positive and negative pairs of values; the positive values correspond to the open configuration shown, while the negative values correspond to the crossed closure.

In Section 1.10 the concept of transmission angle was discussed in connection with the subject of mechanical advantage. With Fig. 2.19 we see that this angle is given by the equation

$$\gamma = \pm \cos^{-1} \frac{r_3^2 + r_4^2 - s^2}{2r_3 r_4} \tag{2.33}$$

2.9 COMPLEX-ALGEBRA SOLUTIONS OF PLANAR VECTOR EQUATIONS

In planar problems it is often desirable to express a vector by specifying its magnitude and direction in *polar notation:*

$$\mathbf{R} = R \angle \theta \tag{a}$$

In Fig. 2.20*a*, the two-dimensional vector

$$\mathbf{R} = R^x \hat{\mathbf{i}} + R^y \hat{\mathbf{j}} \tag{2.34}$$

has two rectangular components of magnitudes

$$R^x = R \cos\theta \quad \text{and} \quad R^y = R \sin\theta \tag{2.35}$$

with

$$R = \sqrt{(R^x)^2 + (R^y)^2} \quad \text{and} \quad \theta = \tan^{-1} \frac{R^y}{R^x} \tag{2.36}$$

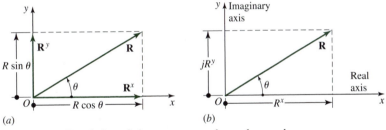

Figure 2.20 Correlation of planar vectors and complex numbers.

Note that we have made the arbitrary choice here of accepting the positive square root for the magnitude R when calculating from the components of **R**. Therefore we must be careful to interpret the signs of R^x and R^y individually when deciding upon the quadrant of θ. Note that *θ is defined as the angle from the positive x axis to the positive end of vector* **R**, *measured about the origin of the vector, and is positive when measured counterclockwise.*

EXAMPLE 2.3

Express the vectors $\mathbf{A} = 10\angle 30°$ and $\mathbf{B} = 8\angle -15°$ in rectangular notation* and find their sum.

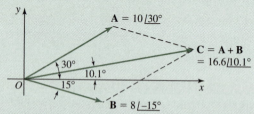

Figure 2.21 Example 2.3.

SOLUTION

The vectors are shown in Fig. 2.21 and are

$$\mathbf{A} = 10\cos 30°\hat{\mathbf{i}} + 10\sin 30°\hat{\mathbf{j}} = 8.66\hat{\mathbf{i}} + 5.00\hat{\mathbf{j}} \qquad \textit{Ans.}$$

$$\mathbf{B} = 8\cos(-15°)\hat{\mathbf{i}} + 8\sin(-15°)\hat{\mathbf{j}} = 7.73\hat{\mathbf{i}} - 2.07\hat{\mathbf{j}} \qquad \textit{Ans.}$$

$$\mathbf{C} = \mathbf{A} + \mathbf{B} = (8.66 + 7.73)\hat{\mathbf{i}} + (5.00 - 2.07)\hat{\mathbf{j}}$$

$$= 16.39\hat{\mathbf{i}} + 2.93\hat{\mathbf{j}}$$

The magnitude of the resultant is found from Eq. (2.36),

$$C = \sqrt{16.39^2 + 2.93^2} = 16.6$$

as is the angle,

$$\theta = \tan^{-1} \frac{2.93}{16.39} = 10.1°$$

The final result in planar notation is

$$\mathbf{C} = 16.6\angle 10.1° \qquad \textit{Ans.}$$

---

*Many calculators are equipped to perform polar-rectangular and rectangular-polar conversions directly.

2.10 COMPLEX POLAR ALGEBRA

Another way of treating two-dimensional vector problems analytically makes use of complex algebra. Although complex numbers are not vectors, they can be used to represent vectors in a plane by choosing an origin and real and imaginary axes. In two-dimensional kinematics problems, these axes can conveniently be chosen coincident with the $x_1 y_1$ axes of the absolute coordinate system.

As shown in Fig. 2.20b, the location of any point in the plane can be specified either by its absolute-position vector or by its corresponding real and imaginary coordinates,

$$\mathbf{R} = R^x + jR^y \tag{2.37}$$

where the operator j is defined as the unit imaginary number,

$$j = \sqrt{-1} \tag{2.38}$$

The real usefulness of complex numbers in planar analysis stems from the ease with which they can be switched to polar form. Employing complex rectangular notation for the vector $\mathbf{R}$, we can write

$$\mathbf{R} = R\angle\theta = R\cos\theta + jR\sin\theta \tag{2.39}$$

But using the well-known Euler equation from trigonometry,

$$e^{\pm j\theta} = \cos\theta \pm j\sin\theta \tag{2.40}$$

we can also write $\mathbf{R}$ in complex polar form as

$$\mathbf{R} = Re^{j\theta} \tag{2.41}$$

where the magnitude and direction of the vector appear explicitly. As we will see in the next two chapters, expression of a vector in this form is especially useful when differentiation is required.

Some familiarity with useful manipulation techniques for vectors written in complex polar forms can be gained by solving the four cases of the loop-closure equation again. Writing Eq. (2.16) in complex polar form, we have

$$Ce^{j\theta_C} = Ae^{j\theta_A} + Be^{j\theta_B} \tag{2.42}$$

In case 1 the two unknowns are C and θ_C. We begin the solution by separating the real and imaginary parts of the equation. By substituting Euler's equation (2.40), we obtain

$$C(\cos\theta_C + j\sin\theta_C) = A(\cos\theta_A + j\sin\theta_A) + B(\cos\theta_B + j\sin\theta_B) \tag{a}$$

Upon equating the real terms and the imaginary terms separately, we obtain two real equations corresponding to the horizontal and vertical components of the two-dimensional vector equation:

$$C \cos \theta_C = A \cos \theta_A + B \cos \theta_B \tag{b}$$

$$C \sin \theta_C = A \sin \theta_A + B \sin \theta_B \tag{c}$$

By squaring and adding these two equations, θ_C is eliminated and a solution is found for C:

$$C = \sqrt{A^2 + B^2 + 2AB \cos(\theta_B - \theta_A)} \tag{2.43}$$

The positive square root is chosen arbitrarily; the negative square root would yield a negative solution for C with θ_C differing by $180°$. The angle θ_C is found from

$$\theta_C = \tan^{-1} \frac{A \sin \theta_A + B \sin \theta_B}{A \cos \theta_A + B \cos \theta_B} \tag{2.44}$$

where the signs of the numerator and denominator must be considered separately in determining the proper quadrant of θ_C. Only a single solution is found for case 1, as previously illustrated in Fig. 2.10.

The graphical solution to case 2 was shown in Fig. 2.11. The two unknowns are A and θ_B. One convenient way of solving in complex polar form is first to divide Eq. (2.42) by $e^{j\theta_A}$:

$$Ce^{j(\theta_C - \theta_A)} = A + Be^{j(\theta_B - \theta_A)} \tag{d}$$

Comparing this equation with Fig. 2.22, we see that division by the complex polar form of a unit vector $e^{j\theta_A}$ has the effect of rotating the real and imaginary axes by the angle $-\theta_A$ such that the real axis lies along the vector **A**. We can now use Euler's equation (2.40) to separate the real and imaginary components,

$$C \cos(\theta_C - \theta_A) = A + B \cos(\theta_B - \theta_A) \tag{e}$$

$$C \sin(\theta_C - \theta_A) = B \sin(\theta_B - \theta_A) \tag{f}$$

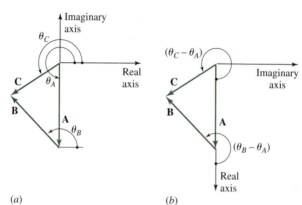

Figure 2.22 Rotation of axes by division of complex polar equations by $e^{-j\theta_A}$: (*a*) original axes; (*b*) rotated axes.

(*a*)   (*b*)

The solutions are then obtained directly from Eqs. (f) and (e), respectively:

$$\theta_B = \theta_A + \sin^{-1} \frac{C \sin(\theta_C - \theta_A)}{B} \tag{2.45}$$

$$A = C \cos(\theta_C - \theta_A) - B \cos(\theta_B - \theta_A) \tag{2.46}$$

The solutions shown by Eqs. (2.45) and (2.46) are intentionally shown in this order because it is noted that Eq. (2.46) cannot be evaluated numerically until after θ_B is found. We also note that the arc sine term in Eq. (2.45) is double-valued and therefore case 2 has two distinct solutions, A, θ_B and A', θ'_B; these are shown in Fig. 2.11.

For case 3 the two unknowns of Eq. (2.42) are the two magnitudes A and B. The graphical solution to this case was shown in Fig. 2.12. Again we begin by aligning the real axis along vector **A** by dividing Eq. (2.42) by $e^{j\theta_A}$ as was done to obtain Eq. (d) before. Separating the real and imaginary components leads again to Eqs. (e) and (f) where we note that the vector **A**, now real, has been totally eliminated from Eq. (f). The solution for B is now easily found:

$$B = C \frac{\sin(\theta_C - \theta_A)}{\sin(\theta_B - \theta_A)} \tag{2.47}$$

The solution for the other unknown magnitude A is found in completely analogous fashion. Dividing Eq. (2.42) by $e^{j\theta_B}$ aligns the real axis along vector **B**. That equation is then separated into real and imaginary parts and yields

$$A = C \frac{\sin(\theta_C - \theta_B)}{\sin(\theta_A - \theta_B)} \tag{2.48}$$

As before, case 3 yields a unique solution.

Case 4 has the two angles θ_A *and* θ_B as unknowns. The graphical solution was shown in Fig. 2.13. In this case we align the real axis along vector **C**:

$$\mathbf{C} = A e^{j(\theta_A - \theta_C)} + B e^{j(\theta_B - \theta_C)} \tag{g}$$

Using Euler's equation to separate components and then rearranging terms, we obtain

$$A \cos(\theta_A - \theta_C) = C - B \cos(\theta_B - \theta_C) \tag{h}$$

$$A \sin(\theta_A - \theta_C) = -B \sin(\theta_B - \theta_C) \tag{i}$$

Squaring both equations and adding results in

$$A^2 = C^2 + B^2 - 2BC \cos(\theta_B - \theta_C) \tag{j}$$

We recognize this as the law of cosines for the vector triangle. It can be solved for θ_B:

$$\theta_B = \theta_C \mp \cos^{-1} \frac{C^2 + B^2 - A^2}{2BC} \tag{2.49}$$

Putting C on the other side of Eq. (h) before squaring and adding results in another form of the law of cosines, from which

$$\theta_A = \theta_C \pm \cos^{-1}\frac{C^2 + A^2 - B^2}{2AC} \tag{2.50}$$

The plus-or-minus signs in these two equations are a reminder that the arc cosines are each double-valued and therefore θ_B and θ_A each have two solutions. These two pairs of angles can be paired naturally together as θ_A, θ_B and θ'_A, θ'_B under the restriction of Eq. (i) above. Thus case 4 has two distinct solutions, as shown in Fig. 2.13.

2.11 POSITION ANALYSIS TECHNIQUES

EXAMPLE 2.4

Make a position analysis of the sliding-block linkage of Fig. 2.23 by finding θ_4 and the distance $O_4 A$.

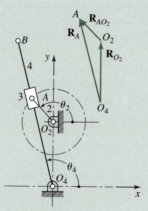

Figure 2.23 A sliding-block linkage, $O_4 O_2 = 9.0$ in, $O_2 A = 4.5$ in, $\theta_2 = 135°$.

SOLUTION

Though you may be able to think of others, here we choose to employ five different approaches, for illustrative purposes, to solve the problem. These are:

(a) Graphic approach
(b) Analytic approach
(c) Complex algebraic approach
(d) Vector algebraic approach
(e) Numeric approach

First, using the vector diagram in Fig. 2.23 we recognize this as a case 1 problem, and so

$$\overset{\text{oo}}{\mathbf{R}_A} = \overset{\sqrt{\sqrt{}}}{\mathbf{R}_{O_2}} + \overset{\sqrt{\sqrt{}}}{\mathbf{R}_{AO_2}} \tag{1}$$

Approach a   Figure 2.23 was originally drawn at a space scale of 6 in/in. By direct measurement of the figure we find

$$\theta_4 = 105.3° \qquad\qquad \textit{Ans.}$$

$$O_4 A = 2.08(6) = 12.48 \text{ in} \qquad\qquad \textit{Ans.}$$

Approach b   For case 1 we employ Eqs. (2.43) and (2.44); thus

$$R_A = \left[R_{O_2}^2 + R_{AO_2}^2 + 2R_{O_2}R_{AO_2}\cos(\theta_2 - 90°) \right]^{1/2}$$

$$= [9^2 + 4.5^2 + 2(9)(4.5)\cos(135° - 90°)]^{1/2}$$

$$= 12.59 \text{ in} \qquad\qquad \textit{Ans.}$$

$$\theta_4 = \tan^{-1} \frac{R_{O_2}\sin 90° + R_{AO_2}\sin\theta_2}{R_{O_2}\cos 90° + R_{AO_2}\cos\theta_2}$$

$$= \tan^{-1} \frac{9\sin 90° + 4.5\sin 135°}{9\cos 90° + 4.5\cos 135°} = \tan^{-1}\left(\frac{12.182}{-3.182} \right)$$

$$\theta_4 = -75.36°$$

However, the calculator used here does not recognize that the negative sign in the denominator indicates an angle in the second quadrant. Therefore

$$\theta_4 = 180° - 75.36° = 104.64° \qquad\qquad \textit{Ans.}$$

Approach c   The terms for Eq. (1) are

$$\mathbf{R}_A = R_A \angle \theta_4 = R_A \cos\theta_4 + jR_A \sin\theta_4$$

$$\mathbf{R}_{O_2} = 9\angle 90° = 9\cos 90° + j9\sin 90° = 0 + j9 \text{ in}$$

$$\mathbf{R}_{AO_2} = 4.5\angle 135° = 4.5\cos 135° + j4.5\sin 135°$$

$$= -3.182 + j3.182 \text{ in}$$

Substituting these into Eq. (1) yields

$$\mathbf{R}_A = (0 + j9) + (-3.182 + j3.182) = -3.182 + j12.182 \text{ in}$$

Thus

$$R_A = [(-3.182)^2 + (12.182)^2]^{1/2} = 12.59 \text{ in} \qquad\qquad \textit{Ans.}$$

and

$$\theta_4 = \tan^{-1} \frac{y}{x} = \tan^{-1} \frac{12.182}{-3.182} = 104.64° \qquad \textit{Ans.}$$

However, note the comment above regarding arc tangent results from the calculator.

Approach d   Here we employ a scientific hand-calculator, which will add and subtract complex numbers in rectangular notation and will convert from rectangular to polar notation or vice versa. Thus

$$\mathbf{R}_{O_2} = 9\angle 90° = 0 + j9 \text{ in}$$

$$\mathbf{R}_{AO_2} = 4.5\angle 135° = -3.182 + j3.182 \text{ in}$$

$$\mathbf{R}_A = \mathbf{R}_{O_2} + \mathbf{R}_{AO_2} = (0 + j9) + (-3.182 + j3.182)$$

$$= -3.182 + j12.182 = 12.59\angle 104.64° \text{ in} \qquad \textit{Ans.}$$

Approach e   Some scientific hand-calculators permit equations containing complex numbers to be solved in either rectangular or polar form or both with the results displayed in either mode. Using this calculator, we enter

$$\mathbf{R}_A = 9\angle 90° + 4.5\angle 135° = 12.59\angle 104.64° \text{ in} \qquad \textit{Ans.}$$

EXAMPLE 2.5

Perform a position analysis of the four-bar linkage shown in Fig. 2.24 by finding θ_3 and θ_4 for the two positions shown.

SOLUTION

For the open position we refer to Fig. 2.24b and observe that

$$\mathbf{R}_A = 0.140\angle 150° \text{ m}, \quad \mathbf{R}_{O_4} = 0.600\angle 0° \text{ m}$$

Therefore

$$\mathbf{S} = \mathbf{R}_{O_4} - \mathbf{R}_A = 0.600\angle 0° - 0.140\angle 150° = 0.725\angle -5.54° \text{ m}$$

Referring again to Fig. 2.24b, we note that the vectors in the triangle ABO_4 are related by the equation

$$\overset{\sqrt{\sqrt{}}}{\underset{\mathbf{S}}{}} \quad \overset{\sqrt{}_O}{} \quad \overset{\sqrt{}_O}{} = \mathbf{R}_{BA} - \mathbf{R}_{BO_4} \qquad (a)$$

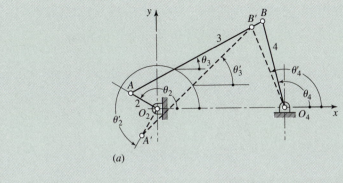

(a)

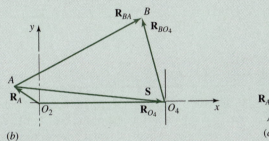

(b)

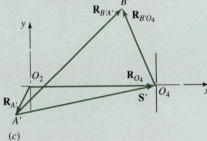

(c)

Figure 2.24 Crank-and-rocker linkage shown in two different input positions. $O_2 O_4 = 600$ mm, $O_2 A = 140$ mm, $AB = 690$ mm, $O_4 B = 400$ mm, $\theta_2 = 150°$, $\theta_2' = 240°$.

There are two unknown directions in this equation, and so we identify this as case 4. Using Eq. (2.49) and substituting **S** for **C**, R_{BA} for A, $-R_{BO_4}$ for B, θ_S for θ_C, and θ_4 for θ_B gives

$$\theta_4 = \theta_S \mp \cos^{-1} \frac{S^2 + R_{BO_4}^2 - R_{BA}^2}{2SR_{BO_4}}$$

$$= -5.54° \mp \cos^{-1} \frac{(0.725)^2 + (-0.400)^2 - (0.690)^2}{2(0.725)(-0.400)}$$

$$= -5.54° \mp 111.18° = -116.72° \quad \text{or} \quad +105.64° \qquad Ans.$$

Note that we could have substituted $R_{BO_4} = +0.400$ m and we would have obtained $\theta_4 + 180°$ for the final result.

Next, using Eq. (2.50) and substituting θ_3 for θ_A gives

$$\theta_3 = \theta_S \pm \cos^{-1} \frac{S^2 + R_{BA}^2 - R_{BO_4}^2}{2SR_{BA}}$$

$$= -5.54° \pm \cos^{-1} \frac{(0.725)^2 + (0.690)^2 - (-0.400)^2}{2(0.725)(0.690)}$$

$$= -5.54° \pm 32.72° = 27.19° \qquad Ans.$$

We follow the same procedure for the other input angle. Using Fig. 2.24c yields

$$\mathbf{S}' = \mathbf{R}_{O_4} - \mathbf{R}_{A'} = 0.600\angle 0° - 0.140\angle 240° = 0.681\angle 10.26° \text{ m}$$

$$\theta'_4 = 10.26° \mp \cos^{-1} \frac{(0.681)^2 + (-0.400)^2 - (0.690)^2}{2(0.681)(-0.400)}$$

$$= 10.26° \mp 105.73° = -95.47° \text{ or } 115.99° \qquad \textit{Ans.}$$

$$\theta'_3 = 10.26° \pm \cos^{-1} \frac{(0.681)^2 + (0.690)^2 - (-0.400)^2}{2(0.681)(0.690)}$$

$$= 10.26° \pm 33.92° = 44.18° \qquad \textit{Ans.}$$

2.12 THE CHACE SOLUTIONS TO PLANAR VECTOR EQUATIONS

As we saw in the last section, the algebra involved in solving even simple planar vector equations can become cumbersome. Chace took advantage of the brevity of vector notation in obtaining explicit closed-form solutions to both two- and three-dimensional vector equations.[2] In this section we will study his solutions for planar equations in terms of the four cases of the loop-closure equation.

We again recall Eq. (2.16), the typical planar vector equation. In terms of magnitudes and unit vectors it can be written

$$C\hat{\mathbf{C}} = A\hat{\mathbf{A}} + B\hat{\mathbf{B}} \qquad (2.51)$$

and it may contain two unknowns consisting of two magnitudes, two directions, or one magnitude and one direction.

Case 1 is the situation where the magnitude and direction of the same vector, say, C and $\hat{\mathbf{C}}$, form the two unknowns. The method of solution for this case was shown in Example 2.2. The general form of the solution is

$$\mathbf{C} = (\mathbf{A} \cdot \hat{\mathbf{i}} + \mathbf{B} \cdot \hat{\mathbf{i}})\hat{\mathbf{i}} + (\mathbf{A} \cdot \hat{\mathbf{j}} + \mathbf{B} \cdot \hat{\mathbf{j}})\hat{\mathbf{j}} \qquad (2.52)$$

For case 2, the unknowns are the magnitude of one vector and the direction of another, say, A and $\hat{\mathbf{B}}$. The Chace approach for this case consists of eliminating one of the unknowns by taking the dot product of every vector with a new vector chosen so that one of the unknowns is eliminated. We can eliminate the vector **A** by taking the dot product of every term of the equation with $\hat{\mathbf{A}} \times \hat{\mathbf{k}}$.

$$\mathbf{C} \cdot (\hat{\mathbf{A}} \times \hat{\mathbf{k}}) = A\hat{\mathbf{A}} \cdot (\hat{\mathbf{A}} \times \hat{\mathbf{k}}) + B\hat{\mathbf{B}} \cdot (\hat{\mathbf{A}} \times \hat{\mathbf{k}}) \qquad (a)$$

Thus, because $\hat{\mathbf{A}} \times \hat{\mathbf{k}}$ is perpendicular to $\hat{\mathbf{A}}$, $\hat{\mathbf{A}} \cdot (\hat{\mathbf{A}} \times \hat{\mathbf{k}}) = 0$:

$$\mathbf{C} \cdot (\hat{\mathbf{A}} \times \hat{\mathbf{k}}) = B\hat{\mathbf{B}} \cdot (\hat{\mathbf{A}} \times \hat{\mathbf{k}}) \qquad (b)$$

Now, from the definition of the dot product of two vectors,

$$\mathbf{P} \cdot \mathbf{Q} = PQ \cos \phi$$

we note that

$$B\hat{\mathbf{B}} \cdot (\hat{\mathbf{A}} \times \hat{\mathbf{k}}) = B \cos \phi \qquad (c)$$

where ϕ is the angle between the vectors $\hat{\mathbf{B}}$ and $(\hat{\mathbf{A}} \times \hat{\mathbf{k}})$. Thus

$$\cos \phi = \hat{\mathbf{B}} \cdot (\hat{\mathbf{A}} \times \hat{\mathbf{k}}) \qquad (d)$$

The vectors $\hat{\mathbf{A}}$ and $\hat{\mathbf{A}} \times \hat{\mathbf{k}}$ are perpendicular to each other; hence we are free to choose another coordinate system $\hat{\boldsymbol{\lambda}}\hat{\boldsymbol{\mu}}$ having the directions $\hat{\boldsymbol{\lambda}} = \hat{\mathbf{A}} \times \hat{\mathbf{k}}$ and $\hat{\boldsymbol{\mu}} = \hat{\mathbf{A}}$. In this reference system, the unknown unit vector $\hat{\mathbf{B}}$ can be written as

$$\hat{\mathbf{B}} = \cos \phi (\hat{\mathbf{A}} \times \hat{\mathbf{k}}) + \sin \phi \hat{\mathbf{A}} \qquad (e)$$

If we now substitute Eq. (d) into (b) and solve for $\cos \phi$, we obtain

$$\cos \phi = \frac{\mathbf{C} \cdot (\hat{\mathbf{A}} \times \hat{\mathbf{k}})}{B} \qquad (f)$$

Then

$$\sin \phi = \pm \sqrt{1 - \cos^2 \phi} = \pm \frac{1}{B} \sqrt{B^2 - [\mathbf{C} \cdot (\hat{\mathbf{A}} \times \hat{\mathbf{k}})]^2} \qquad (g)$$

Substituting Eqs. (f) and (g) into (e) and multiplying both sides by the known magnitude B gives

$$\mathbf{B} = [\mathbf{C} \cdot (\hat{\mathbf{A}} \times \hat{\mathbf{k}})](\hat{\mathbf{A}} \times \hat{\mathbf{k}}) \pm \sqrt{B^2 - [\mathbf{C} \cdot (\hat{\mathbf{A}} \times \hat{\mathbf{k}})]^2} \hat{\mathbf{A}} \qquad (2.53)$$

To obtain the vector $\mathbf{A}$ we may wish to use Eq. (2.51) directly and perform the vector subtraction. Alternatively, if we substitute Eq. (2.53) and rearrange, we obtain

$$\mathbf{A} = \mathbf{C} - [\mathbf{C} \cdot (\hat{\mathbf{A}} \times \hat{\mathbf{k}})](\hat{\mathbf{A}} \times \hat{\mathbf{k}}) \mp \sqrt{B^2 - [\mathbf{C} \cdot (\hat{\mathbf{A}} \times \hat{\mathbf{k}})]^2} \hat{\mathbf{A}} \qquad (h)$$

The first two terms of this equation can be simplified as shown in Fig. 2.25a. The $\hat{\mathbf{A}} \times \hat{\mathbf{k}}$ direction is located 90° clockwise from the $\hat{\mathbf{A}}$ direction. The magnitude $\mathbf{C} \cdot (\hat{\mathbf{A}} \times \hat{\mathbf{k}})$ is the projection of $\mathbf{C}$ in the $\hat{\mathbf{A}} \times \hat{\mathbf{k}}$ direction. Therefore, when $[\mathbf{C} \cdot (\hat{\mathbf{A}} \times \hat{\mathbf{k}})](\hat{\mathbf{A}} \times \hat{\mathbf{k}})$ is subtracted from $\mathbf{C}$, the result is a vector of magnitude $\mathbf{C} \cdot \hat{\mathbf{A}}$ in the $\hat{\mathbf{A}}$ direction. With this substitution, Eq. (h) becomes

$$\mathbf{A} = \left[\mathbf{C} \cdot \hat{\mathbf{A}} \mp \sqrt{B^2 - [\mathbf{C} \cdot (\hat{\mathbf{A}} \times \hat{\mathbf{k}})]^2} \right] \hat{\mathbf{A}} \qquad (2.54)$$

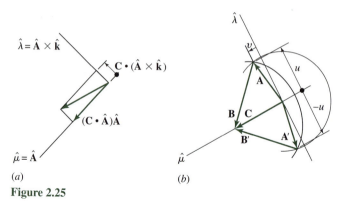

(a)   (b)

Figure 2.25

In case 3 the magnitudes of two different vectors—say, A and B—are unknown. We begin the solution by eliminating $\mathbf{B}$ from Eq. (2.51)

$$\mathbf{C} \cdot (\hat{\mathbf{B}} \times \hat{\mathbf{k}}) = A\hat{\mathbf{A}} \cdot (\hat{\mathbf{B}} \times \hat{\mathbf{k}}) + B\hat{\mathbf{B}} \cdot (\hat{\mathbf{B}} \times \hat{\mathbf{k}})$$

Thus, because $\hat{\mathbf{B}} \times \hat{\mathbf{k}}$ is perpendicular to $\hat{\mathbf{B}}$, we have $\hat{\mathbf{B}} \cdot (\hat{\mathbf{B}} \times \hat{\mathbf{k}}) = 0$; hence, we obtain the magnitude

$$A = \frac{\mathbf{C} \cdot (\hat{\mathbf{B}} \times \hat{\mathbf{k}})}{\hat{\mathbf{A}} \cdot (\hat{\mathbf{B}} \times \hat{\mathbf{k}})} \tag{2.55}$$

In a similar manner, we obtain the unknown magnitude B:

$$B = \frac{\mathbf{C} \cdot (\hat{\mathbf{A}} \times \hat{\mathbf{k}})}{\hat{\mathbf{B}} \cdot (\hat{\mathbf{A}} \times \hat{\mathbf{k}})} \tag{2.56}$$

Finally, in case 4 the unknowns are the directions of two different vectors—say, $\hat{\mathbf{A}}$ and $\hat{\mathbf{B}}$. This case is illustrated in Fig. 2.25b, where the vector $\mathbf{C}$ and the two magnitudes A and B are given. The problem is solved by finding the points of intersection of two circles of radii A and B. We begin by defining a new coordinate system $\hat{\lambda}\hat{\mu}$ whose axes are directed so that $\hat{\lambda} = \hat{\mathbf{C}} \times \hat{\mathbf{k}}$ and $\hat{\mu} = \hat{\mathbf{C}}$, as shown in the figure. If the coordinates of one of the points of intersection in the $\hat{\lambda}\hat{\mu}$ coordinate system are designated as u and v, then

$$\mathbf{A} = u\hat{\lambda} + v\hat{\mu} \quad \text{and} \quad \mathbf{B} = -u\hat{\lambda} + (C - v)\hat{\mu} \tag{i}$$

The equation of the circle of radius A is

$$u^2 + v^2 = A^2 \tag{j}$$

The circle of radius B has the equation

$$u^2 + (v - C)^2 = B^2$$

or

$$u^2 + v^2 - 2Cv + C^2 = B^2 \qquad (k)$$

Subtracting Eq. (k) from (j) and solving for v yields

$$v = \frac{A^2 - B^2 + C^2}{2C} \qquad (l)$$

Substituting this into Eq. (j) and solving for u gives

$$u = \pm \sqrt{A^2 - \left(\frac{A^2 - B^2 + C^2}{2C}\right)^2} \qquad (m)$$

The final step is to substitute these values of u and v into Eqs. (i) and to replace $\hat{\boldsymbol{\lambda}}$ and $\hat{\boldsymbol{\mu}}$ according to their definitions. The results are

$$\mathbf{A} = \pm \sqrt{A^2 - \left(\frac{A^2 - B^2 + C^2}{2C}\right)^2}\,(\hat{\mathbf{C}} \times \hat{\mathbf{k}}) + \frac{A^2 - B^2 + C^2}{2C}\hat{\mathbf{C}} \qquad (2.57)$$

$$\mathbf{B} = \mp \sqrt{A^2 - \left(\frac{A^2 - B^2 + C^2}{2C}\right)^2}\,(\hat{\mathbf{C}} \times \hat{\mathbf{k}}) + \frac{B^2 - A^2 + C^2}{2C}\hat{\mathbf{C}} \qquad (2.58)$$

To apply the Chace equations, begin with the slider-crank linkage of Fig. 2.14a. From Fig. 2.14b, the loop-closure equation is written in the form

$$R_B\hat{\mathbf{R}}_B = R_A\hat{\mathbf{R}}_A + R_{BA}\hat{\mathbf{R}}_{BA} \qquad (n)$$

With θ_2 given, the unknowns in Eq. (n) are the magnitude R_B and the direction $\hat{\mathbf{R}}_{BA}$. The solution corresponds to case 2 and is found by making appropriate substitutions in Eqs. (2.53) and (2.54):

$$\mathbf{R}_{BA} = -[\mathbf{R}_A \cdot (\hat{\mathbf{R}}_B \times \hat{\mathbf{k}})](\hat{\mathbf{R}}_B \times \hat{\mathbf{k}}) + \left\{R_{BA}^2 - [\mathbf{R}_A \cdot (\hat{\mathbf{R}}_B \times \hat{\mathbf{k}})]^2\right\}^{1/2}\hat{\mathbf{R}}_B \qquad (2.59)$$

$$\mathbf{R}_B = \left(\mathbf{R}_A \cdot \hat{\mathbf{R}}_B + \left\{R_{BA}^2 - [\mathbf{R}_A \cdot (\hat{\mathbf{R}}_B \times \hat{\mathbf{k}})]^2\right\}^{1/2}\right)\hat{\mathbf{R}}_B \qquad (2.60)$$

EXAMPLE 2.6

Use the Chace equations to find the position of the slider of Fig. 2.14 with $R_A = 25$ mm, $R_{BA} = 75$ mm, and $\theta_2 = 150°$.

SOLUTION

Placing the given information in vector form yields

$$\mathbf{R}_A = 25\angle 150° = -27.7\hat{\mathbf{i}} + 12.5\hat{\mathbf{j}} \text{ mm}, \qquad R_{BA} = 75 \text{ mm}, \qquad \hat{\mathbf{R}}_B = \hat{\mathbf{i}}, \qquad \hat{\mathbf{R}}_B \times \hat{\mathbf{k}} = -\hat{\mathbf{j}}$$

Substituting these values into Eq. (2.59) gives

$$\mathbf{R}_{BA} = -[(-21.7\hat{\mathbf{i}} + 12.5\hat{\mathbf{j}}) \cdot (-\hat{\mathbf{j}})](-\hat{\mathbf{j}}) + \{(75)^2 - [(-21.7\hat{\mathbf{i}} + 12.5\hat{\mathbf{j}}) \cdot (-\hat{\mathbf{j}})]^2\}^{1/2}$$

$$= -12.5\hat{\mathbf{j}} + [(75)^2 - (-12.5)^2]^{1/2}\,\hat{\mathbf{i}} = 73.9\hat{\mathbf{i}} - 12.5\hat{\mathbf{j}}\ \text{mm}$$

And so

$$\mathbf{R}_{BA} = 75\angle{-9.60°} = 75\angle{350.4°}\ \text{mm} \hspace{3cm} \textit{Ans.}$$

Using Eq. (2.60) next yields

$$\mathbf{R}_B = \{(-21.7\hat{\mathbf{i}} + 12.5\hat{\mathbf{j}}) \cdot \hat{\mathbf{i}} + [(75)^2 - [(-21.7\hat{\mathbf{i}} + 12.5\hat{\mathbf{j}}) \cdot (-\hat{\mathbf{j}})]^2]^{1/2}\}\hat{\mathbf{i}}$$

$$= 50.2\hat{\mathbf{i}}\ \text{mm}$$

$$\hspace{11cm} \textit{Ans.}$$

2.13 COUPLER-CURVE GENERATION

In Chapter 1, Fig. 1.18, we learned of the limitless variety of useful coupler curves capable of being generated by the four-bar linkage. These curves are quite easy to obtain graphically, but computer-generated curves can be obtained more rapidly and are easier to vary in order to obtain the desired curve characteristics. Here we present the basic equations but omit the computer programming details required for screen display.

Applying the Chace approach to the four-bar linkage of Fig. 2.15, we first form the vector

$$\mathbf{S} = \mathbf{R}_C - \mathbf{R}_A$$

Then we also note that

$$\mathbf{S} = R_{BA}\hat{\mathbf{R}}_{BA} - R_{BC}\hat{\mathbf{R}}_{BC}$$

where the two directions $\hat{\mathbf{R}}_{BA}$ and $\hat{\mathbf{R}}_{BC}$ are the unknowns. This is case 4, and the solutions are given by Eqs. (2.57) and (2.58). Substituting gives

$$\hat{\mathbf{R}}_{BC} = \pm\left[R_{BA}^2 - \left(\frac{R_{BA}^2 - R_{BC}^2 + S^2}{2S}\right)^2\right]^{1/2}(\hat{\mathbf{S}} \times \hat{\mathbf{k}}) + \frac{R_{BC}^2 - R_{BA}^2 + S^2}{2S}\hat{\mathbf{S}} \quad (2.61)$$

$$\hat{\mathbf{R}}_{BA} = \pm\left[R_{BA}^2 - \left(\frac{R_{BA}^2 - R_{BC}^2 + S^2}{2S}\right)^2\right]^{1/2}(\hat{\mathbf{S}} \times \hat{\mathbf{k}}) + \frac{R_{BA}^2 - R_{BC}^2 + S^2}{2S}\hat{\mathbf{S}} \quad (2.62)$$

The upper set of signs gives the solution for the crossed linkage; the lower set thus applies to the open linkage of Figs. 2.15a and 2.15b.

In Fig. 2.15, point P represents a coupler point which generates a curve called the coupler curve when crank 2 is moved or rotated. From Fig. 2.15b we see that

$$\mathbf{R}_P = R_P e^{j\theta_6} = R_A e^{j\theta_2} + R_{PA} e^{j(\theta_3 + \alpha)} \tag{2.63}$$

We recognize this as a case 1 vector equation because R_p and θ_6 are two unknowns. The solutions can be found directly by applying Eqs. (2.43) and (2.44):

$$R_p = \left[R_A^2 + R_{PA}^2 + 2R_A R_{PA} \cos(\theta_3 + \alpha - \theta_2) \right] \tag{2.64}$$

$$\theta_6 = \tan^{-1} \frac{R_A \sin\theta_2 + R_{PA} \sin(\theta_3 + \alpha)}{R_A \cos\theta_2 + R_{PA} \cos(\theta_3 + \alpha)} \tag{2.65}$$

Note that both of these equations give double values coming from the double values for θ_3 and corresponding to the two closures of the linkage.

EXAMPLE 2.7

Calculate and plot points on the coupler curve of a four-bar linkage having the following dimensions: $R_C = 200$ mm, $R_A = 100$ mm, $R_{BC} = 300$ mm, $R_{PA} = 150$ mm, $\alpha = -45°$. This notation corresponds to that of Fig. 2.15b.

SOLUTION

For each crank angle value θ_2, the transmission angle γ is evaluated from Eq. (2.33). Next, Eqs. (2.29) and (2.31) or Eq. (2.50) are applied to give θ_3. Finally, the coupler-point positions are calculated from Eqs. (2.64) and (2.65). The solutions for crank angles from 0° to 90° are displayed in Table 2.3, and the full coupler curve is shown in Fig. 2.26. Only one of the two solutions is calculated and plotted.

TABLE 2.3 Coupler-Curve Points Obtained for Example 2.7

θ_2, deg	γ, deg	θ_3, deg	R_P, mm	θ_6, deg	R_P^x, mm	R_P^y, mm
0.0	18.2	110.5	212	40.1	162	136
10.0	18.9	99.4	232	36.9	186	139
20.0	20.9	87.8	245	33.7	204	136
30.0	23.9	77.5	250	31.5	213	131
40.0	27.4	69.2	248	30.5	213	126
50.0	31.3	62.9	241	30.7	207	123
60.0	35.2	58.3	230	31.7	196	121
70.0	39.2	55.1	218	33.5	182	120
80.0	43.1	53.0	204	35.8	166	119
90.0	46.9	51.8	190	38.3	149	118

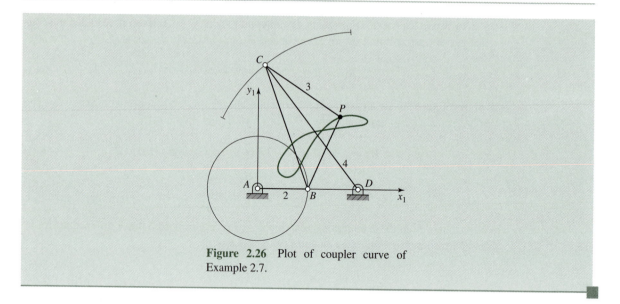

Figure 2.26 Plot of coupler curve of Example 2.7.

2.14 DISPLACEMENT OF A MOVING POINT

We have concerned ourselves so far with only a single instantaneous position of a point, but because we wish to study motion, we must be concerned with the relationship between a succession of positions.

In Fig. 2.27 a particle, originally at point P, is moving along the path shown and, some time later, arrives at the position P'. The *displacement* of the point $\mathbf{\Delta R}_P$ during the time interval is defined as the *net change in position*

$$\mathbf{\Delta R}_P = \mathbf{R}'_P - \mathbf{R}_P \tag{2.66}$$

Displacement is a vector quantity having the magnitude and direction of the vector from point P to point P'.

It is important to note that the displacement $\mathbf{\Delta R}_P$ is the net change in position and does not depend on the particular path taken between points P and P'. Its magnitude is *not*

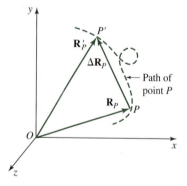

Figure 2.27 Displacement of a moving point.

necessarily equal to the length of the path (the distance traveled), and direction is *not* necessarily along the tangent to the path, although both these are true when the displacement is infinitesimally small. Knowledge of the path actually traveled from P to P' is not even necessary to find the displacement vector as long as the initial and final positions are known.

2.15 DISPLACEMENT DIFFERENCE BETWEEN TWO POINTS

In this section we consider the difference in the displacements of two moving points. In particular we are concerned with the case where the two moving points are both particles of the same rigid body. The situation is shown in Fig. 2.28, where rigid body 2 moves from an initial position defined by $x_2 y_2 z_2$ to a later position defined by $x_2' y_2' z_2'$.

From Eq. (2.6), the position difference between the two points P and Q of body 2 at the initial instant is

$$\mathbf{R}_{PQ} = \mathbf{R}_P - \mathbf{R}_Q \qquad (a)$$

After the displacement of body 2, the two points are located at P' and Q'. At that time the position difference is

$$\mathbf{R}'_{PQ} = \mathbf{R}'_P - \mathbf{R}'_Q \qquad (b)$$

During the time interval of the movement the two points have undergone individual displacements of $\mathbf{\Delta R}_P$ and $\mathbf{\Delta R}_Q$, respectively.

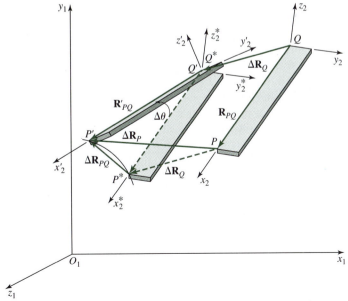

Figure 2.28 Displacement difference between two points on the same rigid body.

As the name implies, the *displacement difference* between the two points is defined as the net difference between their respective displacements and is given the symbol $\mathbf{\Delta R}_{PQ}$

$$\mathbf{\Delta R}_{PQ} = \mathbf{\Delta R}_P - \mathbf{\Delta R}_Q \tag{2.67}$$

Note that this equation corresponds to the vector triangle PP^*P' in Fig. 2.28. As stated in the previous section, the displacement depends only on the net change in position and not on the path by which it was achieved. Thus, no matter how the body containing points P and Q was *actually* displaced, we are free to visualize the path as we choose. Equation (2.67) leads us to visualize the displacement as having taken place in two stages. First, the body translates (slides without rotation) from $x_2 y_2 z_2$ to $x_2^* y_2^* z_2^*$; during this movement all particles, including P and Q, have the same displacement $\mathbf{\Delta R}_Q$. Next we visualize the body as rotating about point Q' through the angle $\Delta\theta$ to the final position $x_2' y_2' z_2'$.

By manipulating Eq. (2.67) we can obtain a different interpretation,

$$\mathbf{\Delta R}_{PQ} = (\mathbf{R}'_P - \mathbf{R}_P) - (\mathbf{R}'_Q - \mathbf{R}_Q)$$
$$= (\mathbf{R}'_P - \mathbf{R}'_Q) - (\mathbf{R}_P - \mathbf{R}_Q) \tag{c}$$

and then, from Eqs. (*a*) and (*b*),

$$\mathbf{\Delta R}_{PQ} = \mathbf{R}'_{PQ} - \mathbf{R}_{PQ} \tag{2.68}$$

This equation corresponds to the vector triangle $Q'P^*P'$ in Fig. 2.28 and shows that the displacement difference, defined as the difference between two displacements, is equal to the net change between the position-difference vectors.

In either interpretation we are illustrating *Euler's theorem,* which states that *any displacement of a rigid body is equivalent to the sum of a net translation of one point (Q) and a net rotation of the body about that point.* We also see that only the rotation contributes to the displacement difference between two points on the same rigid body; that is, *there is no difference between the displacements of any two points of the same rigid body as the result of a translation.* (See Section 2.16 below for the definition of the term *translation.*)

In view of the above discussion, we can visualize the displacement difference $\mathbf{\Delta R}_{PQ}$ as the displacement that would be seen for point P by a moving observer who travels along, always staying coincident with point Q *but not turning* with the moving body—that is, always using the absolute coordinate axes $x_1 y_1 z_1$ to measure direction. It is important to understand the difference between the interpretation of an observer moving with point Q but not rotating and the case of the observer *on* the moving body. To an observer *on* body 2, both points P and Q would appear stationary; neither would be seen to have a displacement because they do not move relative to the observer, and the displacement difference seen by such an observer would be zero.

2.16 ROTATI

Using the concept of displacement difference between two points of the same rigid body, we are now able to define translation and rotation.

Translation is defined as *a state of motion of a body for which the displacement difference between any two points P and Q of the body is zero* or, from the displacement-

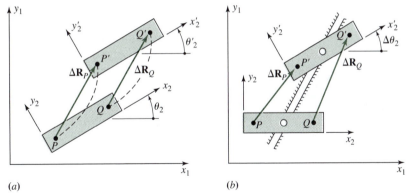

(a)   (b)

Figure 2.29 (a) Translation: $\mathbf{\Delta R}_P = \mathbf{\Delta R}_Q$, $\Delta\theta_2 = 0$; (b) rotation $\mathbf{\Delta R}_P \neq \mathbf{\Delta R}_Q$, $\Delta\theta_2 \neq 0$.

difference equation (2.68),

$$\mathbf{\Delta R}_{PQ} = \mathbf{\Delta R}_P - \mathbf{\Delta R}_Q = 0$$

$$\mathbf{\Delta R}_P = \mathbf{\Delta R}_Q$$

(2.69)

which states that *the displacements of any two points of the body are equal.* Rotation is a state of motion of the body for which different points of the body exhibit different displacements.

Figure 2.29a illustrates a situation where the body has moved along a curved path from position $x_2 y_2$ to position $x_2' y_2'$. In spite of the fact that the point paths are curved,* $\mathbf{\Delta R}_P$ is still equal to $\mathbf{\Delta R}_Q$ and the body has undergone a translation. Note that in translation the point paths described by any two points on the body are identical and there is no change of angular orientation between the moving coordinate system and the observer's coordinate system; that is, $\Delta\theta_2 = \theta_2' - \theta_2 = 0$.

In Fig. 2.29b the center point of the moving body is *constrained* to move along a straight-line path. Yet, as it does so, the body rotates so that $\Delta\theta_2 = \theta_2' - \theta_2 \neq 0$ and the displacements $\mathbf{\Delta R}_P$ and $\mathbf{\Delta R}_Q$ are not equal. Even though there is no obvious point on the body about which it has rotated, the coordinate system $x_2' y_2'$ has changed angular orientation relative to $x_1 y_1$ and the body is said to have undergone a rotation. Note that the point paths described by P and Q are not equal.

We see from these two examples that rotation or translation of a body cannot be defined from the motion of a single point. These are characteristic motions of a body or a coordinate system. It is improper to speak of "rotation of a point" because there is no meaning for angular orientation of a point. It is also improper to associate the terms "rotation" and "translation" with the rectilinear or curvilinear characteristics of a single point path. Although it does not matter which points of the body are chosen, the motion of two or more points must be compared before meaningful definitions exist for these terms.

---

*Translation in which the point paths are not straight lines is referred to as curvilinear translation.

2.17 APPARENT DISPLACEMENT

We have already observed that the displacement of a moving point does not depend on the particular path traveled. However, because displacement is computed from the position vectors of the endpoints of the path, a knowledge of the coordinate system of the observer is essential.

In Fig. 2.30 we identify three bodies. The first, body 1, is a fixed or stationary body containing the absolute reference system $x_1 y_1 z_1$. The second is a moving body, which we call body 2, and we fix to it the reference system $x_2 y_2 z_2$. Thus $x_2 y_2 z_2$ is a moving system. We then identify the third as body 3 and permit it to move with respect to body 2.

We also identify two observers. Designate HE as an observer fixed to body 1, the stationary system. Designate SHE as another observer on body 2, fixed to the moving system $x_2 y_2 z_2$. Thus, you might say, HE is on the ground observing while SHE is going for a ride on body 2.

Now consider a particle, or passenger, P_3, situated on body 3 and moving along a known path on body 2. HE and SHE are both observing the motion of particle P_3 and we want to discover what they see.

We must define another point P_2, which is rigidly attached to body 2 and which is instantaneously coincident with P_3.

Now let body 2 and $x_2 y_2 z_2$ be displaced to a new position $x_2' y_2' z_2'$. While this motion is taking place, let P_3 move to another position on body 2 which we identify as P_3'. But P_2, because it is fixed to body 2, moves with 2 and is now found in the new position identified as P_2'. SHE, on body 2, reports the motion of the passenger or particle P_3 as the vector $\mathbf{\Delta R}_{P_3/2}$, which is read as follows: the displacement of P_3 as it appears to an observer on body 2. This is called the apparent-displacement vector. Note that SHE sees no motion of P_2 because it appears to be stationary. Therefore, $\mathbf{\Delta R}_{P_2/2} = 0$.

However, HE, on the stationary body, reports the displacement of particle P_3 by the vector $\mathbf{\Delta R}_{P_3}$. Note that HE also reports the displacement of P_2 by the vector $\mathbf{\Delta R}_{P_2}$. From the vector triangle in Fig. 2.30 we see that the two observers' observations are related by the *apparent-displacement equation,*

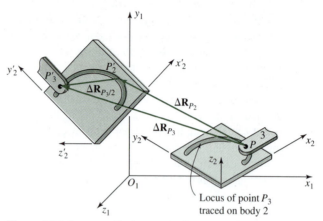

Figure 2.30 Apparent displacement of a point.

$$\Delta \mathbf{R}_{P_3} = \Delta \mathbf{R}_{P_2} + \Delta \mathbf{R}_{P_3/2} \tag{2.70}$$

We can take this equation as the definition of the apparent-displacement vector, although it is important also to understand the physical concepts involved. Notice that the apparent-displacement vector relates the absolute displacements of two *coincident points* which are particles of *different moving bodies*. Notice also that there is no restriction on the actual location of the observer moving with coordinate system 2, only that SHE be fixed in that coordinate system so that SHE senses no displacement for point P_2.

One primary use of the apparent displacement is to determine an absolute displacement. It is not uncommon in machines to find a point such as P_3 which is constrained to move along a known slot or path or guideway defined by the shape of another moving link 2. In such cases it may be much more convenient to measure or calculate $\Delta \mathbf{R}_{P_2}$ and $\Delta \mathbf{R}_{P_3/2}$ and to use Eq. (2.70) than to measure the absolute displacement $\Delta \mathbf{R}_{P_3}$ directly.

2.18 ABSOLUTE DISPLACEMENT

In reflecting on the definition and concept of the apparent-displacement vector, we conclude that the absolute displacement of a moving point $\Delta \mathbf{R}_{P_3/1}$ is the special case of an apparent displacement where the observer is fixed in the absolute coordinate system. As explained for the position vector, the notation is often abbreviated to read $\Delta \mathbf{R}_{P_3}$ or just $\Delta \mathbf{R}_P$ and an absolute observer is implied whenever not noted explicitly.

Perhaps a better physical understanding of apparent displacement can be achieved by relating it to absolute displacement. Imagine an automobile P_3 traveling along a roadway and under observation by an absolute observer some distance off to one side. Consider how this observer visually senses the motion of the car. Although HE may not be conscious of all of the following steps, the contention here is that the observer first imagines a point P_1, *coincident* with P_3, which HE defines in his mind as stationary; HE may relate to a fixed point of the roadway or nearby tree, for example. HE then compares his later observations of the car P_3 with those of P_1 to sense displacement. Notice that HE does not compare with his own location but with the initially coincident point P_1. In this instance the apparent-displacement equation becomes an identity:

$$\Delta \mathbf{R}_{P_3} = \Delta \mathbf{R}_{P_1}^{0} + \Delta \mathbf{R}_{P_3/1}$$

NOTES

[1.]   See, for example, C. R. Mischke, **Mathematical Model Building,** Iowa State University Press, Ames, 1980, p. 86.

[2.]   M. A. Chace, Vector Analysis of Linkages, *J. Eng. Ind., ASME Trans.,* ser. B, vol. 55, no. 3, pp. 289–297, August 1963.

PROBLEMS*

2.1 Describe and sketch the locus of a point A which moves according to the equations $R_A^x = at\cos(2\pi t)$, $R_A^y = at\sin(2\pi t)$, $R_A^z = 0$.

2.2 Find the position difference from point P to point Q on the curve $y = x^2 + x - 16$, where $R_P^x = 2$ and $R_Q^x = 4$.

2.3 The path of a moving point is defined by the equation $y = 2x^2 - 28$. Find the position difference from point P to point Q if $R_P^x = 4$ and $R_Q^x = -3$.

2.4 The path of a moving point P is defined by the equation $y = 60 - x^3/3$. What is the displacement of the point if its motion begins when $R_P^x = 0$ and ends when $R_P^x = 3$?

2.5 If point A moves on the locus of Problem 2.1, find its displacement from $t = 2$ to $t = 2.5$.

2.6 The position of a point is given by the equation $\mathbf{R} = 100e^{j2\pi t}$. What is the path of the point? Determine the displacement of the point from $t = 0.10$ to $t = 0.40$.

2.7 The equation $\mathbf{R} = (t^2 + 4)e^{-j\pi t/10}$ defines the position of a point. In which direction is the position vector rotating? Where is the point located when $t = 0$? What is the next value t can have if the direction of the position vector is to be the same as it is when $t = 0$? What is the displacement from the first position of the point to the second?

2.8 The location of a point is defined by the equation $\mathbf{R} = (4t + 2)e^{j\pi t^2/30}$, where t is time in seconds. Motion of the point is initiated when $t = 0$. What is the displacement during the first 3 s? Find the change in angular orientation of the position vector during the same time interval.

2.9 Link 2 in the figure rotates according to the equation $\theta = \pi t/4$. Block 3 slides outward on link 2 according to the equation $r = t^2 + 2$. What is the absolute displacement $\Delta\mathbf{R}_{P_3}$ from $t = 1$ to $t = 2$? What is the apparent displacement $\Delta\mathbf{R}_{P_3/2}$?

2.10 A wheel with center at O rolls without slipping so that its center is displaced 10 in to the right. What is the displacement of point P on the periphery during this interval?

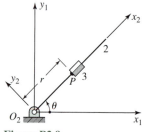

Figure P2.9

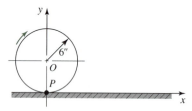

Figure P2.10 Rolling wheel.

2.11 A point Q moves from A to B along link 3 while link 2 rotates from $\theta_2 = 30°$ to $\theta_2' = 120°$. Find the absolute displacement of Q.

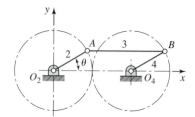

Figure P2.11 $R_{AO_2} = R_{BO_4} = 3$ in, $R_{BA} = R_{O_4O_2} = 6$ in.

2.12 The linkage shown is driven by moving the sliding block 2. Write the loop-closure equation. Solve analytically for the position of sliding block 4. Check the result graphically for the position where $\phi = -45°$.

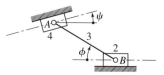

Figure P2.12 $R_{AB} = 200$ mm, $\psi = 15°$.

2.13 The offset slider-crank mechanism is driven by rotating crank 2. Write the loop-closure equation.

Solve for the position of the slider 4 as a function of θ_2.

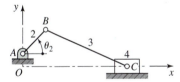

Figure P2.13 $R_{AO} = 1$ in, $R_{BA} = 2.5$ in, $R_{CB} = 7$ in.

2.14 Write a calculator program to find the sum of any number of two-dimensional vectors expressed in mixed rectangular or polar forms. The result should be obtainable in either form with the magnitude and angle of the polar form having only positive values.

2.15 Write a computer program to plot the coupler curve of any crank-rocker or double-crank form of the four-bar linkage. The program should accept four link lengths and either rectangular or polar coordinates of the coupler point relative to the coupler.

2.16 For each linkage shown in the figure, find the path of point P: (a) inverted slider-crank mechanism; (b) second inversion of the slider-crank mechanism; (c) straight-line mechanism; (d) drag-link mechanism.

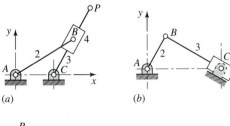

(a)

(b)

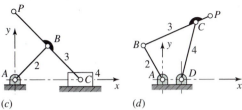

(c)

(d)

Figure P2.16 (a) $R_{CA} = 2$ in, $R_{BA} = 3.5$ in, $R_{PC} = 4$ in; (b) $R_{CA} = 40$ mm, $R_{BA} = 20$ mm, $R_{PB} = 65$ mm; (c) $R_{BA} = R_{CB} = R_{PB} = 25$ mm; (d) $R_{DA} = 1$ in, $R_{BA} = 2$ in, $R_{CC} = R_{DC} = 3$ in, $R_{PB} = 4$ in.

2.17 Using the offset slider-crank mechanism of Fig. 2.15, find the crank angles corresponding to the extreme values of the transmission angle.

2.18 In Section 1.10 it is pointed out that the transmission angle reaches an extreme value for the four-bar linkage when the crank lies on the line between the fixed pivots. Referring to Fig. 2.19, this means that γ reaches a maximum or minimum when crank 2 is coincident with the line $O_2 O_4$. Show, analytically, that this statement is true.

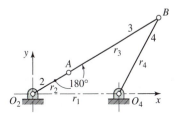

Figure P2.19 A limit position of a four-bar linkage.

2.19 The figure illustrates a crank-and-rocker linkage in the first of its two limit positions. In a limit position, points O_2, A, and B lie on a straight line; that is, links 2 and 3 form a straight line. The two limit positions of a crank-rocker describe the extreme positions of the rocking angle. Suppose that such a linkage has $r_1 = 400$ mm, $r_2 = 200$ mm, $r_3 = 500$ mm, and $r_4 = 400$ mm.

(a) Find θ_2 and θ_4 corresponding to each limit position.

(b) What is the total rocking angle of link 4?

(c) What are the transmission angles at the extremes?

2.20 A double-rocker mechanism has a dead-center position and may also have a limit position (see Problem 2.19). These positions occur when links 3 and 4 in the figure lie along a straight line. In the dead-center position the transmission angle is 180° and the mechanism is locked. The designer must either avoid such positions or provide the external force, such as a spring, to unlock the linkage. Suppose, for the linkage shown in the figure, that $r_1 = 14$ in,

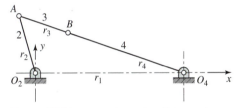

Figure P2.20 A double-rocker linkage shown in its dead-center position.

$r_2 = 5.5$ in, $r_3 = 5$ in, and $r_4 = 12$ in. Find θ_2 and θ_4 corresponding to the dead-center position. Is there a limit position?

2.21 The figure shows a slider-crank mechanism that has an offset e and that is placed in one of its limiting positions. By changing the offset e, it is possible to cause the angle that crank 2 makes in traversing between the two limiting positions to vary in such a manner that the driving or forward stroke of the slider takes place over a larger angle than the angle used for the return stroke. Such a linkage is then called a quick-return mechanism. The problem here is to develop a formula for the crank angle traversed during the forward stroke and also develop a similar formula for the angle traversed during the return stroke. The ratio of these two angles would then constitute a time ratio of the drive to return strokes. Also, determine which direction the crank should rotate.

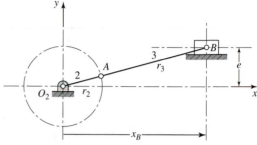

Figure P2.21 An offset slider-crank mechanism shown in one of its limiting positions.

3 Velocity

3.1 DEFINITION OF VELOCITY

In Fig. 3.1 a moving point is first observed at location P defined by the absolute position vector $\mathbf{R}_P$. After a short time interval Δt, its location is observed to have changed to P', defined by $\mathbf{R}'_P$. From Eq. (2.66) we recall that the displacement during this time interval is defined as

$$\Delta\mathbf{R}_P = \mathbf{R}'_P - \mathbf{R}_P$$

The *average velocity* of the point during the time interval Δt is $\Delta\mathbf{R}_P/\Delta t$. The *instantaneous velocity* (hereafter called simply *velocity*) is defined by the limit of this ratio for an infinitesimally small time interval and is given by

$$\mathbf{V}_P = \lim_{\Delta t \to 0} \frac{\Delta\mathbf{R}_P}{\Delta t} = \frac{d\mathbf{R}_P}{dt} \tag{3.1}$$

Because $\Delta\mathbf{R}_P$ is a vector, there are two convergences in taking this limit, the magnitude and the direction. Therefore the velocity of a point is a vector quantity equal to the time rate of change of its position. Like the position and displacement vectors, the velocity vector is defined for a specific point; "velocity" should not be applied to a line, coordinate system, volume, or other collection of points, because the velocity of each point may differ.

We recall that the position vectors $\mathbf{R}_P$ and $\mathbf{R}'_P$ depend on the location and orientation of the observer's coordinate system for their definitions. The displacement vector $\Delta\mathbf{R}_P$ and the velocity vector $\mathbf{V}_P$, on the other hand, are independent of the initial location of the coordinate system or the observer's location within the coordinate system. However, the velocity vector $\mathbf{V}_P$ does depend critically on the motion, if any, of the observer or the

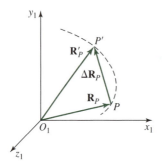

Figure 3.1 Displacement of a moving particle.

coordinate system during the time interval; it is for this reason that the observer is assumed to be stationary within the coordinate system. If the coordinate system involved is the absolute coordinate system, the velocity is referred to as an *absolute velocity* and is denoted by $\mathbf{V}_{P/1}$ or simply $\mathbf{V}_P$. This is consistent with the notation used for absolute displacement.

3.2 ROTATION OF A RIGID BODY

When a rigid body translates, as we saw in Section 2.16, the motion of any particular particle is equal to the motion of every other particle of the same body. When the body rotates, however, two arbitrarily chosen particles P and Q do not undergo the same motion and a coordinate system attached to the body does not remain parallel to its initial orientation; that is, the body undergoes some angular displacement $\Delta\theta$.

Angular displacements were not treated in detail in Chapter 2 because, in general, they cannot be treated as vectors. The reason is that they do not obey the usual laws of vector addition; if several gross angular displacements in three dimensions are undergone in succession, the result depends on the order in which they take place.

To illustrate, consider the rectangle $ABCO$ in Fig. 3.2a. The rectangular body is first rotated by $-90°$ about the y axis and then rotated by $+90°$ about the x axis. The final position of the body is seen to be in the yz plane. In Fig. 3.2b the body occupies the same starting position and is again rotated about the same axes, through the same angles, and in the same directions; however, the first rotation is about the x axis and the second is about the y axis. The order of the rotations is reversed, and the final position of the rectangle is now seen to be in the zx plane rather than the yz plane, as it was before. Because this characteristic does not correspond to the commutative law of vector addition, three-dimensional angular displacements cannot be treated as vectors.

Angular displacements that occur about the same axis or parallel axes, on the other hand, do follow the commutative law. Also, infinitesimally small angular displacements are commutative. To avoid confusion we will treat all finite angular displacements as scalar quantities. However, we will have occasion to treat infinitesimal angular displacements as vectors.

In Fig. 3.3 we recall the definition of the displacement difference between two points, P and Q, both attached to the same rigid body. As pointed out in Section 2.16, the displacement-difference vector is entirely attributable to the rotation of the body; there is no displacement difference between points in a body undergoing a translation. We reached this conclusion by picturing the displacement as occurring in two steps. First the body was

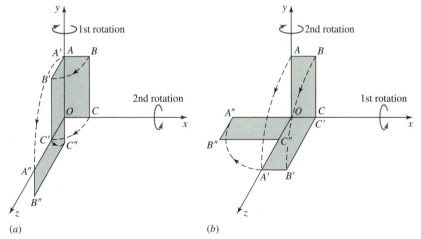

Figure 3.2 Angular displacements cannot be added vectorially because the result depends on the order in which they are added.

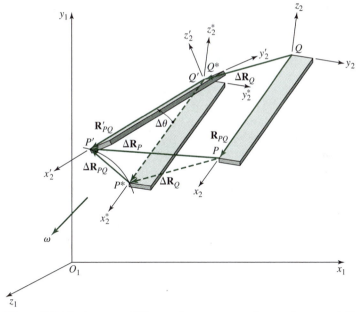

Figure 3.3 Displacement difference between two points on the same rigid link.

assumed to translate through the displacement $\mathbf{\Delta R}_Q$ to the position $x_2^* y_2^* z_2^*$. Next the body was rotated about point Q^* to the position $x_2' y_2' z_2'$.

Another way to picture the displacement difference $\mathbf{\Delta R}_{PQ}$ is to conceive of a moving coordinate system whose origin travels along with point Q but whose axes remain parallel to the absolute axes $x_1 y_1 z_1$. Note that this coordinate system does not rotate. An observer in this moving coordinate system observes no motion for point Q because it remains at the

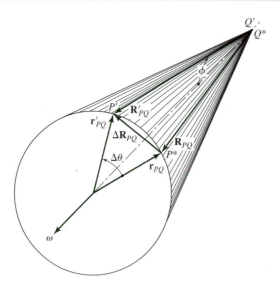

Figure 3.4 Displacement difference $\Delta\mathbf{R}_{PQ}$ as seen by a translating observer.

origin of HER coordinate system. For the displacement of point P, SHE will observe the displacement difference vector $\Delta\mathbf{R}_{PQ}$. It seems to such an observer that point Q remains fixed and that the body rotates about this fixed point as shown in Fig. 3.4.

No matter whether the observer is in the ground coordinate system or in the moving coordinate system described, the body appears to rotate through some total angle $\Delta\theta$ in its displacement from $x_2 y_2 z_2$ to $x_2' y_2' z_2'$. If we take the point of view of the fixed observer, the location of the axis of rotation is not obvious. As seen by the translating observer, the axis passes through the apparently stationary point Q; all points in the body appear to travel in circular paths about this axis, and any line in the body whose direction is normal to this axis appears to undergo an identical angular displacement of $\Delta\theta$.

The *angular velocity* of a rotating body is now defined as a vector quantity $\boldsymbol{\omega}$ having a direction along the instantaneous axis of rotation. The magnitude of the angular velocity is defined as the time rate of change of the angular orientation of any line in the body whose direction is normal to the axis of rotation. If we designate the angular displacement of any of these lines as $\Delta\theta$ and the time interval as Δt, the magnitude of the angular velocity vector $\boldsymbol{\omega}$ is

$$\omega = \lim_{\Delta t \to 0} \frac{\Delta\theta}{\Delta t} = \frac{d\theta}{dt} \tag{3.2}$$

Because we have agreed to treat counterclockwise rotations as positive, the sense of the $\boldsymbol{\omega}$ vector along the axis of rotation is in accordance with the right-hand rule.

3.3 VELOCITY DIFFERENCE BETWEEN POINTS OF A RIGID BODY

Figure 3.5*a* shows another view of the same rigid-body displacement pictured in Fig. 3.3. This is the view seen by an observer in the absolute coordinate system looking directly along the axis of rotation of the moving body, from the tip end of the $\boldsymbol{\omega}$ vector. In this view,

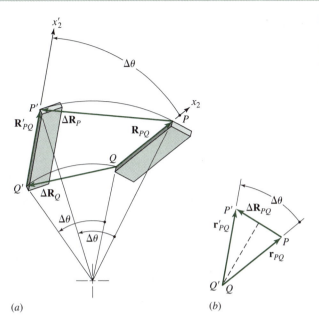

Figure 3.5 (*a*) True view of angular displacements of Fig. 3.3. (*b*) Vector subtraction to form displacement difference $\mathbf{\Delta R}_{PQ}$.

(*a*)

(*b*)

the angular displacement $\Delta\theta$ is observed in true size, and *all* lines in the body rotate through this same angle during the displacement. The displacement vectors and the position difference vectors shown are not necessarily seen in true size; they may appear fore-shortened under this viewing angle.

Figure 3.5*b* shows the same rigid-body rotation from the same viewing angle, but this time from the point of view of the translating observer. Thus this figure corresponds to the base of the cone shown in Fig. 3.4. We note that the two vectors labeled $\mathbf{r}_{PQ}$ and $\mathbf{r}'_{PQ}$ are the foreshortened views of $\mathbf{R}_{PQ}$ and $\mathbf{R}'_{PQ}$, and from Fig. 3.4 we observe that their magnitudes are

$$r_{PQ} = r'_{PQ} = R_{PQ}\sin\phi \tag{a}$$

where ϕ is the constant angle from the angular velocity vector $\boldsymbol{\omega}$ to the rotating position-difference vector $\mathbf{R}_{PQ}$ as it traverses the cone.

Looking again at Fig. 3.5*b*, we see that it can also be interpreted as a scale drawing corresponding to Eq. (2.68). It shows that the displacement different vector $\mathbf{\Delta R}_{PQ}$ is equal to the vector change in the absolute position difference $\mathbf{R}_{PQ}$ produced during the displacement

$$\mathbf{\Delta R}_{PQ} = \mathbf{R}'_{PQ} - \mathbf{R}_{PQ} \tag{b}$$

We are now ready to calculate the magnitude of the displacement-difference vector $\mathbf{\Delta R}_{PQ}$. In Fig. 3.5*b*, where it appears in true size, we construct its perpendicular bisector, from which we see that

$$\Delta R_{PQ} = 2r_{PQ}\sin\frac{\Delta\theta}{2} \tag{c}$$

and, from Eq. (a),

$$\Delta R_{PQ} = 2(R_{PQ}\sin\phi)\sin\frac{\Delta\theta}{2} \qquad (d)$$

If we now limit ourselves to small motions, the sine of the angular displacement can be approximated by the angle itself:

$$\Delta R_{PQ} = 2(R_{PQ}\sin\phi)\frac{\Delta\theta}{2} = \Delta\theta R_{PQ}\sin\phi \qquad (e)$$

Dividing by the small time increment Δt, noting that the magnitude R_{PQ} and the angle ϕ are constant during the interval, and taking the limit, we get

$$\lim_{\Delta t\to 0}\left(\frac{\Delta R_{PQ}}{\Delta t}\right) = \lim_{\Delta t\to 0}\left(\frac{\Delta\theta}{\Delta t}\right)R_{PQ}\sin\phi = \omega R_{PQ}\sin\phi \qquad (f)$$

Upon recalling the definition of ϕ as the angle between the $\boldsymbol{\omega}$ and $\mathbf{R}_{PQ}$ vectors, we can restore the vector attributes of the above equation by recognizing it as the form of a cross product. Thus

$$\lim_{\Delta t\to 0}\left(\frac{\mathbf{\Delta R}_{PQ}}{\Delta t}\right) = \frac{d\mathbf{R}_{PQ}}{dt} = \boldsymbol{\omega}\times\mathbf{R}_{PQ} \qquad (g)$$

This form is so important and so useful that it is given its own name and symbol; it is called the *velocity-difference* vector and is denoted $\mathbf{V}_{PQ}$:

$$\mathbf{V}_{PQ} = \frac{d\mathbf{R}_{PQ}}{dt} = \boldsymbol{\omega}\times\mathbf{R}_{PQ} \qquad (3.3)$$

Let us now recall the displacement-difference equation (2.67),

$$\mathbf{\Delta R}_P = \mathbf{\Delta R}_Q + \mathbf{\Delta R}_{PQ} \qquad (h)$$

Dividing this equation by Δt and taking the limit gives

$$\lim_{\Delta t\to 0}\left(\frac{\mathbf{\Delta R}_P}{\Delta t}\right) = \lim_{\Delta t\to 0}\left(\frac{\mathbf{\Delta R}_Q}{\Delta t}\right) + \lim_{\Delta t\to 0}\left(\frac{\mathbf{\Delta R}_{PQ}}{\Delta t}\right) \qquad (i)$$

which, by Eqs. (3.1) and (3.3), becomes

$$\mathbf{V}_P = \mathbf{V}_Q + \mathbf{V}_{PQ} \qquad (3.4)$$

This extremely important equation is called the *velocity-difference equation;* together with Eq. (3.3), it forms one of the primary bases of all velocity-analysis techniques. Equation (3.4) can be written for any two points with no restriction. However, as will be recognized by reviewing the above derivation, Eq. (3.3) should not be applied to any arbitrary pair of points. *This form is valid only if the two points are both attached to the same rigid body.* This restriction* can perhaps be better remembered if all subscripts are written explicitly:

$$\mathbf{V}_{P_2 Q_2} = \boldsymbol{\omega}_2 \times \mathbf{R}_{P_2 Q_2} \tag{j}$$

but in the interest of brevity, the link-number subscripts are often suppressed. Note that the link-number subscripts are the same throughout Eq. (j). If a mistaken attempt is made to apply Eq. (3.3) when points P and Q are not part of the same link, the error should be discovered because it will not be clear which $\boldsymbol{\omega}$ vector should be used.

3.4 GRAPHIC METHODS; VELOCITY POLYGONS

One major approach to velocity analysis is graphical. As seen in graphical position analysis, it is primarily of use in two-dimensional problems when only a single position requires solution. The major advantages are that a solution can be achieved quickly and that visualization of, and insight into, the problem are enhanced by the graphical approach.

As a first example of graphical velocity analysis, let us consider the two-dimensional motion of the unconstrained link shown in Fig. 3.6a. Suppose that we know the velocities of points A and B and wish to determine the velocity of point C and the angular velocity of the link. We assume that a scale diagram, Fig. 3.6a, has already been drawn of the link at the instant considered—that is, that a position analysis has been completed and that position difference vectors can be measured from the diagram.

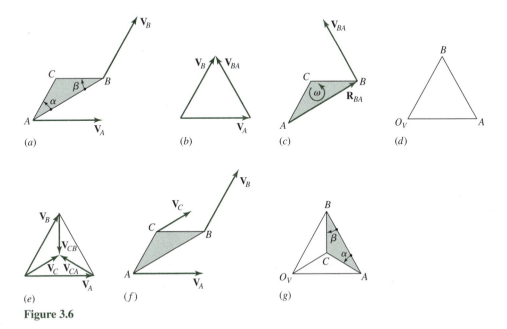

Figure 3.6

---

*More precisely, the restriction is the requirement that the distance R_{PQ} must remain constant. However, in application, the above wording fits most real situations.

Next we consider the velocity difference equation (3.4) relating points A and B,

$$\overset{\sqrt{\sqrt{}} \quad \sqrt{\sqrt{}} \quad oo}{\mathbf{V}_B = \mathbf{V}_A + \mathbf{V}_{BA}} \qquad (a)$$

where the two unknowns are the magnitude and direction of the velocity-difference vector $\mathbf{V}_{BA}$, as indicated above this symbol in the equation. Figure 3.6b shows the graphical solution to the equation. After choosing a scale to represent velocity vectors, the vectors $\mathbf{V}_A$ and $\mathbf{V}_B$ are both drawn to scale, starting from a common origin and in the given directions. The vector spanning the termini of $\mathbf{V}_A$ and $\mathbf{V}_B$ is the velocity-difference vector $\mathbf{V}_{BA}$ and is correct, within graphical accuracy, in both magnitude and direction.

The angular velocity $\boldsymbol{\omega}$ for the link can now be found from Eq. (3.3):

$$\mathbf{V}_{BA} = \boldsymbol{\omega} \times \mathbf{R}_{BA} \qquad (b)$$

Because the link is in planar motion, the $\boldsymbol{\omega}$ vector lies perpendicular to the plane of motion, that is, perpendicular to the vectors $\mathbf{V}_{BA}$ and $\mathbf{R}_{BA}$. Therefore, considering the magnitudes in the above equation,

$$V_{BA} = \omega R_{BA}$$

or

$$\omega = \frac{V_{BA}}{R_{BA}} \qquad (c)$$

The numerical magnitude of ω can therefore be found by scaling V_{BA} from Fig. 3.6b and R_{BA} from Fig. 3.6a, being careful to apply the scale factors for units properly; it is usual practice to evaluate ω in units of radians per second.

The magnitude of ω is not a complete solution for the angular velocity vector; the direction must also be determined. As observed above, the ω vector is perpendicular to the plane of the link itself because the motion is planar. However, this does not say whether ω is directed into or out of the plane of the figure. This is determined as shown in Fig. 3.6c. Taking the point of view of a translating observer, moving with point A but not rotating, we can picture the link as rotating about point A. The velocity difference $\mathbf{V}_{BA}$ is the only velocity seen by such an observer. Therefore, interpreting $\mathbf{V}_{BA}$ to indicate the direction of rotation of point B about point A, we find the direction of ω, counterclockwise in this example. Although it is not strict vector notation, it is common practice in two-dimensional problems to indicate the final solution in the form $\omega = 15$ rad/s ccw, which indicates both magnitude and direction.

The practice of constructing vector diagrams using thick black lines, such as in Fig. 3.6b, makes them easy to read; but when the diagram is the graphical solution of an equation, it is not very accurate. For this reason it is customary to construct the graphical solution with thin sharp lines, made with a hard drawing pencil, as shown in Fig. 3.6d. The solution is started by choosing a scale and a point labeled O_V to represent zero velocity. Absolute velocities such as $\mathbf{V}_A$ and $\mathbf{V}_B$ are constructed with their origins at O_V, and their termini are labeled as points A and B. The line *from A to B* then represents the velocity difference $\mathbf{V}_{BA}$. As we continue, it will be seen that these labels at the vertices are sufficient to determine the precise notation of all velocity differences represented by lines in the diagram. Notice, for example, that $\mathbf{V}_{BA}$ is represented by the vector *to point B from point A*. With this labeling convention, no arrowheads or additional notation are necessary and do not clutter the diagram. Such a diagram is called a *velocity polygon* and, as we will see, adds considerable convenience to the graphical solution technique.

A danger of this convention, however, is that the analyst will begin to think of the technique as a series of graphical "tricks" and lose sight of the fact that each line drawn can and should be fully justified by a corresponding vector equation. The graphics are merely a convenient solution technique, not a substitute for a sound theoretical basis.

Returning to Fig. 3.6c, it may have appeared as coincidence that the vector $\mathbf{V}_{BA}$ was perpendicular to $\mathbf{R}_{BA}$. Looking back to Eq. (b), however, we see that it was a necessary outcome, resulting from the cross product with the $\boldsymbol{\omega}$ vector. We will take advantage of this property in the next step.

Now that $\boldsymbol{\omega}$ has been found, let us determine the absolute velocity of point C. We can relate this by two velocity-difference equations to the absolute velocities of both points A and B.

$$\overset{\text{oo}}{\phantom{\mathbf{V}_C}} \overset{\surd\surd}{} \overset{\text{o}\surd}{\phantom{\mathbf{V}_A}} \overset{\surd\surd}{} \overset{\text{o}\surd}{\phantom{\mathbf{V}_{CA}}}$$
$$\mathbf{V}_C = \mathbf{V}_A + \mathbf{V}_{CA} = \mathbf{V}_B + \mathbf{V}_{CB} \qquad\qquad (d)$$

Because points A, B, and C are all on the same rigid link, each of the velocity difference vectors, $\mathbf{V}_{CA}$ and $\mathbf{V}_{CB}$, is of the form $\boldsymbol{\omega} \times \mathbf{R}$, using $\mathbf{R}_{CA}$ and $\mathbf{R}_{CB}$, respectively. As a result, $\mathbf{V}_{CA}$ is perpendicular to $\mathbf{R}_{CA}$ and $\mathbf{V}_{CB}$ is perpendicular to $\mathbf{R}_{CB}$. The directions of these two terms are therefore indicated as known in Eq. (d).

Because $\boldsymbol{\omega}$ has already been determined, it is easy to calculate the magnitudes of $\mathbf{V}_{CA}$ and $\mathbf{V}_{CB}$ by using a formula like Eq. (c); however, we will assume that this is not done. Instead, we form the graphical solution to Eq. (d). This equation states that a vector that is perpendicular to $\mathbf{R}_{CA}$ must be added to $\mathbf{V}_A$ and that the result will equal the sum of $\mathbf{V}_B$ and a vector perpendicular to $\mathbf{R}_{CB}$. The solution is illustrated in Fig. 3.6e. In practice the solution is continued on the same diagram as Fig. 3.6d and results in Fig. 3.6g. A line perpendicular to $\mathbf{R}_{CA}$ (representing $\mathbf{V}_{CA}$) is drawn starting at point A (representing addition to $\mathbf{V}_A$); similarly, a line is drawn perpendicular to $\mathbf{R}_{CB}$ starting at point B. The point of intersection of these two lines is labeled C and represents the solution to Eq. (d). The line from O_V to point C now represents the absolute velocity $\mathbf{V}_C$. This velocity can be transferred back to the link and interpreted as $\mathbf{V}_C$ in both magnitude and direction as shown in Fig. 3.6f.

In seeing the shading and the labeled angles α and β in Fig. 3.6g and 3.6a, we are led to investigate whether the two triangles labeled ABC in each of these figures are similar in shape, as they appear to be. In reviewing the construction steps we see that indeed they are, because the velocity-difference vectors $\mathbf{V}_{BA}$, $\mathbf{V}_{CA}$, and $\mathbf{V}_{CB}$ are perpendicular to the respective position-difference vectors, $\mathbf{R}_{BA}$, $\mathbf{R}_{CA}$, and $\mathbf{R}_{CB}$. This property would be true no matter what the shape of the moving link; a similarly shaped figure would appear in the velocity polygon. Its sides are always scaled up or down by a factor equal to the angular velocity of the link, and it is always rotated by 90° in the direction of the angular velocity. The properties result from the fact that each velocity-difference vector between two points on the link is of the form of a cross product of the same $\boldsymbol{\omega}$ vector with the corresponding position-difference vector. This similarly shaped figure in the velocity polygon is commonly referred to as the velocity image of the link, and any moving link will have a corresponding *velocity image* in the velocity polygon.

If the concept of the velocity image had been known initially, the solution process could have been speeded up considerably. Once the solution has progressed to the state of Fig. 3.6d, the velocity-image points A and B are known. One can use these two points as the base of a triangle similar to the link shape and label the image point C directly, without writing Eq. (d). Care must be taken not to allow the triangle to be flipped over between the position diagram and the velocity image; but the solution can proceed quickly, accurately, and naturally, resulting in Fig. 3.6g. Here again the caution is repeated that all steps in the solution are based on strictly derived vector equations and are not tricks. It is very wise to continue to write the corresponding vector equations until one is thoroughly familiar with the procedure.

To increase familiarity with graphical velocity-analysis techniques, we analyze two typical example problems.

EXAMPLE 3.1

The four-bar linkage, shown to scale in Fig. 3.7a with all necessary dimensions, is driven by crank 2 at a constant angular velocity of $\omega_2 = 900$ rev/min ccw. Find the instantaneous velocities of points E and F and the angular velocities of links 3 and 4 at the position shown.

SOLUTION

To obtain a graphical solution, we first calculate the angular velocity of link 2 in radians per second. This is

$$\omega_2 = \left(900 \, \frac{\text{rev}}{\text{min}} \right) \left(2\pi \, \frac{\text{rad}}{\text{rev}} \right) \left(\frac{1 \, \text{min}}{60 \, \text{s}} \right) = 94.2 \, \text{rad/s ccw} \tag{1}$$

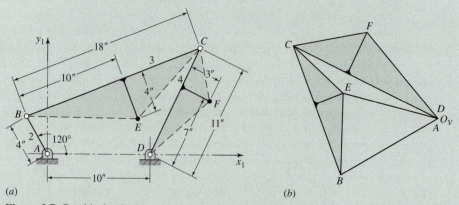

(a)   (b)

Figure 3.7 Graphical velocity analysis of a four-bar linkage. Example 3.1: (a) scale diagram; (b) velocity polygon.

Then we notice that the point A remains fixed and calculate the velocity of point B:

$$\mathbf{V}_B = \overset{0}{\cancel{\mathbf{V}_A}} + \mathbf{V}_{BA} = \boldsymbol{\omega}_2 \times \mathbf{R}_{BA}$$

$$V_B = (94.2 \, \text{rad/s}) \left(\frac{4}{12} \, \text{ft} \right) = 31.4 \, \text{ft/s} \tag{2}$$

We note that the form $\boldsymbol{\omega} \times \mathbf{R}$ was used for the velocity difference, not for the absolute velocity $\mathbf{V}_B$ directly. In Fig. 3.7b we choose the point O_V and a velocity scale factor. We note that the image point A is coincident with O_V and construct the line AB perpendicular to $\mathbf{R}_{BA}$ and toward the left because of the counterclockwise direction of $\boldsymbol{\omega}_2$; this line represents $\mathbf{V}_{BA}$.

If we attempt at this time to write an equation directly for the velocity of point E, we find by counting the unknowns that it cannot be solved yet. Therefore, we next write two equations for the velocity of point C. Because the velocities of points C_3 and C_4 must be equal (links 3 and 4 are pinned together at C), we obtain

$$\mathbf{V}_C = \mathbf{V}_B + \mathbf{V}_{CB} = \overset{0}{\cancel{\mathbf{V}_D}} + \mathbf{V}_{CD} \tag{3}$$

We now construct two lines in the velocity polygon: The line BC is drawn from B perpendicular to $\mathbf{R}_{CB}$, and the line DC is drawn from D (coincident with O_V because $\mathbf{V}_D = 0$) perpendicular to $\mathbf{R}_{CD}$. We label the point of intersection as point C. When the lengths of these lines are scaled, we find that $V_{CB} = 38.4 \, \text{ft/s}$ and $V_C = V_{CD} = 45.5 \, \text{ft/s}$. The angular velocities of links 3 and 4 can now be found:

$$\omega_3 = \frac{V_{CB}}{R_{CB}} = \frac{38.4 \, \text{ft/s}}{18/12 \, \text{ft}} = 25.6 \, \text{rad/s ccw} \qquad \qquad Ans. \tag{4}$$

$$\omega_4 = \frac{V_{CD}}{R_{CD}} = \frac{45.5 \, \text{ft/s}}{11/12 \, \text{ft}} = 49.6 \, \text{rad/s ccw} \qquad \qquad Ans. \tag{5}$$

where the directions of $\boldsymbol{\omega}_3$ and $\boldsymbol{\omega}_4$ were found by the technique illustrated in Fig. 3.6c.

There are now several methods of finding the velocity of point E; that is, $\mathbf{V}_E$. In one method we measure $\mathbf{R}_{EB}$ from the scale drawing of Fig. 3.7a; and then, because points B and E are both attached to link 3, we can calculate*

$$V_{EB} = \omega_3 R_{EB} = (25.6 \, \text{rad/s}) \left(\frac{10.8}{12} \, \text{ft} \right) = 23.0 \, \text{ft/s} \tag{6}$$

We can now construct the line BE in the velocity polygon, drawn to the proper scale and perpendicular to $\mathbf{R}_{EB}$, thus solving** the velocity-difference equation

$$\mathbf{V}_E = \mathbf{V}_B + \mathbf{V}_{EB} \tag{7}$$

The result is

$$V_E = 27.6 \, \text{ft/s} \qquad \qquad Ans.$$

as scaled from the velocity polygon.

---

*There is no restriction in our derivation which requires that $\mathbf{R}_{EB}$ lie along the material portion of link 3 in order to use Eq. (6).

**Note that numerical values should not be substituted into Eq. (7) directly: This equation requires vector addition, not scalar, and this is precisely the purpose of constructing the velocity polygon.

Alternatively, $\mathbf{V}_E$ can be obtained from

$$\mathbf{V}_E = \mathbf{V}_C + \mathbf{V}_{EC} \tag{8}$$

by an identical procedure to that used for Eq. (7). This solution would produce the triangle $O_V EC$ in the velocity polygon.

Suppose we wish to find $\mathbf{V}_E$ without the intermediate step of calculating $\boldsymbol{\omega}_3$. In this case we write Eqs. (7) and (8) simultaneously.

$$\overset{\text{oo}}{\mathbf{V}_E} = \overset{\sqrt{}\sqrt{}}{\mathbf{V}_B} + \overset{\text{o}\sqrt{}}{\mathbf{V}_{EB}} = \overset{\sqrt{}\sqrt{}}{\mathbf{V}_C} + \overset{\text{o}\sqrt{}}{\mathbf{V}_{EC}} \tag{9}$$

Drawing lines EB (perpendicular to $\mathbf{R}_{EB}$) and EC (perpendicular to $\mathbf{R}_{EC}$) in the velocity polygon, we find their intersection and so solve Eq. (9).

Perhaps the easiest method of solving for $\mathbf{V}_E$, however, is to take advantage of the concept of the velocity image of link 3. Recognizing that the velocity-image points B and C have already been found, we can construct the triangle BEC in the velocity polygon, similar in shape to the triangle BEC in the scale diagram of link 3. This locates point E in the velocity polygon and thus gives a solution for $\mathbf{V}_E$.

The velocity $\mathbf{V}_F$ can also be found by any of the above methods using points C, D, and F of link 4. The result is

$$V_F = 31.8 \text{ ft/s} \qquad\qquad \textit{Ans.}$$

EXAMPLE 3.2

The offset slider-crank mechanism of Fig. 3.8a is driven by slider 4 at a speed of $V_C = 10$ m/s to the left at the phase shown. Determine the instantaneous velocity of point D and the angular velocities of links 2 and 3.

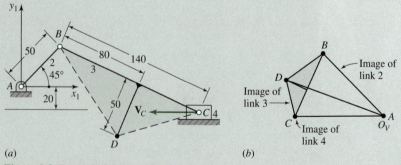

(a)   (b)

Figure 3.8 Example 3.2: (a) scale diagram of offset slider-crank mechanism (dimensions in millimeters); (b) velocity polygon.

SOLUTION

The velocity scale and pole O_V are chosen and $\mathbf{V}_C$ is drawn, thus locating point C as shown in Fig. 3.8b. Simultaneous equations are then written for the velocity of point B,

$$
\overset{\text{oo}}{\mathbf{V}_B} = \overset{\surd\surd}{\mathbf{V}_C} + \overset{\text{o}\surd}{\mathbf{V}_{BC}} = \overset{0\ \text{o}\surd}{\mathbf{V}_A} + \overset{\text{o}\surd}{\mathbf{V}_{BA}} \tag{10}
$$

and solved for the location of point B in the velocity polygon.

Having found points B and C, we can construct the velocity image of link 3 as shown to locate point D; we then scale the line O_VD, which gives

$$
V_D = 12.0 \text{ m/s} \qquad\qquad\qquad Ans.
$$

The angular velocities of links 2 and 3, respectively, are

$$
\omega_2 = \frac{V_{BA}}{R_{BA}} = \frac{10.0 \text{ m/s}}{0.05 \text{ m}} = 200 \text{ rad/s ccw} \qquad\qquad Ans. \quad (11)
$$

$$
\omega_3 = \frac{V_{BC}}{R_{BC}} = \frac{7.5 \text{ m/s}}{0.14 \text{ m}} = 53.6 \text{ rad/s cw} \qquad\qquad Ans. \quad (12)
$$

In this second example problem, Fig. 3.8b, the velocity image of each link is indicated in this polygon. If the analysis of any problem is carried through completely, there will be a velocity image for each link of the mechanism. The following points are true, in general, and can be verified in the above examples.

1. The velocity image of each link is a scale reproduction of the shape of the link in the velocity polygon.
2. The velocity image of each link is rotated $90°$ in the direction of the angular velocity of the link.
3. The letters identifying the vertices of each link are the same as those in the velocity polygon and progress around the velocity image in the same order and in the same angular direction as around the link.
4. The ratio of the size of the velocity image of a link to the size of the link itself is equal to the magnitude of the angular velocity of the link. In general, it is not the same for different links in the same mechanism.
5. The velocity of all points on a translating link are equal, and the angular velocity is zero. Therefore, the velocity image of a link which is translating shrinks to a single point in the velocity polygon.
6. The point O_V in the velocity polygon is the image of all points with zero absolute velocity; it is the velocity image of the fixed link.
7. The absolute velocity of any point on any link is represented by the line from O_V to the image of the point. The velocity difference vector between any two points, say P and Q, is represented by the line from image point P to image point Q.

3.5 APPARENT VELOCITY OF A POINT IN A MOVING COORDINATE SYSTEM

In analyzing the velocities of various machine components, we frequently encounter problems in which it is convenient to describe how a point moves with respect to another moving link but not at all convenient to describe the absolute motion of the point. An example of this occurs when a rotating link contains a slot along which another link is constrained to slide. With the motion of the link containing the slot and the relative sliding motion taking place in the slot as known quantities, we may wish to find the absolute motion of the sliding member. It was for problems of this type that the apparent-displacement vector was defined in Section 2.17, and we now wish to extend this concept to velocity.

In Fig. 3.9 we recall the definition of the apparent-displacement vector. A rigid link having some general motion carries a coordinate system $x_2 y_2 z_2$ attached to it. At a certain time t, the coordinate system lies at $x_2' y_2' z_2'$. All points of link 2 move with the coordinate system.

Also, during the same time interval, another point P_3 of another link 3 is constrained in some manner to move along a known path with respect to link 2. In Fig. 3.9 this constraint is depicted as a slot carrying a pin from link 3; the center of the pin is the point P_3. Although it is pictured in this way, the constraint may occur in a variety of different forms. *The only assumption here is that the path which the moving point P_3 traces in coordinate system $x_2 y_2 z_2$, that is, the locus of the tip of the apparent-position vector* $\mathbf{R}_{P_3/2}$, *is known.*

Recalling the apparent-displacement equation (2.70),

$$\Delta \mathbf{R}_{P_3} = \Delta \mathbf{R}_{P_2} + \Delta \mathbf{R}_{P_3/2}$$

we divide by Δt and take the limit,

$$\lim_{\Delta t \to 0} \left(\frac{\Delta \mathbf{R}_{P_3}}{\Delta t} \right) = \lim_{\Delta t \to 0} \left(\frac{\Delta \mathbf{R}_{P_2}}{\Delta t} \right) + \lim_{\Delta t \to 0} \left(\frac{\Delta \mathbf{R}_{P_3/2}}{\Delta t} \right) \qquad (a)$$

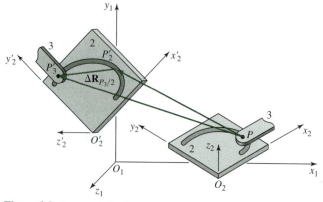

Figure 3.9 Apparent displacement.

We now define the *apparent-velocity* vector as

$$\mathbf{V}_{P_3/2} = \lim_{\Delta t \to 0} \left(\frac{\Delta \mathbf{R}_{P_3/2}}{\Delta t} \right) = \frac{d\mathbf{R}_{P_3/2}}{dt} \qquad (3.5)$$

and, in the limit, Eq. (*a*) becomes

$$\mathbf{V}_{P_3} = \mathbf{V}_{P_2} + \mathbf{V}_{P_3/2} \qquad (3.6)$$

called the *apparent-velocity* equation.

We note from its definition, Eq. (3.5), that the apparent velocity resembles the absolute velocity except that it comes from the apparent displacement rather than the absolute displacement. Thus, in concept, it is the velocity of the moving point P_3 *as it would appear to an observer attached to the moving link* 2 and making observations in coordinate system $x_2y_2z_2$. This concept accounts for its name. We also note that the absolute velocity is a special case of the apparent velocity where the observer happens to be fixed to the $x_1y_1z_1$ coordinate system.

We can get further insight into the nature of the apparent-velocity vector by studying Fig. 3.10. This figure shows the view of the moving point P_3 as it would be seen by the moving observer. To HER, the path traced on link 2 appears stationary and the moving point moves along this path from P_3 to P_3'. Working in this coordinate system, suppose we locate the point C as the center of curvature of the path of point P_3. For small distances from P_2 the path follows the circular arc $P_3 P_3'$ with center C and radius of curvature ρ. We now define the unit-vector tangent to the path $\hat{\tau}$ with positive sense in the direction of positive movement. The plane defined by this tangent vector $\hat{\tau}$ and the center of curvature C is called the *osculating plane*. If we choose a preferred side of this plane as the positive side and denote it by the positive $\hat{v}$ unit vector, we can complete a right-hand Cartesian coordinate system by defining the unit vector normal to the path

$$\hat{\rho} = \hat{\tau} \times \hat{v} \qquad (3.7)$$

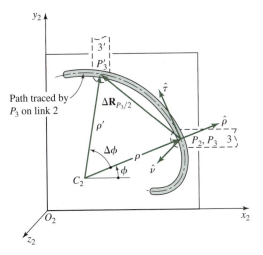

Figure 3.10 Apparent displacement of point P_3 as seen by an observer on link 2.

This coordinate system moves with its origin tracking the motion of point P_3. However, it rotates with the radius-of-curvature vector (through the angle $\Delta\phi$) as the motion progresses, *not* the same rotation as either link 2 or link 3. Note that if positive movement had been chosen in the opposite direction along this curve, the sense of the $\hat{\boldsymbol{\tau}}$ vector and then the $\hat{\boldsymbol{\rho}}$ vector would have been reversed. This would mean that the radius of curvature ρ would have had a negative value; however, the unit vector $\hat{\boldsymbol{\rho}}$ would still have come from Eq. (3.7) rather than from the sense of the radius of curvature vector. Similarly, the angle $\Delta\phi$ would then have been clockwise (from the positive $\hat{\boldsymbol{v}}$ side of the plane) and would have had a negative value.

We now define the scalar Δs as the distance along the curve from P_3 to P_3' and note that $\boldsymbol{\Delta R}_{P_3/2}$ is the chord of the same arc. However, for very short Δt, the magnitude of the chord and the arc distance approach equality. Therefore,

$$\lim_{\Delta t \to 0} \left(\frac{\boldsymbol{\Delta R}_{P_3/2}}{\Delta s} \right) = \frac{d\mathbf{R}_{P_3/2}}{ds} = \hat{\boldsymbol{\tau}} \tag{3.8}$$

Here both $\boldsymbol{\Delta R}_{P_3/2}$ and Δs are considered to be functions of time. Therefore, from Eq. (3.5),

$$\mathbf{V}_{P_3/2} = \lim_{\Delta t \to 0} \left(\frac{\boldsymbol{\Delta R}_{P_3/2}}{\Delta s} \frac{\Delta s}{\Delta t} \right) = \frac{d\mathbf{R}_{P_3/2}}{ds} \frac{ds}{dt} = \frac{ds}{dt}\hat{\boldsymbol{\tau}} \tag{3.9}$$

There are two important conclusions from this result: The magnitude of the apparent velocity is equal to the speed with which the point P_3 progresses along the path, and the apparent-velocity vector is always tangent to the path traced by the point in the coordinate system of the observer. The first of these two results is seldom useful in the solution of problems, although it is an important concept. The second result is extremely useful, because the apparent path traced can often be visualized from the nature of the constraints and thus the direction of the apparent-velocity vector becomes known. Note that only the tangent to the path needs to be determined in this chapter; the radius of curvature ρ is not needed until we attempt acceleration analysis in the next chapter.

EXAMPLE 3.3

Shown in Fig. 3.11a is an inversion of the slider-crank mechanism. Link 2, the crank, is driven at an angular velocity of 36 rad/s cw. Link 3 slides on link 4 and is pivoted to the crank at A. Find the angular velocity of link 4.

SOLUTION

We first calculate the velocity of point A:

$$\mathbf{V}_A = \mathbf{V}_E^{\,0} + \mathbf{V}_{AE} = \boldsymbol{\omega}_2 \times \mathbf{R}_{AE}$$

$$V_A = (36 \text{ rad/s})\left(\frac{3}{12} \text{ ft}\right) = 9 \text{ ft/s} \tag{1}$$

and we plot this from the pole O_V to locate point A in the velocity polygon (see Fig. 3.11b).

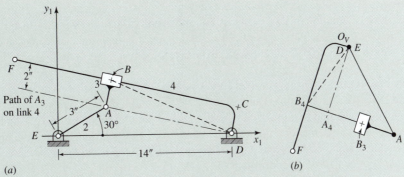

Figure 3.11 Example 3.3: (*a*) inverted slider-crank mechanism; (*b*) velocity polygon.

Next we distinguish two different points B_3 and B_4 at the location of sliding. Point B_3 is part of link 3, and point B_4 is attached to link 4, but at the instant shown the two are coincident. Note that, as seen by an observer on link 4, point B_3 seems to slide along link 4, thus defining a straight path along the line CF. Thus we can write the apparent-velocity equation

$$\mathbf{V}_{B_3} = \mathbf{V}_{B_4} + \mathbf{V}_{B_3/4} \tag{2}$$

When point B_3 is related to A and point B_4 is related to D by velocity differences, expansion of Eq. (2) gives

$$
\overset{\sqrt{\sqrt{}}}{\mathbf{V}_A} + \overset{\text{o}\sqrt{}}{\mathbf{V}_{B_3A}} = \overset{0}{\cancel{\mathbf{V}_D}} + \overset{\text{o}\sqrt{}}{\mathbf{V}_{B_4D}} + \overset{\text{o}\sqrt{}}{\mathbf{V}_{B_3/4}} \tag{3}
$$

where $\mathbf{V}_{B_3A}$ is perpendicular to $\mathbf{R}_{BA}$, $\mathbf{V}_{B_4D}$ is perpendicular to $\mathbf{R}_{BD}$ (shown dashed), and $\mathbf{V}_{B_3/4}$ has a direction defined by the tangent to the path of sliding at B.

Although Eq. (3) appears to have three unknowns, if we note that $\mathbf{V}_{B_3A}$ and $\mathbf{V}_{B_3/4}$ have identical directions, the equation can be rearranged as

$$
\overset{\sqrt{\sqrt{}}}{\mathbf{V}_A} + \overset{\text{o}\sqrt{}}{\overbrace{(\mathbf{V}_{B_3A} - \mathbf{V}_{B_3/4})}} = \overset{\text{o}\sqrt{}}{\mathbf{V}_{B_4D}} \tag{4}
$$

and the difference shown in parentheses can be treated as a single vector of known direction. The equation is thereby reduced to two unknowns and can be solved graphically to locate point B_4 in the velocity polygon.

The magnitude $\mathbf{R}_{BD}$ can be computed or measured from the diagram, and $\mathbf{V}_{B_4D}$ can be scaled from the velocity polygon (the dashed line from O_V to B_4). Therefore,

$$\omega_4 = \frac{V_{B_4D}}{R_{BD}} = \frac{7.3 \text{ ft/s}}{11.6/12 \text{ ft}} = 7.55 \text{ rad/s ccw} \qquad \textit{Ans.} \tag{5}$$

Although the problem as stated is now complete, the velocity polygon has been extended to show the images of links 2, 3, and 4. In doing so it was necessary to note that since links 3 and

4 always remain perpendicular to each other, they must rotate at the same rate. Thus $\omega_3 = \omega_4$. This allowed the calculation of $\mathbf{V}_{BA} = \omega_3 \times \mathbf{R}_{BA}$ and the plotting of the velocity-image point B_3. We also note that the velocity images of links 3 and 4 are of comparable size because $\omega_3 = \omega_4$. However, they have quite a different scale than the velocity image of link 2, the line $O_V A$, because ω_2 is a larger angular velocity.

Another approach to the same problem avoids the need to combine terms as in Eq. (4). If we consider an observer riding on link 4 and ask what SHE would see for the path of point A in HER coordinate system, we find that this path is a straight line parallel to the line CF, as indicated in Fig. 3.11a. Let us now define one point of this path as A_4. At the instant shown, the point A_4 is located coincident with points A_2 and A_3. However, A_4 *does not move with the pin; it is attached to link 4 and rotates with the path around the fixed point D.* Because we can identify the path traced by A_2 and A_3 on link 4, we can write the apparent-velocity equation

$$\mathbf{V}_{A_2} = \mathbf{V}_{A_4} + \mathbf{V}_{A_2/4} \tag{6}$$

and because point A_4 is part of link 4, we have

$$\mathbf{V}_{A_4} = \cancel{\mathbf{V}_D}^{0} + \mathbf{V}_{A_4 D} \tag{7}$$

Substituting Eq. (6) into (7) gives

$$\overset{\sqrt{\sqrt{}}}{\mathbf{V}_{A_2}} = \overset{o\sqrt{}}{\mathbf{V}_{A_4 D}} + \overset{o\sqrt{}}{\mathbf{V}_{A_2/4}} \tag{8}$$

where $\mathbf{V}_{A_4 D}$ is perpendicular to $\mathbf{R}_{AD}$ and $\mathbf{V}_{A_2/4}$ is tangent to the path. Solving this equation locates image point A_4 in the velocity polygon and allows a solution for $\omega_4 = V_{A_4 D}/R_{AD}$. The remainder of the velocity polygon can then be found as shown above.

It would, however, indicate a wrong concept to attempt to use the equation

$$\mathbf{V}_{A_4} = \mathbf{V}_{A_2} + \mathbf{V}_{A_4/2}$$

rather than Eq. (6) because the *path* traced by point A_4 in a coordinate system attached to link 2 *is not known.**

Another insight into the nature and use of the apparent-velocity equation is provided by the following example.

EXAMPLE 3.4

An airplane is traveling at a speed of 300 km/h on a circular path of radius 5 km with center at C, as shown in Fig. 3.12. A rocket 30 km away from the plane is traveling on a straight course at a speed of 2000 km/h. Determine the velocity of the rocket as observed by the pilot of the plane.

---

*Although the use of this equation would suggest a faulty understanding, it still yields a correct solution. If the corresponding path were found, it would be tangent to the path used for point A_2 on link 4. Since the *tangents* to the two paths are the same even though the paths are not, the two solutions would both yield the same correct numeric result. This is not true in acceleration analysis, Chapter 4; therefore, the concept should be studied and this "backward" use should be avoided.

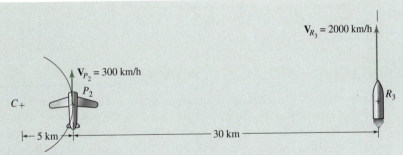

Figure 3.12 Example 3.4.

SOLUTION

Because the plane is on a circular course, the point C_2, attached to the coordinate system of the plane but coincident with C, has no motion. Therefore the angular velocity of the plane is

$$\omega_2 = \frac{V_{PC}}{R_{PC}} = \frac{V_P}{R_{PC}} = \frac{300 \text{ km/h}}{5 \text{ km}} = 60 \text{ rad/h ccw} \qquad (9)$$

The posed question obviously requires the calculation of the apparent velocity $\mathbf{V}_{R_3/2}$, but this equation applies *only between coincident points*. Therefore, we define another point R_2, attached to the rotating coordinate system of the plane but located coincident with the rocket R_3 at the instant shown. As part of the plane, the velocity of this point is

$$\mathbf{V}_{R_2} = \mathbf{V}_P + \omega_2 \times \mathbf{R}_{RP} = 300 \frac{\text{km}}{\text{h}} + \left(60 \frac{\text{rad}}{\text{h}}\right)(30 \text{ km}) = 2100 \text{ km/h} \qquad (10)$$

where the values are added algebraically because the vectors are parallel. The apparent velocity can now be calculated:

$$\mathbf{V}_{R_3/2} = \mathbf{V}_{R_3} - \mathbf{V}_{R_2}$$
$$V_{R_3/2} = 2000 \text{ km/h} - 2100 \text{ km/h} = -100 \text{ km/h} \qquad \textit{Ans.} \quad (11)$$

Thus, as seen by the pilot of the plane, the rocket appears to be backing up at a speed of 100 km/h. This result becomes better understood as we consider the motion of point R_2. This point is being treated as *attached to the plane* and, therefore, seems stationary to the pilot. Yet, in the absolute coordinate system, this point is traveling faster than the rocket; the rocket is not keeping up with this point and therefore appears to the pilot to be backing up.

3.6 APPARENT ANGULAR VELOCITY

When two rigid bodies rotate with different angular velocities, the vector difference between the two is defined as the *apparent angular velocity*. Thus

$$\omega_{3/2} = \omega_3 - \omega_2 \qquad (3.10)$$

which can also be written as

$$\omega_3 = \omega_2 + \omega_{3/2} \qquad (3.11)$$

It will be seen that $\omega_{3/2}$ is the angular velocity of body 3 as it would appear to an observer attached to, and rotating with, body 2. Compare this equation with Eq. (3.6) for the apparent velocity of a point.

3.7 DIRECT CONTACT AND ROLLING CONTACT

Two elements of a mechanism which are in direct contact with each other have relative motion that may or may not involve sliding between the links at the point of direct contact. In the cam-and-follower system shown in Fig. 3.13a, the cam, link 2, drives the follower, link 3, by direct contact. We see that if slip were not possible between links 2 and 3 at point P, the triangle PAB would form a truss; therefore, sliding as well as rotation must take place between the links.

Let us distinguish between the two points P_2 attached to link 2 and P_3 attached to link 3. They are coincident points, both located at P at the instant shown; therefore we can write the apparent-velocity equation

$$\mathbf{V}_{P_3/2} = \mathbf{V}_{P_3} - \mathbf{V}_{P_2} \tag{3.12}$$

If the two absolute velocities $\mathbf{V}_{P_3}$ and $\mathbf{V}_{P_2}$ were both known, they could be subtracted to find $\mathbf{V}_{P_3/2}$. Components could then be taken along directions defined by the common normal and common tangent to the surfaces at the point of direct contact. The components of $\mathbf{V}_{P_3}$ and $\mathbf{V}_{P_2}$ along the common normal must be equal, and this component of $\mathbf{V}_{P_3/2}$ must be zero. Otherwise, either the two links would separate or they would interfere, both contrary to our basic assumption that contact persists. The total apparent velocity $\mathbf{V}_{P_3/2}$ *must therefore lie along the common tangent* and is the velocity of the relative sliding motion within the direct-contact interface. The velocity polygon for this system is shown in Fig. 3.13b.

It is possible in other mechanisms for there to be direct contact between links without slip between the links. The cam-follower system of Fig. 3.14, for example, might have high friction between the roller, link 3, and the cam surface, link 2, and restrain the wheel to roll against the cam without slip. Henceforth we will restrict the use of the term *rolling contact* to situations where *no slip* takes place. By "no slip" we imply that the apparent "slipping" velocity of Eq. (3.12) is zero:

$$\mathbf{V}_{P_3/2} = 0 \tag{3.13}$$

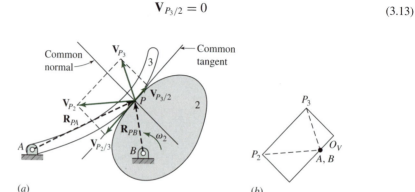

(a)   (b)

Figure 3.13 Apparent sliding velocity at a point of direct contact.

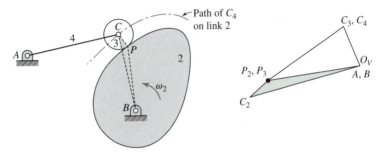

Figure 3.14 Cam-follower system with rolling contact between links 2 and 3.

This equation is sometimes referred to as the rolling-contact condition for velocity. By Eq. (3.12) it can also be written as

$$\mathbf{V}_{P_3} = \mathbf{V}_{P_2} \tag{3.14}$$

which says that *the absolute velocities of two points in rolling contact are equal.*

The graphical solution of the problem of Fig. 3.14 is also shown in the figure. Given ω_2, the velocity difference $\mathbf{V}_{P_2B}$ can be calculated and plotted, thus locating point P_2 in the velocity polygon. Using Eq. (3.13), the rolling contact condition, we also label this point P_3. Next, writing simultaneous equations for $\mathbf{V}_C$, using $\mathbf{V}_{CP_3}$ and $\mathbf{V}_{CA}$, we can find the velocity-image point C. Then ω_3 and ω_4 can be found from $\mathbf{V}_{CP}$ and $\mathbf{V}_{CA}$, respectively.

Another approach to the solution of the same problem involves defining a fictitious point C_2 which is located instantaneously coincident with points C_3 and C_4 but which is understood to be attached to, and move with, link 2, as shown by the shaded triangle BPC_2. When the velocity-image concept is used for link 2, the velocity-image point C_2 can be located. Noticing that point C_4 (and C_3) traces a known path on link 2, we can write and solve the apparent-velocity equation involving $\mathbf{V}_{C_4/2}$, thus obtaining the velocity $\mathbf{V}_{C_4}$ (and ω_4, if desired) without dealing with the point of direct contact. This second approach would be necessary if we had not assumed rolling contact (no-slip) at P.

3.8 SYSTEMATIC STRATEGY FOR VELOCITY ANALYSIS

Review of the preceding sections and their example problems will show that we have now developed enough tools for dealing with those situations that normally arise in the analysis of rigid-body mechanical systems. It will also be noticed that the word "relative" velocity has been carefully avoided. Instead we note that whenever the desire for using "relative" velocity arises, there are always two points whose velocities are to be "related"; also, these two points are attached either to the same or to two different rigid bodies. Therefore, we can organize all situations into the four cases shown in Table 3.1. In this table we can see that when the two points are separated by a distance, only the velocity difference equation is appropriate for use and two points on the same link should be used. When it is desirable to switch to points of another link, then coincident points should be chosen and the apparent velocity equation should be used.

It will be noted that even the notation has been made different to continually remind us that these are two totally different situations and the formulae are not interchangeable

TABLE 3.1 "Relative" Velocity Equations

Points Are	Coincident	Separated
On same body	*Trivial Case:* $\mathbf{V}_P = \mathbf{V}_Q$.	*Velocity Difference;* $\mathbf{V}_P = \mathbf{V}_Q + \mathbf{V}_{PQ}$ $\mathbf{V}_{PQ} = \boldsymbol{\omega}_j \times \mathbf{R}_{PQ}$
On different bodies	*Apparent Velocity;* $\mathbf{V}_{P_i} = \mathbf{V}_{P_j} + \mathbf{V}_{P_i/j}$ where path $P_{i/j}$ is known.	*Too general; use two steps.*
	Rolling Contact Velocity $\mathbf{V}_{P_i} = \mathbf{V}_{P_j}$ and $\mathbf{V}_{P_i/j} = 0$	

between the two. We should not try to use an "$\boldsymbol{\omega} \times \mathbf{R}$" formula when using the apparent velocity, but will not find a useful $\boldsymbol{\omega}$ or $\mathbf{R}$ if we try. Similarly, when using the velocity difference, there will be no question of which $\boldsymbol{\omega}$ to use because only one link will pertain. Even further advantages will become clear when we study accelerations in Chapter 4.

Careful review of Examples 3.1 through 3.4 show how the suggested strategy is applied to problems of velocity analysis.

3.9 ANALYTIC METHODS

Next let us consider the simple algebraic methods of velocity analysis. For some types of linkages the final numerical solution is often more convenient by computer usage, especially when the mechanisms must be analyzed for all phases of their motions.

The slider-crank mechanism shown in Fig. 3.15 is the engine mechanism and, for this reason, we designate the crank radius as r and the connecting-rod length as l. From the geometry of the figure we see that

$$r \sin \theta = l \sin \phi \tag{a}$$

$$x = r \cos \theta + l \cos \phi \tag{b}$$

Now solve Eqs. (a) and (b) simultaneously to eliminate ϕ and obtain x. This gives

$$x = r \cos \theta + l \left(1 - \frac{r^2}{l^2} \sin^2 \theta \right)^{1/2} \tag{3.15}$$

Differentiating Eq. (b) gives the slider (piston) velocity as

$$\dot{x} = -r\omega \left(\sin \theta + \frac{r \sin 2\theta}{2l \cos \phi} \right) \tag{c}$$

or, if ϕ is eliminated,

$$\dot{x} = -r\omega \left\{ \sin \theta + \frac{r \sin 2\theta}{2l[1 - (r^2/l^2)\sin^2 \theta]^{1/2}} \right\} \tag{3.16}$$

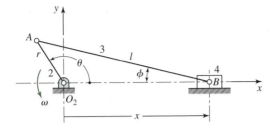

Figure 3.15 The slider-crank mechanism.

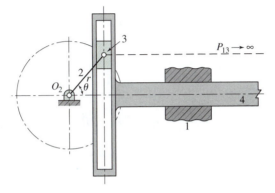

Figure 3.16 The Scotch-yoke linkage.

The *Scotch yoke* shown in Fig. 3.16 is an interesting variation of the slider-crank mechanism. Here, link 3 rotates about a point called an instant center located at infinity (see Section 3.13). This has the effect of a connecting rod of infinite length, and the second terms of Eqs. (3.15) and (3.16) become zero. Hence, for the Scotch yoke we have

$$x = r\cos\theta \quad \text{and} \quad \dot{x} = -r\omega\sin\theta \tag{3.17}$$

Thus the slider moves with simple harmonic motion. It is for this reason that the deviation of the kinematics of the slider-crank motion from simple harmonic motion is sometimes said to be due to "the angularity of the connecting rod."

3.10 COMPLEX-ALGEBRA METHODS

We recall from Section 2.10 that complex algebra provides an alternative algebraic formulation for two-dimensional kinematics problems. As we saw, the complex-algebra formulation provides the advantage of increased accuracy over the graphical methods, and it is amenable to solution by digital computer at a large number of positions once the program is written. On the other hand, the solution of the loop-closure equation for its unknown position variables is a nonlinear problem and can lead to tedious algebraic manipulations. Fortunately, as we will see, the extension of the complex-algebra approach to velocity analysis leads to a set of *linear* equations, and solution is quite straightforward.

Recalling the complex polar form of a two-dimensional vector from Eq. (2.41),

$$\mathbf{R} = Re^{j\theta}$$

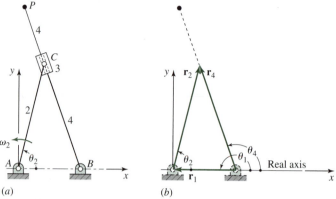

Figure 3.17 Inverted slider-crank mechanism.

we find the general form of its time derivative,

$$\dot{\mathbf{R}} = \frac{d\mathbf{R}}{dt} = \dot{R}e^{j\theta} + j\dot{\theta}Re^{j\theta} \tag{3.18}$$

where $\dot{R}$ and $\dot{\theta}$ denote the time rates of change of the magnitude and angle of $\mathbf{R}$, respectively. We will see in the following examples that the first term of this equation usually represents an apparent velocity and the second term often represents a velocity difference. The methods illustrated in these examples were developed by Raven. Although the original work[1] gives methods applicable to both planar and spatial mechanisms, only the planar aspects are shown here.

To illustrate Raven's approach, let us analyze the inversion of the slider-crank mechanism shown in Fig. 3.17a. We will consider link 2, the driver, to have a known angular position θ_2 and a known angular velocity ω_2 at the instant considered. We wish to derive expressions for the angular velocity of link 4 and the absolute velocity of point P.

To simplify the notation in this example we will use the symbolism shown in Fig. 3.17b for the position-difference vectors; thus $\mathbf{R}_{AB}$ is denoted $\mathbf{r}_1$, $\mathbf{R}_{C_2A}$ is denoted $\mathbf{r}_2$, and $\mathbf{R}_{C_4B}$ is denoted $\mathbf{r}_4$. In terms of these symbols, the loop-closure equation is

$$\mathbf{r}_1 + \mathbf{r}_2 = \mathbf{r}_4 \tag{a}$$

where the vector $\mathbf{r}_1$ has a constant magnitude and direction.* The vector $\mathbf{r}_2$ has constant magnitude and its direction θ_2 varies, but is the input angle. We assume that θ_2 is known or, more precisely, that other unknowns will be solved for as functions of θ_2. The vector $\mathbf{r}_4$ has unknown magnitude and direction.

Recognizing this as case 1 (Section 2.7), we obtain the position solution from Eqs. (2.43) and (2.44):

$$r_4 = \sqrt{r_1^2 + r_2^2 - 2r_1r_2\cos\theta_2} \tag{b}$$

$$\theta_4 = \tan^{-1}\left(\frac{r_2\sin\theta_2}{r_2\cos\theta_2 - r_1}\right) \tag{c}$$

---

*Note particularly that the angle of $\mathbf{r}_1$ is $\theta_1 = 180°$, not zero.

The velocity solution is initiated by differentiating the loop-closure equation (a) with respect to time. Applying the general form, Eq. (3.18), to each term of this equation in turn and keeping in mind that r_1, θ_1, and r_2 are constants, we obtain

$$j\dot{\theta}_2 r_2 e^{j\theta_2} = \dot{r}_4 e^{j\theta_4} + j\dot{\theta}_4 r_4 e^{j\theta_4} \qquad (d)$$

Because $\dot{\theta}_2$ and $\dot{\theta}_4$ are the same as ω_2 and ω_4, respectively, and because we recognize that

$$\dot{\theta}_2 r_2 = V_{C_2}, \qquad \dot{r}_4 = V_{C_2/4}, \qquad \dot{\theta}_4 r_4 = V_{C_4}$$

we see that Eq. (d) is, in fact, the complex polar form of the apparent-velocity equation

$$\mathbf{V}_{C_2} = \mathbf{V}_{C_4} + \mathbf{V}_{C_2/4}$$

(This is pointed out for comparison only and is not a necessary step in the solution process.)

The velocity solution is performed by using Euler's formula, Eq. (2.40), to separate Eq. (d) into its real and imaginary components. This gives

$$-\omega_2 r_2 \sin\theta_2 = \dot{r}_4 \cos\theta_4 - \omega_4 r_4 \sin\theta_4 \qquad (e)$$

$$\omega_2 r_2 \cos\theta_2 = \dot{r}_4 \sin\theta_4 + \omega_4 r_4 \cos\theta_4 \qquad (f)$$

Solving these two equations simultaneously for the two unknowns $\dot{r}_4$ and ω_4 yields

$$\dot{r}_4 = \omega_2 r_2 \sin(\theta_4 - \theta_2) \qquad (3.19)$$

$$\omega_4 = \omega_2 \frac{r_2}{r_4} \cos(\theta_4 - \theta_2) \qquad (3.20)$$

Although the variables r_4 and θ_4 could be substituted from Eqs. (b) and (c) to reduce these results to functions of θ_2 and ω_2 alone, the above forms are considered sufficient because in writing a computer program, numeric values are normally found first for r_4 and θ_4 while performing the position analysis, and these numeric values can then be used in finding $\dot{r}_4$ and ω_4 at each phase angle θ_2.

To find the velocity of point P we write

$$\mathbf{R}_P = R_{PB} e^{j\theta_4} \qquad (g)$$

and we use Eq. (3.18) to differentiate with respect to time, remembering that R_{PB} is a constant length. This yields

$$\mathbf{V}_P = j\omega_4 R_{PB} e^{j\theta_4} \qquad (h)$$

which, upon substituting from Eq. (3.20), becomes

$$\mathbf{V}_P = j\omega_2 R_{PB} \frac{r_2}{r_4} \cos(\theta_4 - \theta_2) e^{j\theta_4} \qquad (3.21)$$

The horizontal and vertical components are

$$V_P^x = -\omega_2 R_{PB} \frac{r_2}{r_4} \cos(\theta_4 - \theta_2)\sin\theta_4 \qquad (i)$$

$$V_P^y = -\omega_2 R_{PB} \frac{r_2}{r_4} \cos(\theta_4 - \theta_2)\cos\theta_4 \qquad (j)$$

As another illustration of Raven's approach, consider the following example problem.

EXAMPLE 3.5

Develop an equation for the relationship between the angular velocities of the input and output cranks of the four-bar linkage shown in Fig. 3.18*a*.

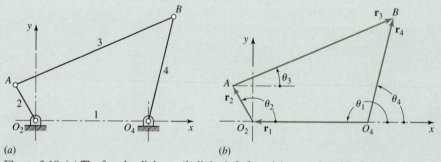

(*a*)   (*b*)

Figure 3.18 (*a*) The four-bar linkage; (*b*) links 1, 2, 3, and 4 are replaced by vectors.

SOLUTION

First, replace each link of Fig. 3.18*a* by a vector as shown in Fig. 3.18*b*. Then, using the loop closure, we have the vector equation

$$\mathbf{r}_1 + \mathbf{r}_2 + \mathbf{r}_3 - \mathbf{r}_4 = 0 \qquad (1)$$

Now replace each vector in Eq. (1) with a complex number in polar form. The result is

$$r_1 e^{j\theta_1} + r_2 e^{j\theta_2} + r_3 e^{j\theta_3} - r_4 e^{j\theta_4} = 0 \qquad (2)$$

Because link 1 is the fixed link, the first time derivative of Eq. (2) is

$$jr_2\dot{\theta}_2 e^{j\theta_2} + jr_3\dot{\theta}_3 e^{j\theta_3} - jr_4\dot{\theta}_4 e^{j\theta_4} = 0 \qquad (3)$$

Now we transform Eq. (3) into rectangular form and separate the real and the imaginary terms. Noting that $\omega_2 = \dot\theta_2$, $\omega_3 = \dot\theta_3$, and $\omega_4 = \dot\theta_4$, we have

$$r_2\omega_2 \cos\theta_2 + r_3\omega_3 \cos\theta_3 - r_4\omega_4 \cos\theta_4 = 0$$

$$r_2\omega_2 \sin\theta_2 + r_3\omega_3 \sin\theta_3 - r_4\omega_4 \sin\theta_4 = 0$$

(4)

With ω_2 given, ω_3 and ω_4 are the unknowns. Solving Eqs. (4) simultaneously yields

$$\omega_3 = \frac{r_2 \sin(\theta_2 - \theta_4)}{r_3 \sin(\theta_4 - \theta_3)}\omega_2 \quad \text{and} \quad \omega_4 = \frac{r_2 \sin(\theta_2 - \theta_3)}{r_4 \sin(\theta_4 - \theta_3)}\omega_2 \qquad \textit{Ans.} \quad (3.22)$$

Note that in both the above problems, the simultaneous equations that were solved were linear equations. This was not a coincidence but is true in all velocity solutions. It results from the fact that the general equation (3.18) is linear in the velocity variables. When real and imaginary components are taken, the *coefficients* may become complicated, but the *equations* remain linear with respect to the velocity unknowns. Therefore, their solution is straightforward.

Another symptom of the linearity of velocity relationships is seen by recalling that in the graphical velocity solutions of previous sections it was possible to pick an arbitrary scale factor for a velocity polygon. If the input speed of a mechanism is doubled, the scale factor of the velocity polygon could be doubled and the same polygon would be valid. This is another indication of linear equations.

It is also worth noting that Eqs. (3.22) both include $\sin(\theta_4 - \theta_3)$ in their denominators. In general, any velocity-analysis problem will have similar denominators in the solutions for each of the velocity unknowns; these denominators are the determinant of the matrix of coefficients of the unknowns of the linear equations, as will be recognized by recalling Cramer's rule. In the four-bar linkage it can be seen from Fig. 2.15 that $(\theta_4 - \theta_3)$ is the transmission angle. When the transmission angle becomes small, the ratio of the output to the input velocity becomes very large and difficulty results.

3.11 THE METHOD OF KINEMATIC COEFFICIENTS

A useful method that provides geometric insight into the motion of a linkage is to differentiate the x and y components of the loop-closure equation with respect to the input position variable, rather than differentiating directly with respect to time. This analytical approach is referred to as the method of kinematic coefficients. The numeric values of the first-order kinematic coefficients can also be checked with the graphical approach of finding the locations of the instantaneous centers of zero velocity. This method is illustrated here by again solving the four-bar linkage problem that was presented in Example 3.1 and the offset slider-crank mechanism that was presented in Example 3.2.

EXAMPLE 3.6

The four-bar linkage shown in Figure 3.7a is driven by crank 2 at a constant angular velocity of $\omega_2 = 900$ rev/min ccw. Find the angular velocities of the coupler link and the output link and the instantaneous velocities of points E and F for the position shown.

SOLUTION

First, replace each link of Fig. 3.7a by a vector (see Fig. 3.18b). Then the vector loop-closure equation can be written as

$$\mathbf{r}_1 + \mathbf{r}_2 + \mathbf{r}_3 - \mathbf{r}_4 = 0 \tag{1}$$

The two scalar equations are

$$r_1 \cos\theta_1 + r_2 \cos\theta_2 + r_3 \cos\theta_3 - r_4 \cos\theta_4 = 0$$
$$r_1 \sin\theta_1 + r_2 \sin\theta_2 + r_3 \sin\theta_3 - r_4 \sin\theta_4 = 0 \tag{2}$$

Because the input is the joint displacement θ_2, the method of kinematic coefficients implies differentiating Eqs. (2) with respect to θ_2. Rearranging the result gives

$$-r_3 \sin\theta_3\theta_3' + r_4 \sin\theta_4\theta_4' = r_2 \sin\theta_2$$

and

$$r_3 \cos\theta_3\theta_3' - r_4 \cos\theta_4\theta_4' = -r_2 \cos\theta_2 \tag{3}$$

where

$$\theta_3' = \frac{d\theta_3}{d\theta_2} \quad \text{and} \quad \theta_4' = \frac{d\theta_4}{d\theta_2} \tag{4}$$

are referred to as the *first-order kinematic coefficients* of links 3 and 4, respectively. Symbolic forms for the kinematic coefficients can be obtained by using Cramer's rule. Writing Eqs. (3) in matrix form gives

$$\begin{bmatrix} -r_3 \sin\theta_3 & r_4 \sin\theta_4 \\ r_3 \cos\theta_3 & -r_4 \cos\theta_4 \end{bmatrix} \begin{bmatrix} \theta_3' \\ \theta_4' \end{bmatrix} = \begin{bmatrix} r_2 \sin\theta_2 \\ -r_2 \cos\theta_2 \end{bmatrix} \tag{5}$$

The determinant of the (2×2) coefficient matrix can be written as

$$\Delta = r_3 r_4 \sin(\theta_3 - \theta_4) \tag{6}$$

and provides geometric insight into special positions of the mechanism (see Example 3.5, Section 3.10). For example, when this determinant tends toward zero, the kinematic coefficients tend toward infinity. Note that the determinant is zero (that is, the transmission angle $\theta_4 - \theta_3$ is zero) when (i) $\theta_3 = \theta_4$ or when (ii) $\theta_3 = \theta_4 \pm 180°$; that is, when links 3 and 4 are aligned.

From Eq. (5), the first-order kinematic coefficients of links 3 and 4 are

$$\theta_3' = \frac{-r_2 \sin(\theta_2 - \theta_4)}{r_3 \sin(\theta_3 - \theta_4)} \quad \text{and} \quad \theta_4' = \frac{-r_2 \sin(\theta_2 - \theta_3)}{r_4 \sin(\theta_3 - \theta_4)} \tag{7}$$

The angular velocities of links 3 and 4, obtained from the chain rule, are

$$\omega_3 = \theta_3' \omega_2 \quad \text{and} \quad \omega_4 = \theta_4' \omega_2 \tag{8}$$

Substituting Eqs. (7) into Eqs. (8) gives the same results as Eqs. (3.22). For the given link dimensions and the specified input angle θ_2, the angular positions θ_3 and θ_4 are known from the position analysis. The answers can be shown to be $\theta_3 = 20.92°$ and $\theta_4 = 64.05°$. Substituting the known data into Eqs. (7), the first-order kinematic coefficients of links 3 and 4 are

$$\theta_3' = +0.2718 \text{ rad/rad} \quad \text{and} \quad \theta_4' = +0.5265 \text{ rad/rad} \tag{9}$$

Substituting these values and the input angular velocity into Eqs. (8), the angular velocities of links 3 and 4 are

$$\omega_3 = 25.6 \text{ rad/s ccw} \quad \text{and} \quad \omega_4 = 49.6 \text{ rad/s ccw} \qquad \textit{Ans.} \tag{10}$$

These answers agree with the results obtained from the velocity polygon method; see Eqs. (4) and (5) of Example 3.1.

The position of point E with respect to the ground pivot A (see Fig. 3.7a) can be written as

$$\mathbf{r}_E = \mathbf{r}_2 + \mathbf{r}_{EB} \tag{11}$$

The x and y components of this vector equation are

$$x_E = r_2 \cos\theta_2 + r_{EB} \cos(\theta_3 - \phi)$$
$$y_E = r_2 \sin\theta_2 + r_{EB} \sin(\theta_3 - \phi) \tag{12}$$

where $\phi = \tan^{-1}(4''/10'') = 21.80°$.

Differentiating Eqs. (12) with respect to the input joint displacement θ_2, the first-order kinematic coefficients for point E are

$$x_E' = -r_2 \sin\theta_2 - r_{EB} \sin(\theta_3 - \phi)\theta_3'$$
$$y_E' = r_2 \cos\theta_2 + r_{EB} \cos(\theta_3 - \phi)\theta_3' \tag{13}$$

Substituting the known values into these equations, the first-order kinematic coefficients are

$$x_E' = -3.4191 \text{ in/rad} \quad \text{and} \quad y_E' = +0.9269 \text{ in/rad} \tag{14}$$

The velocity of point E can be written from the chain rule as

$$\mathbf{V}_E = (x'_E\hat{\mathbf{i}} + y'_E\hat{\mathbf{j}})\omega_2 \tag{15}$$

Substituting Eqs. (14) and the input angular velocity into this equation, the velocity of point E is

$$\mathbf{V}_E = -26.84\hat{\mathbf{i}} + 7.28\hat{\mathbf{j}} \text{ ft/s} \qquad\qquad Ans.$$

Therefore, the instantaneous speed of point E is $V_E = 27.8$ ft/s, which agrees closely with the result obtained from the graphical method in Example 3.1 ($V_E = 27.6$ ft/s).

The velocity of point F can be obtained in a similar manner, and the result is

$$\mathbf{V}_F = -20.60\hat{\mathbf{i}} + 23.82\hat{\mathbf{j}} \text{ ft/s} \qquad\qquad Ans.$$

Therefore, the instantaneous speed of point F is $V_F = 31.49$ ft/s, which agrees closely with the result obtained in Example 3.1 ($V_F = 31.8$ ft/s).

EXAMPLE 3.7

The offset slider-crank linkage shown in Fig. 3.8 is driven by slider 4 at a speed of $V_C = 10$ m/s to the left at the position shown. Determine the angular velocity of links 2 and 3 and the instantaneous velocity of point D.

SOLUTION

The loop-closure equation for the offset slider-crank linkage is

$$\mathbf{r}_1 + \mathbf{r}_2 + \mathbf{r}_3 - \mathbf{r}_4 = 0 \tag{1}$$

The two scalar equations are

$$r_1 \cos\theta_1 + r_2 \cos\theta_2 + r_3 \cos\theta_3 - r_4 \cos\theta_4 = 0$$
$$r_1 \sin\theta_1 + r_2 \sin\theta_2 + r_3 \sin\theta_3 - r_4 \sin\theta_4 = 0 \tag{2}$$

Because the input is the linear displacement of link 4 then differentiating Eqs. (2) with respect to r_4 gives

$$-r_2 \sin\theta_2 \theta'_2 - r_3 \sin\theta_3 \theta'_3 = \cos\theta_4$$
$$r_2 \cos\theta_2 \theta'_2 + r_3 \cos\theta_3 \theta'_3 = \sin\theta_4 \tag{3}$$

where now

$$\theta'_2 = \frac{d\theta_2}{dr_4} \quad \text{and} \quad \theta'_3 = \frac{d\theta_3}{dr_4} \tag{4}$$

are the first-order kinematic coefficients of links 2 and 3. Symbolic equations for the first-order kinematic coefficients are obtained by using Cramer's rule. Writing Eqs. (3) in matrix form gives

$$
\begin{bmatrix} -r_2 \sin \theta_2 & -r_3 \sin \theta_3 \\ r_2 \cos \theta_2 & r_3 \cos \theta_3 \end{bmatrix} \begin{bmatrix} \theta_2' \\ \theta_3' \end{bmatrix} = \begin{bmatrix} \cos \theta_4 \\ \sin \theta_4 \end{bmatrix}
\tag{5}
$$

The determinant of the (2×2) coefficient matrix in Eq. (5) can be written as

$$
\Delta = r_2 r_3 \sin(\theta_3 - \theta_2)
\tag{6}
$$

Note that the determinant is zero when (i) $\theta_2 = \theta_3$ or when (ii) $\theta_2 = \theta_3 \pm 180°$—that is, when links 2 and 3 are fully extended or aligned over each other.

From Eq. (5), the first-order kinematic coefficients of links 2 and 3 can be written as

$$
\theta_2' = \frac{\cos(\theta_3 - \theta_4)}{r_2 \sin(\theta_3 - \theta_2)} \quad \text{and} \quad \theta_3' = \frac{-\cos(\theta_2 - \theta_4)}{r_3 \sin(\theta_3 - \theta_2)}
\tag{7}
$$

For the given link dimensions and the given input position $r_4 = 164$ mm, the angular positions of links 2 and 3 are $\theta_2 = 45°$ and $\theta_3 = 337°$. Substituting these values into Eqs. (7), the first-order kinematic coefficients of links 2 and 3 are

$$
\theta_2' = -19.77 \text{ rad/m} \quad \text{and} \quad \theta_3' = 5.436 \text{ rad/m}
\tag{8}
$$

The angular velocities of links 2 and 3 can be written from the chain rule as

$$
\omega_2 = \theta_2' \dot{r}_4 \quad \text{and} \quad \omega_3 = \theta_3' \dot{r}_4
\tag{9}
$$

where the linear velocity of the slider is $\dot{r}_4 = -10$ m/s. Substituting Eqs. (8) and the velocity of the slider into Eqs. (9), the angular velocities of the two links are

$$
\omega_2 = 197.7 \text{ rad/s ccw} \quad \text{and} \quad \omega_3 = 54.4 \text{ rad/s cw} \qquad\qquad Ans.
$$

These answers agree closely with the results obtained from the velocity polygon method in Example 3.2 ($\omega_2 = 200$ rad/s ccw and $\omega_3 = 53.6$ rad/s cw).

The position of point D with respect to the ground pivot A (see Fig. 3.8a) can be written as

$$
\mathbf{r}_D = \mathbf{r}_2 + \mathbf{r}_{DB}
\tag{10}
$$

The x and y components of this vector equation are

$$
x_D = r_2 \cos \theta_2 + r_{DB} \cos(\theta_3 - \mu)
\tag{11a}
$$

$$
y_D = r_2 \sin \theta_2 + r_{DB} \sin(\theta_3 - \mu)
\tag{11b}
$$

where $\mu = \tan^{-1}(50 \text{ mm}/80 \text{ mm}) = -32.01°$.

Differentiating Eqs. (11) with respect to the input displacement r_4, the first-order kinematic coefficients for point D are

$$x'_D = -r_2 \sin\theta_2 \theta'_2 - r_{DB} \sin(\theta_3 - \mu)\theta'_3 \qquad (12a)$$

$$y'_D = r_2 \cos\theta_2 \theta'_2 + r_{DB} \cos(\theta_3 - \mu)\theta'_3 \qquad (12b)$$

Substituting the known values into these equations, the first-order kinematic coefficients are

$$x'_D = +1.121 \text{ m/m} \quad \text{and} \quad y'_D = -0.4071 \text{ m/m} \qquad (13)$$

The velocity of the point D can be written as

$$\mathbf{V}_D = (x'_D \hat{\mathbf{i}} + y'_D \hat{\mathbf{j}})\dot{r}_4 \qquad (14)$$

Substituting Eqs. (13) and the input velocity $\dot{r}_4 = -10$ m/s into this equation, the velocity of point D is

$$\mathbf{V}_D = -11.21\hat{\mathbf{i}} + 4.07\hat{\mathbf{j}} \text{ m/s} \qquad Ans. \quad (15)$$

Therefore, the instantaneous speed of point D is 11.92 m/s, which agrees closely with the result obtained from the velocity polygon method in Example 3.2 ($V_D = 12$ m/s). The method of kinematic coefficients also provides geometric insight into the motion of mechanisms that have links in rolling contact. The following examples help to illustrate this.

EXAMPLE 3.8

For the mechanism shown in Fig. 3.19, the wheel is rolling without slipping on the ground. The radius of the wheel is 10 cm. The length of the input link 2 is 1.5 times the radius of the wheel. The distance from pin B to the center of the wheel G is one-half the radius of the wheel, and the distance from pin O_2 to the point of contact C is twice the radius of the wheel. For the position shown in the figure, that is, with $\theta_2 = 90°$, determine:

(a) The first-order kinematic coefficients of links 3 and 4
(b) The angular velocities of links 3 and 4 if the input link is rotating with a constant angular velocity of 10 rad/s ccw
(c) The velocity of the center of the wheel

SOLUTION

The loop-closure equation for the mechanism is

$$\mathbf{r}_2 + \mathbf{r}_3 + \mathbf{r}_4 + \mathbf{r}_7 - \mathbf{r}_9 = 0 \qquad (1)$$

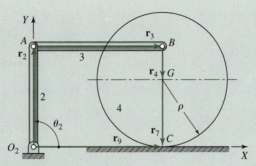

Figure 3.19 Example 3.8.

The horizontal and vertical component scalar equations are

$$r_2 \cos \theta_2 + r_3 \cos \theta_3 + r_4 \cos \theta_4 + r_7 \cos \theta_7 - r_9 \cos \theta_9 = 0 \tag{2a}$$

$$r_2 \sin \theta_2 + r_3 \sin \theta_3 + r_4 \sin \theta_4 + r_7 \sin \theta_7 - r_9 \sin \theta_9 = 0 \tag{2b}$$

Because the input is the angular displacement of link 2, we differentiate Eqs. (2) with respect to θ_2; and, by substituting $\theta_9 = 0°$, we obtain

$$-r_2 \sin \theta_2 - r_3 \sin \theta_3 \theta_3' - r_4 \sin \theta_4 \theta_4' - r_9' = 0 \tag{3a}$$

and

$$r_2 \cos \theta_2 + r_3 \cos \theta_3 \theta_3' + r_4 \cos \theta_4 \theta_4' = 0 \tag{3b}$$

where we have taken notice that both r_7 and θ_7 are constants and where

$$\theta_3' = \frac{d\theta_3}{d\theta_2}, \quad \theta_4' = \frac{d\theta_4}{d\theta_2}, \quad \text{and} \quad r_9' = \frac{dr_9}{d\theta_2} \tag{4}$$

are the first-order kinematic coefficients of links 3 and 4 and the position of the contact point.

We see that the rotation of the wheel, θ_4, and the change in the distance r_9 are not independent because the wheel is rolling on the ground without slipping. This rolling contact (no slip) condition can be written as

$$-\Delta r_9 = \rho(\Delta \theta_4 - \Delta \theta_7) \tag{5}$$

where ρ is the radius of the wheel, $\theta_7 = 270°$ is constant for all positions of the wheel, and $\Delta \theta_7 = 0$; the factor $(\Delta \theta_4 - \Delta \theta_7)$ is the small angular displacement from the position of link 4 shown with respect to the vertical orientation of vector 7. The negative sign on the left of this equation is due to the decreasing magnitude of r_9 for a positive change in the input $\Delta \theta_2$.

In the limit, for infinitesimal displacements, the constraint Eq. (5) can be written in terms of first-order kinematic coefficients as

$$-r_9' = \rho \theta_4' \tag{6}$$

Substituting Eq. (6) into Eqs. (3) and writing the resulting equations in matrix form gives

$$\begin{bmatrix} -r_3 \sin\theta_3 & -r_4 \sin\theta_4 + \rho \\ +r_3 \cos\theta_3 & +r_4 \cos\theta_4 \end{bmatrix} \begin{bmatrix} \theta_3' \\ \theta_4' \end{bmatrix} = \begin{bmatrix} +r_2 \sin\theta_2 \\ -r_2 \cos\theta_2 \end{bmatrix} \tag{7}$$

The determinant of the (2×2) coefficient matrix in Eq. (7) can be written as

$$\Delta = r_3[r_4 \sin(\theta_4 - \theta_3) - \rho \cos\theta_3] \tag{8}$$

From Eq. (7), using Cramer's rule, the symbolic equations for the first-order kinematic coefficients of links 4 and 5 are

$$\theta_3' = \frac{r_2[\rho \cos\theta_2 - r_4 \sin(\theta_4 - \theta_2)]}{\Delta} \tag{9a}$$

$$\theta_4' = \frac{r_2 r_3 \sin(\theta_3 - \theta_2)}{\Delta} \tag{9b}$$

For the position shown in the figure we have $\theta_2 = 90°$, $\theta_3 = 0°$, and $\theta_4 = -90°$. Therefore Eq. (8) can be written as

$$\Delta = -r_3(r_4 + \rho) \tag{10}$$

Also, Eqs. (9) can now be shown to yield

$$\theta_3' = 0 \quad \text{and} \quad \theta_4' = +\frac{r_2}{r_4 + \rho} = +1 \qquad Ans. \tag{11}$$

Note that these results are in agreement with our intuition; in this position, link 3 is not rotating and link 4 is rotating at the same angular velocity as the input link. The angular velocity of link 3 is

$$\omega_3 = \theta_3' \omega_2 = 0 \qquad Ans.$$

and the angular velocity of the wheel, link 4, is

$$\omega_4 = \theta_4' \omega_2 = 10 \text{ rad/s ccw} \qquad Ans.$$

The velocity of the center of the wheel, point G, is

$$V_G = r_9' \omega_2 \tag{12}$$

Substituting Eq. (11) into Eq. (6) gives $r_9' = -\rho = -10$ cm/rad, and substituting this into Eq. (12) gives

$$V_G = r_9' \omega_2 = (-10 \text{ cm/rad})(10 \text{ rad/s}) = -100 \text{ cm/s} \qquad Ans.$$

This result can be verified by direct calculation by noting that

$$\mathbf{V}_G = \boldsymbol{\omega}_4 \times (\rho \hat{\mathbf{j}}) = 10\hat{\mathbf{k}} \times 10\hat{\mathbf{j}} = -100\hat{\mathbf{i}} \text{ cm/s} \qquad Ans.$$

EXAMPLE 3.9

For the mechanism shown in Fig. 3.20, gear 3 is rolling without slip on input gear 2. The radius of gear 3 is half the radius of gear 2, the length of link 5 is three times the radius of the input gear, and the distance from pin A to pin B is half the radius of gear 3. For the position shown—that is, with link 5 horizontal—determine (i) the first-order kinematic coefficients of links 3, 4, and 5 and (ii) the angular velocities of links 3, 4, and 5 if the input gear is rotating with a constant angular velocity of 10 rad/s ccw.

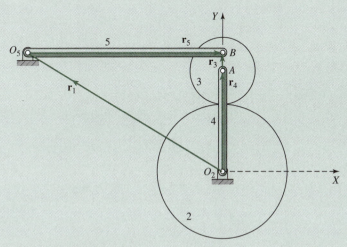

Figure 3.20 Example 3.9.

SOLUTION

The loop-closure equation for the mechanism is

$$\mathbf{r}_4 + \mathbf{r}_3 - \mathbf{r}_5 + \mathbf{r}_1 = 0 \tag{1}$$

The two scalar equations are

$$r_4 \cos \theta_4 + r_3 \cos \theta_3 - r_5 \cos \theta_5 + r_1 \cos \theta_1 = 0 \tag{2a}$$

$$r_4 \sin \theta_4 + r_3 \sin \theta_3 - r_5 \sin \theta_5 + r_1 \sin \theta_1 = 0 \tag{2b}$$

Because the input is the angular displacement of gear 2, differentiating Eqs. (2) with respect to θ_2 gives

$$-r_4 \sin \theta_4 \theta_4' - r_3 \sin \theta_3 \theta_3' + r_5 \sin \theta_5 \theta_5' = 0 \tag{3a}$$

$$r_4 \cos \theta_4 \theta_4' + r_3 \cos \theta_3 \theta_3' - r_5 \cos \theta_5 \theta_5' = 0 \tag{3b}$$

where

$$\theta_3' = \frac{d\theta_3}{d\theta_2}, \quad \theta_4' = \frac{d\theta_4}{d\theta_2}, \quad \text{and} \quad \theta_5' = \frac{d\theta_5}{d\theta_2} \tag{4}$$

are the first-order kinematic coefficients of links 3, 4, and 5.

Note that the joint variables θ_2, θ_3, and θ_4 are not independent because gear 3 is constrained to roll on gear 2. The rolling contact condition can be written to show that the arc distances passed during a small movement are equal on the two surfaces:

$$\rho_2(\Delta\theta_2 - \Delta\theta_4) = -\rho_3(\Delta\theta_3 - \Delta\theta_4) \tag{5a}$$

where $\Delta\theta_2$, $\Delta\theta_3$, and $\Delta\theta_4$ are finite small angular displacements of links 2, 3, and 4 from the position shown in the figure, and the negative sign arises from the change in the direction of the angular displacements.

This constraint equation can be divided by a small change in the input $\Delta\theta_2$, and the limit for small increments taken; this yields the equivalent constraint equation written in terms of first-order kinematic coefficients:

$$\rho_2(\theta_2' - \theta_4') = -\rho_3(\theta_3' - \theta_4') \tag{5b}$$

Because the input is gear 2, then, by definition, $\theta_2' = 1$; and by substituting $\rho_2 = 2\rho_3$, Eq. (5b) can be written as

$$\theta_3' = 3\theta_4' - 2 \tag{6}$$

Substituting Eq. (6) into Eqs. (3) and writing the resulting equations in matrix form gives

$$\begin{bmatrix} -r_4 \sin\theta_4 - 3r_3 \sin\theta_3 & r_5 \sin\theta_5 \\ r_4 \cos\theta_4 + 3r_3 \cos\theta_3 & -r_5 \cos\theta_5 \end{bmatrix} \begin{bmatrix} \theta_4' \\ \theta_5' \end{bmatrix} = \begin{bmatrix} -2r_3 \sin\theta_3 \\ +2r_3 \cos\theta_3 \end{bmatrix} \tag{7}$$

The determinant of the (2×2) coefficient matrix in Eq. (7) can be written as

$$\Delta = r_5[r_4 \sin(\theta_4 - \theta_5) + 3r_3 \sin(\theta_3 - \theta_5)] \tag{8}$$

From Eq. (7), using Cramer's rule, the symbolic equations for the first-order kinematic coefficients of links 4 and 5 are

$$\theta_4' = \frac{2r_3 r_5 \sin(\theta_3 - \theta_5)}{\Delta} \tag{9a}$$

$$\theta_5' = \frac{2r_3 r_4 \sin(\theta_3 - \theta_4)}{\Delta} \tag{9b}$$

For the position shown in the figure, we have $\theta_3 = 90°$, $\theta_4 = 90°$, and $\theta_5 = 0°$; therefore, Eq. (8) can be written as

$$\Delta = r_5(r_4 + 3r_3) \tag{10}$$

Also, Eqs. (9) reduce to

$$\theta_4' = +\frac{2}{9} \quad \text{and} \quad \theta_5' = 0 \tag{11}$$

It is interesting to note that link 5 is not rotating in this position; that is, its angular velocity is $\omega_5 = 0$. By substituting Eqs. (11) into Eq. (6), the first-order kinematic coefficient of gear 3 becomes

$$\theta_3' = -\frac{4}{3} \tag{12}$$

The angular velocity of link j can now be written as

$$\omega_j = \theta_j' \omega_2 \tag{13}$$

Therefore, by substituting Eqs. (11) and (12) into Eq. (13) and setting $\omega_2 = 10$ rad/s ccw, the angular velocities of gear 3 and links 4 and 5 become

$$\omega_3 = -13.33 \text{ rad/s}, \quad \omega_4 = +2.22 \text{ rad/s}, \quad \text{and} \quad \omega_5 = 0 \qquad Ans. \tag{14}$$

where the negative sign indicates the clockwise direction and the positive sign indicates the counterclockwise direction.

Note that kinematic coefficients are a function of *position only;* that is, they are not directly a function of time. Also, note that the units of the first-order kinematic coefficients depend on the specified input and the variable under consideration. The units may be nondimensional (rad/rad or length/length), length (length/rad), or the reciprocal of length (rad/length). Table 3.2 summarizes the first-order kinematic coefficients that are related to link j of a mechanism (that is, the vector $\mathbf{r}_j$) having (i) unknown angle θ_j and/or (ii) unknown magnitude r_j.

TABLE 3.2 Summary of First-Order Kinematic Coefficients

	Variable of Interest Angle θ_j (use symbol θ_j' for kinematic coefficient regardless of input)	Variable of Interest Magnitude r_j (use symbol r_j' for kinematic coefficient regardless of input)
Input $\psi = $ angle θ_i	$\omega_j = \theta_j' \dot{\psi}$ $\theta_j' = \dfrac{d\theta_j}{d\psi}$ (dimensionless, rad/rad)	$\dot{r}_j = r_j' \dot{\psi}$ $r_j' = \dfrac{dr_j}{d\psi}$ (length, length/rad)
Input $\psi = $ magnitude r_i	$\omega_j = \theta_j' \dot{\psi}$ $\theta_j' = \dfrac{d\theta_j}{d\psi}$ (1/length, rad/length)	$\dot{r}_j = r_j' \dot{\psi}$ $r_j' = \dfrac{dr_j}{d\psi}$ (dimensionless, length/length)

3.12 THE VECTOR METHOD

Chace's approach to position analysis was discussed in Section 2.12. It will be shown here how this approach is applied to the velocity analysis of linkages. The method is illustrated by again solving the inverted slider-crank mechanism of Fig. 3.17.

The procedure begins by writing the loop-closure equation

$$\mathbf{r}_1 + \mathbf{r}_2 = \mathbf{r}_4 \tag{a}$$

The velocity relationships are found by differentiating this equation with respect to time. The derivative of a typical term becomes

$$\dot{\mathbf{R}} = \frac{d}{dt}(R\hat{\mathbf{R}}) = \dot{R}\hat{\mathbf{R}} + R\dot{\hat{\mathbf{R}}} \tag{b}$$

However, because $\hat{\mathbf{R}}$ is of constant length and because it usually rotates with one of the links, $\dot{\hat{\mathbf{R}}}$ can be expressed as

$$\dot{\hat{\mathbf{R}}} = \boldsymbol{\omega} \times \hat{\mathbf{R}} = \omega(\hat{\mathbf{k}} \times \hat{\mathbf{R}}) \tag{3.23}$$

from which Eq. (b) becomes

$$\dot{\mathbf{R}} = \dot{R}\hat{\mathbf{R}} + \omega R(\hat{\mathbf{k}} \times \hat{\mathbf{R}}) \tag{3.24}$$

Using this general form and recognizing that the magnitudes r_1 and r_2 and the direction $\hat{\mathbf{r}}_1$ are constant, we can take the time derivative of the loop-closure equation (a). This yields

$$\omega_2 r_2(\hat{\mathbf{k}} \times \hat{\mathbf{r}}_2) = \dot{r}_4\hat{\mathbf{r}}_4 + \omega_4 r_4(\hat{\mathbf{k}} \times \hat{\mathbf{r}}_4) \tag{c}$$

Because it is assumed that r_4 and $\hat{\mathbf{r}}_4$ would be known from a previous position analysis, perhaps following the Chace approach of Section 2.10, and because ω_2 is a known input speed, the only two unknowns of this equation are the velocities $\dot{r}_4$ and ω_4.

Rather than taking components of Eq. (c) in the horizontal and vertical directions, which would lead to two simultaneous equations in two unknowns, Chace's approach leads to the elimination of one unknown by careful choice of the directions along which components are taken. In Eq. (c), for example, we note that the unit vector $\hat{\mathbf{r}}_4$ is perpendicular to $\hat{\mathbf{k}} \times \hat{\mathbf{r}}_4$, and therefore

$$\hat{\mathbf{r}}_4 \cdot (\hat{\mathbf{k}} \times \hat{\mathbf{r}}_4) = 0 \tag{d}$$

We take advantage of this fact to eliminate the unknown $\dot{r}_4$. Taking the dot product of each term of Eq. (c) with $(\hat{\mathbf{k}} \times \hat{\mathbf{r}}_4)$, we obtain

$$\omega_2 r_2(\hat{\mathbf{k}} \times \hat{\mathbf{r}}_2) \cdot (\hat{\mathbf{k}} \times \hat{\mathbf{r}}_4) = \omega_4 r_4$$

from which we solve for ω_4:

$$\omega_4 = \omega_2 \frac{r_2}{r_4}(\hat{\mathbf{k}} \times \hat{\mathbf{r}}_2) \cdot (\hat{\mathbf{k}} \times \hat{\mathbf{r}}_4) \tag{e}$$

Similarly, we can take the dot product of Eq. (c) with the unit vector $\hat{\mathbf{r}}_4$ and thus eliminate ω_4. This gives

$$\dot{r}_4 = \omega_2 r_2 (\hat{\mathbf{k}} \times \hat{\mathbf{r}}_2) \cdot \hat{\mathbf{r}}_4 \qquad (f)$$

That these solutions are indeed the same as those obtained by Raven's method can be shown quite simply. From Eq. (e) we can write

$$\hat{\mathbf{k}} \times \hat{\mathbf{r}}_2 = \begin{vmatrix} \hat{\mathbf{i}} & \hat{\mathbf{j}} & \hat{\mathbf{k}} \\ 0 & 0 & 1 \\ \cos\theta_2 & \sin\theta_2 & 0 \end{vmatrix} = -\sin\theta_2\hat{\mathbf{i}} + \cos\theta_2\hat{\mathbf{j}}$$

and, similarly,

$$\hat{\mathbf{k}} \times \hat{\mathbf{r}}_4 = -\sin\theta_4\hat{\mathbf{i}} + \cos\theta_4\hat{\mathbf{j}}$$

Then

$$(\hat{\mathbf{k}} \times \hat{\mathbf{r}}_2) \cdot (\hat{\mathbf{k}} \times \hat{\mathbf{r}}_4) = (-\sin\theta_2\hat{\mathbf{i}} + \cos\theta_2\hat{\mathbf{j}}) \cdot (-\sin\theta_4\hat{\mathbf{i}} + \cos\theta_4\hat{\mathbf{j}})$$

$$= \sin\theta_2\sin\theta_4 + \cos\theta_2\cos\theta_4 = \cos(\theta_4 - \theta_2) \qquad (g)$$

and, in a similar manner,

$$(\hat{\mathbf{k}} \times \hat{\mathbf{r}}_2) \cdot \hat{\mathbf{r}}_4 = \sin(\theta_4 - \theta_2) \qquad (h)$$

When the terms of Eqs. (g) and (h) are substituted into Eqs. (e) and (f), the results match identically with those obtained from Raven's approach, Eqs. (3.19) and (3.20).

3.13 INSTANTANEOUS CENTER OF VELOCITY

One of the more interesting concepts in kinematics is that of an instantaneous velocity axis for a pair of rigid bodies that move with respect to one another. In particular, we shall find that an axis exists which is common to both bodies and about which either body can be considered as rotating with respect to the other.

Because our study of these axes is restricted to planar motions,* each axis is perpendicular to the plane of the motion. We shall refer to them as *instant centers of velocity* or *velocity poles*. These instant centers are regarded as pairs of coincident points, one attached to each body, about which one body has an apparent rotational velocity, but no translation velocity, with respect to the other. This property is true only instantaneously, and a new pair of coincident points becomes the instant center at the next instant. It is not correct, therefore, to speak of an instant center as the center of rotation, because it is generally not located at the center of curvature of the apparent point path which a point of one body generates with respect to the coordinate system of the other. Even with this restriction, however, we will find that instant centers contribute substantially to understanding the kinematics of planar motion.

---

*For three-dimensional motion, this axis is referred to as the *instantaneous screw axis*. The classic work covering its properties is R. S. Ball, *A Treatise on the Theory of Screws,* Cambridge University Press, Cambridge, 1900.

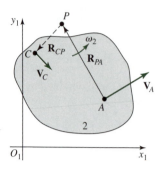

Figure 3.21

The instantaneous center of velocity is defined as *the instantaneous location of a pair of coincident points of two different rigid bodies for which the absolute velocities of the two points are equal.* It may also be defined as the location of a pair of coincident points of two different rigid bodies for which the apparent velocity of one of the points is zero as seen by an observer on the other body.

Let us consider a rigid body 2 having some general motion with respect to the $x_1 y_1$ plane; the motion might be translation, rotation, or a combination of both. As shown in Fig. 3.21, suppose that some point A of the body has a known velocity $\mathbf{V}_A$ and that the body has a known angular velocity ω_2. With these two quantities known, the velocity of any other point of the body can be found from the velocity-difference equation. Suppose we define a point P, for example, whose position difference $\mathbf{R}_{PA}$ from the point A is chosen to be

$$\mathbf{R}_{PA} = \frac{\omega_2 \times \mathbf{V}_A}{\omega_2^2} \tag{3.25}$$

Because of the cross product we see that the point P is located on the perpendicular to $\mathbf{V}_A$ and the vector $\mathbf{R}_{PA}$ is rotated from the direction of $\mathbf{V}_A$ in the direction of ω_2, as shown in Fig. 3.21. The length of $\mathbf{R}_{PA}$ can be calculated from the above equation, and point P can be located. We see that its velocity is

$$\mathbf{V}_P = \mathbf{V}_A + \mathbf{V}_{PA} = \mathbf{V}_A + \omega_2 \times \mathbf{R}_{PA} = \mathbf{V}_A + \omega_2 \times \frac{\omega_2 \times \mathbf{V}_A}{\omega_2^2}$$

But upon replacing this triple product by a vector identity we get

$$\mathbf{V}_P = \mathbf{V}_A + \frac{\overbrace{(\omega_2 \cdot \mathbf{V}_A)}^{0}\omega_2 - \overbrace{(\omega_2 \cdot \omega_2)}^{\omega_2^2} \mathbf{V}_A}{\omega_2^2} = \mathbf{V}_A - \mathbf{V}_A = 0 \tag{a}$$

Because the absolute velocity of the particular point P chosen is zero, the same as the velocity of the coincident point of the fixed link, this point P is the instant center between links 1 and 2.

The velocity of any third point C of the moving body can now be found,

$$\mathbf{V}_C = \cancel{\mathbf{V}_P}^{0} + \mathbf{V}_{CP} = \omega_2 \times \mathbf{R}_{CP} \tag{b}$$

as shown in Fig. 3.21.

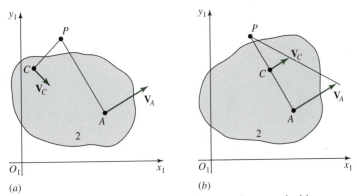

Figure 3.22 Locating an instant center from two known velocities.

The instant center can be located more easily when the absolute velocities of two points are given. In Fig. 3.22a, suppose that points A and C have known velocities $\mathbf{V}_A$ and $\mathbf{V}_C$. Perpendiculars to $\mathbf{V}_A$ and $\mathbf{V}_C$ intersect at P, the instant center. Figure 3.22b shows how to locate the instant center P when the points A, C, and P happen to fall on the same straight line.

The instant center between two bodies, in general, is not a stationary point. It changes its location with respect to both bodies as the motion progresses and describes a path or locus on each. These paths of the instant centers, called centrodes, will be discussed in Section 3.21.

Because we have adopted the convention of numbering the links of a mechanism, it is convenient to designate an instant center by using the numbers of the two links associated with it. Thus P_{32} identifies the instant centers between links 3 and 2. This same center could be identified as P_{23}, because the order of the numbers has no significance. A mechanism has as many instant centers as there are ways of pairing the link numbers. Thus the number of instant centers in an n-link mechanism is

$$N = \frac{n(n-1)}{2} \tag{3.26}$$

3.14 THE ARONHOLD–KENNEDY THEOREM OF THREE CENTERS

According to Eq. (3.26), the number of instant centers in a four-bar linkage is six. As shown in Fig. 3.23a, we can identify four of them by inspection; we see that the four pins can be identified as instant centers P_{12}, P_{23}, P_{34}, and P_{14}, because each satisfies the definition. Point P_{23}, for example, is a point of link 2 about which link 3 appears to rotate; it is a point of link 3 which has no apparent velocity as seen from link 2; it is a pair of coincident points of links 2 and 3 which has the same absolute velocities.

A good method of keeping track of which instant centers have been found is to space the link numbers around the perimeter of a circle, as shown in Fig. 3.23b. Then, as each instant center is identified, a line is drawn connecting the corresponding pair of link numbers. Figure 3.23b shows that P_{12}, P_{23}, P_{34}, and P_{14} have been found; it also shows missing

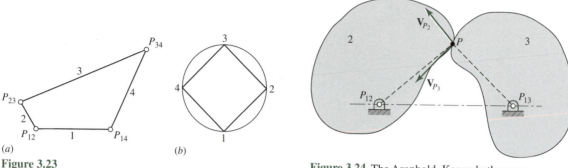

Figure 3.23

Figure 3.24 The Aronhold–Kennedy theorem.

lines where P_{13} and P_{24} have not been located. These two instant centers cannot be found simply by applying the definition visually.

After finding as many of the instant centers as possible by inspection—that is, by locating points which obviously fit the definition—others are located by applying the Aronhold–Kennedy theorem (often just called Kennedy's theorem*) of three centers. This theorem states that *the three instant centers shared by three rigid bodies in relative motion to one another (whether or not they are connected) all lie on the same straight line.*

The theorem can be proven by contradiction, as shown in Fig. 3.24. Link 1 is a stationary frame, and instant center P_{12} is located where link 2 is pin-connected to it. Similarly, P_{13} is located at the pin connecting links 1 and 3. The shapes of links 2 and 3 are arbitrary. The Aronhold–Kennedy theorem states that the three instant centers P_{12}, P_{13}, and P_{23} must all lie on the same straight line, the line connecting the two pins. Let us suppose that this were not true; in fact let us suppose that P_{23} were located at the point labeled P in Fig. 3.24. Then the velocity of P as a point of link 2 would have the direction $\mathbf{V}_{P2}$, perpendicular to $\mathbf{R}_{PP_{12}}$. But the velocity of P as a point of link 3 would have the direction $\mathbf{V}_{P3}$, perpendicular to $\mathbf{R}_{PP_{13}}$. The directions are inconsistent with the definition that an instant center must have equal absolute velocities as a part of either link. The point P chosen therefore cannot be the instant center P_{23}. This same contradiction in the directions of $\mathbf{V}_{P_2}$ and $\mathbf{V}_{P_3}$ occurs for any location chosen for point P unless it is chosen on the straight line through P_{12} and P_{13}.

3.15 LOCATING INSTANT CENTERS OF VELOCITY

In the last two sections we have considered several methods of locating instant centers of velocity. They can often be located by inspecting the figure of a mechanism and visually seeking out a point that fits the definition, such as a pin-joint center. Also, after some instant centers are found, others can be found from them by using the theorem of three centers. Section 3.13 demonstrated that an instant center between a moving body and the fixed link can be found if the directions of the absolute velocities of two points of the body are known or if the absolute velocity of one point and the angular velocity of the body are known. The purpose of this section is to expand this list of techniques and to present examples.

*This theorem is named after its two independent discoverers, Aronhold (1872), and Kennedy (1886). It is known as the Aronhold theorem in German-speaking countries and is called Kennedy's theorem in English-speaking countries.

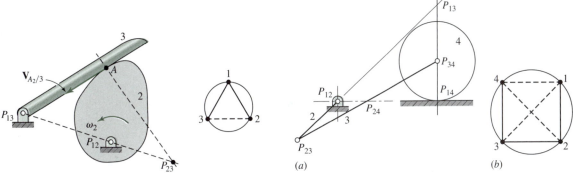

Figure 3.25 Instant centers of a disk cam with a flat-faced follower.

Figure 3.26 Instant center at a point of rolling contact.

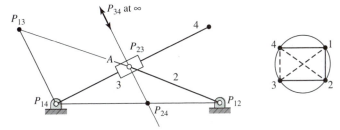

Figure 3.27 Instant centers of an inverted slider-crank mechanism.

Consider the cam-follower system of Fig. 3.25. The instant centers P_{12} and P_{13} can be located, by inspection, at the two pin centers. However, the remaining instant center, P_{23}, is not as obvious. According to the Aronhold–Kennedy theorem, it must lie on the straight line connecting P_{12} and P_{13}, but where on this line? After some reflection we see that the direction of the apparent velocity $\mathbf{V}_{A_2/3}$ must be along the common tangent to the two moving links at the point of contact; and, as seen by an observer on link 3, this velocity must appear as a result of the apparent rotation of body 2 about the instant center P_{23}. Therefore, P_{23} must lie on the line that is perpendicular to $\mathbf{V}_{A_2/3}$. This line now locates P_{23} as shown. The concept illustrated in this example should be remembered because it is often useful in locating the instant centers of mechanisms involving direct contact.

A special case of direct contact, as we have seen before, is rolling contact with no slip. Considering the mechanism of Fig. 3.26, we can immediately locate the instant centers P_{12}, P_{23}, and P_{34}. If the contact between links 1 and 4 involves any slippage, we can only say that instant center P_{14} is located on the vertical line through the point of contact. However, if we also know that there is no slippage—that is, *if there is rolling contact—then the instant center is located at the point of contact*. This is also a general principle, as can be seen by comparing the definition of rolling contact, Eq. (3.14), and the definition of an instant center; they are equivalent.

Another special case of direct contact is evident between links 3 and 4 in Fig 3.27. In this case there is an apparent (slip) velocity $\mathbf{V}_{A_3/4}$ between points A of links 3 and 4, but there *is no apparent rotation between the links*. Here, as in Fig. 3.25, the instant center P_{34} lies along a common perpendicular to the known line of sliding, but now it is located infinitely far away, in the direction defined by this perpendicular line. This infinite distance can be shown by considering the kinematic inversion of the mechanism in which link 4 becomes stationary. Writing Eq. (3.25) for the inverted mechanism, we see that

$$\mathbf{R}_{P_{34}A} = \frac{\boldsymbol{\omega}_{3/4} \times \mathbf{V}_{A_3/4}}{\omega_{3/4}^2} = \hat{\boldsymbol{\omega}}_{3/4} \times \left(\frac{\mathbf{V}_{A_3/4}}{\omega_{3/4}}\right) = \infty \tag{3.27}$$

The direction stated earlier is confirmed by the cross product of Eq. (3.27). We also see that, because there is no relative rotation between links 3 and 4, the denominator is zero and the distance to P_{34} is infinite. The other instant centers of Fig. 3.27 are found by inspection or by the Aronhold–Kennedy theorem. Notice in this figure how the line through P_{14} and P_{34} (at infinity) was used in locating P_{13}.

One more example will be presented to again illustrate the above principles.

EXAMPLE 3.10

Locate all the instant centers of the mechanism of Fig. 3.28 assuming rolling contact between links 1 and 2.

SOLUTION

The instant centers P_{13}, P_{34}, and P_{15}, being pinned joints, are located by inspection. Also, P_{12} is located at the point of rolling contact. The instant center P_{24} may possibly be noticed by the fact that this is the center of the apparent rotation between links 2 and 4; if not, it can be located by drawing perpendicular lines to the directions of the apparent velocities at two of the corners of link 4. One line for the instant center P_{25} comes from noticing the direction of slipping between links 2 and 5; the other comes from the line $P_{12}P_{15}$. After these, all other instant centers can be found by repeated applications of the theorem of three centers.

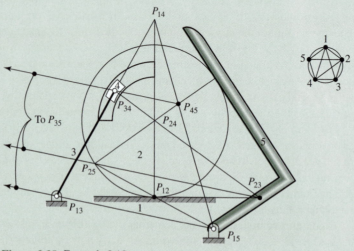

Figure 3.28 Example 3.10.

It should be noted before closing this section that in the above examples the locations of all instant centers were found without having to specify the actual operating speed of the mechanism. This is another indication of the linearity of the equations relating velocities, as pointed out in Section 3.9. *For a single-degree-of-freedom mechanism, the locations of all instant centers are uniquely determined by the geometry alone and do not depend on the operating speed.*

3.16 VELOCITY ANALYSIS USING INSTANT CENTERS

The properties of instant centers also provide a simple graphical approach for the velocity analysis of planar-motion mechanisms.

EXAMPLE 3.11

In Fig. 3.29a we assume that the angular velocity ω_2 of crank 2 is given, and we wish to find the velocities of points B, D, and E at the instant shown.

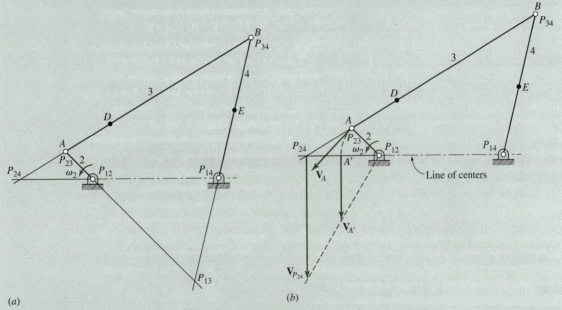

(a)   (b)

Figure 3.29 Example 3.11: Graphical velocity determination by the instant-center method.

SOLUTION

Consider the straight line defined by the instant centers P_{12}, P_{14}, and P_{24}. This must be a straight line according to the Kennedy–Aronhold theorem and is called the line of centers. According to its definition, P_{24} is common to both links 2 and 4 and has equal absolute velocities in each.

First consider instant center P_{24} as a point of link 2. The velocity $\mathbf{V}_A$ can be found from ω_2 using the velocity-difference equation about P_{12}, and the velocity of P_{24} can be found from it; the graphical

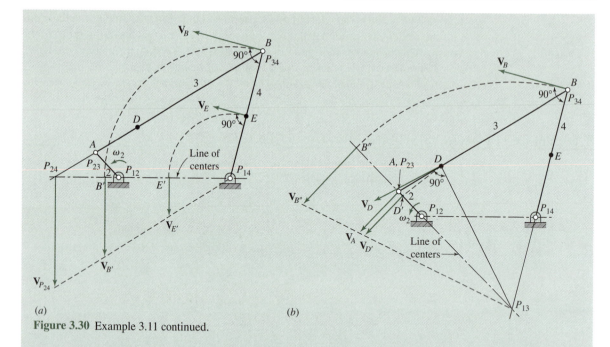

(a)   (b)

Figure 3.30 Example 3.11 continued.

construction is shown in Fig. 3.29b. When point A' of link 2 is located on the line of centers at an equal distance from P_{12}, its absolute velocity $V_{A'}$ is equal in magnitude to V_A. Now the magnitude of $\mathbf{V}_{P_{24}}$ can be found* by constructing a line from P_{12} through the terminus of $\mathbf{V}_{A'}$ as shown.

Next consider P_{24} as a point of link 4 rotating about P_{14}. Knowing $\mathbf{V}_{P_{24}}$, we can find the velocity of any other point of link 4, such as B' or E' (see Fig. 3.30a), by using the reverse construction. Since B' and E' were chosen to have the same radii from P_{14} as B and E, their velocities have magnitudes equal to those of $\mathbf{V}_B$ and $\mathbf{V}_E$, respectively, and these can be laid out in their proper directions as shown in Fig. 3.30a.

To obtain $\mathbf{V}_D$ we note that D is in link 3; the known velocity ω_2 (or $\mathbf{V}_A$) is for link 2, and the reference link is link 1. Therefore, a new line of centers $P_{12}P_{13}P_{23}$ is chosen, as shown in Fig. 3.30b. Using ω_2 and P_{12}, we find the absolute velocity of the common instant center P_{23}. Here this step is trivial, because $\mathbf{V}_{P_{23}} = \mathbf{V}_A$. Locating point D' on the new line of centers, we find $\mathbf{V}_{D'}$ as shown and use its magnitude to find the desired velocity $\mathbf{V}_D$. Note that, according to the definition, the instant center P_{13}, as part of link 3, has zero velocity at this instant. Because B can also be considered a point of link 3, its velocity can be found in a similar manner by finding $\mathbf{V}_{B''}$ as shown.

The *line-of-centers* method of velocity analysis using instant centers can be summarized as follows:

1. Identify the three link numbers associated with the given velocity and the velocity to be found. One of these is usually link 1, because usually absolute velocity information is given and requested.

---

*Note that $\mathbf{V}_{P_{24}}$ could have been found directly from its velocity difference from P_{12}. This construction was shown to illustrate the principle of the graphical method.

2. Locate the three instant centers defined by the links of step 1 and draw the line of centers.
3. Find the velocity of the common instant center by treating it as a point of the link whose velocity is given.
4. With the velocity of the common instant center known, consider it as a point of the link whose velocity is to be found. The velocity of any other point in that link can now be found.

Another example will illustrate the procedure and will show how to treat instant centers located at infinity.

EXAMPLE 3.12

Only some of the links of the device shown in Fig. 3.31 can be seen; others are enclosed in a housing, but it is known that the instant center P_{25} has the location shown. Find the angular velocity of the crank ω_2 which is necessary to produce a velocity $\mathbf{V}_C$ of 10 m/s to the right.

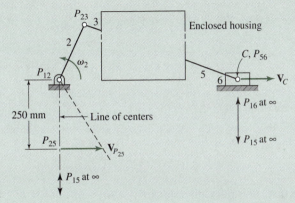

Figure 3.31 Example 3.12.

SOLUTION

Because we are given $\mathbf{V}_{C_5/1}$ and want $\omega_{2/1}$, we need to use the instant centers P_{15}, P_{12}, and P_{25}. After locating P_{25}, P_{56}, and P_{16} by inspection and applying the theorem of three centers, we locate P_{15} at infinity as shown. We now draw the line of centers $P_{12}P_{25}P_{15}$.

Considering P_{25} as a part of link 5, we wish to find its velocity from the given $\mathbf{V}_C$. We have difficulty in locating a point C' on the line of centers at the same radius from P_{15} as C because P_{15} is at infinity. How can we proceed?

Recalling the discussion of Section 3.15 and Eq. (3.27), we see that because P_{15} is at infinity, the relative motion between links 5 and 1 is translation and $\omega_{5/1} = 0$. Because this is true, *every* point of link 5 has the same absolute velocity, including $\mathbf{V}_{P_{25}} = \mathbf{V}_C$. Thus we lay out $\mathbf{V}_{P_{25}}$ on the figure.

Next we treat P_{25} as a point of link 2, rotating about P_{15}, and solve for the angular velocity of link 2:

$$\omega_2 = \frac{V_{P_{25}}}{R_{P_{25}P_{12}}} = \frac{10 \text{ m/s}}{0.25 \text{ m}} = 40 \text{ rad/s ccw} \qquad\qquad Ans.$$

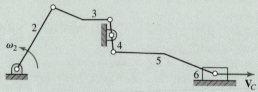

Figure 3.32

Noticing the apparent paradox between the directions of $\mathbf{V}_C$ and ω_2, we may speculate on the validity of our solution. This would be resolved, however, by opening the enclosed housing and discovering the linkage shown in Fig. 3.32.

3.17 THE ANGULAR-VELOCITY-RATIO THEOREM

In Fig. 3.33, P_{24} is the instant center common to links 2 and 4. Its absolute velocity $\mathbf{V}_{P_{24}}$ is the same whether P_{24} is considered as a point of link 2 or of link 4. Considering it each way, we can write

$$\mathbf{V}_{P_{24}} = \mathbf{V}_{P_{12}}^{0} + \omega_{2/1} \times \mathbf{R}_{P_{24}P_{12}} = \mathbf{V}_{P_{14}}^{0} + \omega_{4/1} \times \mathbf{R}_{P_{24}P_{14}} \qquad (a)$$

where $\omega_{2/1}$ and $\omega_{4/1}$ are the same as ω_2 and ω_4, respectively, but the additional subscript has been written to emphasize the presence of the third link (the frame).

Considering the magnitudes only, we can rearrange Eq. (a) to read

$$\frac{\omega_{4/1}}{\omega_{2/1}} = \frac{R_{P_{24}P_{12}}}{R_{P_{24}P_{14}}} \qquad (b)$$

This system illustrates the *angular-velocity-ratio theorem*. The theorem states that *the angular-velocity ratio of any two bodies in planar motion with respect to a third body is inversely proportional to the segments into which the common instant center cuts the line of centers.* Written in general notation for the motion of bodies j and k with respect to body i, the equation is

$$\frac{\omega_{k/i}}{\omega_{j/i}} = \frac{R_{P_{jk}P_{ij}}}{R_{P_{jk}P_{ik}}} \qquad (3.28)$$

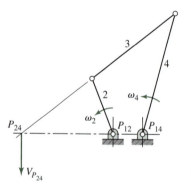

Figure 3.33 The angular-velocity-ratio theorem.

Picking an arbitrary positive direction along the line of centers, you should prove for yourself that the angular-velocity ratio is positive when the common instant center falls outside the other two centers and negative when it falls between them.

3.18 RELATIONSHIPS BETWEEN FIRST-ORDER KINEMATIC COEFFICIENTS AND INSTANT CENTERS

The first-order kinematic coefficients (see Section 3.11) can be expressed in terms of the location of the instantaneous centers of velocity. For the four-bar linkage, the angular velocity of link j can be written from the chain rule as

$$\omega_j = \theta'_j \omega_i \tag{3.29}$$

where ω_i is the angular velocity of the input link. Therefore, the first-order kinematic coefficient of link j can be written as

$$\theta'_j = \frac{\omega_j}{\omega_i}$$

Consistent with Eq. (3.28), the first-order kinematic coefficient of link j can be written as

$$\theta'_j = \frac{R_{P_{ij}P_{1i}}}{R_{P_{ij}P_{1j}}} \tag{3.30}$$

where P_{1i} and P_{1j} are the absolute instant centers of the input link i and link j, respectively, and P_{ij} is the relative instant center of links i and j.

For the four-bar linkage in Example 3.11, the first-order kinematic coefficients of link 3 and 4 can be written from Eqs. (3.30) as

$$\theta'_3 = \frac{R_{P_{23}P_{12}}}{R_{P_{23}P_{13}}} \quad \text{and} \quad \theta'_4 = \frac{R_{P_{24}P_{12}}}{R_{P_{24}P_{14}}}$$

From the scaled drawing of the four-bar linkage (see Fig. 3.29a), we measure $R_{P_{23}P_{13}} = 15$ in and $R_{P_{24}P_{12}} = 11.2$ in. Therefore, the first-order kinematic coefficients of the two links are

$$\theta'_3 = +\frac{4 \text{ in}}{15 \text{ in}} = +0.267 \quad \text{and} \quad \theta'_4 = +\frac{11.2 \text{ in}}{21.2 \text{ in}} = +0.528$$

Substituting these values and the input angular velocity $\omega_2 = 94.2$ rad/s ccw into Eqs. (3.29), the angular velocity of the coupler link and the output link, respectively, are

$$\omega_3 = 25.15 \text{ rad/s ccw} \quad \text{and} \quad \omega_4 = 49.74 \text{ rad/s ccw}$$

These answers agree closely with the results obtained from the velocity polygon method of Example 3.1 ($\omega_3 = 25.6$ rad/s ccw and $\omega_4 = 49.6$ rad/s ccw).

For a slider-crank linkage with the slider regarded as the input (denoted as link i), the angular velocity of link j can be written from the chain rule as

$$\theta'_j = \frac{\omega_j}{\dot{r}_i} \tag{3.31}$$

where $\dot{r}_i$ is the velocity of the slider. When $R_{P_{ij}P_{1i}}$ becomes infinite, Eq. (3.30) for the first-order kinematic coefficient of link j becomes

$$\theta'_j = \frac{1}{R_{P_{ij}P_{1j}}} \tag{3.32}$$

Therefore, for the offset slider-crank linkage in Example 3.12, the first-order kinematic coefficients of links 2 and 3 can be written from Eq. (3.32) as

$$\theta'_2 = \frac{1}{R_{P_{24}P_{12}}} \quad \text{and} \quad \theta'_3 = \frac{1}{R_{P_{34}P_{13}}}$$

Because we are using velocity of the input, slider 4, as positive to the right, we have $\hat{\tau} = \hat{i}$ for the positive input direction and we are required to choose $\hat{\rho} = \hat{\tau} \times \hat{k} = -\hat{j}$ (downward) as the positive direction along the line of centers.* From the scaled drawing of the offset slider-crank linkage (Fig. 3.8a), therefore, we measure $R_{P_{24}P_{12}} = -51$ mm and $R_{P_{34}P_{13}} = 185$ mm and we find that the kinematic coefficients are

$$\theta'_2 = \frac{1}{-0.051 \text{ m}} = -19.6 \text{ rad/m} \quad \text{and} \quad \theta'_3 = \frac{1}{+0.185 \text{ m}} = +5.41 \text{ rad/m}$$

Substituting these values and the input velocity, $\dot{r}_4 = -10$ m/s, into Eqs. (3.31), the angular velocities of links 2 and 3, respectively, are

$$\omega_2 = 196 \text{ rad/s ccw} \quad \text{and} \quad \omega_3 = 54.1 \text{ rad/s cw}$$

These answers agree closely with the results obtained from the velocity polygon method in Example 3.2 ($\omega_2 = 200$ rad/s ccw and $\omega_3 = 53.6$ rad/s cw). However, because graphical measurements are used for positions of the instant centers, results are probably still not as accurate as those of Example 3.7 ($\omega_2 = 197.7$ rad/s ccw and $\omega_3 = 54.4$ rad/s cw).

The velocity of a point, say P, fixed in a link of a mechanism can be written as

$$\mathbf{V}_P = V_P \hat{\tau}_P \tag{3.33a}$$

or as

$$\mathbf{V}_P = (x'_P \hat{i} + y'_P \hat{j})\dot{\psi} \tag{3.33b}$$

where $\dot{\psi}$ is the generalized input velocity to the mechanism. The magnitude of the velocity, commonly referred to as the speed, can be written from the chain rule as

$$V_P = r'_P \dot{\psi} \tag{3.34a}$$

---

*Note that this sign convention is the same as saying that positive input motion of slider 4 is the same as a positive (counterclockwise) rotation about instant center P_{14} now located at infinity in the negative $\hat{\rho}$ direction.

where the first-order kinematic coefficient is defined as

$$r'_P = \pm\sqrt{x'^2_P + y'^2_P} \tag{3.34b}$$

Here the sign convention is as follows: We use the positive sign if the instantaneous change in the input position is positive, and we use the negative sign if the instantaneous change in the input position is negative.

Rearranging Eq. (3.33a), the unit tangent vector to the point trajectory can be written as

$$\hat{\tau}_P = \frac{\mathbf{V}_P}{V_P} \tag{3.35}$$

Then substituting Eqs. (3.33b) and (3.34a) into this relation gives

$$\hat{\tau}_P = \left(\frac{x'_P}{r'_P}\right)\hat{\mathbf{i}} + \left(\frac{y'_P}{r'_P}\right)\hat{\mathbf{j}} \tag{3.36}$$

Consistent with Eq. (3.7), the unit normal vector to the point trajectory can now be written as $\hat{\rho}_P = \hat{\tau}_P \times \mathbf{k}$. Substituting Eq. (3.36) into this relation, the unit normal vector can be written

$$\hat{\rho}_P = \left(\frac{y'_P}{r'_P}\right)\hat{\mathbf{i}} + \left(\frac{-x'_P}{r'_P}\right)\hat{\mathbf{j}} \tag{3.37}$$

3.19 FREUDENSTEIN'S THEOREM

In the analysis and design of linkages it is often important to know the phases of the linkage at which the extreme values of the output velocity occur or, more precisely, the phases at which the ratio of the output and input velocities reaches its extremes.

The earliest work in determining extreme values is apparently that of Krause,[2] who stated that the velocity ratio ω_4/ω_2 of the drag-link mechanism (Fig. 3.34) reaches an extreme value when the connecting rod and follower, links 3 and 4, become perpendicular to each other. Rosenauer, however, showed that this is not strictly true.[3] Following Krause, Freudenstein developed a simple graphical method for determining the phases of the four-bar linkage at which the extreme values of the velocity do occur.[4]

Figure 3.34 The drag-link mechanism.

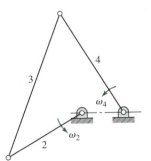

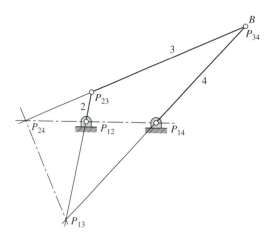

Figure 3.35 The collineation axis.

Freudenstein's theorem makes use of the line connecting instant centers P_{13} and P_{24} (Fig. 3.35), called the *collineation axis*. The theorem states that at *an extreme of the output to input angular velocity ratio of a four-bar linkage, the collineation axis is perpendicular to the coupler link.*[5]

Using the angular-velocity-ratio theorem, Eq. (3.28), we write

$$\frac{\omega_4}{\omega_2} = \frac{R_{P_{24}P_{12}}}{R_{P_{24}P_{12}} + R_{P_{12}P_{14}}}$$

Because $R_{P_{12}P_{14}}$ is the fixed length of the frame link, the extremes of the velocity ratio occur when $R_{P_{24}P_{12}}$ is either a maximum or a minimum. Such positions may occur on either or both sides of P_{12}. Thus the problem reduces to finding the geometry of the linkage for which $R_{P_{24}P_{12}}$ is an extremum.

During motion of the linkage, P_{24} travels along the line $P_{12}P_{14}$ as seen by the theorem of three centers; but at an extreme value of the velocity ratio, P_{24} must instantaneously be at rest (its direction of travel on this line must be reversing). This occurs when the velocity of P_{24}, considered as a point of link 3, is directed along the coupler link. This will be true only when the coupler link is perpendicular to the collineation axis, because P_{13} is the instant center of link 3.

An inversion of the theorem (treating link 2 as fixed) states that an *extreme value of the velocity ratio ω_3/ω_2 of a four-bar linkage occurs when the collineation axis is perpendicular to the follower (link 4).*

3.20 INDICES OF MERIT; MECHANICAL ADVANTAGE

In this section we will study some of the various ratios, angles, and other parameters of mechanisms that tell us whether a mechanism is a good one or a poor one. Many such parameters have been defined by various authors over the years, and there is no common agreement on a single "index of merit" for all mechanisms. Yet the many used have a number of features in common, including the fact that most can be related to the velocity ratios of the mechanism and, therefore, can be determined solely from the geometry of the mechanism. In addition, most depend on some knowledge of the application of the mechanism, especially of which are the input and output links. It is often desirable in the analysis or synthesis of mechanisms to plot these indices of merit for a revolution of the input crank and to notice in particular their minimum and maximum values when evaluating the design of the mechanism or its suitability for a given application.

In Section 3.17 we learned that the ratio of the angular velocity of the output link to the input link of a mechanism is inversely proportional to the segments into which the common instant center cuts the line of centers. Thus, in the four-bar linkage of Fig. 3.36, if links 2 and 4 are the input and output links, respectively, then

$$\frac{\omega_4}{\omega_2} = \frac{R_{PA}}{R_{PD}}$$

is the equation for the output- to input-velocity ratio. We also learned in Section 3.19 that the extremes of this ratio occur when the collineation axis is perpendicular to the coupler, link 3.

If we now assume that the linkage of Fig. 3.36 has no friction or inertia forces during its operation or that these are negligible compared with the input torque T_2, applied to link 2, and the output torque T_4, the resisting load torque on link 4, then we can derive a relation between T_2 and T_4. Because friction and inertia forces are negligible, the input power applied to link 2 is the negative of the power applied to link 4 by the load; hence

$$T_2\omega_2 = -T_4\omega_4 \qquad (a)$$

or

$$\frac{T_4}{T_2} = -\frac{\omega_2}{\omega_4} = -\frac{R_{PD}}{R_{PA}} \qquad (3.38)$$

The *mechanical advantage* of a mechanism is the instantaneous ratio of the output force (torque) to the input force (torque). Here we see that the mechanical advantage is the negative reciprocal of the velocity ratio. Either can be used as an index of merit in judging a mechanism's ability to transmit force or power.

The mechanism is redrawn in Fig. 3.37 at the position where links 2 and 3 are on the same straight line. At this position, R_{PA} and ω_4 are passing through zero; hence an extreme

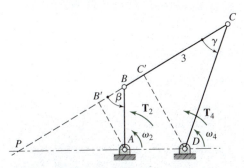

Figure 3.36 Four-bar linkage.

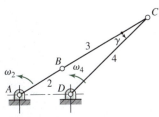

Figure 3.37 Four-bar linkage in toggle.

value (infinity) of the mechanical advantage is obtained. A mechanism in this phase is said to be *in toggle*. Such toggle positions are often used to produce a high mechanical advantage; an example is the clamping mechanism of Fig. 2.8.

Proceeding further, we construct $B'A$ and $C'D$ perpendicular to the line PBC in Fig. 3.36. Also, we assign labels β and γ to the acute angles made by the coupler, or its extension and the input and output links, respectively. Then, by similar triangles,

$$\frac{R_{PD}}{R_{PA}} = \frac{R_{C'D}}{R_{B'A}} = \frac{R_{CD} \sin \gamma}{R_{BA} \sin \beta} \qquad (b)$$

Then, using Eq. (3.38), we see that another expression for mechanical advantage is

$$\frac{T_4}{T_2} = -\frac{\omega_2}{\omega_4} = -\frac{R_{CD} \sin \gamma}{R_{BA} \sin \beta} \qquad (3.39)$$

This equation shows that the mechanical advantage is infinite whenever the angle β is 0 or 180°—that is, whenever the mechanism is in toggle.

In Sections 1.10 and 2.8 we defined the angle γ between the coupler and the follower link as the *transmission angle*. This angle is also often used as an index of merit for a four-bar linkage. Equation (3.39) shows that the mechanical advantage diminishes when the transmission angle is much less than a right angle. If the transmission angle becomes too small, the mechanical advantage becomes small and even a very small amount of friction may cause a mechanism to lock or jam. To avoid this, a common rule of thumb is that a four-bar linkage should not be used in a region where the transmission angle is less than, say, 45° or 50°. The best four-bar linkage, based on the quality of its force transmission, will have a transmission angle that deviates from 90° by the smallest amount.

In other mechanisms—for example, meshing gear teeth or a cam-follower system—the pressure angle is used as an index of merit. The pressure angle is defined as the acute angle between the direction of the output force and the direction of the velocity of the point where the output force is applied. Pressure angles are discussed more thoroughly in Chapters 5 and 6. In the four-bar linkage, the pressure angle is the complement of the transmission angle.

Another index of merit which has been proposed[6] is the determinant of the coefficients of the simultaneous equations relating the dependent velocities of a mechanism. In Example 3.5, for example, we saw that the dependent velocities of a four-bar linkage are related by

$$(r_3 \sin \theta_3)\omega_3 - (r_4 \sin \theta_4)\omega_4 = -(r_2 \sin \theta_2)\omega_2$$

$$(r_3 \cos \theta_3)\omega_3 - (r_4 \cos \theta_4)\omega_4 = -(r_2 \cos \theta_2)\omega_2$$

The determinant of the coefficients is

$$\Delta = \begin{vmatrix} r_3 \sin \theta_3 & -r_4 \sin \theta_4 \\ r_3 \cos \theta_3 & -r_4 \cos \theta_4 \end{vmatrix} = r_3 r_4 \sin(\theta_4 - \theta_3)$$

As is obvious from Cramer's rule, the solutions for the dependent velocities, in this case ω_3 and ω_4, must include this determinant in the denominator. This is borne out in the solution of the four-bar linkage, Eqs. (3.22). Although the form of this determinant changes for different mechanisms, such a determinant can always be defined and always appears in the denominators of all dependent velocity solutions.

When the determinant becomes small, the mechanical advantage also becomes small and the usefulness of the mechanism is reduced in such regions. We have not seen it yet, but it is also true that this same determinant appears in the denominator of the dependent accelerations (see Section 4.10) and all other quantities which require taking derivatives of the loop-closure equation. If this determinant is small, the mechanism will function poorly in all respects—force transmission, motion transformation, sensitivity to manufacturing errors, and so on.

3.21 CENTRODES

We noted in Section 3.13 that the location of an instant center of velocity is defined only instantaneously and changes as the mechanism moves. When the changing locations of an instant center are found for all possible phases of the mechanism, they describe a curve or locus, called a *centrode*.* In Fig. 3.38, the instant center P_{13} is located at the intersection of the extension of links 2 and 4. As the linkage is moved through all possible positions, P_{13} traces out the curve called the *fixed centrode* on link 1.

Figure 3.39 shows the inversion of the same linkage in which link 3 is fixed and link 1 is movable. When this inversion is moved through all possible positions, P_{13} traces a

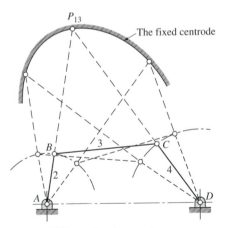

Figure 3.38 The fixed centrode.

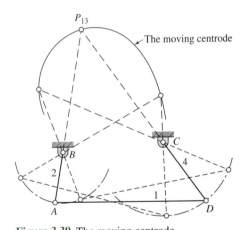

Figure 3.39 The moving centrode.

*Opinion seems about equally divided on whether these loci should be termed *centrodes* or *polodes*. Generally, those preferring the name *instant center* call them *centrodes* and those who use the word *pole* call them *polodes*. The French name *roulettes* has also been applied. The three-dimensional equivalents are ruled surfaces and are referred to as *axodes*.

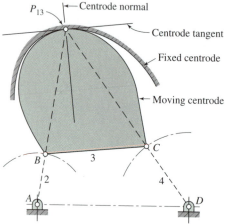

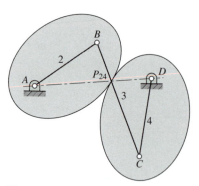

Figure 3.40 Rolling contact between centrodes.

Figure 3.41

different curve on link 3. For the original linkage, with link 1 fixed, this is the curve traced by P_{13} on the coordinate system of the moving link 3; it is called the *moving centrode*.

Figure 3.40 shows the moving centrode, attached to link 3, and the fixed centrode, attached to link 1. It is imagined here that links 1 and 3 have been machined to the actual shapes of the respective centrodes and that links 2 and 4 have been removed entirely. If the moving centrode is now permitted to roll on the fixed centrode without slip, link 3 will have exactly the same motion as it had in the original linkage. This remarkable property, which stems from the fact that a point of rolling contact is an instant center, turns out to be quite useful in the synthesis of linkages.

We can restate this property as follows: *The plane motion of one rigid body with respect to another is completely equivalent to the rolling motion of one centrode on the other.* The instantaneous point of rolling contact is the instant center, as shown in Fig. 3.40. Also shown are the common tangent to the two centrodes and the common normal, called the *centrode tangent* and the *centrode normal;* they are sometimes used as the axes of a coordinate system for developing equations for a coupler curve or other properties of the motion.

The centrodes of Fig. 3.40 were generated by the instant center P_{13} on links 1 and 3. Another set of centrodes, both moving, is generated on links 2 and 4 when instant center P_{24} is considered. Figure 3.41 shows these as two ellipses for the case of a crossed double-crank linkage with equal cranks. These two centrodes roll upon each other and describe the identical motion between links 2 and 4 which would result from the operation of the original four-bar linkage. This construction can be used as the basis for the development of a pair of elliptical gears.

NOTES

1. F. H. Raven, Velocity and Acceleration Analysis of Plane and Space Mechanisms by Means of Independent-Position Equations, *J. Appl. Mech., ASME Trans.,* ser. E, vol. 80, pp. 1–6, 1958.

2.   R. Krause, Die Doppelkurbel und Ihre Geschwindigkeitsgrenzen, *Maschinenbau/Getriebetechnik,* vol. 18, pp. 37–41, 1939; Zur Synthese der Doppelkurbel, *Maschinenbau/Getriebetechnik,* vol. 18, pp. 93–94, 1939.

3.   N. Rosenauer, Synthesis of Drag-Link Mechanisms for Producing Nonuniform Rotational Motion with Prescribed Reduction Ratio Limits, *Aust. J. Appl. Sci.,* vol. 8, pp. 1–6, 1957.

4.   F. Freudenstein, On the Maximum and Minimum Velocities and Accelerations in Four-Link Mechanisms, *Trans. ASME,* vol. 78, pp. 779–787, 1956.

5.   A. S. Hall, Jr. contributed a rigorous proof of this theorem in an appendix to Freudenstein's paper.

6.   J. Denavit et al., Velocity, Acceleration, and Static Force Analysis of Spatial Linkages, *J. Appl. Mech., ASME Trans.,* vol. 87, ser. E, no. 4, pp. 903–910, 1965.

Problems*

3.1   The position vector of a point is given by the equation $\mathbf{R} = 100e^{j\pi t}$, where R is in inches. Find the velocity of the point at $t = 0.40$ s.

3.2   The equation $R = (t^2 + 4)e^{j\pi t/10}$ defines the path of a particle. If R is in meters, find the velocity of the particle at $t = 20$ s.

3.3   If automobile A is traveling south at 55 mi/h and automobile B north $60°$ east at 40 mi/h, what is the velocity difference between B and A? What is the apparent velocity of B to the driver of A?

3.4   In the figure, wheel 2 rotates at 600 rev/min and drives wheel 3 without slipping. Find the velocity difference between points B and A.

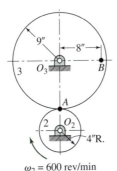

$\omega_2 = 600$ rev/min

Figure P3.4

3.5   Two points A and B, located along the radius of a wheel (see figure), have speeds of 80 and 140 m/s, respectively. The distance between the points is $R_{BA} = 300$ mm.
(a) What is the diameter of the wheel?
(b) Find $\mathbf{V}_{AB}$, $\mathbf{V}_{BA}$, and the angular velocity of the wheel.

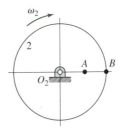

Figure P3.5

3.6   A plane leaves point B and flies east at 350 mi/h. Simultaneously, at point A, 200 miles southeast (see figure), a plane leaves and flies northeast at 390 mi/h.
(a) How close will the planes come to each other if they fly at the same altitude?

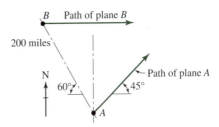

Figure P3.6 $R_{AB} = 200$ miles.

*When assigning problems, the instructor may wish to specify the method of solution to be used, because a variety of approaches are presented in the text.

(b) If they both leave at 6:00 p.m., at what time will this occur?

3.7 To the data of Problem 3.6, add a wind of 30 mi/h from the west.
(a) If A flies the same heading, what is its new path?
(b) What change does the wind make in the results of Problem 3.6?

3.8 The velocity of point B on the linkage shown in the figure is 40 m/s. Find the velocity of point A and the angular velocity of link 3.

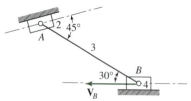

Figure P3.8 $R_{AB} = 400$ mm.

3.9 The mechanism shown in the figure is driven by link 2 at $\omega_2 = 45$ rad/s ccw. Find the angular velocities of links 3 and 4.

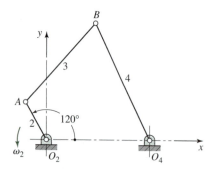

Figure P3.9 $R_{AO_2} = 4$ in, $R_{BA} = 10$ in, $R_{O_4O_2} = 10$ in, $R_{BO_4} = 12$ in.

3.10 Crank 2 of the push-link mechanism shown in the figure is driven at $\omega_2 = 60$ rad/s cw. Find the velocities of points B and C and the angular velocities of links 3 and 4.

3.11 Find the velocity of point C on link 4 of the mechanism shown in the figure if crank 2 is driven at $\omega_2 = 48$ rad/s ccw. What is the angular velocity of link 3?

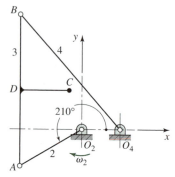

Figure P3.10 $R_{AO_2} = 150$ mm, $R_{BA} = 300$ mm, $R_{O_4O_2} = 75$ mm, $R_{BO_4} = 300$ mm, $R_{DA} = 150$ mm, $R_{CD} = 100$ mm.

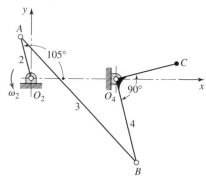

Figure P3.11 $R_{AO_2} = 8$ in, $R_{BA} = 32$ in, $R_{O_4O_2} = 16$ in, $R_{BO_4} = 16$ in.

3.12 The figure illustrates a parallel-bar linkage, in which opposite links have equal lengths. For this linkage, show that ω_3 is always zero and that $\omega_4 = \omega_2$. How would you describe the motion of link 4 with respect to link 2?

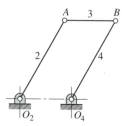

Figure P3.12

3.13 The figure illustrates the antiparallel or crossed-bar linkage. If link 2 is driven at $\omega_2 = 1$ rad/s ccw, find the velocities of points C and D.

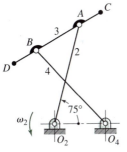

Figure P3.13 $R_{AO_2} = R_{BO_4} =$
300 mm, $R_{BA} = R_{O_4O_2} = 150$ mm,
$R_{CA} = R_{DB} = 75$ mm.

3.14 Find the velocity of point C of the linkage shown in the figure assuming that link 2 has an angular velocity of 60 rad/s ccw. Also find the angular velocities of links 3 and 4.

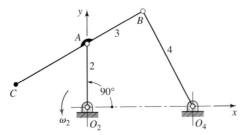

Figure P3.14 $R_{AO_2} = R_{BA} = 6$ in,
$R_{O_4O_2} = R_{BO_4} = 10$ in, $R_{CA} = 8$ in.

3.15 The inversion of the slider-crank mechanism shown in the figure is driven by link 2 at $\omega_2 = 60$ rad/s ccw. Find the velocity of point B and the angular velocities of links 3 and 4.

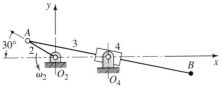

Figure P3.15 $R_{AO_2} = 75$ mm, $R_{BA} = 400$ mm,
$R_{O_4O_2} = 125$ mm.

3.16 Find the velocity of the coupler point C and the angular velocities of links 3 and 4 of the mechanism shown if crank 2 has an angular velocity of 30 rad/s cw.

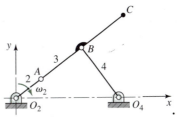

Figure P3.16 $R_{AO_2} = 3$ in,
$R_{BA} = R_{CB} = 5$ in, $R_{O_4O_2} = 10$ in,
$R_{BO_4} = 6$ in.

3.17 Link 2 of the linkage shown in the figure has an angular velocity of 10 rad/s ccw. Find the angular velocity of link 6 and the velocities of points B, C, and D.

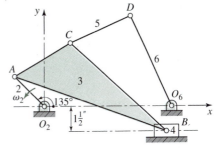

Figure P3.17 $R_{AO_2} = 2.5$ in, $R_{BA} = 10$ in,
$R_{CB} = 8$ in, $R_{CA} = R_{DC} = 4$ in, $R_{O_6O_2} =$
8 in, $R_{DO_6} = 6$ in.

3.18 The angular velocity of link 2 of the drag-link mechanism shown in the figure is 16 rad/s cw. Plot a polar velocity diagram for the velocity of point B for all crank positions. Check the positions of maximum and minimum velocities by using Freudenstein's theorem.

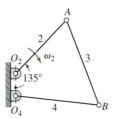

Figure P3.18 $R_{AO_2} = 350$ mm,
$R_{BA} = 425$ mm, $R_{O_4O_2} = 100$ mm,
$R_{BO_4} = 400$ mm.

3.19 Link 2 of the mechanism shown in the figure is driven at $\omega_2 = 36$ rad/s cw. Find the angular velocity of link 3 and the velocity of point B.

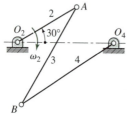

Figure P3.19 $R_{AO_2} = 5$ in,
$R_{BA} = R_{BO_4} = 8$ in, $R_{O_4O_2} = 7$ in.

3.20 Find the velocity of point C and the angular velocity of link 3 of the push-link mechanism shown in the figure. Link 2 is the driver and rotates at 8 rad/s ccw.

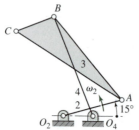

Figure P3.20 $R_{AO_2} = 150$ mm,
$R_{BA} = R_{BO_4} = 250$ mm,
$R_{O_4O_2} = 75$ mm, $R_{CA} = 300$ mm,
$R_{CB} = 100$ mm.

3.21 Link 2 of the mechanism shown in the figure has an angular velocity of 56 rad/s ccw. Find the velocity of point C.

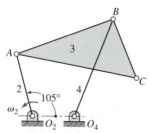

Figure P3.21 $R_{BA} = R_{BO_4} = 250$ mm,
$R_{O_4O_2} = 100$ mm, $R_{CA} = 300$ mm.

3.22 Find the velocities of points B, C, and D of the double-slider mechanism shown in the figure if crank 2 rotates at 42 rad/s cw.

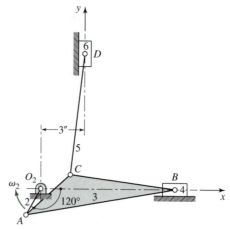

Figure P3.22 $R_{AO_2} = 2$ in, $R_{BA} = 10$ in,
$R_{CA} = 4$ in, $R_{CB} = 7$ in, $R_{DC} = 8$ in.

3.23 The figure shows the mechanism used in a two-cylinder 60° V engine consisting, in part, of an articulated connecting rod. Crank 2 rotates at 2000 rev/min cw. Find the velocities of points B, C, and D.

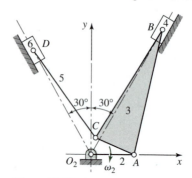

Figure P3.23 $R_{AO_2} = 2$ in, $R_{BA} = R_{CB} = 6$ in, $R_{CA} = 2$ in, $R_{DC} = 5$ in.

3.24 Make a complete velocity analysis of the linkage shown in the figure given that $\omega_2 = 24$ rad/s cw. What is the absolute velocity of point B? What is its apparent velocity to an observer moving with link 4?

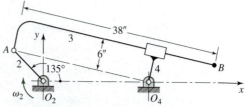

Figure P3.24 $R_{AO_2} = 8$ in, $R_{O_4O_2} = 20$ in.

3.25 Find $\mathbf{V}_B$ for the linkage shown in the figure if $V_A = 1$ ft/s.

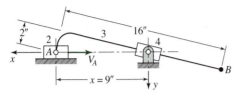

Figure P3.25

3.26 The figure shows a variation of the Scotch-yoke mechanism. It is driven by crank 2 at $\omega_2 = 36$ rad/s ccw. Find the velocity of the crosshead, link 4.

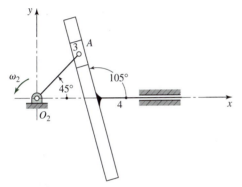

Figure P3.26 $R_{AO_2} = 250$ mm.

3.27 Make a complete velocity analysis of the linkage shown in the figure for $\omega_2 = 72$ rad/s ccw.

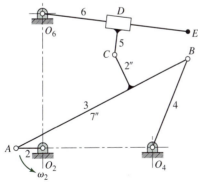

Figure P3.27 $R_{AO_2} = R_{DC} = 1.5$ in,
$R_{BA} = 10.5$ in, $R_{O_4O_2} = 6$ in, $R_{BO_4} = 5$ in,
$R_{O_6O_2} = 7$ in, $R_{EO_6} = 8$ in.

3.28 Slotted links 2 and 3 are driven independently at $\omega_2 = 30$ rad/s cw and $\omega_3 = 20$ rad/s cw, respec-

tively. Find the absolute velocity of the center of the pin P_4 carried in the two slots.

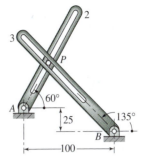

Figure P3.28 Dimensions in millimeters.

3.29 The mechanism shown is driven such that $V_C = 10$ in/s to the right. Rolling contact is assumed between links 1 and 2, but slip is possible between links 2 and 3. Determine the angular velocity of link 3.

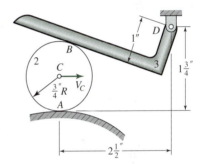

Figure P3.29

3.30 The circular cam shown is driven at an angular velocity of $\omega_2 = 15$ rad/s cw. There is rolling contact between the cam and the roller, link 3. Find the angular velocity of the oscillating follower, link 4.

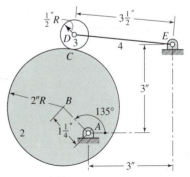

Figure P3.30

3.31 The mechanism shown is driven by link 2 at 10 rad/s ccw. There is rolling contact at point F. Determine the velocity of points E and G and the angular velocities of links 3, 4, 5, and 6.

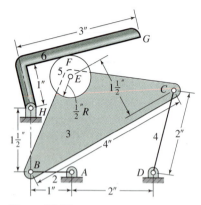

Figure P3.31

3.32 The figure shows the schematic diagram for a two-piston pump. It is driven by a circular eccentric, link 2, at $\omega_2 = 25$ rad/s ccw. Find the velocities of the two pistons, links 6 and 7.

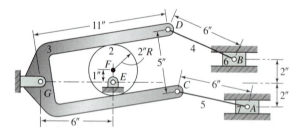

Figure P3.32

3.33 The epicyclic gear train shown is driven by the arm, link 2, at $\omega_2 = 10$ rad/s cw. Determine the angular velocity of the output shaft, attached to gear 3.

3.34 The diagram shows a planar schematic approximation of an automotive front suspension. The roll center is the term used by the industry to describe the point about which the auto body seems to rotate with

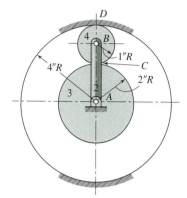

Figure P3.33

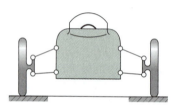

Figure P3.34

respect to the ground. The assumption is made that there is pivoting but no slip between the tires and the road. After making a sketch, use the concepts of instant centers to find a technique to locate the roll center.

3.35 Locate all instant centers for the linkage of Problem 3.22.

3.36 Locate all instant centers for the mechanism of Problem 3.25.

3.37 Locate all instant centers for the mechanism of Problem 3.26.

3.38 Locate all instant centers for the mechanism of Problem 3.27.

3.39 Locate all instant centers for the mechanism of Problem 3.28.

3.40 Locate all instant centers for the mechanism of Problem 3.29.

4 | Acceleration

4.1 DEFINITION OF ACCELERATION

In Fig. 4.1a a moving point is first observed at location P, where it has a velocity of $\mathbf{V}_P$. After a short time interval Δt, the point is observed to have moved along some path to a new location P', and its velocity has changed to $\mathbf{V}'_p$, which may differ from $\mathbf{V}_P$ in both magnitude and direction. We can evaluate the change in velocity $\mathbf{\Delta V}_P$ as shown in Fig. 4.1b:

$$\mathbf{\Delta V}_P = \mathbf{V}'_p - \mathbf{V}_P$$

The *average acceleration* of the point P during the time interval is $\mathbf{\Delta V}_P / \Delta t$. The *instantaneous acceleration* (hereafter called simply the *acceleration*) of point P is defined as the time rate of change of its velocity—that is, the limit of the average acceleration for an infinitesimally small time interval:

$$\mathbf{A}_P = \lim_{\Delta t \to 0} \left(\frac{\mathbf{\Delta V}_P}{\Delta t} \right) = \frac{d\mathbf{V}_P}{dt} = \frac{d^2 \mathbf{R}_P}{dt^2} \tag{4.1}$$

Because velocity is a vector quantity, $\mathbf{\Delta V}_P$ and the acceleration $\mathbf{A}_P$ are also vector quantities and have both magnitude and direction. Also, like velocity, the acceleration vector is properly defined only for a point; the term should not be applied to a line, coordinate system, volume, or other collection of points, because the accelerations of the several points involved may differ.

Like velocity, the acceleration of a moving point will appear differently to different observers. Acceleration does not depend on the actual location of the observer but does

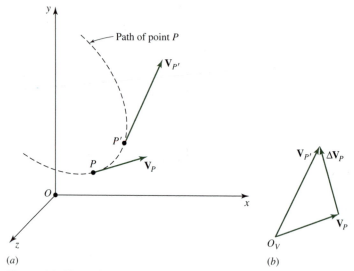

Figure 4.1 Change in velocity of a moving point.

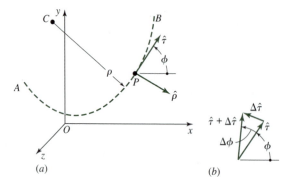

Figure 4.2 The motion of point P generates the space path AB.

depend critically on the observer's motion or, more precisely, on the motion of the observer's coordinate system. If the acceleration is sensed by an observer in the absolute coordinate system, it is referred to as an *absolute acceleration* and is denoted by the symbol $\mathbf{A}_{P/1}$ or simply $\mathbf{A}_P$, which is consistent with the notation used for position, displacement, and velocity.

Next, in Fig. 4.2a, we designate the velocity of point P, moving along the path AB, as

$$\mathbf{V}_P = \dot{s}\hat{\boldsymbol{\tau}} \tag{a}$$

where $\dot{s}$ is the instantaneous speed of P along the path. As we did in Section 3.5, we define $\hat{\boldsymbol{\tau}}$ as a unit vector tangent to the path of P and with positive sense in the direction of positive motion for s. Identifying the osculating plane defined by $\hat{\boldsymbol{\tau}}$ and the center of curvature C and designating its preferred positive side by the unit vector $\hat{\boldsymbol{v}}$, we complete the right-handed vector triad $\hat{\boldsymbol{\rho}}\hat{\boldsymbol{\tau}}\hat{\boldsymbol{v}}$ by defining $\hat{\boldsymbol{\rho}} = \hat{\boldsymbol{\tau}} \times \hat{\boldsymbol{v}}$ as was done in Eq. (3.7). Thus $\hat{\boldsymbol{\rho}}$ and $\hat{\boldsymbol{\tau}}$ are normal and tangent, respectively, to the path at the instantaneous position of P. Differentiating

Eq. (*a*) to obtain the acceleration yields

$$\mathbf{A}_P = \dot{s}\dot{\hat{\boldsymbol{\tau}}} + \ddot{s}\hat{\boldsymbol{\tau}} \tag{b}$$

The initial term of this equation requires additional clarification.

In Fig. 4.2*a*, let ϕ represent the inclination angle of $\hat{\boldsymbol{\tau}}$ to any axis selected in the osculating plane defined by $\hat{\boldsymbol{\rho}}$, $\hat{\boldsymbol{\tau}}$, and point C, the instantaneous center of curvature of the path at P. As P moves along the path AB, both $\hat{\boldsymbol{\tau}}$ and $\hat{\boldsymbol{\rho}}$ are functions of ϕ. In Fig. 4.2*b*, the unit vectors $\hat{\boldsymbol{\tau}}$ and $\hat{\boldsymbol{\tau}} + \boldsymbol{\Delta}\hat{\boldsymbol{\tau}}$ have been transferred to the same origin. Then the limit of the ratio $\boldsymbol{\Delta}\hat{\boldsymbol{\tau}}/\Delta\phi$ is

$$\frac{d\hat{\boldsymbol{\tau}}}{d\phi} = \lim_{\Delta\phi \to 0}\left(\frac{\boldsymbol{\Delta}\hat{\boldsymbol{\tau}}}{\Delta\phi}\right) = \lim_{\Delta\phi \to 0}\left(\frac{2\sin(\Delta\phi/2)(-\hat{\boldsymbol{\rho}})}{\Delta\phi}\right) = -\hat{\boldsymbol{\rho}} \tag{c}$$

The first term of Eq. (*b*) can now be arranged in the form

$$\dot{s}\dot{\hat{\boldsymbol{\tau}}} = \frac{ds}{dt}\frac{d\hat{\boldsymbol{\tau}}}{d\phi}\frac{d\phi}{ds}\frac{ds}{dt} \tag{d}$$

The term $d\phi/ds$ is the rate with respect to the change in distance s along the path with which the inclination of the tangent to the path changes. This is called the *curvature*, and its reciprocal is the *radius of curvature* ρ. Thus

$$\frac{d\phi}{ds} = \frac{1}{\rho} \tag{e}$$

With the help of Eq. (*c*), Eq. (*d*) can now be written as

$$\dot{s}\dot{\hat{\boldsymbol{\tau}}} = -\frac{\dot{s}^2}{\rho}\hat{\boldsymbol{\rho}} \tag{f}$$

Hence Eq. (*b*) becomes

$$\mathbf{A}_P = -\frac{\dot{s}^2}{\rho}\hat{\boldsymbol{\rho}} + \ddot{s}\hat{\boldsymbol{\tau}} \tag{4.2}$$

Thus the acceleration vector $\mathbf{A}_P$ has two perpendicular components: a normal component of magnitude $\dot{s}^2/\rho$ (normal because it is oriented in the negative $\hat{\boldsymbol{\rho}}$ direction*) and a tangential component of magnitude $\ddot{s}$ in the $\hat{\boldsymbol{\tau}}$ direction. Thus Eq. (4.2) can be written as

$$\mathbf{A}_P = \mathbf{A}_P^n + \mathbf{A}_P^t \tag{4.3}$$

where the superscripts are n for the normal direction and t for the tangential direction.

---

*The reader should verify that this normal term is always oriented toward the center of curvature no matter which orientations are chosen as positive for $\hat{\boldsymbol{\tau}}$ and $\hat{\boldsymbol{v}}$.

4.2 ANGULAR ACCELERATION

We have seen that the acceleration of a point is a vector quantity having a magnitude and a direction. But a point has no dimensions, so we cannot speak of the angular acceleration of a point. Thus, *angular acceleration deals with the motion of a rigid body;* whereas *rectilinear acceleration,* or just plain acceleration, *deals with the motion of a point.*

Suppose, at one instant in time, a rigid body has an angular velocity of ω and an instant later, an angular velocity of ω'. The difference,

$$\Delta\omega = \omega' - \omega \tag{a}$$

is also a vector quantity. The angular velocities ω and ω' may have magnitudes that differ from one another as well as different directions. Thus we define *angular acceleration* as *the time rate of change of the angular velocity of a rigid body* and designate it by the symbol α; that is,

$$\alpha = \lim_{\Delta t \to 0} \left(\frac{\Delta\omega}{\Delta t} \right) = \frac{d\omega}{dt} = \dot{\omega} \tag{4.4}$$

As is the case with $\Delta\omega$, there is no reason to believe that α has a direction along either ω or ω'; it may have an entirely new direction.

We further note that the angular acceleration vector α applies to the absolute rotation of the entire rigid body and hence may be subscripted by the number of the coordinate system of the rigid body—for example, α_2 or $\alpha_{2/1}$.

4.3 ACCELERATION DIFFERENCE BETWEEN
POINTS OF A RIGID BODY

In Section 3.3 the velocity difference between two points in a rigid body moving with both translation and rotation was determined. It was found that the velocity of a point in a rigid body could be obtained as the sum of the velocity of *any* reference point of the body plus a rotational component, called the velocity difference, due to the angular velocity ω of the body about the reference point. Thus, the velocity of any point P in a rigid body can be obtained from the equation

$$\mathbf{V}_P = \mathbf{V}_Q + \mathbf{V}_{PQ} \tag{a}$$

where $\mathbf{V}_Q$ is the velocity of the reference point and $\mathbf{V}_{PQ}$ is the velocity difference and is given by the equation

$$\mathbf{V}_{PQ} = \omega \times \mathbf{R}_{PQ} \tag{b}$$

Here ω is the angular velocity of the body and $\mathbf{R}_{PQ}$ is the position difference vector that defines the position P with respect to the reference point Q. Thus we can write

$$\mathbf{V}_P = \mathbf{V}_Q + \omega \times \mathbf{R}_{PQ} \tag{c}$$

We employ a similar nomenclature in Fig. 4.3 and specify that a reference point Q of a rigid body has an acceleration $\mathbf{A}_Q$, and that the body has an angular acceleration $\boldsymbol{\alpha}$ in addition to its angular velocity $\boldsymbol{\omega}$ as shown. Note that $\boldsymbol{\alpha}$ does not, generally, have the same direction as $\boldsymbol{\omega}$. The acceleration of point P is obtained by taking the derivative of Eq. (c):

$$\dot{\mathbf{V}}_P = \dot{\mathbf{V}}_Q + \boldsymbol{\omega} \times \dot{\mathbf{R}}_{PQ} + \dot{\boldsymbol{\omega}} \times \mathbf{R}_{PQ} \qquad (d)$$

However, $\dot{\mathbf{V}}_P = \mathbf{A}_P$, $\dot{\mathbf{V}}_Q = \mathbf{A}_Q$, $\dot{\mathbf{R}}_{PQ} = \boldsymbol{\omega} \times \mathbf{R}_{PQ}$, and $\dot{\boldsymbol{\omega}} = \boldsymbol{\alpha}$; therefore Eq. ($d$) can be written as

$$\mathbf{A}_P = \mathbf{A}_Q + \boldsymbol{\omega} \times (\boldsymbol{\omega} \times \mathbf{R}_{PQ}) + \boldsymbol{\alpha} \times \mathbf{R}_{PQ} \qquad (4.5)$$

In this expression $\mathbf{A}_Q$ is the acceleration of the reference and is one component of the total acceleration of P. The other two components are due to the rotation of the body. In order to picture the directions of these, let us first study them in terms of a two-dimensional problem.

In Fig. 4.4, let P and Q be two points in a rigid body that has a combined motion of translation and rotation in the ground $x_1 y_1$ reference plane. Define, also, a moving $x_2 y_2$ system with origin at Q, but restrict this system to only translation; thus, x_2 must remain parallel to x_1. As given quantities we specify the velocity and acceleration of reference point Q. We also specify the angular velocity $\boldsymbol{\omega}$ and the angular acceleration $\boldsymbol{\alpha}$ of the rigid body. For plane motion these angular rates can be treated as scalars because the corresponding vectors always have axes perpendicular to the plane of motion. They may have different senses, however, because scalar quantities can be either positive or negative.

The location of point P in Fig. 4.4 can now be specified by the position-difference equation

$$\mathbf{R}_P = \mathbf{R}_Q + \mathbf{R}_{PQ} \qquad (e)$$

This equation can also be written in the alternative form,

$$\mathbf{R}_P = \mathbf{R}_Q + R_{PQ} \angle \theta \qquad (f)$$

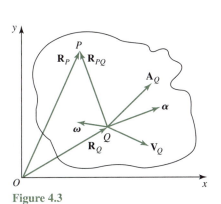

Figure 4.3

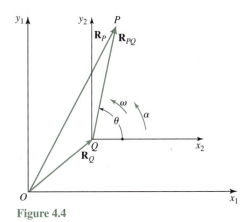

Figure 4.4

or, in complex polar form it can be written as

$$\mathbf{R}_P = \mathbf{R}_Q + R_{PQ}e^{j\theta} \tag{g}$$

The first derivative of Eq. (g) gives the velocity of point P and can be written as

$$\mathbf{V}_P = \mathbf{V}_Q + j\omega R_{PQ}e^{j\theta} \tag{h}$$

Note that this is the complex polar form of Eq. (c) for plane motion where the second term on the right-hand side corresponds to the velocity difference $\mathbf{V}_{PQ}$. Its magnitude ωR_{PQ} and its direction $je^{j\theta}$ is perpendicular to $\mathbf{R}_{PQ}$ in the sense of ω as shown in Fig. 4.5.

The acceleration of point P is given by the derivative of Eq. (h), and this can be written as

$$\mathbf{A}_P = \mathbf{A}_Q - \omega^2 R_{PQ}e^{j\theta} + j\alpha R_{PQ}e^{j\theta} \tag{i}$$

The second and third components of Eq. (i) correspond exactly with the second and third components in Eq. (4.5). The term $j\alpha R_{PQ}e^{j\theta}$ is due to the angular acceleration of the body. The magnitude is αR_{PQ} and the direction is $je^{j\theta}$, which is in the same direction as the velocity difference $\mathbf{V}_{PQ}$. Note that P traces out a circle in its motion relative to the translating reference Q. Because the component $j\alpha R_{PQ}e^{j\theta}$ is perpendicular to the position-difference vector $\mathbf{R}_{PQ}$ and hence tangent to the circle, it is convenient to call it the *tangential* component and designate its magnitude by the equation

$$A^t_{PQ} = \alpha R_{PQ} \tag{4.6}$$

The second component of acceleration in Eq. (i) is called the *normal* or the *centripetal* component. For two-dimensional motion, its magnitude is

$$\omega^2 R_{PQ} = \frac{V^2_{PQ}}{R_{PQ}}$$

and its direction $-e^{j\theta}$ is opposite to the position-difference vector $\mathbf{R}_{PQ}$. This component

Figure 4.5 $A^t_{PQ} = \alpha R_{PQ}$; $V_{PQ} = \omega R_{PQ}$.

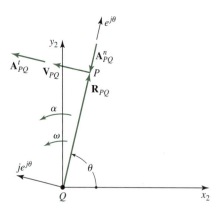

is also shown in Fig. 4.5. The superscript n is used to denote the normal component as follows:

$$A^n_{PQ} = \omega^2 R_{PQ} = \frac{V^2_{PQ}}{R_{PQ}} \tag{4.7}$$

Let us now examine the last two terms of Eq. (4.5) again, but this time in three-dimensional space. According to the definition of the cross product, the term

$$\mathbf{A}^t_{PQ} = \boldsymbol{\alpha} \times \mathbf{R}_{PQ} \tag{4.8}$$

is perpendicular to the plane containing $\boldsymbol{\alpha}$ and $\mathbf{R}_{PQ}$ with a sense according to the right-hand rule. Because of $\boldsymbol{\alpha}$ we visualize P as accelerating around a circle, as shown in Fig. 4.6. The plane of this circle is normal to the plane containing $\boldsymbol{\alpha}$ and $\mathbf{R}_{PQ}$. Using the definition of the cross product, we find that the magnitude of this tangential acceleration component is

$$|\boldsymbol{\alpha} \times \mathbf{R}_{PQ}| = \alpha R_{PQ} \sin \theta$$

where $R_{PQ} \sin \theta$, as shown, is the radius of the circle.

The direction of the normal acceleration component

$$\mathbf{A}^n_{PQ} = \boldsymbol{\omega} \times (\boldsymbol{\omega} \times \mathbf{R}_{PQ}) \tag{4.9}$$

is shown in Fig. 4.7. This component is in the plane containing $\boldsymbol{\omega}$ and $\mathbf{R}_{PQ}$, and it is perpendicular to $\boldsymbol{\omega}$. The magnitude is

$$|\boldsymbol{\omega} \times (\boldsymbol{\omega} \times \mathbf{R}_{PQ})| = \omega^2 R_{PQ} \sin \phi = \frac{V^2_{PQ}}{R_{PQ} \sin \phi}$$

where $R_{PQ} \sin \phi$ is the radius of the circle in Fig. 4.7.

Again we emphasize that $\boldsymbol{\alpha}$ and $\boldsymbol{\omega}$ do not generally have the same directions in three-dimensional space.

Let us now summarize the results of this section. The acceleration of a point fixed in a rigid body can be found from the sum of three components. The first of these is the acceleration of a reference point (point Q in Fig. 4.3). This is only one component of the acceleration, and its value depends on the motion, if any, of the particular reference point

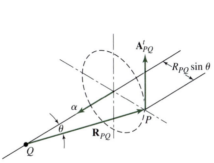

Figure 4.6 $\mathbf{A}^t_{PQ} = \boldsymbol{\alpha} \times \mathbf{R}_{PQ}$.

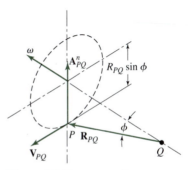

Figure 4.7 $\mathbf{V}_{PQ} = \boldsymbol{\omega} \times \mathbf{R}_{PQ}$; $\mathbf{A}^n_{PQ} = \boldsymbol{\omega} \times (\boldsymbol{\omega} \times \mathbf{R}_{PQ})$.

selected. There are two additional components of acceleration due to the rotation of the body. One of these is the normal component and comes solely from the angular velocity. The other is the tangential component, and it is due to the rate of change of the angular velocity of the body.

Equation (4.5) can now be written as

$$\mathbf{A}_P = \mathbf{A}_Q + \mathbf{A}_{PQ} \tag{4.10}$$

which is called the *acceleration-difference equation*. It is also convenient to designate the components of the acceleration difference as

$$\mathbf{A}_{PQ} = \mathbf{A}^n_{PQ} + \mathbf{A}^t_{PQ} \tag{4.11}$$

The acceleration-difference equation can also be solved by most of the methods that were used in Chapter 3.

EXAMPLE 4.1

For the four-bar linkage shown in Fig. 4.8a, find the acceleration of points A and B and the angular acceleration of links 3 and 4. Crank 2 has a constant angular velocity of 200 rad/s ccw.

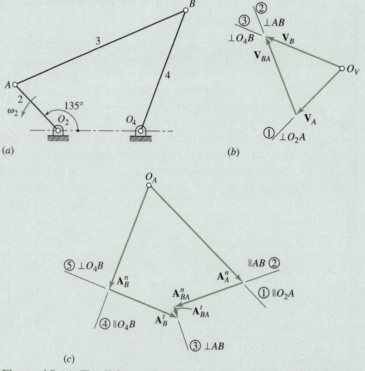

Figure 4.8 (a) The linkage; $R_{AO_2} = 6$ in, $R_{BA} = 18$ in, $R_{BO_4} = 12$ in, $R_{O_4O_2} = 8$ in. (b) The velocity polygon. (c) The acceleration polygon.

SOLUTION

Figures 4.8b and 4.8c show the velocity and acceleration polygons, respectively, with the circled numbers indicating the order of the steps in the construction; the method of finding the directions of each vector is also indicated, using one symbol to indicate parallelism and another to indicate perpendicularity. The velocity polygon must, of course, be found first, because the angular velocities of links 3 and 4 are needed in the acceleration analysis.

The velocity of point A is

$$V_A = \omega_2 R_{AO_2} = (200 \text{ rad/s})(6/12 \text{ ft}) = 100 \text{ ft/s}$$

Now the velocity polygon (Fig. 4.8b) can be drawn. From the polygon we obtain

$$V_{BA} = 128 \text{ ft/s} \quad \text{and} \quad V_B = 129 \text{ ft/s}$$

and, therefore, the angular velocities of links 3 and 4, respectively, are

$$\omega_3 = \frac{V_{BA}}{R_{BA}} = \frac{128 \text{ ft/s}}{18/12 \text{ ft}} = 85.3 \text{ rad/s ccw}$$

$$\omega_4 = \frac{V_{BO_4}}{R_{BO_4}} = \frac{129 \text{ ft/s}}{12/12 \text{ ft}} = 129 \text{ rad/s ccw}$$

where the directions are obtained from an examination of the velocity polygon.

The next step is to write the acceleration-difference equation in terms of its components. Thus, because

$$\mathbf{A}_B = \mathbf{A}_A + \mathbf{A}_{BA}$$

we have

$$\overset{\surd\surd}{\mathbf{A}^n_{BO_4}} + \overset{o\surd}{\mathbf{A}^t_{BO_4}} = \overset{\surd\surd}{\mathbf{A}^n_{AO_2}} + \overset{0}{\cancel{\mathbf{A}^t_{AO_2}}} + \overset{\surd\surd}{\mathbf{A}^n_{BA}} + \overset{o\surd}{\mathbf{A}^t_{BA}}$$

Note that there are two unknowns, the magnitudes of the two tangential-component vectors. We now have enough information to calculate some of the components as follows:

$$A^t_{AO_2} = \alpha_2 R_{AO_2} = 0$$

$$A^n_{AO_2} = \omega_2^2 R_{AO_2} = (200 \text{ rad/s})^2(6/12 \text{ ft}) = 20\,000 \text{ ft/s}^2 = A_A \qquad \textit{Ans.}$$

$$A^n_{BA} = \frac{V^2_{BA}}{R_{BA}} = \frac{(128 \text{ ft/s})^2}{18/12 \text{ ft}} = 10\,923 \text{ ft/s}^2$$

$$A^n_{BO_4} = \frac{V^2_{BO_4}}{R_{BO_4}} = \frac{(129 \text{ ft/s})^2}{(12/12 \text{ ft})} = 16\,641 \text{ ft/s}^2$$

Beginning with the right-hand side of the equation, the acceleration polygon is drawn by constructing $\mathbf{A}_{AO_2}^n$, then $\mathbf{A}_{BA}^n$, and then $\mathbf{A}_{BA}^t$, temporarily of indefinite length because its magnitude is unknown. Beginning again at the acceleration pole O_A and using the left-hand side of the equation, we now construct $\mathbf{A}_{BO_4}^n$, and then $\mathbf{A}_{BO_4}^t$ with magnitude also unknown. When the lines for the two unknowns intersect, this completes the polygon; we label it as shown and measure the following results:

$$A_{BO_4}^t = 11\,600 \text{ ft/s}^2, \quad A_{BA}^t = 1\,550 \text{ ft/s}^2, \quad \text{and} \quad A_B = 20\,600 \text{ ft/s}^2 \qquad \textit{Ans.}$$

The angular accelerations are then computed as follows:

$$\alpha_3 = \frac{A_{BA}^t}{R_{BA}} = \frac{1550 \text{ ft/s}^2}{18/12 \text{ ft}} = 1\,033 \text{ rad/s}^2 \text{ cw} \qquad \textit{Ans.}$$

$$\alpha_4 = \frac{A_{BO_4}^t}{R_{BO_4}} = \frac{11\,600 \text{ ft/s}^2}{12/12 \text{ ft}} = 11\,600 \text{ rad/s}^2 \text{ cw} \qquad \textit{Ans.}$$

EXAMPLE 4.2

Solve Example 4.1 using the direct analytical approach.

SOLUTION

The first step is to perform a position analysis of the linkage. Because this procedure has been demonstrated in earlier chapters, we omit the analysis here and display only the results (see Fig. 4.9).

In vector form, the position vectors corresponding to the links are as follows:

$$\mathbf{R}_{AO_2} = \left(\frac{6}{12}\right) \text{ft} \angle 135° = -0.353\,55\hat{\mathbf{i}} + 0.353\,55\hat{\mathbf{j}} \text{ ft}$$

$$\mathbf{R}_{BA} = \left(\frac{18}{12}\right) \text{ft} \angle 22.4° = 1.386\,81\hat{\mathbf{i}} + 0.571\,61\hat{\mathbf{j}} \text{ ft}$$

$$\mathbf{R}_{BO_4} = \left(\frac{12}{12}\right) \text{ft} \angle 68.4° = 0.368\,13\hat{\mathbf{i}} + 0.929\,78\hat{\mathbf{j}} \text{ ft}$$

The results of a velocity analysis are

$$\omega_2 = 200\hat{\mathbf{k}} \text{ rad/s}, \quad \omega_3 = 84.253\hat{\mathbf{k}} \text{ rad/s}, \quad \text{and} \quad \omega_4 = 129.393\hat{\mathbf{k}} \text{ rad/s}$$

The known acceleration components are computed next as

$$\mathbf{A}_{AO_2}^n = \mathbf{A}_A = \omega_2 \times (\omega_2 \times \mathbf{R}_{AO_2}) = 14\,142\hat{\mathbf{i}} - 14\,142\hat{\mathbf{j}} \text{ ft/s}^2 \qquad \textit{Ans.} \quad (1)$$

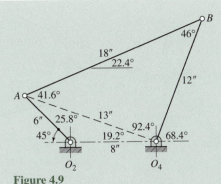

Figure 4.9

$$\mathbf{A}^n_{BA} = \boldsymbol{\omega}_3 \times (\boldsymbol{\omega}_3 \times \mathbf{R}_{BA}) = -9\,844\hat{\mathbf{i}} - 4\,058\hat{\mathbf{j}} \text{ ft/s}^2 \tag{2}$$

$$\mathbf{A}^n_{BO_4} = \boldsymbol{\omega}_4 \times (\boldsymbol{\omega}_4 \times \mathbf{R}_{BO_4}) = -6\,163\hat{\mathbf{i}} - 15\,567\hat{\mathbf{j}} \text{ ft/s}^2 \tag{3}$$

Though $\boldsymbol{\alpha}_3$ and $\boldsymbol{\alpha}_4$ are both unknown, we can incorporate them into the solution in the following manner:

$$\mathbf{A}^t_{BA} = \boldsymbol{\alpha}_3 \times \mathbf{R}_{BA} = \begin{vmatrix} \hat{\mathbf{i}} & \hat{\mathbf{j}} & \hat{\mathbf{k}} \\ 0 & 0 & \alpha_3 \\ 1.38681 & 0.57161 & 0 \end{vmatrix} = -0.571\,61\alpha_3\hat{\mathbf{i}} + 1.386\,81\alpha_3\hat{\mathbf{j}} \text{ ft/s}^2 \tag{4}$$

$$\mathbf{A}^t_{BO_4} = \boldsymbol{\alpha}_4 \times \mathbf{R}_{BO_4} = \begin{vmatrix} \hat{\mathbf{i}} & \hat{\mathbf{j}} & \hat{\mathbf{k}} \\ 0 & 0 & \alpha_4 \\ 0.36813 & 0.92978 & 0 \end{vmatrix} = -0.929\,78\alpha_4\hat{\mathbf{i}} + 0.368\,13\alpha_4\hat{\mathbf{j}} \text{ ft/s}^2 \tag{5}$$

Writing the acceleration-difference equation for point B and noting that $\mathbf{A}^t_{AO_2} = 0$ gives

$$\mathbf{A}^n_{BO_4} + \mathbf{A}^t_{BO_4} = \mathbf{A}^n_{AO_2} + \mathbf{A}^n_{BA} + \mathbf{A}^t_{BA} \tag{6}$$

If we now substitute Eqs. (1) through (5) into (6) and separate the $\hat{\mathbf{i}}$ and $\hat{\mathbf{j}}$ components, we obtain the following pair of simultaneous equations:

$$0.571\,61\alpha_3 - 0.929\,78\alpha_4 = 10\,461 \tag{7}$$

$$-1.386\,81\alpha_3 + 0.368\,13\alpha_4 = -2\,633 \tag{8}$$

When these equations are solved simultaneously, the results are found to be

$$\boldsymbol{\alpha}_3 = -1300\hat{\mathbf{k}} \text{ rad/s}^2 \quad \text{and} \quad \boldsymbol{\alpha}_4 = -12\,050\hat{\mathbf{k}} \text{ rad/s}^2 \qquad \textit{Ans.}$$

4.4 ACCELERATION POLYGONS

The acceleration image of a link is obtained in much the same manner as a velocity image. Figure 4.10a is a slider-crank mechanism in which links 2 and 3 have been given triangular shapes in order to illustrate their images. Figure 4.10b shows the velocity images, and

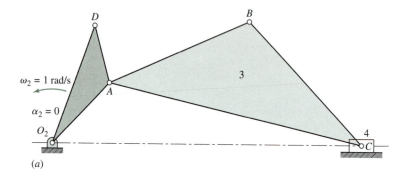

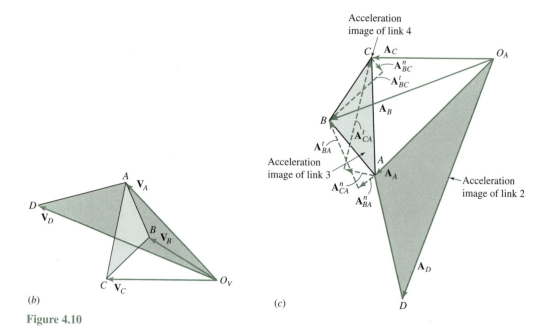

Figure 4.10

the acceleration images are shown in Fig. 4.10c. The angular acceleration of the crank (link 2) is zero, and notice that the corresponding acceleration image is turned 180° from the orientation of the link itself. On the other hand, notice that link 3 has a counterclockwise angular acceleration and that its image is oriented less than 180° from the orientation of the link itself. Thus the orientation of the acceleration image depends on the angular acceleration of the link in question. It can be shown from the geometry of the figure that the orientation of an acceleration image is given by the equation

$$\delta = 180° - \tan^{-1} \frac{\alpha}{\omega^2} \tag{4.12}$$

where δ is the angle in degrees, measured in the (positive) counterclockwise direction, from the orientation of the link itself to its acceleration image.

The acceleration image is used in exactly the same manner as the velocity image. Note particularly that the acceleration image is formed by the *total acceleration difference* vectors, not the component vectors. We also note the following properties of the acceleration image:

1. The acceleration image of each rigid link is a scale reproduction of the shape of the link in the acceleration polygon.
2. The letters identifying the vertices of each link are the same as those in the acceleration polygon, and they progress around the acceleration image in the same order and in the same angular direction as around the link itself.
3. The point O_A in the acceleration polygon is the image of all points with zero absolute acceleration. It is the acceleration image of the fixed link.
4. The absolute acceleration of any point on any link is represented by the line from O_A to the image of the point in the acceleration polygon. The acceleration difference between two points, say P and Q, is represented by the line from acceleration image point P to acceleration image point Q.

EXAMPLE 4.3

The four-bar linkage shown in Fig. 4.11 was analyzed for velocities in Example 3.1. The velocity polygon is shown in Fig. 3.7*b*. Assuming that the angular velocity of link 2 is a constant 900 rev/min = 94.2 rad/s ccw, determine the angular accelerations of links 3 and 4 and the absolute accelerations of points E and F.

SOLUTION

Since the angular acceleration of link 2 is zero, there remains only the normal component of the acceleration of B, and hence

$$A_B = A_{BA}^n = \omega_2^2 R_{BA} = (94.2 \text{ rad/s})^2(4/12 \text{ ft}) = 2\,958 \text{ ft/s}^2$$

Point O_A and an acceleration scale are chosen and $\mathbf{A}_B$ is constructed (opposite in direction to the vector $\mathbf{R}_{BA}$) to locate the acceleration image point B, as shown in Fig. 4.12.

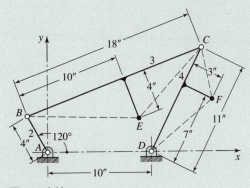

Figure 4.11

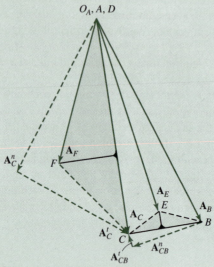

Figure 4.12 Acceleration polygon.

Next, the acceleration-difference equation is written to relate point C to point B on the acceleration polygon. Thus

$$\mathbf{A}_C = \mathbf{A}_B + \mathbf{A}^n_{CB} + \mathbf{A}^t_{CB} = \mathbf{A}^n_{CD} + \mathbf{A}^t_{CD} \tag{1}$$

Using information scaled from the velocity polygon found in Example 3.7, we calculate the magnitude of the two normal components of Eq. (1). Thus

$$A^n_{CB} = \frac{V^2_{CB}}{R_{CB}} = \frac{(38.4\ \text{ft/s})^2}{18/12\ \text{ft}} = 983\ \text{ft/s}^2$$

$$A^n_{CD} = \frac{V^2_{CD}}{R_{CD}} = \frac{(45.5\ \text{ft/s})^2}{11/12\ \text{ft}} = 2\,258\ \text{ft/s}^2$$

These two normal components are constructed with directions opposite to the vectors $\mathbf{R}_{CB}$ and $\mathbf{R}_{CD}$, respectively. As required by Eq. (1), they are added to the acceleration polygon originating from points B and D, respectively, and are shown by dashed lines in Fig. 4.12. Perpendicular dashed lines are then drawn through the termini of these two normal components; these represent the addition of the two tangential components A^t_{CB} and A^t_{CD}, as required to complete Eq. (1). Their intersection is labeled as acceleration-image point C.

The angular accelerations of links 3 and 4 are now found from the two tangential components which are scaled from the acceleration polygon

$$\alpha_3 = \frac{A^t_{CB}}{R_{CB}} = \frac{170\ \text{ft/s}^2}{18/12\ \text{ft}} = 113\ \text{rad/s}^2\ \text{ccw} \qquad \textit{Ans.}$$

$$\alpha_4 = \frac{A^t_{CD}}{R_{CD}} = \frac{1670\ \text{ft/s}^2}{11/12\ \text{ft}} = 1\,822\ \text{rad/s}^2\ \text{cw} \qquad \textit{Ans.}$$

There are several approaches for finding the acceleration of point E. One method is to relate it through acceleration-difference equations to points B and C, which are also on link 3. Thus

$$\mathbf{A}_E = \mathbf{A}_B + \mathbf{A}_{EB}^n + \mathbf{A}_{EB}^t = \mathbf{A}_C + \mathbf{A}_{EC}^n + \mathbf{A}_{EC}^t \qquad (2)$$

The solution of these equations can follow the same methods used for Eq. (1) if desired. A second approach is to use the value of α_3, now known, to calculate one or both of the tangential components in Eq. (2). Probably the easiest approach, however, and the one used in Fig. 4.12, is to form the acceleration-image triangle BCE for link 3, using $\mathbf{A}_{CB}$ and the shape of link 3 as the basis.* Any of these approaches leads to the location of the acceleration-image point E shown in Fig. 4.12. The absolute acceleration of point E is then scaled and found to be

$$A_E = 2\,580 \text{ ft/s}^2 \qquad\qquad Ans.$$

Any of these approaches can also be used to find the absolute acceleration of point F. The result is found to be

$$A_F = 1\,960 \text{ ft/s}^2 \qquad\qquad Ans.$$

4.5 APPARENT ACCELERATION OF A POINT IN A MOVING COORDINATE SYSTEM

In Section 3.5 we found it helpful to develop the apparent-velocity equation for situations where it was convenient to describe the path along which a point moves relative to another moving link but where it was not convenient to describe the absolute motion of the same point. Let us now investigate the acceleration of such a point.

To review, Fig. 4.13 illustrates a point P_3 of link 3 that moves along a known path (the slot) relative to the moving reference frame $x_2 y_2 z_2$. Point P_2 is fixed to the moving link 2 and is instantaneously coincident with P_3. The problem is to find an equation relating the accelerations of points P_3 and P_2 in terms of meaningful parameters that can be calculated (or measured) in a typical mechanical system.

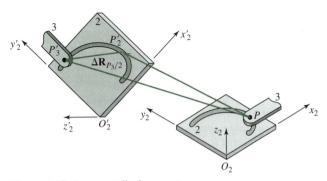

Figure 4.13 Apparent displacement.

---

*We must be extremely careful that the shape of the acceleration-image is not "flipped over" with respect to the original shape. A convenient test is to notice that, for link 3 of this example, since the labels BEC are in clockwise order for the original link shape, they are still clockwise in the velocity and the acceleration-images.

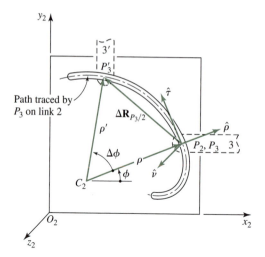

Figure 4.14 Apparent displacement of point P_3 as seen by an observer on link 2.

In Fig. 4.14 we recall how the same situation would be perceived by a moving observer attached to link 2. To HER the path of P_3, the slot, would appear stationary and the point P_3 would appear to move along its tangent with the apparent velocity $\mathbf{V}_{P_3/2}$.

In Section 3.5 we defined another moving coordinate system $\hat{\rho}\hat{\tau}\hat{v}$, where $\hat{\tau}$ is defined as the unit tangent vector to the path of P in the direction of positive movement, and $\hat{v}$ is normal to the plane containing $\hat{\tau}$ and C and is directed positive to the preferred side of the plane. A third unit vector $\hat{\rho}$ is then found from Eq. (3.8), $\hat{\rho} = \hat{\tau} \times \hat{v}$, thus completing a right-hand Cartesian coordinate system. After defining s as the scalar arc distance measuring the travel of P_3 along its curved path, we derived Eq. (3.9) for the apparent velocity

$$\mathbf{V}_{P_3/2} = \frac{ds}{dt}\hat{\tau} \tag{a}$$

Consider the rotation of the radius of curvature vector $\boldsymbol{\rho}$; it sweeps through some small angle $\Delta\phi$ as P_3 travels the small arc distance Δs during a short time interval Δt. The small angle and arc distance are related by

$$\Delta\phi = \frac{\Delta s}{\rho} \tag{b}$$

Notice here that the center of curvature C can be either along the positive or the negative extension of $\hat{\rho}$. Therefore the radius of curvature ρ, measured from C to P, can have either a positive or a negative value according to the sense of $\hat{\rho}$. Also, this implies that the angle $\Delta\phi$ is positive when counterclockwise as seen from positive $\hat{v}$.

Dividing Eq. (b) by Δt and taking the limit for infinitesimally small Δt, we find

$$\frac{d\phi}{dt} = \frac{1}{\rho}\frac{ds}{dt} = \frac{V_{P_3/2}}{\rho} \tag{c}$$

This is the angular rate at which the radius of curvature vector $\boldsymbol{\rho}$ (and also $\hat{\tau}$) appears to rotate as seen by a moving observer in coordinate system 2 as point P_3 moves along its path. We can give this rotational speed its proper vector properties as an apparent angular velocity by noticing that the axis of this rotation is parallel to $\hat{\mathbf{v}}$. Thus we define

$$\dot{\boldsymbol{\varphi}} = \frac{d\phi}{dt}\hat{\mathbf{v}} = \frac{V_{P_3/2}}{\rho}\hat{\mathbf{v}} \tag{d}$$

Next we seek to find the time derivative of the unit vector $\hat{\boldsymbol{\tau}}$ so that we can differentiate Eq. (a). Because $\hat{\boldsymbol{\tau}}$ is a unit vector, its length does not change; however, it does have a derivative due to its change in direction, its rotation. In the absolute coordinate system, $\hat{\boldsymbol{\tau}}$ is subject to the rotation $\dot{\boldsymbol{\varphi}}$ and also to the angular velocity $\boldsymbol{\omega}$, with which the moving coordinate system 2 is rotating. Therefore,

$$\frac{d\hat{\boldsymbol{\tau}}}{dt} = (\boldsymbol{\omega} + \dot{\boldsymbol{\varphi}}) \times \hat{\boldsymbol{\tau}} = \boldsymbol{\omega} \times \hat{\boldsymbol{\tau}} + \dot{\boldsymbol{\varphi}} \times \hat{\boldsymbol{\tau}} \tag{e}$$

Substituting Eq. (d) into this equation gives

$$\frac{d\hat{\boldsymbol{\tau}}}{dt} = \boldsymbol{\omega} \times \hat{\boldsymbol{\tau}} + \frac{V_{P_3/2}}{\rho}\hat{\mathbf{v}} \times \hat{\boldsymbol{\tau}} = \boldsymbol{\omega} \times \hat{\boldsymbol{\tau}} - \frac{V_{P_3/2}}{\rho}\hat{\boldsymbol{\rho}} \tag{f}$$

Now, taking the time derivative of Eq. (a) and using Eq. (f), we find that

$$\frac{d\mathbf{V}_{P_3/2}}{dt} = \frac{ds}{dt}\frac{d\hat{\boldsymbol{\tau}}}{dt} + \frac{d^2s}{dt^2}\hat{\boldsymbol{\tau}} = \frac{ds}{dt}\hat{\boldsymbol{\omega}} \times \hat{\boldsymbol{\tau}} - \frac{ds}{dt}\frac{V_{P_3/2}}{\rho}\hat{\boldsymbol{\rho}} + \frac{d^2s}{dt^2}\hat{\boldsymbol{\tau}}$$

Finally, using Eqs. (a) and (c), this equation reduces to

$$\frac{d\mathbf{V}_{P_3/2}}{dt} = \boldsymbol{\omega} \times \mathbf{V}_{P_3/2} - \frac{V_{P_3/2}^2}{\rho}\hat{\boldsymbol{\rho}} + \frac{d^2s}{dt^2}\hat{\boldsymbol{\tau}} \tag{g}$$

Notice that the three terms in the above equation are *not* defined as the *apparent-acceleration* components. To be consistent, the term apparent acceleration should include only those components *that would be seen by an observer attached to the moving coordinate system*. The above equation is derived in the absolute coordinate system and includes the rotational effect of $\boldsymbol{\omega}$, which would not be sensed by the moving observer. The apparent acceleration, which is given the notation $\mathbf{A}_{P_3/2}$, can easily be found, however, by setting $\boldsymbol{\omega}$ to zero in Eq. (f). This gives the two remaining components,

$$\mathbf{A}_{P_3/2} = \mathbf{A}_{P_3/2}^n + \mathbf{A}_{P_3/2}^t \tag{4.13}$$

where

$$\mathbf{A}_{P_3/2}^n = -\frac{V_{P_3/2}^2}{\rho}\hat{\boldsymbol{\rho}} \tag{4.14}$$

is called the *normal component,* indicating that it is always normal to the path and is always directed from point *P*—that is, toward the center of curvature (the $-\hat{\boldsymbol{\rho}}$ direction when ρ is positive, or the $+\hat{\boldsymbol{\rho}}$ when ρ is negative)—while

$$\mathbf{A}_{P_3/2}^t = \frac{d^2s}{dt^2}\hat{\boldsymbol{\tau}} \tag{4.15}$$

is called the *tangential component,* indicating that it is always tangent to the path (along the $\hat{\tau}$ direction, but it may be either positive or negative).

Next we note that the radius of curvature vector ρ rotates due to both ω and $\dot{\varphi}$. Therefore, its derivative is*

$$\frac{d\rho}{dt} = (\omega + \dot{\varphi}) \times \rho = \omega \times \rho + \mathbf{V}_{P_3/2} \tag{h}$$

We can now write the position equation from Fig. 4.14,

$$\mathbf{R}_{P_3} = \mathbf{R}_{C_2} + \rho$$

and with the help of Eq. (h) its time derivative can be taken**:

$$\mathbf{V}_{P_3} = \mathbf{V}_{C_2} + \omega \times \rho + \mathbf{V}_{P_3/2} \tag{i}$$

Differentiating this equation again with respect to time, we obtain

$$\mathbf{A}_{P_3} = \mathbf{A}_{C_2} + \alpha \times \rho + \omega \times \frac{d\rho}{dt} + \frac{d\mathbf{V}_{P_3/2}}{dt}$$

and, with the help of Eqs. (g) and (h), this becomes

$$\mathbf{A}_{P_3} = \mathbf{A}_{C_2} + \alpha \times \rho + \omega \times (\omega \times \rho) + 2\omega \times \mathbf{V}_{P_3/2} - \frac{V_{P_3/2}^2}{\rho}\hat{\rho} + \frac{d^2 s}{dt^2}\hat{\tau} \tag{j}$$

We recognize the first three terms of this equation as the components of $\mathbf{A}_{P_2}$ [see Eq. (4.5)] and identify the last two terms as the components of the apparent acceleration $\mathbf{A}_{P_3/2}$ [see Eqs. (4.13)–(4.15)]. We therefore define a symbol for the remaining term,

$$\mathbf{A}_{P_3 P_2}^c = 2\omega \times \mathbf{V}_{P_3/2} \tag{4.16}$$

This term is called the *Coriolis component of acceleration.* We note that it is one term of the total apparent acceleration equation. However, unlike the components of $\mathbf{A}_{P_3/2}$, it is not sensed by an observer attached to the moving coordinate system 2. Still, it is a necessary term in Eq. (j) and it is a part of the difference between $\mathbf{A}_{P_3}$ and $\mathbf{A}_{P_2}$ sensed by an absolute observer.

With the definition of Eq. (4.16), Eq. (j) can be written in the following form, called the *apparent-acceleration equation,*

$$\mathbf{A}_{P_3} = \mathbf{A}_{P_2} + \mathbf{A}_{P_3 P_2}^c + \mathbf{A}_{P_3/2}^n + \mathbf{A}_{P_3/2}^t \tag{4.17}$$

where the definitions of the individual components are given in Eqs. (4.14) through (4.16).

---

*Note that the magnitude of ρ is treated as constant in the vicinity of point P due to its definition. It is not truly constant, but it is at a stationary value; its second derivative is nonzero, but its first derivative is zero at the instant considered.

**The first two terms of Eq. (h) are equal to $\mathbf{V}_{P_2}$; thus this is equivalent to the apparent-velocity equation. Note, however, that even though $\rho = \mathbf{R}_{P_2 C_2}$, their derivatives are not equal; they do not rotate at the same rate. Thus, some of the terms of the next equation would be missed if the apparent-velocity equation were differentiated instead.

It is extremely important to recognize certain features of Eq. (4.17):

1. It serves the objectives of this section because it relates the accelerations of two *coincident points on different links* in a meaningful way.
2. There is only *one unknown* among the three new components defined. The Coriolis and normal components can be calculated from Eqs. (4.16) and (4.14) from velocity information; they do not contribute any new unknowns. The tangential component $A^t_{P_3/2}$, however, will almost always have an unknown magnitude in application, because d^2s/dt^2 will not be known.
3. It is important to notice the dependence of Eq. (4.17) on the ability to recognize in each application the point path that P_3 traces on coordinate system 2. This path is the basis for the axes for the normal and tangential components and is also necessary for determination of ρ for Eq. (4.14).

Finally, a word of warning: *The path described by P_3 on link 2 is not necessarily the same as the path described by P_2 on link 3.* In Fig. 4.14 the path of P_3 on link 2 is clear; it is the curved slot. The path of P_2 on link 3 is not at all clear. As a result, there is a natural "right" and "wrong" way to write the apparent-acceleration equation for that situation. The equation

$$\mathbf{A}_{P_2} = \mathbf{A}_{P_3} + \mathbf{A}^c_{P_2 P_3} + \mathbf{A}^n_{P_2/3} + \mathbf{A}^t_{P_2/3}$$

is a perfectly valid equation, but *useless* because ρ is not known for the normal component. Note also that $\mathbf{A}^c_{P_3 P_2}$ makes use of $\boldsymbol{\omega}_2$ while $\mathbf{A}^c_{P_2 P_3}$ would make use of $\boldsymbol{\omega}_3$. *We must be extremely careful to write the appropriate equation for each application, recognizing which path is known.*

EXAMPLE 4.4

In Fig. 4.15a block 3 slides outward on link 2 at a uniform rate of 30 m/s, while link 2 is rotating at a constant angular velocity of 50 rad/s ccw. Determine the absolute acceleration of point A of the block.

SOLUTION

We first calculate the absolute acceleration of the coincident point A_2, immediately under the block but attached to link 2:

$$\mathbf{A}_{A_2} = \overset{0}{\cancel{\mathbf{A}}_{O_2}} + \mathbf{A}^n_{A_2 O_2} + \overset{0}{\cancel{\mathbf{A}}^t_{A_2 O_2}}$$

$$A^n_{A_2 O_2} = \omega_2^2 R_{A_2 O_2} = (50 \text{ rad/s})^2 (0.5 \text{ m}) = 1\,250 \text{ m/s}^2$$

We plot this, determining the acceleration-image point A_2. Next, we recognize that point A_3 is constrained to travel only along the axis of link 2. This provides a path for which we can write the apparent-acceleration equation

$$\mathbf{A}_{A_3} = \mathbf{A}_{A_2} + \mathbf{A}^c_{A_3 A_2} + \mathbf{A}^n_{A_3/2} + \mathbf{A}^t_{A_3/2}$$

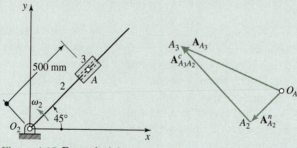

Figure 4.15 Example 4.4.

The terms for this equation are computed as follows and graphically added in the acceleration polygon:

$$A^c_{A_3 A_2} = 2\omega_2 V_{A_3/2} = 2(50 \text{ rad/s})(30 \text{ m/s}) = 3\,000 \text{ m/s}^2$$

$$A^n_{A_3/2} = \frac{V^2_{A_3/2}}{\rho} = \frac{(30 \text{ m/s})^2}{\infty} = 0$$

$$A^t_{A_3/2} = \frac{d^2 s}{dt^2} = 0 \qquad \text{(uniform rate along path)}$$

This locates the acceleration-image point A_3, and the result is

$$A_{A_3} = 3\,250 \text{ m/s}^2 \qquad\qquad\qquad Ans.$$

EXAMPLE 4.5

Perform an acceleration analysis of the linkage shown in Fig. 4.16 for the constant input speed of $\omega_2 = 18$ rad/s cw.

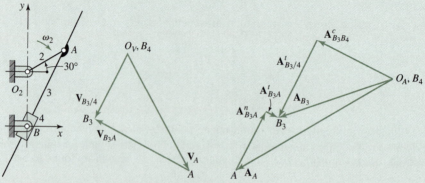

Figure 4.16 Example 4.5; $R_{AO_2} = 8$ in, $R_{BO_2} = 10$ in.

SOLUTION

A complete velocity analysis is performed first, as shown in the figure. This yields

$$V_A = 12.0 \text{ ft/s}, \quad V_{B_3A} = 10.0 \text{ ft/s}, \quad \text{and} \quad V_{B_3/4} = 6.7 \text{ ft/s}$$

$$\omega_3 = \omega_4 = 7.67 \text{ rad/s cw}$$

To solve the accelerations we first find

$$A_A = \overset{0}{\cancel{A^t_{O_2}}} + A^n_{AO_2} + \overset{0}{\cancel{A^t_{AO_2}}}$$

$$A^n_{AO_2} = \omega_2^2 R_{AO_2} = (18 \text{ rad/s})^2 (8/12 \text{ ft}) = 216.0 \text{ ft/s}^2 \qquad \textit{Ans.}$$

and plot this to locate the acceleration image point A. Next we write the acceleration-difference equation

$$\overset{oo}{\mathbf{A}_{B_3}} = \overset{\sqrt{\sqrt{}}}{\mathbf{A}_A} + \overset{\sqrt{\sqrt{}}}{\mathbf{A}^n_{B_3A}} + \overset{o\sqrt{}}{\mathbf{A}^t_{B_3A}} \tag{1}$$

$$A^n_{B_3A} = \frac{V_{B_3A}^2}{R_{BA}} = \frac{(10.0 \text{ ft/s})^2}{(15.6/12) \text{ ft}} = 76.9 \text{ ft/s}^2$$

The term $\mathbf{A}^n_{B_3A}$ is directed from B toward A and is added to the acceleration polygon as shown. The term $\mathbf{A}^t_{B_3A}$ has unknown magnitude but is perpendicular to $\mathbf{R}_{BA}$.

Because Eq. (1) has three unknowns, it cannot be solved by itself. Therefore we seek a second equation for $\mathbf{A}_{B_3}$. Consider the view of an observer located on link 4; SHE would see point B_3 moving on a straight-line path along the centerline of the block. Using this path, we can write the apparent-acceleration equation

$$\mathbf{A}_{B_3} = \overset{0}{\cancel{\mathbf{A}_{B_4}}} + \mathbf{A}^c_{B_3B_4} + \mathbf{A}^n_{B_3/4} + \mathbf{A}^t_{B_3/4} \tag{2}$$

Because point B_4 is pinned to the ground link, it has zero acceleration. The other components of Eq. (2) are

$$A^c_{B_3B_4} = 2\omega_4 V_{B_3/4} = 2(7.67 \text{ rad/s})(6.7 \text{ ft/s}) = 103 \text{ ft/s}^2$$

$$A^n_{B_3/4} = \frac{V_{B_3/4}^2}{\rho} = \frac{(6.5 \text{ ft/s})^2}{\infty} = 0$$

The Coriolis component is added to the acceleration polygon, originating at point $B_4(O_A)$ as shown. Finally, $\mathbf{A}^t_{B_3/4}$, whose magnitude is unknown, is graphically added to this in the direction defined by the path tangent. It crosses the unknown line from $\mathbf{A}^t_{B_3A}$, Eq. (1), locating the acceleration-image point B_3. When the polygon is scaled, the results are found to be

$$A^t_{B_3/4} = 103 \text{ ft/s}^2, \qquad A^t_{B_3A} = 17 \text{ ft/s}^2, \qquad A_{B_3} = 145 \text{ ft/s}^2 \qquad \textit{Ans.}$$

The angular accelerations of links 3 and 4 are

$$\alpha_4 = \alpha_3 = \frac{A^t_{B_3A}}{R_{BA}} = \frac{17 \text{ ft/s}^2}{15.6/12 \text{ ft}} = 13.1 \text{ rad/s}^2 \text{ ccw} \qquad \text{Ans.}$$

We note that in this example the path of B_3 on link 4 and the path of B_4 on link 3 can both be visualized and either could have been used in deciding the approach. However, even though B_4 is pinned to ground (link 1), the path of point B_3 on link 1 is not known. Therefore, the term $A^n_{B_3/1}$ cannot be calculated directly.

EXAMPLE 4.6

The velocity analysis of the inverted slider-crank mechanisms of Fig. 4.17 was performed in Example 3.3 (Fig. 3.11). Determine the angular acceleration of link 4 if link 2 is driven at constant speed.

SOLUTION

From Example 3.3 we recall that

$$V_{A_2} = 9 \text{ ft/s}, \qquad V_{AD} = 7.17 \text{ ft/s}, \qquad V_{A_2/4} = 5.48 \text{ ft/s}$$

$$\omega_2 = 36 \text{ rad/s cw}, \qquad \omega_3 = \omega_4 = 7.55 \text{ rad/s ccw}$$

To analyze for accelerations, we start by writing

$$\mathbf{A}_{A_2} = \overset{0}{\cancel{\mathbf{A}^t_E}} + \mathbf{A}^n_{AE} + \overset{0}{\cancel{\mathbf{A}^t_{AE}}}$$

$$A^n_{AE} = \omega_2^2 R_{AE} = (36 \text{ rad/s})^2(3/12 \text{ ft}) = 324 \text{ ft/s}^2$$

and plot this as shown in the figure.

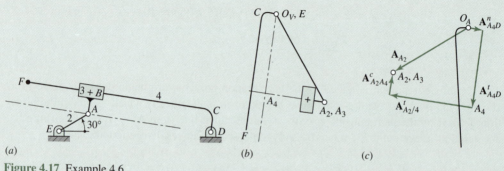

(a)   *(b)*   *(c)*

Figure 4.17 Example 4.6.

Next we notice that point A_2 travels along the straight-line path shown as seen by an observer on link 4. Knowing this path, we write

$$\mathbf{A}_{A_2} = \mathbf{A}_{A_4} + \mathbf{A}_{A_2 A_4}^{c} + \overset{0}{\mathbf{A}_{A_2/4}^{n}} + \mathbf{A}_{A_2/4}^{t} \tag{3}$$

$$A_{A_2 A_4}^{c} = 2\omega_4 V_{A_2/4} = 2(7.55\ \text{rad/s})(5.48\ \text{ft/s}) = 82.7\ \text{ft/s} \tag{4}$$

and $A_{A_2/4}^{n} = 0$ because $\rho = \infty$. In Eq. (3), $\mathbf{A}_{A_4}$ was marked as having only one unknown because

$$\mathbf{A}_{A_4} = \overset{0}{\mathbf{A}_{D}} + \mathbf{A}_{A_4 D}^{n} + \mathbf{A}_{A_4 D}^{t}$$

$$A_{A_4 D}^{n} = \frac{V_{A_4 D}^{2}}{R_{A_4 D}} = \frac{(7.17\ \text{ft/s})^{2}}{11.5/12\ \text{ft}} = 53.6\ \text{ft/s}^2 \tag{5}$$

The term $\mathbf{A}_{A_4 D}^{n}$ is added from O_A followed by a line of unknown length for $\mathbf{A}_{A_4 D}^{t}$. Because the image point A_4 is not yet known, the terms $\mathbf{A}_{A_2 A_4}^{c}$ and $\mathbf{A}_{A_2/4}^{t}$ cannot be added as directed by Eq. (3). However, these two terms can be transferred to the other side of Eq. (3) and graphically subtracted from image point A_2, thus completing the acceleration polygon. The angular acceleration of link 4 can now be found:

$$\alpha_4 = \frac{A_{A_4 D}^{t}}{R_{AD}} = \frac{279.9\ \text{ft/s}^2}{11.5/12\ \text{ft}} = 292\ \text{rad/s}^2\ \text{ccw} \qquad\qquad \textit{Ans.}$$

This need to subtract vectors is common in acceleration problems involving the Coriolis component and should be studied carefully. Note that the opposite equation involving $\mathbf{A}_{A_4/2}$ cannot be used, because ρ and therefore $A_{A_4/2}^{n}$ would become an additional (third) unknown.

4.6 APPARENT ANGULAR ACCELERATION

Although it is seldom useful, completeness suggests that we should also define the term *apparent angular acceleration*. When two rigid bodies rotate with different angular accelerations, the vector difference between them is defined as the apparent angular acceleration,

$$\boldsymbol{\alpha}_{3/2} = \boldsymbol{\alpha}_3 - \boldsymbol{\alpha}_2$$

The apparent angular acceleration equation can also be written as

$$\boldsymbol{\alpha}_3 = \boldsymbol{\alpha}_2 + \boldsymbol{\alpha}_{3/2} \tag{4.18}$$

It will be seen that $\boldsymbol{\alpha}_{3/2}$ is the angular acceleration of body 3 as it would appear to an observer attached to, and rotating with, body 2.

4.7 DIRECT CONTACT AND ROLLING CONTACT

We recall from Section 3.7 that the relative motion between two bodies in direct contact can be of two different kinds; there may be an apparent slipping velocity between the bodies, or there may be no such slip. We defined the term *rolling contact* to imply that no slip is in progress and developed the rolling contact condition, Eq. (3.13), to indicate that the apparent velocity at such a point is zero. Here we intend to investigate the apparent acceleration at a point of rolling contact.

Consider the case of a circular wheel in rolling contact with another straight link, as shown in Fig. 4.18. Although this is admittedly a very simplified case, the arguments made and the conclusions reached are completely general and apply to any rolling-contact situation, no matter what the shapes of the two bodies or whether either is the ground link. To keep this clear in our minds, the ground link has been numbered 2 for this example.

Once the acceleration $\mathbf{A}_C$ of the center point of the wheel is given, the pole O_A can be chosen and the acceleration polygon can be started by plotting $\mathbf{A}_C$. In relating the accelerations of points P_3 and P_2 at the rolling contact point, however, we are dealing with two coincident points of different bodies. Therefore, it is appropriate to use the apparent acceleration equation. To do this we must identify a path that one of these points traces on the other body. The path* that point P_3 traces on link 2 is shown in the figure. Although the precise shape of the path depends on the shapes of the two contacting links, it will always have a cusp at the point of rolling contact and the tangent to this cusp-shaped path will always be normal to the surfaces that are in contact.

Because this path is known, we are free to write the apparent acceleration equation:

$$\mathbf{A}_{P_3} = \mathbf{A}_{P_2} + \mathbf{A}^c_{P_3 P_2} + \mathbf{A}^n_{P_3/2} + \mathbf{A}^t_{P_3/2}$$

In evaluating the components, we keep in mind the rolling contact velocity condition—that is, $\mathbf{V}_{P_3/2} = 0$. Then

$$\mathbf{A}^c_{P_3 P_2} = 2\boldsymbol{\omega}_2 \times \mathbf{V}_{P_3/2} = 0 \quad \text{and} \quad A^n_{P_3/2} = \frac{V^2_{P_3/2}}{\rho} = 0$$

Thus only one component of the apparent acceleration, $\mathbf{A}^t_{P_3/2}$, may be nonzero.

Because of possible confusion in calling this nonzero term a tangential component (tangent to the cusp-shaped path) while its direction is *normal* to the rolling surfaces, we will adopt a new superscript and refer to it as the *rolling contact acceleration* $\mathbf{A}^r_{P_3/2}$. Thus,

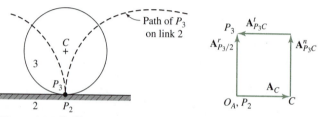

Figure 4.18 Rolling contact.

---

*This particular curve is called a *cycloid*.

for rolling contact, the apparent acceleration equation becomes

$$\mathbf{A}_{P_3} = \mathbf{A}_{P_2} + \mathbf{A}^r_{P_3/2} \tag{4.19}$$

and the term $\mathbf{A}^r_{P_3/2}$ is known always to have a line of action normal to the surfaces at the point of rolling contact.

EXAMPLE 4.7

Given the scaled drawing and velocity analysis of the direct contact circular cam with oscillating flat-faced follower system shown in Fig. 4.19a, determine the angular acceleration of the follower, link 3, at the instant shown. The angular velocity and angular acceleration of the cam at the instant depicted are $\omega_2 = 10$ rad/s cw and $\alpha_2 = 25$ rad/s^2 cw, respectively.

SOLUTION

The velocity polygon of the cam-and-follower system is shown in Fig. 4.19c.

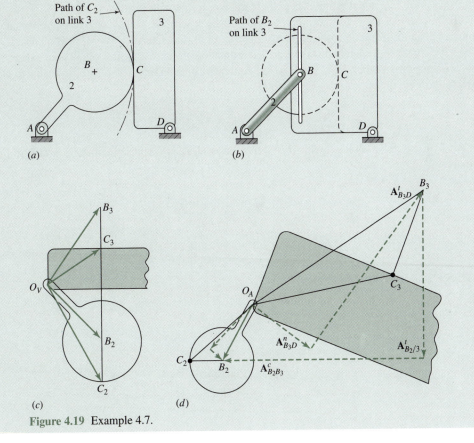

Figure 4.19 Example 4.7.

The acceleration polygon (see Fig. 4.19d) is begun by calculating and plotting the acceleration of point B_2 relative to point A:

$$\mathbf{A}_{B_2} = \overset{0}{\cancel{\mathbf{A}_A}} + \mathbf{A}^n_{B_2A} + \mathbf{A}^t_{B_2A}$$

$$A^n_{B_2A} = \omega_2^2 R_{B_2A} = (10 \text{ rad/s})^2(3 \text{ in}) = 300 \text{ in/s}^2$$

$$A^t_{B_2A} = \alpha_2 R_{B_2A} = (25 \text{ rad/s}^2)(3 \text{ in}) = 75 \text{ in/s}^2$$

These are plotted as shown in the figure, and then the acceleration image-point C_2 is found by constructing the acceleration image of the triangle AB_2C_2.

Proceeding as in Section 3.7, we would next like to find the velocity of the point C_3. If we attempt this approach to acceleration analysis, however, the equation is

$$\mathbf{A}_{C_2} = \mathbf{A}_{C_3} + \mathbf{A}^c_{C_2C_3} + \mathbf{A}^n_{C_2/3} + \mathbf{A}^t_{C_2/3}$$

and is based on the path which point C_2 traces on link 3, shown in Fig. 4.19a. However, this approach is useless here because the radius of curvature of this path is not known and, therefore, $A^n_{C_2/3}$ cannot be calculated. To avoid this problem, we take a different approach; we look for another pair of coincident points where the curvature of the path is known.

If we consider the path traced by point B_2 on the (extended) link 3, we see that it remains a constant distance from the surface; the path is a straight line. Our new approach is better visualized if we consider the mechanism of Fig. 4.19b and note that it has motion equivalent to the original. Having thus visualized a "slot" in the equivalent mechanism for a path, it becomes clear how to proceed; the appropriate equation is

$$\mathbf{A}_{B_2} = \mathbf{A}_{B_3} + \mathbf{A}^c_{B_2B_3} + \mathbf{A}^n_{B_2/3} + \mathbf{A}^t_{B_2/3} \tag{1}$$

where B_3 is a point coincident with B_2 but attached to link 3. Because the path is a straight line, we obtain

$$A^n_{B_2/3} = \frac{V^2_{B_2/3}}{\rho} = \frac{(50 \text{ in/s})^2}{\infty} = 0$$

$$A^c_{B_2B_3} = 2\omega_3 V_{B_2/3} = 2(10 \text{ rad/s})(50 \text{ in/s}) = 1\,000 \text{ in/s}^2$$

The acceleration of B_3 can be found from

$$\mathbf{A}_{B_3} = \overset{0}{\cancel{\mathbf{A}_D}} + \mathbf{A}^n_{B_3D} + \mathbf{A}^t_{B_3D} \tag{2}$$

where

$$A^n_{B_3D} = \frac{V^2_{B_3D}}{R_{B_3D}} = \frac{(35.4 \text{ in/s})^2}{3.58 \text{ in}} = 350 \text{ in/s}^2$$

Substituting Eq. (2) into Eq. (1) and rearranging terms, we arrive at an equation with only two unknowns:

$$\overset{\sqrt{}\sqrt{}}{\mathbf{A}_{B_2}} - \overset{\sqrt{}\sqrt{}}{\mathbf{A}^c_{B_2 B_3}} - \overset{0\sqrt{}}{\mathbf{A}^t_{B_2/3}} = \overset{\sqrt{}\sqrt{}}{\mathbf{A}^n_{B_3 D}} + \overset{0\sqrt{}}{\mathbf{A}^t_{B_3 D}} \tag{3}$$

This equation is solved graphically as shown in Fig. 4.19d. Once the image point B_3 has been found, the image point C_3 is easily found by constructing the acceleration image of the triangle DB_3C_3, all on link 3. Figure 4.19d has been extended to show the complete acceleration images of links 2 and 3 to aid visualization and to illustrate once again that there is no obvious relation between the final locations of image points C_2 and C_3, as suggested by Eq. (2).

Finally, the angular acceleration of link 3 is found as follows:

$$\alpha_3 = \frac{A^t_{B_3 D}}{R_{B_3 D}} = \frac{938 \text{ in/s}^2}{3.58 \text{ in}} = 262 \text{ rad/s}^2 \text{ cw} \qquad\qquad Ans.$$

4.8 SYSTEMATIC STRATEGY FOR ACCELERATION ANALYSIS

Review of the preceding sections and their example problems will show that we have now developed enough tools for dealing with those situations that normally arise in the acceleration analysis of rigid-body mechanical systems. It will also be noticed that the word "relative" acceleration has been carefully avoided. Instead we note that, as with velocity analysis, whenever the desire for using "relative" acceleration arises, there are always two points whose accelerations are to be "related"; also, these two points are attached either to the same or to two different rigid bodies. Therefore, as was done for velocity analysis, we can organize all situations into the four cases shown in Table 4.1.

In this table we see that, when the two points are separated by a distance, only the acceleration difference equation is appropriate for use, and two points on the *same link* should be used. When it is desirable to switch to a point of another link, then *coincident points* should be chosen and the apparent acceleration equation should be used. The path of one of these points on the other link is then needed.

It will be noted that even the notation has been made different to continually remind us that these are two totally different situations and the formulae are not interchangeable between the two. We should not try to use an "$\alpha \times \mathbf{R}$" formula when using the apparent acceleration, but will not find a useful α or $\mathbf{R}$ if we try. Similarly, when using the acceleration difference, there is no question of which α to use because only one link pertains. The two questions that always arise with respect to apparent acceleration are: (a) When should I include the Coriolis term and when do I only use normal and tangential components? and (b) Which ω should I use in the Coriolis term? The answer to the first question is easy: Whenever we use the apparent acceleration equation, the Coriolis term should be included; if it should not be there, such as when the "path" is not rotating, it should be included anyway and it will calculate to zero. The answer to the second question is also easy: Whenever we use the apparent acceleration equation, we must always visualize a path which the

TABLE 4.1 "Relative" Acceleration Equations

Points are	Coincident	Separated
On same body	*Trivial Case:* $\mathbf{A}_P = \mathbf{A}_Q$	*Acceleration Difference;* $\mathbf{A}_P = \mathbf{A}_Q + \mathbf{A}^n_{PQ} + \mathbf{A}^t_{PQ}$ $\mathbf{A}^n_{PQ} = \boldsymbol{\omega} \times (\boldsymbol{\omega} \times \mathbf{R}_{PQ})$ $\mathbf{A}^t_{PQ} = \boldsymbol{\alpha} \times \mathbf{R}_{PQ}$
On different bodies	*Apparent Acceleration;* $\mathbf{A}_{P_i} = \mathbf{A}_{P_j} + \mathbf{A}^c_{P_i P_j} + \mathbf{A}^n_{P_i/j} + \mathbf{A}^t_{P_i/j}$ where path $P_{i/j}$ is known, and $\mathbf{A}^c_{P_i P_j} = 2\boldsymbol{\omega}_j \times \mathbf{V}_{P_i/j}$ $\mathbf{A}^n_{P_i/j} = -\dfrac{V^2_{P_i/j}}{\rho}\hat{\boldsymbol{\rho}}$ $\mathbf{A}^t_{P_i/j} = \dfrac{d^2 s}{dt^2}\hat{\boldsymbol{\tau}}$	*Too general; use two steps.*
	Rolling Contact Acceleration: $\mathbf{A}_{P_i} = \mathbf{A}_{P_j} + \mathbf{A}^r_{P_i/j}$ where path $\mathbf{A}^r_{P_i/j}$ is normal to surfaces at point of contact.	

point P_i makes on link j; then $\boldsymbol{\omega}_j$ for the link that contains the path is used. As pointed out in Section 4.7 and in Table 4.1, rolling contact acceleration is a special case of apparent acceleration.

Careful review of Examples 4.1 through 4.7 show how the suggested strategy is applied to a variety of problems of acceleration analysis.

4.9 ANALYTIC METHODS

In this section we shall continue some of the analytic approaches that we commenced in Chapter 3. One mechanism studied there was the centered slider-crank linkage of Fig. 4.20.

The acceleration of the slider can be obtained by differentiating Eq. (*c*) of Section 3.9. The result, after considerable manipulation, is

$$\ddot{x} = -r\alpha \left(\sin\theta + \frac{r\sin 2\theta}{2l\cos\phi} \right) - r\omega^2 \left(\cos\theta + \frac{r\cos 2\theta}{l\cos\phi} + \frac{r^3 \sin^2 2\theta}{4l^3 \cos^3 \phi} \right) \tag{4.20}$$

This is an involved expression, and it becomes even more involved when the substitution

$$\cos\phi = \sqrt{1 - \left(\frac{r}{l}\sin\theta\right)^2} \tag{4.21}$$

is made. An approximate expression is sometimes used for the acceleration of the slider. The last term on the right in Eq. (4.20) can be neglected for large (l/r) ratios. Also, for

Figure 4.20

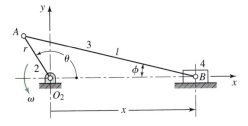

large values of (l/r), $\cos\phi$ is near unity. By using these approximations, Eq. (4.20) can be written as

$$\ddot{x} = -r\alpha\left(\sin\theta + \frac{r}{2l}\sin 2\theta\right) - r\omega^2\left(\cos\theta + \frac{r}{l}\cos 2\theta\right) \tag{4.22}$$

These expressions are still complicated; and if a computer is to be used in the analysis, Raven's method is easier to use.

4.10 COMPLEX-ALGEBRA METHODS

Let us now see how Raven's method is extended for the analysis of the accelerations. The general method is illustrated here for the offset slider-crank linkage shown in Fig. 4.21.

In complex polar form the loop-closure equation is

$$r_2 e^{j\theta_2} + r_3 e^{j\theta_3} - r_1 e^{-j(\pi/2)} - r_4 e^{j0} = 0 \tag{a}$$

If we separate this equation into its real and the imaginary components, we obtain the two position equations:

$$r_2\cos\theta_2 + r_3\cos\theta_3 - r_4 = 0 \tag{b}$$

$$r_2\sin\theta_2 + r_3\sin\theta_3 + r_1 = 0 \tag{c}$$

With r_1, r_2, r_3, and θ_2 given, these can be solved for the two position unknowns. In this example, Eq. (c) can be solved for θ_3. The result is

$$\theta_3 = \sin^{-1}\left(\frac{r_1 + r_2\sin\theta_2}{r_3}\right) \tag{4.23}$$

and then Eq. (b) can be solved for r_4. For this we obtain

$$r_4 = r_2\cos\theta_2 + r_3\cos\theta_3 \tag{4.24}$$

Differentiating Eq. (a) once with respect to time gives the velocity equation, which is really the equation of the velocity polygon in exponential form. After replacing $\dot{\theta}$ with ω, the result is

$$jr_2\omega_2 e^{j\theta_2} + jr_3\omega_3 e^{j\theta_3} - \dot{r}_4 e^{j0} = 0 \tag{d}$$

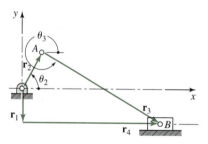

Figure 4.21 Offset slider-crank mechanism.

Raven's method, as we have seen, consists of making the trigonometric transformation and separating the result into real and imaginary terms as was done in obtaining Eqs. (*b*) and (*c*) from Eq. (*a*). When this procedure is carried out for the velocity equation in this example, Eq. (*d*), we find

$$\omega_3 = -\frac{r_2 \cos \theta_2}{r_3 \cos \theta_3} \omega_2 \tag{4.25}$$

and

$$\dot{r}_4 = -r_2 \sin \theta_2 \omega_2 - r_3 \sin \theta_3 \omega_3 \tag{4.26}$$

We use the very same approach to obtain the acceleration equations by first differentiating Eq. (*d*). Separating the result into its real and imaginary parts gives two equations that can be solved for the two acceleration unknowns. The acceleration results for this example are

$$\alpha_3 = \frac{r_2 \sin \theta_2 \omega_2^2 - r_2 \cos \theta_2 \alpha_2 + r_3 \sin \theta_3 \omega_3^2}{r_3 \cos \theta_3} \tag{4.27}$$

$$\ddot{r}_4 = -r_2 \cos \theta_2 \omega_2^2 - r_2 \sin \theta_2 \alpha_2 - r_3 \cos \theta_3 \omega_3^2 - r_3 \sin \theta_3 \alpha_3 \tag{4.28}$$

If the input angular velocity is constant; that is, if $\alpha_2 = 0$, then the acceleration results are

$$\alpha_3 = -\frac{r_2 \sin \theta_2 \omega_2^2 + r_3 \sin \theta_3 \omega_3^2}{r_3 \cos \theta_3} \tag{4.29}$$

$$\ddot{r}_4 = -r_2 \cos \theta_2 \omega_2^2 - r_3 \sin \theta_3 \alpha_3 - r_3 \cos \theta_3 \omega_3^2 \tag{4.30}$$

The same procedure, when carried out for the four-bar linkage, gives results that can be used for computer solutions of both the crank-and-rocker linkage and the drag-link mechanism. Unfortunately, they cannot be used for other four-bar linkages unless an arrangement is built into the program to cause the solution to stop when an extreme position is reached. If, for the four-bar linkage, we write

$$\mathbf{r}_1 + \mathbf{r}_2 + \mathbf{r}_3 - \mathbf{r}_4 = 0 \tag{e}$$

where the subscripts are the link numbers and where link 2 is the driver having a constant input angular velocity, then the acceleration relations which are obtained by Raven's method are

$$\alpha_3 = \frac{r_2 \cos(\theta_2 - \theta_4)\omega_2^2 + r_3 \cos(\theta_3 - \theta_4)\omega_3^2 - r_4\omega_4^2}{r_3 \sin(\theta_4 - \theta_3)} \tag{4.31}$$

$$\alpha_4 = \frac{r_2 \cos(\theta_2 - \theta_3)\omega_2^2 - r_4 \cos(\theta_3 - \theta_4)\omega_4^2 + r_3\omega_3^2}{r_4 \sin(\theta_4 - \theta_3)} \tag{4.32}$$

4.11 THE METHOD OF KINEMATIC COEFFICIENTS

Again we employ the four-bar linkage as an example of this method of solution. The angular acceleration of links 3 and 4 can be obtained by differentiating Eqs. (3) in Example 3.6, with respect to the input position θ_2; that is,

$$-r_3 \cos\theta_3\theta_3'^2 - r_3 \sin\theta_3\theta_3'' + r_4 \cos\theta_4\theta_4'^2 + r_4 \sin\theta_4\theta_4'' = r_2 \cos\theta_2$$
$$-r_3 \sin\theta_3\theta_3'^2 + r_3 \cos\theta_3\theta_3'' + r_4 \sin\theta_4\theta_4'^2 - r_4 \cos\theta_4\theta_4'' = r_2 \sin\theta_2 \tag{4.33}$$

where by definition

$$\theta_3'' = \frac{d^2\theta_3}{d\theta_2^2} \quad \text{and} \quad \theta_4'' = \frac{d^2\theta_4}{d\theta_2^2} \tag{4.34}$$

and are referred to as the *second-order kinematic coefficients* of links 3 and 4. Writing Eqs. (4.33) in matrix form gives

$$\begin{bmatrix} -r_3 \sin\theta_3 & r_4 \sin\theta_4 \\ r_3 \cos\theta_3 & -r_4 \cos\theta_4 \end{bmatrix} \begin{bmatrix} \theta_3'' \\ \theta_4'' \end{bmatrix} = \begin{bmatrix} r_2 \cos\theta_2 + r_3 \cos\theta_3\theta_3'^2 - r_4 \cos\theta_4\theta_4'^2 \\ r_2 \sin\theta_2 + r_3 \sin\theta_3\theta_3'^2 - r_4 \sin\theta_4\theta_4'^2 \end{bmatrix} \tag{4.35}$$

Note that the right-hand side of this matrix equation is completely known (that is, the position and velocity analysis have been completed), and for convenience we let

$$B_1 = r_2 \cos\theta_2 + r_3 \cos\theta_3\theta_3'^2 - r_4 \cos\theta_4\theta_4'^2$$
$$B_2 = r_2 \sin\theta_2 + r_3 \sin\theta_3\theta_3'^2 - r_4 \sin\theta_4\theta_4'^2 \tag{4.36}$$

Also, note that the (2×2) coefficient matrix on the left-hand side of Eq. (4.35) is the same as the (2×2) coefficient matrix in the velocity analysis*; see Eq. (5) in Example 3.6. This is not a coincidence. The coefficient matrix is the same for all positional derivatives—that is, first-order, second-order, third-order kinematic coefficients, and so on. Therefore, this matrix provides a built-in check of all higher-order differentiation. The determinant of the coefficient matrix is as given by Eq. (6) in Example 3.6; that is,

$$\Delta = r_3 r_4 \sin(\theta_3 - \theta_4) \tag{4.37}$$

---

*This coefficient matrix is called the *Jacobian* of the system.

TABLE 4.2 Summary of the Second-Order Kinematic Coefficients

	Variable of Interest Angle θ_j (use symbol θ_j'' for kinematic coefficient regardless of input)	Variable of Interest Magnitude r_j (use symbol r_j'' for kinematic coefficient regardless of input)
Input $\psi = $ angle θ_i	$\alpha_j = \theta_j'' \dot\psi^2 + \theta_j' \ddot\psi$	$\ddot r_j = r_j'' \dot\psi^2 + r_j' \ddot\psi$
	$\theta_j' = \dfrac{d\theta_j}{d\psi}$ (dimensionless)	$r_j' = \dfrac{dr_j}{d\psi}$ (length)
	$\theta_j'' = \dfrac{d^2\theta_j}{d\psi^2}$ (dimensionless)	$r_j'' = \dfrac{d^2 r_j}{d\psi^2}$ (length)
Input $\psi = $ magnitude r_i	$\alpha_j = \theta_j'' \dot\psi^2 + \theta_j' \ddot\psi$	$\ddot r_j = r_j'' \dot\psi^2 + r_j' \ddot\psi$
	$\theta_j' = \dfrac{d\theta_j}{d\psi}$ (1/length)	$r_j' = \dfrac{dr_j}{d\psi}$ (dimensionless)
	$\theta_j'' = \dfrac{d^2\theta_j}{d\psi^2}$ (1/length)^2	$r_j'' = \dfrac{d^2 r_j}{d\psi^2}$ (1/length)

The second-order kinematic coefficients for links 3 and 4 can be obtained from Eq. (4.35) using Cramer's rule and the results are as

$$\theta_3'' = \frac{-B_1 \cos\theta_4 - B_2 \sin\theta_4}{r_3 \sin(\theta_3 - \theta_4)} \quad \text{and} \quad \theta_4'' = \frac{-B_1 \cos\theta_3 - B_2 \sin\theta_3}{r_4 \sin(\theta_3 - \theta_4)} \tag{4.38}$$

Note that the second-order kinematic coefficients, for the four-bar linkage, are nondimensional. The angular accelerations of links 3 and 4, obtained from the chain rule, are

$$\alpha_3 = \theta_3'' \omega_2^2 + \theta_3' \alpha_2 \quad \text{and} \quad \alpha_4 = \theta_4'' \omega_2^2 + \theta_4' \alpha_2 \tag{4.39}$$

Table 4.2 summarizes the second-order kinematic coefficients that are related to link j of a planar mechanism (that is, vector $\mathbf{r}_j$) having (i) variable angular displacement θ_j and/or (ii) variable magnitude r_j.

EXAMPLE 4.8

To illustrate the method of kinematic coefficients for determining the angular acceleration of a link and the rectilinear acceleration of a point fixed to a link, we will revisit Example 4.3. Recall that link 2 of that four-bar linkage is driven at a constant angular velocity of 94.2 rad/s ccw; the problem is to determine the angular acceleration of links 3 and 4 and the absolute acceleration of points E and F.

SOLUTION

Recall that the first-order kinematic coefficients, from Eq. (9) in Example 3.6, are

$$\theta_3' = +0.271\,8 \text{ rad/rad} \quad \text{and} \quad \theta_4' = +0.526\,5 \text{ rad/rad} \tag{1}$$

Substituting Eqs. (1) and the known data into Eqs. (4.36) and substituting the results into Eq. (4.38), we obtain the second-order kinematic coefficients of links 3 and 4:

$$\theta_3'' = +0.013\,1 \quad \text{and} \quad \theta_4'' = -0.203\,0 \tag{2}$$

Then substituting Eqs. (1) and (2) and the angular velocity and acceleration of the input link into Eqs. (4.39), we obtain the angular accelerations of links 3 and 4:

$$\alpha_3 = 116 \text{ rad/s}^2 \text{ ccw} \quad \text{and} \quad \alpha_4 = 1\,802 \text{ rad/s}^2 \text{ cw} \qquad \textit{Ans.}$$

These answers are in good agreement with the answers of Example 4.3; that is, $\alpha_3 = 113$ rad/s^2 ccw and $\alpha_4 = 1\,822$ rad/s^2 cw.

To obtain the acceleration of point E, we differentiate Eqs. (14) in Example 3.6 with respect to the input angular displacement θ_2. Therefore, the second-order kinematic coefficients for point E are

$$x_E'' = -r_2 \cos\theta_2 - r_{EB} \cos(\theta_3 - \phi)\theta_3'^2 - r_{EB} \sin(\theta_3 - \phi)\theta_3''$$
$$y_E'' = -r_2 \sin\theta_2 - r_{EB} \sin(\theta_3 - \phi)\theta_3'^2 + r_{EB} \cos(\theta_3 - \phi)\theta_3'' \tag{3}$$

Substituting Eqs. (1) and (2) and the known data into Eqs. (3) gives

$$x_E'' = +1.206\,7 \text{ in/rad}^2 \quad \text{and} \quad y_E'' = -3.310\,8 \text{ in/rad}^2 \tag{4}$$

Differentiating the velocity of point E, from Eq. (16) in Example 3.6, with respect to time, the acceleration of point E can be written as

$$\mathbf{A}_E = \left(x_E'' \omega_2^2 + x_E' \alpha_2\right)\hat{\mathbf{i}} + \left(y_E'' \omega_2^2 + y_E' \alpha_2\right)\hat{\mathbf{j}} \tag{5}$$

Recall that the first-order kinematic coefficients for point E, from Eq. (15) in Example 3.6, are

$$x_E' = -3.419\,1 \text{ in/rad} \quad \text{and} \quad y_E' = +0.926\,9 \text{ in/rad} \tag{6}$$

Substituting Eqs. (4) and (6) and the angular velocity and acceleration of the input link into Eq. (5) gives

$$\mathbf{A}_E = 892.32\hat{\mathbf{i}} - 2\,448.24\hat{\mathbf{j}} \text{ ft/s}^2 \qquad \textit{Ans.}$$

Therefore, the magnitude of the acceleration of point E is $A_E = 2\,606$ ft/s^2. This result is in good agreement with the answer in Example 4.3; that is, $A_E = 2\,580$ ft/s^2.

The acceleration of point F can be obtained in a similar manner. The first-order and second-order kinematic coefficients for point F are

$$x'_F = -2.626\,2 \text{ in/rad} \quad \text{and} \quad y'_F = +3.068\,6 \text{ in/rad}$$
$$x''_F = -0.586\,0 \text{ in/rad}^2 \quad \text{and} \quad y''_F = -2.551\,7 \text{ in/rad}^2 \tag{7}$$

Substituting Eqs. (7) and the angular velocity and acceleration of the input link into the equation for the acceleration of point F, namely,

$$\mathbf{A}_F = \left(x''_F \omega_2^2 + x'_F \alpha_2\right)\hat{\mathbf{i}} + \left(y''_F \omega_2^2 + y'_F \alpha_2\right)\hat{\mathbf{j}}$$

we obtain

$$\mathbf{A}_F = -433\hat{\mathbf{i}} - 1\,887\hat{\mathbf{j}} \text{ ft/s}^2 \qquad\qquad Ans.$$

Therefore, the magnitude of the acceleration of point F is $A_F = 1\,936 \text{ ft/s}^2$. This result is in good agreement with the answer of Example 4.3; that is, $A_F = 1\,960 \text{ ft/s}^2$.

EXAMPLE 4.9

Link 4 of the offset slider-crank linkage, shown in Fig. 3.8, is driven at a constant speed of 10 m/s to the left at the position shown. Determine the angular acceleration of links 2 and 3 and the instantaneous acceleration of point D.

SOLUTION

The angular acceleration of links 3 and 4 is obtained by differentiating Eqs. (3) in Example 3.7, with respect to the input displacement r_4; that is,

$$-r_2 \cos\theta_2\theta_2'^2 - r_2 \sin\theta_2\theta_2'' - r_3 \cos\theta_3\theta_3'^2 - r_3 \sin\theta_3\theta_3'' = 0$$
$$-r_2 \sin\theta_2\theta_2'^2 + r_2 \cos\theta_2\theta_2'' - r_3 \sin\theta_3\theta_3'^2 + r_3 \cos\theta_3\theta_3'' = 0 \tag{1}$$

where the second-order kinematic coefficients of links 2 and 3 are defined as

$$\theta_2'' = \frac{d^2\theta_2}{dr_4^2} \quad \text{and} \quad \theta_3'' = \frac{d^2\theta_3}{dr_4^2}$$

We recall the first-order kinematic coefficients, from Eq. (8) in Example 3.7; they are

$$\theta_2' = -19.77 \text{ rad/m} \quad \text{and} \quad \theta_3' = +5.436 \text{ rad/m} \tag{2}$$

Substituting Eqs. (2) and the known data into Eqs. (1), we find the second-order kinematic coefficients of links 2 and 3 to be

$$\theta_2'' = -246.25 \text{ rad/m}^2 \quad \text{and} \quad \theta_3'' = +162.73 \text{ rad/m}^2 \tag{3}$$

The angular acceleration of links 2 and 3 are now written from Table 4.2 as

$$\alpha_2 = \theta_2'' \dot{r}_4^2 + \theta_2' \ddot{r}_4 \quad \text{and} \quad \alpha_3 = \theta_3'' \dot{r}_4^2 + \theta_3' \ddot{r}_4 \tag{4}$$

Substituting Eqs. (2) and (3) and the constant input speed $\dot{r}_4 = -10$ m/s into Eq. (4), we find the angular accelerations of links 2 and 3:

$$\alpha_2 = 24\,625 \text{ rad/s}^2 \text{ cw} \quad \text{and} \quad \alpha_3 = 16\,273 \text{ rad/s}^2 \text{ ccw} \qquad Ans.$$

To obtain the acceleration of point D, we differentiate Eqs. (12) in Example 3.7 with respect to the input displacement r_4. The second-order kinematic coefficients for point D are

$$x_D'' = -r_2 \cos\theta_2 \theta_2'^2 - r_2 \sin\theta_2 \theta_2'' - r_{DB} \cos(\theta_3 - \mu)\theta_3'^2 - r_{DB} \sin(\theta_3 - \mu)\theta_3''$$
$$y_D'' = -r_2 \sin\theta_2 \theta_2'^2 + r_2 \cos\theta_2 \theta_2'' - r_{DB} \sin(\theta_3 - \mu)\theta_3'^2 + r_{DB} \cos(\theta_3 - \mu)\theta_3'' \tag{5}$$

Substituting Eqs. (2) and (3) and the known data into Eqs. (5), we find

$$x_D'' = +6.52 \text{ m/m}^2 \quad \text{and} \quad y_D'' = -12.31 \text{ m/m}^2 \tag{6}$$

Differentiating the velocity of point D, Eq. (14) in Example 3.7, with respect to time, the acceleration of point D is now written as

$$\mathbf{A}_D = \left(x_D'' \dot{r}_4^2 + x_D' \ddot{r}_4\right)\hat{\mathbf{i}} + \left(y_D'' \dot{r}_4^2 + y_D' \ddot{r}_4\right)\hat{\mathbf{j}} \tag{7}$$

Next we recall the first-order kinematic coefficients for point D, from Eq. (13) in Example 3.7, as

$$x_D' = +1.121 \text{ m/m} \quad \text{and} \quad y_D' = -0.407\,1 \text{ m/m} \tag{8}$$

Substituting Eqs. (8) and the speed and acceleration of the input link into Eq. (7), we get

$$\mathbf{A}_D = 652\hat{\mathbf{i}} - 1\,231\hat{\mathbf{j}} \text{ m/s}^2 \qquad Ans.$$

Therefore, the magnitude of the acceleration of point D is $A_D = 1\,393$ m/s^2.

4.12 THE CHACE SOLUTIONS

Here we employ the four-bar linkage as an example of Chace's method of solution. First, we write the loop-closure equation

$$\mathbf{r}_1 + \mathbf{r}_2 + \mathbf{r}_3 - \mathbf{r}_4 = 0 \tag{a}$$

Here, again, the subscripts designate the link numbers. In using Chace's approach, we first differentiate Eq. (a) to get the velocities. Because r_2, r_3, and r_4 are constants for the four-bar linkage, this yields

$$\boldsymbol{\omega}_2 \times \mathbf{r}_2 + \boldsymbol{\omega}_3 \times \mathbf{r}_3 - \boldsymbol{\omega}_4 \times \mathbf{r}_4 = 0 \tag{b}$$

In this example we are taking link 2 as the driver, ω_2 is a known constant, and $\alpha_2 = 0$. Now, in a plane-motion mechanism all the angular velocities are in the $\hat{\mathbf{k}}$ direction. For this reason we can write Eq. (b) as

$$r_2\omega_2(\hat{\mathbf{k}} \times \hat{\mathbf{r}}_2) + r_3\omega_3(\hat{\mathbf{k}} \times \hat{\mathbf{r}}_3) - r_4\omega_4(\hat{\mathbf{k}} \times \hat{\mathbf{r}}_4) = 0 \tag{c}$$

To solve this equation, we take the dot product with $\hat{\mathbf{r}}_4$ and then with $\hat{\mathbf{r}}_3$. Each of these operations reduces Eq. (c) to a scalar equation with only a single unknown. Thus, dotting through with $\hat{\mathbf{r}}_4$ because

$$r_4\omega_4(\hat{\mathbf{k}} \times \hat{\mathbf{r}}_4) \cdot \hat{\mathbf{r}}_4 = 0$$

gives

$$r_2\omega_2(\hat{\mathbf{k}} \times \hat{\mathbf{r}}_2) \cdot \hat{\mathbf{r}}_4 + r_3\omega_3(\hat{\mathbf{k}} \times \hat{\mathbf{r}}_3) \cdot \hat{\mathbf{r}}_4 = 0 \tag{d}$$

Solving Eq. (d) for ω_3 then yields

$$\omega_3 = -\frac{r_2\omega_2}{r_3} \frac{(\hat{\mathbf{k}} \times \hat{\mathbf{r}}_2) \cdot \hat{\mathbf{r}}_4}{(\hat{\mathbf{k}} \times \hat{\mathbf{r}}_3) \cdot \hat{\mathbf{r}}_4} \tag{4.40}$$

We then dot Eq. (c) through with $\hat{\mathbf{r}}_3$ in a similar manner and solve the result for ω_4. Thus we obtain

$$\omega_4 = \frac{r_2\omega_2}{r_4} \frac{(\hat{\mathbf{k}} \times \hat{\mathbf{r}}_2) \cdot \hat{\mathbf{r}}_3}{(\hat{\mathbf{k}} \times \hat{\mathbf{r}}_4) \cdot \hat{\mathbf{r}}_3} \tag{4.41}$$

Next, we differentiate Eq. (b) to get the acceleration relations, remembering that the input angular acceleration α_2 equals 0. Thus, in vector form, we have

$$\boldsymbol{\omega}_2 \times (\boldsymbol{\omega}_2 \times \mathbf{r}_2) + \boldsymbol{\omega}_3 \times (\boldsymbol{\omega}_3 \times \mathbf{r}_3) + \dot{\boldsymbol{\omega}}_3 \times \mathbf{r}_3 - \boldsymbol{\omega}_4 \times (\boldsymbol{\omega}_4 \times \mathbf{r}_4) - \dot{\boldsymbol{\omega}}_4 \times \mathbf{r}_4 = 0 \tag{e}$$

The unknowns are $\dot{\omega}_3$ and $\dot{\omega}_4$, which we now replace with α_3 and α_4. For plane motion we can write Eq. (e), as before, in the form

$$-r_2\omega_2^2\hat{\mathbf{r}}_2 - r_3\omega_3^2\hat{\mathbf{r}}_3 + r_3\alpha_3(\hat{\mathbf{k}} \times \hat{\mathbf{r}}_3) + r_4\omega_4^2\hat{\mathbf{r}}_4 - r_4\alpha_4(\hat{\mathbf{k}} \times \hat{\mathbf{r}}_4) = 0 \tag{f}$$

Let us now dot this equation through with the unit vector $\hat{\mathbf{r}}_3$. Then

$$r_3\alpha_3(\hat{\mathbf{k}} \times \hat{\mathbf{r}}_3) \cdot \hat{\mathbf{r}}_3 = 0$$

because $\hat{\mathbf{k}} \times \hat{\mathbf{r}}_3$ is perpendicular to $\hat{\mathbf{r}}_3$. Also

$$r_3\omega_3^2\hat{\mathbf{r}}_3 \cdot \hat{\mathbf{r}}_3 = r_3\omega_3^2$$

Therefore, Eq. (f) becomes

$$-r_2\omega_2^2\hat{\mathbf{r}}_2 \cdot \hat{\mathbf{r}}_3 - r_3\omega_3^2 + r_4\omega_4^2\hat{\mathbf{r}}_4 \cdot \hat{\mathbf{r}}_3 - r_3\alpha_4(\hat{\mathbf{k}} \times \hat{\mathbf{r}}_4) \cdot \hat{\mathbf{r}}_3 = 0 \qquad (g)$$

The only unknown in this equation is α_4. Because it is a scalar equation, we can solve the equation and obtain

$$\alpha_4 = \frac{r_2\omega_2^2\hat{\mathbf{r}}_2 \cdot \hat{\mathbf{r}}_3 + r_3\omega_3^2 - r_4\omega_4^2\hat{\mathbf{r}}_4 \cdot \hat{\mathbf{r}}_3}{-r_4(\hat{\mathbf{k}} \times \hat{\mathbf{r}}_4) \cdot \hat{\mathbf{r}}_3} \qquad (4.42)$$

By taking the dot product of Eq. (f) throughout with the unit vector $\hat{\mathbf{r}}_4$ and using a similar procedure, the angular acceleration of link 3 is found to be

$$\alpha_3 = \frac{r_2\omega_2^2\hat{\mathbf{r}}_2 \cdot \hat{\mathbf{r}}_4 + r_3\omega_3^2\hat{\mathbf{r}}_3 \cdot \hat{\mathbf{r}}_4 - r_4\omega_4^2}{r_3(\hat{\mathbf{k}} \times \hat{\mathbf{r}}_3) \cdot \hat{\mathbf{r}}_4} \qquad (4.43)$$

You should prove for yourself that Eqs. (4.42) and (4.43) are identical to Eqs. (4.32) and (4.31), respectively, found by Raven's method.

4.13 THE INSTANT CENTER OF ACCELERATION

While it is of little use in analysis, it is desirable to define the *instant center of acceleration,* or *acceleration pole,* for a planar-motion mechanism, if only to avoid the implication that the instant center of velocity is also the instant center of acceleration. The instant center of acceleration is defined as *the instantaneous location of a pair of coincident points of two different rigid bodies where the absolute accelerations of the two points are equal.* If we consider a fixed and a moving body, the instantaneous center of acceleration is the point of the moving body which has zero absolute acceleration at the instant considered.

In Fig. 4.22a, let P be the instant center of acceleration, a point of zero absolute acceleration whose location is unknown. Assume that another point A of the moving plane has a known acceleration $\mathbf{A}_A$ and that $\boldsymbol{\omega}$ and $\boldsymbol{\alpha}$ of the moving plane are known. The acceleration-difference equation can then be written as

$$\mathbf{A}_P = \mathbf{A}_A - \omega^2\mathbf{R}_{PA} + \boldsymbol{\alpha} \times \mathbf{R}_{PA} = 0 \qquad (a)$$

Figure 4.22 The instantaneous center of acceleration.

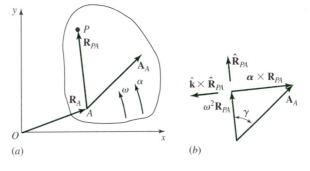

(a) (b)

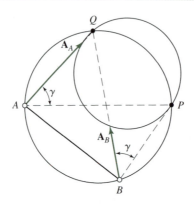

Figure 4.23 The four-circle method of locating the instantaneous center of acceleration P.

Solving for $\mathbf{A}_A$ gives

$$\mathbf{A}_A = \omega^2 R_{PA} \hat{\mathbf{R}}_{PA} - \alpha R_{PA}(\hat{\mathbf{k}} \times \hat{\mathbf{R}}_{PA}) \qquad (b)$$

Now, because $\hat{\mathbf{R}}_{PA}$ is perpendicular to $\hat{\mathbf{k}} \times \hat{\mathbf{R}}_{PA}$, the two terms on the right of Eq. (b) are the rectangular components of $\mathbf{A}_A$, as shown in Fig. 4.22b. From this figure we can solve for the magnitude and direction of $\hat{\mathbf{R}}_{PA}$:

$$\gamma = \tan^{-1} \frac{\alpha}{\omega^2} \qquad (4.44a)$$

$$R_{PA} = \frac{A_A}{\sqrt{\omega^4 + \alpha^2}} = \frac{A_A \cos \gamma}{\omega^2} \qquad (4.44b)$$

Equation (4.44b) states that the distance R_{PA} from point A to the instant center of acceleration can be found from the magnitude of the acceleration A_A of any point of the moving plane. Because the denominator ω^2 is always positive, the angle γ, measured from the direction of $\mathbf{A}_A$, is always acute, but can be either positive or negative depending on the sign of α.

There are many graphical methods of locating the instant center of acceleration.[1] Here, without proof, we present one method. In Fig. 4.23 we are given points A and B and their absolute accelerations $\mathbf{A}_A$ and $\mathbf{A}_B$. We extend $\mathbf{A}_A$ and $\mathbf{A}_B$ until they intersect at Q; then we construct a circle through points A, B, and Q. Now we draw another circle through the termini of $\mathbf{A}_A$ and $\mathbf{A}_B$ and point Q. The intersection of the two circles locates point P, the instant center of acceleration.

4.14 THE EULER–SAVARY EQUATION[2]

In Section 4.5 we developed the apparent-acceleration equation (4.17). Then, in the examples that followed, we found that it was very important to carefully choose a point whose apparent path was known so that the radius of curvature of the path, needed for the normal component in Eq. (4.14), could be found by inspection. This need to know the radius of curvature of the path often dictates the method of approach to such a problem, and sometimes even requires the visualization of an equivalent mechanism. It would be convenient if any arbitrary point could be chosen and the radius of curvature of its path could be calculated. In planar mechanisms this can be done by the methods presented below.

When two rigid bodies move relative to each other with planar motion, any arbitrarily chosen point A of one describes a path or locus relative to a coordinate system fixed to the other. At any given instant there is a point A', attached to the other body, which is the center of curvature of the locus of A. If we take the kinematic inversion of this motion, A' also describes a locus relative to the body containing A, and it so happens that A is the center of curvature of this locus. Each point therefore acts as the center of curvature of the path traced by the other, and the two points are called *conjugates* of each other. The distance between these two conjugate points is the radius of curvature of either locus.

Figure 4.24 shows two circles with centers at C and C'. Let us think of the circle with center C' as the fixed centrode and think of the circle with center C as the moving centrode of two bodies experiencing some particular relative planar motion. In actuality, the fixed centrode need not be fixed but is attached to the body that contains the path whose curvature is sought. Also, it is not necessary that the two centrodes be circles; we are interested only in instantaneous values and, for convenience, we will think of the centrodes as circles matching the curvatures of the two actual centrodes in the region near their point of contact P. As pointed out in Section 3.21, when the bodies containing the two centrodes move relative to each other, the centrodes appear to roll against each other without slip. Their point of contact P, of course, is the instant center of velocity. Because of these properties, we can think of the two circular centrodes as actually representing the shapes of the two moving bodies if this helps in visualizing the motion.

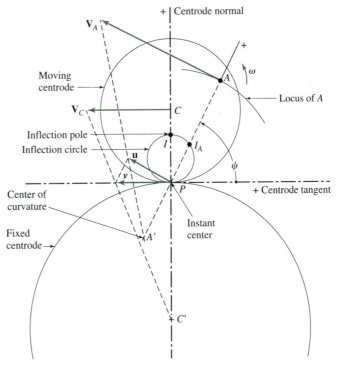

Figure 4.24 The Hartmann construction.

If the moving centrode has some angular velocity ω relative to the fixed centrode, the instantaneous velocity* of point C is

$$V_C = \omega R_{CP} \tag{a}$$

Similarly, the arbitrary point A, whose conjugate point A' we wish to find, has a velocity of

$$V_A = \omega R_{AP} \tag{b}$$

As the motion progresses, the point of contact of the two centrodes, and therefore the location of the instant center P, moves along both centrodes with some velocity v. As shown in Fig. 4.24, v can be found by connecting a straight line from the terminus of V_C to the point C'. Alternatively, its size can be found from

$$v = \frac{R_{PC'}}{R_{CC'}} V_C \tag{c}$$

A graphical construction for A', the center of curvature of the locus of point A, is shown in Fig. 4.24 and is called the *Hartmann constuction*. First the component u of the instant center's velocity v is found as that component parallel to $\mathbf{V}_A$ or perpendicular to $\mathbf{R}_{AP}$. Then the intersection of the line AP and a line connecting the termini of the velocities $\mathbf{V}_A$ and u gives the location of the conjugate point A'. The radius of curvature ρ of the locus of point A is $\rho = R_{AA'}$.

An analytical expression for locating point A' would also be desirable and can be derived from the Hartmann construction. The magnitude of the velocity u is given by

$$u = v \sin \psi \tag{d}$$

where ψ is the positive counterclockwise angle measured from the centrode tangent to the line of $\mathbf{R}_{AP}$. Then, noticing the similar triangles in Fig. 4.24, we can write

$$u = \frac{R_{PA'}}{R_{AA'}} V_A \tag{e}$$

Now, equating the expressions of Eqs. (d) and (e) and substituting from Eqs. (a), (b), and (c) gives

$$u = \frac{R_{PC'} R_{CP}}{R_{CC'}} \omega \sin \psi = \frac{R_{PA'} R_{AP}}{R_{AA'}} \omega \tag{f}$$

Dividing by $\omega \sin \psi$ and inverting, we obtain

$$\frac{R_{AA'}}{R_{AP} R_{PA'}} \sin \psi = \frac{R_{CC'}}{R_{CP} R_{PC'}} = \frac{\omega}{v} \tag{g}$$

Next, upon noticing that $\mathbf{R}_{AA'} = \mathbf{R}_{AP} - \mathbf{R}_{A'P}$ and $\mathbf{R}_{CC'} = \mathbf{R}_{CP} - \mathbf{R}_{C'P}$, we can reduce this equation to the form

$$\left(\frac{1}{R_{AP}} - \frac{1}{R_{A'P}} \right) \sin \psi = \left(\frac{1}{R_{CP}} - \frac{1}{R_{C'P}} \right) \tag{4.45}$$

---

*All velocities used in this section are actually apparent velocities, relative to the coordinate system of the fixed centrode; they are written as absolute velocities to simplify the notation.

This important equation is one form of the *Euler–Savary equation*. Once the radii of curvature of the two centrodes R_{CP} and $R_{C'P}$ are known, this equation can be used to determine the positions of the two conjugate points A and A' relative to the instant center P.

Before continuing, an explanation of the sign convention is important. In using the Euler–Savary equation, we may arbitrarily choose a positive sense for the centrode tangent; the positive centrode normal is then 90° counterclockwise from it. This establishes a positive direction for the line CC' which may be used in assigning appropriate signs to R_{CP} and $R_{C'P}$. Similarly, an arbitrary positive direction can be chosen for the line AA'. The angle ψ is then taken as positive counterclockwise from the positive centrode tangent to the positive sense of line AA'. The sense of line AA' also gives the appropriate signs for R_{AP} and $R_{A'P}$ for Eq. (4.45).

There is a major drawback to the above form of the Euler–Savary equation in that the radii of curvature of both centrodes, R_{CP} and $R_{C'P}$, must be found. Usually they are not known, any more than the curvature of the locus itself was known. However, this difficulty can be overcome by seeking another form of the equation.

Let us consider the particular point labeled I in Fig. 4.24. This point is located on the centrode normal at a location defined by

$$\frac{1}{R_{IP}} = \frac{1}{R_{CP}} - \frac{1}{R_{C'P}} \qquad (h)$$

If this particular point is chosen for A in Eq. (4.45), we find that its conjugate point I' must be located at infinity. The radius of curvature of the path of point I is infinite, and the locus of I therefore has an inflection point at I. The point I is called the *inflection pole*.

Let us now consider whether there are any other points A of the moving body which also have infinite radii of curvature at the instant considered. If so, then for each of these points $R_{I_A P} = \infty$ and, from Eqs. (4.45) and (h),

$$R_{I_A P} = R_{IP} \sin \psi \qquad (4.46)$$

This equation defines a circle called the *inflection circle,* whose diameter is R_{IP}, as shown in Fig. 4.24. Every point on this circle has its conjugate point at infinity, and each therefore has an infinite radius of curvature at the instant shown.

Now, with the help of Eq. (4.46), the Euler–Savary equation can be written in the form

$$\frac{1}{R_{AP}} - \frac{1}{R_{A'P}} = \frac{1}{R_{I_A P}} \qquad (4.47)$$

Also, after some further manipulation, this can be put in the form

$$\rho = R_{AA'} = \frac{R_{AP}^2}{R_{AI_A}} \qquad (4.48)$$

Either of these two forms of the Euler–Savary equation, Eqs. (4.47) and (4.48), is more useful in practice than Eq. (4.45), because they do not require knowledge of the curvatures of the two centrodes. They do require finding the inflection circle, but it will be shown how this can be done in the following example.

EXAMPLE 4.10

Find the inflection circle for the motion of the coupler of the slider-crank linkage of Fig. 4.25 and determine the instantaneous radius of curvature of the path of the coupler point C.

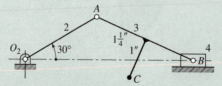

Figure 4.25 Example 4.10: $R_{AO_2} = 2$ in, $R_{BA} = 21.5$ in.

SOLUTION

We begin in Fig. 4.26 by locating the instant center P at the intersection of the line $O_2 A$ and a line through B perpendicular to its direction of travel. Points B and P must both lie on the inflection circle by definition; hence we need only one additional point to construct the circle.

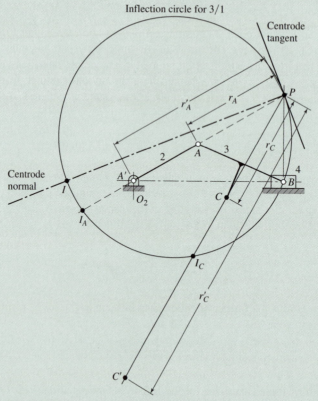

Figure 4.26 Example 4.10.

The center of curvature A is, of course, at O_2, which we now call A'. Taking the positive sense of the line AP as being downward and to the left, we measure $R_{AP} = 2.64$ in and $R_{AA'} = -2$ in. Then, substituting into Eq. (4.48), we obtain

$$R_{AI_A} = \frac{R_{AP}^2}{R_{AA'}} = \frac{2.64^2}{-2.00} = -3.48 \text{ in}$$

With this, we lay off 3.48 in from A to locate I_A, a third point on the inflection circle. The circle can now be constructed through the three points B, P, and I_A, and its diameter can be found:

$$R_{IP} = 6.28 \text{ in} \qquad\qquad Ans.$$

The centrode normal and centrode tangent can also be drawn, if desired, as shown in the figure.

Next, drawing the ray R_{CI_c} and taking its positive sense as downward and to the left, we can measure $R_{CP} = 3.1$ in and $R_{CI_c} = -1.75$ in. Substituting these into Eq. (4.48), we find the instantaneous radius of curvature of the path of point C:

$$\rho = R_{CC'} = \frac{R_{CP}^2}{R_{CI_c}} = \frac{3.1^2}{-1.75} = -5.49 \text{ in} \qquad\qquad Ans.$$

where the negative sign indicates that C' is below C on the line $C'I_CCP$.

4.15 THE BOBILLIER CONSTRUCTIONS

The Hartmann construction provides one graphical method of finding the conjugate point and the radius of curvature of the path of a moving point, but it requires knowledge of the curvature of the fixed and moving centrodes. It would be desirable to have *graphical* methods of obtaining the inflection circle and the conjugate of a given point without requiring the curvature of the centrodes. Such graphical solutions are presented in this section and are called the *Bobillier constructions*.

To understand these constructions, consider the inflection circle and the centrode normal N and centrode tangent T shown in Fig. 4.27. Let us select any two points A and B of the moving body which are not on a straight line through P. Now, by using the Euler–Savary equation, we can find the two corresponding conjugate points A' and B'. The intersection of the lines AB and $A'B'$ is labeled Q. Then, the straight line drawn through P and Q is called the *collineation axis*. This axis applies only to the two lines AA' and BB' and so is said to belong to these two rays; also, the point Q will be located differently on the collineation axis if another set of points A and B is chosen on the same rays. Nevertheless, there is a unique relationship between the collineation axis and the two rays used to define it. This relationship is expressed in *Bobillier's theorem*, which states that *the angle from the centrode tangent to one of these rays is the negative of the angle from the collineation axis to the other ray*.

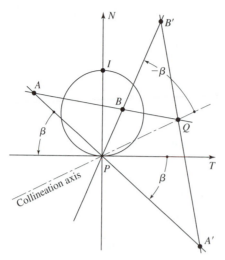

Figure 4.27 The Bobillier theorem.

In applying the Euler–Savary equation to a planar mechanism, we can usually find two pairs of conjugate points by inspection, and from these we wish to determine the inflection circle graphically. For example, a four-bar linkage with a crank O_2A and a follower O_4B has A and O_2 as one set of conjugate points and B and O_4 as the other, when we are interested in the motion of the coupler relative to the frame. Given these two pairs of conjugate points, how do we use the Bobillier theorem to find the inflection circle?

In Fig. 4.28a, let A and A' and B and B' represent the known pairs of conjugate points. Rays constructed through each pair intersect at P, the instant center of velocity, giving one point on the inflection circle. Point Q is located next by the intersection of a ray through A and B with a ray through A' and B'. Then the collineation axis can be drawn as the line PQ.

The next step is shown in Fig. 4.28b. Drawing a straight line through P parallel to $A'B'$, we identify the point W as the intersection of this line with the line AB. Now, through W we draw a second line parallel to the collineation axis. This line intersects AA' at I_A and BB' at I_B, the two additional points on the inflection circle for which we are searching.

We could now construct the circle through the three points I_A, I_B, and P, but there is an easier way. Remembering that a triangle inscribed in a semicircle is a right triangle having the diameter as its hypotenuse, we erect a perpendicular to AP at I_A and another perpendicular to BP at I_B. The intersection of these two perpendiculars gives point I, the inflection pole, as shown in Fig. 4.28c. Because PI is the diameter, the inflection circle, the centrode normal N, and the centrode tangent T can all be easily constructed.

To show that this construction satisfies the Bobillier theorem, note that the arc from P to I_A is inscribed by the angle that I_AP makes with the centrode tangent. But this same arc is also inscribed by the angle PI_BI_A. Therefore these two angles are equal. But the line I_BI_A was originally constructed parallel to the collineation axis. Therefore, the line PI_B also makes the same angle β with the colliineation axis.

Our final problem is to learn how to use the Bobillier theorem to find the conjugate of another arbitrary point, say C, when the inflection circle is given. In Fig. 4.29 we join C

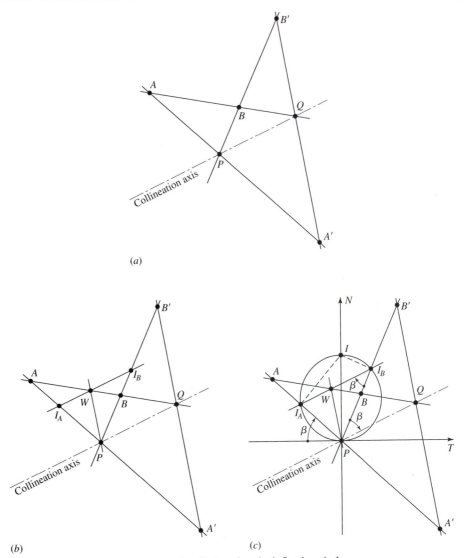

Figure 4.28 The Bobillier construction for locating the inflection circle.

with the instant center P and locate the intersection point I_C with the inflection circle. This ray serves as one of the two necessary to locate the collineation axis. For the other we may as well use the centrode normal, because I and its conjugate point I', at infinity, are both known. For these two rays the collineation axis is a line through P parallel to the line $I_C I$, as we learned in Fig. 4.28c. The balance of the construction is similar to that of Fig. 4.27. Point Q is located by the intersection of a line through I and C with the collineation axis. Then a line through Q and I' at infinity intersects the ray PC at C', the conjugate point for C.

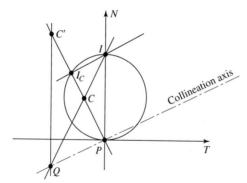

Figure 4.29 The Bobillier construction for locating the conjugate point C'.

EXAMPLE 4.11

Use the Bobillier theorem to find the center of curvature of the coupler curve of point C of the four-bar linkage shown in Fig. 4.30.

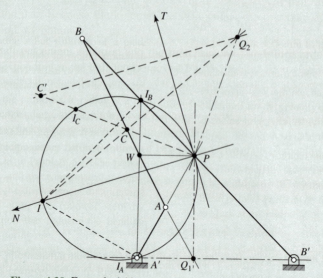

Figure 4.30 Example 4.11.

SOLUTION

Locate the instant center P at the intersection of AA' and BB'; also locate Q_1 at the intersection of AB and $A'B'$. PQ_1 is the first collineation axis. Through P draw a line parallel to $A'B'$ to locate W on AB. Through W draw a line parallel to PQ_1 to locate I_A on AA' and I_B on BB'. Then, through I_A draw a perpendicular to AA' and through I_B draw a perpendicular to BB'. These perpendiculars intersect at the inflection pole I and define the inflection circle, the centrode normal N, and the centrode tangent T.

To obtain the conjugate point of C, draw the ray PC and locate I_C on the inflection circle. The second collineation axis PQ_2, belonging to the pair of rays PC and PI, is a line through P parallel to a line (not shown) from I to I_C. The point Q_2 is obtained as the intersection of this collineation axis and a line IC. Now, through Q_2 draw a line parallel to the centrode normal; its intersection with the ray PC yields C', the center of curvature of the path of C.

4.16 RADIUS OF CURVATURE OF A POINT TRAJECTORY USING KINEMATIC COEFFICIENTS

The radius of curvature of a point trajectory (say point P), at the instant considered, can be written from Eqs. (4.2) and (4.3) as

$$\rho = \frac{V_P^2}{A_P^n} \tag{a}$$

From Eq. (3.34a), the speed of the point can be written as

$$V_P = r_P' \dot{\psi} \tag{b}$$

and, from Eq. (4.14), the normal component of the acceleration of point P can be written as

$$A_P^n = -\mathbf{A}_P \cdot \hat{\boldsymbol{\rho}} \tag{c}$$

where the unit normal vector to the point trajectory at the position considered [see Eq. (3.37)] is

$$\hat{\boldsymbol{\rho}} = \left(\frac{y_P'}{r_P'}\right)\hat{\mathbf{i}} + \left(\frac{-x_P'}{r_P'}\right)\hat{\mathbf{j}} \tag{d}$$

Taking the time derivative of Eq. (3.33b), the acceleration of point P can be written as

$$\mathbf{A}_P = (x_P'' \dot{\psi}^2 + x_P' \ddot{\psi})\hat{\mathbf{i}} + (y_P'' \dot{\psi}^2 + y_P' \ddot{\psi})\hat{\mathbf{j}} \tag{4.49}$$

Substituting Eqs. (d) and (4.49) into Eq. (c) gives

$$A_P^n = \left(\frac{x_P' y_P'' - y_P' x_P''}{r_P'}\right)\dot{\psi}^2 \tag{e}$$

Then substituting Eqs. (b) and (e) into Eq. (a), the radius of curvature of the point trajectory at the position considered can be written as

$$\rho = \frac{r_P'^3}{x_P' y_P'' - y_P' x_P''} \tag{4.50}$$

The sign convention is as follows: If the unit normal vector to the point trajectory points away from the center of curvature of the path, then the radius of curvature has a positive value. If the unit normal vector to the point trajectory points toward the center of curvature of the path, then the radius of curvature has a negative value.

The coordinates of the center of curvature of the point trajectory, at the position under investigation, can be written as

$$x_C = x_P - \rho\left(\frac{y'_P}{r'_P}\right) \quad \text{and} \quad y_C = y_P + \rho\left(\frac{x'_P}{r'_P}\right) \tag{4.51}$$

4.17 THE CUBIC OF STATIONARY CURVATURE

Consider a point on the coupler of a planar four-bar linkage that generates a path relative to the frame whose radius of curvature, at the instant considered, is ρ. For most cases, because the coupler curve is of sixth order, this radius of curvature changes continuously as the point moves. In certain situations, however, the path will have stationary curvature, which means that

$$\frac{d\rho}{ds} = 0 \tag{a}$$

where s is the increment traveled along the path. The locus of all points on the coupler or moving plane which have stationary curvature at the instant considered is called the *cubic of stationary curvature* or sometimes the *circling-point curve*. It should be noted that stationary curvature does not necessarily mean constant curvature, but rather that the continually varying radius of curvature is passing through a maximum or minimum.

Here we will present a fast and simple graphical method for obtaining the cubic of stationary curvature, as described by Hain.[3] In Fig. 4.31 we have a four-bar linkage $A'ABB'$,

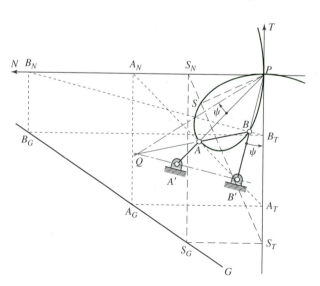

Figure 4.31 The cubic of stationary curvature.

with A' and B' the frame pivots. Then points A and B have stationary curvature—in fact, constant curvature about centers at A' and B'; hence, A and B lie on the cubic.

The first step of the construction is to obtain the centrode normal and centrode tangent. Because the inflection circle is not needed, we locate the collineation axis PQ as shown and draw the centrode tangent T at the angle ψ from the line PA' to the collineation axis. This construction follows directly from Bobillier's theorem. We also construct the centrode normal N. At this point it may be convenient to reorient the drawing on the working surface so that the T-square or horizontal lies along the centrode normal.

Next we construct a line through A perpendicular to PA and another line through B perpendicular to PB. These lines intersect the centrode normal and centrode tangent at A_N, A_T and B_N, B_T, respectively, as shown in Fig. 4.31. Now we draw the two rectangles $PA_NA_GA_T$ and $PB_NB_GB_T$; the points A_G and B_G define an auxiliary line G that we will use to obtain other points on the cubic.

Next we choose any point S_G on the line G. A ray parallel to N locates S_T, and another ray parallel to T locates S_N. Connecting S_T with S_N and drawing a perpendicular to this line through P locates point S, another point on the cubic of stationary curvature. We now repeat this process as often as desired by choosing different points on G, and we draw the cubic as a smooth curve through all the points S obtained.

Note that the cubic of stationary curvature has two tangents at P, the *centrode-normal tangent* and the *centrode-tangent tangent*. The radius of curvature of the cubic at these tangents is obtained as follows: Extend G to intersect T at G_T and N at G_N (not shown). Then, half the distance PG_T is the radius of curvature of the cubic at the centrode-normal tangent, and half the distance PG_N is the radius of curvature of the cubic at the centrode-tangent tangent.

A point with interesting properties occurs at the intersection of the cubic of stationary curvature with the inflection circle; this point is called *Ball's point*. A point of the coupler coincident with Ball's point describes a path that is approximately a straight line because it is located at an inflection point of its path and has stationary curvature.

The equation of the cubic of stationary curvature[4] is

$$\frac{1}{M \sin \psi} + \frac{1}{N \cos \psi} - \frac{1}{r} = 0 \qquad (4.52)$$

where r is the distance from the instant center to the point on the cubic, measured at an angle ψ from the centrode tangent. The constants M and N are found by using any two points known to lie on the cubic, such as points A and B of Fig. 4.31. It so happens[5] that M and N are, respectively, the diameters PG_T and PG_N of the circles centered on the centrode tangent and centrode normal whose radii represent the two curvatures of the cubic at the instant center.

NOTES

[1.]   N. Rosenauer and A. H. Willis, *Kinematics of Mechanisms,* Associated General Publications, Sydney, Australia, 1953, pp. 145–156; republished by Dover, New York, 1967; K. Hain (translated by T. P. Goodman et al.) *Applied Kinematics,* 2nd ed., McGraw-Hill, New York, 1967, pp. 149–158.

2. The most important and most useful references on this subject are Rosenauer and Willis, *Kinematics of Mechanisms,* Chapter 4; A. E. R. de Jonge, A Brief Account of Modern Kinematics, *Trans. ASME,* vol. 65, 1943, pp. 663–683; R. S. Hartenberg and J. Denavit, *Kinematic Synthesis of Linkages,* McGraw-Hill, New York, 1964, Chapter 7; A. S. Hall, Jr., *Kinematics and Linkage Design,* Prentice-Hall, Englewood Cliffs, NJ, 1961, Chapter 5 (this book is a real classic on the theory of mechanisms and contains many useful examples); Hain, *Applied Kinematics,* Chapter 4.

3. Hain, *Applied Kinematics,* pp. 498–502.

4. For a derivation of this equation see A. S. Hall, Jr., *Kinematics and Linkage Design,* Prentice-Hall, 1961, p. 98, or R. S. Hartenberg and J. Denavit, *Kinematic Synthesis of Linkages,* 1965, p. 206.

5. D. C. Tao, *Applied Linkage Synthesis,* Addison-Wesley, Reading, MA, 1964, p. 111.

PROBLEMS*

4.1 The position vector of a point is defined by the equation

$$\mathbf{R} = \left(4t - \frac{t^3}{3}\right)\hat{\mathbf{i}} + 10\hat{\mathbf{j}}$$

where R is in inches and t is in seconds. Find the acceleration of the point at $t = 2s$.

4.2 Find the acceleration at $t = 3$ s of a point which moves according to the equation

$$\mathbf{R} = \left(t^2 - \frac{t^3}{6}\right)\hat{\mathbf{i}} + \frac{t^3}{3}\hat{\mathbf{j}}$$

The units are meters and seconds.

4.3 The path of a point is described by the equation

$$\mathbf{R} = (t^2 + 4)e^{-j\pi t/10}$$

where R is in millimeters and t is in seconds. For $t = 20$ s, find the unit tangent vector for the path, the normal and tangential components of the point's absolute acceleration, and the radius of curvature of the path.

4.4 The motion of a point is described by the equations

$$x = 4t \cos \pi t^3 \quad \text{and} \quad y = \frac{t^3 \sin 2\pi t}{6}$$

where x and y are in feet and t is in seconds. Find the acceleration of the point at $t = 1.40$ s.

4.5 Link 2 in the figure has an angular velocity of $\omega_2 = 120$ rad/s ccw and an angular acceleration of $4\,800$ rad/s² ccw at the instant shown. Determine the absolute acceleration of point A.

Figure P4.5 $R_{AO_2} = 500$ mm.

4.6 Line 2 is rotating clockwise as shown in the figure. Find its angular velocity and acceleration and the acceleration of its midpoint C.

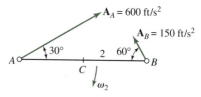

Figure P4.6 $R_{BA} = 20$ in.

4.7 For the data given in the figure, find the velocity and acceleration of points B and C.

---

*When assigning problems, the instructor may wish to specify the method of solution to be used because a variety of approaches are provided in the text.

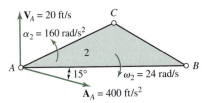

Figure P4.7 $R_{BA} = 16$ in, $R_{CA} = 10$ in, $R_{CB} = 8$ in.

4.8   For the straight-line mechanism shown in the figure, $\omega_2 = 20$ rad/s cw and $\alpha_2 = 140$ rad/s^2 cw. Determine the velocity and acceleration of point B and the angular acceleration of link 3.

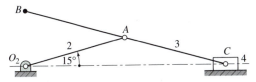

Figure P4.8 $R_{AO_2} = R_{CA} = R_{BA} = 100$ mm.

4.9   In the figure above, the slider 4 is moving to the left with a constant velocity of 20 m/s^2. Find the angular velocity and acceleration of link 2.

4.10 Solve Problem 3.8, for the acceleration of point A and the angular acceleration of link 3.

4.11 For Problem 3.9, find the angular accelerations of links 3 and 4.

4.12 For Problem 3.10, find the acceleration of point C and the angular accelerations of links 3 and 4.

4.13 For Problem 3.11, find the acceleration of point C and the angular accelerations of links 3 and 4.

4.14 Using the data of Problem 3.13, solve for the accelerations of points C and D and the angular acceleration of link 4.

4.15 For Problem 3.14, find the acceleration of point C and the angular acceleration of link 4.

4.16 Solve Problem 3.16 for the acceleration of point C and the angular acceleration of link 4.

4.17 For Problem 3.17, find the acceleration of point B and the angular accelerations of links 3 and 6.

4.18 For the data of Problem 3.18, what angular acceleration must be given to link 2 for the position shown to make the angular acceleration of link 4 zero?

4.19 For the data of Problem 3.19, what angular acceleration must be given to link 2 for the angular accel-

eration of link 4 to be 100 rad/s^2 cw at the instant shown?

4.20 Solve Problem 3.20 for the acceleration of point C and the angular acceleration of link 3.

4.21 For Problem 3.21, find the acceleration of point C and the angular acceleration of link 3.

4.22 Find the acceleration of points B and D of Problem 3.22.

4.23 Find the accelerations of points B and D of Problem 3.23.

4.24 to 4.30 The nomenclature for this group of problems is shown in the figure, and the dimensions and data are given in the accompanying table. For each problem, determine θ_3, θ_4, ω_3, ω_4, α_3, and α_4. The angular velocity ω_2 is constant for each problem, and a negative sign is used to indicate the clockwise direction. The dimensions of even-numbered problems are given in inches, and odd-numbered problems are given in millimeters.

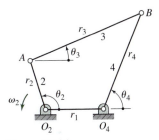

Figure P4.24

TABLE P4.24 TO P4.30

Problem	r_1	r_2	r_3	r_4	θ_2, deg	ω_2, rad/s
4.24	4	6	9	10	240	1
4.25	100	150	250	250	−45	56
4.26	14	4	14	10	0	10
4.27	250	100	500	400	70	−6
4.28	8	2	10	6	40	12
4.29	400	125	300	300	210	−18
4.30	16	5	12	12	315	−18

4.31 Crank 2 of the system shown in the figure has a speed of 60 rev/min ccw. Find the velocity and acceleration of point B and the angular velocity and acceleration of link 4.

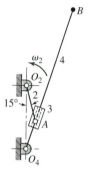

Figure P4.31 $R_{O_4 O_2} = 12$ in,
$R_{A O_2} = 7$ in, $R_{B O_4} = 28$ in.

4.32 The mechanism shown in the figure is a marine steering gear called *Rapson's slide*. $O_2 B$ is the tiller, and AC is the actuating rod. If the velocity of AC is 10 ft/min to the left, find the angular acceleration of the tiller.

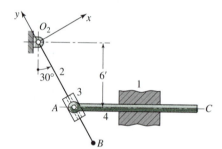

Figure P4.32

4.33 Determine the acceleration of link 4 of Problem 3.26.

4.34 For Problem 3.27, find the acceleration of point E.

4.35 Find the acceleration of point B and the angular acceleration of link 4 of Problem 3.24.

4.36 For Problem 3.25, find the acceleration of point B and the angular acceleration of link 3.

4.37 Assuming that both links 2 and 3 of Problem 3.28 are rotating at constant speed, find the acceleration of point P_4.

4.38 Solve Problem 3.32, for the accelerations of points A and B.

4.39 For Problem 3.33, determine the acceleration of point C_4 and the angular acceleration of link 3 if crank 2 is given an angular acceleration of 2 rad/s² ccw.

4.40 Determine the angular accelerations of links 3 and 4 of Problem 3.30.

4.41 For Problem 3.31, determine the acceleration of point G and the angular accelerations of links 5 and 6.

4.42 Find the inflection circle for motion of the coupler of the double-slider mechanism shown in the figure. Select several points on the centrode normal and find their conjugate points. Plot portions of the paths of these points to demonstrate for yourself that the conjugates are indeed the centers of curvature.

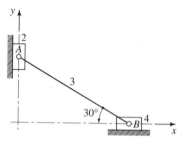

Figure P4.42 $R_{BA} = 125$ mm.

4.43 Find the inflection circle for motion of the coupler relative to the frame of the linkage shown in the figure. Find the center of curvature of the coupler curve of point C and generate a portion of the path of C to verify your findings.

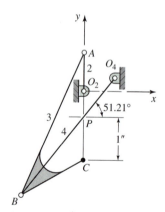

Figure P4.43 $R_{CA} = 2.5$ in,
$R_{A O_2} = 0.9$ in, $R_{B O_4} = 3.5$ in,
$R_{P O_4} = 1.17$ in.

4.44 For the motion of the coupler relative to the frame, find the inflection circle, the centrode normal, the centrode tangent, and the centers of curvature of points C and D of the linkage of Problem 3.13. Choose points on the coupler coincident with the instant center and inflection pole and plot their paths.

4.45 On 18- by 24-in paper, draw the linkage shown in the figure in full size, placing A' at 6 in from the lower edge and 7 in from the right edge. Better utilization of the paper is obtained by tilting the frame through about 15° as shown.

(a) Find the inflection circle.

(b) Draw the cubic of stationary curvature.

(c) Choose a coupler point C coincident with the cubic and plot a portion of its coupler curve in the vicinity of the cubic.

(d) Find the conjugate point C'. Draw a circle through C with center at C' and complete this circle with the actual path of C.

(e) Find Ball's point. Locate a point D on the coupler at Ball's point and plot a portion of its path. Compare the result with a straight line.

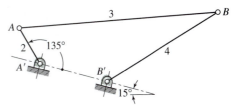

Figure P4.45 $R_{AA'} = 1$ in, $R_{BA} = 5$ in, $R_{B'A'} = 1.75$ in, $R_{BB'} = 3.25$ in.

PART 2

Design of Mechanisms

5 Cam Design

5.1 INTRODUCTION

In the previous chapters we have been learning how to analyze the kinematic characteristics of a given mechanism. We were given the design of a mechanism and we studied ways to determine its mobility, its position, its velocity, and its acceleration, and we even discussed its suitability for given types of tasks. However, we have said little about how the mechanism was designed—that is, how the sizes and shapes of its links are chosen by the designer.

The next several chapters will introduce this *design* point of view as it relates to mechanisms. We will find ourselves looking more at individual types of machine components, and learning when and why such components are used and how they are sized. In this chapter, devoted to the design of cams, for example, we will assume that we know the task to be accomplished. However, we will not know, but will look for techniques to help discover, the size and shape of the cam to perform this task.

Of course, there is the creative process of deciding whether we should use a cam in the first place, or rather a gear train, or a linkage, or some other idea. This question often cannot be answered on the basis of scientific principles alone; it requires experience and imagination and involves such factors as economics, marketability, reliability, maintenance, esthetics, ergonomics, ability to manufacture, and suitability to the task. These aspects are not well-studied by a general scientific approach; they require human judgment of factors that are often not easily reduced to numbers or formulae. There is usually not a single "right" answer, and these questions cannot be answered by this or any other text or reference book.

On the other hand, this is not to say that there is no place for a general science-based approach in design situations. Most mechanical design is based on repetitive analysis. Therefore, in this chapter and in future chapters, we will use the principles of analysis presented in the previous chapters. Also, we will use the governing analysis equations to help in our choice of part sizes and shapes and to help us assess the quality of our design as we proceed. The coming several chapters will still be based on the laws of mechanics. The primary shift for Part 2 of this book is that the component dimensions will often be the unknowns of the problem, while the input and output speeds, for example, may be given information. In this chapter we will discover how to determine a cam contour which will deliver a specified motion.

5.2 CLASSIFICATION OF CAMS AND FOLLOWERS

A *cam* is a mechanical element used to drive another element, called the *follower,* through a specified motion by direct contact. Cam-and-follower mechanisms are simple and inexpensive, have few moving parts, and occupy a very small space. Furthermore, follower motions having almost any desired characteristics are not difficult to design. For these reasons, cam mechanisms are used extensively in modern machinery.

The versatility and flexibility in the design of cam systems are among their more attractive features, yet this also leads to a wide variety of shapes and forms and the need for terminology to distinguish them.

Cams are classified according to their basic shapes. Figure 5.1 illustrates four different types of cams:

(a) A *plate cam,* also called a *disk cam* or a *radial cam*
(b) A wedge cam
(c) A *cylindric cam* or *barrel cam*
(d) An *end cam* or *face cam*

The least common of these in practical applications is the wedge cam because of its need for a reciprocating motion rather than a continuous input motion. By far the most common is the plate cam. For this reason, most of the remainder of this chapter specifically addresses plate cams, although the concepts presented pertain universally.

Cam systems can also be classified according to the basic shape of the follower. Figure 5.2 shows plate cams acting with four different types of followers:

(a) A *knife-edge* follower
(b) A *flat-face* follower
(c) A *roller* follower
(d) A *spherical-face* or *curved-shoe* follower

Notice that the follower face is usually chosen to have a simple geometric shape and the motion is achieved by careful design of the shape of the cam to mate with it. This is not always the case; and examples of *inverse cams,* where the output element is machined to a complex shape, can be found.

Another method of classifying cams is according to the characteristic output motion allowed between the follower and the frame. Thus, some cams have *reciprocating* (translating) followers, as in Figs. 5.1a, 5.1b, and 5.1d and Figs. 5.2a and 5.2b, while others have *oscillating* (rotating) followers, as in Fig. 5.1c and Figs. 5.2c and 5.2d. Further classification of reciprocating followers distinguishes whether the centerline of the follower stem relative to the center of the cam is *offset,* as in Fig. 5.2a, or *radial,* as in Fig. 5.2b.

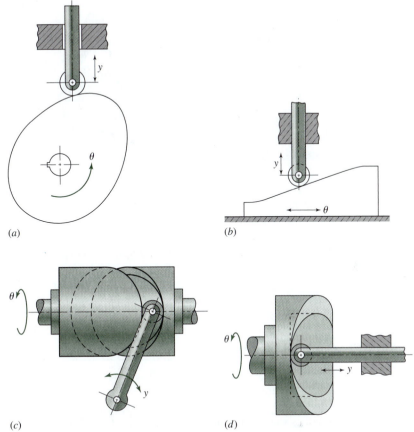

(a)   (b)   (c)   (d)

Figure 5.1 Types of cams: (*a*) plate cam; (*b*) wedge cam; (*c*) barrel cam; (*d*) face cam.

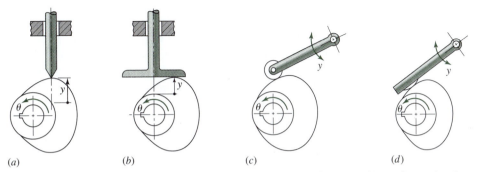

(a)   (b)   (c)   (d)

Figure 5.2 Plate cams with (*a*) an offset reciprocating knife-edge follower; (*b*) a reciprocating flat-face follower; (*c*) an oscillating roller follower; and (*d*) an oscillating curved-shoe follower.

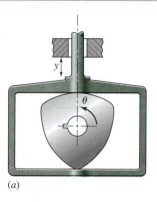 

Figure 5.3 (*a*) A constant-breadth cam with a reciprocating flat-face follower. (*b*) Conjugate cams with an oscillating roller follower.

(*a*)   (*b*)

In all cam systems the designer must ensure that the follower maintains contact with the cam at all times. This can be accomplished by depending on gravity, by the inclusion of a suitable spring, or by a mechanical constraint. In Fig. 5.1*c*, the follower is constrained by the groove. Figure 5.3*a* shows an example of a *constant-breadth* cam, where two contact points between the cam and the follower provide the constraint. Mechanical constraint can also be introduced by employing *dual* or *conjugate* cams in an arrangement like that illustrated in Fig. 5.3*b*. Here each cam has its own roller, but the rollers are mounted on a common follower.

5.3 DISPLACEMENT DIAGRAMS

In spite of the wide variety of cam types used and their differences in form, they also have certain features in common which allow a systematic approach to their design. Usually a cam system is a single-degree-of-freedom device. It is driven by a known input motion, usually a shaft that rotates at constant speed, and it is intended to produce a certain desired periodic output motion for the follower.

In order to investigate the design of cams in general, we will denote the known input motion by $\theta(t)$ and the output motion by y. Reviewing Figs. 5.1 to 5.3 will demonstrate the definitions of y and θ for various types of cams. These figures also show that y is a translational distance for a reciprocating follower but is an angle for an oscillating follower.

During the rotation of the cam through one cycle of input motion, the follower executes a series of events as shown in graphical form in the *displacement diagram* of Fig. 5.4. In such a diagram the abscissa represents one cycle of the input motion θ (one revolution of the cam) and is drawn to any convenient scale. The ordinate represents the follower travel y and for a reciprocating follower is usually drawn at full scale to help in the layout of the cam. On a displacement diagram it is possible to identify a portion of the graph called the *rise,* where the motion of the follower is away from the cam center. The maximum rise is called the *lift.* Portions of the cycle during which the follower is at rest are referred to as *dwells,* and the *return* is the portion in which the motion of the follower is toward the cam center.

Many of the essential features of a displacement diagram, such as the total lift or the placement and duration of dwells, are usually dictated by the requirements of the application. There are, however, many possible choices of follower motions that might be used for the rise and return, and some are preferable to others depending on the situation. One of the

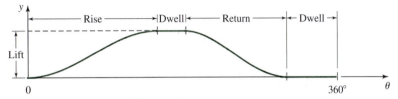

Figure 5.4 Displacement diagram for a cam.

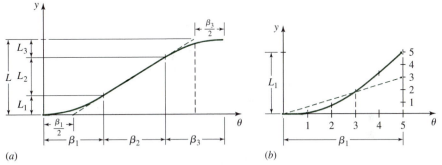

(a)   (b)

Figure 5.5 Parabolic motion displacement diagram: (a) interfaces with uniform motion; (b) graphical construction.

key steps in the design of a cam is the choice of suitable forms for these motions. Once the motions have been chosen—that is, once the exact relationship is set between the input θ and the output y—the displacement diagram can be constructed precisely and is a graphical representation of the functional relationship

$$y = y(\theta)$$

This equation has stored in it the exact nature of the shape of the final cam, the necessary information for its layout and manufacture, and also the important characteristics that determine the quality of its dynamic performance. Before looking further at these topics, however, we will display graphical methods of constructing the displacement diagrams for various rise and return motions.

The displacement diagram for *uniform motion* is a straight line with a constant slope. Thus, for constant input speed, the velocity of the follower is also constant. This motion is not useful for the full lift because of the corners produced at the boundaries with other sections of the displacement diagram. It is often used, however, between other curve sections that eliminate the corners.

The displacement diagram for a *modified uniform motion* is illustrated in Fig. 5.5a. The central portion of the diagram, subtended by the cam angle β_2 and the lift L_2, is uniform motion. The ends, angles β_1 and β_3 and corresponding lifts L_1 and L_3, are shaped to deliver *parabolic motion* to the follower. Soon we shall learn that this produces constant acceleration of the follower. The diagram shows how to match the slopes of the parabolic motion with that of the uniform motion. With β_1, β_2, β_3, and the total lift (or rise) L known, the individual lifts L_1, L_2, and L_3 are found by locating the midpoints of the β_1 and β_3 sections and constructing a straight line as shown. Figure 5.5b illustrates a graphical construction for a parabola to be fit within a given rectangular boundary defined by L_1 and β_1. The abscissa and ordinate are first divided into a convenient but equal number of divisions and numbered as shown. The construction of each point of the parabola then follows that shown by dashed lines for point 3.

In the graphical layout of an actual cam, a great many divisions must be employed to obtain good accuracy. At the same time, the drawing is often made to a large scale, perhaps 10 times full size. However, for clarity in reading, the figures in this chapter are shown with a minimum number of points to define the curves and illustrate the graphic techniques.

The displacement diagram for *simple harmonic motion* is shown in Fig. 5.6. The graphical construction makes use of a semicircle having a diameter equal to the rise L. The semicircle and abscissa are divided into an equal number of parts, and the construction then follows that shown by dashed lines for point 2.

Cycloidal motion obtains its name from the geometric curve called a cycloid. As shown on the left-hand side of Fig. 5.7, a circle of radius $L/(2\pi)$, where L is the total rise, will make exactly one revolution when rolling along the ordinate from the origin to $y = L$.

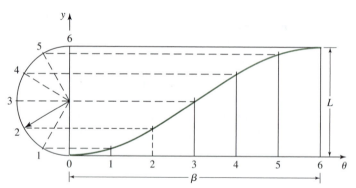

Figure 5.6 Simple harmonic motion displacement diagram; graphical construction.

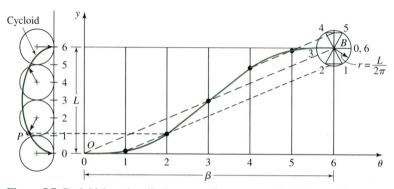

Figure 5.7 Cycloidal motion displacement diagram; graphical construction.

A point P of the circle, originally located at the origin, traces a cycloid as shown. As the circle rolls without slip at a constant rate, the graph of the point's vertical position y versus rotation angle gives the displacement diagram shown at the right of the figure. We find it much more convenient for graphical purposes to draw the circle only once, using point B as a center. After dividing the circle and the abscissa into an equal number of parts and numbering them as shown, we can project each point of the circle horizontally until it intersects the ordinate; next, from the ordinate, we project parallel to the diagonal OB to obtain the corresponding point on the displacement diagram.

5.4 GRAPHICAL LAYOUT OF CAM PROFILES

Let us now examine the problem of determining the exact shape of a cam surface required to deliver a specified follower motion. We assume here that the required motion has been completely defined—graphically, analytically, or numerically—as discussed in later sections. Thus a complete displacement diagram can be drawn to scale for the entire cam rotation. The problem now is to lay out the proper cam shape to achieve the follower motion represented by this displacement diagram.

We illustrate the procedure using the case of a plate cam as shown in Fig. 5.8. Let us first note some additional nomenclature shown in this figure.

The *trace point* is a theoretical point of the follower; it corresponds to the tip of a fictitious knife-edge follower. It is located at the center of a roller follower or along the surface of a flat-face follower.

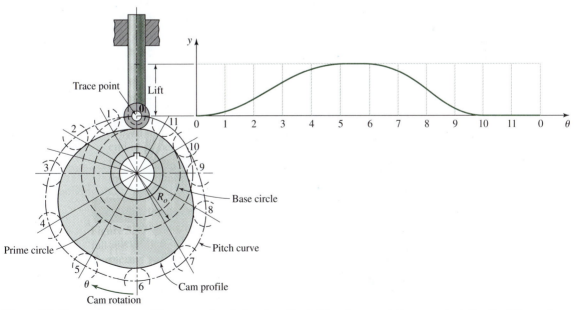

Figure 5.8 Cam nomenclature. The cam surface is developed by holding the cam stationary and rotating the follower from station 0 through stations 1, 2, 3, etc.

The *pitch curve* is the locus generated by the trace point as the follower moves relative to the cam. For a knife-edge follower, the pitch curve and cam surface are identical. For a roller follower they are separated by the radius of the roller.

The *prime circle* is the smallest circle that can be drawn with center at the cam rotation axis and tangent to the pitch curve. The radius of this circle is denoted as R_0.

The *base circle* is the smallest circle centered on the cam rotation axis and tangent to the cam surface. For a roller follower it is smaller than the prime circle by the radius of the roller, and for a flat-face follower it is identical with the prime circle.

In constructing the cam profile, we employ the principle of kinematic inversion. We imagine the sheet of paper on which we are working to be fixed to the cam, and we allow the follower to appear to rotate *opposite to the actual direction of cam rotation.* As shown in Fig. 5.8, we divide the prime circle into a number of segments and assign station numbers to the boundaries of these segments. Dividing the displacement-diagram abscissa into corresponding segments, we transfer distances, by means of dividers, from the displacement diagram directly onto the cam layout to locate the corresponding positions of the trace point. The smooth curve through these points is the pitch curve. For the case of a roller-follower, as in this example, we simply draw the roller in its proper position at each station and then construct the cam profile as a smooth curve tangent to all these roller positions.

Figure 5.9 shows how the method of construction must be modified for an offset roller-follower. We begin by constructing an *offset circle,* using a radius equal to the amount of

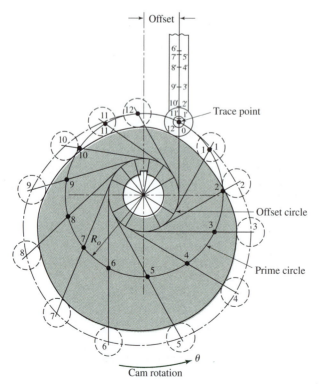

Figure 5.9 Graphical layout of a plate cam profile with an offset reciprocating roller follower.

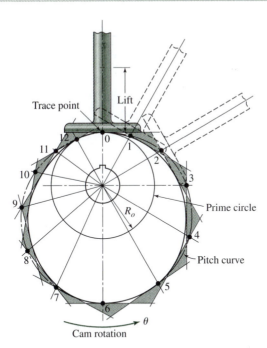

Figure 5.10 Graphical layout of a plate cam profile with a reciprocating flat-face follower.

the offset. After identifying station numbers around the prime circle, the centerline of the follower is constructed for each station, making it tangent to the offset circle. The roller centers for each station are now established by transferring distances from the displacement diagram directly to these follower centerlines, always measuring positive outward from the prime circle. An alternative procedure is to identify the points $0'$, $1'$, $2'$, and so on, on a single follower centerline and then to rotate them about the cam center to the corresponding follower centerline positions. In either case, the roller circles can be drawn next and a smooth curve tangent to all roller circles is the required cam profile.

Figure 5.10 shows the construction for a plate cam with a reciprocating flat-face follower. The pitch curve is constructed by using a method similar to that used for the roller follower in Fig. 5.8. A line representing the flat face of the follower is then constructed in each position. The cam profile is a smooth curve drawn tangent to all the follower positions. It may be helpful to extend each straight line representing a position of the follower face to form a series of triangles. If these triangles are lightly shaded, as suggested in the illustration, it may be easier to draw the cam profile inside all the shaded triangles and tangent to the inner sides of the triangles.

Figure 5.11 shows the layout of the profile of a plate cam with an oscillating roller follower. In this case we must rotate the fixed pivot center of the follower opposite the direction of cam rotation to develop the cam profile. To perform this inversion, first a circle is drawn about the camshaft center through the fixed pivot of the follower. This circle is then divided and given station numbers to correspond to the displacement diagram. Next arcs are drawn about each of these centers, all with equal radii corresponding to the length of the follower.

In the case of an oscillating follower, the ordinate values of the displacement diagram represent angular movements of the follower. If the vertical scale of the displacement

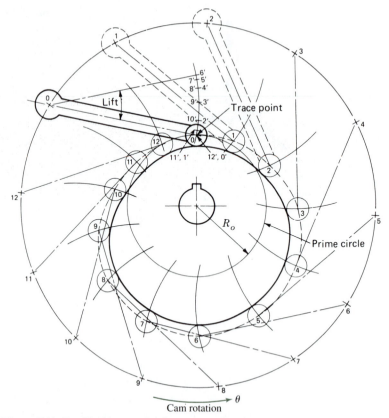

Figure 5.11 Graphical layout of a plate cam profile with an oscillating roller follower.

diagram is properly chosen initially, however, and if the total lift of the follower is a reasonably small angle, ordinate distances of the displacement diagram at each station can be transferred directly to the corresponding arc traveled by the roller by using dividers and measuring positive outward along the arc from the prime circle to locate the center of the roller for that station. Finally, circles representing the roller positions are drawn at each station, and the cam profile is constructed as a smooth curve tangent to each of these roller positions.

From the examples presented in this section, it should be clear that each different type of cam-and-follower system requires its own method of construction to determine the cam profile graphically from the displacement diagram. The examples presented here are not intended to be exhaustive of those possible, but they illustrate the general approach. They should also serve to illustrate and reinforce the discussion of the previous section; it should now be clear that much of the detailed shape of the cam itself results directly from the shape of the displacement diagram. Although different types of cams and followers have different shapes for the same displacement diagram, once a few parameters (such as prime-circle radius) are chosen to determine the size of a cam, the remainder of its shape results directly from the motion requirements specified in the displacement diagram.

5.5 KINEMATIC COEFFICIENTS OF THE FOLLOWER MOTION

We have seen that the displacement diagram is plotted with the follower motion y as the ordinate and the cam input motion θ as the abscissa no matter what the type of the cam or follower. The displacement diagram is, therefore, a graph representing some mathematical function relating the input and output motions of the cam system. In general terms, this relationship is

$$y = y(\theta) \tag{5.1}$$

Additional graphs can be plotted representing the derivatives of y with respect to the input position θ; that is, the kinematic coefficients of the follower motion. The first-order kinematic coefficient is denoted as

$$y'(\theta) = \frac{dy}{d\theta} \tag{5.2}$$

and represents the slope of the displacement diagram at each input position θ. The first-order kinematic coefficient, although it may now seem of little practical value, is a measure of the "steepness" of the displacement diagram. We will find in later sections that it is closely related to the mechanical advantage of the cam system and manifests itself in such things as the pressure angle (see Section 5.10). If we consider a wedge cam (see Fig. 5.1b) with a knife-edge follower, the displacement diagram itself is of the same shape as the corresponding cam. Here we can visualize that difficulties will occur if the cam is too steep—that is, if the first-order kinematic coefficient y' has too high a value.

The second-order kinematic coefficient (that is, the second derivative of y with respect to the input position θ) is also significant. The second-order kinematic coefficient is denoted as

$$y''(\theta) = \frac{d^2 y}{d\theta^2} \tag{5.3}$$

Although it is not quite as easy to visualize, the second-order kinematic coefficient is very closely related to the radius of curvature of the cam at various points along its profile. Because there is an inverse relationship (that is, as y'' becomes large, the radius of curvature becomes small), then if the second-order kinematic coefficient becomes infinite, the cam profile at such a position becomes pointed. This is a highly unsatisfactory condition from the point of view of contact stresses between the cam and the follower surfaces.

The third-order kinematic coefficient denoted as

$$y'''(\theta) = \frac{d^3 y}{d\theta^3} \tag{5.4}$$

can also be plotted if desired. Although it is not easy to describe geometrically, this is the rate of change of y'' with respect to the input position θ. We will see below that the third-order kinematic coefficient should also be controlled when choosing the detailed shape of the displacement diagram.

EXAMPLE 5.1

Derive equations to describe the displacement diagram of a cam that rises with parabolic motion from a dwell to another dwell such that the total lift is L and the total cam rotation angle is β. Plot the displacement diagram and the first-, second-, and third-order kinematic coefficients, with respect to cam rotation.

SOLUTION

As illustrated in Fig. 5.5a, two parabolas will be required, meeting at an inflection point taken here at midrange. For the first half of the motion we choose the general equation of a parabola; that is,

$$y = A\theta^2 + B\theta + C \tag{a}$$

The first three derivatives of Eq. (a), with respect to the input position θ, are

$$y' = 2A\theta + B \tag{b}$$

$$y'' = 2A \tag{c}$$

$$y''' = 0 \tag{d}$$

To match the position and the slope with those of the preceding dwell properly, at $\theta = 0$ we have $y(0) = y'(0) = 0$. Thus, Eqs. (a) and (b) show that $B = C = 0$. Looking next at the inflection point, at $\theta = \beta/2$ we want $y = L/2$; therefore, Eq. (a) gives

$$A = \frac{2L}{\beta^2}$$

Therefore, the displacement equation for the first half of the parabolic motion can be written as

$$y = 2L\left(\frac{\theta}{\beta}\right)^2 \tag{5.5}$$

Differentiating this equation with respect to the input θ, the first-, second-, and third-order kinematic coefficients, respectively, are

$$y' = \frac{4L}{\beta}\left(\frac{\theta}{\beta}\right) \tag{5.6}$$

$$y'' = \frac{4L}{\beta^2} \tag{5.7}$$

$$y''' = 0 \tag{5.8}$$

The maximum values for the first-order kinematic coefficient (that is, the maximum slope) occurs at the inflection point, where $\theta = \beta/2$. Substituting this value into Eq. (5.6), the maximum value for the first-order kinematic coefficient is

$$y'_{max} = \frac{2L}{\beta} \tag{5.9}$$

For the second half of the parabolic motion we return to the general equations (a) through (d) for a parabola. Substituting the conditions that, at $\theta = \beta$, $y = L$, and $y' = 0$ into Eqs. (a) and (b) gives

$$L = A\beta^2 + B\beta + C \tag{e}$$

$$0 = 2A\beta + B \tag{f}$$

Because the slope must match that of the first parabola at $\theta = \beta/2$, then from Eqs. (b) and (5.9) we have

$$\frac{2L}{\beta} = 2A\frac{\beta}{2} + B \tag{g}$$

Solving Eqs. (e), (f), and (g) simultaneously gives

$$A = -\frac{2L}{\beta^2}, \qquad B = \frac{4L}{\beta}, \qquad C = -L \tag{h}$$

Substituting these constraints into Eq. (a), the displacement equation for the second half of the parabolic motion can be written as

$$y = L\left[1 - 2\left(1 - \frac{\theta}{\beta}\right)^2\right] \tag{5.10}$$

Also, substituting Eqs. (h) into Eqs. (b), (c), and (d), the first-, second-, and third-order kinematic coefficients for the second half of the parabolic motion, respectively, are

$$y' = \frac{4L}{\beta}\left(1 - \frac{\theta}{\beta}\right) \tag{5.11}$$

$$y'' = -\frac{4L}{\beta^2} \tag{5.12}$$

$$y''' = 0 \tag{5.13}$$

The displacement diagram and the first three derivatives for this example of parabolic motion are shown in Fig. 5.12.

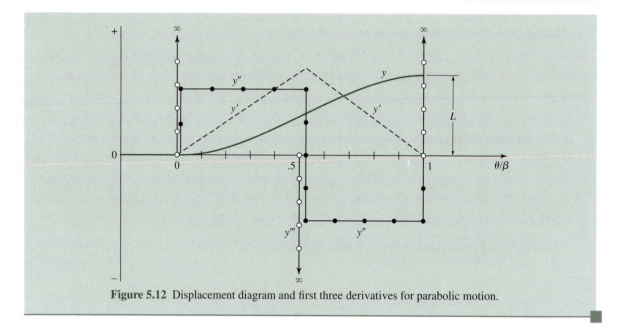

Figure 5.12 Displacement diagram and first three derivatives for parabolic motion.

The above discussion relates to the kinematic coefficients of the follower motion. These coefficients are derivatives with respect to the input position θ and relate to the geometry of the cam system. Let us now consider the derivatives of the follower motions with respect to time. First, we assume that the time history of the input motion $\theta(t)$ is known. The angular velocity $\omega = d\theta/dt$, the angular acceleration $\alpha = d^2\theta/dt^2$, and the next derivative (often called angular jerk or second angular acceleration) $\dot{\alpha} = d^3\theta/dt^3$ are also assumed to be known. Usually, a plate cam is driven by a constant-speed shaft. In this case, ω is a known constant, $\theta = \omega t$, and $\alpha = \dot{\alpha} = 0$. During start-up of the cam system, however, this is not the case, and we will consider the more general situation first.

From the general equation of the displacement diagram

$$y = y(\theta) \quad \text{and} \quad \theta = \theta(t)$$

We can therefore differentiate to find the time derivatives of the follower motion. The velocity of the follower, for example, is given by

$$\dot{y} = \frac{dy}{dt} = \left(\frac{dy}{d\theta}\right)\left(\frac{d\theta}{dt}\right) \tag{5.14a}$$

which can be written as

$$\dot{y} = y'\omega \tag{5.14b}$$

Similarly, the acceleration and the jerk of the follower can be written, respectively, as

$$\ddot{y} = \frac{d^2y}{dt^2} = y''\omega^2 + y'\alpha \tag{5.15}$$

and

$$\dddot{y} = \frac{d^3 y}{dt^3} = y'''\omega^3 + 3y''\omega\alpha + y'\dot{\alpha} \tag{5.16}$$

When the camshaft speed is constant, then Eqs. (5.14) through (5.16) reduce to

$$\dot{y} = y'\omega, \qquad \ddot{y} = y''\omega^2, \qquad \dddot{y} = y'''\omega^3 \tag{5.17}$$

For this reason, it has become somewhat common to refer to the graphs of the kinematic coefficients y', y'', y''', such as those shown in Fig. 5.12, as the "velocity," "acceleration," and "jerk" curves for a given motion. These would be appropriate names for a constant-speed cam only, and then only when scaled by ω, ω^2, and ω^3, respectively.* However, it is helpful to use these names for the kinematic coefficients when considering the physical implications of a certain choice of displacement diagram. For the parabolic motion of Fig. 5.12, for example, the "velocity" of the follower rises linearly to a maximum and then decreases to zero. The "acceleration" of the follower is zero during the initial dwell and changes abruptly to a constant positive value upon beginning the rise. There are two more abrupt changes in "acceleration" of the follower, at the midpoint and end of the rise. At each of the abrupt changes of "acceleration," the "jerk" of the follower becomes infinite.

5.6 HIGH-SPEED CAMS

Continuing with our discussion of parabolic motion, let us consider briefly the implications of the "acceleration" curve of Fig. 5.12 on the dynamic performance of the cam system. Any real follower must, of course, have some mass and, when multiplied by acceleration, will exert an inertia force (see Chapter 15). Therefore, the "acceleration" curve of Fig. 5.12 can also be thought of as indicating the inertia force of the follower, which, in turn, must be felt at the follower bearings and at the contact point with the cam surface. An "acceleration" curve with abrupt changes, such as parabolic motion, will exert abruptly changing contact stresses at the bearings and on the cam surface and lead to noise, surface wear, and eventual failure. Thus it is very important in choosing a displacement diagram to ensure that the first- and second-order kinematic coefficients (i.e., the "velocity" and "acceleration" curves) are continuous—that is, that they contain no step changes.

Sometimes in low-speed applications compromises are made with the velocity and acceleration relationships. It is sometimes simpler to employ a reverse procedure and design the cam shape first, obtaining the displacement diagram as a second step. Such cams are often composed of some combination of curves such as straight lines and circular arcs, which are readily produced by machine tools. Two examples are the *circle-arc cam* and the *tangent cam* of Fig. 5.13. The design approach is by iteration. A trial cam is designed and its kinematic characteristics computed. The process is then repeated until a cam with acceptable characteristics is obtained. Points A, B, C, and D of the circle-arc cam and the tangent cam are points of tangency or blending points. It is worth noting, as with the

---

*Accepting the word "velocity" literally, for example, leads to consternation when it is discovered that for a plate cam with a reciprocating follower, the units of y' are length per radian. Multiplying these units by radians per second, the units of ω, give units of length per second for $\dot{y}$, however.

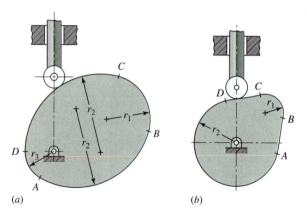

Figure 5.13 (*a*) Circle-arc cam. (*b*) Tangent cam.

parabolic-motion example before, that the acceleration changes abruptly at each of the blending points because of the instantaneous change in radius of curvature.

Although cams with discontinuous acceleration characteristics are sometimes found in low-speed applications, such cams are certain to exhibit major problems if the speed is raised. For any high-speed cam application, it is extremely important that not only the displacement and "velocity" curves but also the "acceleration" curve be made continuous for the entire motion cycle. No discontinuities should be allowed at the boundaries of different sections of the cam.

As shown by Eq. (5.17), the importance of continuous derivatives becomes more serious as the camshaft speed is increased. The higher the speed, the greater the need for smooth curves. At very high speeds it might also be desirable to require that jerk, which is related to rate of change of force, and perhaps even higher derivatives, be made continuous as well. In most applications, however, this is not necessary.

There is no simple answer as to how high a speed one must have before considering the application to require high-speed design techniques. The answer depends not only on the mass of the follower but also on the stiffness of the return spring, the materials used, the flexibility of the follower, and many other factors.[1] Further analysis techniques on cam dynamics are presented in Chapter 20. Still, with the methods presented below, it is not difficult to achieve continuous derivative displacement diagrams. Therefore, it is recommended that this be undertaken as standard practice. Parabolic-motion cams are no easier to manufacture than cycloidal-motion cams, for example, and there is no good reason for their use. The circle-arc cam and the tangent cam are easier to produce; but with modern machining methods, cutting of more complex cam shapes is not expensive.

5.7 STANDARD CAM MOTIONS

Example 5.1 gave a detailed derivation of the equations for parabolic motion and its derivatives. Then, in Section 5.6, reasons were provided for avoiding the use of parabolic motion in high-speed cam systems. The purpose of this section is to present equations for a number of standard types of displacement curves that can be used to address most high-speed cam-motion requirements. The derivations parallel those of Example 5.1 and are not presented in this text.

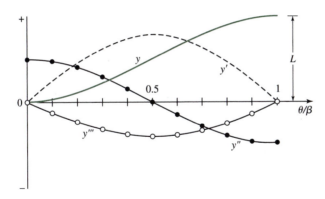

Figure 5.14 Displacement diagram and derivatives for full-rise simple harmonic motion, Eqs. (5.18).

The displacement equation and the first-, second-, and third-order kinematic coefficients for full-rise simple harmonic motion are

$$y = \frac{L}{2}\left(1 - \cos\frac{\pi\theta}{\beta}\right) \tag{5.18a}$$

$$y' = \frac{\pi L}{2\beta}\sin\frac{\pi\theta}{\beta} \tag{5.18b}$$

$$y'' = \frac{\pi^2 L}{2\beta^2}\cos\frac{\pi\theta}{\beta} \tag{5.18c}$$

$$y''' = -\frac{\pi^3 L}{2\beta^3}\sin\frac{\pi\theta}{\beta} \tag{5.18d}$$

The displacement diagram and the first-, second-, and third-order kinematic coefficients are shown in Fig. 5.14. Unlike parabolic motion, simple harmonic motion shows no discontinuity at the inflection point.

The displacement equation and the first-, second-, and third-order kinematic coefficients for full-rise cycloidal motion are

$$y = L\left(\frac{\theta}{\beta} - \frac{1}{2\pi}\sin\frac{2\pi\theta}{\beta}\right) \tag{5.19a}$$

$$y' = \frac{L}{\beta}\left(1 - \cos\frac{2\pi\theta}{\beta}\right) \tag{5.19b}$$

$$y'' = \frac{2\pi L}{\beta^2}\sin\frac{2\pi\theta}{\beta} \tag{5.19c}$$

$$y''' = \frac{4\pi^2 L}{\beta^3}\cos\frac{2\pi\theta}{\beta} \tag{5.19d}$$

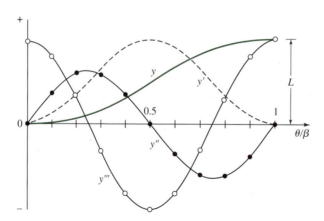

Figure 5.15 Displacement diagram and derivatives for full-rise cycloidal motion, Eqs. (5.19).

The displacement diagram and the first-, second-, and third-order kinematic coefficients are shown in Fig. 5.15.

Unlike the simple harmonic motion curves, the cycloidal rise motion has zero values at the beginning and end of its span, for all derivatives including the second-order kinematic coefficient (or "acceleration" curve). The peak "velocity," "acceleration," and "jerk" values, however, are higher than those for simple harmonic motion.

The displacement equation and the first-, second-, and third-order kinematic coefficients for a rise motion formed from an eighth-order polynomial are

$$y = L\left[6.09755\left(\frac{\theta}{\beta}\right)^3 - 20.78040\left(\frac{\theta}{\beta}\right)^5 + 26.73155\left(\frac{\theta}{\beta}\right)^6\right.$$
$$\left. - 13.60965\left(\frac{\theta}{\beta}\right)^7 + 2.56095\left(\frac{\theta}{\beta}\right)^8\right] \tag{5.20a}$$

$$y' = \frac{L}{\beta}\left[18.29265\left(\frac{\theta}{\beta}\right)^2 - 103.90200\left(\frac{\theta}{\beta}\right)^4 + 160.38930\left(\frac{\theta}{\beta}\right)^5\right.$$
$$\left. - 95.26755\left(\frac{\theta}{\beta}\right)^6 + 20.48760\left(\frac{\theta}{\beta}\right)^7\right] \tag{5.20b}$$

$$y'' = \frac{L}{\beta^2}\left[36.58530\left(\frac{\theta}{\beta}\right) - 415.60800\left(\frac{\theta}{\beta}\right)^3 + 801.94650\left(\frac{\theta}{\beta}\right)^4\right.$$
$$\left. - 571.60530\left(\frac{\theta}{\beta}\right)^5 + 143.41320\left(\frac{\theta}{\beta}\right)^6\right] \tag{5.20c}$$

$$y''' = \frac{L}{\beta^3}\left[36.58530 - 1246.82400\left(\frac{\theta}{\beta}\right)^2 + 3207.78600\left(\frac{\theta}{\beta}\right)^3\right.$$
$$\left. - 2858.02650\left(\frac{\theta}{\beta}\right)^4 + 860.47920\left(\frac{\theta}{\beta}\right)^5\right] \tag{5.20d}$$

The displacement diagram and the first-, second-, and third-order kinematic coefficients for a rise motion formed from this eighth-order polynomial are shown in Fig. 5.16. Equations (5.20) have seemingly strange coefficients for the polynomial because they have been

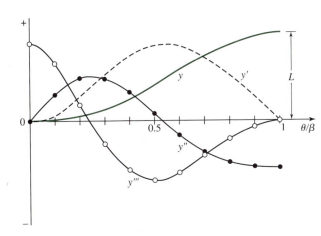

Figure 5.16 Displacement diagram and derivatives for full-rise eighth-order polynomial motion, Eqs. (5.20).

specially derived to have many "nice" properties.[2] Among these, Fig. 5.16 shows that several of the kinematic coefficients are zero at both ends of the range but that the "acceleration" characteristics are nonsymmetric while the peak values of "acceleration" are kept as small as possible.

Polynomial displacement equations of much higher order and meeting many more conditions than those presented here are also in common use. Automated procedures for determining the coefficients have been developed by Stoddart,[3] who also shows how the choice of coefficients can be made to compensate for elastic deformation of the follower system under dynamic conditions. Such cams are referred to as *polydyne cams*.

The displacement diagrams of simple harmonic, cycloidal, and eighth-order polynomial motion look quite similar at first glance. Each rises through a lift of L in a total cam rotation angle of β, and each begins and ends with a horizontal slope. For these reasons they are all referred to as *full-rise* motions. However, their "acceleration" curves are quite different. Simple harmonic motion has nonzero "acceleration" at the two ends of the range; cycloidal motion has zero "acceleration" at both boundaries; and eighth-order polynomial motion has one zero and one nonzero "acceleration" at its two ends. This variety provides the selection necessary when matching these curves with neighboring curves of different types.

Full-return motions of the same three types are shown in Figs. 5.17 through 5.19.

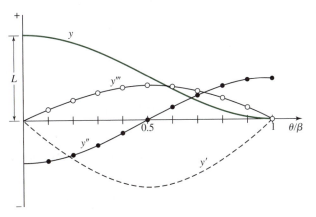

Figure 5.17 Displacement diagram and derivatives for full-return simple harmonic motion, Eqs. (5.21).

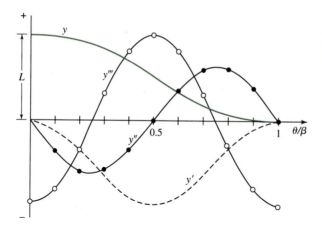

Figure 5.18 Displacement diagram and derivatives for full-return cycloidal motion, Eqs. (5.22).

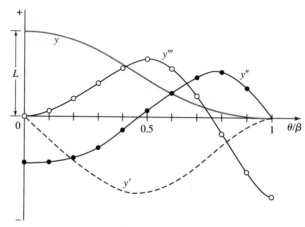

Figure 5.19 Displacement diagram and derivatives for the full-return eighth-order polynomial motion, Eqs. (5.23).

The displacement equation and the first-, second-, and third-order kinematic coefficients for full-return simple harmonic motion are

$$y = \frac{L}{2}\left(1 + \cos\frac{\pi\theta}{\beta}\right) \qquad (5.21a)$$

$$y' = -\frac{\pi L}{2\beta}\sin\frac{\pi\theta}{\beta} \qquad (5.21b)$$

$$y'' = -\frac{\pi^2 L}{2\beta^2}\cos\frac{\pi\theta}{\beta} \qquad (5.21c)$$

$$y''' = \frac{\pi^3 L}{2\beta^2}\sin\frac{\pi\theta}{\beta} \qquad (5.21d)$$

For full-return cycloidal motion, the displacement equation and the first-, second-, and third-order kinematic coefficients are

$$y = L\left(1 - \frac{\theta}{\beta} + \frac{1}{2\pi}\sin\frac{2\pi\theta}{\beta}\right) \qquad (5.22a)$$

$$y' = -\frac{L}{\beta}\left(1 - \cos\frac{2\pi\theta}{\beta}\right) \tag{5.22b}$$

$$y'' = -\frac{2\pi L}{\beta^2}\sin\frac{2\pi\theta}{\beta} \tag{5.22c}$$

$$y''' = -\frac{4\pi^2 L}{\beta^3}\cos\frac{2\pi\theta}{\beta} \tag{5.22d}$$

For the eighth-order-polynomial full-return motion, the displacement equation and the first-, second-, and third-order kinematic coefficients are

$$y = L\left[1.00000 - 2.63415\left(\frac{\theta}{\beta}\right)^2 + 2.78055\left(\frac{\theta}{\beta}\right)^5 + 3.17060\left(\frac{\theta}{\beta}\right)^6 \right.$$
$$\left. - 6.87795\left(\frac{\theta}{\beta}\right)^7 + 2.56095\left(\frac{\theta}{\beta}\right)^8\right] \tag{5.23a}$$

$$y' = -\frac{L}{\beta}\left[5.26830\frac{\theta}{\beta} - 13.90275\left(\frac{\theta}{\beta}\right)^4 - 19.02360\left(\frac{\theta}{\beta}\right)^5 \right.$$
$$\left. + 48.14565\left(\frac{\theta}{\beta}\right)^6 - 20.48760\left(\frac{\theta}{\beta}\right)^7\right] \tag{5.23b}$$

$$y'' = -\frac{L}{\beta^2}\left[5.26830 - 55.61100\left(\frac{\theta}{\beta}\right)^3 - 95.11800\left(\frac{\theta}{\beta}\right)^4 \right.$$
$$\left. + 288.87390\left(\frac{\theta}{\beta}\right)^5 - 143.41320\left(\frac{\theta}{\beta}\right)^6\right] \tag{5.23c}$$

$$y''' = \frac{L}{\beta^3}\left[166.83300\left(\frac{\theta}{\beta}\right)^2 + 380.47200\left(\frac{\theta}{\beta}\right)^3 - 1444.36950\left(\frac{\theta}{\beta}\right)^4 \right.$$
$$\left. + 860.47920\left(\frac{\theta}{\beta}\right)^5\right] \tag{5.23d}$$

In addition to the full-rise and full-return motions presented above, it is often useful to have a selection of standard *half-rise* or *half-return* motions available. These are curves for which one boundary has a nonzero slope and can be used to blend with uniform motion. The displacement diagrams and the first-, second-, and third-order kinematic coefficients for simple harmonic half-rise motions, sometimes called *half-harmonics,* are shown in Fig. 5.20. The equations corresponding to Fig. 5.20a are

$$y = L\left(1 - \cos\frac{\pi\theta}{2\beta}\right) \tag{5.24a}$$

$$y' = \frac{\pi L}{2\beta}\sin\frac{\pi\theta}{2\beta} \tag{5.24b}$$

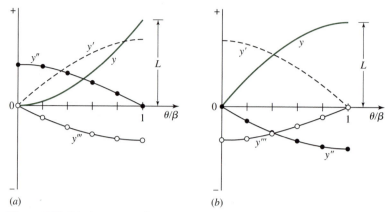

(a)   (b)

Figure 5.20 Displacement diagram and derivatives for half-harmonic rise motions: (a) Eqs. (5.24); (b) Eqs. (5.25).

$$y'' = \frac{\pi^2 L}{4\beta^2} \cos \frac{\pi\theta}{2\beta} \tag{5.24c}$$

$$y''' = -\frac{\pi^3 L}{8\beta^3} \sin \frac{\pi\theta}{2\beta} \tag{5.24d}$$

The displacement equation and the first-, second-, and third-order kinematic coefficients corresponding to the half-harmonics of Fig. 5.20b are

$$y = L \sin \frac{\pi\theta}{2\beta} \tag{5.25a}$$

$$y' = \frac{\pi L}{2\beta} \cos \frac{\pi\theta}{2\beta} \tag{5.25b}$$

$$y'' = -\frac{\pi^2 L}{4\beta^2} \sin \frac{\pi\theta}{2\beta} \tag{5.25c}$$

$$y''' = -\frac{\pi^3 L}{8\beta^3} \cos \frac{\pi\theta}{2\beta} \tag{5.25d}$$

The half-harmonic curves for half-return motions are shown in Fig. 5.21. The equations corresponding to Fig. 5.21a are

$$y = L \cos \frac{\pi\theta}{2\beta} \tag{5.26a}$$

$$y' = -\frac{\pi L}{2\beta} \sin \frac{\pi\theta}{2\beta} \tag{5.26b}$$

$$y'' = -\frac{\pi^2 L}{4\beta^2} \cos \frac{\pi\theta}{2\beta} \tag{5.26c}$$

$$y''' = \frac{\pi^3 L}{8\beta^3} \sin \frac{\pi\theta}{2\beta} \tag{5.26d}$$

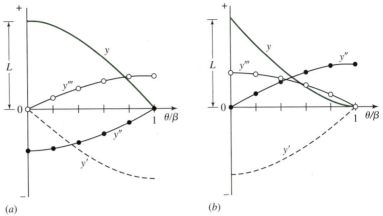

Figure 5.21 Displacement diagram and derivatives for half-harmonic return motions: (*a*) Eqs. (5.26); (*b*) Eqs. (5.27).

The equations corresponding to Fig. 5.21*b* are

$$y = L \left(1 - \sin \frac{\pi \theta}{2\beta} \right) \tag{5.27a}$$

$$y' = -\frac{\pi L}{2\beta} \cos \frac{\pi \theta}{2\beta} \tag{5.27b}$$

$$y'' = \frac{\pi^2 L}{4\beta^2} \sin \frac{\pi \theta}{2\beta} \tag{5.27c}$$

$$y''' = \frac{\pi^3 L}{8\beta^3} \cos \frac{\pi \theta}{2\beta} \tag{5.27d}$$

Besides the half-harmonics, half-cycloidal motions are also useful, because their "accelerations" are zero at both boundaries. The displacement diagrams and first-, second-, and third-order kinematic coefficients for half-cycloidal half-rise motions are shown in Fig. 5.22. The equations corresponding to Fig. 5.22*a* are

$$y = L \left(\frac{\theta}{\beta} - \frac{1}{\pi} \sin \frac{\pi \theta}{\beta} \right) \tag{5.28a}$$

$$y' = \frac{L}{\beta} \left(1 - \cos \frac{\pi \theta}{\beta} \right) \tag{5.28b}$$

$$y'' = \frac{\pi L}{\beta^2} \sin \frac{\pi \theta}{\beta} \tag{5.28c}$$

$$y''' = \frac{\pi^2 L}{\beta^3} \cos \frac{\pi \theta}{\beta} \tag{5.28d}$$

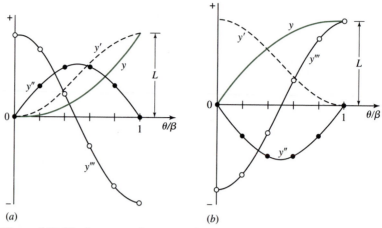

Figure 5.22 Displacement diagram and derivatives for half-cycloidal rise motions: (*a*) Eqs. (5.28); (*b*) Eqs. (5.29).

The equations corresponding to Fig. 5.22*b* are

$$y = L\left(\frac{\theta}{\beta} + \frac{1}{\pi}\sin\frac{\pi\theta}{\beta}\right) \tag{5.29a}$$

$$y' = \frac{L}{\beta}\left(1 + \cos\frac{\pi\theta}{\beta}\right) \tag{5.29b}$$

$$y'' = -\frac{\pi L}{\beta^2}\sin\frac{\pi\theta}{\beta} \tag{5.29c}$$

$$y''' = -\frac{\pi^2 L}{\beta^3}\cos\frac{\pi\theta}{\beta} \tag{5.29d}$$

The half-cycloidal curves for half-return motions are shown in Fig. 5.23. The equations corresponding to Fig. 5.23*a* are

$$y = L\left(1 - \frac{\theta}{\beta} + \frac{1}{\pi}\sin\frac{\pi\theta}{\beta}\right) \tag{5.30a}$$

$$y' = -\frac{L}{\beta}\left(1 - \cos\frac{\pi\theta}{\beta}\right) \tag{5.30b}$$

$$y'' = -\frac{\pi L}{\beta^2}\sin\frac{\pi\theta}{\beta} \tag{5.30c}$$

$$y''' = -\frac{\pi^2 L}{\beta^3}\cos\frac{\pi\theta}{\beta} \tag{5.30d}$$

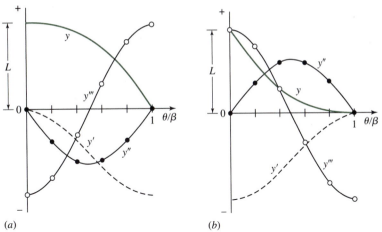

(a)   (b)

Figure 5.23 Displacement diagram and derivatives for half-cycloidal return motions: (a) Eqs. (5.30); (b) Eqs. (5.31).

The equations corresponding to Fig. 5.23b are

$$y = L\left(1 - \frac{\theta}{\beta} - \frac{1}{\pi}\sin\frac{\pi\theta}{\beta}\right) \tag{5.31a}$$

$$y' = -\frac{L}{\beta}\left(1 + \cos\frac{\pi\theta}{\beta}\right) \tag{5.31b}$$

$$y'' = \frac{\pi L}{\beta^2}\sin\frac{\pi\theta}{\beta} \tag{5.31c}$$

$$y''' = \frac{\pi^2 L}{\beta^3}\cos\frac{\pi\theta}{\beta} \tag{5.31d}$$

We shall soon see how the graphs and equations presented in this section can greatly reduce the analytical effort involved in designing the full displacement diagram for a high-speed cam. First, however, we should note several important features of the graphs of Figs. 5.14 through 5.23.

Each graph shows only one section of a full displacement diagram; the total lift for that section is labeled L for each, and the total cam travel is labeled β. The abscissa of each graph is normalized so that the ratio (θ/β) ranges from zero at the left end to unity at the right end $(\theta = \beta)$.

The scales used in plotting the graphs are not shown but are consistent for all full-rise and full-return curves and for all half-rise and half-return curves. Thus, in judging the suitability of one curve compared with another, the magnitudes of the "accelerations," for example, can be compared. For this reason, when other factors are equivalent, simple harmonic motion should be used when possible in order to keep "accelerations" small.

Finally, it should be noted that the standard cam motions presented in this section do not form an exhaustive set; cams with good dynamic characteristics can also be formed from a wide variety of other possible motion curves. The set presented here, however, is sufficiently complete for most practical applications.

5.8 MATCHING DERIVATIVES OF DISPLACEMENT DIAGRAMS

In the previous section, a great many equations were presented which might be used to represent the different segments of the displacement diagram of a cam. In this section we will study how they can be joined together to form the motion specification for a complete cam. The procedure is one of solving for proper values of L and β for each segment so that:

1. The motion requirements of the particular application are met.
2. The displacement diagram, as well as the diagrams of the first- and second-order kinematic coefficients, is continuous across the boundaries of the segments. The diagram of the third-order kinematic coefficient may be allowed discontinuities if necessary, but must not become infinite; that is, the "acceleration" curve may contain corners but not discontinuities.
3. The maximum magnitudes of the "velocity" and "acceleration" peaks are kept as low as possible consistent with the above conditions.

The procedure may best be understood through an example.

EXAMPLE 5.2

A plate cam with a reciprocating follower is to be driven by a constant-speed motor at 150 rev/min. The follower is to start from a dwell, accelerate to a uniform velocity of 25 in/s, maintain this velocity for 1.25 in of rise, decelerate to the top of the lift, return, and then dwell for 0.10 s. The total lift is to be 3.0 in. Determine the complete specifications of the displacement diagram.

SOLUTION

The speed of the input shaft is

$$\omega = 150 \text{ rev/min} = 15.70796 \text{ rad/s} \tag{1}$$

Using Eq. (5.14b), the first-order kinematic coefficient (that is, the slope of the uniform velocity segment) is

$$y' = \frac{\dot{y}}{\omega} = \frac{25 \text{ in/s}}{15.70796 \text{ rad/s}} = 1.59155 \text{ in/rad} \tag{2}$$

Because this is held constant for 1.25 in of rise, the cam rotation in this segment is*

$$\beta_2 = \frac{L_2}{y'} = \frac{1.25 \text{ in}}{1.59155 \text{ in/rad}} = 0.78540 \text{ rad} = 45.00° \tag{3}$$

Similarly, from Eq. (1), the cam rotation during the final dwell is

$$\beta_5 = 0.10 \text{ s} (15.70796 \text{ rad/s}) = 1.570796 \text{ rad} = 90.00° \tag{4}$$

---

*Note that several digits of accuracy higher than usual are used in this example. Any inaccuracy in the L and β values results in discontinuities in the smoothness of derivatives at the junctures of the segments and discontinuities in force as previously explained.

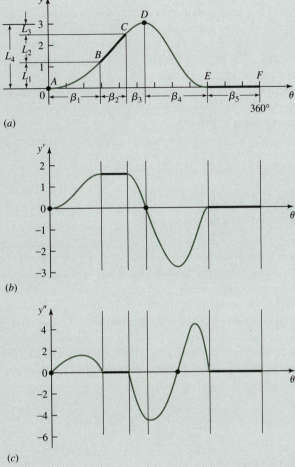

Figure 5.24 Example 5.2: (*a*) displacement diagram, in;
(*b*) velocity diagram, in/rad; (*c*) acceleration diagram, in/rad^2.

From these results and the given information, we can sketch the beginnings of the displacement diagram, not necessarily working to scale, but in order to visualize the motion requirements. This gives the general shapes shown by the heavy curves of Fig. 5.24*a*. The lighter sections of the displacement curve are not yet accurately known, but can be sketched by lightly outlining a smooth curve for visualization. Working from this curve, we can also sketch the general nature of the derivative curves. From the slope of the displacement diagram we sketch the "velocity" curve (see Fig. 5.24*b*), and from the slope of this curve we sketch the "acceleration" curve (see Fig. 5.24*c*). At this time no attempt is made to produce accurate curves drawn to scale, only to provide an idea of the desired curve shapes.

Now using the sketches of Fig. 5.24, we compare the desired motion curve with the various standard curves of Figs. 5.14 through 5.23 in order to choose an appropriate set of equations for each segment of the cam. In the segment AB, for example, we find that Fig. 5.22a is the only standard motion curve available with half-rise characteristics, an appropriate slope curve, and the necessary zero "acceleration" at both ends of the segment. Thus we choose the half-cycloidal rise motion of Eq. (5.28) for that portion of the cam. There are two sets of choices possible for the segments CD and DE. One set might be the choice of Fig. 5.22b matched with Fig. 5.18. However, in order to keep the peak "accelerations" low and to keep the "jerk" curves as smooth as possible, we will choose Fig. 5.20b matched with Fig. 5.19. Thus for segment CD we will use the half-harmonic rise motion of Eq. (5.25), and for segment DE we will choose the eighth-order-polynomial return motion of Eq. (5.23).

Choosing the motion curve types, however, is not sufficient to fully specify the motion characteristics. We must also find values for the unknown parameters of the motion equations; these are L_1, L_3, β_1, β_3, and β_4. We do this by equating the kinematic coefficients at each nonzero boundary. For example, to match the "velocities" at point B we must equate the first-order kinematic coefficient from Eq. (5.28b) at $\theta/\beta = 1$ (that is, at the right end) with the first-order kinematic coefficient of the BC segment; that is,

$$y_B' = \frac{2L_1}{\beta_1} = \frac{L_2}{\beta_2} = \frac{1.25 \text{ in}}{0.78540 \text{ rad}} = 1.59155 \text{ in/rad}$$

or

$$L_1 = 0.79577\beta_1 \tag{5}$$

Similarly, to match the "velocities" at point C, we equate the first-order kinematic coefficient of the BC segment with the first-order kinematic coefficient of Eq. (5.25b) at $\theta/\beta = 0$ (that is, at its left end). This gives

$$y_C' = \frac{L_2}{\beta_2} = \frac{\pi L_3}{2\beta_3} = 1.59155 \text{ in/rad}$$

or

$$L_3 = 1.01321\beta_3 \tag{6}$$

In order to match the "accelerations" (that is, the curvatures) at point D, we equate the second-order kinematic coefficient from Eq. (5.25c) at $\theta/\beta = 1$ (that is, at its right end) with the second-order kinematic coefficient from Eq. (5.23c) at $\theta/\beta = 0$ (that is, at its left end). This gives

$$y_D'' = -\frac{\pi^2 L_3}{4\beta_3^2} = -5.26830\frac{L_4}{\beta_4^2}$$

where $L_4 = 3$ in is the total lift. Substituting Eq. (6) and the total lift into this result, along with rearranging, gives

$$\beta_3 = 0.15818\beta_4^2 \tag{7}$$

Finally, for geometric compatibility, we have the constraints

$$L_1 + L_3 = L_4 - L_2 = 1.750 \text{ in} \tag{8}$$

and

$$\beta_1 + \beta_3 + \beta_4 = 2\pi - \beta_2 - \beta_5 = 3.92699 \text{ rad} \tag{9}$$

Solving the five equations—that is, Eqs. (5) through (9)—simultaneously for the five unknowns L_1, L_3, β_1, β_3, and β_4 provides the proper values of the remaining parameters. In summary, the results are

$$
\begin{aligned}
L_1 &= 1.1831 \text{ in} & \beta_1 &= 1.48674 \text{ rad} = 85.184° \\
L_2 &= 1.2500 \text{ in} & \beta_2 &= 0.78540 \text{ rad} = 45.000° \\
L_3 &= 0.5669 \text{ in} & \beta_3 &= 0.55951 \text{ rad} = 32.058° \\
L_4 &= 3.0000 \text{ in} & \beta_4 &= 1.88074 \text{ rad} = 107.758° \\
L_5 &= 0.0000 \text{ in} & \beta_5 &= 1.57080 \text{ rad} = 90.000°
\end{aligned}
$$

At this time an accurate layout of the displacement diagram and, if desired, the kinematic coefficients can be made to replace the original sketches. The curves of Fig. 5.24 have been drawn to scale using these values.

5.9 PLATE CAM WITH RECIPROCATING FLAT-FACE FOLLOWER

Once the displacement diagram of a cam system has been completely determined, as described in Section 5.8, the layout of the actual cam shape can be made, as shown in Section 5.4. In laying out the cam, however, we recall the need for a few more parameters, depending on the type of cam and follower—for example, the prime-circle radius, any offset distance, roller radius, and so on. Also, as we will see, each different type of cam can be subject to certain further problems unless these remaining parameters are properly chosen.

In this section we study the problems that may be encountered in the design of a plate cam with a reciprocating flat-face follower. The geometric parameters of such a system that may yet be chosen are the prime-circle radius R_0, the offset (or eccentricity) ε of the follower stem, and the minimum width of the follower face.

Figure 5.25 shows the layout of a plate cam with a radial reciprocating flat-face follower. In this case the displacement chosen was a cycloidal rise of $L = 100$ mm during $\beta_1 = 90°$ of cam rotation, followed by a cycloidal return during the remaining $\beta_2 = 270°$ of cam rotation. The layout procedure of Fig. 5.10 was followed to develop the cam shape, and a prime-circle radius of $R_0 = 25$ mm was used. Obviously, there is a problem because the cam profile intersects itself. In machining, part of the cam shape would be lost and during operation the intended cycloidal motion would not be achieved. Such a cam is said to be *undercut*.

Why did undercutting occur in this example, and how can it be avoided? It resulted from attempting to achieve too great a lift in too little cam rotation with too small a cam.

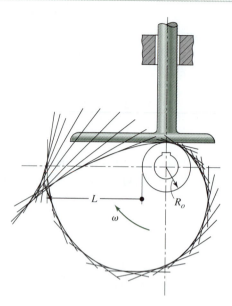

Figure 5.25 Undercut plate-cam profile layout with flat-face follower.

One possibility is to decrease the desired lift L or to increase the cam rotation angle β_1 to avoid the problem. However, this is not possible while still achieving the original design objectives. Another solution is to use the same displacement characteristics but to increase the prime-circle radius R_0 to avoid undercutting. This does produce a larger cam, but with sufficient increase it will overcome the undercutting problem.

The minimum increase in R_0 to eliminate undercutting can be found by developing an equation for the radius of curvature of the cam profile. We start by writing the loop-closure equation using the vectors shown in Fig. 5.26. Using complex polar notation, the loop-closure equation is

$$r e^{j(\theta+\alpha)} + j\rho = j(R_0 + y) + s \tag{a}$$

Here we have carefully chosen the vectors so that point C is the instantaneous center of curvature and ρ is the radius of curvature corresponding to the current contact point. The line along the vector $\mathbf{u}$ which separates the angles θ and α is fixed on the cam and is horizontal for the cam position $\theta = 0$.

Separating Eq. (a) into real and imaginary parts, respectively, gives

$$r \cos(\theta + \alpha) = s \tag{b}$$

$$r \sin(\theta + \alpha) + \rho = R_0 + y \tag{c}$$

Because the center of curvature C is stationary on the surface of the cam, the magnitudes of r, α, and ρ do not change for small changes in cam rotation*; that is,

$$\frac{dr}{d\theta} = \frac{d\alpha}{d\theta} = \frac{d\rho}{d\theta} = 0$$

---

*The values of r, α, and ρ are not constant but are currently at stationary values; their higher derivatives are nonzero.

Figure 5.26

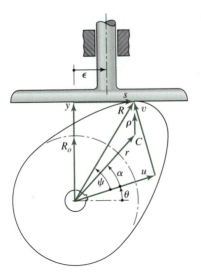

Therefore, differentiating Eq. (a) with respect to θ gives

$$jre^{j(\theta+\alpha)} = jy' + s' \tag{d}$$

where the first-order kinematic coefficient $s' = ds/d\theta$. Separating Eq. (d) into real and imaginary parts, the first-order kinematic coefficients are

$$s' = -r\sin(\theta + \alpha) \tag{e}$$

$$y' = r\cos(\theta + \alpha) \tag{f}$$

Equating Eqs. (b) and (f), the location of the trace point along the surface of the follower can be written as

$$s = y' \tag{5.32}$$

Differentiating this equation with respect to the cam rotation angle θ, we find its first-order kinematic coefficient

$$s' = y'' \tag{g}$$

Substituting Eq. (g) into Eq. (e) and then substituting the result into Eq. (c), the radius of curvature of the cam profile can be written as

$$\rho = R_0 + y + y'' \tag{5.33}$$

We should note carefully the importance of Eq. (5.33); it states that the radius of curvature of the cam can be obtained for each value of the cam rotation angle θ directly from the displacement equations, *without laying out the cam profile*. All that is needed is the radius of the prime circle R_0 and values for the displacement y and the second-order kinematic coefficient y''.

We can use Eq. (5.33) to select a value for R_0 which will avoid undercutting. When undercutting occurs, the radius of curvature of the cam profile switches sign from positive to negative. If we are on the verge of undercutting, the cam will come to a point and the radius of curvature will be zero for some value of the input angle θ. We can choose R_0 large enough that this is never the case. In fact, to avoid high contact stresses, we may wish to ensure that ρ is everywhere larger than some specified value ρ_{min}. Then, from Eq. (5.33), we must require that

$$\rho = R_0 + y + y'' > \rho_{min}$$

Because R_0 and y are always positive, the critical situation occurs where the second-order kinematic coefficient y'' has its largest negative value. Denoting this minimum value of y'' as y''_{min} and remembering that y corresponds to the same position, defined by cam angle θ, we have the condition

$$R_0 > \rho_{min} - y - y''_{min} \tag{5.34}$$

which must be satisfied. This can easily be checked once the displacement equations have been established, and an appropriate value of R_0 can be chosen before the cam layout is attempted.

Returning now to Eq. (5.32), we see from Fig. 5.26 that this can also be of value. The equation states that the distance of travel of the point of contact on either side of the cam rotation center corresponds precisely with the plot of the first-order kinematic coefficient. Thus the minimum face width for the flat-face follower must extend at least y'_{max} to the right and $-y'_{min}$ to the left of the cam center to maintain contact; that is,

$$\text{Face width} > y'_{max} - y'_{min} \tag{5.35}$$

EXAMPLE 5.3

Assuming that the displacement characteristics found in Example 5.2 are to be achieved by a plate cam with a reciprocating flat-face follower, determine the minimum face width and the minimum prime-circle radius to ensure that the radius of curvature of the cam is everywhere greater than 0.25 in.

SOLUTION

From Fig. 5.24b we see that the maximum "velocity" (that is, the maximum value of the first-order kinematic coefficient) occurs in the segment BC and is

$$y'_{max} = \frac{L_2}{\beta_2} = \frac{1.25 \text{ in}}{0.785 \text{ rad}} = 1.59 \text{ in/rad} \tag{1}$$

The minimum "velocity" occurs in segment DE at approximately $\theta/\beta_4 = 0.5$. From Eq. (5.23b), the minimum value of the first-order kinematic coefficient is

$$y'_{min} = -2.81 \text{ in/rad} \tag{2}$$

Substituting Eqs. (1) and (2) into Eq. (5.35), the minimum face width is

$$\text{Face width} > 1.59 + 2.81 = 4.40 \text{ in} \qquad\qquad \textit{Ans.}$$

Therefore, the follower would be positioned with 1.59 in to the right and 2.81 in to the left of the cam rotation axis, and some appropriate additional allowance would be added on each side.

The maximum negative "acceleration" occurs at point D. The minimum value of the second-order kinematic coefficients can be obtained from Eq. (5.25c) at $\theta/\beta_3 = 1$; that is,

$$y''_{\min} = -\frac{\pi^2 L_3}{4\beta_3^2} = -\frac{\pi^2(0.567 \text{ in})}{4(0.559 \text{ rad})^2} = -4.47 \text{ in/rad}^2$$

Substituting this result and the known parameters into Eq. (5.34), the minimum prime-circle radius is

$$R_0 > 0.25 - (-4.47) - 3.00 = 1.72 \text{ in} \qquad\qquad \textit{Ans.}$$

From this calculation we would choose the actual prime-circle radius as, say, $R_0 = 1.75$ in.

We can see that the eccentricity of the flat-face follower stem does not affect the geometry of the cam. This eccentricity is usually chosen to relieve high bending stresses in the follower.

Looking again at Fig. 5.26, we can write another loop-closure equation; that is,

$$ue^{j\theta} + ve^{j(\theta+\pi/2)} = j(R_0 + y) + s$$

where u and v denote the coordinates of the contact point in a coordinate system attached to the cam. Dividing this equation by $e^{j\theta}$ gives

$$u + jv = j(R_0 + y)e^{-j\theta} + se^{-j\theta}$$

Using Eq. (5.32), the real and imaginary parts of this equation can be written as

$$u = (R_0 + y)\sin\theta + y'\cos\theta \qquad\qquad (5.36a)$$

$$v = (R_0 + y)\cos\theta - y'\sin\theta \qquad\qquad (5.36b)$$

These two equations give the coordinates of the cam profile and provide an alternative to the graphical layout procedure of Fig. 5.10. They can be used to generate a table of numeric rectangular coordinate data from which the cam can be machined. Polar-coordinate equations for this same curve are

$$R = \sqrt{(R_0 + y)^2 + (y')^2} \qquad\qquad (5.37a)$$

and

$$\psi = \left(\frac{\pi}{2} - \theta\right) - \tan^{-1}\frac{y'}{R_0 + y} \qquad\qquad (5.37b)$$

5.10 PLATE CAM WITH RECIPROCATING ROLLER FOLLOWER

Figure 5.27 shows a plate cam with a reciprocating roller follower. We see that three geometric parameters remain to be chosen after the displacement diagram is completed and before the cam layout can be accomplished. These three parameters are: the radius of the prime circle R_0, the eccentricity ε, and the radius of the roller R_r. There are also two potential problems to be considered when choosing these parameters. One problem is undercutting and the other is an improper pressure angle.

Pressure angle is the name used for the angle between the axis of the follower stem and the line of the force exerted by the cam onto the roller follower, the normal to the pitch curve through the trace point. The pressure angle is labeled ϕ in Figure 5.27. Only the component of force along the line of motion of the follower is useful in overcoming the output load; the perpendicular component should be kept low to reduce sliding friction between the follower and its guideway. Too high a pressure angle increases the deleterious effect of friction and may cause the translating follower to chatter or perhaps even to jam. Cam pressure angles of up to about $30°$ to $35°$ are about the largest that can be used without causing difficulties.

In Fig. 5.27 we see that the normal to the pitch curve intersects the horizontal axis at point P_{24}—that is, at the instantaneous center of velocity between cam 2 and follower 4. Because the follower is translating, all points of the follower have velocities equal to that of P_{24}. This velocity must also be equal to the velocity of the coincident point of link 2; that is,

$$V_{P_{24}} = \dot{y} = \omega R_{P_{24}O_2}$$

Dividing this equation by ω, the first-order kinematic coefficient is

$$y' = \frac{\dot{y}}{\omega} = R_{P_{24}O_2}$$

Figure 5.27

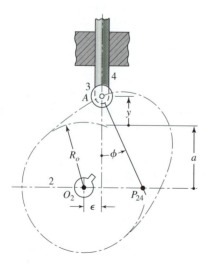

This coefficient can also be expressed in terms of the eccentricity and the pressure angle as

$$y' = \varepsilon + (a + y)\tan\phi \qquad (a)$$

where, as shown in Fig. 5.27, the vertical distance from the cam axis to the prime circle is

$$a = \sqrt{R_o^2 - \varepsilon^2} \qquad (b)$$

Substituting this into Eq. (a) and rearranging, the pressure angle can be written as

$$\phi = \tan^{-1}\left(\frac{y' - \varepsilon}{\sqrt{R_0^2 - \varepsilon^2} + y}\right) \qquad (5.38)$$

From this equation we observe that, after the displacement equations and the first-order kinematic coefficients have been determined, the two parameters, R_0 and ε, can be adjusted to obtain a suitable pressure angle. We also notice that the pressure angle is continuously changing as the cam rotates, and therefore we are interested in studying its extreme values.

Let us first consider the effect of the eccentricity. From Eq. (5.38), we observe that increasing ε either increases or decreases the magnitude of the numerator, depending on the sign of the first-order kinematic coefficient y'. Thus a small eccentricity ε can be used to reduce the pressure angle ϕ during the rise motion when y' is positive, but only at the expense of an increased pressure angle during the return motion when y' is negative. Still, because the magnitudes of the forces are usually greater during rise, it is common practice to offset the follower to take advantage of this reduction in pressure angle.

A much more significant effect can be made in reducing the pressure angle by increasing the prime-circle radius R_0. In order to study this effect, let us take the conservative approach and assume that there is no eccentricity (that is, that $\varepsilon = 0$). Equation (5.38) then reduces to

$$\phi = \tan^{-1}\left(\frac{y'}{R_0 + y}\right) \qquad (5.39)$$

To find the extremum values of the pressure angle, it is possible to differentiate this equation with respect to the cam rotation angle and equate it to zero, thus finding the values of the rotation angle θ that yield the maximum and minimum pressure angles. This is a tedious mathematical process, however, and can be avoided by using the nomogram of Fig. 5.28. This nomogram was produced by searching out on a digital computer the maximum value of ϕ from Eq. (5.39) for each of the standard full-rise motion curves of Section 5.7. With the nomogram it is possible to use the known values of L and β for each segment of the displacement diagram and to read directly the maximum pressure angle occurring in that segment for a particular choice of R_0. Alternatively, a desired maximum pressure angle can be chosen and an appropriate value of R_0 can be found. The process is best illustrated by an example.

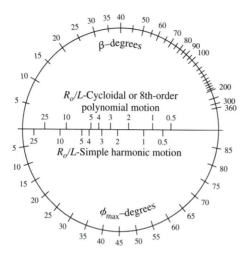

Figure 5.28 Nomogram relating maximum pressure angle ϕ_{max} to prime-circle radius R_0, lift L, and active cam angle β for radial roller/follower cams with simple harmonic, cycloidal, and eighth-order-polynomial full-rise or full-return motion.

EXAMPLE 5.4

Assuming that the displacement characteristics of Example 5.2 are to be achieved by a plate cam with a reciprocating radial roller follower, determine the minimum prime-circle radius which ensures that the pressure angle is everywhere less than 30°.

SOLUTION

Each segment of the displacement diagram can be checked in succession using the nomogram of Fig. 5.28.

For segment AB of Fig. 5.24, we have half-cycloidal motion with $L_1 = 1.184$ in and $\beta_1 = 85.18°$. Because this is a half-rise curve, whereas Fig. 5.28 is only good for full-rise curves, it is necessary to double both L_1 and β_1, thus pretending that the curve is full-rise. This gives $L = 2.37$ in and $\beta = 170°$. Next, connecting a straight line from $\beta = 170°$ to $\phi_{max} = 30°$, we read from the upper scale on the center axis of the nomogram a value of $R_0/L = 0.75$, from which

$$R_0 = 0.75(2.37 \text{ in}) = 1.78 \text{ in}$$

The segment BC need not be checked, because its maximum pressure angle will occur at the boundary B and cannot be greater than that for segment AB.

The segment CD has half-harmonic motion with $L_3 = 0.566$ in and $\beta_3 = 32.05°$. Again, because this is a half-rise curve, these values are doubled and $L = 1.13$ in and $\beta = 64°$ are used instead. Then, from the nomogram, we find $R_0/L = 2.45$, from which

$$R_0 = 2.45(1.13 \text{ in}) = 2.77 \text{ in}$$

Here we must be careful, however. This value is the radius of a fictitious prime circle for which the horizontal axis of our doubled "full-rise" harmonic curve would have $y = 0$. It is not the R_0

we seek, because our full-harmonic curve has a nonzero y value at its base of

$$y = 3.00 - 1.13 = 1.87 \text{ in}$$

The appropriate value for R_0 for this situation is

$$R_0 = 2.77 - 1.87 = 0.90 \text{ in}$$

Next we check the segment DE, which has eighth-order polynomial motion with $L_4 = 3.00$ in and $\beta_4 = 108°$. Because this is a full-return motion curve with $y = 0$ at its base, no adjustments are necessary for use of the nomogram. We find $R_0/L = 1.3$ and

$$R_0 = 1.3(3.00 \text{ in}) = 3.90 \text{ in}$$

In order to ensure that the pressure angle does not exceed $30°$ throughout all segments of the cam, we must chose the prime-circle radius to be at least as large as the maximum of these predicted values. Remembering the inability to read the nomogram with great precision, we might choose a larger value, such as

$$R_0 = 4.00 \text{ in} \qquad\qquad\qquad Ans.$$

Now that a final value has been chosen, we can use Fig. 5.28 again to find the actual maximum pressure angle in each segment of the motion:

AB:
$$\frac{R_0}{L} = \frac{4.00}{2.37} = 1.69, \qquad \phi_{max} = 18°$$

CD:
$$\frac{R_0}{L} = \frac{5.87}{1.13} = 5.19, \qquad \phi_{max} = 14°$$

DE:
$$\frac{R_0}{L} = \frac{4.00}{3.00} = 1.33, \qquad \phi_{max} = 29°$$

Even though the prime circle has been sized to give a satisfactory pressure angle, the follower may still not complete the desired motion; if the curvature of the pitch curve is too sharp, the cam profile may be "undercut." Figure 5.29a shows a portion of a cam pitch curve and two cam profiles generated by two different-size rollers. The cam profile generated by the larger roller is undercut and intersects itself. The result, after machining, is a pointed cam that does not produce the desired motion. It is also clear from the same figure that a smaller roller moving on the same pitch curve generates a satisfactory cam profile. Similarly, if the prime circle and thus the cam size is increased enough, the larger roller will also operate satisfactorily.

In Fig. 5.29b we see that the cam profile will be pointed when the roller radius R_r is equal to the radius of curvature of the pitch curve. Therefore, to achieve some chosen minimum value ρ_{min} for the minimum radius of curvature of the cam profile, the radius of

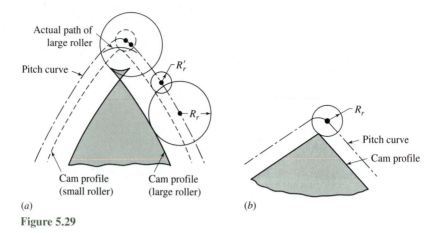

Actual path of
large roller

Pitch curve

R'_r

R_r

Cam profile
(small roller)

Cam profile
(large roller)

R_r

Pitch curve

Cam profile

(a)

(b)

Figure 5.29

curvature of the pitch curve must be greater than this value by the radius of the roller; that is,

$$\rho_{\text{pitch}} = \rho + R_r \qquad (c)$$

Now, for the case of a radial (non-offset) roller follower, the polar coordinates of the pitch curve are θ and

$$R = R_0 + y \qquad (d)$$

From any standard text on differential calculus we can find the general expression for radius of curvature of a curve in polar coordinates. Therefore, the equation for the radius of curvature of the pitch curve can be written as

$$\rho_{\text{pitch}} = \rho + R_r = \frac{[(R_0 + y)^2 + (y')^2]^{3/2}}{(R_0 + y)^2 + 2(y')^2 - (R_0 + y)y''} \qquad (5.40)$$

As before, it is technically possible to differentiate this expression with respect to the cam rotation θ and to seek out the minimum value of ρ for a particular choice of the displacement equation y and a particular prime-circle radius R_0. However, because this would be an extremely tedious calculation to repeat for each new cam design, the minimum radius of curvature has been sought out by digital computer program for each of the standard cam motions of Section 5.7. The results are presented graphically in Figs. 5.30 through 5.34. Each of these figures shows graphs of $(\rho_{\text{min}} + R_r)/R_0$ versus β for one type of standard-motion curves with various ratios of R_0/L. Because we have already chosen the displacement diagram and have found a suitable value of R_0, each segment of the cam can now be checked to find its minimum radius of curvature.

Saving even more effort, it is not necessary to check those segments of the cam where the second-order kinematic coefficient y'' remains positive throughout the segment, such as the half-rise motions of Eqs. (5.24) and (5.28) or the half-return motions of Eqs. (5.27) and (5.31). Assuming that the "acceleration" curve has been made continuous, the minimum radius of curvature of the cam cannot occur in these segments; Eq. (5.40) yields $\rho_{\text{min}} = R_0 - R_r$ for each of these.

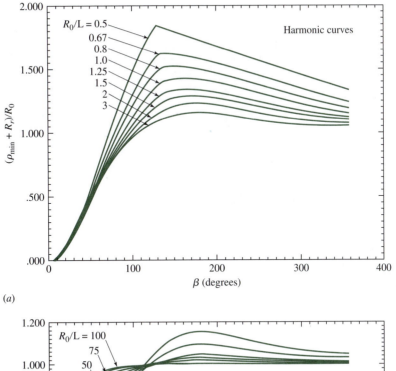

(a)

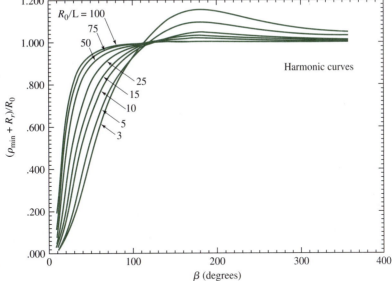

(b)

Figure 5.30 Minimum radius-of-curvature charts for radial roller follower cam with full-rise or full-return simple harmonic motion, Eqs. (5.18) and (5.21). (From M. A. Ganter and J. J. Uicker, Jr., *J. Mech. Des., ASME Trans.,* ser. B, vol. 101, no. 3, pp. 465–470, 1979, by permission.)

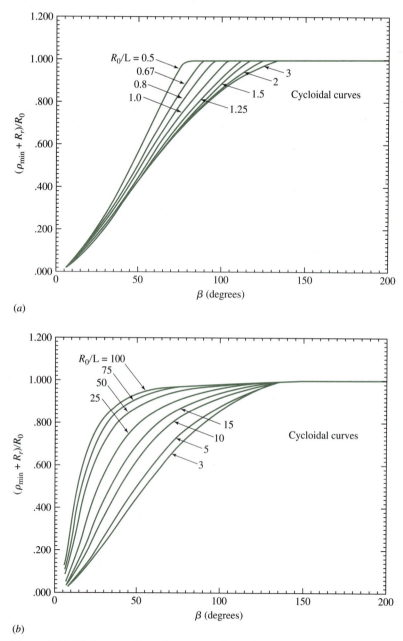

Figure 5.31 Minimum radius-of-curvature charts for radial roller follower cam with full-rise or full-return cycloidal motion, Eqs. (5.19) and (5.22). (From M. A. Ganter and J. J. Uicker, Jr., *J. Mech. Des., ASME Trans.,* ser. B, vol. 101, no. 3, pp. 465–470, 1979, by permission.)

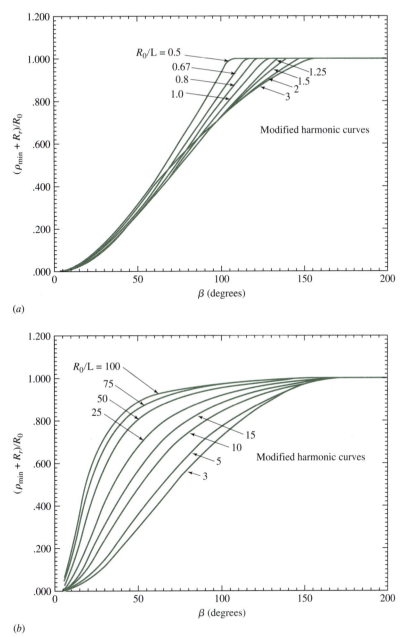

(a)

(b)

Figure 5.32 Minimum radius-of-curvature charts for radial roller follower cam with full-rise or full-return eighth-order polynomial motion, Eqs. (5.20) and (5.23). (From M. A. Ganter and J. J. Uicker, Jr., *J. Mech. Des., ASME Trans.,* ser B, vol. 101, no. 3, pp. 465–470, 1979, by permission.)

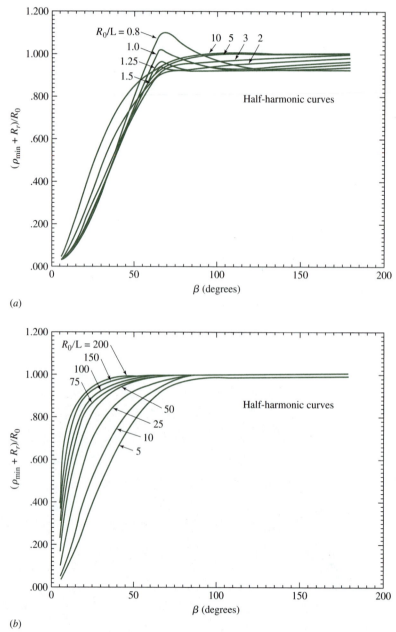

Figure 5.33 Minimum radius-of-curvature charts for radial roller follower cam with half-harmonic motion. Eqs. (5.25) and (5.26). (From M. A. Ganter and J. J. Uicker, Jr., *J. Mech. Des., ASME Trans.,* ser. B, vol. 101, no. 3, pp. 465–470, 1979, by permission.)

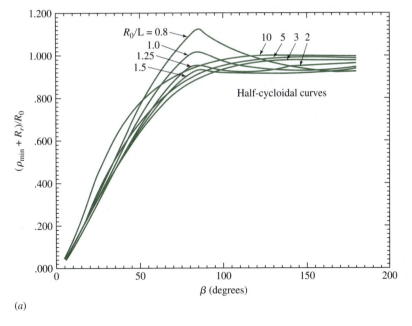

(a)

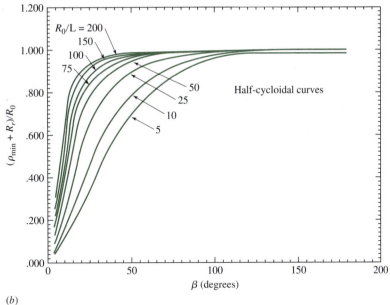

(b)

Figure 5.34 Minimum radius-of-curvature charts for radial roller follower cam with half-cycloidal motion. Eqs. (5.29) and (5.30). (From M. A. Ganter and J. J. Uicker, Jr., *J. Mech. Des., ASME Trans.,* ser. B, vol. 101, no. 3, pp. 465–470, 1979, by permission.)

EXAMPLE 5.5

Assuming that the displacement characteristics of Example 5.2 are to be achieved by a plate cam with a reciprocating roller follower, determine the minimum radius of curvature of the cam profile if a prime-circle radius of $R_0 = 4.00$ in (found in Example 5.4) and a roller radius of $R_r = 0.50$ in are used.

SOLUTION

For the segment AB of Fig. 5.24, we have no need to check, because the second-order kinematic coefficient y'' is positive throughout the segment.

For the segment CD we have half-harmonic motion with $L_3 = 0.566$ in and $\beta_3 = 32°$, from which we find

$$\frac{R_0}{L} = \frac{4.00 + (L_1 + L_2)}{L_3} = \frac{6.43 \text{ in}}{0.57 \text{ in}} = 11.3$$

where R_0 was adjusted by $(L_1 + L_2)$ because the graphs of Fig. 5.33 were plotted for $y = 0$ at the base of the segment. Now, using Fig. 5.33b, we find $(\rho_{\min} + R_r)/R_0 = 0.66$, and therefore

$$\rho_{\min} = 0.66R_0 - R_r = 0.66(6.43) - 0.50 = 3.74 \text{ in}$$

where, again, the adjusted value of R_0 was used.

For the segment DE we have eighth-order polynomial motion with $L_4 = 3.00$ in and $\beta_4 = 108°$, from which $R_0/L = 1.33$. Using Fig. 5.32a, we find $(\rho_{\min} + R_r)/R_0 = 1.00$, and

$$\rho_{\min} = 1.00R_0 - R_r = 1.00(4.00) - 0.50 = 3.50 \text{ in}$$

Choosing the smaller value, the minimum radius of curvature of the entire cam profile is taken as

$$\rho_{\min} = 3.50 \text{ in} \qquad\qquad Ans.$$

The rectangular coordinates of the cam profile of a plate cam with a reciprocating roller follower are given by

$$u = \left(\sqrt{R_0^2 - \varepsilon^2} + y\right)\sin\theta + \varepsilon\cos\theta + R_r\sin(\phi - \theta) \tag{5.41a}$$

$$v = \left(\sqrt{R_0^2 - \varepsilon^2} + y\right)\cos\theta - \varepsilon\sin\theta - R_r\cos(\phi - \theta) \tag{5.41b}$$

where ϕ is the pressure angle given by Eq. (5.38). The polar coordinates are

$$R = \sqrt{\left(\sqrt{R_0^2 - \varepsilon^2} + y - R_r\cos\phi\right)^2 + (\varepsilon + R_r\sin\phi)^2} \tag{5.42a}$$

$$\psi = -\theta + \tan^{-1}\left(\frac{\sqrt{R_0^2 - \varepsilon^2} - R_r\cos\phi}{\varepsilon + R_r\sin\phi}\right) \tag{5.42b}$$

In this section and the previous section, we have considered the problems that result from improper choice of prime-circle radius for a plate cam with a reciprocating follower. Although the equations are different for oscillating followers or other types of cams, a similar approach can be used to guard against undercutting[4] and severe pressure angles.[5] Similar equations can also be developed for cam profile data.[6] A good survey of the literature has been compiled by Chen.[7]

EXAMPLE 5.6

A plate cam with a reciprocating radial (that is, with eccentricity of $\varepsilon = 0$) roller follower is to be designed such that the lift of the follower will be

$$y = 0.5(1 - \cos 2\theta) \text{ in} \tag{1}$$

The prime-circle radius is to be $R_0 = 1.50$ in, the roller radius is $R_r = 0.50$ in, and the cam is to rotate counterclockwise. For the cam rotation angle $\theta = 30°$, determine:

(a) The coordinates of the point of contact between the cam and the follower in the moving coordinate system
(b) The radius of curvature of the cam profile
(c) The pressure angle of the cam

SOLUTION

The coordinates of the roller center in the moving coordinate system attached to the cam can be written as

$$u = X \cos \theta + Y \sin \theta \tag{2}$$

$$v = -X \sin \theta + Y \cos \theta \tag{3}$$

where X and Y are the coordinates of the follower center in the fixed coordinate system. In this example, the roller follower is radial therefore

$$X = 0 \quad \text{and} \quad Y = R_0 + y \tag{4}$$

Substituting Eqs. (4) into Eqs. (2) and (3), the coordinates of the follower center (the pitch curve) in the moving coordinates of the cam are

$$u = (R_0 + y)\sin \theta \tag{5}$$

$$v = (R_0 + y)\cos \theta \tag{6}$$

Differentiating Eqs. (5) and (6) with respect to the rotation of the cam, the first-order kinematic coefficients of the follower center are

$$u' = (R_0 + y)\cos \theta + y' \sin \theta \tag{7}$$

$$v' = -(R_0 + y)\sin \theta + y' \cos \theta \tag{8}$$

The coordinates of the point of contact between the cam and the follower, in the moving coordinate system, can be written as

$$u_{\text{cam}} = u + R_r \left(\frac{v'}{w'}\right) \tag{5.43a}$$

$$v_{\text{cam}} = v - R_r \left(\frac{u'}{w'}\right) \tag{5.43b}$$

where

$$w' = +\sqrt{u'^2 + v'^2} \tag{5.43c}$$

Differentiating Eqs. (7) and (8) with respect to the position of the cam, the second-order kinematic coefficients of the follower center are

$$u'' = -(R_0 + y)\sin\theta + 2y'\cos\theta + y''\sin\theta \tag{9}$$
$$v'' = -(R_0 + y)\cos\theta - 2y'\sin\theta + y''\cos\theta \tag{10}$$

The radius of curvature of the pitch curve can be written, from Eq. (5.40), as

$$\rho_{\text{pitch}} = \frac{w'^3}{u'v'' - v'u''} \tag{5.44}$$

Notice that positive rotation of the cam causes the point on the pitch curve to increment in the clockwise direction around the cam in the moving uv coordinate system. This defines the positive sense of the unit tangent vector $\hat{\tau}_{\text{pitch}}$ for the pitch curve in the moving coordinate system, and the unit normal vector is then given by $\hat{\rho}_{\text{pitch}} = \hat{k} \times \hat{\tau}_{\text{pitch}}$ in the moving coordinate system. Because this points toward the center of curvature of the pitch curve, the radius of curvature is expected to yield a negative value.

Substituting Eqs. (7) and (8) into Eq. (5.43c) gives

$$w' = +\sqrt{(R_0 + y)^2 + y'^2} \tag{11}$$

Differentiating Eq. (1) with respect to the position of the cam for this specific example, the first- and second-order kinematic coefficients (that is, the slope and the change of the slope) of the lift curve are

$$y' = \sin 2\theta \quad \text{and} \quad y'' = 2\cos 2\theta \tag{12}$$

Substituting Eqs. (7)–(11) into Eq. (5.44), the radius of curvature of the pitch curve can be written as

$$\rho_{\text{pitch}} = \frac{+[(R_0 + y)^2 + y'^2]^{3/2}}{-(R_0 + y)^2 - 2y'^2 + (R_0 + y)y''} \tag{13}$$

that, except for the negative denominator (explained above), exactly matches Eq. (5.40).

At the cam rotation angle $\theta = 30°$, the lift of the follower from Eq. (1) is

$$y = 0.25 \text{ in} \tag{14}$$

The first- and second-order kinematic coefficients of the lift curve, from Eqs. (12), are

$$y' = 0.866 \text{ in/rad} \quad \text{and} \quad y'' = 1.00 \text{ in/rad} \tag{15}$$

Substituting the radius of the prime circle and Eq. (14) into Eq. (4), the Y-coordinate of the follower center is

$$Y = 1.75 \text{ in} \tag{16}$$

Substituting this value and the known dimensions into Eqs. (5) and (6), the coordinates of the follower center in the moving coordinate system are

$$u = 0.8750 \text{ in} \quad \text{and} \quad v = 1.5155 \text{ in} \tag{17}$$

Substituting Eqs. (15) and the given geometry into Eqs. (7)–(10), the first- and second-order kinematic coefficients of the follower center are

$$u' = 1.7500 \cos 30° + 0.8660 \sin 30° = 1.9486 \text{ in/rad} \tag{18}$$

$$v' = -1.7500 \sin 30° + 0.8660 \cos 30° = -0.1250 \text{ in/rad} \tag{19}$$

$$u'' = -1.7500 \sin 30° + 1.7321 \cos 30° + \sin 30° = 1.1250 \text{ in/rad}^2 \tag{20}$$

$$v'' = -1.7500 \cos 30° - 1.7321 \sin 30° + \cos 30° = -1.5155 \text{ in/rad}^2 \tag{21}$$

Also, from Eq. (11), we have

$$w' = +\sqrt{1.7500^2 + 0.8660^2} = 1.9526 \text{ in/rad} \tag{22}$$

Substituting Eqs. (17)–(19) and (22) and the given dimensions into Eqs. (5.43a) and (5.43b), the coordinates of the point of contact, in the moving coordinate system, are

$$u_{\text{cam}} = 0.8750 + 0.5 \left(\frac{-0.1250}{1.9525} \right) = +0.8430 \text{ in} \qquad Ans.$$

$$v_{\text{cam}} = 1.5155 - 0.5 \left(\frac{1.9486}{1.9526} \right) = +1.0166 \text{ in} \qquad Ans.$$

Substituting the given geometry and Eqs. (18)–(22) into Eq. (13) or Eq. (5.44), the radius of curvature of the pitch curve is

$$\rho_{\text{pitch}} = \frac{1.9526^3}{(1.9486)(-1.5155) - (-0.1250)(1.1250)} = -2.647 \text{ in}$$

The negative sign indicates that the unit normal vector to the pitch curve, through the follower center, points in the negative $\hat{\boldsymbol{\rho}}_{pitch}$ direction, away from the center of curvature as explained above. Therefore, the radius of curvature of the cam profile is

$$\rho_{cam} = \rho_{pitch} + R_r = -2.647 + 0.500 = -2.147 \text{ in} \qquad Ans.$$

The pressure angle of the cam can be written from Fig. (5.27) as

$$\cos\phi = (-\hat{\boldsymbol{\rho}}_{pitch}) \cdot \left(\frac{\mathbf{V}_A}{V_A}\right) \qquad (5.45a)$$

where $(-\hat{\boldsymbol{\rho}}_{pitch})$ is the outward unit normal to the pitch curve and $\mathbf{V}_A$ is the absolute velocity of the follower center at the cam rotation angle θ. The outward unit normal vector can be expressed in terms of the first-order kinematic coefficients by Eq. (3.37) as

$$\hat{\boldsymbol{\rho}}_{pitch} = \left(\frac{v'}{w'}\right)\hat{\mathbf{i}} + \left(\frac{-u'}{w'}\right)\hat{\mathbf{j}}$$

Substituting this equation into Eq. (5.45a) gives

$$\cos\phi = \left[\left(-\frac{v'}{w'}\right)\hat{\mathbf{i}} + \left(\frac{u'}{w'}\right)\hat{\mathbf{j}}\right] \cdot [\sin\theta\hat{\mathbf{i}} + \cos\theta\hat{\mathbf{j}}]$$

which can be written as

$$\cos\phi = \left(-\frac{v'}{w'}\right)\sin\theta + \left(\frac{u'}{w'}\right)\cos\theta \qquad (5.45b)$$

Substituting Eqs. (18), (19), and (22), the pressure angle of the cam for this specific example (at the cam rotation angle $\theta = 30°$) is

$$\phi = 26.33° \qquad Ans.$$

EXAMPLE 5.7

A plate cam with an oscillating roller follower is to be designed such that the lift of the follower will be

$$y = 0.5(1 - \cos 2\theta) \text{ rad} \qquad (1)$$

The prime-circle radius is $R_0 = 1.50$ in, the roller radius is $R_r = 0.50$ in, the distance between the center of the camshaft and the follower pivot is $R_{MO} = r_1 = 3.00$ in, the length of the follower arm is $R_{CM} = r_2 = 2.598$ in, and the cam is to rotate counterclockwise. For the cam rotation angle of $\theta = 45°$, determine:

(a) The coordinates of the point of contact between the cam and the follower
(b) The radius of curvature of the cam profile
(c) The pressure angle of the cam

SOLUTION

The coordinates of the trace point C in the moving coordinate system of the cam are

$$u = X \cos\theta + Y \sin\theta \tag{2}$$

$$v = -X \sin\theta + Y \cos\theta \tag{3}$$

The coordinates of the trace point C in the fixed coordinate system are

$$X = r_1 + r_2 \cos\theta_2 \tag{4}$$

$$Y = r_2 \sin\theta_2 \tag{5}$$

where $r_1 = 3$ in, $r_2 = 2.598$ in, and θ_2 is the angle of rotation of the follower arm.

$$\theta_2 = 5\pi/6 - 0.5(1 - \cos 2\theta) \text{ rad} \tag{6}$$

Substituting Eqs. (4) and (5) into Eqs. (2) and (3), the coordinates of the trace point C in the moving system are

$$u = (r_2 \sin\theta_2)\sin\theta + (r_1 + r_2 \cos\theta_2)\cos\theta$$

$$v = (r_2 \sin\theta_2)\cos\theta - (r_1 + r_2 \cos\theta_2)\sin\theta$$

which can be written as

$$u = r_1 \cos\theta + r_2 \cos(\theta_2 - \theta) \tag{7}$$

$$v = -r_1 \sin\theta + r_2 \sin(\theta_2 - \theta) \tag{8}$$

Differentiating Eqs. (7) and (8) with respect to the rotation angle of the cam θ, the first and second-order kinematic coefficients of the trace point C are

$$u' = -r_1 \sin\theta - r_2 \sin(\theta_2 - \theta)[\theta_2' - 1] \tag{9}$$

$$v' = -r_1 \cos\theta + r_2 \cos(\theta_2 - \theta)[\theta_2' - 1] \tag{10}$$

$$u'' = -r_1 \cos\theta - r_2 \cos(\theta_2 - \theta)[\theta_2' - 1]^2 - r_2\theta_2'' \sin(\theta_2 - \theta) \tag{11}$$

$$v'' = +r_1 \sin\theta - r_2 \sin(\theta_2 - \theta)[\theta_2' - 1]^2 + r_2\theta_2'' \cos(\theta_2 - \theta) \tag{12}$$

where the first- and second-order kinematic coefficients of the follower arm, from Eq. (6), are

$$\theta_2' = -\sin 2\theta \quad \text{and} \quad \theta_2'' = -2\cos 2\theta \tag{13}$$

The coordinates of the point of contact between the cam and the follower, in the moving coordinate system, are

$$u_{cam} = u + R_r \left(\frac{v'}{w'} \right) \tag{14}$$

$$v_{cam} = v - R_r \left(\frac{u'}{w'} \right) \tag{15}$$

where

$$w' = +\sqrt{u'^2 + v'^2} \tag{16}$$

At the cam rotation angle $\theta = 45°$, from Eqs. (6) and (13) we have

$$r_2 = 2.598 \text{ in}, \quad \theta_2 = 2.118 \text{ rad} = 121.35°, \quad \theta_2' = -1, \quad \text{and} \quad \theta_2'' = 0 \tag{17}$$

Substituting Eqs. (17) into Eqs. (7) and (8), the coordinates of the trace point C are

$$u = 3 \cos 45° + 2.598 \cos(121.35° - 45°) = 2.734 \text{ in} \qquad Ans. \tag{18}$$
$$v = -3 \sin 45° + 2.598 \sin(121.35° - 45°) = 0.403 \text{ in} \qquad Ans. \tag{19}$$

Substituting Eqs. (17) and the given geometry into Eqs. (9) and (10), the first-order kinematic coefficients of the trace point are

$$u' = -3 \sin 45° - 2.598 \sin(121.35° - 45°)[-2] = 2.928 \text{ in/rad} \tag{20}$$
$$v' = -3 \cos 45° + 2.598 \cos(121.35° - 45°)[-2] = -3.347 \text{ in/rad} \tag{21}$$

and from Eq. (16) we have

$$w' = +\sqrt{2.928^2 + 3.347^2} = 4.447 \text{ in/rad} \tag{22}$$

Then substituting Eqs. (18)–(22) into Eqs. (14) and (15), the coordinates of the contact point between the cam and the follower, in the moving coordinate system, are

$$u_{cam} = 2.734 + 0.5 \left(\frac{-3.347}{4.447} \right) = 2.358 \text{ in} \qquad Ans.$$

$$v_{cam} = 0.403 - 0.5 \left(\frac{2.928}{4.447} \right) = 0.074 \text{ in} \qquad Ans.$$

The pressure angle of the cam can be written from Eq. (5.38) as

$$\cos \phi = (-\hat{\boldsymbol{\rho}}_{pitch}) \cdot \left(\frac{\mathbf{V}_{FC}}{V_{FC}} \right) \tag{23}$$

where

$$\frac{\mathbf{V}_{FC}}{V_{FC}} = \cos\beta\hat{\mathbf{i}} - \sin\beta\hat{\mathbf{j}}$$

and the angle

$$\beta = (30° + 28.65°) - 45° = 13.65° \tag{24}$$

Therefore, Eq. (23) can be written in terms of the kinematic coefficients as

$$\cos\phi = \left[\left(-\frac{v'}{w'}\right)\hat{\mathbf{i}} + \left(\frac{u'}{w'}\right)\hat{\mathbf{j}}\right] \cdot [\cos\beta\hat{\mathbf{i}} - \sin\beta\hat{\mathbf{j}}]$$

or as

$$\cos\phi = \left(-\frac{v'}{w'}\right)\cos\beta - \left(\frac{u'}{w'}\right)\sin\beta$$

Substituting Eqs. (20)–(22) and (24) into this equation, the pressure angle (at the cam rotation angle $\theta = 45°$) is*

$$\phi = 54.83° \qquad\qquad Ans.$$

Substituting Eqs. (17) and the given geometry into Eqs. (11) and (12), the second-order kinematic coefficients of the follower center are

$$u'' = -3\cos 45° - 2.598\cos(121.35° - 45°)[-2]^2 = -4.573 \text{ in/rad}^2 \tag{25}$$

$$v'' = +3\sin 45° - 2.598\sin(121.35° - 45°)[-2]^2 = -7.977 \text{ in/rad}^2 \tag{26}$$

The radius of curvature of the pitch curve can be written, as it was in Eq. (5.44), as

$$\rho_{\text{pitch}} = \frac{w'^3}{u'v'' - v'u''}$$

Substituting Eqs. (20)–(22), (25), and (26) into this equation, the radius of curvature is

$$\rho_{\text{pitch}} = \frac{4.447^3}{(2.928)(-7.977) - (-3.347)(-4.573)} = -2.275 \text{ in}$$

The negative sign indicates that the unit normal vector to the pitch curve, through the follower center, points in the negative $\hat{\rho}_{\text{pitch}}$ direction, away from the center of curvature as explained in Example 5.6 above. Therefore, the radius of curvature of the cam profile is

$$\rho_{\text{cam}} = \rho_{\text{pitch}} + R_r = -2.275 + 0.500 = -1.775 \text{ in} \qquad\qquad Ans.$$

---

*This is a high value for the pressure angle (that is, $\phi \geq 30°$ can provide difficulties); therefore, design changes should be considered.

EXAMPLE 5.8

A plate cam with a reciprocating flat-face follower is to be designed such that the lift of the follower will be

$$y = 2.5\theta^2 - \theta^3 \text{ in} \tag{1}$$

where the cam rotation angle θ is in radians. The prime-circle radius $R_0 = 1.50$ in and the cam is to rotate counterclockwise. For the cam rotation angle $\theta = 60° = \pi/3$ rad, determine

(a) The coordinates of the trace point in the moving coordinate system
(b) The radius of curvature of the cam profile

SOLUTION

The coordinates of the trace point C, in the moving coordinate system, are

$$u = X \cos\theta + Y \sin\theta \tag{2}$$

$$v = -X \sin\theta + Y \cos\theta \tag{3}$$

The coordinates of the trace point in the fixed coordinate system are

$$X = y' \quad \text{and} \quad Y = R_0 + y \tag{4}$$

Substituting Eqs. (4) into Eqs. (2) and (3), the coordinates of the trace point in the moving coordinate system are

$$u = (R_0 + y)\sin\theta + y' \cos\theta \tag{5}$$

$$v = (R_0 + y)\cos\theta - y' \sin\theta \tag{6}$$

Differentiating these two equations with respect to the input rotation of the cam, the first- and second-order kinematic coefficients of the trace point can be written as

$$u' = (R_0 + y + y'')\cos\theta \tag{7}$$

$$v' = -(R_0 + y + y'')\sin\theta \tag{8}$$

$$u'' = (y' + y''')\cos\theta - (R_0 + y + y'')\sin\theta \tag{9}$$

$$v'' = -(y' + y''')\sin\theta - (R_0 + y + y'')\cos\theta \tag{10}$$

and

$$w' = +(R_0 + y + y'') \tag{11}$$

From Eqs. (1), the first-, second-, and third-order kinematic coefficients of the lift curve are

$$y' = 5\theta - 3\theta^2 \text{ in/rad}, \quad y'' = 5 - 6\theta \text{ in/rad}^2, \quad \text{and} \quad y''' = -6 \text{ in/rad}^3 \tag{12}$$

The radius of curvature of the cam profile can be written, from Eq. (5.44), as

$$\rho_{\text{cam}} = \frac{w'^3}{u'v'' - v'u''}$$

Substituting Eqs. (8)–(11) into this equation and simplifying, the radius of curvature is

$$\rho_{\text{cam}} = -w' = -(R_0 + y + y'') \tag{13}$$

The pressure angle of the cam for a flat-face follower can be obtained as in Eq. (5.45b) and is

$$\cos \phi = \left(-\frac{v'}{w'}\right) \sin \theta + \left(\frac{u'}{w'}\right) \cos \theta$$

Substituting Eqs. (7), (8), and (11) into this equation gives $\cos \phi = 1$, which says that the pressure angle of the cam for a flat-face follower is always zero.

At the cam rotation angle $\theta = 60° = 1.047$ rad, the lift of the follower from Eq. (1) is

$$y = 1.593 \text{ in} \tag{14}$$

The first-, second-, and third-order kinematic coefficients of the lift curve, from Eqs. (12), are

$$y' = \frac{\pi}{3}\left(5 - \frac{\pi}{3}\right) = 4.139 \text{ in/rad}, \quad y'' = 5 - 2\pi = -1.283 \text{ in/rad}^2, \quad \text{and} \quad y''' = -6 \text{ in/rad}^3 \tag{15}$$

Substituting the radius of the prime circle and Eqs. (14) and (15) into Eqs. (4), the coordinates of the trace point in the fixed system are

$$X = 4.139 \text{ in} \quad \text{and} \quad Y = 3.093 \text{ in}$$

Substituting these values and the known geometry into Eqs. (5) and (6), the coordinates of the trace point in the moving coordinate system are

$$u = 2.037 \text{ in} \quad \text{and} \quad v = 2.658 \text{ in} \qquad \qquad \textit{Ans.}$$

Substituting Eqs. (14) and (15) and the known geometry into Eq. (13), the radius of curvature of the cam profile is

$$\rho_{\text{cam}} = -1.810 \text{ in} \qquad \qquad \textit{Ans.}$$

NOTES

1. A good analysis of this subject is presented in D. Tesar and G. K. Matthew, *The Dynamic Synthesis, Analysis, and Design of Modeled Cam Systems,* Heath, Lexington, MA, 1976.

2. M. Kloomak and R. V. Muffley, Plate Cam Design—with Emphasis on Dynamic Effects, *Prod. Eng.,* vol. 26, no. 2, 1955.

3. D. A. Stoddart, Polydyne Cam Design, *Mach. Design,* vol. 25, no. 1, pp. 121–135; vol. 25, no. 2, pp. 146–154; vol. 25, no. 3, pp. 149–164, 1953.

4. M. Kloomak and R. V. Mufley, Plate Cam Design: Radius of Curvature, *Prod. Eng.,* vol. 26, no. 9, pp. 186–201, 1955.

5. M. Kloomak and R. V. Mufley, Plate Cam Design: Pressure Angle Analysis, *Prod. Eng.,* vol. 26, no. 5, pp. 155–171, 1955.

6. See, for example, the excellent book by S. Molian, *The Design of Cam Mechanisms and Linkages,* Constable, London, 1968.

7. F. Y. Chen, A Survey of the State of the Art of Cam System Dynamics, *Mech. Mach. Theory,* vol. 12, no. 3, pp. 201–224, 1977.

PROBLEMS

5.1   The reciprocating radial roller follower of a plate cam is to rise 2 in with simple harmonic motion in 180° of cam rotation and return with simple harmonic motion in the remaining 180°. If the roller radius is 0.375 in and the prime-circle radius is 2 in, construct the displacement diagram, the pitch curve, and the cam profile for clockwise cam rotation.

5.2   A plate cam with a reciprocating flat-face follower has the same motion as in Problem 5.1. The prime-circle radius is 2 in, and the cam rotates counterclockwise. Construct the displacement diagram and the cam profile, offsetting the follower stem by 0.75 in in the direction that reduces the bending stress in the follower during rise.

5.3   Construct the displacement diagram and the cam profile for a plate cam with an oscillating radial flat-face follower that rises through 30° with cycloidal motion in 150° of counterclockwise cam rotation, then dwells for 30°, returns with cycloidal motion in 120°, and dwells for 60°. Determine the necessary length for the follower face, allowing 5-mm clearance at each end. The prime-circle radius is 30 mm, and the follower pivot is 120 mm to the right.

5.4   A plate cam with an oscillating roller follower is to produce the same motion as in Problem 5.3. The prime-circle radius is 60 mm, the roller radius is 10 mm, the length of the follower is 100 mm, and it is pivoted at 125 mm to the left of the cam rotation axis. The cam rotation is clockwise. Determine the maximum pressure angle.

5.5   For a full-rise simple harmonic motion, write the equations for the velocity and the jerk at the midpoint of the motion. Also, determine the acceleration at the beginning and the end of the motion.

5.6   For a full-rise cycloidal motion, determine the values of θ for which the acceleration is maximum and minimum. What is the formula for the acceleration at these points? Find the equations for the velocity and the jerk at the midpoint of the motion.

5.7   A plate cam with a reciprocating follower is to rotate clockwise at 400 rev/min. The follower is to dwell for 60° of cam rotation, after which it is to rise to a lift of 2.5 in. During 1 in of its return stroke, it must have a constant velocity of 40 in/s. Recommend standard cam motions from Section 5.7 to be used for high-speed operation and determine the corresponding lifts and cam rotation angles for each segment of the cam.

5.8   Repeat Problem 5.7 except with a dwell that is to be for 20° of cam rotation.

5.9   If the cam of Problem 5.7 is driven at constant speed, determine the time of the dwell and the maximum and minimum velocity and acceleration of the follower for the cam cycle.

5.10 A plate cam with an oscillating follower is to rise through 20° in 60° of cam rotation, dwell for 45°, then rise through an additional 20°, return, and dwell for 60° of cam rotation. Assuming high-speed operation, recommend standard cam motions from Section 5.7 to be used, and determine the lifts and cam-rotation angles for each segment of the cam.

5.11 Determine the maximum velocity and acceleration of the follower for Problem 5.10, assuming that the cam is driven at a constant speed of 600 rev/min.

5.12 The boundary conditions for a polynomial cam motion are as follows: for $\theta = 0$, $y = 0$, and $y' = 0$; for $\theta = \beta$, $y = L$, and $y' = 0$. Determine the appropriate displacement equation and the first three derivatives of this equation with respect to the cam rotation angle. Sketch the corresponding diagrams.

5.13 Determine the minimum face width using 0.1-in allowances at each end, and determine the minimum radius of curvature for the cam of Problem 5.2.

5.14 Determine the maximum pressure angle and the minimum radius of curvature for the cam of Problem 5.1.

5.15 A radial reciprocating flat-face follower is to have the motion described in Problem 5.7. Determine the minimum prime-circle radius if the radius of curvature of the cam is not to be less than 0.5 in. Using this prime-circle radius, what is the minimum length of the follower face using allowances of 0.15 in on each side?

5.16 Graphically construct the cam profile of Problem 5.15 for clockwise cam rotation.

5.17 A radial reciprocating roller follower is to have the motion described in Problem 5.7. Using a prime-circle radius of 20 in, determine the maximum pressure angle and the maximum roller radius that can be used without producing undercutting.

5.18 Graphically construct the cam profile of Problem 5.17 using a roller radius of 0.75 in. The cam rotation is to be clockwise.

5.19 A plate cam rotates at 300 rev/min and drives a reciprocating radial roller follower through a full rise of 75 mm in of cam rotation. Find the minimum radius of the prime-circle if simple harmonic motion is used and the pressure angle is not to exceed 25°. Find the maximum acceleration of the follower.

5.20 Repeat Problem 5.19 except that the motion is cycloidal.

5.21 Repeat Problem 5.19 except that the motion is eighth-order polynomial.

5.22 Using a roller diameter of 20 mm, 180° determine whether the cam of Problem 5.19 will be undercut.

5.23 Equations (5.36) and (5.37) describe the profile of a plate cam with a reciprocating flat-face follower. If such a cam is to be cut on a milling machine with cutter radius R_c, determine similar equations for the center of the cutter.

5.24 Write computer programs for each of the displacement equations of Section 5.7.

5.25 Write a computer program to plot the cam profile for Problem 5.2.

6 | Spur Gears

Gears are machine elements used to transmit rotary motion between two shafts, usually with a constant speed ratio. The *pinion* is a name given to the smaller of the two mating gears; the larger is often called the *gear* or the *wheel*. In this chapter we will discuss the case where the axes of the two shafts are parallel and the teeth are straight and parallel to the axes of rotation of the shafts; such gears are called *spur gears*. A pair of spur gears in mesh is shown in Fig. 6.1.

6.1 TERMINOLOGY AND DEFINITIONS

The terminology of gear teeth is illustrated in Fig. 6.2, where most of the following definitions are shown:

The *pitch circle* is a theoretical circle on which all calculations are usually based. The pitch circles of a pair of mating gears are tangent to each other, and it is these pitch circles that were pictured as rolling without slip in earlier chapters.

The *diametral pitch P* is the number of teeth on the gear per inch of its pitch diameter. The diametral pitch can be found from the equation

$$P = \frac{N}{2R} \tag{6.1}$$

where N is the number of teeth, R is the pitch circle radius in inches, and P has units of teeth/inch. Note that the diametral pitch cannot be directly measured on the gear itself.

Figure 6.1 Pair of spur gears in mesh. (Courtesy of Gleason Works, Rochester, NY.)

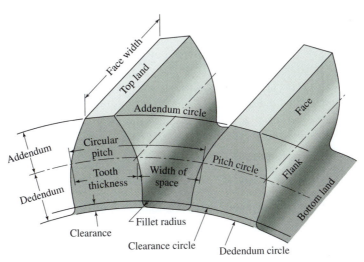

Figure 6.2 Terminology.

Figure 6.3 Tooth sizes for various diametral pitches. (Courtesy of Gleason Cutting Tools Corp., Loves Park, IL.)

Note also that as the diametral pitch values become larger, the teeth become smaller; this is shown clearly in Fig. 6.3.

The *module m* is the ratio of the pitch diameter to the number of teeth

$$m = \frac{2R}{N} \tag{6.2}$$

where R is now in mm and module has units of mm/tooth. The module is the usual unit indicating tooth size in SI while diametral pitch is used in US customary units.

The *circular pitch p* is the distance from one tooth to the adjacent tooth measured along the pitch circle. Therefore it can be found from

$$p = \frac{2\pi R}{N} \tag{6.3}$$

The circular pitch is related to the previous definitions, depending on the units, by

$$p = \frac{\pi}{P} = \pi m \tag{6.4}$$

The *addendum a* is the radial distance between the pitch circle and the top land of each tooth.

The *dedendum d* is the radial distance from the pitch circle to the bottom land of each tooth.

The *whole depth* is the sum of the addendum and dedendum.

The *clearance c* is the amount by which the dedendum of a gear exceeds the addendum of the mating gear.

The *backlash* is the amount by which the width of a tooth space exceeds the thickness of the engaging tooth measured along the pitch circles.

6.2 FUNDAMENTAL LAW OF TOOTHED GEARING

Mating gear teeth acting against each other to produce rotary motion are similar to a cam and follower. When the tooth profiles (or cam and follower profiles) are shaped so as to produce a constant angular velocity ratio between the two shafts during meshing, then the two mating surfaces are said to be *conjugate*. It is possible to specify an arbitrary profile for one tooth and then to find a profile for the mating tooth so that the two surfaces are conjugate. One possible choice for such conjugate solutions is the *involute* profile, which, with few exceptions, is in universal use for gear teeth.

The action of a single pair of mating gear teeth as they pass through their entire phase of action must be such that the ratio of the angular velocity of the driven gear to that of the driving gear; that is, *the first-order kinematic coefficient must remain constant*. This is the fundamental criterion that governs the choice of the tooth profiles. If this were not true in gearing, very serious vibration and impact problems would result, even at low speeds.

In Section 3.17 we learned that the angular velocity ratio theorem states that the first-order kinematic coefficient of any mechanism is inversely proportional to the segments into which the common instant center cuts the line of centers. In Fig. 6.4 two profiles are in contact at point T; let profile 2 represent the driver and 3 be driven. The normal to the profiles at the point of contact T intersects the line of centers $O_2 O_3$ at the instant center P.

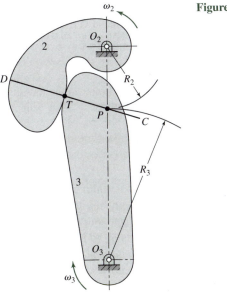

Figure 6.4

In gearing, P is called the *pitch point* and the normal to the surfaces CD is called the *line of action*. Designating the pitch circle radii of the two gear profiles as R_2 and R_3, from Eq. (3.28) we see

$$\frac{\omega_2}{\omega_3} = \frac{R_3}{R_2} \tag{6.5}$$

This equation is frequently used to define what is called the fundamental law of gearing, which states that, as gears go through their mesh, *the pitch point must remain stationary on the line of centers* for the speed ratio to remain constant. This means that the line of action for every new instantaneous point of contact must always pass through the stationary pitch point P. Thus the problem of finding a conjugate profile is to find for a given shape a mating shape that satisfies the fundamental law of gearing.

It should not be assumed that just any shape or profile is satisfactory just because a conjugate profile can be found. Even though theoretically conjugate curves might be found, the practical problems of reproducing these curves on steel gear blanks, or other materials, while using existing machinery still exist. In addition the sensitivity of the law of gearing to small dimensional changes of the shaft center distance due either to misalignment or to large forces must also be considered. Finally, the tooth profile selected must be one that can be reproduced quickly and economically in very large quantities. A major portion of this chapter is devoted to illustrating how the involute curve profile fulfills these requirements.

6.3 INVOLUTE PROPERTIES

An *involute* curve is the path generated by a tracing point on a cord as the cord is unwrapped from a cylinder called the *base cylinder*. This is shown in Fig. 6.5, where T is the tracing point. Note that the cord AT is normal to the involute at T, and the distance AT is the instantaneous value of the radius of curvature. As the involute is generated from its origin T_0 to T_1, the radius of curvature varies continuously; it is zero at T_0 and increases continuously to T_1. Thus the cord is the generating line, and it is always normal to the involute.

If the two mating tooth profiles both have the shapes of involute curves, the condition that the pitch point P remain stationary is satisfied. This is illustrated in Fig. 6.6, where two gear blanks with fixed centers O_2 and O_3 are shown having base cylinders with respective

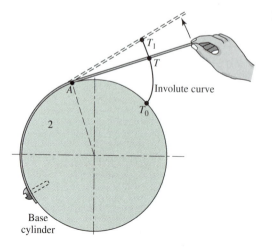

Figure 6.5 Involute curve.

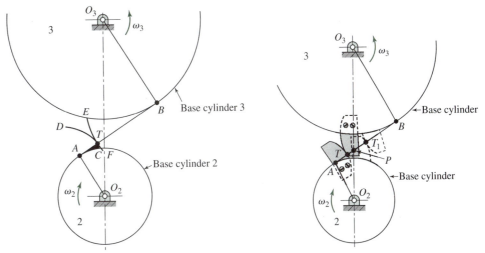

Figure 6.6 Conjugate involute curves. **Figure 6.7** Involute action.

radii of O_2A and O_3B. We now imagine that a cord is wound clockwise abound the base cylinder of gear 2, pulled tightly between points A and B, and wound counterclockwise around the base cylinder of gear 3. If now the two base cylinders are rotated in opposite directions so as to keep the cord tight, a tracing point T traces out the involutes EF on gear 2 and GH on gear 3. The involutes thus generated simultaneously by the single tracing point T are conjugate profiles.

Next imagine that the involutes of Fig. 6.6 are scribed on plates and that the plates are cut along the scribed curves and then bolted to the respective cylinders in the same positions. The result is shown in Fig. 6.7. The cord can now be removed and, if gear 2 is moved clockwise, gear 3 is caused to move counterclockwise by the camlike action of the two curved plates. The path of contact is the line AB formerly occupied by the cord. Because the line AB is the generating line for each involute, it is normal to both profiles at all points of contact. Also, it always occupies the same position because it is always tangent to both base cylinders. Therefore point P is the pitch point. Point P does not move; therefore the involute curves are conjugate curves and satisfy the fundamental law of gearing.

6.4 INTERCHANGEABLE GEARS; AGMA STANDARDS

A *tooth system* is the name given to a *standard** that specifies the relationships between addendum, dedendum, clearance, tooth thickness, and fillet radius to attain interchangeability of gears of all tooth numbers but of the same pressure angle and diametral pitch or module. We

---

*Standards are defined by the American Gear Manufacturers Association (AGMA) and the American National Standards Institute (ANSI). The AGMA standards may be quoted or extracted in their entirety, provided that an appropriate credit line is included—for example, "Extracted from AGMA Information Sheet—Strength of Spur, Helical, Herringbone, and Bevel Gear Teeth (AGMA 225.01) with permission of the publisher, the American Gear Manufacturers Association, 1500 King Street, Suite 201, Alexandria, VA 22314." These standards have been used extensively in this chapter and in the chapters that follow.

TABLE 6.1 Standard Tooth Sizes

Standard Diametral Pitches U.S. Customary, teeth/in	
Coarse	$1, 1\frac{1}{4}, 1\frac{1}{2}, 1\frac{3}{4}, 2, 2\frac{1}{4}, 2\frac{1}{2}, 3, 4, 6, 8, 10, 12, 16$
Fine	20, 24, 32, 40, 48, 64, 80, 96, 120, 150, 200

Standard Modules SI, mm/tooth	
Preferred	1, 1.25, 1.5, 2, 2.5, 3, 4, 5, 6, 8, 10, 12, 16, 20, 25, 32, 40, 50
Next choice	1.125, 1.375, 1.75, 2.25, 2.75, 3.5, 4.5, 5.5, 7, 9, 11, 14, 18, 22, 28, 36, 45

TABLE 6.2 Standard Tooth Systems for Spur Gears

System	Pressure Angle, ϕ (deg)	Addendum, a	Dedendum, d
Full depth	20°	$1/P$ or $1m$	$1.25/P$ or $1.25m$
Full depth	22.5°	$1/P$ or $1m$	$1.25/P$ or $1.25m$
Full depth	25°	$1/P$ or $1m$	$1.25/P$ or $1.25m$
Stub teeth	20°	$0.8/P$ or $0.8m$	$1/P$ or $1m$

should be aware of the advantages and disadvantages so that we can choose the best gears for a given design and have a basis for comparison if we depart from a standard tooth profile.

In order for a pair of spur gears to properly mesh, they must share the same pressure angle and the same tooth size as specified by the choice of diametral pitch or module. The numbers of teeth and the pitch diameters of the two gears in mesh need not match, but are chosen to give the desired speed ratio as shown in Eq. (6.5).

The sizes of the teeth used are chosen by selecting the diametral pitch or module. Standard cutters are generally available for the sizes listed in Table 6.1. Once the diametral pitch or module is chosen, the remaining dimensions of the tooth are set by the standards shown in Table 6.2. Tables 6.1 and 6.2 contain the standards for the spur gears most in use today.

Let us illustrate the design choices by an example:

EXAMPLE 6.1

Two parallel shafts, separated by a distance of 3.5 in, are to be connected by a gear set so that the output shaft rotates at 40% of the speed of the input shaft. Design a gearset to fit this situation.

SOLUTION

From the given information we have $R_2 + R_3 = 3.5$ in, and from Eq. (6.5) we find $\omega_3/\omega_2 = R_2/R_3 = 0.40$. We solve these equations simultaneously to find $R_2 = 1.0$ in and $R_3 = 2.5$ in. Next we choose the size of the teeth by tentatively picking $P = 10$ teeth/in, and from Eq. (6.1) we find the numbers of teeth on the two gears to be $N_2 = 2PR_2 = 20$ teeth and $N_3 = 2PR_3 = 50$ teeth. This choice of P for tooth size must later be checked for possible undercutting, as we will study in Section 6.7, and for strength and wear of the teeth.[1]

6.5 FUNDAMENTALS OF GEAR-TOOTH ACTION

In order to illustrate the fundamentals we now proceed, step by step, through the actual graphical layout of a pair of spur gears. The dimensions used are those of Example 6.1 above assuming standard 20° full-depth involute tooth form as specified in Table 6.2. The various steps in correct order are illustrated in Figs. 6.8 and 6.9 and are as follows:

STEP 1   Calculate the two pitch circle radii, R_2 and R_3, as in Example 6.1 and draw the two pitch circles tangent to each other, identifying O_2 and O_3 as the two shaft centers (Fig. 6.8).

STEP 2   Draw the common tangent to the pitch circles perpendicular to the line of centers and through the pitch point P (Fig. 6.8). Draw the *line of action* at an angle equal to the *pressure angle* $\phi = 20°$ from the common tangent. This line of action corresponds to the generating line discussed in Section 6.3; it is always normal to the involute curves and always passes through the pitch point. It is called the line of action or pressure line because, assuming no friction during operation, the resultant tooth force acts along this line.

STEP 3   Through the centers of the two gears, draw the two perpendiculars O_2A and O_3B to the line of action (Fig. 6.8). Draw the two *base circles* with radii of $r_2 = O_2A$ and $r_3 = O_3B$; these correspond to the base cylinders of Section 6.3.

STEP 4   From Table 6.2, continuing with $P = 10$ teeth/in, the addendum for the gears is found to be

$$a = \frac{1}{P} = \frac{1}{10} = 0.10 \text{ in}$$

Adding this to each of the pitch circle radii, draw the two addendum circles that define the top lands of the teeth on each gear. Carefully identify and label point C where the addendum circle of gear 3 intersects the line of action (Fig. 6.9). Similarly, identify and label

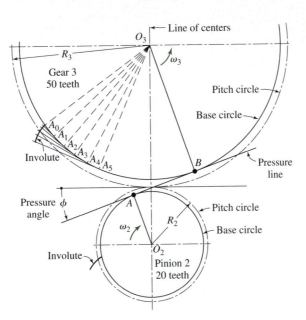

Figure 6.8 Partial gear pair layout.

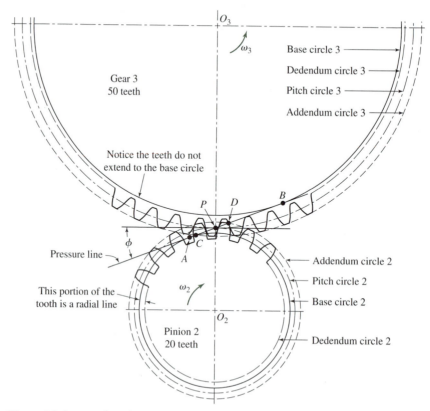

Figure 6.9 Layout of a pair of spur gears.

point D where the addendum circle of gear 2 intersects the line of action. Visualizing the rotation of the two gears in the directions shown, we see that contact is not possible before point C because the teeth of gear 3 are not of sufficient height; thus C is the first point of contact between this pair of teeth. Similarly, the teeth of gear 2 are too short to allow further contact after reaching point D; thus contact between one or more pairs of mating teeth continues between C and D and then ceases.

Steps 1 through 4 are critical for verifying the choice of any gear pair. We will continue with the diagram shown in Fig. 6.9 when we check for interference, undercutting, and contact ratio in later sections. However, in order to complete our visualization of gear tooth action, let us first proceed to the construction of the complete involute tooth shapes as shown in Fig. 6.8.

STEP 5 From Table 6.2 the dedendum for each gear is found to be

$$d = \frac{1.25}{P} = \frac{1.25}{10} = 0.125 \text{ in}$$

Subtracting this from each of the pitch circle radii, draw the two dedendum circles that define the bottom lands of the teeth on each gear (Fig. 6.9). Note that the dedendum circles often lie quite close to the base circles; however, they have distinctly different meanings.

In this example, the dedendum circle of gear 3 is larger than its base circle; however, the dedendum circle of the pinion 2 is smaller than its base circle.

STEP 6   Generate an involute curve on each base circle as illustrated for gear 3 in Fig. 6.8. This is done by first dividing a portion of the base circle into a series of equal small parts A_0, A_1, A_2, and so on. Next the radial lines O_3A_0, O_3A_1, O_3A_2, and so on, are constructed, and tangents to the base circle are drawn perpendicular to each of these. The involute begins at A_0. The second point is obtained by striking an arc with center A_1 and radius A_0A_1 up to the tangent line through A_1. The next point is found by striking a similar arc with center at A_2, and so on. This construction is continued until the involute curve is generated far enough to meet the addendum circle of gear 3. If the dedendum circle lies inside of the base circle, as is true for pinion 2 of this example, then, except for the fillet, the curve is extended inward to the dedendum circle by a radial line; this portion of the curve is not involute.

STEP 7   Using cardboard or, preferably, a sheet of clear plastic cut a template for the involute curve and mark on it the center point of the corresponding gear. Notice that two templates are needed because the involute curves are different for gears 2 and 3.

STEP 8   Calculate the circular pitch using Eq. (6.4):

$$p = \frac{\pi}{P} = \frac{\pi}{10 \text{ teeth/in}} = 0.314\ 16 \text{ in/tooth}$$

This distance from one tooth to the next is now marked along the pitch circle and the template is used to draw the involute portion of each tooth (Fig. 6.9). The width of a tooth and of a tooth space are each equal to half of the circular pitch or $(0.314\ 16)/2 = 0.157\ 08$ in. These distances are marked along the pitch circle, and the same template is turned over and used to draw the opposite sides of the teeth. The portion of the tooth space between the clearance and the dedendum may be used for a fillet radius. The top and bottom lands are now drawn as circular arcs along the addendum and dedendum circles to complete the tooth profiles. The same process is performed on the other gear using the other template.

Remember that steps 5 through 8 are not necessary for the proper design of a gear set. They are only included here to help us to visualize the relation between real tooth shapes and the theoretical properties of the involute curve.

Involute Rack   We may imagine a *rack* as a spur gear having an infinitely large pitch diameter. Therefore, in theory, a rack has an infinite number of teeth and its base circle is located an infinite distance from the pitch point. For involute teeth, the curves on the sides of the teeth become straight lines making an angle with the line of centers equal to the pressure angle. The addendum and dedendum distances are the same as shown in Table 6.2. Figure 6.10 shows an involute rack in mesh with the pinion of the previous example.

Base Pitch   Corresponding sides of involute teeth are parallel curves. The *base pitch* is the constant and fundamental distance between these curves—that is, the distance from one tooth to the next, measured along the common normal to the tooth profiles which is the line of action (Fig. 6.10). The base pitch p_b and the circular pitch p are related as follows:

$$p_b = p \cos \phi \qquad (6.6)$$

The base pitch is a much more fundamental measurement as we will see next.

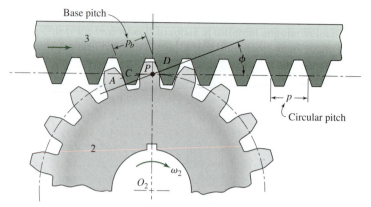

Figure 6.10 Involute pinion and rack.

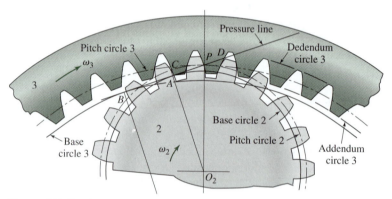

Figure 6.11 Involute pinion and internal gear.

Internal Gear   Figure 6.11 depicts the pinion of the preceding example in mesh with an *internal,* or *annular,* gear. With internal contact, both centers are on the same side of the pitch point. Thus the positions of the addendum and dedendum circles of an internal gear are reversed with respect to the pitch circle; the addendum circle of the internal gear lies *inside* the pitch circle while the dedendum circle lies *outside* the pitch circle. The base circle lies inside the pitch circle as with an external gear, but is now near the addendum circle. Otherwise, Fig. 6.11 is constructed the same as was Fig. 6.9.

6.6 THE MANUFACTURE OF GEAR TEETH

There are many ways of manufacturing the teeth of gears; for example, they can be made by sand casting, shell molding, investment casting, permanent-mold casting, die casting, or centrifugal casting. They can be formed by the powder-metallurgy process, or a single bar of aluminum can be formed by extrusion and then sliced into gears. Gears that carry large loads in comparison with their sizes are usually made of steel and are cut with either *form*

Figure 6.12 Manufacture of gear teeth by a form cutter. (*a*) A single-tooth involute hob. (*b*) Machining of a single tooth space. (Courtesy of Gleason Works, Rochester, NY.)

cutters or *generating cutters*. In form cutting, the cutter is of the exact shape of the tooth space. For generating, a tool having a shape different from the tooth space is moved relative to the gear blank to obtain the proper shape for the teeth.

Probably the oldest method of cutting gear teeth is *milling*. A form milling cutter corresponding to the shape of the tooth space, such as that shown in Fig. 6.12*a*, is used to machine one tooth space at a time, as shown in Fig. 6.12*b*, after which the gear is indexed through one circular pitch to the next position. Theoretically, with this method, a different cutter is required for each gear to be cut because, for example, the shape of the tooth space in a 25-tooth gear is different from the shape of the tooth space in, say, a 24-tooth gear. Actually, the change in tooth space shape is not too great, and eight form cutters can be used to cut any gear in the range from 12 teeth to a rack with reasonable accuracy. Of course, a separate set of form cutters is required for each pitch.

Shaping is a highly favored method of *generating* gear teeth. The cutting tool may be either a rack cutter or a pinion cutter. The operation is explained by reference to Fig. 6.13. For shaping, the reciprocating cutter is first fed into the gear blank until the pitch circles are tangent. Then, after each cutting stroke, the gear blank and the cutter roll slightly on their pitch circles. When the blank and cutter have rolled by a total distance equal to the circular pitch, one tooth has been generated and the cutting continues with the next tooth until all teeth have been cut. Shaping of an internal gear with a pinion cutter is shown in Fig. 6.14.

Hobbing is another method of generating gear teeth which is quite similar to shaping them with a rack cutter. However, hobbing is done with a special tool called a *hob,* which is a cylindrical cutter with one or more helical threads quite like a screw-thread tap; the threads have straight sides like a rack. A number of different gear hobs are displayed in Fig. 6.15. A view of the hobbing of a gear is shown in Fig. 6.16. The hob and the gear blank are both rotated continuously at the proper angular velocity ratio, and the hob is fed slowly across the face of the blank to cut the full thickness of the teeth.

Following the cutting process, grinding, lapping, shaving, and burnishing are often used as final finishing processes when tooth profiles of very good accuracy and surface finish are desired.

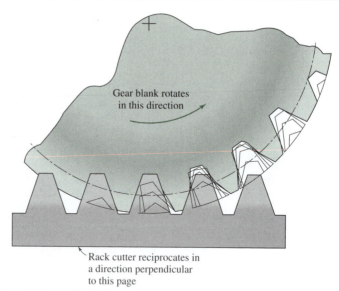

Gear blank rotates
in this direction

Rack cutter reciprocates in
a direction perpendicular
to this page

Figure 6.13 Shaping of involute teeth with a rack cutter.

Figure 6.14 Shaping of an
internal gear with a pinion cutter.
(Courtesy of Gleason Works,
Rochester, NY.)

Figure 6.15 A variety of
involute gear hobs. (Courtesy of
Gleason Works, Rochester, NY.)

Figure 6.16 The hobbing of a gear. (Courtesy of Gleason Works, Rochester, NY.)

6.7 INTERFERENCE AND UNDERCUTTING

Figure 6.17 shows the pitch circles of the same gears used for discussion in Section 6.5. Let us assume that the pinion is the driver and that it is rotating clockwise.

We saw in Section 6.5 that, for involute teeth, contact always takes place along the line of action AB. Contact begins at point C where the addendum circle of the driven gear crosses the line of action. Thus initial contact is on the tip of the driven gear tooth and on the flank of the pinion tooth.

As the pinion tooth drives the gear tooth, both approach the pitch point P. Near the pitch point, contact slides *up* the flank of the pinion tooth and *down* the face of the gear tooth. At the pitch point, contact is at the pitch circles; note that P is the instant center, and therefore the motion must be rolling with no slip at that point. Note also that this is the only location where the motion can be true rolling.

As the teeth recede from the pitch point, the point of contact continues to travel in the same direction as before along the line of action. Contact continues to slide *up* the face of the pinion tooth and *down* the flank of the gear tooth. The last point of contact occurs at the tip of the pinion and the flank of the gear tooth, at the intersection D of the line of action and the addendum circle of the pinion.

The *approach* phase of the motion is the period between the initial contact and the pitch point. The *angles of approach* are the angles through which the two gears rotate as the point of contact progresses from C to P. However, reflecting on the unwrapping cord analogy of Fig. 6.6, we see that the distance CP is equal to a length of cord unwrapped from the base circle of the pinion during the approach phase of the motion. Similarly, an equal

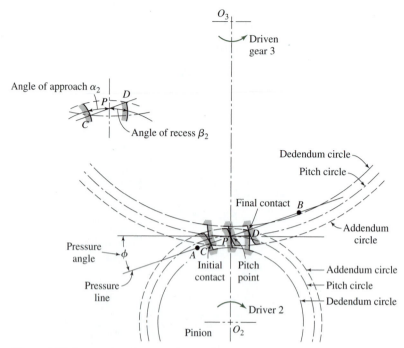

Figure 6.17 Approach and recess phases of gear tooth action.

amount of cord has wrapped onto the gear during that same phase. Thus the angles of approach for the pinion and the gear, in radians, must be

$$\alpha_2 = \frac{CP}{r_2} \quad \text{and} \quad \alpha_3 = \frac{CP}{r_3} \tag{6.7}$$

The *recess* phase of the motion is the period during which contact progresses from the pitch point P to final contact at point D. The *angles of recess* are the angles through which the two gears rotate as the point of contact progresses from P to D. Again, from the un-wrapping cord analogy, we find these angles, in radians, to be

$$\beta_2 = \frac{PD}{r_2} \quad \text{and} \quad \beta_3 = \frac{PD}{r_3} \tag{6.8}$$

If the teeth were to come into contact such that they are not conjugate, this is called *interference* (consider Fig. 6.18). Illustrated here are two 16-tooth $14\frac{1}{2}°$ pressure angle gears with full-depth involute teeth.* The driver, gear 2, turns clockwise. As with previous figures, the points labeled A and B indicate the points of tangency of the line of action with the two base circles, while the points labeled C and D indicate the initial and final points of contact. Now notice that the points C and D are *outside* of points A and B. This indicates interference.

---

*Such gears came from an older standard and are now obsolete. They are chosen here only to illustrate an example of interference.

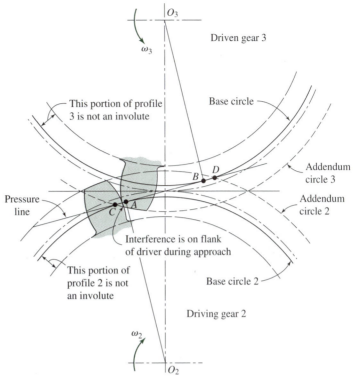

Figure 6.18 Interference in gear tooth action.

The interference is explained as follows. Contact begins when the tip of the driven gear 3 contacts the flank of the driving tooth. In this case the flank of the driving tooth first makes contact with the driven tooth at point C, and this occurs *before* the involute portion of the driving tooth comes within range. In other words, contact occurs before the two teeth become tangent. The actual effect is that the tip of the driven gear interferes with and digs out the nontangent flank of the driver.

In this example a similar effect occurs again as the teeth leave contact. Contact should end at or before point B. Because, for this example, it does not end until point D, the effect is for the tip of the driving tooth to interfere with and dig out the nontangent flank of the driven tooth.

When gear teeth are produced by a generating process, interference is automatically eliminated because the cutting tool removes the interfering portion of the flank. This effect is called *undercutting;* if undercutting is at all pronounced, the undercut tooth can be considerably weakened. Thus the effect of eliminating interference by a generation process is merely to substitute another problem for the original.

The importance of the problem of teeth that have been weakened by undercutting cannot be overemphasized. Of course, interference can be eliminated by using more teeth on the gears. However, if the gears are to transmit a given amount of power, more teeth can be used only by increasing the pitch diameter. This makes the gears larger, which is seldom desirable. It also increases the pitch-line velocity, which makes the gears noisier and somewhat reduces the power transmission, although not in direct proportion. In general, however, the use of more teeth to eliminate interference or undercutting is seldom an acceptable solution.

Another method of reducing interference and the resulting undercutting is to employ a larger pressure angle. The larger pressure angle creates smaller base circles, so that a greater portion of the tooth profile has an involute shape. In effect, this means that fewer teeth can be used; as a result, gears with larger pressure angle are often smaller.

Of course, the use of standard gears is far less expensive than manufacturing specially made nonstandard gears. However, as shown in Table 6.2, gears with larger pressure angles can be found without deviating from the standards.

One more way to eliminate or reduce interference is to use gears with shorter teeth. If the addendum distance is reduced, then points C and D move inward. One way to do this is to purchase standard gears and then grind the tops of the teeth to a new addendum distance. This, of course, makes the gears nonstandard and causes concern about repair or replacement, but it can be effective in eliminating interference. Again, careful study of Table 6.2 shows that this is possible by use of the 20° stub teeth gear standard.

6.8 CONTACT RATIO

The zone of action of meshing gear teeth is shown in Fig. 6.19, where tooth contact begins and ends at the intersections of the two addendum circles with the line of action. As always, initial contact occurs at C and final contact at D. Tooth profiles drawn through these points intersect the base circle at points c and d. Thinking back to our analogy of the unwrapping cord of Fig. 6.6, the linear distance CD, measured along the line of action, is equal to the arc length cd, measured along the base circle.

Consider a situation in which the arc length cd, or distance CD, is exactly equal to the base pitch p_b of Eq. (6.6). This means that one tooth and its space spans the entire arc cd. In other words, when a tooth is just beginning contact at C, the tooth ahead of it is just ending its contact at D. Therefore, during the tooth action from C to D there is exactly one pair of teeth in contact.

Next, consider a situation for which the arc length cd, or distance CD, is greater than the base pitch, but not very much greater, say $cd = 1.2p_b$. This means that when one pair of teeth is just entering contact at C, the previous pair, already in contact, has not yet reached D. Thus, for a short time, there are two pairs of teeth in contact, one in the vicinity

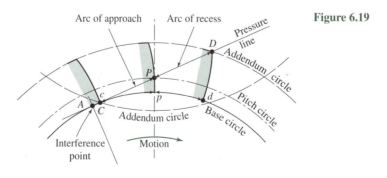

Figure 6.19

of C and the other nearing D. As meshing proceeds, the previous pair reaches D and ceases contact, leaving only a single pair of teeth in contact again, until the procedure repeats itself with the next pair of teeth.

Because of the nature of this tooth action, with one, two, or even more pairs of teeth in contact simultaneously, it is convenient to define the term contact ratio m_c as

$$m_c = \frac{CD}{p_b} \tag{6.9}$$

This is a value for which the next lower integer indicates the average number of pairs of teeth in contact.

The distance CD is quite convenient to measure if we are working graphically by making a drawing like Fig. 6.20 or Fig. 6.9. However, the distances CP and PD can also be found analytically. From triangles O_3BC and O_3BP we can write

$$CP = \sqrt{(R_3 + a)^2 - (R_3 \cos\phi)^2} - R_3 \sin\phi \tag{6.10}$$

Similarly, from triangles O_2AD and O_2AP we have

$$PD = \sqrt{(R_2 + a)^2 - (R_2 \cos\phi)^2} - R_2 \sin\phi \tag{6.11}$$

The contact ratio is then obtained by substituting the sum of Eqs. (6.10) and (6.11) into Eq. (6.9).

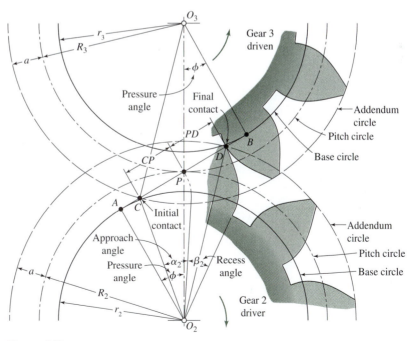

Figure 6.20

We should note, however, that Eqs. (6.10) and (6.11) are only valid for the conditions where

$$CP \leq R_2 \sin \phi \quad \text{and} \quad PD \leq R_3 \sin \phi \qquad (6.12)$$

because proper contact cannot begin before point A or end after point B. If either of these inequalities fails, then the gear teeth have interference and undercutting will result.

6.9 VARYING THE CENTER DISTANCE

Figure 6.21a illustrates a pair of meshing gears having $20°$ involute full-depth teeth. Because both sides of the teeth are in contact, the center distance O_2O_3 cannot be reduced without jamming or deforming the teeth. However, Fig. 6.21b shows the same pair of gears, but mounted with a slightly increased distance between the shaft centers O_2O_3, as might happen through the accumulation of tolerances of surrounding parts. Clearance, or *backlash,* now exists between the teeth, as shown.

When the center distance is increased, the base circles of the two gears do not change; they are fundamental to the shapes of the gears once manufactured. However, review of Fig. 6.6 shows that the same involute tooth shapes still touch as conjugate curves and the fundamental law of gearing is still satisfied. However, the larger center distance results in

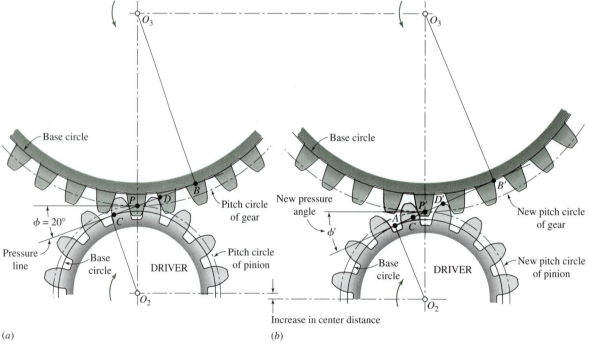

(a)   (b)

Figure 6.21 Effect of increased center distance on the action of involute gearing; mounting the gears at (a) normal center distance and (b) increased center distance.

an increase of the pressure angle and larger pitch circles passing through a new adjusted pitch point.

In Fig. 6.21b we can see that the triangles $O_2A'P'$ and $O_3B'P'$ are still similar to each other, though they are both modified by the change in pressure angle. Also, the distances O_2A' and O_3B' are the base circle radii and have not changed. Therefore, the ratio of the new pitch radii, O_2P' and O_3P', and the new velocity ratio remain the same as the original design.

Another effect, observable in Fig. 6.21, of increasing the center distance is the shortening of the path of contact. The original path of contact CD in Fig. 6.21a is shortened to $C'D'$ in Fig. 6.21b. The contact ratio, Eq. (6.9), is also reduced as the path of contact $C'D'$ is shortened. Because a contact ratio of less than unity would imply periods during which no teeth would be in contact at all, the center distance must not be larger than that corresponding to a contact ratio of unity.

6.10 INVOLUTOMETRY

The study of the geometry of the involute curve is called *involutometry*. In Fig. 6.22 a base circle with center at O is used to generate the involute BC. AT is the generating line, ρ is the instantaneous radius of curvature of the involute, and r is the radius to any point T on the curve. If we designate the radius of the base circle as r_b, the line AT has the same length as the arc AB and so

$$\rho = r_b(\alpha + \varphi) \tag{a}$$

where α is the angle between the origin of the involute and the radius AT, and φ is the angle between the radius of the base circle OA and the radius AT. Because OAT is a right triangle,

$$\rho = r_b \tan \varphi \tag{6.13}$$

Figure 6.22

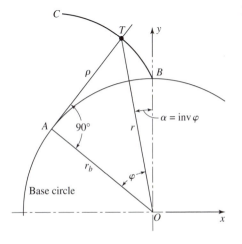

Solving Eqs. (*a*) and (6.13) simultaneously to eliminate ρ and r_b gives

$$\alpha = \tan \varphi - \varphi$$

which can be written

$$\text{inv } \varphi = \tan \varphi - \varphi \tag{6.14}$$

and defines the involute function. The angle φ in this equation is the variable involute pressure angle, and it must be specified in radians. Once φ is known, inv φ can readily be determined. The inverse problem, when inv φ is given and φ is to be found, is more difficult. One approach is to expand Eq. (6.14) in an infinite series and to employ the first several terms to obtain a numerical approximation. Another approach is to use a root finding technique.[2] Here we include Appendix Table 6 in which the value of φ is tabulated and can be found directly, in degrees.

Referring again to Fig. 6.22, we see that

$$r = \frac{r_b}{\cos \varphi} \tag{6.15}$$

To illustrate the use of the relations obtained above, let us determine the tooth dimensions of Fig. 6.23. Here, only the portion of the tooth extending above the base circle has been drawn, and the thickness of the tooth t_p at the pitch circle (point A), equal to half of the circular pitch, is given. The problem is to determine the tooth thickness at some other point, say point T. The various quantities shown in Fig. 6.23 are identified as follows:

r_b = radius of base circle
r_p = radius of pitch circle

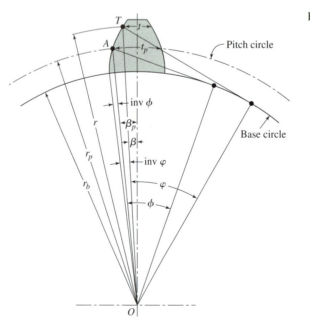

Figure 6.23

r = radius at which tooth thickness is to be determined
t_p = arc tooth thickness at pitch circle
t = arc tooth thickness to be determined
ϕ = pressure angle corresponding to pitch radius r_p
φ = pressure angle corresponding to point T
β_p = angular half-tooth thickness at pitch circle
β = angular half-tooth thickness at point T

The half-tooth thicknesses at points A and T are

$$\frac{t_p}{2} = \beta_p r_p, \qquad \frac{t}{2} = \beta r \tag{b}$$

so that

$$\beta_p = \frac{t_p}{2r_p}, \qquad \beta = \frac{t}{2r} \tag{c}$$

Now we can write

$$\text{inv}\,\varphi - \text{inv}\,\phi = \beta_p - \beta = \frac{t_p}{2r_p} - \frac{t}{2r} \tag{d}$$

The tooth thickness at point T is obtained by solving Eq. (d) for t:

$$t = 2r\left(\frac{t_p}{2r_p} + \text{inv}\,\phi - \text{inv}\,\varphi\right) \tag{6.16}$$

EXAMPLE 6.2

A gear has 20° involute teeth, cut full-depth, with a diametral pitch of 2 teeth/in, and 22 teeth. Find the thickness of the teeth at the base circle and at the addendum circle.

SOLUTION

By the equations of Section 6.1 and Table 6.2 we find $a = 0.500$ in, $d = 0.625$ in, $r_p = R = 5.500$ in, and $p = 1.571$ in. The radius of the base circle is found from Eq. (6.15):

$$r_b = r_p \cos\phi = 5.500 \cos 20° = 5.168 \text{ in}$$

The thickness of the tooth at the pitch circle is

$$t_p = \frac{p}{2} = \frac{1.571}{2} = 0.785 \text{ in}$$

Converting the tooth pressure angle of 20° to radians gives $\phi = 0.349$ rad. Then

$$\text{inv}\,\phi = \tan 0.349 - 0.349 = 0.014\,9 \text{ rad}$$

At the base circle $\varphi_b = 0$, so inv $\varphi_b = 0$. By Eq. (6.16) the tooth thickness at the base circle is

$$t_b = 2r_b \left[\frac{t_p}{2r_p} + \text{inv } \phi - \text{inv } \varphi \right] = 2(5.168) \left[\frac{0.785}{2(5.500)} + 0.014\,9 - 0 \right] = 0.892 \text{ in} \quad \textit{Ans.}$$

The radius of the addendum circle is $r_a = 6.000$ in. The involute pressure angle corresponding to this radius, from Eq. (6.15), is

$$\varphi = \cos^{-1} \left(\frac{r_b}{r} \right) = \cos^{-1} \left(\frac{5.168}{6.000} \right) = 0.533 \text{ rad}$$

Thus

$$\text{inv } \varphi = \tan 0.533 - 0.533 = 0.056\,9 \text{ rad}$$

and Eq. (6.16) gives the tooth thickness at the addendum circle as

$$t_a = 2r_a \left[\frac{t_p}{2r_p} + \text{inv } \phi - \text{inv } \varphi \right] = 2(6.000) \left[\frac{0.785}{2(5.500)} + 0.014\,9 - 0.056\,9 \right] = 0.352 \text{ in} \quad \textit{Ans.}$$

6.11 NONSTANDARD GEAR TEETH

In this section we examine the effects obtained by modifying such things as pressure angle, tooth depth, addendum, or center distance. Some of these modifications do not eliminate interchangeability; all of them are made with the intent of obtaining improved performance.

The designer is often under great pressure to produce gear designs that are small and yet will transmit large amounts of power. Consider, for example, a gearset that must have a 4:1 velocity ratio. If the smallest pinion that will carry the load has a pitch diameter of 2 in, the mating gear will have a pitch diameter of 8 in, making the overall space required for the two gears more than 10 in. On the other hand, if the pitch diameter of the pinion can be reduced by only $\frac{1}{4}$ in the pitch diameter of the gear is reduced by a full 1 in and the overall size of the gearset is reduced by $1\frac{1}{4}$ in. This reduction assumes considerable importance when it is realized that the sizes of associated machine elements, such as shafts, bearings, and enclosures, are also reduced. If a tooth of a certain pitch is required to carry the load, the only method of decreasing the pinion diameter is to use fewer teeth. However, we have already seen that problems involving interference, undercutting, and contact ratio are encountered when the tooth numbers are made too small. Thus the three principal reasons for employing nonstandard gears are to eliminate undercutting, to prevent interference, and to maintain a reasonable contact ratio. It should be noted too that, if a pair of gears is manufactured of the same material, the pinion is the weaker and is subject to greater wear because its teeth are in contact a greater portion of the time. Then, any undercutting weakens the tooth that is already the weaker of the two. Thus, another objective of nonstandard gears is to gain a better balance of strength between the pinion and the gear.

As an involute curve is generated from its base circle, its radius of curvature becomes larger and larger. Near the base circle the radius of curvature is quite small, being exactly zero at the base circle. Contact near this region of sharp curvature should be avoided if possible because of the difficulty of obtaining good cutting accuracy in areas of small curvature, and because the contact stresses are likely to be very high. Nonstandard gears present the opportunity of designing to avoid these sensitive areas.

Clearance Modifications

A larger fillet radius at the root of the tooth increases the fatigue strength of the tooth and provides extra depth for shaving the tooth profile. Because interchangeability is not lost, the clearance is sometimes increased to $0.300/P$ or $0.400/P$ to obtain space for a larger fillet radius.

Center-Distance Modifications

When gears of low tooth numbers are to be paired with each other, or with larger gears, reduction in interference and improvement in the contact ratio can be obtained by increasing the center distance. Although such a system changes the tooth proportions and the pressure angle of the gears, the resulting tooth shapes can be generated with rack cutters (or hobs) of standard pressure angles or with standard pinion shapers. Before introducing this system, it will be of value to develop additional relations about the geometry of gears.

The first new relation is for finding the thickness of a tooth that is cut by a rack cutter (or hob) when the pitch line of the rack cutter has been displaced or offset a distance e from the pitch circle of the gear being cut. What we are doing here is moving the rack cutter away from the center of the gear being cut. Stated another way, suppose the rack cutter does not cut as deeply into the gear blank and the teeth are not cut to full depth. This produces teeth that are thicker than the standard, and this thickness will now be found. Figure 6.24a shows the problem, and Fig. 6.24b shows the solution. The increase of tooth thickness at the pitch circle is $2e \tan \phi$, so that

$$t = 2e \tan \phi + \frac{p}{2} \tag{6.17}$$

where ϕ is the pressure angle of the rack cutter and t is the thickness of the gear tooth on its own pitch circle.

Now suppose that two gears of different tooth numbers have both been cut with the cutter offset from their pitch circles as in the previous paragraph. Because the teeth of both

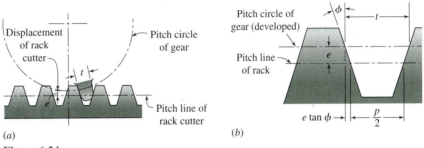

(a)   (b)

Figure 6.24

have been cut with offset cutters, they will mate at a new pressure angle and with new pitch circles and consequently new center distances. The word *new* is used here in the sense of being nonstandard. Our problem is to determine the radii of these new pitch circles and the value of the new pressure angle.

In the following notation the word *standard* refers to values that would have been obtained had the usual, or standard, systems been employed to obtain the dimensions:

ϕ = pressure angle of rack generating cutter
ϕ' = new pressure angle at which gears will mate
R_2 = standard pitch radius of pinion
R_2' = new pitch radius of pinion when meshing with given gear
R_3 = standard pitch radius of gear
R_3' = new pitch radius of gear when meshing with given pinion
t_2 = thickness of pinion tooth at standard pitch radius R_2
t_2' = thickness of pinion tooth at new pitch radius R_2'
t_3 = thickness of gear tooth at standard pitch radius R_3
t_3' = thickness of gear tooth at new pitch radius R_3'

From Eq. (6.16),

$$t_2' = 2R_2' \left(\frac{t_2}{2R_2} + \operatorname{inv} \phi - \operatorname{inv} \phi' \right) \tag{a}$$

$$t_3' = 2R_3' \left(\frac{t_3}{2R_3} + \operatorname{inv} \phi - \operatorname{inv} \phi' \right) \tag{b}$$

The sum of these two thicknesses must be the same as the circular pitch, or, from Eq. (6.3),

$$t_2' + t_3' = p = \frac{2\pi R_2'}{N_2} \tag{c}$$

The pitch diameters of a pair of mating gears are proportional to their tooth numbers, and so

$$R_3 = \frac{N_3}{N_2} R_2 \quad \text{and} \quad R_3' = \frac{N_3}{N_2} R_2' \tag{d}$$

Substituting Eqs. (a), (b), and (d) into Eq. (c) and rearranging gives

$$\operatorname{inv} \phi' = \frac{N_2(t_2' + t_3') - 2\pi R_2}{2R_2(N_2 + N_3)} + \operatorname{inv} \phi \tag{6.18}$$

Equation (6.18) gives the pressure angle ϕ' at which a pair of gears will operate when the tooth thicknesses on their standard pitch circles are modified to t_2' and t_3'.

It has been demonstrated that although the base circle of a gear is fundamental to its shape and fixed once the gear is generated, gears have no pitch circles until a pair of them are brought into contact. Bringing a pair of gears into contact creates a pair of pitch circles that are tangent to each other at the pitch point. Throughout this discussion, the idea of a pair of so-called standard pitch circles has been used in order to define a certain point on the involute curves. These standard pitch circles, we have seen, are the ones that would come into existence when the gears are paired *if the gears are not modified from the standard dimensions*. On the other hand, the base circles are fixed circles that are not changed by tooth modifications. The base circle remains the same whether the tooth dimensions are changed or not; so we can determine the base circle radius using either the standard pitch circle or the new pitch circle. Thus, from Eq. (6.15),

$$R_2 \cos \phi = R_2' \cos \phi'$$

or

$$R_2' = \frac{R_2 \cos \phi}{\cos \phi'} \tag{6.19}$$

Similarly, for the gear,

$$R_3' = \frac{R_3 \cos \phi}{\cos \phi'} \tag{6.20}$$

These equations give the values of the actual pitch radii when the two gears with modified teeth are brought into mesh without backlash. The new center distance is, of course, the sum of these radii.

All the necessary relations have now been developed to create nonstandard gears with changes in the center distance. The use of these relations is best illustrated by an example. Figure 6.25 is a drawing of a 20° pressure angle, 1 tooth/in diametral pitch, 12-tooth pinion generated with a rack cutter with a standard clearance of $0.250/P$. In the 20° full-depth

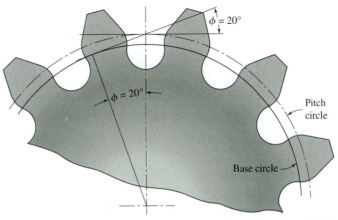

Figure 6.25 Standard 20°, 1-tooth/in diametral pitch, 12-tooth full-depth involute gear showing undercut.

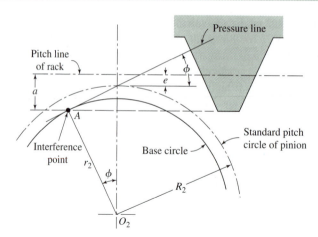

Figure 6.26 Offset of a rack cutter to cause its addendum line to pass through the interference point.

system, interference is severe whenever the number of teeth is less than 14. The resulting undercutting is evident in the drawing. If this pinion were mated with a standard 40-tooth gear, the contact ratio would be 1.41, which can easily be verified by Eq. (6.9).

In an attempt to eliminate the undercutting, improve the tooth action, and increase the contact ratio, let the 12-tooth pinion be cut from a larger blank. Then let the resulting pinion be paired again with the 40-tooth standard gear and let us determine the degree of improvement. If we designate the pinion as subscript 2 and designate the gear as 3, the following values are found:

$$P = 1 \text{ tooth/in}, \qquad \phi = 20°, \qquad R_2 = 6 \text{ in}, \qquad R_3 = 20 \text{ in}$$

$$p = 3.142 \text{ in/tooth}, \qquad t_3 = 1.571 \text{ in}, \qquad N_2 = 12 \text{ teeth}, \qquad N_3 = 40 \text{ teeth}$$

We shall offset the rack cutter so that its addendum passes through the interference point A of the pinion—that is, the point of tangency of the 20° line of action and the base circle—as shown in Fig. 6.26. From Eq. (6.15) we have

$$r_2 = R_2 \cos \phi \tag{e}$$

Then, from Fig. 6.26,

$$e = a + r_2 \cos \phi - R_2 \tag{f}$$

Substituting Eq. (e) into Eq. (f) gives

$$e = a + R_2 \cos^2 \phi - R_2 = a - R_2 \sin^2 \phi \tag{6.21}$$

For a standard rack, from Table 6.2, the addendum is $a = 1/P$; so, for this example, $a = 1.0$ in.

The offset to be used is

$$e = 1.0 - 6.0 \sin^2 20° = 0.298 \text{ in}$$

Then, using Eq. (6.17) for the thickness of the pinion tooth at its 6 in pitch circle, we get

$$t = 2e \tan \phi + \frac{p}{2} = 2(0.298)\tan 20° + \frac{3.142}{2} = 1.788 \text{ in}$$

The pressure angle at which these (and only these) gears will operate is found from Eq. (6.18)

$$\text{inv } \phi' = \frac{N_2(t_2' + t_3') - 2\pi R_2}{2R_2(N_2 + N_3)} + \text{inv } \phi$$

$$= \frac{12(1.788 + 1.571) - 2\pi(6.0)}{2(6.0)(12 + 40)} + \text{inv } 20°$$

$$= 0.019 \, 08 \text{ rad}$$

From Appendix Table 6, $\phi' = 21.65°$.

Using Eqs. (6.19) and (6.20), the new pitch radii are found to be

$$R_2' = \frac{R_2 \cos \phi}{\cos \phi'} = \frac{6.0 \cos 20°}{\cos 21.65°} = 6.066 \text{ in}$$

$$R_3' = \frac{R_3 \cos \phi}{\cos \phi'} = \frac{20.0 \cos 20°}{\cos 21.65°} = 20.220 \text{ in}$$

So the new center distance is

$$R_2' + R_3' = 6.066 + 20.220 = 26.286 \text{ in}$$

Notice that the center distance has not increased as much as the offset of the rack cutter. Initially a clearance of $0.25/P$ was specified, making the standard dedendums equal to $1.25/P$ according to Table 6.2. So the root radii of the two gears are

$$\text{Root radius of pinion} = \quad 6.298 - 1.250 = \quad 5.048 \text{ in}$$

$$\text{Root radius of gear} \quad = 20.000 - 1.250 = \underline{18.750 \text{ in}}$$

$$\text{Sum of root radii} \qquad\qquad\qquad = 23.798 \text{ in}$$

The difference between this sum and the center distance is the working depth plus twice the clearance. Because the clearance is 0.25 in for each gear, the working depth is

$$\text{Working depth} = 26.286 - 23.798 - 2(0.250) = 1.988 \text{ in}$$

The outside radius of each gear is the sum of the root radius, the clearance, and the working depth.

$$\text{Outside radius of pinion} = 5.048 + 0.250 + 1.988 = 7.286 \text{ in}$$

$$\text{Outside radius of gear} \quad = 18.750 + 0.250 + 1.988 = 20.988 \text{ in}$$

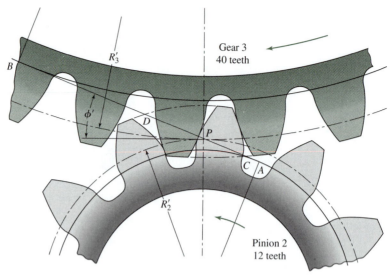

Figure 6.27

The result is shown in Fig. 6.27, and the pinion is seen to have a stronger looking form than the one of Fig. 6.25. Undercutting has been completely eliminated. The contact ratio can be obtained from Eqs. (6.9) through (6.11). The following quantities are needed:

$$\text{Outside radius of pinion} = R_2' + a = 7.286 \text{ in}$$

$$\text{Outside radius of gear} \quad = R_3' + a = 20.988 \text{ in}$$

$$r_2 = R_2 \cos \phi = 6.000 \cos 20° = 5.638 \text{ in}$$

$$r_3 = R_3 \cos \phi = 20.000 \cos 20° = 18.794 \text{ in}$$

$$p_b = p \cos \phi = 3.1416 \cos 20° = 2.952 \text{ in/tooth}$$

Therefore, for Eqs. (6.10) and (6.11) we have

$$CP = \sqrt{(R_3' + a)^2 - r_3^2} - R_3' \sin \phi$$

$$= \sqrt{(20.988)^2 - (18.794)^2} - 20.220 \sin 21.65°$$

$$= 1.883 \text{ in}$$

$$PD = \sqrt{(R_2' + a)^2 - r_2^2} - R_2' \sin \phi$$

$$= \sqrt{(7.286)^2 - (5.638)^2} - 6.066 \sin 21.65°$$

$$= 2.377 \text{ in}$$

and, finally, from Eq. (6.9), the contact ratio is

$$m_c = \frac{CP + PD}{p_b} = \frac{1.883 + 2.377}{2.952} = 1.443 \text{ teeth avg.}$$

Thus, the contact ratio has increased only slightly. The modification, however, is justified because of the elimination of undercutting which results in a very substantial improvement in the strength of the teeth.

Long-and-Short-Addendum Systems   It often happens in the design of machinery that the center distance between a pair of gears is fixed by some other design consideration or feature of the machine. In such a case, modifications to obtain improved performance cannot be made by varying the center distance.

In the previous section we saw that improved action and tooth shape can be obtained by backing out the rack cutter from the gear blank during forming of the teeth. The effect of this withdrawal is to create the active tooth profile farther away from the base circle. Examination of Fig. 6.27 reveals that more dedendum could be used on the gear (not the pinion) before the interference point is reached. If the rack cutter is advanced into the gear blank by a distance equal to the offset from the pinion blank, more of the gear dedendum will be used and at the same time the center distance will not be changed. This is called the *long-and-short-addendum system*.

In the long-and-short-addendum system there is no change in the pitch circles and consequently none in the pressure angle. The effect is to move the contact region away from the pinion center toward the gear center, thus shortening the approach action and lengthening the recess action.

The characteristics of the long-and-short-addendum system can be explained by reference to Fig. 6.28. Figure 6.28a illustrates a conventional (standard) set of gears having a dedendum equal to the addendum plus the clearance. Interference exists, and the tip of the gear tooth will have to be relieved as shown or the pinion will be undercut. This is so because the addendum circle crosses the line of action at C, outside of the tangency or interference point A; hence, the distance AC is a measure of the degree of interference.

To eliminate the undercutting or interference, the pinion addendum may be enlarged, as in Fig. 6.28b until the addendum circle of the pinion passes through the interference point (point B) of the gear. In this manner we shall be using all of the gear-tooth profile. The same whole depth may be retained; hence the dedendum of the pinion may be reduced by the same amount that the addendum is increased. This means that we must also lengthen the gear dedendum and shorten the dedendum of the mating pinion. With these changes the path of contact is the line CD of Fig. 6.27b. It is longer than the path AD of Fig. 6.28a, and so the contact ratio is higher. Notice too that the base circles, the pitch circles, the pressure angle, and the center distance have not changed. Both gears can be cut with standard cutters by advancing the cutter into the gear blank, for this modification, by a distance equal to the amount of withdrawal from the pinion blank. Finally, note that the blanks from which the pinion and gear are cut must now be of different diameters than the standard blanks.

The tooth dimensions for the long-and-short-addendum system can be determined by using the equations developed in the previous sections.

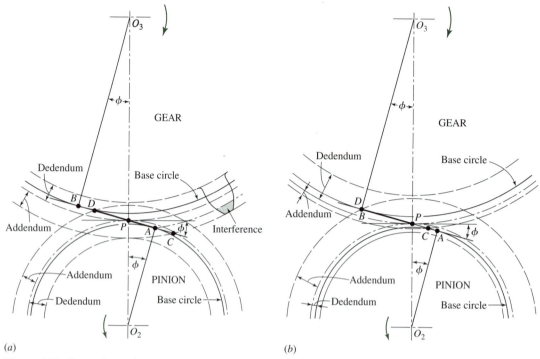

Figure 6.28 Comparison of standard gears and gears cut by the long-and-short-addendum system: (*a*) gear and pinion with standard addendum and dedendum; (*b*) gear and pinion with long-and-short addendum.

A less obvious advantage of the long-and-short-addendum system is that more recess action than approach action is obtained. The approach action of gear teeth is analogous to pushing a piece of chalk across a blackboard; the chalk screeches. But when the chalk is pulled across a blackboard, analogous to the recess action, it glides smoothly. Thus, recess action is always preferable because of the smoothness and the lower frictional forces.

NOTES

[1.]   The strength and wear of gears are covered in texts such as J. E. Shigley and C. R. Mischke, *Mechanical Engineering Design,* 6th ed., McGraw-Hill, 2001.

[2.]   See, for example, C. R. Mischke, *Mathematical Model Building,* Iowa State University Press, Ames, Iowa, 1980, Chapter 4.

PROBLEMS

6.1   Find the diametral pitch of a pair of gears having 32 and 84 teeth, respectively, whose center distance is 3.625 in.

6.2   Find the number of teeth and the circular pitch of a 6-in pitch diameter gear whose diametral pitch is 9 teeth/in.

6.3   Determine the module of a pair of gears having 18 and 40 teeth, respectively, whose center distance is 58 mm.

6.4   Find the number of teeth and the circular pitch of a gear whose pitch diameter is 200 mm if the module is 8 mm/tooth.

6.5 Find the diametral pitch and the pitch diameter of a 40-tooth gear whose circular pitch is 3.50 in/tooth.

6.6 The pitch diameters of a pair of mating gears are 3.50 in and 8.25 in, respectively. If the diametral pitch is 16 teeth/in, how many teeth are there on each gear?

6.7 Find the module and the pitch diameter of a gear whose circular pitch is 40 mm/tooth if the gear has 36 teeth.

6.8 The pitch diameters of a pair of gears are 60 mm and 100 mm, respectively. If their module is 2.5 mm/tooth, how many teeth are there on each gear?

6.9 What is the diameter of a 33-tooth gear if its circular pitch is 0.875 in/tooth?

6.10 A shaft carries a 30-tooth, 3-teeth/in diametral pitch gear that drives another gear at a speed of 480 rev/min. How fast does the 30-tooth gear rotate if the shaft center distance is 9 in?

6.11 Two gears having an angular velocity ratio of 3:1 are mounted on shafts whose centers are 136 mm apart. If the module of the gears is 4 mm/tooth, how many teeth are there on each gear?

6.12 A gear having a module of 4 mm/tooth and 21 teeth drives another gear at a speed of 240 rev/min. How fast is the 21-tooth gear rotating if the shaft center distance is 156 mm?

6.13 A 4-tooth/in diametral pitch, 24-tooth pinion is to drive a 36-tooth gear. The gears are cut on the 20° full-depth involute system. Find and tabulate the addendum, dedendum, clearance, circular pitch, base pitch, tooth thickness, base circle radii, length of paths of approach and recess, and contact ratio.

6.14 A 5-tooth/in diametral pitch, 15-tooth pinion is to mate with a 30-tooth internal gear. The gears are 20° full-depth involute. Make a drawing of the gears showing several teeth on each gear. Can these gears be assembled in a radial direction? If not, what remedy should be used?

6.15 A $2\frac{1}{2}$-tooth/in diametral pitch 17-tooth pinion and a 50-tooth gear are paired. The gears are cut on the 20° full-depth involute system. Find the angles of approach and recess of each gear and the contact ratio.

6.16 A gearset with a module of 5 mm/tooth has involute teeth with 22.5° pressure angle, and has 19 and 31 teeth, respectively. (Such gears came from an older standard and are now obsolete.) They have 1.0m for the addendum and 1.35m for the dedendum. (In SI, tooth sizes are given in modules, m, and $a = 1.0m$ means 1 module, not 1 meter.) Tabulate the addendum, dedendum, clearance, circular pitch, base pitch, tooth thickness, base circle radius, and contact ratio.

6.17 A gear with a module of 8 mm/tooth and 22 teeth is in mesh with a rack; the pressure angle is 25°. The addendum and dedendum are 1.0m and 1.25m, respectively. (In SI, tooth sizes are given in modules, m, and $a = 1.0m$ means 1 module, not 1 meter.) Find the lengths of the paths of approach and recess and determine the contact ratio.

6.18 Repeat Problem 6.15 using the 25° full-depth system.

6.19 Draw a 2-tooth/in diametral pitch, 26-tooth, 20° full-depth involute gear in mesh with a rack.
  (*a*) Find the lengths of the paths of approach and recess and the contact ratio.
  (*b*) Draw a second rack in mesh with the same gear but offset $\frac{1}{8}$ in away from the gear center. Determine the new contact ratio. Has the pressure angle changed?

6.20 through 6.24 Shaper gear cutters have the advantage that they can be used for either external or internal gears and also that only a small amount of runout is necessary at the end of the stroke. The generating action of a pinion shaper cutter can easily be simulated by employing a sheet of clear plastic. The figure illustrates one tooth of a 16-tooth pinion cutter with 20° pressure angle as it can be cut from a plastic sheet. To construct the cutter, lay out the tooth on a sheet of drawing paper. Be sure to include the clearance at the top of the tooth. Draw radial lines through the pitch circle spaced at distances equal to one fourth of the tooth thickness as shown in the figure. Now fasten the sheet of plastic to the drawing and scribe the cutout, the pitch circle, and the radial lines onto the sheet. The sheet is then removed and the tooth outline trimmed with a razor blade. A small piece of fine sandpaper should then be used to remove any burrs.

To generate a gear with the cutter, only the pitch circle and the addendum circle need be drawn. Divide the pitch circle into spaces equal to those used on the template and construct radial lines through them. The tooth outlines are then obtained by rolling the template pitch circle upon that of the gear and drawing the cutter tooth lightly for each position. The resulting generated tooth upon the gear will be evident. The following problems all employ a standard 1-tooth/in diametral pitch full depth template

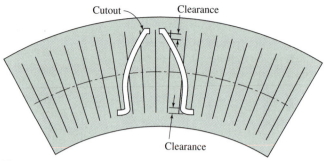

Figure P6.20

constructed as described above. In each case you should generate a few teeth and estimate the amount of undercutting.

TABLE P6.20–P6.24

Problem Number:	6.20	6.21	6.22	6.23	6.24
Number of Teeth:	10	12	14	20	36

6.25 A 10-mm/tooth module gear has 17 teeth, a 20° pressure angle, an addendum of $1.0m$, and a dedendum of $1.25m$. (In SI, tooth sizes are given in modules, m, and $a = 1.0m$ means 1 module, not 1 meter.) Find the thickness of the teeth at the base circle and at the addendum circle. What is the pressure angle corresponding to the addendum circle?

6.26 A 15-tooth pinion has $1\frac{1}{2}$-tooth/in diametral pitch 20° full-depth teeth. Calculate the thickness of the teeth at the base circle. What are the tooth thickness and the pressure angle at the addendum circle?

6.27 A tooth is 0.785 in thick at a pitch circle radius of 8 in and a pressure angle of 25°. What is the thickness at the base circle?

6.28 A tooth is 1.571 in thick at the pitch radius of 16 in and a pressure angle of 20°. At what radius does the tooth become pointed?

6.29 A 25° involute, 12-tooth/in diametral pitch pinion has 18 teeth. Calculate the tooth thickness at the base circle. What are the tooth thickness and pressure angle at the addendum circle?

6.30 A nonstandard 10-tooth 8-tooth/in diametral pitch pinion is to be cut with a $22\frac{1}{2}°$ pressure angle. (Such gears came from an older standard and are now obsolete.) What maximum addendum can be used before the teeth become pointed?

6.31 The accuracy of cutting gear teeth can be measured by fitting hardened and ground pins in diametrically opposite tooth spaces and measuring the distance over the pins. For a 10-tooth/in diametral pitch 20° full-depth involute system 96 tooth gear:
  (a) Calculate the pin diameter that will contact the teeth at the pitch lines if there is to be no backlash.
  (b) What should be the distance measured over the pins if the gears are cut accurately?

6.32 A set of interchangeable gears with a 4-tooth/in diametral pitch is cut on the 20° full-depth involute system. The gears have tooth numbers of 24, 32, 48, and 96. For each gear, calculate the radius of curvature of the tooth profile at the pitch circle and at the addendum circle.

6.33 Calculate the contact ratio of a 17-tooth pinion that drives a 73-tooth gear. The gears are 96-tooth/in diametral pitch and cut on the 20° fine pitch system.

6.34 A 25° pressure angle 11-tooth pinion is to drive a 23-tooth gear. The gears have a diametral pitch of 8 teeth/in and have stub teeth. What is the contact ratio?

6.35 A 22-tooth pinion mates with a 42-tooth gear. The gears are full depth, have a diametral pitch of 16 teeth/in, and are cut with a $17\frac{1}{2}°$ pressure angle. (Such gears came from an older standard and are now obsolete.) Find the contact ratio.

6.36 The center distance of two 24-tooth, 20° pressure angle, full-depth involute spur gears with diametral pitch of 2 teeth/in is increased by 0.125 in over the standard distance. At what pressure angle do the gears operate?

6.37 The center distance of two 18-tooth, 25° pressure angle, full-depth involute spur gears with diametral pitch of 3 teeth/in is increased by 0.0625 in over the standard distance. At what pressure angle do the gears operate?

6.38 A pair of mating gears have 24-teeth/in diametral pitch and are generated on the 20° full-depth involute system. If the tooth numbers are 15 and 50, what maximum addendums may they have if interference is not to occur?

6.39 A set of gears is cut with a $4\frac{1}{2}$-in/tooth circular pitch and a $17\frac{1}{2}^{\circ}$ pressure angle. (Such gears came from an older standard and are now obsolete.) If the gear has 240 teeth, what maximum addendum may it have in order to avoid interference?

6.40 Using the method described for Problems P6.20 through P6.24, cut a 1-tooth/in diametral pitch 20° pressure angle full-depth involute rack tooth from a sheet of clear plastic. Use a nonstandard clearance of $0.35/P$ in order to obtain a stronger fillet. This template can be used to simulate the generating action of a hob. Now, using the variable-center-distance system, generate an 11-tooth pinion to mesh with a 25-tooth gear without interference. Record the values found for center distance, pitch radii, pressure angle, gear blank diameters, cutter offset, and contact ratio. Note that more than one satisfactory solution exists.

6.41 Using the template cut in Problem 6.40 generate an 11-tooth pinion to mesh with a 44-tooth gear with the long-and-short-addendum system. Determine and record suitable values for gear and pinion addendum and dedendum and for the cutter offset and contact ratio. Compare the contact ratio with that of standard gears.

6.42 A pair of spur gears with 9 and 36 teeth are to be cut with a 20° full-depth cutter with diametral pitch of 3 teeth/in.

  (*a*) Determine the amount that the addendum of the gear must be decreased in order to avoid interference.

  (*b*) If the addendum of the pinion is increased by the same amount, determine the contact ratio.

6.43 A standard 20° pressure angle full-depth 1-tooth/in diametral pitch 20-tooth pinion drives a 48-tooth gear. The speed of the pinion is 500 rev/min. Using the position of the point of contact along the line of action as the abscissa, plot a curve showing the sliding velocity at all points of contact. Notice that the sliding velocity changes sign when the point of contact passes through the pitch point.

7

Helical Gears

When rotational motion is to be transmitted between parallel shafts, engineers often prefer to use spur gears because they are easier to design and often more economical to manufacture. However, sometimes the design requirements are such that helical gears are a better choice. This is especially true when the loads are heavy, the speeds are high, or the noise level must be kept low.

7.1 PARALLEL-AXIS HELICAL GEARS

The shape of the tooth of a helical gear is illustrated in Fig. 7.1. If a piece of paper is cut into the shape of a parallelogram and wrapped around a cylinder, the angular edge of the paper wraps into a helix. The cylinder plays the same role as the base cylinder of Chapter 6. If the paper is then unwound, each point on the angular edge generates an involute curve as was shown in Section 6.3 for spur gears. The surface obtained when every point on the angular edge of the paper generates an involute is called an *involute helicoid*. If we imagine the strip of paper as unwrapping from a base cylinder on one gear and wrapping up onto the base cylinder of another, then a line on this strip of paper generates two involute helicoids meshing as two tangent tooth shapes.

The initial contact of spur gear teeth, as we saw in the previous chapter, is a line extending across the face of the tooth. The initial contact of helical gear teeth starts as a point and changes into a line as the teeth come into more engagement; in helical gears, however, the line is diagonal across the face of the tooth. It is this gradual engagement of the teeth and the smooth transfer of load from one tooth to another that give helical gears the ability to quietly transmit heavy loads at high speeds.

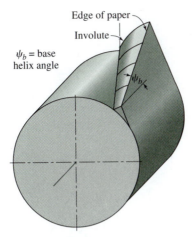

Figure 7.1 An involute helicoid.

7.2 HELICAL GEAR TOOTH RELATIONS

As shown in Fig. 7.2, to mesh properly, two parallel shaft helical gears must have equal pitches and equal helix angles, but must be of opposite hand.

Figure 7.3 represents a portion of the top view of a helical rack. Lines AB and CD are the centerlines of two adjacent helical teeth taken on the pitch plane. The angle ψ is the helix angle and is measured at the pitch diameter unless otherwise specified. The distance AC, in the plane of rotation of the gear, is the *transverse circular pitch* p_t. The distance AE is the *normal circular pitch* p_n and is related to the transverse circular pitch as follows:

$$p_n = p_t \cos \psi \qquad (7.1)$$

The distance AD is called axial pitch p_x and is

$$p_x = \frac{p_t}{\tan \psi} \qquad (7.2)$$

Because $p_n P_n = \pi$, the normal diametral pitch is

$$P_n = \frac{P_t}{\cos \psi} \qquad (7.3)$$

where P_t is the transverse diametral pitch.

Because of the angularity of the teeth, we must define two different pressure angles, the *normal pressure angle* ϕ_n and the *transverse pressure angle* ϕ_t, both shown in Fig. 7.3. They are related by

$$\tan \phi_n = \tan \phi_t \cos \psi \qquad (7.4)$$

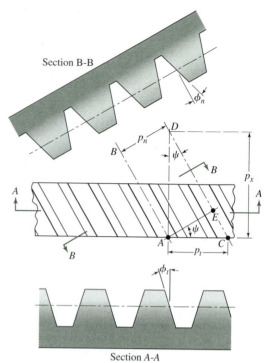

Figure 7.2 Pair of helical gears in mesh. Note the opposite hand of the two gears. (Courtesy of Gleason Works, Rochester, NY.)

Figure 7.3 Helical-gear-tooth relations.

In applying these equations it is convenient to remember that all the equations and relations that are valid for spur gears apply equally for the transverse plane of a helical gear.

A better picture of the tooth relations can be obtained by examination of Fig. 7.4. In order to obtain the geometric relations, a helical gear has been cut by the oblique plane AA at an angle ψ to a right section. For convenience, only the pitch cylinder of radius R is shown. The figure shows that the intersection of the AA plane and the pitch cylinder is an ellipse whose radius at the pitch point P is R_e. This is called the *equivalent pitch radius*, and it is the radius of curvature of the pitch surface in the normal cross section. For the condition that $\psi = 0$, this radius of curvature is $R_e = R$. If we imagine the angle ψ to be gradually increased from 0 to 90°, we see that R_e begins at a value of $R_e = R$ and increases until, when $\psi = 90°$, $R_e = \infty$.

It is shown on pages 294–295[1] that

$$R_e = \frac{R}{\cos^2 \psi} \tag{7.5}$$

where R is the pitch radius of the helical gear and R_e is the pitch radius of an equivalent spur gear. This equivalence is taken on the normal section of the helical gear.

Let us define the number of teeth on the helical gear as N and on the equivalent spur gear as N_e. Then

$$N_e = 2R_e P_n \tag{d}$$

Figure 7.4

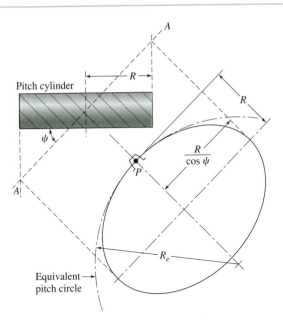

Using Eq. (7.3), we can write Eq. (*d*) as

$$N_e = 2\frac{R}{\cos^2 \psi}\frac{P_t}{\cos \psi} = \frac{N}{\cos^3 \psi} \tag{7.6}$$

7.3 HELICAL GEAR TOOTH PROPORTIONS

Except for fine pitch (diametral pitch of 20 teeth/in and finer), there is no generally accepted standard for the proportions of helical gear teeth.

In determining the tooth proportions for helical gears, it is necessary to consider the manner in which the teeth are formed. If the helical gear is to be hobbed, then tooth proportions are calculated in a plane normal to the tooth. As a general guide, tooth proportions are then often based on a normal pressure angle of $\phi_n = 20°$. Most of the proportions shown in Table 6.2 can then be used. The tooth proportions are calculated by using the normal diametral pitch P_n. These proportions are suitable for helix angles from 0° to 30°, and all helix angles can be cut with the same hob. Of course the normal diametral pitch of the hob and the gear must be the same.

If the gear is to be cut by a shaper, an alternative set of tooth proportions is used based on a transverse pressure angle of $\phi_t = 20°$ and the transverse diametral pitch P_t. For these gears the helix angles are generally restricted to 15°, 23°, 30°, or 45°; helix angles of greater than 45° are not recommended. The normal diametral pitches must be used to compute the tooth dimensions; the proportions shown in Table 6.2 are usually satisfactory. If the shaper method is used, however, the same cutter cannot be used to cut both spur and helical gears.

7.4 CONTACT OF HELICAL GEAR TEETH

For spur gears, contact between meshing teeth occurs along a line that is parallel to their axes of rotation. As shown in Fig. 7.5, contact between meshing helical gear teeth occurs along a diagonal line.

Several kinds of contact ratios are used in evaluating the performance of helical gearsets. The *transverse contact ratio* is designated by m_t and is the contact ratio in the transverse plane. It is obtained exactly as was m_c for spur gears.

The *normal contact ratio* m_n is the contact ratio in the normal section. It is also found exactly as was m_c for spur gears, but the equivalent spur gears must be used in the determination. The base helix angle ψ_b and the pitch helix angle ψ, for helical gears, are related by

$$\tan \psi_b = \tan \psi \cos \phi \tag{7.7}$$

Then the transverse and normal contact ratios are related by

$$m_n = \frac{m_t}{\cos \psi_b} \tag{7.8}$$

The *axial contact ratio* m_x, also called the *face contact ratio,* is the ratio of the face width of the gear to the axial pitch, found from Eq. (7.2). It is given by

$$m_x = \frac{F}{p_x} = \frac{F \tan \psi}{p_t} \tag{7.9}$$

where F is the face width of the helical gear. It can be seen from Fig. 7.5 that this contact ratio is greater than unity when another tooth is beginning contact solely because of the helix angle of the teeth before the previous tooth contact has finished. Note that this face contact ratio, also called *overlap,* has no parallel for spur gears and that, because of the helix angle, this face contact ratio can be made greater than unity for helical gears by the choice of face width in spite of the choice of tooth size. If the face width is made greater than the axial pitch, continuous contact of at least one tooth is assured. This means that fewer teeth may be used on helical pinions than on spur pinions. The overlapping action also results in smoother operation of the gears. Note also that the face contact ratio depends solely on the geometry of a single gear, while the transverse and normal contact ratios depend upon the geometry of a pair of mating gears.

The *total contact ratio* is the sum of the face contact ratio m_x and the transverse contact ratio m_t. In a sense this sum gives the average total number of teeth in contact.

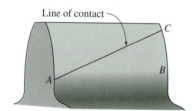

Line of contact

Figure 7.5 When contact of another tooth is just beginning at *A*, contact at the other end of the tooth has already progressed from *B* to *C*.

7.5 REPLACING SPUR GEARS WITH HELICAL GEARS

Because of their ability to carry heavy loads with little noise, it is sometimes desirable to replace a pair of spur gears by parallel shaft helical gears even though the cost may be slightly more. An example will illustrate the calculations.

EXAMPLE 7.1

A pair of 20° full-depth involute spur gears with 32 and 80 teeth, diametral pitch of 16 teeth/in, and face width of 0.75 in are to be replaced by helical gears. The same hob used for the spur gears is to be used for the helical gears. The center distance and the angular velocity ratio must remain the same. The helix angle is to be as small as possible, and the overlap is to be 1.5 or greater. Determine the helix angle, the numbers of teeth, and the face width of the new helical gears.

SOLUTION

From the spur gear data and Eq. (6.1) the center distance is

$$R_2 + R_3 = \frac{N_2 + N_3}{2P} = \frac{32 + 80}{2(16)} = 3.5 \text{ in} \tag{1}$$

From Eq. (6.5) the first-order kinematic coefficient, the angular velocity ratio, is

$$|\theta'_{3/2}| = \left|\frac{\omega_3}{\omega_2}\right| = \frac{R_2}{R_3} = \frac{N_2}{N_3} = \frac{32}{80} = 0.4 \tag{2}$$

Because the same hob is to be used, P_n for the helical gears must also be 16 teeth/in.

$$R_2 + R_3 = \frac{N_2 + N_3}{2P_n \cos \psi} = \frac{N_2 + N_3}{2(16)\cos \psi} = 3.5 \text{ in}$$

or

$$\cos \psi = \frac{N_2 + N_3}{112} \tag{3}$$

which implies that $N_2 + N_3$ must be less than 112 while Eq. (2) requires that their ratio remain 0.4. Because $N_2 = 31$ teeth does not give an integer solution for N_3, the next smallest integer solution (which gives the smallest nonzero helix angle ψ) is $N_2 = 30$ and $N_3 = 75$ teeth, giving a helix angle of $\psi = 20.364°$. The transverse circular pitch is

$$p_t = \frac{\pi}{P_n \cos \psi} = \frac{\pi}{16 \cos 20.364°} = 0.209 \text{ in/tooth}$$

for which Eq. (7.9) shows a face width of $F = 0.846$ in. Unfortunately, the space available will not allow this increase in face width. Therefore this solution is not acceptable.

The next solution is $N_2 = 28$ and $N_3 = 70$ teeth giving a helix angle of $\psi = 28.955°$. The transverse circular pitch is $p_t = \pi/(P_n \cos \psi) = 0.224$ in/tooth and the face width is $F = 0.608$ in. This is an acceptable solution with face width of less than the original spur gears.

7.6 HERRINGBONE GEARS

Double-helical or *herringbone gears* comprise teeth having a right- and a left-hand helix cut on the same gear blank as illustrated schematically in Fig. 7.6. One of the primary disadvantages of the single helical gear is the existence of axial thrust loads that must be accounted for in the design of the bearings. In addition, the desire to obtain a good overlap without an excessively large face width may lead to the use of a comparatively larger helix angle, thus producing higher axial thrust loads. These thrust loads are eliminated by the herringbone configuration because the axial force of the right-hand half is balanced by that of the left-hand half. Thus, with the absence of thrust reactions, helix angles are usually larger for herringbone gears than for single-helical gears. However, one of the members of a herringbone gearset should always be mounted with some axial play or float to accommodate slight tooth errors and mounting allowances.

For the efficient transmission of large power at high speeds, herringbone gears are almost universally employed.

Figure 7.6 Schematic drawing of the pitch cylinder of a herringbone gear.

7.7 CROSSED-AXIS HELICAL GEARS

Crossed helical or *spiral gears* are sometimes used when the shaft centerlines are neither parallel nor intersecting. These are essentially nonenveloping worm gears (see Chapter 9) because the gear blanks have a cylindrical form with the two cylinder axes skew to each other.

The tooth action of crossed-axis helical gears is quite different from that of parallel-axis helical gears. The teeth of crossed helical gears have only *point contact*. In addition, there is much greater sliding action along the tooth surfaces than for parallel-axis helical gears. For these reasons they are chosen only to carry small loads. Because of the point contact, however, they need not be mounted accurately; either the center distance or the shaft angle may be varied slightly without affecting the amount of contact.

There is no difference between crossed helical gears and other helical gears until they are mounted in mesh. They are manufactured in the same way. Two meshing crossed helical gears usually have the same hand; that is, a right-hand driver goes with a right-hand

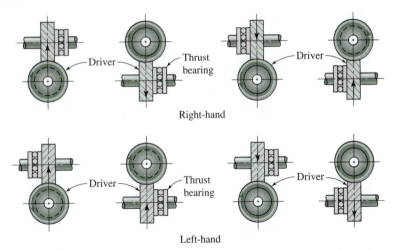

Figure 7.7 Thrust, rotation, and hand relations for crossed-helical gearing. (Courtesy of Boston Gear Works, Inc., North Quincy, MA.)

driven gear. The relation between thrust, hand, and rotation for crossed helical gears is shown in Fig. 7.7.

For crossed helical gears to mesh properly, they must share the same normal pitch. When tooth sizes are specified, the normal pitch should always be used. The reason for this is that when different helix angles are used for the driver and the driven gear, the transverse pitches are not the same. The relation between the shaft and helix angles is

$$\Sigma = \psi_2 \pm \psi_3 \tag{7.10}$$

where Σ is the shaft angle. The positive sign is used when both helix angles are of the same hand, and the negative sign is used when they are of opposite hand. Opposite hand crossed helical gears are used when the shaft angle Σ is small. The first-order kinematic coefficient, the angular velocity ratio between the shafts, is

$$|\theta'_{3/2}| = \left| \frac{\omega_3}{\omega_2} \right| = \frac{N_2}{N_3} = \frac{R_2 \cos \psi_2}{R_3 \cos \psi_3} \tag{7.11}$$

Crossed helical gears have the least sliding at the point of contact when the two helix angles are equal. If the two helix angles are not equal, the largest helix angle should be used with the driver if both gears have the same hand.

There is no widely accepted standard for crossed-axis helical gear tooth proportions. Many different proportions give good tooth action. Because the teeth are in point contact, an effort should be made to obtain a contact ratio of 2 or more. For this reason, crossed-axis helical gears are usually cut with a low pressure angle and a deep tooth. The tooth proportions shown in Table 7.1 are representative of good design. The driver tooth numbers shown are the minimum required to avoid undercut. The driven gear should have 20 teeth or more if a contact ratio of 2 is to be obtained.

TABLE 7.1 Tooth Proportions for Crossed-Axis Helical Gears[a]

Driver			
Helix Angle, ψ_2	Minimum Tooth Number, N_2	Driven: Helix Angle, ψ_3	Both: Normal Pressure Angle, ϕ_n
45°	20	45°	14.5°
60°	9	30°	17.5°
75°	4	15°	19.5°
86°	1	4°	20°

[a]Normal diametral pitch $P_n = 1$ teeth/in; working depth $= 2.400$ in; whole depth $= 2.650$ in; addendum $a = 1.200$ in.

To illustrate the calculations for a pair of crossed-axis helical gears, consider the following example:

EXAMPLE 7.2

Two shafts at an angle of 60° are to have a velocity ratio of 1 : 1.5. The center distance between the shafts is 8.63 in. Design a pair of crossed-axis helical gears for this application.

SOLUTION

Choosing $\psi_2 = 35°$ for the pinion, Eq. (7.10) gives $\psi_3 = 25°$. Then Eq. (7.11) shows

$$|\theta'_{3/2}| = \left|\frac{\omega_3}{\omega_2}\right| = \frac{R_2 \cos 35°}{R_3 \cos 25°} = \frac{1}{1.5}$$

This, along with the center distance, $R_2 + R_3 = 8.63$ in, gives $R_2 = 3.663$ in and $R_3 = 4.967$ in. Choosing a normal diametral pitch of $P_n = 6$ teeth/in, the numbers of teeth on the two gears are found to be

$$N_2 = 2P_n R_2 \cos \psi_2 = 2(6)(3.663)\cos 35° = 36 \text{ teeth}$$

and

$$N_3 = 2P_n R_3 \cos \psi_3 = 2(6)(4.967)\cos 25° = 54 \text{ teeth}$$

NOTE

[1.]   The equation of an ellipse with its center at the origin of an xy coordinate system with a and b as its semimajor and semiminor axes, respectively, is

$$\frac{x^2}{a^2} + \frac{y^2}{b^2} = 1 \qquad (a)$$

Also the formula for radius of curvature is

$$\rho = \frac{[1 + (dy/dx)^2]^{3/2}}{d^2y/dx^2} \qquad (b)$$

By using these two equations, it is not difficult to find the radius of curvature corresponding to $x = 0$, $y = b$. The result is

$$\rho = a^2/b \qquad (c)$$

Then, referring to Fig. 7.3, we substitute $a = R/\cos \psi$ and $b = R$ into Eq. (c) and obtain Eq. (7.5).

PROBLEMS

7.1 A pair of parallel-axis helical gears has 14.5° normal pressure angle, diametral pitch of 6 teeth/in, and 45° helix angle. The pinion has 15 teeth, and the gear has 24 teeth. Calculate the transverse and normal circular pitch, the normal diametral pitch, the pitch diameters, and the equivalent tooth numbers.

7.2 A set of parallel-axis helical gears are cut with a 20° normal pressure angle and a 30° helix angle. They have diametral pitch of 16 teeth/in and have 16 and 40 teeth, respectively. Find the transverse pressure angle, the normal circular pitch, the axial pitch, and the pitch radii of the equivalent spur gears.

7.3 A parallel-axis helical gearset is made with a 20° transverse pressure angle and a 35° helix angle. The gears have diametral pitch of 10 teeth/in and have 15 and 25 teeth, respectively. If the face width is 0.75 in, calculate the base helix angle and the axial contact ratio.

7.4 A set of helical gears is to be cut for parallel shafts whose center distance is to be about 3.5 in to give a velocity ratio of approximately 1.80. The gears are to be cut with a standard 20° pressure angle hob whose diametral pitch is 8 teeth/in. Using a helix angle of 30°, determine the transverse values of the diametral and circular pitches and the tooth numbers, pitch radii, and center distance.

7.5 A 16-tooth helical pinion is to run at 1800 rev/min and drive a helical gear on a parallel shaft at 400 rev/min. The centers of the shafts are to be spaced 11.0 in apart. Using a helix angle of 23° and a pressure angle of 20°, determine the values for the tooth numbers, pitch diameters, normal circular and diametral pitch, and face width.

7.6 The catalog description of a pair of helical gears is as follows: 14.5° normal pressure angle, 45° helix angle, diametral pitch of 8 teeth/in, 1.0 in face width, and normal diametral pitch of 11.31 teeth/in. The pinion has 12 teeth and a 1.500-in pitch diameter, and the gear has 32 teeth and a 4.000-in pitch diameter. Both gears have full-depth teeth, and they may be purchased either right- or left-handed. If a right-hand pinion and left-hand gear are placed in mesh, find the transverse contact ratio, the normal contact ratio, the axial contact ratio, and the total contact ratio.

7.7 In a medium-sized truck transmission a 22-tooth clutch-stem gear meshes continuously with a 41-tooth countershaft gear. The data show normal diametral pitch of 7.6 teeth/in, 18.5° normal pressure angle, 23.5° helix angle, and a 1.12-in face width. The clutch-stem gear is cut with a left-hand helix, and the countershaft gear is cut with a right-hand helix. Determine the normal and total contact ratio if the teeth are cut full-depth with respect to the normal diametral pitch.

7.8 A helical pinion is right-hand, has 12 teeth, has a 60° helix angle, and is to drive another gear at a velocity ratio of 3.0. The shafts are at a 90° angle, and the normal diametral pitch of the gears is 8 teeth/in. Find the helix angle and the number of teeth on the mating gear. What is the shaft center distance?

7.9 A right-hand helical pinion is to drive a gear at a shaft angle of 90°. The pinion has 6 teeth and a 75° helix angle and is to drive the gear at a velocity ratio of 6.5. The normal diametral pitch of the gear is 12 teeth/in. Calculate the helix angle and the number of teeth on the mating gear. Also determine the pitch diameter of each gear.

7.10 Gear 2 in the figure is to rotate clockwise and drive gear 3 counterclockwise at a velocity ratio of 2. Use a normal diametral pitch of 5 teeth/in, a shaft center distance of about 10 in, and the same helix angle for both gears. Find the tooth numbers, the helix angles, and the exact shaft center distance.

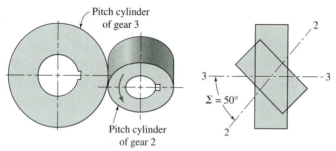

Figure P7.10

8 Bevel Gears

When rotational motion is to be transmitted between shafts whose axes intersect, some form of bevel gears are usually used. Bevel gears have pitch surfaces that are cones, with their cone axes matching the two shaft rotation axes as shown in Fig. 8.1. The gears are mounted so that the apexes of the two pitch cones are coincident with the point of intersection of the shaft axes. These pitch cones roll together without slipping.

Although bevel gears are often made for an angle of 90° between the shafts, they can be designed for almost any angle. When the shaft angle is other than 90° the gears are called *angular bevel gears*. For the special case where the shaft angle is 90° and both gears are of equal size, such bevel gears are called *miter gears*. A pair of miter gears is shown in Fig 8.2.

8.1 STRAIGHT-TOOTH BEVEL GEARS

For straight-tooth bevel gears, the true shape of a tooth is obtained by taking a spherical section through the tooth, where the center of the sphere is at the common apex, as shown in Fig. 8.1. As the radius of the sphere increases, the same number of teeth is projected onto a larger surface; therefore, the size of the teeth increases as larger spherical sections are taken. We have seen that the action and contact conditions for spur gear teeth may be viewed in a plane taken at right angles to the axes of the spur gears. For bevel gear teeth, the action and contact conditions should properly be viewed on a spherical surface (instead of a plane). We can even think of spur gears as a special case of bevel gears in which the spherical radius is infinite, thus producing the plane surface on which the tooth action is viewed. Figure 8.3 is typical of many straight-tooth bevel gear sets.

It is standard practice to specify the pitch diameters of a bevel gear at the large end of the teeth. In Fig. 8.4 the pitch cones of a pair of bevel gears are drawn and the pitch radii are given as R_2 and R_3, respectively, for the pinion and the gear. The cone angles γ_2 and γ_3

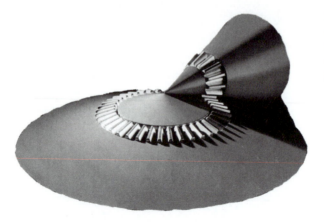

Figure 8.1 The pitch surfaces of bevel gears are cones that have only rolling contact. (Courtesy of Gleason Works, Rochester, NY.)

Figure 8.2 A pair of miter gears in mesh. (Courtesy of Gleason Works, Rochester, NY.)

Figure 8.3 A pair of straight-tooth bevel gears. (Courtesy of Gleason Works, Rochester, NY.)

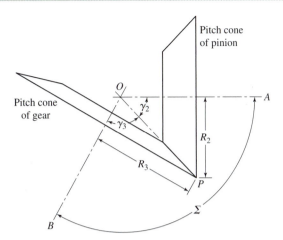

Figure 8.4 Pitch cones of bevel gears.

are defined as the pitch angles, and their sum is equal to the shaft angle Σ. The first-order kinematic coefficient, the angular velocity ratio between the shafts, is obtained in the same manner as for spur gears and is

$$|\theta'_{3/2}| = \left| \frac{\omega_3}{\omega_2} \right| = \frac{R_2}{R_3} = \frac{N_2}{N_3} \tag{8.1}$$

In the kinematic design of bevel gears the tooth numbers of the two gears and the shaft angle are usually given, and the corresponding pitch angles are to be determined. Although they can be found graphically, the analytical approach gives more exact values. From Fig. 8.4 the distance OP may be written as

$$OP = \frac{R_2}{\sin \gamma_2} = \frac{R_3}{\sin \gamma_3} \tag{a}$$

so that

$$\sin \gamma_2 = \frac{R_2}{R_3} \sin \gamma_3 = \frac{R_2}{R_3} \sin(\Sigma - \gamma_2) \tag{b}$$

or

$$\sin \gamma_2 = \frac{R_2}{R_3} (\sin \Sigma \cos \gamma_2 - \cos \Sigma \sin \gamma_2) \tag{c}$$

Dividing both sides of this equation by $\cos \gamma_2$ and rearranging gives

$$\tan \gamma_2 = \frac{\sin \Sigma}{(R_3/R_2) + \cos \Sigma} = \frac{\sin \Sigma}{(N_3/N_2) + \cos \Sigma} \tag{8.2}$$

Similarly,

$$\tan \gamma_3 = \frac{\sin \Sigma}{(N_2/N_3) + \cos \Sigma} \tag{8.3}$$

For a shaft angle of $\Sigma = 90°$ the above expressions reduce to

$$\tan \gamma_2 = \frac{N_2}{N_3} \tag{8.4}$$

and

$$\tan \gamma_3 = \frac{N_3}{N_2} \tag{8.5}$$

The projection of bevel gear teeth onto the surface of a sphere would indeed be a difficult and time-consuming task. Fortunately, an approximation is available which reduces the problem to that of ordinary spur gears. This approximation is called *Tredgold's approximation;* and as long as the gear has eight or more teeth, it is accurate enough for practical purposes. It is in almost universal use, and the terminology of bevel gear teeth has evolved around it.

In using Tredgold's method, a *back cone* is formed of elements that are perpendicular to the elements of the pitch cone at the large end of the teeth. This is shown in Fig. 8.5. The length of a back-cone element is called the back-cone radius. Now an equivalent spur gear is constructed whose pitch radius R_e is equal to the back-cone radius. Thus, from a pair of bevel gears, by using Tredgold's approximation, we can obtain a pair of equivalent spur gears that are then used to define the tooth profiles. They can also be used to determine the tooth action and the contact conditions exactly as was done for ordinary spur gears, and the results will correspond closely to those for the bevel gears.

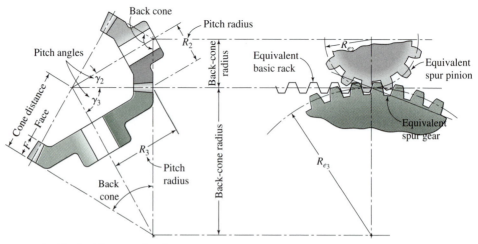

Figure 8.5 Tredgold's approximation.

From the geometry of Fig. 8.5, the equivalent pitch radii are

$$R_{e2} = \frac{R_2}{\cos \gamma_2}, \qquad R_{e3} = \frac{R_3}{\cos \gamma_3} \tag{8.6}$$

The number of teeth on one of the equivalent spur gears is

$$N_e = \frac{2\pi R_e}{p} \tag{8.7}$$

where p is the circular pitch of the bevel gear measured at the large end of the teeth. Usually the equivalent spur gears will *not* have an integral number of teeth.

8.2 TOOTH PROPORTIONS FOR BEVEL GEARS

Practically all straight-tooth bevel gears manufactured today use a 20° pressure angle. It is not necessary to use an interchangeable tooth form because bevel gears cannot be interchanged anyway. For this reason the long-and-short-addendum system, described in Section 6.11, is used. The proportions are tabulated in Table 8.1.

Bevel gears are usually mounted on the outboard side of the bearings because the shaft axes intersect, and this means that the effect of shaft deflection is to pull the small end of the teeth away from mesh, causing the larger end to take more of the load. Thus the load across the tooth is variable; for this reason, it is desirable to design a fairly short tooth. As shown in Table 8.1, the face width is usually limited to about one-third of the cone distance. We note also that a short face width simplifies the tooling problems in cutting bevel gear teeth.

TABLE 8.1 Tooth Proportions for 20° Straight-Tooth Bevel Gears

Item	Formula			
Working depth	$h_k = 2.0/P$			
Clearance	$c = 0.188/P + 0.002$ in			
Addendum of gear	$a_G = \dfrac{0.540}{P} + \dfrac{0.460}{P(m_{90})^2}$			
Gear ratio	$m_G = N_G/N_P$			
Equivalent 90° ratio	$m_{90} = \begin{cases} m_G & \text{when } \Sigma = 90° \\ \sqrt{m_G \dfrac{\cos \gamma_P}{\cos \gamma_G}} & \text{when } \Sigma \neq 90° \end{cases}$			
Face width	$F = \dfrac{1}{3}$ or $F = \dfrac{10}{P}$, whichever is smaller			
Minimum number of teeth	Pinion: 16	15	14	13
	Gear: 16	17	20	30

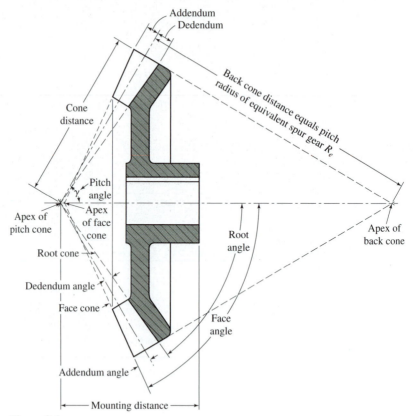

Figure 8.6

Figure 8.6 defines additional terminology characteristic of bevel gears. Note that a constant clearance is maintained by making the elements of the face cone parallel to the elements of the root cone of the mating gear. This explains why the face cone apex is not coincident with the pitch cone apex in Fig. 8.6. This permits a larger fillet than would otherwise be possible.

8.3 CROWN AND FACE GEARS

If the pitch angle of one of a pair of bevel gears is made equal to 90°, the pitch cone becomes a flat surface and the resulting gear is called a *crown gear*. Figure 8.7 shows a crown gear in mesh with a bevel pinion. Notice that a crown gear is the counterpart of a rack in spur gearing. The back cone of a crown gear is a cylinder, and the resulting involute teeth have straight sides, as indicated in Fig. 8.5.

A pseudo-bevel gearset can be obtained by using a cylindrical spur gear for a pinion in mesh with a gear having a planar pitch surface (similar to a crown gear) called a *face gear*. When the axis of the pinion and gear intersect, the face gear is called *on-center;* when the axes do not intersect, the face gear is called *off-center*.

To understand how a spur pinion, with a cylindrical rather than conical pitch surface, can properly mesh with a face gear we must consider how the face gear is formed; it is

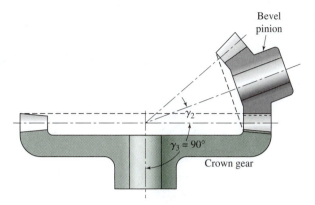

Bevel
pinion

Figure 8.7 A crown gear and bevel pinion.

γ_2

$\gamma_3 = 90°$

Crown gear

generated by a reciprocating cutter that is a replica of the spur pinion. Because the cutter and the gear blank are rotated as if in mesh, the resulting face gear is conjugate to the cutter and, therefore, to the spur pinion. The face width of the teeth on the face gear must be held quite short, however; otherwise the top land will become pointed.

Face gears are not capable of carrying heavy loads, but, because the axial mounting position of the pinion is not critical, they are sometimes more suitable for angular drives than bevel gears.

8.4 SPIRAL BEVEL GEARS

Straight-tooth bevel gears are easy to design and simple to manufacture and give very good results in service if they are mounted accurately and positively. As in the case of spur gears, however, they become noisy at higher pitch-line velocities. In such cases it is often good design practice to use *spiral bevel gears,* which are the bevel counterparts of helical gears. Figure 8.8 shows a mating pair of spiral bevel gears, and it can be seen that the pitch surfaces and the nature of contact are the same as for straight-tooth bevel gears except for the differences brought about by the spiral shaped teeth.

Spiral bevel gear teeth are conjugate to a basic crown rack, which can be generated as shown in Fig. 8.9 by using a circular cutter. The spiral angle ψ is measured at the mean radius of the gear. As with helical gears, spiral bevel gears give much smoother tooth action than straight-tooth bevel gears and hence are useful where high speeds are encountered. In order to obtain true spiral tooth action, the face contact ratio should be at least 1.25.

Pressure angles used with spiral bevel gears are generally $14\frac{1}{2}°$ to 20°, while the spiral angle is about 30° or 35°. As far as tooth action is concerned, the hand of the spiral may be either right or left; it makes no difference. However, looseness in the bearings might result in the teeth's jamming or separating, depending on the direction of rotation and the hand of the spiral. Because jamming of the teeth would do the most damage, the hand of the spiral should be such that the teeth tend to separate.

Zerol Bevel Gears   The Zerol bevel gear is a patented gear which has curved teeth but a zero-degree spiral angle. An example is shown in Fig. 8.10. It has no advantage in tooth action over the straight-tooth bevel gear and is designed simply to take advantage of the cutting machinery used for cutting spiral bevel gears.

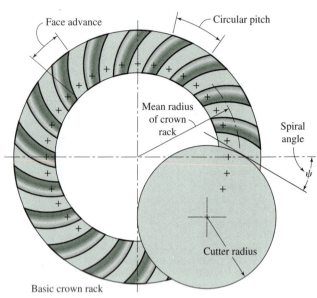

Figure 8.8 Spiral bevel gears. (Courtesy of Gleason Works, Rochester, NY.)

Figure 8.9 Cutting spiral bevel gear teeth on a basic crown rack.

Figure 8.10 Zerol bevel gears. (Courtesy of Gleason Works, Rochester, NY.)

8.5 HYPOID GEARS

It is frequently desirable, as in the case of rear-wheel drive automotive differential applications, to have a gearset similar to bevel gears but where the shafts do not intersect. Such gears are called *hypoid gears* because, as shown in Fig. 8.11, their pitch surfaces are hyperboloids of revolution. Figure 8.12 shows a pair of hypoid gears in mesh. The tooth action between these gears is a combination of rolling and sliding along a straight line and has much in common with that of worm gears (see Chapter 9).

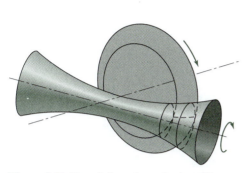

Figure 8.11 The pitch surfaces for hypoid gears are hyperboloids of revolution.

Figure 8.12 Hypoid gears. (Courtesy of Gleason Works, Rochester, NY.)

PROBLEMS

8.1 A pair of straight-tooth bevel gears is to be manufactured for a shaft angle of 90°. If the driver is to have 18 teeth and the velocity ratio is to be 3:1, what are the pitch angles?

8.2 A pair of straight-tooth bevel gears has a velocity ratio of 1.5 and a shaft angle of 75°. What are the pitch angles?

8.3 A pair of straight-tooth bevel gears is to be mounted at a shaft angle of 120°. The pinion and gear are to have 15 teeth and 33 teeth, respectively. What are the pitch angles?

8.4 A pair of straight-tooth bevel gears with diametral pitch of 2 teeth/in have 19 teeth and 28 teeth, respectively. The shaft angle is 90°. Determine the pitch diameters, pitch angles, addendum, dedendum, face width, and pitch diameters of the equivalent spur gears.

8.5 A pair of straight-tooth bevel gears with diametral pitch of 8 teeth/in have 17 teeth and 28 teeth, respectively, and a shaft angle of 105°. For each gear, calculate the pitch diameter, pitch angle, addendum, dedendum, face width, and the equivalent tooth numbers. Make a sketch of the two gears in mesh. Use the standard tooth proportions as for a 90° shaft angle.

9 Worms and Worm Gears

9.1 BASICS

A *worm* is a machine member having a screw-like thread, and worm teeth are frequently spoken of as threads. A worm meshes with a conjugate gear-like member called a *worm wheel* or a *worm gear*. Figure 9.1 shows a worm and worm gear in an application. These gears are used with nonintersecting shafts that are usually at a shaft angle of 90°, but there is no reason why shaft angles other than 90° cannot be used if a design demands it.

Worms in common use have from one to four teeth and are said to be *single-threaded, double-threaded,* and so on. As we will see, there is no definite relation between the number of teeth and the pitch diameter of a worm. The number of teeth on a worm gear is usually much higher and, therefore, the angular velocity of the worm gear is usually much lower than that of the worm. In fact, often, one primary type of application for a worm and worm gear is in order to obtain a very large angular velocity reduction—that is, a very low first-order kinematic coefficient or angular velocity ratio. In keeping with the low velocity ratio, the worm gear is usually the driven member of the pair and the worm is usually the driving member.

A worm gear, unlike a spur or helical gear, has a face that is made concave so that it partially wraps around, or envelops, the worm, as shown in Fig. 9.2. Worms are sometimes designed with a cylindrical pitch surface, or they may have an hourglass shape, such that the worm also wraps around or partially encloses the worm gear. If the enveloping worm gear is mated with a cylindrical worm, the set is said to be *single-enveloping*. When the worm is hourglass-shaped, the worm and worm gearset is said to be *double-enveloping* because each member partially wraps around the other; such a worm is sometimes called a Hindley worm. The nomenclature of a single-enveloping worm and worm gearset is shown in Fig. 9.2.

Figure 9.1 A single-enveloping worm and worm gearset. (Courtesy of Gleason Works, Rochester, NY.)

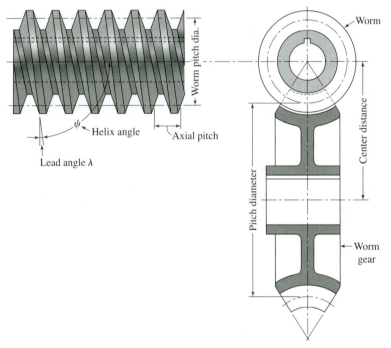

Figure 9.2 Nomenclature of a single-enveloping worm and worm gearset.

A worm and worm gear combination is similar to a pair of mating crossed-helical gears except that the worm gear partially envelops the worm. For this reason they have line contact instead of the point contact found in crossed-helical gears and are thus able to transmit more power. When a double-enveloping worm and worm gearset is used, even more power can be transmitted, at least in theory, because contact is distributed over an area on the tooth surfaces.

In a single-enveloping worm and worm gearset it makes no difference whether the worm rotates on its own axis and drives the gear by a screwing action or whether the worm is translating along its axis and drives the worm gear through rack action. The resulting motion and contact are the same. For this reason a single-enveloping worm need not be accurately mounted along its shaft axis. However, the worm gear should be accurately mounted along its rotation axis; otherwise its pitch surface is not properly aligned with the worm axis. In a double-enveloping worm and worm gearset, both members are throated and therefore both must be accurately mounted in all directions in order to obtain correct contact.

A mating worm and worm gear with a 90° shaft angle have the same hand of helix, but the helix angles are usually very different. The helix angle on the worm is usually quite large (at least for one or two teeth) and quite small on the worm gear. On the worm, the *lead angle* is the complement of the helix angle, as shown in Fig. 9.2. Because of this, it is customary to specify the lead angle for the worm and specify the helix angle for the worm gear. This is convenient because the two angles are equal for a 90° shaft angle.

In specifying the pitch of a worm and worm gearset, it is usual to specify the axial pitch of the worm and the circular pitch of the worm gear. These are equal if the shaft angle is 90°. It is common to employ even fractions, such as $\frac{1}{4}$, $\frac{3}{8}$, $\frac{1}{2}$, $\frac{3}{4}$, 1, $1\frac{1}{4}$ in/tooth, and so on, for the circular pitch of the worm gear; there is no reason, however, why the AGMA standard diametral pitches used for spur gears should not also be used for worm gears.

The pitch radius of a worm gear is found the same as for a spur gear:

$$R_3 = \frac{N_3 p}{2\pi} \tag{9.1}$$

where all values are defined the same as for spur gears, but refer to the parameters of the worm gear.

The pitch radius of the worm may have any value, but it should be the same as that of the hob used to cut the worm gear teeth. AGMA recommends the following relation between the pitch radius of the worm and the center distance:

$$R_2 = \frac{(R_2 + R_3)^{0.875}}{4.4} \tag{9.2}$$

where the quantity $(R_2 + R_3)$ is the center distance. This equation gives proportions that result in good power capacity. The AGMA standard also states that the denominator of Eq. (9.2) may vary from 3.4 to 6.0 without appreciably affecting the power capacity. Equation (9.2) is not required, however; other proportions will also serve well and, in fact, power capacity may not always be the primary consideration. However, there are a lot of variables in worm gear design, and the equation is helpful in obtaining trial dimensions.

The *lead* of a worm has the same meaning as for a screw thread and is the axial distance through which a point on the helix will move when the worm is turned through one revolution. Thus, in equation form,

$$l = p_x N_2 \tag{9.3}$$

where l is the lead, p_x is the axial pitch, and N_2 is the number of teeth (threads) on the worm. The lead and the *lead angle* are related as follows:

$$\lambda = \tan^{-1}\left(\frac{l}{2\pi R_2}\right) \tag{9.4}$$

where λ is the lead angle, as shown in Fig. 9.2.

The teeth on a worm are usually cut in a milling machine or on a lathe. Worm gear teeth are most often produced by hobbing. Except for clearance at the top of the hob teeth, the worm should be an exact duplicate of the hob in order to obtain conjugate action. This also means that, where possible, the worm should be designed using the dimensions of existing hobs.

The pressure angles used on worms and worm-gear sets vary widely and should depend approximately on the value of the lead angle. Good tooth action is obtained if the pressure angle is made large enough to eliminate undercutting of the worm gear tooth on the side at which the contact ends. Recommended values are shown in Table 9.1.

A satisfactory tooth depth that has about the right relation to the lead angle is obtained by making the depth a proportion of the normal circular pitch. Using an addendum of $1/P = p_n/\pi$, as for full-depth spur gears, we obtain the following proportions for worms and worm gears:

$$\text{Addendum} = 1.000/P = 0.3183 p_n$$

$$\text{Whole depth} = 2.000/P = 0.6366 p_n$$

$$\text{Clearance} = 0.200/P = 0.0637 p_n$$

The face width of the worm gear should be obtained as shown in Fig. 9.3. This makes the face of the worm gear equal to the length of a tangent to the worm pitch circle between its points of intersection with the addendum circle.

TABLE 9.1 Recommended Pressure Angles for Worm and Worm Gearsets

Lead Angle, λ	Pressure Angle, ϕ
0°–16°	14.5°
16°–25°	20°
25°–35°	25°
35°–45°	30°

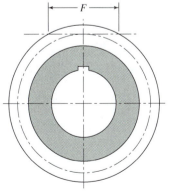

Figure 9.3 Face width of a worm gear.

PROBLEMS

9.1   A worm having 4 teeth and a lead of 1.0 in drives a worm gear at a velocity ratio of 7.5. Determine the pitch diameters of the worm and worm gear for a center distance of 1.75 in.

9.2   Specify a suitable worm and worm gear combination for a velocity ratio of 60 and a center distance of 6.50 in using an axial pitch of 0.500 in/tooth.

9.3   A triple-threaded worm drives a worm gear having 40 teeth. The axial pitch is 1.25 in and the pitch diameter of the worm is 1.75 in. Calculate the lead and lead angle of the worm. Find the helix angle and pitch diameter of the worm gear.

9.4   A triple-threaded worm with a lead angle of 20° and an axial pitch of 0.400 in/tooth drives a worm gear with a velocity reduction of 15 to 1. Determine for the worm gear: (*a*) the number of teeth, (*b*) the pitch diameter, and (*c*) the helix angle. (*d*) Determine the pitch diameter of the worm. (*e*) Compute the center distance.

10 | Mechanism Trains

Mechanisms arranged in series or parallel combinations so that the driven member of one mechanism is the driver for another mechanism are called *mechanism trains*. With certain exceptions, to be explored, the analysis of such trains can proceed in serial fashion by using the methods developed in the previous chapters.

10.1 PARALLEL-AXIS GEAR TRAINS

In Chapter 3 we learned that the *kinematic coefficient* is the term used to describe the ratio of the angular velocity of the driven member to that of the driving member. Thus, for example, in a four-bar linkage with link 2 as the driving or input member and link 4 as the driven or output member, we had

$$\theta'_{4/2} = \frac{\omega_4}{\omega_2} = \frac{d\theta_4/dt}{d\theta_2/dt} = \frac{d\theta_4}{d\theta_2} \tag{a}$$

In this chapter where we deal largely with gear trains it is convenient to deal with speeds, which are commonly expressed in revolutions per minute (rev/min) and, in a few cases, in revolutions per second (rev/s). Thus, for gearing, we prefer to write Eq. (a) as

$$\theta'_{L/F} = \frac{\omega_L}{\omega_F} = \frac{d\theta_L/dt}{d\theta_F/dt} = \frac{d\theta_L}{d\theta_F} \tag{10.1}$$

where ω_L is the speed of the *last* gear and ω_F is the speed of the *first* gear in the same train. Usually the last gear is the output and is the driven gear, and the first is the input and driving gear.

The term $\theta'_{L/F}$ shown in Eq. (10.1) is called the *kinematic coefficient,* or the *speed ratio* by some, or the *train value* by others. All three terms are completely appropriate. The equation is often written in the more convenient form

$$\omega_L = \theta'_{L/F}\omega_F \tag{10.2}$$

Next we consider a pinion 2 driving a gear 3. The speed of the driven gear is

$$\omega_3 = \frac{R_2}{R_3}\omega_2 = \frac{N_2}{N_3}\omega_2 \tag{b}$$

where, for each gear, N is the number of teeth, and ω is either the speed in revolutions per minute or the number of revolutions during a certain time interval.

For parallel-shaft gearing, the directions can be kept track of by following the vector sense—that is, by specifying that angular velocity is positive or negative when clockwise or counterclockwise from a chosen side. For parallel-shaft gearing we shall use the following sign convention: If the last gear of a parallel-shaft gear train rotates in the same sense as the first gear, then we take $\theta'_{L/F}$ as positive; if the last gear rotates in the opposite sense to the first gear, then we take $\theta'_{L/F}$ as negative. This approach is not as easy, however, when the gear shafts are not parallel to each other, as in bevel, crossed-helical, or worm gearing. For these reasons it is often simpler to track the directions by visually inspecting a sketch of the train.

The gear train shown in Fig. 10.1 is made up of five gears in series. Using Eq. (b) we find the speed of gear 6 to be

$$\omega_6 = \frac{N_5}{N_6}\frac{N_4}{N_5}\frac{N_2}{N_3}\omega_2 \tag{c}$$

Here we notice that gear 5 is an idler; that is, its tooth numbers cancel in Eq. (c) and hence the only purpose served by gear 5 is to change the direction of rotation of gear 6. We further notice that gears 5, 4, and 2 are drivers, while gears 6, 5, and 3 are driven members. Thus Eq. (10.1) can also be written

$$\theta'_{L/F} = \frac{\text{product of driving tooth numbers}}{\text{product of driven tooth numbers}} \tag{10.3}$$

Note also that pitch radii can be used in Eq. (10.3) just as well as tooth numbers.

Figure 10.1

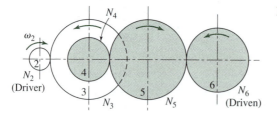

10.2 EXAMPLES OF GEAR TRAINS

In speaking of gear trains it is often convenient to describe one having only one gear on each axis as a *simple* gear train. A *compound* gear train then is one that has two or more gears on one or more axes, like the train in Fig. 10.1.

Figure 10.2 shows an example of a compound gear train. It shows a transmission for a small- or medium-sized truck, which has four speeds forward and one in reverse.

The gear train shown in Fig. 10.3 is composed of bevel, helical, and spur gears. The helical gears are crossed, and so their direction of rotation depends upon their hand.

A *reverted* gear train, like the one shown in Fig. 10.4, is one in which the first and last gears have collinear axes of rotation. This arrangement produces compactness and is used in such applications as speed reducers, clocks (to connect the hour hand to the minute hand), and machine tools. As an exercise, it is suggested that you seek out a suitable set of

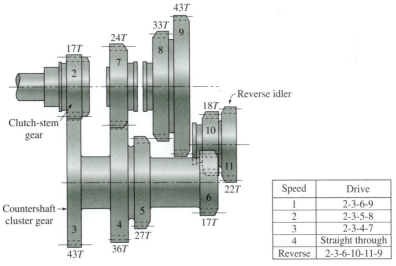

Speed	Drive
1	2-3-6-9
2	2-3-5-8
3	2-3-4-7
4	Straight through
Reverse	2-3-6-10-11-9

Figure 10.2 A truck transmission with gears having diametral pitch of 7 teeth/in and pressure angle of 22.5°.

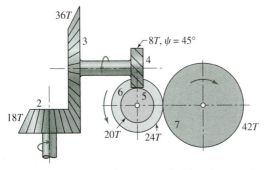

Figure 10.3 A gear train composed of bevel, crossed-helical, and spur gears.

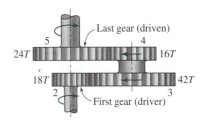

Figure 10.4 A reverted gear train.

diametral pitches for each pair of gears shown in Fig. 10.4 so that the first and last gears will have the same axis of rotation with all gears properly engaged.

10.3 DETERMINING TOOTH NUMBERS

When noticeable power is transmitted through a speed reduction unit, the speed ratio of the last pair of meshing gears is larger than that of the first gear pair because the torque is greater at the low-speed end. In a fixed amount of space, more teeth can be used on lesser pitch gears; hence a greater speed reduction can be obtained at the high-speed end.

Without examining the problem of tooth strength, suppose we wish to use two pairs of gears in a train to obtain an overall kinematic coefficient of $\theta'_{L/F} = 1/12$. Let us also impose the restriction that the tooth numbers must not be less than 15 and that the reduction in the first pair of gears should be about twice that of the second pair. This means that

$$\theta'_{5/2} = \frac{N_4}{N_5} \frac{N_2}{N_3} = \frac{1}{12} \tag{a}$$

where N_2/N_3 is the kinematic coefficient of the first gear pair, and N_4/N_5 is that of the second pair. Because the kinematic coefficient of the first pair should be half that of the second, we obtain

$$\left(\frac{N_4}{N_5}\right)\left(\frac{N_4}{2N_5}\right) = \frac{1}{12} \tag{b}$$

or

$$\frac{N_4}{N_5} = \sqrt{\frac{1}{6}} = 0.408\,248 \tag{c}$$

to six decimal places. The following tooth numbers are seen to be close:

$$\frac{15}{37} \quad \frac{16}{39} \quad \frac{18}{44} \quad \frac{20}{49} \quad \frac{22}{54} \quad \frac{24}{59}$$

Of these, $N_4/N_5 = 20/49$ is the closest approximation, but notice that

$$\theta'_{5/2} = \left(\frac{N_4}{N_5}\right)\left(\frac{N_2}{N_3}\right) = \left(\frac{20}{49}\right)\left(\frac{20}{98}\right) = \frac{400}{4802} = \frac{1}{12.005}$$

which is not quite 1/12. On the other hand, the choice of $N_4/N_5 = 18/44$ gives exactly

$$\theta'_{5/2} = \left(\frac{N_4}{N_5}\right)\left(\frac{N_2}{N_3}\right) = \left(\frac{18}{44}\right)\left(\frac{22}{108}\right) = \frac{396}{4752} = \frac{1}{12}$$

In this case, the reduction in the first gear pair is not exactly twice the reduction in the second gear pair. However, this consideration is usually of only minor importance.

The problem of specifying tooth numbers and the number of pairs of gears to give a kinematic coefficient with a specified degree of accuracy has interested many. Consider, for instance, the problem of specifying a set of gears to have a kinematic coefficient of $\theta'_{L/F} = \pi/10$ accurate to eight decimal places.

10.4 EPICYCLIC GEAR TRAINS

Figure 10.5 shows an elementary epicyclic gear train together with its schematic diagram suggested by Lévai.[1] The train consists of a *central gear* 2 and an *epicyclic gear* 4, which produces epicyclic motion for its points by rolling around the periphery of the central gear. A *crank arm* 3 contains the bearings for the epicyclic gear to maintain the two gears in mesh.

These trains are also called *planetary* or *sun-and-planet* gear trains. In this nomenclature, gear 2 of Fig. 10.5 is called the *sun gear,* gear 4 is called the *planet gear,* and crank 3 is called the *planet carrier.* Figure 10.6 shows the train of Fig. 10.5 with two redundant planet gears added. This produces better force balance; also, adding more planet gears allows lower forces by more force sharing. However, these additional planet gears do not change the kinematic characteristics at all. For this reason we shall generally show only a single planet in the illustrations and problems in this chapter, even though an actual machine would probably be constructed with planets in trios.

The simple epicyclic gear train together with its schematic designation shown in Fig. 10.7 shows how the motion of the planet gear can be transmitted to another central

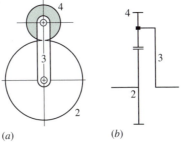

(a)   (b)

Figure 10.5 (*a*) The elementary epicyclic gear train; (*b*) its schematic designation.

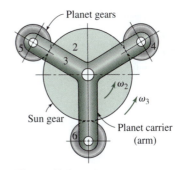

Figure 10.6 A planetary gearset.

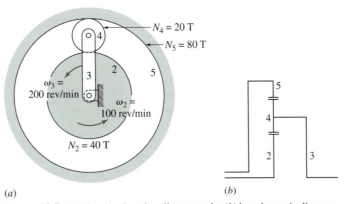

(*a*)   (*b*)

Figure 10.7 (*a*) The simple epicyclic gear train; (*b*) its schematic diagram.

gear. The second central gear in this case is gear 5, an internal gear. Figure 10.8 shows a similar arrangement with the difference that both central gears are external gears. Note, in Fig. 10.8, that the double planet gears are mounted on a single planet shaft and that each planet gear is in mesh with a separate sun gear rotating at a different speed.

In any case, no matter how many planets are used, only one planet carrier or arm may be used. This principle is illustrated in Fig. 10.6, in which redundant planets are used, and in Fig. 10.9, where two planets are used to alter the kinematic performance.

According to Lévai, 12 variations are possible; they are all shown in schematic form in Fig. 10.10 as Lévai arranged them. Those in Fig. 10.10*a* and Fig. 10.10*c* are the simple trains in which the planet gears mesh with both sun gears. The trains shown in Fig. 10.10*b* and Fig. 10.10*d* have planet gear pairs that are partly in mesh with each other and partly in mesh with the sun gears.

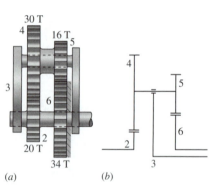

(*a*)   (*b*)

Figure 10.8 A simple epicyclic gear train with double planet gears.

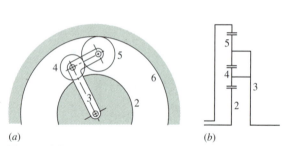

(*a*)   (*b*)

Figure 10.9 An epicyclic gear train with two planet gears.

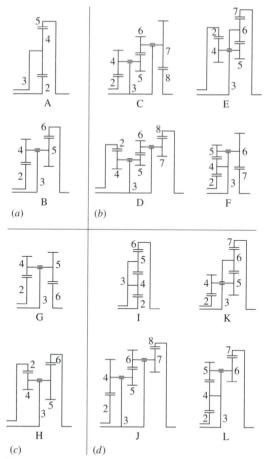

(*a*)   (*b*)

(*c*)   (*d*)

Figure 10.10 All 12 possible epicyclic gear train types according to Lévai. (Reproduced with permission of the author.)

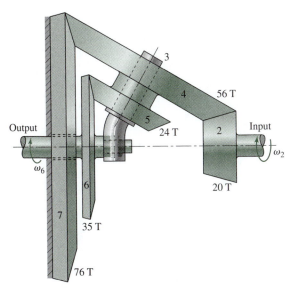

Figure 10.11 Humpage's reduction gear.

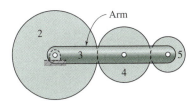

Figure 10.12

10.5 BEVEL GEAR EPICYCLIC TRAINS

The bevel gear train shown in Fig. 10.11 is called *Humpage's reduction gear*. Bevel gear epicyclic trains are used quite frequently, and they are the same as spur gear epicyclic trains except that their axes of rotation are not all on parallel shafts. The train of Fig. 10.11 is, in fact, a double epicyclic train, and the spur gear counterpart of each can be found in Fig. 10.10. We will find in the next section that the analysis of such trains can be done the same as for spur gear trains.

10.6 ANALYSIS OF PLANETARY GEAR TRAINS BY FORMULA

Figure 10.12 shows a planetary gear train composed of a sun gear 2, an arm or planet carrier 3, and planet gears 4 and 5. Using the apparent angular velocity equation, Eq. (3.2), we can write that the angular velocity of gear 2 as it would appear from a coordinate system on the arm 3 is

$$\omega_{2/3} = \omega_2 - \omega_3 \tag{a}$$

Also, the angular velocity of gear 5 as it would appear from arm 3 is

$$\omega_{5/3} = \omega_5 - \omega_3 \tag{b}$$

Dividing Eq. (*b*) by Eq. (*a*) gives

$$\frac{\omega_{5/3}}{\omega_{2/3}} = \frac{\omega_5 - \omega_3}{\omega_2 - \omega_3} \tag{c}$$

Equation (c) expresses the ratio of the apparent angular velocity of gear 5 to that of gear 2 with both taken as they would appear from arm 3. This ratio, which is proportional to the tooth numbers, appears the same whether the arm is rotating or not; it is the first-order kinematic coefficient of the gear train. Therefore we can write

$$\theta'_{5/2} = \frac{\omega_5 - \omega_3}{\omega_2 - \omega_3} \qquad (d)$$

An equation similar to Eq. (d) is all that we need to find the angular velocities in any planetary gear train. It is convenient to express it in the form

$$\theta'_{L/F} = \frac{\omega_L - \omega_A}{\omega_F - \omega_A} \qquad (10.4)$$

where ω_F = angular velocity of the first gear in the train,
  ω_L = angular velocity of the last gear in the train,
  ω_A = angular velocity of the arm.

The following examples illustrate the use of Eq. (10.4).

EXAMPLE 10.1

Figure 10.8 shows a reverted planetary gear train. Gear 2 is fastened to its shaft and is driven at 250 rev/min in a clockwise direction. Gears 4 and 5 are planet gears that are joined but are free to turn on the shaft carried by the arm. Gear 6 is stationary. Find the speed and direction of rotation of the arm.

SOLUTION

We must first decide which gears to designate as the first and last members of the train. Because the speeds of gears 2 and 6 are given, either may be chosen as the first. The choice makes no difference in the results but, once the decision is made, it may not be changed. Here we choose 2 as first; therefore gear 6 is last. Thus

$$\omega_F = \omega_2 = -250 \text{ rev/min} \quad \text{and} \quad \omega_L = \omega_6 = 0 \text{ rev/min}$$

and, according to Eq. (10.3), the first-order kinematic coefficient is

$$\theta'_{L/F} = \theta'_{6/2} = \left(\frac{16}{34}\right)\left(\frac{20}{30}\right) = \frac{16}{51}$$

Choosing counterclockwise as positive and substituting these values into Eq. (10.4) gives

$$\theta'_{L/F} = \frac{16}{51} = \frac{0 - \omega_3}{-250 - \omega_3}$$

$$\omega_A = \omega_3 = 114.3 \text{ rev/min ccw} \qquad \textit{Ans.}$$

EXAMPLE 10.2

In the bevel gear train in Fig. 10.11, the input is to gear 2, and the output is from gear 6, which is connected to the output shaft. The arm 3 turns freely on the output shaft and carries the planets 4 and 5. Gear 7 is fixed to the frame. What is the output shaft speed if gear 2 rotates at 2000 rev/min?

SOLUTION

The problem is solved in two steps. In the first step we consider the train to be made up of gears 2, 4, and 7 and calculate the rotational speed of the arm. Thus

$$\omega_F = \omega_2 = 2000 \text{ rev/min} \quad \text{and} \quad \omega_L = \omega_7 = 0 \text{ rev/min}$$

and, according to Eq. (10.3),

$$\theta'_{L/F} = \theta'_{7/2} = \left(-\frac{56}{76}\right)\left(\frac{20}{56}\right) = -\frac{5}{19}$$

Substituting into Eq. (10.4) and solving for the angular velocity of arm 3 gives

$$\theta'_{L/F} = -\frac{5}{19} = \frac{0 - \omega_3}{2000 - \omega_3}$$

$$\omega_A = \omega_3 = 416.67 \text{ rev/min}$$

Next we consider the train as composed of gears 2, 4, 5, and 6. Then $\omega_F = \omega_2 = 2000$ rev/min, as before, and $\omega_L = \omega_6$, which is to be found. The kinematic coefficient of the train is

$$\theta'_{L/F} = \theta'_{6/2} = \left(\frac{24}{35}\right)\left(-\frac{20}{56}\right) = -\frac{12}{49}$$

Substituting into Eq. (10.4) again and solving for ω_6, with ω_3 now known, gives

$$-\frac{12}{49} = \frac{\omega_L - 416.67}{2000 - 416.67}$$

$$\omega_L = \omega_6 = 28.91 \text{ rev/min} \qquad\qquad Ans.$$

Because the result is positive, we conclude that the output shaft rotates in the same direction as the input shaft 2 with a speed reduction of 2000 : 28.91 or 69.18 : 1.

10.7 TABULAR ANALYSIS OF PLANETARY GEAR TRAINS

Another method of determining the rotations of epicyclic gear trains uses the principle of superposition. The total analysis is carried out by finding the apparent rotations relative to the arm or planet carrier, and then summing with the rotation of the components as if all the gears are fixed to the arm. The process is easily carried out in a tabular procedure.

Figure 10.7 illustrates a planetary gear train composed of a sun gear 2, a planet carrier (arm) 3, a planet gear 4, and an internal gear 5 that is in mesh with the planet gear. Because this gear train has two degrees of freedom, we might reasonably specify the angular velocities of the sun gear and of the arm and wish to determine the angular velocity of the internal gear.

The analysis can be carried out in the following three steps:

1. Consider all gears (including a fixed gear, if any) locked to the arm and allow the arm to rotate with angular velocity ω_A. Tabulate the angular velocities of all components under this condition as also equal to ω_A.
2. Free all constraints of step 1, fix the arm, and allow some gear B (such as the sun gear) to rotate with angular velocity $\omega_{B/A}$. Tabulate the apparent angular velocities of all other gears with respect to the arm as multiples of $\omega_{B/A}$.
3. Add the angular velocities of each gear from steps 1 and 2, and apply the given input velocities in order to find numeric values for ω_A and $\omega_{B/A}$.

EXAMPLE 10.3

As an example of such a solution let us assume the tooth numbers of Fig. 10.7, and let the angular velocities of the sun gear and the arm be $\omega_2 = 100$ rev/min and $\omega_3 = 200$ rev/min, respectively, both in the ccw direction, chosen positive. What is the angular velocity of the internal ring gear 5?

SOLUTION

The solution process described by the three steps explained above is shown in Table 10.1.

TABLE 10.1 Tabular Analysis for Examples 10.3 and 10.4

Step Number	Gear 2	Arm 3	Gear 4	Gear 5
1. Gears fixed to arm	ω_3	ω_3	ω_3	ω_3
2. Arm fixed	$\omega_{2/3}$	0	$(-40/20)\omega_{2/3}$	$(20/80)(-40/20)\omega_{2/3}$
3. Total	$\omega_3 + \omega_{2/3}$	ω_3	$\omega_3 + (-40/20)\omega_{2/3}$	$\omega_3 + (20/80)(-40/20)\omega_{2/3}$

Next, comparing the given input velocities to the bottom row of columns 2 and 3, we see that because $\omega_2 = \omega_3 + \omega_{2/3} = 100$ rev/min and $\omega_3 = 200$ rev/min, then $\omega_{2/3} = -100$ rev/min. Therefore, from the bottom row of column 5 we find $\omega_5 = \omega_3 + (20/80)(-40/20)\omega_{2/3} = 200 + (20/80)(-40/20)(-100) = 250$ rev/min in the same ccw direction as the inputs.

Three more worked examples will help to better understand this tabular method of solution.

EXAMPLE 10.4

What is the angular velocity of the external gear 5 of Fig. 10.7 if gear 2 rotates at 100 rev/min clockwise while arm 3 rotates at 200 rev/min counterclockwise?

SOLUTION

The analysis in Table 10.1 is identical to that done for Example 10.3; however, the input velocities have now changed. Still taking counterclockwise as positive, the bottom row of columns 2 and 3 show that because $\omega_2 = \omega_3 + \omega_{2/3} = -100$ rev/min and $\omega_3 = 200$ rev/min, then $\omega_{2/3} = -300$ rev/min. Therefore, the bottom row of column 5 shows $\omega_5 = \omega_3 + (20/80) \times (-40/20)\omega_{2/3} = 200 + (20/80)(-40/20)(-300) = 350$ rev/min ccw.

EXAMPLE 10.5

The planetary gear train shown in Fig. 10.13 is called Ferguson's paradox.[2] Gear 2 is fixed to the frame. Arm 3 and gears 4 and 5 are free to turn upon the shaft. Gears 2, 4, and 5 have tooth numbers of 100, 101, and 99, respectively, all cut with the same circular pitch (but with slightly different pitch circle radii) so that the planet gear 6 meshes with all of them. Find the angular rotations of gears 4 and 5 when arm 3 is given one counterclockwise turn.

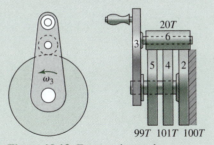

Figure 10.13 Ferguson's paradox.

SOLUTION

The solution process is shown in Table 10.2.

TABLE 10.2 Tabular Analysis for Example 10.5

Step Number	Gear 2	Arm 3	Gear 4	Gear 5	Gear 6
1. Gears fixed to arm	$\Delta\theta_3$	$\Delta\theta_3$	$\Delta\theta_3$	$\Delta\theta_3$	$\Delta\theta_3$
2. Arm fixed	$\Delta\theta_{2/3}$	0	$(-20/101)(-100/20)\Delta\theta_{2/3}$	$(-20/99)(-100/20)\Delta\theta_{2/3}$	$(-100/20)\Delta\theta_{2/3}$
3. Total	$\Delta\theta_3 + \Delta\theta_{2/3}$	$\Delta\theta_3$	$\Delta\theta_3 + (-20/101) \times (-100/20)\Delta\theta_{2/3}$	$\Delta\theta_3 + (-20/99) \times (-100/20)\Delta\theta_{2/3}$	$\Delta\theta_3 + (-100/20) \times \Delta\theta_{2/3}$

It is noticed that angular displacements are shown in the table instead of angular velocities. This comes from the nature of the question asked and recognizing that, during a chosen time interval, $\Delta\theta = \omega\Delta t$ for each of the elements. According to the problem statement we choose the time interval Δt such that for the arm, $\Delta\theta_3 = 1$ rev ccw. Then, in order for gear 2 to remain stationary, column 2 shows that $\Delta\theta_3 + \Delta\theta_{2/3} = 1 + \Delta\theta_{2/3} = 0$ and, therefore, $\Delta\theta_{2/3} = -1$ rev. Finally, from the bottom row of columns 4 and 5 we find that $\Delta\theta_4 = \Delta\theta_3 + (-20/101) \times (-100/20)\Delta\theta_{2/3} = 1 + (-20/101)(-100/20)(-1) = 1/101$ rev ccw and $\Delta\theta_5 = \Delta\theta_3 + (-20/99)(-100/20)\Delta\theta_{2/3} = 1 + (-20/99)(-100/20)(-1) = -1/99$ rev $= 1/99$ rev cw. Thus when arm 3 is turned, gear 4 rotates very slowly in the same direction while gear 5 turns very slowly in the opposite direction.

EXAMPLE 10.6

The overdrive unit shown in Fig. 10.14 is sometimes used to follow a standard automotive transmission to further reduce engine speed. The engine speed (after the transmission) corresponds to the speed of the planet carrier 3, and the drive shaft speed corresponds to that of gear 5; the sun gear 2 is held stationary. Determine the percentage reduction in engine speed obtained when the overdrive is active.

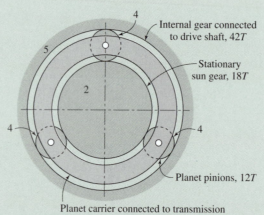

Figure 10.14 Overdrive unit.

SOLUTION

The analysis for this problem is shown in Table 10.3. For gear 2 to remain stationary, column 2 shows that $\omega_3 + \omega_{2/3} = 0$; therefore $\omega_{2/3} = -\omega_3$. Putting this into column 5 shows that $\omega_5 = \omega_3 + (12/42)(-18/12)\omega_{2/3} = \omega_3 + (12/42)(-18/12)(-\omega_3) = 1.429\omega_3$. Thus the percentage reduction in engine speed is $(1.429\omega_3 - 1.0\omega_3)/(1.429\omega_3) = 0.300 = 30\%$.

TABLE 10.3 Tabular Analysis for Example 10.6

Step Number	Gear 2	Arm 3	Gear 4	Gear 5
1. Gears fixed to arm	ω_3	ω_3	ω_3	ω_3
2. Arm fixed	$\omega_{2/3}$	0	$(-18/12)\omega_{2/3}$	$(12/42)(-18/12)\omega_{2/3}$
3. Total	$\omega_3 + \omega_{2/3}$	ω_3	$\omega_3 + (-18/12)\omega_{2/3}$	$\omega_3 + (12/42)(-18/12)\omega_{2/3}$

10.8 ADDERS AND DIFFERENTIALS

Figure 10.15 illustrates a variety of mechanisms used as computing devices. Because each of these is a two-degree-of-freedom mechanism, two input position variables must be defined in order that the positions of the remaining elements of the system are determined. The equation below each of these mechanisms shows that the output variable is a direct measure of the sum of the two input variables. For this reason such a mechanism is referred to as an *adder* or a *summer*.

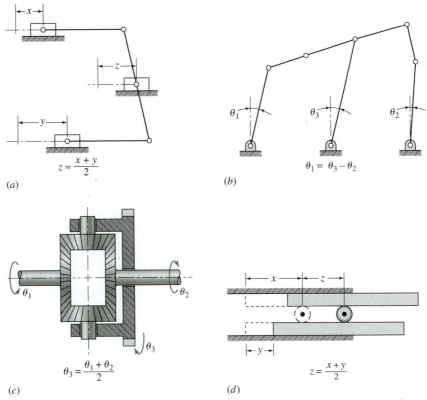

Figure 10.15 Differential mechanisms used for (*a*) adding, (*b*) subtracting, and (*c*, *d*) averaging two quantities.

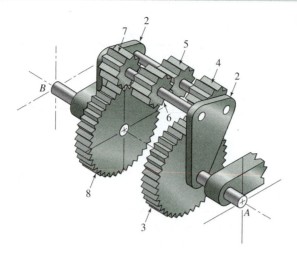

Figure 10.16 A spur gear differential.

The spur gear differential of Fig. 10.16 helps in visualizing its action. If planet carrier 2 is held stationary and gear 3 is turned by some amount $\Delta\theta_3$ then gear 8 turns in the opposite direction by an amount $\Delta\theta_8 = -\Delta\theta_3$. If the planet carrier is also allowed to turn, then $\Delta\theta_8 = \Delta\theta_2 - \Delta\theta_3$. It is for this reason that this two-degree-of-freedom mechanism is called a *differential*. Of course, a better force balance is obtained by employing several sets of planets equally spaced about the sun gears; three sets is usual. Also, you will notice in Fig. 10.16 that planets 4, 5, 6, and 7 are identical; by making planets 4 and 7 longer (thicker), they meet with each other and eliminate the need for planets 5 and 6.

If a spur gear differential were used on the driving axles of an automobile, then shafts *A* and *B* of Fig. 10.16 would drive the right and left wheels, respectively, while arm 2 receives power from a main drive shaft connected to the transmission.

It is interesting that a differential was used in China, long before the invention of the compass, to indicate geographic direction. In Fig. 10.17 each wheel of a carriage drives a vertical shaft through pin wheels. The right-hand shaft drives the upper pin wheel of the differential shown in Fig. 10.18. The left-hand shaft drives the lower pin wheel. When the cart is driven in a straight line, the upper and lower pin wheels rotate at the same speed but in opposite directions. Thus the planet gear turns about its own center, but the axle to which it is mounted remains stationary and so the figure continues to point in the same direction. When the cart makes a turn, one of the pin wheels rotates faster than the other, causing the planet axle to turn just enough to cause the figure to continue to point in the same geographic direction as before.

Figure 10.19 is a schematic drawing of the ordinary bevel gear automotive differential. The drive shaft pinion and the ring gear are normally hypoid gears. The ring gear acts as the planet carrier, and its speed can be calculated as for a simple gear train when the speed of the drive shaft is given. Gears 5 and 6 are connected, respectively, to each rear wheel and, when the car is traveling in a straight line, these two gears rotate in the same direction with exactly the same speed. Thus, for straight-line motion of the car, there is no relative motion between the planet gears and gears 5 and 6. The planet gears, in effect, serve only as keys to transmit motion from the planet carrier to both wheels.

When the vehicle is making a turn, the wheel on the inside of the curve makes fewer revolutions than the wheel with the larger turning radius. Unless this difference in speed is

Figure 10.17 The figure on this chariot continues to point in a constant geographic direction. (Smithsonian Institution, photo P63158-B.)

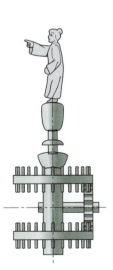

Figure 10.18 The Chinese differential.

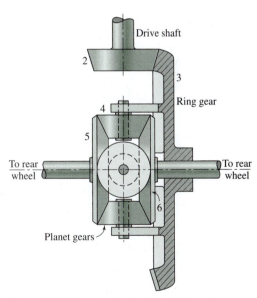

Figure 10.19 Schematic drawing of a bevel gear automotive rear axle differential.

accommodated in some manner, one or both of the tires must slip in order to make the turn. The differential permits the two wheels to rotate at different angular velocities while, at the same time, delivering power to both. During a turn, the planet gears rotate about their own axes, thus permitting gears 5 and 6 to revolve at different angular velocities.

The purpose of the differential is to allow different speeds for the two driving wheels. In the usual differential of a rear-wheel drive passenger car, the torque is divided equally whether the car is traveling in a straight line or on a curve. Sometimes the road conditions are such that the tractive effort developed by the two wheels is unequal. In such a case the total tractive effort is only twice that at the wheel having the least traction, because the differential divides the torque equally. If one wheel happens to be resting on snow or ice, the total tractive effort possible at that wheel is very small because only a small torque is required to cause the wheel to slip. Thus the car sits stationary with one wheel spinning and the other at rest with only trivial tractive effort. If the car is in motion and encounters slippery surfaces, then all traction as well as control is lost!

Limited Slip Differential   It is possible to overcome this disadvantage of the simple bevel gear differential by adding a coupling unit that is sensitive to wheel speeds. The object of such a unit is to cause more of the torque to be directed to the slower moving wheel. Such a combination is then called a *limited slip differential*.

Mitsubishi, for example, utilizes a viscous coupling unit, called a VCU, which is torque-sensitive to wheel speeds. A slight difference in wheel speeds causes slightly more torque to be delivered to the slower moving wheel. A large difference, perhaps caused by the spinning of one wheel on ice, causes a large amount of torque to be delivered to the non-spinning wheel. The arrangement, as used on the rear axle of an automobile, is shown in Fig. 10.20.

Another approach is to employ Coulomb friction, or clutching action, in the coupling. Such a unit, as with the VCU, is engaged whenever a significant difference in wheel speeds occurs.

Of course, it is also possible to design a bevel gear differential that is capable of being locked by the driver whenever dangerous road conditions are encountered. This is equivalent to a solid axle and forces both wheels to move at the same speed. It seems obvious that such a differential should not be locked when the tires are on dry pavement, because of excessive wear caused by tire slipping.

Worm Gear Differential   If gears 3 and 8 in Fig. 10.16 were replaced with worm gears, and planet gears 4 and 7 with mating worm wheels, then the result is a *worm gear differential*. Of course, planet carrier 2 would have to rotate about a new axis perpendicular to the axle AB, because worm and worm wheel axes are at right angles to each other. Such an arrangement can provide the traction of a locked differential or solid axle without the penalty of restricting differential movement between the wheels.

The worm gear differential was invented by Mr. Vernon Gleasman and developed by the Gleason Works as the TORSEN differential, a registered trademark now owned by TK Gleason, Inc. The word TORSEN combines parts of the words "torque-sensing" because the differential can be designed to provide any desired locking value by varying the lead angle of the worm. Figure 10.21 illustrates a TORSEN differential as used on Audi automobiles.

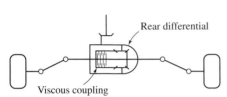

Figure 10.20 Viscous coupling used on the rear axle of the Mitsubishi Galant and Eclipse GSX automobiles.

Figure 10.21 The TORSEN differential used on the drive shaft of Audi automobiles. This is the brainchild of Mr. Vernon Gleasman of Pittsford, New York. (Courtesy of Audi of America, Inc., Troy, MI.)

10.9 ALL WHEEL DRIVE TRAIN

As shown in Fig. 10.22, an all wheel automotive drive train consists of a center differential, geared to the transmission, driving the ring gears on the front and rear axle differentials. Dividing the thrusting force between all four wheels instead of only two is itself an advantage, but it also makes for easier handling on curves and in cross winds.

Dr. Herbert H. Dobbs, Colonel (Ret.), an automotive engineer, states:

One of the major improvements (in automotive design) is antilocking braking. This provides stability and directional control when stopping by ensuring a balanced transfer of momentum from the car to the road through all wheels. As too many have found out, loss of traction at one of the wheels when braking produces unbalanced forces on the car which can throw it out of control.

It is equally important to provide such control during starting and acceleration. As with braking, this is not a problem when a car is driven prudently and driving conditions are good. When driving conditions are not good, the problems quickly become manifold. The antilock braking systems provide the answer for the deceleration portion of the driving cycle, but the devices generally available to help during the remainder of the cycle are much less satisfactory.[3]

An early solution to this problem used by Audi is to lock, by electrical means, the center or the rear differential, or both, when driving conditions deteriorate. Locking only the center differential causes one-half of the power to be delivered to the rear wheels and one-half to the front wheels. If one of the rear wheels, say, rests on slippery ice, the other rear wheel has no traction. But the front wheels still provide traction. So the car has two-wheel drive. If then the rear differential is also locked, the car has three-wheel drive because the rear wheel drive is then 50–50 distributed.

Another solution is to use a limited slip differential as the center differential on an all wheel drive (AWD) vehicle. This then has the effect of distributing most of the driving

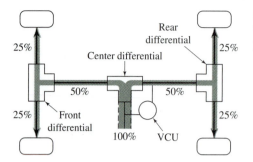

Figure 10.22 All-wheel-drive (AWD) system used on the Mitsubishi Galant, showing the power distribution for straight ahead operation.

torque to the front or rear axle depending on which is moving the slowest. An even better solution is to use limited slip differentials on both the center and the rear differentials.

Unfortunately, both the locking differentials and the limited slip differentials interfere with antilock braking systems. However, they are quite effective during low-speed winter operation.

The most effective solution seems to be the use of TORSEN differentials in an AWD vehicle. Here is what Dr. Dobbs has to say about their use:

> If they are cut to preclude any slip, the TORSEN distributes torque proportional to available traction at the driven wheels under all conditions just like a solid axle does, but it never locks up under any circumstances. Both of the driven wheels are always free to follow the separate paths dictated for them by the vehicle's motion, but are constrained by the balancing gears to stay synchronized with each other. All this adds up to a true "TORque SENsing and proportioning" differential, which of course is where the name came from.
>
> The result, particularly with a high performance front-wheel drive vehicle, is remarkable to say the least. The Army has TORSENS in the High Mobility Multipurpose Wheeled Vehicle (HMMWV) or "Hummer," which is replacing the Jeep. The only machines in production more mobile off-road than this one have tracks, and it is very capable on highway as well. It is fun to drive. The troops love it. Beyond that, Teledyne has an experimental "FAst Attack Vehicle" with TORSENS front, center, and rear. I've driven that machine over 50 mi/h on loose sand washes at Hank Hodges' Nevada Automotive Center, and it handled like it was running on dry pavement. The constant redistribution of torque to where traction was available kept all wheels driving and none digging!

NOTES

[1.] Literature devoted to epicyclic gear trains is indeed scarce. For a comprehensive study in the English language, see Z. L. Lévai, *Theory of Epicyclic Gears and Epicyclic Change-Speed Gears,* Technical University of Building, Civil, and Transport Engineering, Budapest, 1966. This book lists 104 references.

[2.] James Ferguson (1710–1776), Scottish physicist and astronomer, first published this device under the title *The Description and Use of a New Machine Called the Mechanical Paradox,* London, 1764.

[3.] Dr. Herbert H. Dobbs, Rochester Hills, MI, personal communication.

PROBLEMS

10.1 Find the speed and direction of gear 8 in the figure. What is the kinematic coefficient of the train?

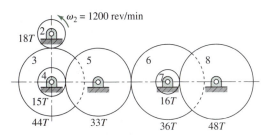

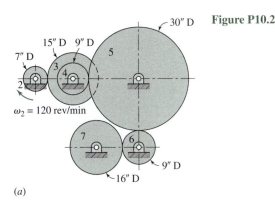

Figure P10.1

10.2 Part (*a*) of the figure gives the pitch diameters of a set of spur gears forming a train. Compute the kinematic

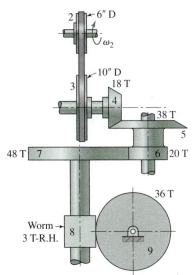

(*a*)

(*b*)

coefficient of the train. Determine the speed and direction of rotation of gears 5 and 7.

10.3 Part (*b*) of Fig. P10.2 shows a gear train consisting of bevel gears, spur gears, and a worm and worm gear. The bevel pinion is mounted on a shaft which is driven by a V-belt on pulleys. If pulley 2 rotates at 1200 rev/min in the direction shown, find the speed and direction of rotation of gear 9.

10.4 Use the truck transmission of Fig. 10.2(*a*) and an input speed of 3000 rev/min to find the drive shaft speed for each forward gear and for reverse gear.

10.5 The figure illustrates the gears in a speed-change gearbox used in machine tool applications. By sliding the cluster gears on shafts *B* and *C*, nine speed changes can be obtained. The problem of the machine tool designer is to select tooth numbers for the various gears so as to produce a reasonable distribution of speeds for the output shaft. The smallest and largest gears are gears 2 and 9, respectively. Using 20 teeth and 45 teeth for these gears, determine a set of suitable tooth numbers for the remaining gears. What are the corresponding speeds of the output shaft? Notice that the problem has many solutions.

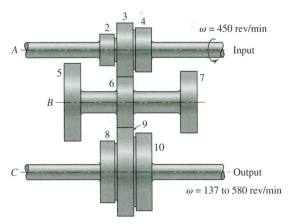

Figure P10.5

10.6 The internal gear (gear 7) in the figure turns at 60 rev/min ccw. What are the speed and direction of rotation of arm 3?

10.7 If the arm in Fig. P10.6 rotates at 300 rev/min ccw, find the speed and direction of rotation of internal gear 7.

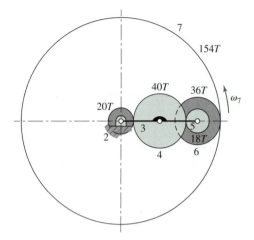

Figure P10.6

10.8 In part (*a*) of the figure, shaft *C* is stationary. If gear 2 rotates at 800 rev/min ccw, what are the speed and direction of rotation of shaft *B*?

10.9 In part (*a*) of Fig. P10.8, consider shaft *B* as stationary. If shaft *C* is driven at 380 rev/min ccw, what are the speed and direction of rotation of shaft *A*?

10.10 In part (*a*) of Fig. P10.8, determine the speed and direction of rotation of shaft *C* if
(*a*) shafts *A* and *B* both rotate at 360 rev/min ccw and
(*b*) shaft *A* rotates at 360 rev/min cw and shaft *B* rotates at 360 rev/min ccw.

10.11 In part (*b*) of Fig. P10.8, gear 2 is connected to the input shaft. If arm 3 is connected to the output shaft, what speed reduction can be obtained? What is the sense of rotation of the output shaft? What changes could be made in the train to produce the opposite sense of rotation for the output shaft?

10.12 The Lévai type-L train shown in Fig. 10.10 has $N_2 = 16$ T, $N_4 = 19$ T, $N_5 = 17$ T, $N_6 = 24$ T, and $N_3 = 95$ T. Internal gear 7 is fixed. Find the speed and direction of rotation of the arm if gear 2 is driven at 100 rev/min cw.

10.13 The Lévai type-A train of Fig. 10.10 has $N_2 = 20$ T and $N_4 = 32$ T.
(*a*) If the module is 6 mm, find the number of teeth on gear 5 and the crank arm radius.
(*b*) If gear 2 is fixed and internal gear 5 rotates at 10 rev/min ccw, find the speed and direction of rotation of the arm.

10.14 The tooth numbers for the automotive differential shown in Fig. 10.15 are $N_2 = 17$ T, $N_3 = 54$ T, $N_4 = 11$ T, and $N_5 = N_6 = 16$ T. The drive shaft turns at 1200 rev/min. What is the speed of the right wheel if it is jacked up and the left wheel is resting on the road surface?

10.15 A vehicle using the differential shown in Fig. 10.15 turns to the right at a speed of 30 mi/hr on a curve of 80-ft radius. Use the same tooth numbers as in Problem 10.14. The tire diameter is 15 in. Use 60 in as the distance between treads.
(*a*) Calculate the speed of each rear wheel.
(*b*) Find the rotational speed of the ring gear.

10.16 The figure shows a possible arrangement of gears in a lathe headstock. Shaft A is driven by a motor at a speed of 720 rev/min. The three pinions can slide along shaft *A* so as to yield the meshes 2 with 5, 3 with 6, or 4 with 8. The gears on shaft *C* can also slide so as to mesh either 7 with 9 or 8 with 10. Shaft *C* is the mandril shaft.

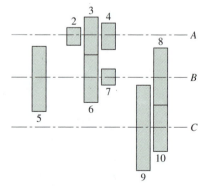

Figure P10.16 $N_2 = 16$ T, $N_3 = 36$ T, $N_4 = 25$ T, $N_5 = 64$ T, $N_6 = 66$ T, $N_7 = 17$ T, $N_8 = 55$ T, $N_9 = 79$ T, $N_{10} = 41$ T.

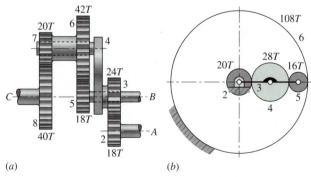

(*a*)   (*b*)

Figure P10.8

(a) Make a table showing all possible gear arrangements, beginning with the slowest speed for shaft C and ending with the highest, and enter in this table the speeds of shafts B and C.

(b) If the gears all have a module of 5 mm, what must be the shaft center distances?

10.17 Shaft A in the figure is the output and is connected to the arm. If shaft B is the input and drives gear 2, what is the speed ratio? Can you identify the Lévai type for this train?

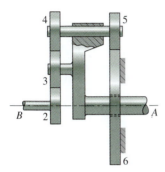

Figure P10.17 $N_2 = 16$ T, $N_3 = 18$ T, $N_4 = 16$ T, $N_5 = 18$ T, $N_6 = 50$ T.

10.18 In Problem 10.17, shaft B rotates at 100 rev/min cw. Find the speed of shaft A and of gears 3 and 4 about their own axes.

10.19 Bevel gear 2 is driven by the engine in the reduction unit shown in the figure. Bevel planets 3 mesh with crown gear 4 and are pivoted on the spider (arm), which is connected to propeller shaft B. Find the percent speed reduction.

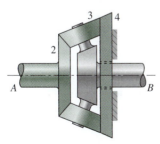

Figure P10.19 A marine reduction differential; $N_2 = 36$ T, $N_3 = 21$ T, $N_4 = 52$ T; crown gear 4 is fixed.

10.20 In the clock mechanism shown in the figure, a pendulum on shaft A drives an anchor (see Fig. 1.9c). The pendulum period is such that one tooth of the 30-T escape wheel on shaft B is released every 2 s, causing shaft B to rotate once every minute. In the figure, note that the second (to the right) 64-T gear is pivoted loosely on shaft D and is connected by a tubular shaft to the hour hand.

(a) Show that the train values are such that the minute hand rotates once every hour and that the hour hand rotates once every 12 hours.

(b) How many turns does the drum on shaft F make every day?

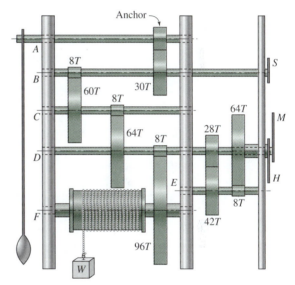

Figure P10.20 Clockwork mechanism.

11 Synthesis of Linkages

In previous chapters we have concentrated primarily on the analysis of mechanisms. By kinematic synthesis we mean the design or the creation of a mechanism to yield a desired set of motion characteristics. Because of the very large number of techniques available, some of which may be quite frustrating, we present here only a few of the more useful approaches to illustrate the applications of the theory.[1]

11.1 TYPE, NUMBER, AND DIMENSIONAL SYNTHESIS

Type synthesis refers to the kind of mechanism selected; it might be a linkage, a geared system, belts and pulleys, or even a cam system. This beginning phase of the total design problem usually involves design factors such as manufacturing processes, materials, safety, space, and economics. The study of kinematics is usually only slightly involved in type synthesis.

Number synthesis deals with the number of links and the number of joints or pairs that are required to obtain a certain mobility (see Section 1.6). Number synthesis is the second step in design following type synthesis.

The third step in design, determining the dimensions of the individual links, is called *dimensional synthesis*. This is the subject of the balance of this chapter.

Extensive bibliographies may be found in K. Hain (translated by H. Kuenzel, T. P. Goodman et al., *Applied Kinematics,* 2nd ed., pp. 639–727, McGraw-Hill, New York, 1967, and F. Freudenstein and G. N. Sandor, Kinematics of Mechanism, in H. A. Rothbart (ed.), *Mechanical Design and Systems Handbook,* 2nd ed., pp. 4-56 to 4-68, McGraw-Hill, New York, 1985.

11.2 FUNCTION GENERATION, PATH GENERATION, AND BODY GUIDANCE

A frequent requirement in design is that of causing an output member to rotate, oscillate, or reciprocate according to a specified function of time or function of the input motion. This is called *function generation*. A simple example is that of synthesizing a four-bar linkage to generate the function $y = f(x)$. In this case, x would represent the motion (crank angle) of the input crank, and the linkage would be designed so that the motion (angle) of the output rocker would approximate the function y. Other examples of function generation are as follows:

1. In a conveyor line the output member of a mechanism must move at the constant velocity of the conveyor while performing some operation—for example, bottle capping, return, pick up the next cap, and repeat the operation.
2. The output member must pause or stop during its motion cycle to provide time for another event. The second event might be a sealing, stapling, or fastening operation of some kind.
3. The output member must rotate at a specified nonuniform velocity function because it is geared to another mechanism that requires such a rotating motion.

A second type of synthesis problem is called path generation. This refers to a problem in which a coupler point is to generate a path having a prescribed shape. Common requirements are that a portion of the path be a circular arc, elliptical, or a straight line. Sometimes it is required that the path cross over itself, as in a figure-of-eight.

The third general class of synthesis problems is called *body guidance*. Here we are interested in moving an object from one position to another. The problem may call for a simple translation or a combination of translation and rotation. In the construction industry, for example, heavy parts such as scoops and bulldozer blades must be moved through a series of prescribed positions.

11.3 TWO-POSITION SYNTHESIS OF SLIDER-CRANK MECHANISMS

The centered slider-crank mechanism of Fig. 11.1*a* has a stroke $B_1 B_2$ equal to twice the crank radius r_2. As shown, the extreme positions of B_1 and B_2, also called limiting positions of the slider, are found by constructing circular arcs through O_2 of length $r_3 - r_2$ and $r_3 + r_2$, respectively.

In general, the centered slider-crank mechanism must have r_3 larger than r_2. However, the special case of $r_1 = r_2$ results in the *isosceles slider-crank mechanism,* in which the slider reciprocates through O_2 and the stroke is four times the crank radius. All points on the coupler of the isosceles slider crank generate elliptic paths. The paths generated by points on the coupler of the slider crank of Fig. 11.1*a* are not elliptical, but they are always symmetrical about the sliding axis $O_2 B$.

The linkage of Fig. 11.1*b* is called the *general* or *offset slider-crank mechanism.* Certain special effects can be obtained by changing the offset distance e. For example, the

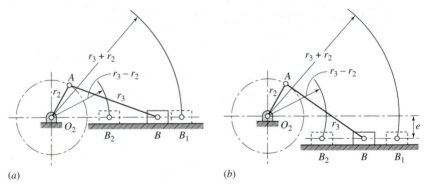

Figure 11.1 (*a*) Centered slider-crank mechanism. (*b*) General or offset slider-crank mechanism.

stroke B_1B_2 is always greater than twice the crank radius. Also, the crank angle required to execute the forward stroke is different from that for the return stroke. This feature can be used to synthesize quick-return mechanisms where a slower working stroke is desired (see Section 1.7). Note from Fig. 11.1*b* that the limiting positions B_1 and B_2 of the slider are found in the same manner as for the centered slider-crank mechanism.

11.4 TWO-POSITION SYNTHESIS OF CRANK-AND-ROCKER MECHANISMS

The limiting positions of the rocker in a crank-and-rocker mechanism are shown as points B_1 and B_2 in Fig. 11.2. Note that these positions are found in the same way as for the slider-crank linkage. Also, note that the crank and the coupler form a single straight line at each extreme position.

In this particular case the crank executes the angle ψ while the rocker moves from B_1 to B_2 through the angle ϕ. Note on the return stroke that the rocker swings from B_2 back to B_1 through the same angle ϕ but the crank moves through the angle $360° − \psi$.

There are many cases in which a crank-and-rocker mechanism is superior to a cam and follower system. Among the advantages over cam systems are smaller forces involved, the elimination of the retaining spring, and the closer clearances because of the use of revolute pairs.

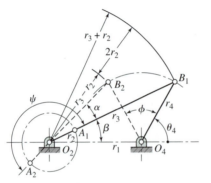

Figure 11.2 The extreme positions of the crank-and-rocker mechanism.

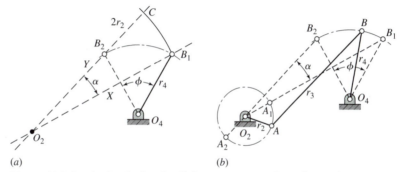

Figure 11.3 Synthesis of a four-bar linkage to generate the rocker angle ϕ.

If $\psi > 180°$ in Fig. 11.2, then $\alpha = \psi - 180°$, where α can be obtained from the equation for the time ratio (see Section 1.7)

$$Q = \frac{180° + \alpha}{180° - \alpha} \qquad (11.1)$$

of the forward and backward motions of the rocker. The first problem that arises in the synthesis of crank-and-rocker linkages is how to obtain the dimensions or geometry that will cause the mechanism to generate a specified output angle ϕ when the time ratio is specified.[2]

To synthesize a crank-and-rocker mechanism for specified values of ϕ and α, locate point O_4 in Fig. 11.3a and choose any desired rocker length r_4. Then draw the two positions $O_4 B_1$ and $O_4 B_2$ of link 4 separated by the angle ϕ as given. Through B_1 construct any line X. Then through B_2 construct the line Y at the given angle α to the line X. The intersection of the these two lines defines the location of the crank pivot O_2. Because line X was originally chosen arbitrarily, there are an infinite number of solutions to this problem.

Next, as shown in Figs. 11.2 and 11.3a, the distance $B_2 C$ is $2r_2$, or twice the crank length. So, we bisect this distance to find r_2. Then the coupler length is $r_3 = O_2 B_1 - r_2$. The completed linkage is shown in Fig. 11.3b.

11.5 CRANK-ROCKER MECHANISMS WITH OPTIMUM TRANSMISSION ANGLE

Brodell and Soni[3] developed an analytic method of synthesizing the crank-rocker linkage in which the time ratio Q equals unity. The design also satisfies the condition

$$\gamma_{\min} = 180° - \gamma_{\max} \qquad (a)$$

where γ is the transmission angle (see Section 1.10).

To develop the method, we use Fig. 11.2 and the law of cosines to write the two equations,

$$\cos(\theta_4 + \phi) = \frac{r_1^2 + r_4^2 - (r_3 - r_2)^2}{2r_1 r_4} \qquad (b)$$

$$\cos \theta_4 = \frac{r_1^2 + r_4^2 - (r_3 + r_2)^2}{2r_1 r_4} \qquad (c)$$

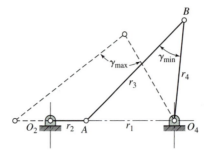

Figure 11.4

Then from Fig. 11.4 we obtain

$$\cos \gamma_{min} = \frac{r_3^2 + r_4^2 - (r_1 - r_2)^2}{2 r_3 r_4} \qquad (d)$$

$$\cos \gamma_{max} = \frac{r_3^2 + r_4^2 - (r_1 + r_2)^2}{2 r_3 r_4} \qquad (e)$$

Equations (a)–(e) are now solved simultaneously; the results are the link-length ratios

$$\frac{r_3}{r_1} = \sqrt{\frac{1 - \cos \phi}{2 \cos^2 \gamma_{min}}} \qquad (11.2)$$

$$\frac{r_4}{r_1} = \sqrt{\frac{1 - (r_3/r_1)^2}{1 - (r_3/r_1)^2 \cos^2 \gamma_{min}}} \qquad (11.3)$$

$$\frac{r_2}{r_1} = \sqrt{\left(\frac{r_3}{r_1}\right)^2 + \left(\frac{r_4}{r_1}\right)^2 - 1} \qquad (11.4)$$

Brodell and Soni plot these results as a design chart, as shown in Fig. 11.5. They state that the transmission angle should be larger than 30° for a good "quality" motion and even larger if high speeds are involved.

The synthesis of a crank-rocker mechanism for optimum transmission angle when the time ratio is not unity is more difficult. An orderly method of accomplishing this is explained by Hall[4] and by Soni.[5] The first step in the approach is illustrated in Fig. 11.6. Here the two points O_2 and O_4 are selected and the points C and C', symmetrical about $O_2 O_4$ and defined by the angles $(\phi/2) - \alpha$ and $\phi/2$, are found. Then, using C as a center and using the distance from C to O_2 as the radius, draw the circular arc that is the locus of B_2. Next, using C' as the center and using the same radius, draw another circle arc which is the locus of B_1.

One of the many possible crank-rocker linkages has been synthesized in Fig. 11.7. To obtain the dimensions, choose any point B_1 on the locus of B_1 and swing an arc about O_4 to locate B_2 on the locus of B_2. With these two points defined, the methods of the preceding section are used to locate points A_1 and A_2 together with the link lengths r_2 and r_3.

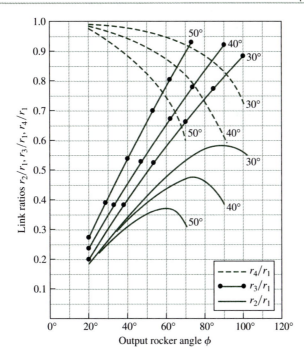

Figure 11.5 The Brodell–Soni chart for the design of the crank-rocker linkage with optimum transmission angle and unity time ratio. The angles shown on the graph are γ_{min}.

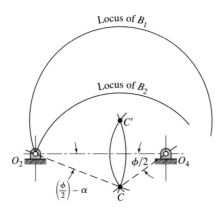

Figure 11.6 Layout showing all possible locations of B_1 and B_2.

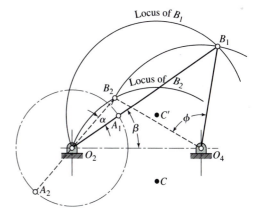

Figure 11.7 Determination of the link lengths for one of the possible crank-rocker mechanisms.

The resulting linkages should always be checked to ensure that crank 2 is capable of rotating through a complete circle.

To obtain a linkage with an optimum transmission angle, choose a variety of points B_1 on the locus of B_1, synthesizing a linkage for each point. Determine the angles $|90° - \gamma_{min}|$ and $|90° - \gamma_{max}|$ for each of these linkages. Then plot these data on a chart using the angle β (Fig. 11.7) as the abscissa to obtain two curves. The mechanism having the best transmission angle is then defined by the low point on one of these curves.

11.6 THREE-POSITION SYNTHESIS

In Fig. 11.8a motion of the input rocker $O_2 A$ through the angle ψ_{12} causes a motion of the output rocker $O_4 B$ through the angle ϕ_{12}. To employ inversion as a technique of synthesis, let us hold $O_4 B$ stationary and permit the remaining links, including the frame, to occupy the same relative positions as in Fig. 11.8a. The result (Fig. 11.8b) is called *inverting on the output rocker*. Note that $A_1 B_1$ is positioned the same in Fig. 11.8a and Fig. 11.8b. Therefore the inversion is made on the $O_4 B_1$ position. Because $O_4 B_1$ is to be fixed, the frame will have to move in order to get the linkage to the $A_2 B_2$ position. In fact, the frame must move *backward* through the angle ϕ_{12}. The second position is therefore $O_2' A_2' B_2' O_4$.

Figure 11.9 illustrates a problem and the synthesized linkage in which it is desired to determine the dimensions of a linkage in which the output lever is to occupy three specified positions corresponding to three given positions of the input lever. In Fig. 11.9 the starting angle of the input lever is θ_2; and ψ_{12}, ψ_{23}, and ψ_{13} are the swing angles, respectively, between the three design positions 1 and 2, 2 and 3, and 1 and 3. Corresponding angles of swing ϕ_{12}, ϕ_{23}, and ϕ_{13} are desired for the output lever. The length of link 4 and the starting position θ_4 are to be determined.

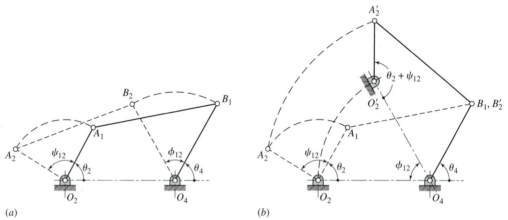

Figure 11.8 (*a*) Rotation of input rocker $O_2 A$ through the angle ψ_{12} causes the output rocker $O_4 B$ to rock through the angle ϕ_{12}. (*b*) Linkage inverted on the $O_4 B$ position.

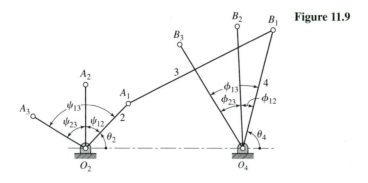

Figure 11.9

Figure 11.10

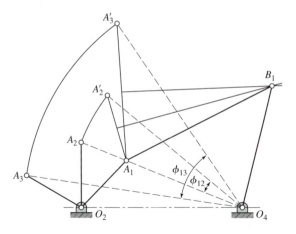

The solution to the problem is illustrated in Fig. 11.10 and is based on inverting the linkage on link 4. First we draw the input rocker O_2A in the three specified positions and locate a desired position for O_4. Because we will invert on link 4 in the first design position, we draw a ray from O_4 to A_2 and rotate it backward through the angle ϕ_{12} to locate A_2'. Similarly, we draw another ray O_4A_3 and rotate it backward through the angle ϕ_{13} to locate A_3'. Because we are inverting on the first design position, A_1 and A_1' are coincident. Now we draw midnormals to the lines $A_1'A_2'$ and $A_2'A_3'$. These intersect at B_1 and define the length of the coupler link 3 and the length and starting position of link 4.

11.7 FOUR-POSITION SYNTHESIS; POINT-POSITION REDUCTION

In the technique called *point-position reduction* the linkage is made symmetrical about the frame centerline O_2O_4 to cause two of the A' points to be coincident. The effect of this is to produce three equivalent A' points through which a circle can be drawn as in three-position synthesis. This technique is best illustrated by an example.

Let us synthesize a linkage to generate the function $y = \log x$ over the interval $10 \le x \le 60$ using an input crank range of $120°$ and an output range of $90°$.

The angle ψ is evaluated for the four design positions from the equation $\psi = ax + b$ and from the boundary conditions $\psi = 0$ when $x = 10$ and $\psi = 120°$ when $x = 60$. This gives $\psi = 2.40x - 24$. The angle ϕ is evaluated in exactly the same manner; thus, we get $\phi = 50y - 115$. The results of this preliminary work are shown in Table 11.1.

For the starting position a choice of four arrangements is shown in Fig. 11.11. In part a the line O_2O_4 bisects both ψ_{12} and ϕ_{12}; and so, if the output member is turned counterclockwise from the O_4B_2 position, then A_1' and A_2' will be coincident and at A_1. The inversion would then be based on the O_4B_1 position. Then A_3 would be rotated through the angle ϕ_{13} about O_4 counterclockwise to A_3', and A_4 would be rotated through the angle ϕ_{14} to A_4'.

In Fig. 11.11b the line O_2O_4 bisects ψ_{23} and ϕ_{23}, while in Fig. 11.11d the angles ψ_{14} and ϕ_{14} are bisected. In obtaining the inversions for each case, great care must be taken to ensure that rotation is made in the correct direction and with the correct angles.

TABLE 11.1[a]

Position	x	ψ, deg	y	ϕ, deg
1	10	0	2.30	0
2	20	24	3.00	35
3	45	94	3.80	75
4	60	120	4.10	90

[a]$\psi_{12} = 24°$ $\phi_{12} = 35°$
$\psi_{23} = 70°$ $\phi_{23} = 40°$
$\psi_{34} = 26°$ $\phi_{34} = 15°$

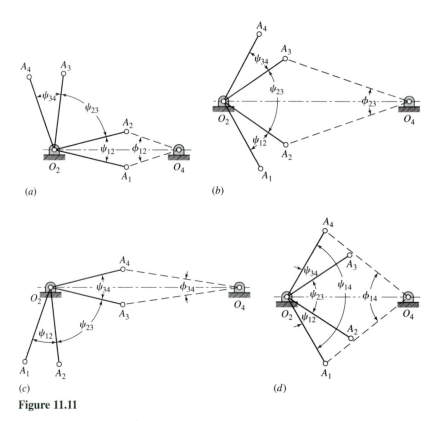

(a)   (b)

(c)   (d)

Figure 11.11

When point-position reduction is used, only the length of the input rocker O_2A can be specified in advance. The distance O_2O_4 is dependent upon the values of ψ and ϕ, as indicated in Fig. 11.11. Note that each synthesis position gives a different value for this distance. This is really quite convenient, because it is not at all unusual to synthesize a linkage that is not workable. When this happens, one of the other arrangements can be tried.

The synthesized linkage is shown in Fig. 11.12. The procedure is exactly the same as that for three positions, except as previously noted. Point B_1 is obtained at the intersection of the midnormals to $A_1'A_3'$ and $A_3'A_4'$. In this example the greatest error is less than 3 percent.

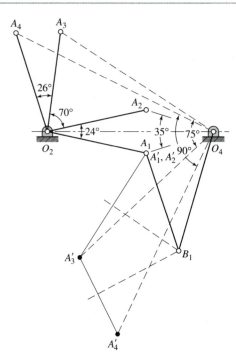

Figure 11.12

11.8 PRECISION POSITIONS; STRUCTURAL ERROR; CHEBYCHEV SPACING

All the synthesis examples we have seen in the preceding sections are of the function generation type. That is, if x is the angular position of the input crank and y is the position of the output link, then we are trying to find the dimensions of a linkage for which the input/output relationship fits a given functional relationship:

$$y = f(x) \tag{a}$$

In general, however, a mechanism has only a limited number of design parameters, a few link lengths, starting values for the input and output angles, and a few more. Therefore, except for very special cases, a linkage synthesis problem usually has no exact solution over its entire range of travel.

In the preceding sections we have chosen to work with two or three or four positions of the linkage, called *precision positions,* and to find a linkage that exactly satisfies the desired function at these few chosen positions. Our implicit assumption is that if the design fits the specifications at these few positions, then it will probably deviate only slightly from the desired function between the precision positions, and that the deviation will be acceptably small. *Structural error* is defined as the theoretical difference between the function produced by the synthesized linkage and the function originally prescribed. For many function generation problems the structural error in a four-bar linkage solution can be held to less than 4 percent. We should note, however, that structural error usually exists even if no *graphical error* were present from a graphical solution process, and even with no *mechanical error* that stems from imperfect manufacturing tolerances.

Of course, the amount of structural error in the solution can be affected by the choice of the precision positions. One of the problems of linkage design is to select a set of precision positions for use in the synthesis procedure which will minimize this structural error.

A very good trial for the spacing of these precision positions is called *Chebychev* spacing. For n precision positions in the range $x_0 \le x \le x_{n+1}$, the *Chebychev* spacing, according to Freudenstein and Sandor,[6] is

$$x_j = \frac{1}{2}(x_{n+1} + x_0) - \frac{1}{2}(x_{n+1} - x_0)\cos\frac{(2j-1)\pi}{2n}, \qquad j = 1, 2, \ldots, n \qquad (11.5)$$

As an example, suppose we wish to devise a linkage to generate the function

$$y = x^{0.8} \qquad (b)$$

over the range $1 \le x \le 3$ using three precision positions. Then, from Eq. (11.5), the three values of x_j are

$$x_1 = \frac{1}{2}(3+1) - \frac{1}{2}(3-1)\cos\frac{(2-1)\pi}{2(3)} = 2 - \cos\frac{\pi}{6} = 1.134$$

$$x_2 = 2 - \cos\frac{3\pi}{6} = 2.000$$

$$x_3 = 2 - \cos\frac{5\pi}{6} = 2.866$$

From Eq. (b), we find the corresponding values of y to be

$$y_1 = 1.106, \qquad y_2 = 1.741, \qquad y_3 = 2.322$$

Chebychev spacing of the precision positions is also easily found using the graphical approach shown in Fig. 11.13. As shown in Fig. 11.13a, a circle is first constructed whose diameter is equal to the range Δx, given by

$$\Delta x = x_{n+1} - x_0 \qquad (c)$$

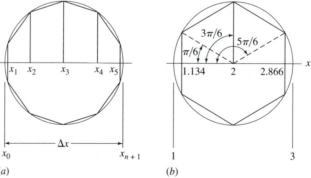

(a)   (b)

Figure 11.13 Graphical determination of Chebychev spacing.

Next we inscribe a regular polygon having $2n$ sides in this circle, with its first side spaced symmetrically about the x axis. Perpendiculars dropped from each jth vertex now intersect the diameter Δx at the precision position value of x_j. Figure 11.13b illustrates the construction for the numerical example before.

It should be noted that Chebychev spacing is a good approximation of precision positions that will reduce structural error in the design; depending on the accuracy requirements of the problem, it may be satisfactory. If additional accuracy is required, then by plotting a curve of structural error versus x we can usually determine visually the adjustments to be made in the choice of precision positions for another trial.

Before closing this section, however, we should note two more problems that can arise to confound the designer in using precision positions for synthesis. These are called *branch defect* and *order defect*. Branch defect refers to a possible completed design that meets all of the prescribed requirements at each of the precision positions, but which cannot be moved continuously between these positions without being taken apart and reassembled. Order defect refers to a developed linkage that can reach all of the precision positions, but not in the desired order.[7]

11.9 THE OVERLAY METHOD

Synthesis of a function generator, say, using the overlay method, is the easiest and quickest of all methods to use. It is not always possible to obtain a solution, and sometimes the accuracy is rather poor. Theoretically, however, one can employ as many precision positions as are desired in the process.

Let us design a function generator to solve the equation

$$y = x^{0.8}, \qquad 1 \le x \le 3 \qquad (a)$$

Suppose we choose six precision positions of the linkage for this example and use uniform spacing of the output rocker. Table 11.2 shows the values of x and y, rounded, and the corresponding angles selected for the input and output rockers.

The first step in the synthesis is shown in Fig. 11.14a. We use a sheet of tracing paper and construct the input rocker O_2A in all its positions. This requires an arbitrary choice for the length of O_2A. Also, on this sheet, we choose another arbitrary length for the coupler AB and draw arcs numbered 1 to 6 using A_1 to A_6, respectively, as centers.

TABLE 11.2

Position	x	ψ, deg	y	ϕ, deg
1	1	0	1	0
2	1.366	22.0	1.284	14.2
3	1.756	45.4	1.568	28.4
4	2.16	69.5	1.852	42.6
5	2.58	94.8	2.136	56.8
6	3.02	121.0	2.420	71.0

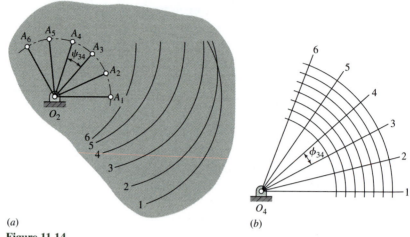

(a)

Figure 11.14

(b)

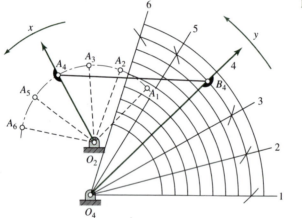

Figure 11.15

Now, on another sheet of paper, we construct the output rocker, whose length is unknown, in all its positions, as shown in Fig. 11.14b. Through O_4 we draw a number of arbitrarily spaced arcs intersecting the lines $O_4 1$, $O_4 2$, and so on; these represent possible lengths of the output rocker.

As the final step we lay the tracing over the drawing and manipulate it in an effort to find a fit. In this case a fit is found, and the result is shown in Fig. 11.15.

11.10 COUPLER-CURVE SYNTHESIS[8]

In this section we use the method of point-position reduction to synthesize a four-bar linkage so that a tracing point on the coupler will trace a previously specified path when the linkage is moved. Then, in sections to follow we will discover that paths having certain characteristics are particularly useful in synthesizing linkages having dwells of the output member for certain periods of rotation of the input member.

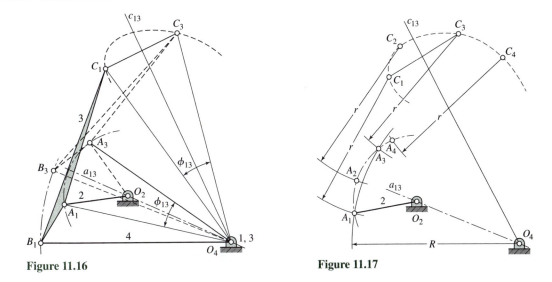

Figure 11.16   **Figure 11.17**

In synthesizing a linkage to generate a path, we can choose up to six precision positions along the path. If the synthesis is successful, the tracing point will pass through each precision position. The final result may or, because of the branch or order defects, may not approximate the desired path.

Two positions of a four-bar linkage are shown in Fig. 11.16. Link 2 is the input member; it is connected at A to coupler 3, containing the tracing point C, and connected to output link 4 at B. Two phases of the linkage are illustrated by the subscripts 1 and 3. Points C_1 and C_3 are two positions of the tracing point on the path to be generated. In this example, C_1 and C_3 have been especially selected so that the midnormal c_{13} passes through O_4. Note, for the selection of points, that the angle $C_1 O_4 C_3$ is the same as the angle $A_1 O_4 A_3$, as indicated in the figure.

The advantage of making these two angles equal is that when the linkage is finally synthesized, the triangles $C_3 A_3 O_4$ and $C_1 A_1 O_4$ are congruent. Thus, if the tracing point is made to pass through C_1 on the path, it will also pass through C_3.

To synthesize a linkage so that the coupler will pass through four precision positions, we locate any four points C_1, C_2, C_3, and C_4 on the desired path (see Fig. 11.17). Choosing C_1 and C_3 say, we first locate O_4 anywhere on the midnormal c_{13}. Then, with O_4 as a center and any radius R, we construct a circular arc. Next, with centers at C_1 and C_3, and any other radius r, we strike arcs to intersect the arc of radius R. These two intersections define points A_1 and A_3 on the input link. We construct the midnormal a_{13} to $A_1 A_3$ and note that it passes through O_4. We locate O_2 anywhere on a_{13}. This provides an opportunity to choose a convenient length for the input rocker. Now we use O_2 as a center and draw the crank circle through A_1 and A_3. Points A_2 and A_4 on this circle are obtained by striking arcs of radius r again about C_2 and C_4. This completes the first phase of the synthesis; we have located O_2 and O_4 relative to the desired path and hence defined the distance $O_2 O_4$. We have also defined the length of the input member and located its positions relative to the four precision points on the path.

Our next task is to locate point B, the point of attachment of the coupler and output member. Any one of the four locations of B can be used; in this example we use the B_1 position.

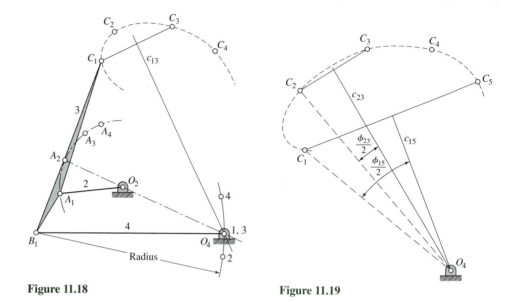

Figure 11.18   **Figure 11.19**

Before beginning the final step, we note that the linkage is now defined. Four arbitrary decisions were made: the location of O_4, the radii R and r, and the location of O_2. Thus an infinite number of solutions are possible.

Referring to Fig. 11.18, locate point 2 by making triangles $C_2A_2O_4$ and C_1A_12 congruent. Locate point 4 by making $C_4A_1O_4$ and C_1A_14 congruent. Points 4, 2, and O_4 lie on a circle whose center is B_1. So B_1 is found at the intersection of the midnormals of O_42 and O_44. Note that the procedure used causes points 1 and 3 to coincide with O_4. With B_1 located, the links can be drawn in place and the mechanism tested to see how well it traces the prescribed path.

To synthesize a linkage to generate a path through five precision points, it is possible to make two point reductions. Begin by choosing five points, C_1 to C_5, on the path to be traced. Choose two pairs of these for reduction purposes. In Fig. 11.19 we choose the pairs C_1C_5 and C_2C_3. Other pairs that could have been chosen are

$$C_1C_5, C_2C_4, \qquad C_1C_5, C_3C_4, \qquad C_1C_4, C_2C_3, \qquad C_2C_5, C_3C_4$$

Construct the perpendicular bisectors c_{23} and c_{15} of the lines connecting each pair. These intersect at point O_4. Note that O_4 can, therefore, be located conveniently by a judicious choice of the pairs to be used as well as by the choice of the positions of the points C_i on the path.

The next step is best performed by using a sheet of tracing paper as an overlay. Secure the tracing paper to the drawing and mark upon it the center O_4, the midnormal c_{23}, and another line from O_4 to C_2. Such an overlay is shown in Fig. 11.20a with the line O_4C_2 designated as O_4C_2'. This defines the angle $\phi_{23}/2$. Now rotate the overlay about the point O_4 until the midnormal coincides with c_{15} and repeat for point C_1. This defines the angle $\phi_{15}/2$ and the corresponding line O_4C_1'.

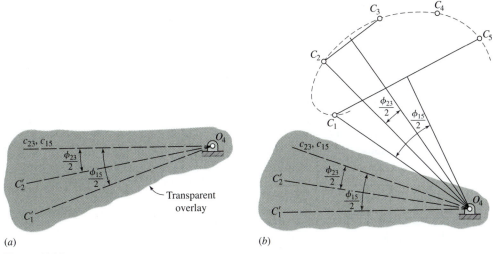

Figure 11.20

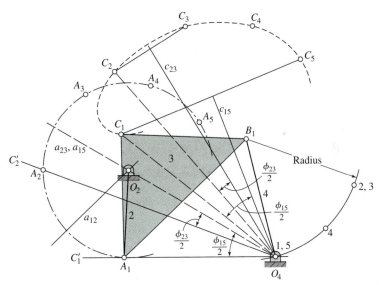

Figure 11.21

Now pin the overlay at point O_4, using a thumbtack, and rotate it until a good position is found. It is helpful to set the compass for some convenient radius r and draw circles about each point C_i. The intersection of these circles with the lines O_4C_1' and O_4C_2' on the overlay, and with each other, will reveal which areas will be profitable to investigate. See Fig. 11.20*b*.

The final steps in the solution are shown in Fig. 11.21. Having located a good position for the overlay, transfer the three lines to the drawing and remove the overlay. Now draw a circle of radius r to intersect O_4C_1' and locate point A_1. Another arc of the same radius r

from point C_2 intersects O_4C_2' at point A_2. With A_1 and A_2 located, draw the midnormal a_{12}; it intersects the midnormal a_{23} at O_2, giving the length of the input rocker. A circle through A_1 about O_2 will contain all the design positions of point A; use the same radius r and locate A_3, A_4, and A_5 on arcs about and C_3, C_4, and C_5.

We have now located everything except point B_1, and this is found as before. A double point 2, 3 exists because of the choice of point O_4 on the midnormal c_{23}. To locate this point, strike an arc from C_1 of radius C_2O_4. Then strike another arc from A_1 of radius A_2O_4. These intersect at point 2, 3. To locate point 4, strike an arc from C_1 of radius C_4O_4, and another from A_1 of radius A_4O_4. Note that point O_4 and the double points 1, 5 are coincident because the synthesis is based on inversion on the O_4B_1 position. Points O_4, 4 and double points 2, 3 lie on a circle whose center is B_1, as shown in Fig. 11.21. The linkage is completed by drawing the coupler link and the follower link in the first design position.

11.11 COGNATE LINKAGES; THE ROBERTS–CHEBYCHEV THEOREM

One of the unusual properties of the planar four-bar linkage is that there is not one but three four-bar linkages that generate the same coupler curve. This was discovered by Roberts[9] in 1875 and by Chebychev in 1878 and hence is known as the Roberts–Chebychev theorem. Though mentioned in an English publication in 1954,[10] it did not appear in the American literature until it was presented, independently and almost simultaneously, by Richard S. Hartenberg and Jacques Denavit of Northwestern University and by Roland T. Hinkle of Michigan State University in 1958.[11]

In Fig. 11.22 let O_1ABO_2 be the original four-bar linkage with a coupler point P attached to AB. The remaining two linkages defined by the Roberts–Chebychev theorem were termed *cognate linkages* by Hartenberg and Denavit. Each of the cognate linkages is shown in Fig. 11.22, one using short dashes for showing the links and the other using long dashes. The construction is evident by observing that there are four similar triangles, each containing the angles α, β, and γ, and three different parallelograms.

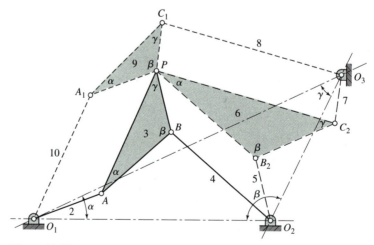

Figure 11.22

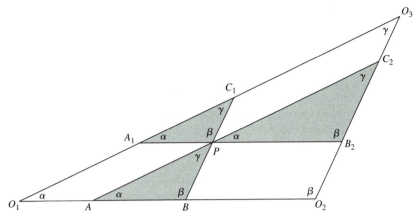

Figure 11.23 The Cayley diagram.

A good way to obtain the dimensions of the two cognate linkages is to imagine that the frame connections O_1, O_2, and O_3 can be unfastened. Then "pull" O_1, O_2, and O_3 away from each other until a straight line is formed by the crank, coupler, and follower of each linkage. If we were to do this for Fig. 11.22, then we would obtain Fig. 11.23. Note that the frame distances are incorrect, but all the movable links are of the correct length and all the angles are correct. Given any four-bar linkage and its coupler point, one can create a drawing similar to Fig. 11.23 and obtain the dimensions of the other two cognate linkages. This approach was discovered by A. Cayley and is called the *Cayley diagram*.[12]

If the tracing point P is on the straight line AB or its extensions, a figure like Fig. 11.23 is of little help because all three linkages are compressed into a single straight line. An example is shown in Fig. 11.24 where O_1ABO_2 is the original linkage having a coupler point P on an extension of AB. To find the cognate linkages, locate O_1 on an extension of O_1O_2 in the same ratio as AB is to BP. Then construct, in order, the parallelograms O_1A_1PA, O_2B_2PB, and $O_3C_1PC_2$.

Hartenberg and Denavit showed that the angular-velocity relations between the links in Fig. 11.22 are

$$\omega_9 = \omega_2 = \omega_7, \qquad \omega_{10} = \omega_3 = \omega_5, \qquad \omega_8 = \omega_4 = \omega_6 \qquad (11.6)$$

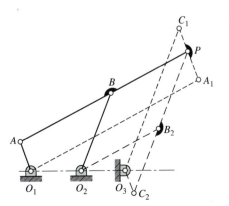

Figure 11.24

They also observed that if crank 2 is driven at a constant angular velocity and if the velocity relationships are to be preserved during generation of the coupler curve, the cognate mechanisms must be driven at variable angular velocities.

11.12 BLOCH'S METHOD OF SYNTHESIS

Sometimes a research paper is published that is a classic in its simplicity and cleverness. Such a paper written by the Russian kinematician Bloch[13] has sparked an entire generation of research. We present the method here more for the additional ideas the method may generate than for its intrinsic value, and also for its historic interest.

In Fig. 11.25 replace the links of a four-bar linkage by position vectors and write the vector equation

$$\mathbf{r}_1 + \mathbf{r}_2 + \mathbf{r}_3 + \mathbf{r}_4 = 0 \tag{a}$$

In complex polar notation Eq. (a) is written as

$$r_1 e^{j\theta_1} + r_2 e^{j\theta_2} + r_3 e^{j\theta_3} + r_4 e^{j\theta_4} = 0 \tag{b}$$

The first and second derivatives of this equation are

$$r_2\omega_2 e^{j\theta_2} + r_3\omega_3 e^{j\theta_3} + r_4\omega_4 e^{j\theta_4} = 0 \tag{c}$$

$$r_2(\alpha_2 + j\omega_2^2)e^{j\theta_2} + r_3(\alpha_3 + j\omega_3^2)e^{j\theta_3} + r_4(\alpha_4 + j\omega_4^2)e^{j\theta_4} = 0 \tag{d}$$

If we now transform Eqs. (b), (c), and (d) back into vector notation, we obtain the simultaneous equations

$$
\begin{array}{cccc}
\mathbf{r}_1 + & \mathbf{r}_2 + & \mathbf{r}_3 + & \mathbf{r}_4 = \mathbf{0} \\
& \omega_2\mathbf{r}_2 + & \omega_3\mathbf{r}_3 + & \omega_4\mathbf{r}_4 = \mathbf{0} \\
(\alpha_2 + j\omega_2^2)\mathbf{r}_2 + & (\alpha_3 + j\omega_3^2)\mathbf{r}_3 + & (\alpha_4 + j\omega_4^2)\mathbf{r}_4 = \mathbf{0}
\end{array}
\tag{e}
$$

This is a set of homogeneous vector equations having complex numbers as coefficients. Bloch specified desired values of all the angular velocities and angular accelerations and then solved the equations for the relative linkage dimensions.

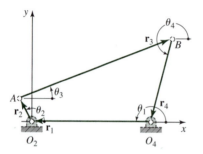

Figure 11.25

Solving Eqs. (*e*) for $\mathbf{r}_2$ gives

$$
\mathbf{r}_2 = \frac{
\begin{vmatrix}
-1 & 1 & 1 \\
0 & \omega_3 & \omega_4 \\
0 & \alpha_3 + j\omega_3^2 & \alpha_4 + j\omega_4^2
\end{vmatrix}
}{
\begin{vmatrix}
1 & 1 & 1 \\
\omega_2 & \omega_3 & \omega_4 \\
\alpha_2 + j\omega_2^2 & \alpha_3 + j\omega_3^2 & \alpha_4 + j\omega_4^2
\end{vmatrix}
}
\tag{f}
$$

Similar expressions can be obtained for $\mathbf{r}_3$ and $\mathbf{r}_4$. It turns out that the denominators for all three expressions—that is, for $\mathbf{r}_2$, $\mathbf{r}_3$, and $\mathbf{r}_4$—are complex numbers and are equal. In division, we divide the magnitudes and subtract the angles. Because these denominators are all alike, the effect of the division would be to change the magnitudes of $\mathbf{r}_2$, $\mathbf{r}_3$, and $\mathbf{r}_4$ by the same factor and to shift all the directions by the same angle. For this reason, we make all the denominators unity; the solutions then give dimensionless vectors for the links. When the determinants are evaluated, we find

$$\mathbf{r}_2 = \omega_4\left(\alpha_3 + j\omega_3^2\right) - \omega_3\left(\alpha_4 + j\omega_4^2\right)$$

$$\mathbf{r}_3 = \omega_2\left(\alpha_4 + j\omega_4^2\right) - \omega_4\left(\alpha_2 + j\omega_2^2\right)$$

$$\mathbf{r}_4 = \omega_3\left(\alpha_2 + j\omega_2^2\right) - \omega_2\left(\alpha_3 + j\omega_3^2\right)$$

$$\mathbf{r}_1 = -\mathbf{r}_2 - \mathbf{r}_3 - \mathbf{r}_4 \tag{11.7}$$

EXAMPLE 11.1

Synthesize a four-bar linkage to give the following values for the angular velocities and accelerations.

$$\omega_2 = 200 \text{ rad/s}, \qquad \omega_3 = 85 \text{ rad/s}, \qquad \omega_4 = 130 \text{ rad/s}$$
$$\alpha_2 = 0 \text{ rad/s}^2, \qquad \alpha_3 = -1000 \text{ rad/s}^2, \qquad \alpha_4 = -1600 \text{ rad/s}^2$$

SOLUTION

Substituting the given values into Eqs. (11.7) gives

$$\mathbf{r}_2 = 130[-1000 + j(85)^2] - 85[-16\,000 + j(130)^2]$$
$$= 1\,230\,000 - j497\,000 = 1\,330\,000\angle{-22°} \text{ units}$$

$$\mathbf{r}_3 = 200[-16\,000 + j(130)^2] - 130[0 + j(200)^2]$$
$$= -3\,200\,000 - j892\,000 = 3\,690\,000\angle{-150.4°} \text{ units}$$

$$\mathbf{r}_4 = 85[0 + j(200)^2] - 200[-1000 + j(85)^2]$$
$$= 200\,000 - j1\,955\,000 = 1\,965\,000\angle{-84.15°} \text{ units}$$

$$\mathbf{r}_1 = -(1\,230\,000 - j497\,000) - (-3\,200\,000 - j1\,820\,000) - (200\,000 + j1\,955\,000)$$
$$= 1\,770\,000 + j362\,000 = 1\,810\,000\angle{11.6°} \text{ units}$$

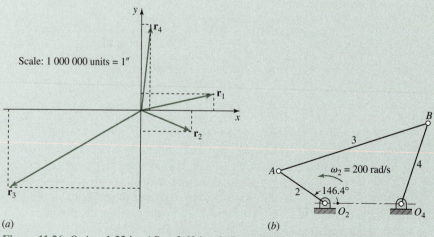

Figure 11.26 $O_2 A = 1.33$ in; $AB = 3.69$ in; $O_4 B = 1.965$ in; $O_2 O_4 = 1.81$ in.

In Fig. 11.26a these four vectors are plotted to a scale of 10^6 units per inch. In order to make $\mathbf{r}_1$ horizontal an in the $-x$ direction, the entire vector system must be rotated counterclockwise $180° - 11.6° = 168.4°$. The resulting linkage can then be constructed by using $\mathbf{r}_1$ for link 1, $\mathbf{r}_2$ for link 2, and so on, as shown in Fig. 11.26b. This mechanism has been dimensioned in inches and, if analyzed, will show that the conditions of the example are fulfilled.

11.13 FREUDENSTEIN'S EQUATION[14]

If Eq. (b) of the preceding section is transformed into complex rectangular form, and if the real and the imaginary components are separated, we obtain the two algebraic equations

$$r_1 \cos\theta_1 + r_2 \cos\theta_2 + r_3 \cos\theta_3 + r_4 \cos\theta_4 = 0 \qquad (a)$$

$$r_1 \sin\theta_1 + r_2 \sin\theta_2 + r_3 \sin\theta_3 + r_4 \sin\theta_4 = 0 \qquad (b)$$

From Fig. 11.25, $\sin\theta_1 = 0$ and $\cos\theta_1 = -1$; therefore

$$-r_1 + r_2 \cos\theta_2 + r_3 \cos\theta_3 + r_4 \cos\theta_4 = 0 \qquad (c)$$

$$r_2 \sin\theta_2 + r_3 \sin\theta_3 + r_4 \sin\theta_4 = 0 \qquad (d)$$

In order to eliminate the coupler angle θ_3 from the equations, we move all terms except those involving r_3 to the right-hand side and square both sides. This gives

$$r_3^2 \cos^2\theta_3 = (r_1 - r_2 \cos\theta_2 - r_4 \cos\theta_4)^2 \qquad (e)$$

$$r_3^2 \sin^2\theta_3 = (-r_2 \sin\theta_2 - r_4 \sin\theta_4)^2 \qquad (f)$$

Now adding the two equations and expanding the right-hand side gives

$$r_3^2 = r_1^2 + r_2^2 + r_4^2 - 2r_1 r_2 \cos\theta_2 - 2r_1 r_4 \cos\theta_4 + 2r_2 r_4 (\cos\theta_2 \cos\theta_4 + \sin\theta_2 \sin\theta_4) \qquad (g)$$

Note that $(\cos \theta_2 \cos \theta_4 + \sin \theta_2 \sin \theta_4) = \cos(\theta_2 - \theta_4)$. If we make this replacement, divide by the factor $2r_2 r_4$, and rearrange again, we obtain

$$\frac{r_3^2 - r_1^2 - r_2^2 - r_4^2}{2r_2 r_4} + \frac{r_1}{r_4} \cos \theta_2 + \frac{r_1}{r_2} \cos \theta_4 = \cos(\theta_2 - \theta_4) \tag{h}$$

Freudenstein writes this equation in the form

$$K_1 \cos \theta_2 + K_2 \cos \theta_4 + K_3 = \cos(\theta_2 - \theta_4) \tag{11.8}$$

where

$$K_1 = \frac{r_1}{r_4} \tag{11.9}$$

$$K_2 = \frac{r_1}{r_2} \tag{11.10}$$

$$K_3 = \frac{r_3^2 - r_1^2 - r_2^2 - r_4^2}{2r_2 r_4} \tag{11.11}$$

We have already learned graphical methods of synthesizing a linkage so that the motion of the output member is coordinated with that of the input member. Freudenstein's equation enables us to perform this same task by analytic means. Thus, suppose we wish the output lever of a four-bar linkage to occupy the positions ϕ_1, ϕ_2, and ϕ_3 corresponding to the angular positions ψ_1, ψ_2, and ψ_3 of the input lever. In Eq. (11.8) we simply replace θ_2 with ψ_i, θ_4 with ϕ_i, and write the equation three times, once for each position. This gives

$$K_1 \cos \psi_1 + K_2 \cos \phi_1 + K_3 = \cos(\psi_1 - \phi_1)$$
$$K_1 \cos \psi_2 + K_2 \cos \phi_2 + K_3 = \cos(\psi_2 - \phi_2) \tag{i}$$
$$K_1 \cos \psi_3 + K_2 \cos \phi_3 + K_3 = \cos(\psi_3 - \phi_3)$$

Equations (i) are solved simultaneously for the three unknowns K_1, K_2, and K_3. Then a length, say r_1, is selected for one of the links and Eqs. (11.9) through (11.11) solved for the dimensions of the other three links. The method is best illustrated by an example.

EXAMPLE 11.2

Synthesize a function generator to solve the equation

$$y = \frac{1}{x} \qquad \text{over the range } 1 \le x \le 2$$

using three precision positions.

SOLUTION

Choosing Chebychev spacing, we find, from Eq. (11.5), the values of x and corresponding values of y to be

$$x_1 = 1.067, \quad y_1 = 0.937$$
$$x_2 = 1.500, \quad y_2 = 0.667$$
$$x_3 = 1.933, \quad y_3 = 0.517$$

We must now choose starting angles for the input and output levers and also total swing angles for each. These are arbitrary decisions and may or may not result in a good linkage design in the sense that the structural errors between the precision points may be large or the transmission angles may be poor. Sometimes, in such a synthesis, it is even found that one of the pivots must be disconnected in order to move from one precision point to another. Generally, some trial-and-error work is necessary to discover the best starting angles and swing angles.

Here, for the input lever, we choose a 30° starting angle and a 90° total swing angle. For the output lever, we choose a starting angle of 240° and again choose a range of 90° total travel. With these choices made, the first and last rows of Table 11.3 can be completed.

TABLE 11.3 Example 11.2: Accuracy Positions

Position	x	ψ, deg	y	ϕ, deg
—	1.000	30.00	1.000	240.00
1	1.067	36.03	0.937	251.34
2	1.500	75.00	0.667	300.00
3	1.933	113.97	0.517	326.94
—	2.000	120.00	0.500	330.00

Next, to obtain the values of ψ and ϕ corresponding to the precision points, we write

$$\psi = ax + b, \quad \phi = cy + d \tag{1}$$

and use the data in the first and last rows of Table 11.3 to evaluate the constants $a, b, c,$ and d. When this is done, we find Eqs. (1) are

$$\psi = 90x - 60, \quad \phi = -180y + 420 \tag{2}$$

These equations can now be used to compute the data for the remaining rows in Table 11.3 and to determine the scales of the input and output levers of the synthesized linkage.

Now take the values of ψ and ϕ from the second, third, and fourth lines of Table 11.3 and substitute them for θ_2 and θ_4 in Eq. (11.8). Repeat this for the third and fourth lines. We then have the three equations

$$K_1 \cos 36.03° + K_2 \cos 251.34° + K_3 = \cos(36.03° - 251.34°)$$

$$K_1 \cos 75.00° + K_2 \cos 300.00° + K_3 = \cos(75.00° - 300.00°) \tag{3}$$

$$K_1 \cos 113.97° + K_2 \cos 326.94° + K_3 = \cos(113.97° - 326.94°)$$

When the trigonometric operations are carried out, we have

$$0.8087K_1 - 0.3200K_2 + K_3 = -0.8160$$

$$0.2588K_1 + 0.5000K_2 + K_3 = -0.7071 \tag{4}$$

$$-0.4062K_1 + 0.8381K_2 + K_3 = -0.8389$$

Upon solving Eqs. (4) we obtain

$$K_1 = 0.4032, \qquad K_2 = 0.4032, \qquad K_3 = -1.0130$$

Using $r_1 = 1.00$ units, we obtain, from Eq. (11.9),

$$r_4 = \frac{r_1}{K_1} = \frac{1.00}{0.4032} = 2.48 \text{ units}$$

Similarly, from Eqs. (11.10) and (11.11), we learn that

$$r_2 = 2.48 \text{ units} \quad \text{and} \quad r_3 = 0.968 \text{ units}$$

The result is the crossed linkage shown in Fig. 11.27.

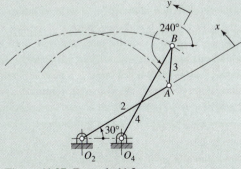

Figure 11.27 Example 11.2.

Freudenstein offers the following suggestions, which are helpful in synthesizing such function generators:

1. The total swing angles of the input and output members should be less than 120°.
2. Avoid the generation of symmetric functions such as $y = x^2$ in the range $-1 \le x \le 1$.
3. Avoid the generation of functions having abrupt changes in slope.

11.14 ANALYTICAL SYNTHESIS USING COMPLEX ALGEBRA

Another very powerful approach to the synthesis of planar linkages takes advantage of the concept of precision positions and the operations available through the use of complex algebra. Basically, as was done with Freudenstein's equation in the previous section, the idea is to write complex algebra equations describing the final linkage in each of its precision positions.

Because the links may not change lengths during the motion, the magnitudes of these complex vectors do not change from one position to the next, but their angles vary. By writing equations at several precision positions, we obtain a set of simultaneous equations that may be solved for the unknown magnitudes and angles.

The method is very flexible and much more general than is shown here. More complete coverage is given in texts such as that by Erdman, Sandor, and Kota.[15] However, the fundamental ideas and some of the operations are illustrated here by an example.

In this example we wish to design a mechanical strip-chart recorder. The concept of the final design is shown in Fig. 11.28. We assume that the signal to be recorded is available as a shaft rotation having a range of $0 \le \phi \le 90°$ clockwise. This rotation is to be converted

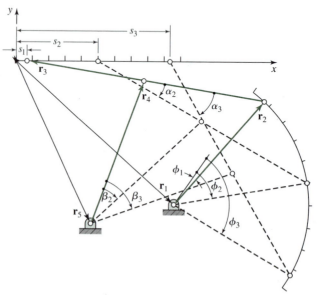

Figure 11.28 Three-position synthesis of chart-recorder linkage using complex algebra.

into a straight-line motion of the recorder pen over a range of $0 \le s \le 4$ in to the right with a linear relationship between changes of ϕ and s.[16]

We choose three accuracy positions for our design approach. Using Chebychev spacing over the range to minimize structural error, and taking counterclockwise rotations as positive, we see that the three accuracy positions are given by Eq. (11.5):

$$\phi_1 = -6° = -0.104\,72 \text{ rad}, \qquad s_1 = 0.267\,95 \text{ in.}$$

$$\phi_2 = -45° = -0.785\,40 \text{ rad}, \qquad s_2 = 2.000\,00 \text{ in.} \qquad (a)$$

$$\phi_3 = -84° = -1.466\,08 \text{ rad}, \qquad s_3 = 3.732\,05 \text{ in.}$$

First we tackle the design of the dyad consisting of the input crank $\mathbf{r}_2$ and the coupler link $\mathbf{r}_3$. Taking these to be complex vectors for the mechanism in its first accuracy position, and designating the unknown position of the fixed pivot by the complex vector $\mathbf{r}_1$, we can write a loop-closure equation at each of the three precision positions:

$$\mathbf{r}_1 + \mathbf{r}_2 + \mathbf{r}_3 = s_1$$

$$\mathbf{r}_1 + \mathbf{r}_2 e^{j(\phi_2 - \phi_1)} + \mathbf{r}_3 e^{j\alpha_2} = s_2 \qquad (b)$$

$$\mathbf{r}_1 + \mathbf{r}_2 e^{j(\phi_3 - \phi_1)} + \mathbf{r}_3 e^{j\alpha_3} = s_3$$

where the angles α_j represent the angular displacements of the coupler link from its first position. Next, by subtracting the first of these equations from each of the others and rearranging, we obtain

$$\left[e^{j(\phi_2 - \phi_1)} - 1\right]\mathbf{r}_2 + [e^{j\alpha_2} - 1]\mathbf{r}_3 = s_2 - s_1$$

$$\left[e^{j(\phi_3 - \phi_1)} - 1\right]\mathbf{r}_2 + [e^{j\alpha_3} - 1]\mathbf{r}_3 = s_3 - s_1 \qquad (c)$$

Here we note that we have two complex equations in two complex unknowns, $\mathbf{r}_2$ and $\mathbf{r}_3$, except that the coupler displacement angles α_j, which appear in the coefficients, are also unknowns. Thus we have more unknowns than equations and are free to specify additional conditions or additional data for the problem. Making estimates based on crude sketches of our contemplated design, therefore, we make the following arbitrary decisions:

$$\alpha_2 = -20° = -0.349\,07 \text{ rad}$$

$$\alpha_3 = -50° = -0.872\,66 \text{ rad} \qquad (d)$$

Collecting the data from Eqs. (a) and (d), substituting into Eqs. (c), and evaluating, we find

$$-(0.222\,85 + 0.629\,32j)\mathbf{r}_2 - (0.060\,31 + 0.342\,02j)\mathbf{r}_3 = 1.732\,05$$

$$-(0.792\,09 + 0.978\,15j)\mathbf{r}_2 - (0.357\,21 + 0.766\,04j)\mathbf{r}_3 = 3.464\,10$$

which we can now solve for the two unknowns:

$$\mathbf{r}_2 = 2.153\,26 + 2.448\,60j = 3.261\angle 48.67° \text{ in} \qquad Ans.$$

$$\mathbf{r}_3 = -5.725\,48 + 0.952\,04j = 5.804\angle 170.56° \text{ in} \qquad Ans.$$

Then, using the first of Eqs. (*b*), we solve for the position of the fixed pivot:

$$\mathbf{r}_1 = s_1 - \mathbf{r}_2 - \mathbf{r}_3$$
$$= 3.840\,17 - 3.400\,64\,j = 5.129\angle{-41.53°} \text{ in} \qquad\qquad Ans.$$

Thus far, we have completed the design of the dyad, which includes the input crank. Before proceeding, we should notice that an identical procedure could have been used for the design of a slider-crank mechanism, or for one dyad of a four-bar linkage used for any path generation or motion generation problem, or for a variety of other applications. Our total design is not yet completed, but we should notice the general applicability of the procedures covered to other linkage synthesis problems.

Continuing with our design of the recording instrument, however, we now need to find the location and dimensions of the dyad, $\mathbf{r}_4$ and $\mathbf{r}_5$ of Fig. 11.28. As shown in the figure, we choose to connect the moving pivot of the output crank at the midpoint of the coupler link in order to minimize its mass and to keep dynamic forces low. Thus we can write another loop-closure equation including the rocker at each of the three precision positions:

$$\mathbf{r}_5 + \mathbf{r}_4 + 0.5\mathbf{r}_3 = s_1$$
$$\mathbf{r}_5 + \mathbf{r}_4 e^{j\beta_2} + 0.5\mathbf{r}_3 e^{j\alpha_2} = s_2 \qquad\qquad (e)$$
$$\mathbf{r}_5 + \mathbf{r}_4 e^{j\beta_3} + 0.5\mathbf{r}_3 e^{j\alpha_3} = s_3$$

Substituting the known data and rearranging these equations we obtain

$$\mathbf{r}_5 + \mathbf{r}_4 + (-3.130\,69 + 0.476\,02\,j) = 0$$
$$\mathbf{r}_5 + (e^{j\beta_2})\mathbf{r}_4 + (-4.527\,25 + 1.426\,52\,j) = 0 \qquad\qquad (f)$$
$$\mathbf{r}_5 + (e^{j\beta_3})\mathbf{r}_4 + (-5.207\,45 + 2.499\,03\,j) = 0$$

These appear to be three simultaneous complex equations in only two complex unknowns, $\mathbf{r}_4$ and $\mathbf{r}_5$, and thus we are not free to choose the rotation angles β_2 and β_3 arbitrarily. In order for Eqs. (*f*) to have consistent nontrivial solutions, it is necessary that the determinant of the matrix of coefficients be zero. Thus β_2 and β_3 must be chosen such that

$$\begin{vmatrix} 1 & 1 & (-3.130\,69 + 0.476\,02\,j) \\ 1 & e^{j\beta_2} & (-4.527\,25 + 1.426\,52\,j) \\ 1 & e^{j\beta_3} & (-5.207\,45 + 2.499\,03\,j) \end{vmatrix} = 0 \qquad\qquad (g)$$

which expands to

$$(-2.076\,76 + 2.023\,01\,j)e^{j\beta_2} + (1.396\,56 - 0.950\,50\,j)e^{j\beta_3}$$
$$+ (0.680\,20 - 1.072\,51\,j) = 0$$

Solving this for $e^{j\beta_2}$ gives

$$e^{j\beta_2} = (0.573\,82 + 0.101\,28\,j)e^{j\beta_3} + (0.426\,19 - 0.101\,28\,j)$$

and equating the real and imaginary parts, respectively, gives

$$\cos \beta_2 = 0.573\,82 \cos \beta_3 - 0.101\,28 \sin \beta_3 + 0.426\,19$$

$$\sin \beta_2 = 0.101\,28 \cos \beta_3 + 0.573\,82 \sin \beta_3 - 0.101\,28$$

(h)

These equations can now be squared and added to eliminate the unknown β_2. The result, after rearrangement, is a single equation in β_3:

$$0.468\,59 \cos \beta_3 - 0.202\,56 \sin \beta_3 - 0.468\,57 = 0 \tag{i}$$

This equation can be solved by substituting the tangent of the half-angle identities

$$x = \tan \frac{\beta_3}{2}$$

$$0.468\,59(1 - x^2) - 0.202\,56(2x) - 0.468\,57(1 + x^2) = 0 \tag{j}$$

This reduces to the quadratic equation

$$-0.937\,16x^2 - 0.405\,12x + 0.000\,02 = 0$$

for which the roots are

$$x = -0.432\,66 \quad \text{or} \quad x = 0.000\,27$$

and, from Eq. (j),

$$\beta_3 = -46.79° \quad \text{or} \quad \beta_3 = 0.03°$$

Guided by our sketch of the desired design, we choose the first of these roots, $\beta_3 = -46.79°$. Then, returning to Eqs. (h), we find the value $\beta_2 = -26.76°$, and finally, from Eqs. (f), we find the final solution:

$$\mathbf{r}_4 = 1.299\,71 + 3.410\,92j = 3.650\angle 69.14° \text{ in} \qquad Ans.$$

$$\mathbf{r}_5 = 1.830\,98 - 3.886\,94j = 4.297\angle{-64.78°} \text{ in} \qquad Ans.$$

Note that this second part of the solution, solving for the rocker $\mathbf{r}_4$, is also a very general approach that could be used to design a crank to go through three given precision positions in a variety of other problems. Although this example presents a specific case, the approach arises repeatedly in linkage design.

Of course, before we finish, we should evaluate the quality of our solution by analysis of the linkage we have designed. This has been done here using the equations of Chapter 2 to find the location of the coupler point for 20 equally spaced crank increments spanning the given range of motion. As expected, there is structural error; the coupler curve of the recording pen tip is not exactly straight, and the displacement increments are not perfectly linear over the range of travel of the pen. However, the solution is quite good; the deviation from a straight line is less than 0.020 in, or 0.5% of the travel, and the linearity between the input crank rotation and coupler point travel is better than 1% of the travel. As expected, the structural error follows a regular pattern and vanishes at the three precision positions. The transmission angle remains larger than 70° throughout the range; thus no problems with force transmission are expected. Although the design might be improved slightly by using additional precision positions, the present solution seems excellent and the additional effort does not seem worthwhile.

11.15 SYNTHESIS OF DWELL MECHANISMS

One of the most interesting uses of coupler curves having straight-line or circle-arc segments is in the synthesis of mechanisms having a substantial dwell during a portion of their operating period. By using segments of coupler curves, it is not difficult to synthesize linkages having a dwell at either or both of the extremes of their motion or at an intermediate point.

In Fig. 11.29a a coupler curve having approximately an elliptical shape is selected from the Hrones and Nelson atlas so that a substantial portion of the curve approximates a circle arc. Connecting link 5 is then given a length equal to the radius of this arc. Thus, in the figure, points D_1, D_2, and D_3 are stationary while coupler point C moves through positions C_1, C_2, and C_3. The length of output link 6 and the location of the frame point O_6 depend upon the desired angle of oscillation of this link. The frame point should also be positioned for optimum transmission angle.

When segments of circular arcs are desired for the coupler curve, an organized method of searching the Hrones and Nelson atlas can be employed. The overlay, shown in Fig. 11.30, is made on a sheet of tracing paper and can be fitted over the paths in the atlas very quickly. It reveals immediately the radius of curvature of the segment, the location of pivot point D, and the swing angle of the connecting link.

Figure 11.29b shows a dwell mechanism employing a slider. A coupler curve having a straight-line segment is used, and the pivot point O_6 is placed on an extension of this line.

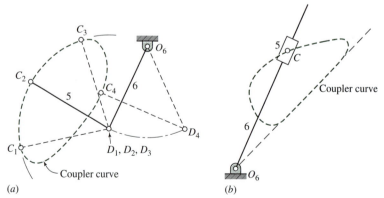

(a)   (b)

Figure 11.29 Synthesis of dwell mechanisms; in each case, the four-bar linkage that generates the coupler curve is not shown. (a) Link 6 dwells while point C travels the circle-arc path $C_1 C_2 C_3$. (b) Link 6 swells while point C travels along the straight portion of the coupler curve.

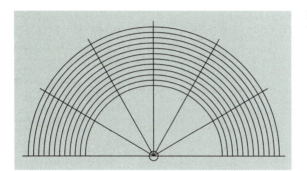

Figure 11.30 Overlay for use with the Hrones and Nelson atlas.

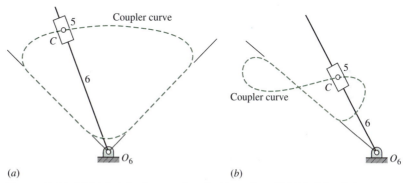

(a)   (b)

Figure 11.31 (a) Link 6 dwells at each end of its swing. (b) Link 6 dwells in the central portion of its swing.

The arrangement shown in Fig. 11.31a has a dwell at both extremes of the motion. A practical arrangement of this mechanism is rather difficult to achieve, however, because link 6 has such a high velocity when the slider is near the pivot O_6.

The slider mechanism of Fig. 11.31b uses a figure-eight coupler curve having a straight-line segment to produce an intermediate dwell linkage. Pivot O_6 must be located on an extension of the straight-line segment, as shown.

11.16 INTERMITTENT ROTARY MOTION

The *Geneva wheel,* or *Maltese cross,* is a camlike mechanism that provides intermittent rotary motion and is widely used in both low-speed and high-speed machinery. Although originally developed as a stop to prevent overwinding of watches, it is now used extensively in automatic machinery, for example, where a spindle, turret, or worktable must be indexed. It is also used in motion-picture projectors to provide the intermittent advance of the film.

A drawing of a six-slot Geneva mechanism is shown in Fig. 11.32. Notice that the centerlines of the slot and crank are mutually perpendicular at engagement and at disengagement. The crank, which usually rotates at a uniform angular velocity, carries a roller to engage with the slots. During one revolution of the crank the Geneva wheel rotates a fractional part of a revolution, the amount of which is dependent upon the number of slots.

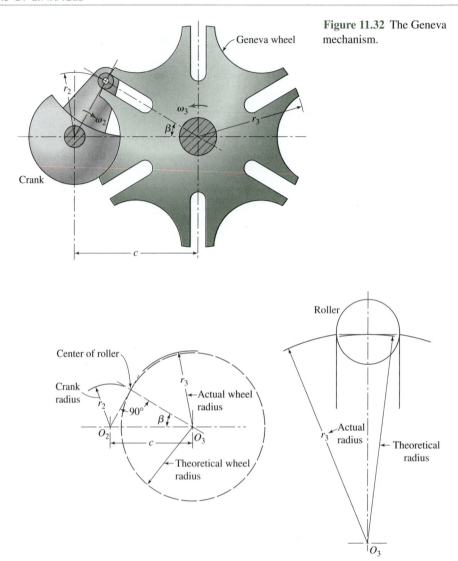

Figure 11.32 The Geneva mechanism.

Figure 11.33 Design of a Geneva wheel.

The circular segment attached to the crank effectively locks the wheel against rotation when the roller is not in engagement and also positions the wheel for correct engagement of the roller with the next slot.

The design of a Geneva mechanism is initiated by specifying the crank radius, the roller diameter, and the number of slots. At least three slots are necessary, but most problems can be solved with wheels having from four to 12 slots. The design procedure is shown in Fig. 11.33. The angle β is half the angle subtended by adjacent slots; that is,

$$\beta = \frac{360°}{2n} \qquad\qquad (a)$$

Figure 11.34

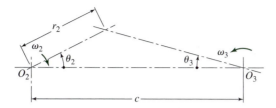

where n is the number of slots in the wheel. Then, defining r_2 as the crank radius, we have

$$c = \frac{r_2}{\sin \beta} \qquad (b)$$

where c is the center distance. Note, too, from Fig. 11.33 that the actual Geneva-wheel radius is more than that which would be obtained by a zero-diameter roller. This is due to the difference between the sine and the tangent of the angle subtended by the roller, measured from the wheel center.

After the roller has entered the slot and is driving the wheel, the geometry is that of Fig. 11.34. Here θ_2 is the crank angle and θ_3 is the wheel angle. They are related trigonometrically by

$$\tan \theta_3 = \frac{\sin \theta_2}{(c/r_2) - \cos \theta_2} \qquad (c)$$

We can determine the angular velocity of the wheel for any value of θ_2 by differentiating Eq. (c) with respect to time. This produces

$$\omega_3 = \omega_2 \frac{(c/r_2)\cos \theta_2 - 1}{1 + (c^2/r_2^2) - 2(c/r_2)\cos \theta_2} \qquad (11.12)$$

The maximum wheel velocity occurs when the crank angle is zero. Substituting $\theta_2 = 0$ therefore gives

$$\omega_3 = \omega_2 \frac{r_2}{c - r_2} \qquad (11.13)$$

The angular acceleration, obtained by differentiating Eq. (11.12) with respect to time, is

$$\alpha_3 = \omega_2^2 \frac{(c/r_2)\sin \theta_2 (1 - c^2/r_2^2)}{[1 + (c/r_2)^2 - 2(c/r_2)\cos \theta_2]^2} \qquad (11.14)$$

The angular acceleration reaches a maximum where

$$\theta_2 = \cos^{-1}\left\{ \pm \sqrt{\left[\frac{1 + (c^2/r_2^2)}{4(c/r_2)} \right]^2 + 2 - \frac{1 + (c/r_2)^2}{4(c/r_2)}} \right\} \qquad (11.15)$$

This occurs when the roller has advanced about 30 percent into the slot.

Several methods have been employed to reduce the wheel acceleration in order to reduce inertia forces and the consequent wear on the sides of the slot. Among these is the idea of using a curved slot. This can reduce the acceleration, but also increases the deceleration and consequently the wear on the other side of the slot.

Another method uses the Hrones–Nelson atlas for synthesis. The idea is to place the roller on the connecting link of a four-bar linkage. During the period in which it drives the wheel, the path of the roller should be curved and should have a low value of acceleration. Figure 11.35 shows one solution and includes the path taken by the roller. This is the path that is sought while leafing through the book.

The inverse Geneva mechanism of Fig. 11.36 enables the wheel to rotate in the same direction as the crank and requires less radial space. The locking device is not shown, but

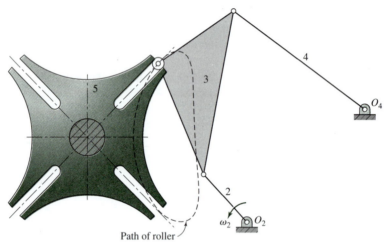

Figure 11.35 Geneva wheel driven by a four-bar linkage synthesized by the Hrones–Nelson atlas. Link 2 is the driving crank.

Figure 11.36 The inverse Geneva mechanism.

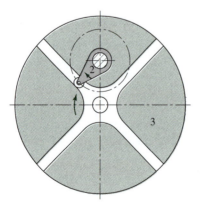

this can be a circular segment attached to the crank, as before, which locks by wiping against a built-up rim on the periphery of the wheel.

NOTES

1. The following are some of the most useful references on kinematic synthesis in the English language: R. Beyer (translated by H. Kuenzel) *Kinematic Synthesis of Linkages,* McGraw-Hill, New York, 1964; A. G. Erdman, G. N. Sandor, and S. Kota, *Mechanism Design: Analysis and Synthesis,* vol. I, 4th ed., Prentice-Hall, Englewood Cliffs, NJ, 2001; R. E. Gustavson, "Linkages," Chapter 41 in J. E. Shigley and C. R. Mischke (eds.), *Standard Handbook of Machine Design,* McGraw-Hill, New York, 1986; R. E. Gustavson, "Linkages," Chapter 3 in J. E. Shigley and C. R. Mischke (eds.); *Mechanisms-A Mechanical Designer's Workbook,* McGraw-Hill, New York, 1990; Hain, *op. cit.;* A. S. Hall, *Kinematics and Linkage Design,* Prentice-Hall, Englewood Cliffs, NJ, 1961; R. S. Hartenberg and J. Denavit, *Kinematic Synthesis of Linkages,* McGraw-Hill, New York, 1964; J. Hirschhorn, *Kinematics and Dynamics of Plane Mechanisms,* McGraw-Hill, New York, 1962; K. H. Hunt, *Kinematic Geometry of Mechanisms,* Oxford University Press, Oxford, 1978; A. H. Soni, *Mechanism Synthesis and Analysis,* McGraw-Hill, New York, 1974; C. H. Suh and C. W. Radcliffe, *Kinematics and Mechanism Design,* Wiley, New York, 1978; D. C. Tao, *Fundamentals of Applied Kinematics,* Addison-Wesley, Reading, MA, 1967.

2. The method described here appears in Hall, *op. cit.,* p. 33, and Soni, *op. cit.,* p. 257. Both Tao, *op. cit.,* p. 241, and Hain, *op. cit.,* p. 317, describe another method that gives different results.

3. R. Joe Brodell and A. H. Soni. "Design of the Crank-Rocker Mechanism with Unit Time Ratio," *J. Mech.,* vol. 5, no. 1, p. 1, 1970.

4. *Op. cit.,* pp. 36–42.

5. *Op. cit.,* p. 258.

6. *Op. cit.,* pp. 14–27.

7. See K. J. Waldron and E. N. Stephensen, Jr., "Elimination of Branch, Grashof, and Order Defects in Path-Angle Generation and Function Generation Synthesis," *J. Mech. Design, Trans. ASME,* vol. 101, no. 3, July 1979, pp. 428–437.

8. The methods presented here were devised by Hain and presented in Hain, *op. cit.,* Chapter 17.

9. By S. Roberts, a mathematician; this was not the same Roberts of the approximate-straight-line generator (Fig. 1.19*b*).

10. P. Grodzinski and E. M'Ewan, "Link Mechanisms in Modern Kinematics," *Proc. Inst. Mech. Eng.,* vol. 168, no. 37, pp. 877–896, 1954.

11. R. S. Hartenberg and J. Denavit, "The Fecund Four-Bar," *Trans. 5th Conf. Mech.,* Purdue University, Lafayette, IN, 1958, p. 194. R. T. Hinkle. "Alternate Four-Bar Linkages," *Prod. Eng.,* vol. 29, p. 4, October 1958.

12. A. Cayley, "On Three-Bar Motion," *Proc. Lond. Math. Soc.,* vol. 7, pp. 136–166, 1876. In Cayley's time a four-bar linkage was described as a three-bar mechanism because the idea of a kinematic chain had not yet been conceived.

13. S. Sch. Bloch, "On the Synthesis of Four-Bar Linkages" (in Russian). *Bull. Acad. Sci. USSR,* pp. 47–54, 1940.

14. Ferdinand Freudenstien, "Approximate Synthesis of Four-Bar Linkages," *Trans. ASME,* vol. 77, no. 6, pp. 853–861, 1955.

15. *Op. cit.,* Chapter 8.

16. This example is adapted from a similar problem solved graphically by Hartenberg and Denavit, *op. cit.,* pp. 244–248 and pp. 274–278.

PROBLEMS

11.1 A function varies from 0 to 10. Find the Chebychev spacing for two, three, four, five, and six precision positions.

11.2 Determine the link lengths of a slider-crank linkage to have a stroke of 600 mm and a time ratio of 1.20.

11.3 Determine a set of link lengths for a slider-crank linkage such that the stoke is 16 in and the time ratio is 1.25.

11.4 The rocker of a crank-rocker linkage is to have a length of 500 mm and swing through a total angle of 45° with a time ratio of 1.25. Determine a suitable set of dimensions for r_1, r_2, and r_3.

11.5 A crank-and-rocker mechanism is to have a rocker 6 ft in length and a rocking angle of 75°. If the time ratio is to be 1.32, what are a suitable set of link lengths for the remaining three links?

11.6 Design a crank and coupler to drive rocker 4 in the figure such that slider 6 will reciprocate through a distance of 16 in with a time ratio of 1.20. Use $a = r_4 = 16$ in and $r_5 = 24$ in with $\mathbf{r}_4$ vertical at midstroke. Record the location of O_2 and dimensions r_2 and r_3.

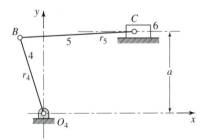

Figure P11.6

11.7 Design a crank and rocker for a six-link mechanism such that the slider in the figure for Problem 11.6 reciprocates through a distance of 800 mm with a time ratio of 1.12, use $a = r_4 = 1\ 200$ mm and $r_5 = 1\ 800$ mm. Locate O_4 such that rocker 4 is vertical when the slider is at midstroke. Find suitable coordinates for O_2 and lengths for r_2 and r_3.

11.8 Design a crank-rocker mechanism with optimum transmission angle, a unit time ratio, and a rocker angle of 45° using a rocker 250 mm in length. Use the chart of Fig. 11.5 and $\gamma_{min} = 50°$.

11.9 The figure shows two positions of a folding seat used in the aisles of buses to accommodate extra passengers. Design a four-bar linkage to support the seat so that it will lock in the open position and fold to a stable closing position along the side of the aisle.

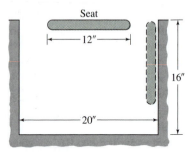

Figure P11.9

11.10 Design a spring-operated four-bar linkage to support a heavy lid like the trunk lid of an automobile. The lid is to swing through an angle of 80° from the closed to the open position. The springs are to be mounted so that the lid will be held closed against a stop, and they should also hold the lid in a stable open position without the use of a stop.

11.11 For part (a) of the figure, synthesize a linkage to move AB from position 1 to position 2 and return.

11.12 For part (b) of the figure synthesize a mechanism to move AB successively through positions 1, 2, and 3.

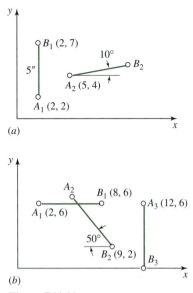

Figure P11.11

11.13 through 11.22* The figure shows a function-generator linkage in which the motion of rocker 2 corresponds to x and the motion of rocker 4 to the function $y = f(x)$. Use four precision points and Chebychev spacing and synthesize a linkage to generate the functions shown in the accompanying table. Plot a curve of the desired function and a curve of the actual function which the linkage generates. Compute the maximum error between them in percent.

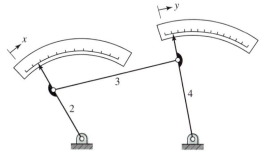

Figure P11.13

Problem Number	Function, $y =$	Range
11.13, 11.23	$\log_{10} x$	$1 \le x \le 2$
11.14, 11.24	$\sin x$	$0 \le x \le \pi/2$
11.15, 11.25	$\tan x$	$0 \le x \le \pi/4$
11.16, 11.26	e^x	$0 \le x \le 1$
11.17, 11.27	$1/x$	$1 \le x \le 2$
11.18, 11.28	$x^{1.5}$	$0 \le x \le 1$
11.19, 11.29	x^2	$0 \le x \le 1$
11.20, 11.30	$x^{2.5}$	$0 \le x \le 1$
11.21, 11.31	x^3	$0 \le x \le 1$
11.22, 11.32	x^2	$-1 \le x \le 1$

11.23 through 11.32 Repeat Problems 11.13 through 11.22 using the overlay method.

11.33 The figure illustrates a coupler curve which can be generated by a four-bar linkage (not shown). Link 5 is to be attached to the coupler point, and link 6 is to be a rotating member with O_6 as the frame connection. In this problem we wish to find a coupler curve from the Hrones and Nelson atlas or by point-position reduction, such that, for an appreciable distance, point C moves through an arc of a circle. Link 5 is then proportioned so that D lies at the center of curvature of this arc. The result is then called a *hesitation motion* because link 6 will hesitate in its rotation for the period during which point C transverse the approximate circle arc. Make a drawing of the complete linkage and plot the velocity-displacement diagram for 360° of displacement of the input link.

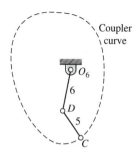

Figure P11.33

11.34 Synthesize a four-bar linkage to obtain a coupler curve having an approximate straight-line segment. Then, using the suggestion included in Fig. 11.29b or 11.31b, synthesize a dwell motion. Using an input crank angular velocity of unity, plot the velocity of rocker 6 versus the input crank displacement.

11.35 Synthesize a dwell mechanism using the idea suggested in Fig. 11.29a and the Hrones and Nelson atlas. Rocker 6 is to have a total angular displacement of 60°. Using this displacement as the abscissa, plot a velocity diagram of the motion of the rocker to illustrate the dwell motion.

---

*Solutions for these problems were among the earliest computer work in kinematic synthesis and results are shown in F. Freudenstein, "Four-bar Function Generators," *Machine Design*, vol. 30, no. 24, pp. 119–123, 1958.

12 | Spatial Mechanisms

12.1 INTRODUCTION

The large majority of mechanisms in use today have *planar* motion. That is, the motions of all points produce paths which lie in parallel planes. This means that all motions can be seen in true size and shape from a single viewing direction, and that graphical methods of solution require only a single view. If the coordinate system is chosen with the x and y axes parallel to the plane(s) of motion, then all z values remain constant and the problem can be solved, either graphically or analytically, with only two dimensional methods.

Although this is usually the case, it is not a necessity, and mechanisms having more general, three-dimensional point paths are called *spatial mechanisms.* Another special category, called *spherical mechanisms,* have point paths which lie on concentric spherical surfaces.

While these definitions were raised in Chapter 1, almost all examples in the previous chapters have dealt only with planar mechanisms. This is justified because of their very wide use in practical situations. Although a few nonplanar mechanisms such as universal shaft couplings and bevel gears have been known for centuries, it is only relatively recently that kinematicians have made substantial progress in developing design procedures for other spatial mechanisms. It is probably not coincidence, given the greater difficulty of the mathematical manipulations, that the emergence of such tools awaited the development and availability of computers.

Although we have concentrated so far on examples with planar motion, a brief review will show that most of the previous theory has been derived in sufficient generality for both planar and spatial motion. Examples have been planar because they can be more easily visualized and require less tedious computations than the three-dimensional case. Still, most of the theory extends directly to spatial mechanisms. This chapter will review some of the previous techniques, showing examples with spatial motion, but will also introduce a few new tools and solution techniques that were not needed for planar motion.

In Section 1.6 we learned that the mobility of a kinematic chain can be obtained from the Kutzbach criterion. The three-dimensional form of this criterion was given in Eq. (1.3),

$$m = 6(n - 1) - 5j_1 - 4j_2 - 3j_3 - 2j_4 - 1j_5 \tag{12.1}$$

where m is the mobility of the mechanism, n is the number of links, and each j_k is the number of joints having k degrees of freedom.

One of the numerical solutions to Eq. (12.1) is $m = 1$, $n = 7$, $j_1 = 7$, $j_2 = j_3 = j_4 = j_5 = 0$. Harrisberger called such a solution a mechanism type[1]; in particular, he called this the $7j_1$ type. Other combinations of j_k's produce other types of mechanisms. For example, with mobility of $m = 1$, the $3j_1 + 2j_2$ type has $n = 5$ links while the $1j_1 + 2j_3$ type has $n = 3$ links.

Each mechanism type contains a finite number of *kinds* of mechanisms; there are as many kinds of mechanisms of each type as there are ways of arranging the different kinds of joints between the links. In Table 1.1 we saw that three of the six lower pairs have one degree of freedom: the revolute, R, the prismatic, P, and the screw, S. Thus, using any seven of these lower pairs, we get 36 kinds of type $7j_1$ mechanisms. All together, Harrisberger lists 435 kinds that satisfy the Kutzbach criterion with mobility of $m = 1$. Not all of these types or kinds are likely to have practical value, however. Consider, for example, the $7j_1$ type with all revolute joints, all connected in series in a single loop.*

For mechanisms having mobility of $m = 1$, Harrisberger has selected nine kinds from two types that appear to him to be useful; these are illustrated in Fig. 12.1. They are all spatial four-bar linkages having four joints with either rotating or sliding input and output members. The designations in the legend, such as *RGCS* for Fig. 12.1f, identify the kinematic pair types (see Table 1.1) beginning with the input link and proceeding through the coupler and output link back to the frame. Thus for the *RGCS*, the input crank is pivoted to the frame by a revolute pair R and to the coupler by a globular pair G, and the motion of the output member is determined by the screw pair S. The freedoms of these pairs, from Table 1.1 are $R = 1$, $G = 3$, $C = 2$, and $S = 1$.

The mechanisms of Figs. 12.1a through 12.1c are described by Harrisberger as type 1 (or type $1j_1 + 3j_2$) mechanisms. The remaining linkages of Fig. 12.1 are described as type 2 or of the $2j_1 + 1j_2 + 1j_3$ type. All have $n = 4$ links and have mobility of $m = 1$.

12.2 EXCEPTIONS IN THE MOBILITY OF MECHANISMS

Curiously enough, the most common and most useful spatial mechanisms that have been discovered date back many years and are exceptions to the Kutzbach criterion. As pointed out in Fig. 1.6, certain geometric conditions sometime occur which are not included in the Kutzbach criterion and lead to such apparent exceptions. As a case in point, consider that any planar mechanism, once constructed, truly exists in three dimensions. Yet a planar four-bar linkage has $n = 4$ and is of type $4j_1$; thus Eq. (12.1) predicts a mobility of $m = 6(4 - 1) - 5(4) = -2$. The special geometric conditions in this case lie in the fact that all revolute axes remain parallel and are all perpendicular to the plane of motion. The

---

*The only example known to the authors for this type is its use in the front landing gear mechanism of the Boeing 727 aircraft.

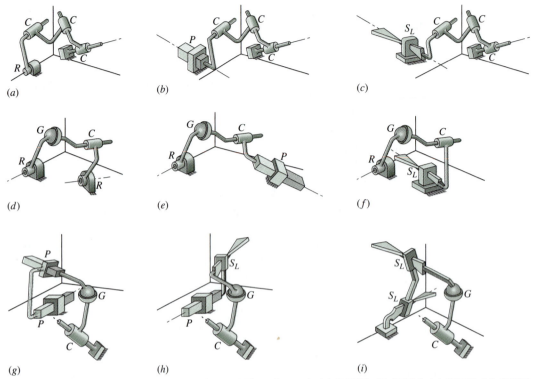

Figure 12.1 Four-bar spatial linkages having mobility of $m = 1$: (a) RCCC, (b) PCCC, (c) SCCC, (d) RGCR, (e) RGCP, (f) RGCS, (g) PPGC, (h) PSGC, (i) SSGC. (From L. Harrisberger, "A Number Synthesis Survey of Three-Dimensional Mechanisms," *J. Eng. Ind., ASME Trans.,* ser. B, vol. 87, no. 2, 1965, published with permission of ASME and the author.)

Kutzbach criterion does not consider these facts and results in error; the true mobility is $m = 1$.

At least three more exceptions to the Kutzbach criterion are also four link *RRRR* mechanisms. Thus, as with the planar four-bar linkage, the Kutzbach criterion predicts $m = -2$, and yet they are truly of mobility $m = 1$. One of these is the spherical four-bar linkage shown in Fig. 12.2. The axes of all four revolutes intersect at the center of a sphere. The links may be regarded as great-circle arcs existing on the surface of the sphere; what would be link lengths are now spherical angles. By properly proportioning these angles, it is possible to design all of the spherical counterparts of the planar four-bar mechanism such as the spherical crank-rocker linkage and the spherical drag-link mechanism. The spherical four-bar linkage is easy to design and manufacture and hence is one of the most useful of all spatial mechanisms. The Hooke or Cardan joint, which is the basis of the universal shaft coupling, is a special case of a spherical four-bar mechanism having input and output cranks that subtend equal spherical angles.

The wobble-plate mechanism, shown in Fig. 12.3, is also a special case. It is another four-link spherical *RRRR* mechanism, which is an exception to the Kutzbach criterion but is movable. Note again that all of the revolute axes intersect at the origin; thus it is a spherical mechanism.

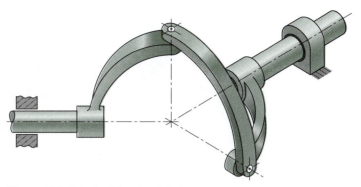

Figure 12.2 Spherical four-bar linkage.

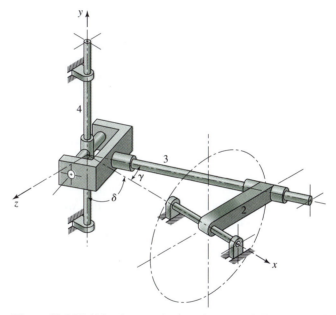

Figure 12.3 Wobble-plate mechanism; input crank 2 rotates and output shaft 4 oscillates. When $\delta = 90°$, the mechanism is called a spherical-slide oscillator. If $\gamma > \delta$, the output shaft rotates.

Bennett's *RRRR* mechanism, shown in Fig. 12.4, is probably one of the most useless of the known spatial linkages. In this mechanism the opposite links are of equal lengths and are twisted by equal amounts. The sines of the twist angles α_1 and α_2 are proportional to the link lengths a_1 and a_2 according to the equation

$$\frac{a_1}{\sin \alpha_1} = \pm \frac{a_2}{\sin \alpha_2}$$

The spatial four-link *RGGR* mechanism, shown in Fig. 12.5, is another important and useful linkage. Because $n = 4$, $j_1 = 2$, and $j_3 = 2$, the Kutzbach criterion of Eq. (12.1)

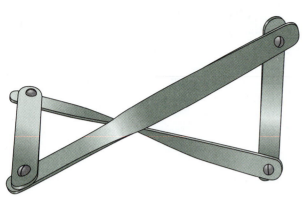

Figure 12.4 Bennett four-link *RRRR* mechanism.

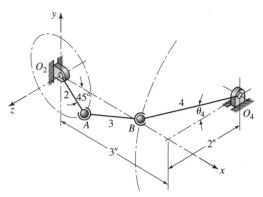

Figure 12.5 Four-link *RGGR* linkage for Example 12.1; $R_{AO_2} = 1$ in, $R_{BA} = 3.5$ in, $R_{BO_4} = 4$ in.

predicts a mobility of $m = 2$. Although this might appear at first glance to be another exception, upon closer examination we find that the second degree of freedom actually exists; it is the freedom of the coupler to spin about its own axis between the two ball joints. Because this degree of freedom does not affect the input–output kinematic relationship, it is called an *idle* freedom. This extra freedom does no harm if the mass of the coupler is distributed along its axis; in fact, it may be an advantage because the rotation of the coupler about its axis may equalize wear on the two ball-and-socket joints. If the mass center lies off-axis, however, then this second freedom is not idle dynamically and can cause quite erratic performance at high speed.

Still other exceptions to the Kutzbach mobility criterion are the Goldberg (not Rube!) five-bar *RRRRR* and the Bricard six-bar *RRRRRR* linkage.[2] Again it is doubtful if these mechanisms have any practical value.

Harrisberger and Soni have sought to identify all spatial linkages having one general constraint, that is, which have mobility of $m = 1$ but for which the Kutzbach criterion predicts $m = 0$.[3] They have identified eight types and 212 kinds and found seven new mechanisms which may have useful applications.

Note that all of the mechanisms mentioned which defy the Kutzbach mobility criterion always predict a mobility less than the actual. This is always the case; the Kutzbach criterion always predicts a lower limit on the mobility, even when the criterion shows exceptions. The reason for this is mentioned in Section 1.6. The argument for the development of the Kutzbach equation came from counting the freedom for motion of all of the bodies before any connections are made less the numbers of these presumably eliminated by connecting various types of joints. Yet, when there are special geometric conditions such as intersections or parallelism between joint axes, the criterion counts each joint as eliminating its own share of freedoms even though two (or more) of them may eliminate the same freedom(s). Thus the exceptions arise from the assumption of independence among the constraints of the joints.

But let us carry this thought one step further. When two (or more) of the constraint conditions eliminate the same motion freedom, the problem is said to have *redundant constraints*. Under these conditions the same redundant constraints both determine how the forces must be shared where the motion freedom is eliminated. Thus, when we come to analyzing the forces, we will find that there are too many constraints (equations) relating the number of unknown forces. The force analysis problem is then said to be *overconstrained,* and we find that there are statically indeterminate forces in the same number as the error in the predicted mobility.

There is an important lesson buried in this argument[4]: Whenever there are redundant constraints on the motion, there are an equal number of statically indeterminate forces in the mechanism. In spite of the higher simplicity in the design equations of planar linkages, for example, we should consider the force effects of these redundant constraints; all of the out-of-plane force and moment components become statically indeterminate. Slight machining tolerance errors or misalignments of axes can cause indeterminate stresses with cyclic loading as the mechanism is operated. What effect will these have on the fatigue life of the members?

On the other hand, as pointed out by Phillips,[5] when motion is only occasional and loads are very high, this might be an ideal design decision. If errors are small, the additional indeterminate forces may be small and such designs are tolerated even though they may seem to exhibit friction effects and wear in the joints. As errors become larger, we may find binding in no uncertain terms.

The very existence of the large number of planar linkages in the world testify that such effects can be lived with by setting tolerances on manufacturing errors, better lubrication, and looser fits between mating joint elements. Still, too few mechanical designers truly understand that the root of the problem is best eliminated by removing the redundant constraints in the first place. Thus, even in planar motion mechanisms, for example, we can make use of ball-and-socket or cylindric joints which, if properly located, can help to relieve statically indeterminate forces and moments.

12.3 THE POSITION-ANALYSIS PROBLEM

Like planar mechanisms, a spatial mechanism is often connected to form one or more closed loops. Thus, following methods similar to those of Section 2.6, loop-closure equations can be written which define the kinematic relationships of the mechanism. A number of different mathematical forms can be used, including vectors,[6] dual numbers, quaternions,[7] and transformation matrices.[8] In vector notation, the closure of a spatial linkage such as the mechanism of Fig. 12.5 can be defined by a loop-closure equation of the form

$$\mathbf{r} + \mathbf{s} + \mathbf{t} + \mathbf{C} = \mathbf{0} \tag{12.2}$$

This equation is called the *vector tetrahedron equation* because the individual vectors can be thought of as defining four of the six edges of a tetrahedron.

The vector tetrahedron equation is three-dimensional and hence can be solved for three scalar unknowns. These can be either magnitudes or angles and can exist in any combination in vectors $\mathbf{r}$, $\mathbf{s}$, and $\mathbf{t}$. The vector $\mathbf{C}$ is the sum of all known vectors in the loop. By using spherical coordinates, each of the vectors $\mathbf{r}$, $\mathbf{s}$, and $\mathbf{t}$ can be expressed as a magnitude and two angles. Vector $\mathbf{r}$, for example, is defined once its magnitude r and two angles θ_r and ϕ_r are known. Thus, in Eq. (12.2), any three of the nine quantities $r, \theta_r, \phi_r, s, \theta_s, \phi_s, t,$

TABLE 12.1 Classification of Solutions of the Vector Tetrahedron Equation

		Known Quantities		
Case Number	Unknowns	Vectors	Scalars	Degree of Polynomial
1	r, θ_r, ϕ_r	$\mathbf{C}$		1
2a	r, θ_r, s	$\mathbf{C}, \hat{\boldsymbol{\omega}}_r, \hat{\mathbf{s}}$	ϕ_r	2
2b	r, θ_r, θ_s	$\mathbf{C}, \hat{\boldsymbol{\omega}}_r, \hat{\boldsymbol{\omega}}_s$	ϕ_r, s, ϕ_s	4
2c	θ_r, ϕ_r, s	$\mathbf{C}, \hat{\mathbf{s}}$	r	2
2d	$\theta_r, \phi_r, \theta_s$	$\mathbf{C}, \hat{\boldsymbol{\omega}}_s$	r, s, ϕ_s	2
3a	r, s, t	$\mathbf{C}, \hat{\mathbf{r}}, \hat{\mathbf{s}}, \hat{\mathbf{t}}$		1
3b	r, s, θ_t	$\mathbf{C}, \hat{\mathbf{r}}, \hat{\mathbf{s}}, \hat{\boldsymbol{\omega}}_t$	t, ϕ_t	2
3c	r, θ_s, θ_t	$\mathbf{C}, \hat{\mathbf{r}}, \hat{\boldsymbol{\omega}}_s, \hat{\boldsymbol{\omega}}_t$	s, ϕ_s, t, ϕ_t	4
3d	$\theta_r, \theta_s, \theta_t$	$\mathbf{C}, \hat{\boldsymbol{\omega}}_r, \hat{\boldsymbol{\omega}}_s, \hat{\boldsymbol{\omega}}_t$	$r, \phi_r, s, \phi_s, t, \phi_t$	8

θ_t, and ϕ_t can be unknowns that must be found from the equation. Chace has solved these nine cases by first reducing each to a polynomial.[9] He classifies the solutions depending upon whether the three unknowns occur in one, two, or three separate vectors, and he tabulates the forms of the solutions as shown in Table 12.1.

In this table, the unit vectors $\hat{\boldsymbol{\omega}}_r$, $\hat{\boldsymbol{\omega}}_s$, and $\hat{\boldsymbol{\omega}}_t$ are known directions of axes about which the known angles ϕ_r, ϕ_s, and ϕ_t are measured. In case 1, vectors $\mathbf{s}$ and $\mathbf{t}$ are not needed, set to zero, and dropped from the equation; the three unknowns are all in vector $\mathbf{r}$. In cases 2a, 2b, 2c, and 2d, vector $\mathbf{t}$ is not needed and dropped; the three unknowns are shared by vectors $\mathbf{r}$ and $\mathbf{s}$. Cases 3a, 3b, 3c and 3d have single unknowns in each of the vectors $\mathbf{r}$, $\mathbf{s}$, and $\mathbf{t}$.

One advantage of the Chace vector tetrahedron solutions is that, since they provide known forms for the solutions of the nine cases, it is easy to write a set of nine subprograms for computer or calculator evaluation. All nine cases have been reduced to explicit closed-form solutions for the unknowns and can therefore be quickly evaluated. Only case 3d, involving the solution of an eighth-order polynomial, must be solved by iterative techniques.

Although the vector tetrahedron equation and its nine case solutions can be used to solve most practical spatial problems, we recall from Section 12.1 that the Kutzbach criterion predicts the existence of up to seven j_1 joints in a single loop mechanism with one degree of freedom. A case such as the seven-link 7R mechanism, for example, would have six unknowns to be found from the loop-closure equation. This is not possible from the vector form of the loop-closure equation, and other tools such as dual-quaternions or transformation matrices must be used instead. Such problems require numerical techniques for their evaluation. Anyone attempting the solution of such equations by hand algebraic techniques quickly appreciates that *position* analysis, not velocity or acceleration, is the most difficult problem in kinematics.

Solving the polynomials of the Chace vector tetrahedron equation turns out to be equivalent to finding the intersections of straight lines or circles with various surfaces of revolution. Such problems can usually be easily and quickly solved by the graphical methods of descriptive geometry. The graphical approach has the additional advantage that the geometric nature of the problem is not concealed in a multiplicity of mathematical operations. However, graphical methods are not nearly as accurate as the analytical approach.

EXAMPLE 12.1

A four-link *RGGR* crank-rocker mechanism is shown in Fig. 12.5. The knowns are the position and plane of rotation of the input crank, the plane of rotation of the output crank, and the dimensions of all four links. Find the positions of all moving links when the input crank is set to $\theta_2 = -45°$ as shown.

GRAPHIC SOLUTION

The position analysis problem consists of finding the positions of the coupler and the rocker, links 3 and 4. If we treat the output link 4 as the vector $\mathbf{R}_{BO_4}$, then its only unknown is the angle θ_4 because the magnitude of the vector and its plane of motion are given. Similarly, if the coupler link 3 is treated as the vector $\mathbf{R}_{BA}$, its magnitude is known, but there exist two unknown spherical coordinate angular direction variables for this vector. The loop-closure equation is

$$(\mathbf{R}_{BA}) + (-\mathbf{R}_{BO_4}) + (\mathbf{R}_{AO_2} - \mathbf{R}_{O_4O_2}) = \mathbf{0} \tag{1}$$

We identify this as case 2*d* in Table 12.1, requiring the solution of a quadratic polynomial and hence yielding two solutions.

This problem is solved graphically by using two orthographic views, the frontal and profile views. If we imagine, in Fig. 12.5, that the coupler is disconnected from the output crank at *B* and allowed to occupy any possible position with respect to *A*, then *B* of link 3 must lie on the surface of a sphere of known radius with center at *A*. With joint *B* still disconnected, the locus of *B* of link 4 is a circle of known radius about O_4 in a plane parallel to the *yz* plane. Therefore, to solve this problem, we need only find the two points of intersection of a circle and a sphere.

The solution is shown in Fig. 12.6 where the subscripts *F* and *P* denote projections on the frontal and profile planes, respectively. First we locate points O_2, *A*, and O_4 in both views. In the profile view we draw a circle of radius $R_{BO_4} = 4$ in about center O_{4P}; this is the locus of point B_4. This circle appears in the frontal view as the vertical line $M_F O_{4F} N_F$. Next, in the frontal view, we construct the outline of a sphere with A_F as center and the coupler length $R_{BA} = 3.5$ in as its radius. If $M_F O_{4F} N_F$ is regarded as the trace of a plane normal to the frontal view plane, the intersection of this plane with the sphere appears as the full circle in the profile view, having diameter $M_P N_P = M_F N_F$. The circular arc of radius R_{BO_4} intersects the circle at two points, yielding two solutions. One of the points is chosen as B_P and is projected back to the frontal view to locate B_F. Links 3 and 4 are drawn in next as the dashed lines shown in the profile and frontal views of Fig. 12.6.

Measuring the *x*, *y*, and *z* projections from the graphic solution, we can write the vectors for each link:

$$\mathbf{R}_{O_4O_2} = 3.00\hat{\mathbf{i}} - 2.00\hat{\mathbf{k}} \text{ in} \qquad\qquad \textit{Ans.}$$

$$\mathbf{R}_{AO_2} = 0.71\hat{\mathbf{i}} - 0.71\hat{\mathbf{j}} \text{ in} \qquad\qquad \textit{Ans.}$$

$$\mathbf{R}_{BA} = 2.30\hat{\mathbf{i}} + 1.95\hat{\mathbf{j}} + 1.77\hat{\mathbf{k}} \text{ in} \qquad\qquad \textit{Ans.}$$

$$\mathbf{R}_{BO_4} = 1.22\hat{\mathbf{j}} + 3.81\hat{\mathbf{k}} \text{ in} \qquad\qquad \textit{Ans.}$$

These components were obtained from a true-size graphical solution; better accuracy would be possible, of course, by making the drawings two or four times actual size.

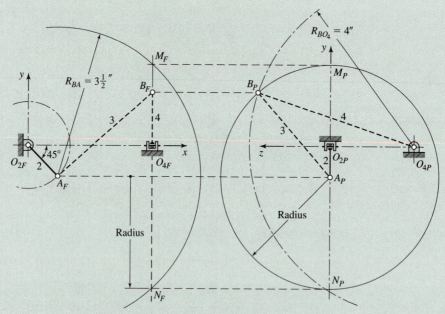

Figure 12.6 Graphical position analysis for Example 12.1.

The given information establishes that

$$\mathbf{R}_{O_4O_2} = 3.0\hat{\mathbf{i}} - 2.0\hat{\mathbf{k}} \text{ in}$$

$$\mathbf{R}_{AO_2} = \cos(-45°) + \sin(-45°) = 0.707\hat{\mathbf{i}} - 0.707\hat{\mathbf{j}} \text{ in}$$

$$\mathbf{R}_{BA} = R_{BA}^x\hat{\mathbf{i}} + R_{BA}^y\hat{\mathbf{j}} + R_{BA}^z\hat{\mathbf{k}} \text{ in}$$

$$\mathbf{R}_{BO_4} = -4.0\sin\theta_4\hat{\mathbf{j}} + 4.0\cos\theta_4\hat{\mathbf{k}} \text{ in}$$

The same vector loop-closure equation as in (1) above needs to be solved. Substituting the given information gives

$$\left(R_{BA}^x\hat{\mathbf{i}} + R_{BA}^y\hat{\mathbf{j}} + R_{BA}^z\hat{\mathbf{k}}\right) + (4.0\sin\theta_4\hat{\mathbf{j}} - 4.0\cos\theta_4\hat{\mathbf{k}}) + (0.707\hat{\mathbf{i}} - 0.707\hat{\mathbf{j}} - 3.0\hat{\mathbf{i}} + 2.0\hat{\mathbf{k}}) = \mathbf{0}$$

On separating this into its $\hat{\mathbf{i}}, \hat{\mathbf{j}}$, and $\hat{\mathbf{k}}$ components and rearranging, we get the following three equations:

$$R_{BA}^x = 2.293 \text{ in}$$

$$R_{BA}^y = -4.0\sin\theta_4 + 0.707 \text{ in}$$

$$R_{BA}^z = 4.0\cos\theta_4 - 2.000 \text{ in}$$

Next we can square and add these and, remembering that $R_{BA} = 3.500$ in, we get

$$(2.293)^2 + (-4.0\sin\theta_4 + 0.707)^2 + (4.0\cos\theta_4 - 2.000)^2 = (3.500)^2$$

$$-5.657\sin\theta_4 - 16.0\cos\theta_4 + 13.507 = 0$$

We now have a single equation with only one unknown, θ_4. However, it contains a mix of sine and cosine terms. A standard technique for dealing with this is to use the identities

$$\sin\theta_4 = \frac{2\tan(\theta_4/2)}{1 + \tan^2(\theta_4/2)} \quad \text{and} \quad \cos\theta_4 = \frac{1 - \tan^2(\theta_4/2)}{1 + \tan^2(\theta_4/2)}$$

Substituting these and multiplying by the common denominator, we get a single quadratic equation in one unknown, $\tan(\theta_4/2)$.

$$-11.314\tan(\theta_4/2) - 16[1 - \tan^2(\theta_4/2)] + 13.507[1 + \tan^2(\theta_4/2)] = 0$$

or

$$29.507\tan^2(\theta_4/2) - 11.314\tan(\theta_4/2) - 2.493 = 0$$

This can now be solved for two roots, which are

$$\tan\frac{\theta_4}{2} = \begin{cases} 0.540, & \theta_4 = 56.7° \\ -0.156, & \theta_4 = -17.8° \end{cases}$$

Of these two, we choose the second as the solution of interest. Back-substituting into the previous conditions, we find numeric values for the four vectors, which, of course, match (within graphical error) those of the graphic solution.

$$\mathbf{R}_{O_4 O_2} = 3.000\hat{\mathbf{i}} - 2.000\hat{\mathbf{k}} \text{ in} \qquad\qquad Ans.$$

$$\mathbf{R}_{A O_2} = 0.707\hat{\mathbf{i}} - 0.707\hat{\mathbf{j}} \text{ in} \qquad\qquad Ans.$$

$$\mathbf{R}_{BA} = 2.293\hat{\mathbf{i}} + 1.929\hat{\mathbf{j}} + 1.809\hat{\mathbf{k}} \text{ in} \qquad\qquad Ans.$$

$$\mathbf{R}_{B O_4} = 1.222\hat{\mathbf{j}} + 3.809\hat{\mathbf{k}} \text{ in} \qquad\qquad Ans.$$

We should note from this example that we have solved for the output angle θ_4 and the current values of the four vectors for the given value of the input crank angle θ_2. However, have we really solved for the positions of all links? No! As pointed out above, we have not found how the coupler link may be rotating about its axis $\mathbf{R}_{BA}$ between the two ball-and-socket joints. This is still unknown and explains how we were able to solve without specifying another input angle for this idle freedom. Depending upon our motivation, the above solution may be sufficient; however, it is important to note that the above vectors are not sufficient to solve for this additional variable.

The four-revolute spherical four-link mechanism shown in Fig. 12.2 is also case *2d* of the vector tetrahedron equation and can be solved in the same manner as the example shown once the position of the input link is given.

12.4 VELOCITY AND ACCELERATION ANALYSES

Once the positions of all members of a spatial mechanism have been found, the velocities and accelerations can be determined by using the methods of Chapters 3 and 4. In planar mechanisms the angular velocity and acceleration vectors were always perpendicular to the plane of motion. This reduced considerably the effort required in the solution process for both graphical and analytical approaches. In spatial problems, however, these vectors may be skew in space. Otherwise, the methods of analysis are the same. The following example will illustrate the differences.

EXAMPLE 12.2

The angular velocity of link 2 of the four-link *RGGR* linkage of Fig. 12.7 is $\omega_2 = 40\hat{k}$ rad/s and is constant. Find the angular velocities and angular accelerations of links 3 and 4 and the velocity and acceleration of point *B* for the position shown.

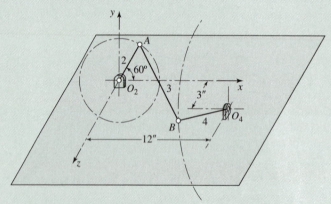

Figure 12.7 Four-link *RGGR* linkage for Example 12.2; $R_{AO_2} = 4$ in, $R_{BA} = 15$ in, $R_{BO_4} = 10$ in.

ANALYTIC SOLUTION

The position analysis follows exactly the procedure shown in Section 12.3 and Example 12.1; the graphical solution is shown in Fig. 12.8. For the crank angle shown, the results are as follows:

$$\mathbf{R}_{O_4O_2} = 12.000\hat{i} + 3.000\hat{k} \text{ in}$$

$$\mathbf{R}_{AO_2} = 2.000\hat{i} + 3.464\hat{j} \text{ in}$$

$$\mathbf{R}_{BA} = 10.000\hat{i} + 2.746\hat{j} + 10.838\hat{k} \text{ in}$$

$$\mathbf{R}_{BO_4} = 6.210\hat{j} + 7.838\hat{k} \text{ in}$$

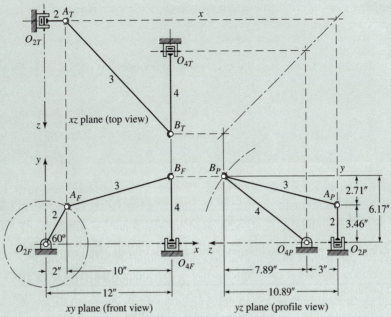

Figure 12.8 Graphical position analysis for Example 12.2.

From the given information and the constraints we see that the angular velocities and accelerations can be written as

$$\boldsymbol{\omega}_2 = 40\hat{\mathbf{k}} \text{ rad/s}, \qquad\qquad \boldsymbol{\alpha}_2 = \mathbf{0}$$

$$\boldsymbol{\omega}_3 = \omega_3^x\hat{\mathbf{i}} + \omega_3^y\hat{\mathbf{j}} + \omega_3^z\hat{\mathbf{k}} \text{ rad/s}, \qquad \boldsymbol{\alpha}_3 = \alpha_3^x\hat{\mathbf{i}} + \alpha_3^y\hat{\mathbf{j}} + \alpha_3^z\hat{\mathbf{k}} \text{ rad/s}^2$$

$$\boldsymbol{\omega}_4 = \omega_4\hat{\mathbf{i}} \text{ rad/s}, \qquad\qquad \boldsymbol{\alpha}_4 = \alpha_4\hat{\mathbf{i}} \text{ rad/s}^2$$

First we find the velocity of point A as the velocity difference from point O_2. Thus

$$\mathbf{V}_A = \boldsymbol{\omega}_2 \times \mathbf{R}_{AO_2} = (40\hat{\mathbf{k}}) \times (2.000\hat{\mathbf{i}} + 3.464\hat{\mathbf{j}})$$

$$= -138.564\hat{\mathbf{i}} + 80.000\hat{\mathbf{j}} \text{ in/s} \tag{1}$$

Similarly, for link 3,

$$\mathbf{V}_{BA} = \left(\omega_3^x\hat{\mathbf{i}} + \omega_3^y\hat{\mathbf{j}} + \omega_3^z\hat{\mathbf{k}}\right) \times (10.000\hat{\mathbf{i}} + 2.746\hat{\mathbf{j}} + 10.838\hat{\mathbf{k}})$$

$$= \left(10.838\omega_3^y - 2.746\omega_3^z\right)\hat{\mathbf{i}} + \left(10.000\omega_3^z - 10.838\omega_3^x\right)\hat{\mathbf{j}}$$

$$+ \left(2.746\omega_3^x - 10.000\omega_3^y\right)\hat{\mathbf{k}} \text{ in/s} \tag{2}$$

and for link 4,

$$\mathbf{V}_B = \boldsymbol{\omega}_4 \times \mathbf{R}_{BO_4} = (\omega_4\hat{\mathbf{i}}) \times (6.210\hat{\mathbf{j}} + 7.838\hat{\mathbf{k}})$$

$$= -7.838\omega_4\hat{\mathbf{j}} + 6.210\omega_4\hat{\mathbf{k}} \text{ in/s} \tag{3}$$

The next step is to substitute these into the velocity difference equation

$$\mathbf{V}_B = \mathbf{V}_A + \mathbf{V}_{BA} \tag{4}$$

Once this is done, we can separate the $\hat{\mathbf{i}}, \hat{\mathbf{j}},$ and $\hat{\mathbf{k}}$ components to obtain these three algebraic equations:

$$10.838\omega_3^y - 2.746\omega_3^z \qquad\qquad = 138.564 \tag{5}$$

$$-10.838\omega_3^x \qquad + 10.000\omega_3^z + 7.838\omega_4 = -80.000 \tag{6}$$

$$2.746\omega_3^x - 10.000\omega_3^y \qquad - 6.210\omega_4 = 0.000 \tag{7}$$

We now have three equations; however, we note that there are four unknowns: ω_3^x, ω_3^y, ω_3^z, and ω_4. This would not normally occur in most problems, but does here because of the idle freedom of the coupler to spin about its own axis. Because this spin does not affect the input–output relationship, we will get the same result for ω_4 no matter what the spin. One way to proceed, therefore, would be to choose one component of $\boldsymbol{\omega}_3$ and give it a value, thus setting the rate of spin, and then solve for the other unknowns. Another approach is to set the rate of spin to zero by requiring that

$$\boldsymbol{\omega}_3 \cdot \mathbf{R}_{BA} = 0$$

$$10.000\omega_3^x + 2.746\omega_3^y + 10.838\omega_3^z = 0 \tag{8}$$

Equations (5) through (8) can now be solved simultaneously for the four unknowns. The result is

$$\boldsymbol{\omega}_3 = -7.692\hat{\mathbf{i}} + 13.704\hat{\mathbf{j}} + 3.625\hat{\mathbf{k}} \text{ rad/s} \qquad\qquad \textit{Ans.}$$

$$\boldsymbol{\omega}_4 = -25.468\hat{\mathbf{i}} \text{ rad/s} \qquad\qquad \textit{Ans.}$$

Substituting into the earlier equation we get

$$\mathbf{V}_B = 199.621\hat{\mathbf{j}} - 158.159\hat{\mathbf{k}} \text{ in/s} \qquad\qquad \textit{Ans.}$$

Turning next to acceleration analysis, we compute the following components:

$$\mathbf{A}_{AO_2}^n = \boldsymbol{\omega}_2 \times (\boldsymbol{\omega}_2 \times \mathbf{R}_{AO_2}) = -3\,200.00\hat{\mathbf{i}} - 5\,542.40\hat{\mathbf{j}} \text{ in/s}^2 \tag{9}$$

$$\mathbf{A}_{AO_2}^t = \boldsymbol{\alpha}_2 \times \mathbf{R}_{AO_2} = \mathbf{0} \tag{10}$$

$$\mathbf{A}_{BA}^n = \boldsymbol{\omega}_3 \times (\boldsymbol{\omega}_3 \times \mathbf{R}_{BA}) = -2\,601.01\hat{\mathbf{i}} - 714.26\hat{\mathbf{j}} - 2\,818.99\hat{\mathbf{k}} \text{ in/s}^2 \tag{11}$$

$$\mathbf{A}_{BA}^t = \boldsymbol{\alpha}_3 \times \mathbf{R}_{BA} = \left(\alpha_3^x\hat{\mathbf{i}} + \alpha_3^y\hat{\mathbf{j}} + \alpha_3^z\hat{\mathbf{k}}\right) \times (10.000\hat{\mathbf{i}} + 2.746\hat{\mathbf{j}} + 10.838\hat{\mathbf{k}})$$

$$= \left(10.838\alpha_3^y - 2.746\alpha_3^z\right)\hat{\mathbf{i}} + \left(10.000\alpha_3^z - 10.838\alpha_3^x\right)\hat{\mathbf{j}}$$

$$+ \left(2.746\alpha_3^x - 10.000\alpha_3^y\right)\hat{\mathbf{k}} \text{ in/s}^2 \tag{12}$$

$$\mathbf{A}_{BO_4}^n = \boldsymbol{\omega}_4 \times (\boldsymbol{\omega}_4 \times \mathbf{R}_{BO_4}) = -4\,028.05\hat{\mathbf{j}} - 5\,084.03\hat{\mathbf{k}} \text{ in/s}^2 \tag{13}$$

$$\mathbf{A}_{BO_4}^t = \boldsymbol{\alpha}_4 \times \mathbf{R}_{BO_4} = -7.838\alpha_4\hat{\mathbf{j}} + 6.210\alpha_4\hat{\mathbf{k}} \text{ in/s}^2 \tag{14}$$

These quantities are now substituted into the acceleration difference equation

$$\mathbf{A}^n_{BO_4} + \mathbf{A}^t_{BO_4} = \mathbf{A}^n_{AO_2} + \mathbf{A}^t_{AO_2} + \mathbf{A}^n_{BA} + \mathbf{A}^t_{BA} \tag{15}$$

and separated into $\hat{\mathbf{i}}$, $\hat{\mathbf{j}}$, and $\hat{\mathbf{k}}$ components. Along with the condition $a_3 \cdot \mathbf{R}_{BA} = 0$ for the spin of the idle freedom, this results in four equations in four unknowns as follows:

$$
\begin{aligned}
10.838\alpha_3^y - 2.746\alpha_3^z && = &\; 5\,801.01 \\
-10.838\alpha_3^x && + 10.000\alpha_3^z + 7.838\alpha_4 = &\; 2\,228.61 \\
2.746\alpha_3^x - 10.000\alpha_3^y && - 6.210\alpha_4 = &\; -2\,265.04 \\
10.000\alpha_3^x + 2.746\alpha_3^y + 10.838\alpha_3^z && = &\; 0.00
\end{aligned}
$$

These are now solved to give the desired accelerations.

$$\boldsymbol{\alpha}_3 = -526.94\hat{\mathbf{i}} + 618.71\hat{\mathbf{j}} + 329.43\hat{\mathbf{k}}\ \text{rad/s}^2 \qquad \textit{Ans.}$$

$$\boldsymbol{\alpha}_4 = -864.59\hat{\mathbf{i}}\ \text{rad/s}^2 \qquad \textit{Ans.}$$

$$\mathbf{A}_B = \mathbf{A}^n_{BO_4} + \mathbf{A}^t_{BO_4} = 2\,748.61\hat{\mathbf{j}} - 10\,453.10\hat{\mathbf{k}}\ \text{in/s}^2 \qquad \textit{Ans.}$$

GRAPHIC SOLUTION

The determination of the velocities and accelerations of a spatial mechanism by graphical means can be conducted in the same manner as for a planar mechanism. However, the position information and also the velocity and acceleration vectors often do not appear in their true lengths in the standard front, top, and profile views, but are foreshortened. This means that spatial motion problems usually require the use of auxiliary views where the vectors do appear in true lengths.

The velocity solution for this example is shown in Fig. 12.9 with notation which corresponds to that used in many works on descriptive geometry. The letters F, T, and P designate the front, top, and profile views, and the numbers 1 and 2 show the first and second auxiliary views, respectively. Points projected into these views bear the subscripts F, T, P, and so on. The steps in obtaining the velocity are as follows:

1. We first construct to scale the front, top, and profile views of the linkage, and designate each point.
2. Next we calculate $\mathbf{V}_A$ as above and place this vector in position with its origin at A in the three views. The velocity $\mathbf{V}_A$ shows in true length in the frontal view. We designate its terminus as a_F and project this point to the top and profile views to find a_T and a_P.
3. The magnitude of the velocity $\mathbf{V}_B$ is unknown, but its direction is perpendicular to $\mathbf{R}_{BO_4}$ and, once the problem is solved, it will show in true size in the profile view. Therefore we construct a line in the profile view that originates at point A_P (the origin of our velocity polygon) and corresponds in direction to that of $\mathbf{V}_B$. We then choose any point d_P and project it to the front and top views to establish the line of action of $\mathbf{V}_B$ in those views.
4. The equation to be solved is

$$\mathbf{V}_B = \mathbf{V}_A + \mathbf{V}_{BA} \tag{16}$$

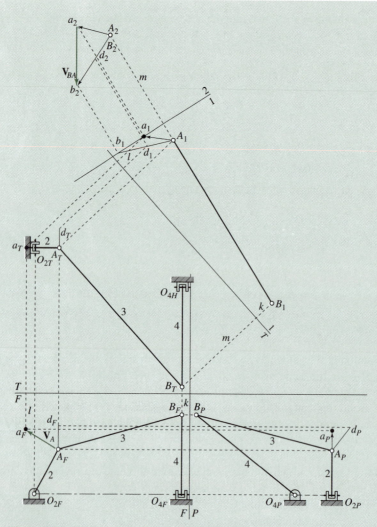

Figure 12.9 Graphical velocity analysis for Example 12.2.

where $\mathbf{V}_A$ and the directions of $\mathbf{V}_B$ and $\mathbf{V}_{BA}$ are known. We note that $\mathbf{V}_{BA}$ must lie in a plane perpendicular (in space) to $\mathbf{R}_{BA}$, but its magnitude is also unknown. The vector $\mathbf{V}_{BA}$ must originate at the terminus of $\mathbf{V}_A$ and must lie in a plane perpendicular to $\mathbf{R}_{BA}$; it terminates by intersecting the line Ad or its extension. To find this plane perpendicular to $\mathbf{R}_{BA}$, we start by constructing the first auxiliary view, which shows vector $\mathbf{R}_{BA}$ in true length; so we construct the edge view of plane 1 parallel to $A_T B_T$ and project $\mathbf{R}_{BA}$ to this plane. In this projection we note that the distances k and l are the same in this first auxiliary view as in the frontal view. The first auxiliary view of AB is $A_1 B_1$ which is true length. We also project points a and d to this view, but the remaining links need not be projected.

5. In this step we construct a second auxiliary view, plane 2, such that the projection of AB upon it is a point. Then all lines drawn parallel to this plane will be perpendicular to $\mathbf{R}_{BA}$. The edge view of such a plane must be perpendicular to A_1B_1 extended. In this example it is convenient to choose this plane so that it contains point a; therefore we construct the edge view of plane 2 through point a_1 perpendicular to A_1B_1 extended. Now we project points A, B, a, and d onto this plane. Note that the distances—m, for example—of points from plane 1 must be the same in the top view and in the second auxiliary view.

6. We now extend the line a_1d_1 until it intersects the edge view of plane 2 at b_1, and we find the projection b_2 of this point in plane 2 where the projector meets the line A_2d_2 extended. Now both points a and b lie in plane 2, and any line drawn in plane 2 is perpendicular to $\mathbf{R}_{BA}$. Therefore the line ab is $\mathbf{V}_{BA}$ and the second auxiliary view of that line shows its true length. The line A_2b_2 is the projection of $\mathbf{V}_B$ on the second auxiliary plane, but not in its true length because point A is not in plane 2.

7. In order to simplify reading of the drawing, step 7 is not shown; if you follow the first six steps carefully, you will have no difficulty with the seventh. We can project the three vectors back to the top, front, and profile views. The velocity $\mathbf{V}_B$ can be measured from the profile view where it appears in true length. The result is

$$\mathbf{V}_B = 200\hat{\mathbf{j}} - 158\hat{\mathbf{k}} \text{ in/s} \qquad\qquad Ans.$$

When all vectors have been projected back to these three views, we can measure their x, y, and z projections directly.

8. If we assume that the idle freedom of spin is not active and, therefore, that ω_3 is perpendicular to $\mathbf{R}_{BA}$ and shows true length in the second auxiliary view,* then the magnitudes of the angular velocities can be found from the equations

$$\omega_3 = \frac{V_{BA}}{R_{BA}} = \frac{242 \text{ in/s}}{15 \text{ in}} = 16.13 \text{ rad/s}$$

$$\omega_4 = \frac{V_{BO_4}}{R_{BO_4}} = \frac{255 \text{ in/s}}{10 \text{ in}} = 25.50 \text{ rad/s}$$

Therefore, we can draw the angular velocity vectors in the views where they appear true length and project them into the views where their vector components can be measured. The results are

$$\omega_3 = -7.69\hat{\mathbf{i}} + 13.71\hat{\mathbf{j}} + 3.63\hat{\mathbf{k}} \text{ rad/s} \qquad\qquad Ans.$$

$$\omega_4 = 25.50\hat{\mathbf{i}} \text{ rad/s} \qquad\qquad Ans.$$

9. The solution to the acceleration problem, also not shown, is obtained in identical manner, using the same two auxiliary planes. The equation to be solved is

$$\mathbf{A}_{BO_4}^n + \mathbf{A}_{BO_4}^t = \mathbf{A}_{AO_2}^n + \mathbf{A}_{AO_2}^t + \mathbf{A}_{BA}^n + \mathbf{A}_{BA}^t \qquad (17)$$

where the vectors $\mathbf{A}_{BO_4}^n$, $\mathbf{A}_{AO_2}^n$, $\mathbf{A}_{AO_2}^t$, and $\mathbf{A}_{BA}^n$ can be found now that the velocity analysis is completed. Also, we see that $\mathbf{A}_{BO_4}^t$ and $\mathbf{A}_{BA}^t$ have known directions. Therefore the

---

*This is the same assumption we used in the analytic solution when we wrote the equation $\omega_3 \cdot \mathbf{R}_{BA} = 0$.

solution can proceed exactly as for the velocity polygon; the only difference in approach is that there are more known vectors. The final results are

$$\alpha_3 = -527\hat{\mathbf{i}} + 619\hat{\mathbf{j}} + 329\hat{\mathbf{k}} \text{ rad/s}^2 \qquad\qquad Ans.$$

$$\alpha_4 = -865\hat{\mathbf{i}} \text{ rad/s}^2 \qquad\qquad Ans.$$

$$\mathbf{A}_B = 2\,750\hat{\mathbf{j}} - 10\,450\hat{\mathbf{k}} \text{ in/s}^2 \qquad\qquad Ans.$$

12.5 THE EULERIAN ANGLES

We saw in Section 3.2 that angular velocity is a vector quantity; hence, like all vector quantities, it has components along any set of rectilinear axes,

$$\boldsymbol{\omega} = \omega^x\hat{\mathbf{i}} + \omega^y\hat{\mathbf{j}} + \omega^z\hat{\mathbf{k}}$$

and it obeys the laws of vector algebra. Unfortunately, we also saw in Fig. 3.2 that three-dimensional angular displacements do not behave as vectors. Their order of summation is not arbitrary; that is, they are not commutative in addition. Therefore they do not follow the rules of vector algebra. The inescapable conclusion is that we cannot find a set of three angles or angular displacements that specify the three-dimensional orientation of a rigid body and that also have ω^x, ω^y, and ω^z as their time derivatives.

To clarify the problem further, we visualize a rigid body rotating in space about a fixed point O that we take as the origin of a grounded or absolute reference frame xyz. We also visualize a moving reference system $x'y'z'$ sharing the same origin, but attached to and moving with the rotating body. The $x'y'z'$ system are called *body-fixed axes*. Our problem here is to find some way to specify the orientation of the body-fixed axes with respect to the absolute reference axes, a method that is convenient for use with three-dimensional finite rotations.

Let us assume that the stationary axes are aligned along unit vector directions $\hat{\mathbf{i}}, \hat{\mathbf{j}}$, and $\hat{\mathbf{k}}$ and that the rotated body-fixed axes have directions denoted by $\hat{\mathbf{i}}', \hat{\mathbf{j}}'$, and $\hat{\mathbf{k}}'$. Then any point in the absolute coordinate system is located by the position vector $\mathbf{R}$, and by $\mathbf{R}'$ in the rotated system. Because both are descriptions of the same point's position, we obtain $\mathbf{R} = \mathbf{R}'$ and

$$R^x\hat{\mathbf{i}} + R^y\hat{\mathbf{j}} + R^z\hat{\mathbf{k}} = R^{x'}\hat{\mathbf{i}}' + R^{y'}\hat{\mathbf{j}}' + R^{z'}\hat{\mathbf{k}}'$$

Now, by taking vector dot products of this equation with $\hat{\mathbf{i}}, \hat{\mathbf{j}}$, and $\hat{\mathbf{k}}$, respectively, we find the transformation equations between the two sets of axes. They are

$$R^x = (\hat{\mathbf{i}} \cdot \hat{\mathbf{i}}')R^{x'} + (\hat{\mathbf{i}} \cdot \hat{\mathbf{j}}')R^{y'} + (\hat{\mathbf{i}} \cdot \hat{\mathbf{k}}')R^{z'}$$

$$R^y = (\hat{\mathbf{j}} \cdot \hat{\mathbf{i}}')R^{x'} + (\hat{\mathbf{j}} \cdot \hat{\mathbf{j}}')R^{y'} + (\hat{\mathbf{j}} \cdot \hat{\mathbf{k}}')R^{z'} \qquad (a)$$

$$R^z = (\hat{\mathbf{k}} \cdot \hat{\mathbf{i}}')R^{x'} + (\hat{\mathbf{k}} \cdot \hat{\mathbf{j}}')R^{y'} + (\hat{\mathbf{k}} \cdot \hat{\mathbf{k}}')R^{z'}$$

But we remember that the dot product of two unit vectors is equal to the cosine of the angle between them. For example, if the angle from $\hat{\mathbf{i}}$ to $\hat{\mathbf{j}}'$ is denoted by $\theta_{ij'}$, then

$$(\hat{\mathbf{i}} \cdot \hat{\mathbf{j}}') = \cos\theta_{ij'} \tag{b}$$

Thus the above equations become

$$R^x = \cos\theta_{ii'}\,R^{x'} + \cos\theta_{ij'}\,R^{y'} + \cos\theta_{ik'}\,R^{z'}$$
$$R^y = \cos\theta_{ji'}\,R^{x'} + \cos\theta_{jj'}\,R^{y'} + \cos\theta_{jk'}\,R^{z'} \tag{12.3}$$
$$R^z = \cos\theta_{ki'}\,R^{x'} + \cos\theta_{kj'}\,R^{y'} + \cos\theta_{kk'}\,R^{z'}$$

or, in matrix notation, this can be written

$$\begin{bmatrix} R^x \\ R^y \\ R^z \end{bmatrix} = \begin{bmatrix} \cos\theta_{ii'} & \cos\theta_{ij'} & \cos\theta_{ik'} \\ \cos\theta_{ji'} & \cos\theta_{jj'} & \cos\theta_{jk'} \\ \cos\theta_{ki'} & \cos\theta_{kj'} & \cos\theta_{kk'} \end{bmatrix} \begin{bmatrix} R^{x'} \\ R^{y'} \\ R^{z'} \end{bmatrix} \tag{12.4}$$

This means of defining the orientation of the $x'y'z'$ axes with respect to the xyz axes uses what are called *direction cosines* as coefficients in a set of *transformation equations* between the two coordinate systems. Although we will have use for these direction cosines, we note the disadvantage that there are nine of them while only three variables can be independent for spatial rotation. The nine direction cosines are not all independent, but are related by six *orthogonality conditions*. We would prefer a technique that had only three independent variables, preferably all angles.

Three angles, called *Eulerian angles,* can be used to specify the orientation of the body-fixed axes with respect to the reference axes as shown in Fig. 12.10. To illustrate Eulerian angles, we begin with the body-fixed axes coincident with the reference axes. We then specify three successive rotations—θ, ϕ, and ψ—which must occur in the specified order and about the specified axes, to move the body-fixed axes into their final orientation.

The first Eulerian angle describes a rotation through the angle θ and is taken positive counterclockwise about the positive z axis as shown in Fig. 12.10a. This rotation goes from

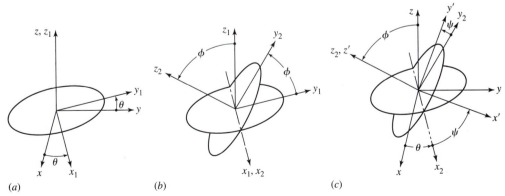

(a)   (b)   (c)

Figure 12.10 Eulerian angles.

the xy axes around by the angle θ to their new x_1y_1 locations shown while z and z_1 remain fixed. The transformation equations for this first rotation are

$$
\begin{bmatrix} R^x \\ R^y \\ R^z \end{bmatrix} = \begin{bmatrix} \cos\theta & -\sin\theta & 0 \\ \sin\theta & \cos\theta & 0 \\ 0 & 0 & 1 \end{bmatrix} \begin{bmatrix} R^{x_1} \\ R^{y_1} \\ R^{z_1} \end{bmatrix}
\tag{12.5}
$$

The second Eulerian angle describes a rotation through the angle ϕ and is taken positive counterclockwise about the positive x_1 axis as shown in Fig. 12.10b. This rotation goes from the displaced y_1z_1 axes around by the angle ϕ to their new y_2z_2 locations shown while x_1 and x_2 remain fixed. The transformation equations for this rotation are

$$
\begin{bmatrix} R^{x_1} \\ R^{y_1} \\ R^{z_1} \end{bmatrix} = \begin{bmatrix} 1 & 0 & 0 \\ 0 & \cos\phi & -\sin\phi \\ 0 & \sin\phi & \cos\phi \end{bmatrix} \begin{bmatrix} R^{x_2} \\ R^{y_2} \\ R^{z_2} \end{bmatrix}
\tag{12.6}
$$

The third Eulerian angle describes a rotation through the angle ψ and is taken positive counterclockwise about the positive z_2 axis as shown in Fig. 12.10c. This rotation goes from the x_2y_2 axes around by the angle ψ to their final $x'y'$ locations shown while z_2 and z' remain fixed. The transformation equations for this third rotation are

$$
\begin{bmatrix} R^{x_2} \\ R^{y_2} \\ R^{z_2} \end{bmatrix} = \begin{bmatrix} \cos\psi & -\sin\psi & 0 \\ \sin\psi & \cos\psi & 0 \\ 0 & 0 & 1 \end{bmatrix} \begin{bmatrix} R^{x'} \\ R^{y'} \\ R^{z'} \end{bmatrix}
\tag{12.7}
$$

Now, substituting Eqs. (12.7) into Eqs. (12.6) and these into Eqs. (12.5), we get the total transformation from the xyz axes to the $x'y'z'$ axes:

$$
\begin{bmatrix} R^x \\ R^y \\ R^z \end{bmatrix} =
$$

$$
\begin{bmatrix} \cos\theta\cos\psi - \sin\theta\cos\phi\sin\psi & -\cos\theta\sin\psi - \sin\theta\cos\phi\cos\psi & \sin\theta\sin\phi \\ \sin\theta\cos\phi + \cos\theta\cos\phi\sin\psi & -\sin\theta\sin\phi + \cos\theta\cos\phi\cos\psi & -\cos\theta\sin\phi \\ \sin\phi\sin\psi & \sin\phi\cos\psi & \cos\phi \end{bmatrix} \begin{bmatrix} R^{x'} \\ R^{y'} \\ R^{z'} \end{bmatrix}
\tag{12.8}
$$

Because the transformation is expressed as a function of only three Eulerian angles—θ, ϕ, and ψ—it represents the same information as Eqs. (12.3) and (12.4), but without the difficulty of having nine variables (direction cosines) related by six (orthogonality) conditions. Thus Eulerian angles have become a useful tool in treating problems involving three-dimensional rotations.

Note, however, that the form of the transformation equations given here depend on using precisely the conventions for the Eulerian angles defined in Fig. 12.10. Unfortunately, there appears to be little agreement among authors on how these angles should be defined. A wide variety of other definitions, differing in the axes about which the rotations are to be measured, or in their order, or in the sign conventions for positive values of the angles, are to be found in other references. The differences are not great, but are sufficient to frustrate easy comparison of the formulae derived.

We must also remember that the time derivatives of the Eulerian angles—$\dot{\theta}$, $\dot{\phi}$, and $\dot{\psi}$—are *not* the components of the angular velocity $\boldsymbol{\omega}$ of the body fixed axes. We see that each of these rotations acts about a different axis and they are in inconsistent coordinate systems. From Fig. 12.10 we can see that the angular velocity of the body fixed axes is

$$\boldsymbol{\omega} = \dot{\theta}\hat{\mathbf{k}} + \dot{\phi}\hat{\mathbf{i}}_1 + \dot{\psi}\hat{\mathbf{k}}'$$

When these different unit vectors are all transformed into the fixed reference directions and then added we get

$$\omega^x = \dot{\phi}\cos\theta + \dot{\psi}\sin\theta\sin\phi \tag{12.9a}$$

$$\omega^y = \dot{\phi}\sin\theta - \dot{\psi}\cos\theta\sin\phi \tag{12.9b}$$

$$\omega^z = \dot{\theta} + \dot{\psi}\cos\phi \tag{12.9c}$$

On the other hand, we can also transform them into the body fixed axes, where we find

$$\omega^{x'} = \dot{\theta}\sin\phi\sin\psi + \dot{\phi}\cos\psi \tag{12.10a}$$

$$\omega^{y'} = \dot{\theta}\sin\phi\cos\psi - \dot{\phi}\sin\psi \tag{12.10b}$$

$$\omega^{z'} = \dot{\theta}\cos\phi + \dot{\psi} \tag{12.10c}$$

12.6 THE DENAVIT–HARTENBERG PARAMETERS

The transformation equations of the previous section dealt only with three-dimensional rotations about a fixed point. Yet the same approach can be generalized to include both translations and rotations, and thus to treat general spatial motions. This has been done, and much literature can be found dealing with these techniques, usually under a title such as "matrix methods." Most modern work on this approach stems from the work of Denavit and Hartenberg, who developed a notation scheme for labeling all single-loop lower pair linkages and also devised a transformation matrix technique for their analysis.[10]

In the Denavit–Hartenberg approach we start by numbering the joints of the linkage, usually starting with the input joint and numbering consecutively around the loop to the output joint. If we assume that the linkage has only a single kinematic loop with only j_1 joints and a mobility of $m = 1$, then the Kutzbach criterion shows that there will be $n = 7$ binary links and $j_1 = 7$ joints numbered from 1 through 7.* Figure 12.11 shows a typical joint of this loop, revolute joint number i in this case, and the two links which it joins.

Next we identify the motion axis of each of the joints, choose a positive orientation on each, and label each as a z_i axis. Even though the z_i axes may be skew in space, it is possible to find a common perpendicular between each consecutive pair and label these as x_i axes such that each x_i axis is the common perpendicular between z_{i-1} and z_i, choosing an arbitrary positive orientation.**

---

*The treatment presented here deals only with j_1 joints; multifreedom lower pairs may be represented as combinations of revolutes and prisms, however, with fictitious links between them.

**When $i = 1$, then $i - 1$ must be taken as n because, in the loop, the last joint is also the joint before the first. Similarly, when $i = n$, then $i + 1$ must be taken as 1.

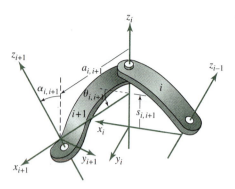

Figure 12.11 Definitions of Denavit–Hartenberg parameters.

Having done this, we can now identify and label y_i axes such that there is a right-hand Cartesian coordinate system $x_i y_i z_i$ associated with each joint of the loop. Careful study of Fig. 12.11 and visualization of the motion allowed by the joint shows that coordinate system $x_i y_i z_i$ remains fixed to the link carrying joint $i - 1$ and joint i, while $x_{i+1} y_{i+1} z_{i+1}$ moves with and remains fixed to the link carrying joint i and joint $i + 1$. Thus we have a basis for assigning numbers to the links that correspond to the numbers of the coordinate system attached to each.

The key to the Denavit–Hartenberg approach comes now in the standard method they defined for determining the shapes of the links. The relative position and orientation of any two consecutive coordinate systems placed as described above can be defined by four parameters, labeled a, α, θ, and s, shown in Fig. 12.11, and defined as follows:

$a_{i,i+1}$ = distance along x_{i+1} from z_i to z_{i+1} with sign taken from the sense of x_{i+1}

$\alpha_{i,i+1}$ = angle from positive z_i to positive z_{i+1} taken positive counterclockwise as seen from positive x_{i+1}

$\theta_{i,i+1}$ = angle from positive x_i to positive x_{i+1} taken positive counterclockwise as seen from positive z_i

$s_{i,i+1}$ = distance along z_i from x_i to x_{i+1} with sign taken from the sense of z_i

We see that, when joint i is a revolute, as depicted in Fig. 12.11, then the $a_{i,i+1}$, $\alpha_{i,i+1}$, and $s_{i,i+1}$ parameters are constants defining the shape of link $i + 1$, but $\theta_{i,i+1}$ is variable; in fact, it serves to measure the joint variable of joint i. When the joint is prismatic, then the $a_{i,i+1}$, $\alpha_{i,i+1}$, and $\theta_{i,i+1}$ parameters are constants and $s_{i,i+1}$ is the pair variable.

EXAMPLE 12.3

Find the Denavit–Hartenberg parameters for the Hooke, or Cardan, universal joint shown in Fig. 12.12.

SOLUTION

From the figure we see that the mechanism has four links and four revolute joints. We choose the axes of the four revolutes and label them as z_1, z_2, z_3, and z_4 as shown. Next we identify and label their common perpendiculars, choose positive orientations for these, and label them x_1, x_2, x_3,

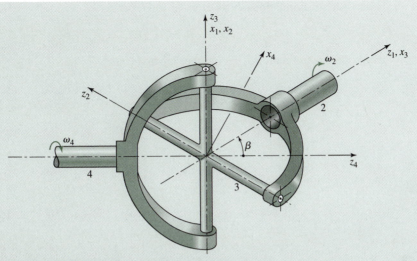

Figure 12.12 The Hooke, or Cardan, universal joint.

and x_4 as shown. Note that, at the instant shown, x_1 and x_3 appear to lie along z_3 and z_1, respectively; this is only a temporary coincidence and will change as the mechanism moves.

Now, following the definitions explained before, we find the values for the parameters:

$$a_{12} = 0, \qquad a_{23} = 0, \qquad a_{34} = 0, \qquad a_{41} = 0 \qquad\qquad Ans.$$

$$\alpha_{12} = 90°, \qquad \alpha_{23} = 90°, \qquad \alpha_{34} = 90°, \qquad \alpha_{41} = \beta \qquad\qquad Ans.$$

$$\theta_{12} = \phi_1, \qquad \theta_{23} = \phi_2, \qquad \theta_{34} = \phi_3, \qquad \theta_{41} = \phi_4 \qquad\qquad Ans.$$

$$s_{12} = 0, \qquad s_{23} = 0, \qquad s_{34} = 0, \qquad s_{41} = 0 \qquad\qquad Ans.$$

We note that all a and s distance parameters are zero; this signifies that the mechanism is spherical with all motion axes intersecting at a common center. Also, we see that all θ parameters are joint variables because this is an *RRRR* linkage. They have been given the symbols ϕ_i rather than numeric values because they are variables; solutions for their values will be found in the next example.

12.7 TRANSFORMATION-MATRIX POSITION ANALYSIS

The Denavit–Hartenberg parameters provide a standard method for measuring the important geometric characteristics of a linkage, but their value does not stop at that. Having standardized the placement of the coordinate systems on each link, Denavit and Hartenberg have also shown that the transformation equations between successive coordinate systems can be written in a standard matrix format that uses these parameters. If we know the position coordinates of some point measured in one of the coordinate systems, say R_{i+1}, then we can find the position coordinates of the same point in the previous coordinate system R_i

by using a transformation matrix $T_{i,i+1}$ as follows:

$$R_i = T_{i,i+1} R_{i+1} \tag{12.11}$$

where the transformation $T_{i,i+1}$ has the standard form*

$$T_{i,i+1} = \begin{bmatrix} \cos\theta_{i,i+1} & -\cos\alpha_{i,i+1}\sin\theta_{i,i+1} & \sin\alpha_{i,i+1}\sin\theta_{i,i+1} & a_{i,i+1}\cos\theta_{i,i+1} \\ \sin\theta_{i,i+1} & \cos\alpha_{i,i+1}\cos\theta_{i,i+1} & -\sin\alpha_{i,i+1}\cos\theta_{i,i+1} & a_{i,i+1}\sin\theta_{i,i+1} \\ 0 & \sin\alpha_{i,i+1} & \cos\alpha_{i,i+1} & s_{i,i+1} \\ 0 & 0 & 0 & 1 \end{bmatrix} \tag{12.12}$$

and each position vector is given by

$$R = \begin{bmatrix} x \\ y \\ z \\ 1 \end{bmatrix} \tag{12.13}$$

Now, by applying Eq. (12.11) recursively from one link to the next, we see that, for an n-link single-loop mechanism,

$$R_1 = T_{12} R_2$$

$$R_1 = T_{12} T_{23} R_3$$

$$R_1 = T_{12} T_{23} T_{34} R_4$$

and, in general,

$$R_1 = T_{12} T_{23} \cdots T_{i-1,i} R_i \tag{12.14}$$

If we agree on a notation in which a product of these T matrices is still denoted as a T matrix,

$$T_{i,j} = T_{i,i+1} T_{i+1,i+2} \cdots T_{j-1,j} \tag{12.15}$$

then Eq. (12.14) becomes

$$R_1 = T_{1,i} R_i \tag{12.16}$$

Finally, because link 1 follows link n at the end of the loop,

$$R_1 = T_{12} T_{23} \cdots T_{n-1,n} T_{n,1} R_1$$

and, because this equation must be true no matter what point we choose for R_1, we see that

$$T_{1,2} T_{2,3} \cdots T_{n-1,n} T_{n,1} = I \tag{12.17}$$

where I is the 4×4 identity transformation matrix.

---

*Note that the first three rows and columns of the T matrix are the direction cosines needed for the rotation between the coordinate systems. The fourth column adds the translation terms for the separation of the origins. The fourth row represents a dummy equation, $1 = 1$, but keeps the T matrix square and invertible.

This important equation is the transformation matrix form of the *loop-closure equation*. Just as the vector tetrahedron equation, Eq. (12.2), states that the *sum of vectors* around a kinematic loop must equal zero for the loop to close, Eq. (12.17) states that the *product of transformation matrices* around a kinematic loop must equal the identity transformation. While the vector sum ensures that the loop returns to its starting location, the transformation matrix product also ensures that the loop returns to its starting angular orientation. This becomes critical, for example, in spherical motion problems and cannot be shown by the vector-tetrahedron equation. The following example illustrates this point.

EXAMPLE 12.4

Analyze the Hooke universal joint of Fig. 12.12 to find equations for the positions of all other joint variables when we are given a value for input shaft angle ϕ_1.

SOLUTION

The Denavit–Hartenberg parameters for this mechanism were found in Example 12.3. Using these we substitute into Eq. (12.12) to find the individual transformation matrices for each link. These are:

$$T_{1,2} = \begin{bmatrix} \cos\phi_1 & 0 & \sin\phi_1 & 0 \\ \sin\phi_1 & 0 & -\cos\phi_1 & 0 \\ 0 & 1 & 0 & 0 \\ 0 & 0 & 0 & 1 \end{bmatrix} \tag{1}$$

$$T_{2,3} = \begin{bmatrix} \cos\phi_2 & 0 & \sin\phi_2 & 0 \\ \sin\phi_2 & 0 & -\cos\phi_2 & 0 \\ 0 & 1 & 0 & 0 \\ 0 & 0 & 0 & 1 \end{bmatrix} \tag{2}$$

$$T_{3,4} = \begin{bmatrix} \cos\phi_3 & 0 & \sin\phi_3 & 0 \\ \sin\phi_3 & 0 & -\cos\phi_3 & 0 \\ 0 & 1 & 0 & 0 \\ 0 & 0 & 0 & 1 \end{bmatrix} \tag{3}$$

$$T_{4,1} = \begin{bmatrix} \cos\phi_4 & -\cos\beta\sin\phi_4 & \sin\beta\sin\phi_4 & 0 \\ \sin\phi_4 & \cos\beta\cos\phi_4 & -\sin\beta\cos\phi_4 & 0 \\ 0 & \sin\beta & \cos\beta & 0 \\ 0 & 0 & 0 & 1 \end{bmatrix} \tag{4}$$

Although we could now use Eq. (12.17) directly, it reduces the number of computations if we first rearrange as follows:

$$T_{1,2}T_{2,3}T_{3,4} = T_{4,1}^{-1} \tag{5}$$

The inverse matrix shown here can easily be found by simply transposing the matrix, that is by switching rows and columns. Therefore, substituting and carrying out the matrix computations,

Eq. (5) becomes

$$\begin{bmatrix} \cos\phi_1\cos\phi_2\cos\phi_3 + \sin\phi_1\sin\phi_3 & \cos\phi_1\sin\phi_2 & -\cos\phi_1\cos\phi_2\sin\phi_3 - \sin\phi_1\cos\phi_3 & 0 \\ \sin\phi_1\cos\phi_2\cos\phi_3 - \cos\phi_1\sin\phi_3 & \sin\phi_1\sin\phi_2 & \sin\phi_1\cos\phi_2\sin\phi_3 + \cos\phi_1\cos\phi_3 & 0 \\ \sin\phi_2\cos\phi_3 & -\cos\phi_2 & \sin\phi_2\sin\phi_3 & 0 \\ 0 & 0 & 0 & 1 \end{bmatrix}$$

$$= \begin{bmatrix} \cos\phi_4 & \sin\phi_4 & 0 & 0 \\ -\cos\beta\sin\phi_4 & \cos\beta\cos\phi_4 & \sin\beta & 0 \\ \sin\beta\sin\phi_4 & -\sin\beta\cos\phi_4 & \cos\beta & 0 \\ 0 & 0 & 0 & 1 \end{bmatrix} \qquad (6)$$

Corresponding rows and columns on both sides of this equation must be equal. Therefore we can equate the ratios of the second-row, second-column elements to the first-row, second-column elements. Rearranging, we get

$$\tan\phi_4 = \frac{\cos\beta}{\tan\phi_1} \qquad \textit{Ans.}$$

Once we have solved for ϕ_4, we can equate the ratios of the third-row, third-column elements to the third-row, first-column elements. This gives

$$\tan\phi_3 = \frac{1}{\tan\beta\sin\phi_4} \qquad \textit{Ans.}$$

Finally, from the elements of the third-row, second-column,

$$\cos\phi_2 = \sin\beta\cos\phi_4 \qquad \textit{Ans.}$$

12.8 MATRIX VELOCITY AND ACCELERATION ANALYSES

The power of the matrix method begins to show in the position analysis method above. However, it is not limited to position analysis. The same standardized approach can be extended to velocity and acceleration analyses. To see this, let us start by noticing that, of the four Denavit–Hartenberg parameters, three are constants describing the link shape, and one is the joint variable. Therefore, in the basic transformation of Eq. (12.12),

$$T = \begin{bmatrix} \cos\theta & -\cos\alpha\sin\theta & \sin\alpha\sin\theta & a\cos\theta \\ \sin\theta & \cos\alpha\cos\theta & -\sin\alpha\cos\theta & a\sin\theta \\ 0 & \sin\alpha & \cos\alpha & s \\ 0 & 0 & 0 & 1 \end{bmatrix} \qquad (12.18)$$

there is only one variable and it might be either of the parameters θ or s depending on the type of joint.

If we consider the case where the joint variable is the angle θ, then the derivative of T with respect to its own variable is

$$\frac{dT}{d\theta} = \begin{bmatrix} -\sin\theta & -\cos\alpha\cos\theta & \sin\alpha\cos\theta & -a\sin\theta \\ \cos\theta & -\cos\alpha\sin\theta & \sin\alpha\sin\theta & a\cos\theta \\ 0 & 0 & 0 & 0 \\ 0 & 0 & 0 & 0 \end{bmatrix} \tag{12.19}$$

On the other hand, if the joint variable is the distance s, as in a prismatic joint, then the derivative is

$$\frac{dT}{ds} = \begin{bmatrix} 0 & 0 & 0 & 0 \\ 0 & 0 & 0 & 0 \\ 0 & 0 & 0 & 1 \\ 0 & 0 & 0 & 0 \end{bmatrix} \tag{12.20}$$

It is interesting that both of these derivatives can be taken by the same formula

$$\frac{dT_{i,i+1}}{d\phi_i} = Q_i T_{i,i+1} \tag{12.21}$$

where we understand that when joint i is a revolute, then $\phi_i = \theta_{i,i+1}$ and we use

$$Q_i = \begin{bmatrix} 0 & -1 & 0 & 0 \\ 1 & 0 & 0 & 0 \\ 0 & 0 & 0 & 0 \\ 0 & 0 & 0 & 0 \end{bmatrix} \tag{12.22}$$

and that when joint i is prismatic, then $\phi_i = s_{i,i+1}$ and we use

$$Q_i = \begin{bmatrix} 0 & 0 & 0 & 0 \\ 0 & 0 & 0 & 0 \\ 0 & 0 & 0 & 1 \\ 0 & 0 & 0 & 0 \end{bmatrix} \tag{12.23}$$

For velocity analysis we need derivatives with respect to time rather than with respect to the joint variables. Therefore, using Eq. (12.21), we find that

$$\frac{dT_{i,i+1}}{dt} = Q_i T_{i,i+1} \dot{\phi}_i \tag{12.24}$$

where $\dot{\phi}_i = d\phi_i/dt$; these are usually not known values, but we will discover next how they can be found. We start by differentiating the loop-closure conditions, Eq. (12.17), with respect to time. Using the chain rule with Eq. (12.24) to differentiate each factor, we get

$$\sum T_{12}T_{23}\cdots T_{i-1,i}\,Q_i\,T_{i,i+1}\cdots T_{n-1,n}T_{n1}\dot{\phi}_i = 0$$

and using the more condensed notation of Eq. (12.15), this becomes

$$\sum T_{1i} Q_i T_{i1} \dot{\phi}_i = 0$$

If we now define the symbol

$$D_i = T_{1i} Q_i T_{i1} \tag{12.25a}$$

and take note that the loop-closure condition shows that this is the same as

$$D_i = T_{1i} Q_i T_{1i}^{-1} \tag{12.25b}$$

then the previous equation becomes

$$\sum D_i \dot{\phi}_i = 0 \tag{12.26}$$

This equation contains all the conditions that the pair variable velocities must obey to fit the mechanism in question and be compatible with each other. Notice that once the position analysis is finished, the D_i matrices can be evaluated from known information. Some of the pair variable velocities will be given as input information, namely the m input variables; all other $\dot{\phi}_i$ values may then be solved for from Eq. (12.26). A continuation of Example 12.4 will make the procedure clear.

EXAMPLE 12.5

Determine the angular velocity of the output shaft for the Hooke universal joint of Example 12.4.

SOLUTION

From the results of the previous example we find the following relationships between the input shaft angle ϕ_1 and the other pair variables:

$$\sin \phi_2 = \frac{\cos \beta}{\sigma}, \qquad \cos \phi_2 = \frac{\sin \beta \sin \phi_1}{\sigma}$$

$$\sin \phi_3 = \sigma, \qquad \cos \phi_3 = \sin \beta \cos \phi_1$$

$$\sin \phi_4 = \frac{\cos \beta \cos \phi_1}{\sigma}, \qquad \cos \phi_4 = \frac{\sin \phi_1}{\sigma}$$

where

$$\sigma = \sqrt{1 - \sin^2 \beta \cos^2 \phi_1}$$

and substituting these into Eq. (12.12) gives the transformation matrices as functions of ϕ_1 alone. These, in turn, can be substituted into Eq. (12.25) to give the derivative operator matrices D_i, which become

$$D_1 = \begin{bmatrix} 0 & -1 & 0 & 0 \\ 1 & 0 & 0 & 0 \\ 0 & 0 & 0 & 0 \\ 0 & 0 & 0 & 0 \end{bmatrix}$$

$$D_2 = \begin{bmatrix} 0 & 0 & -\cos\phi_1 & 0 \\ 0 & 0 & -\sin\phi_1 & 0 \\ \cos\phi_1 & \sin\phi_1 & 0 & 0 \\ 0 & 0 & 0 & 0 \end{bmatrix}$$

$$D_3 = \begin{bmatrix} 0 & \dfrac{\sin\beta\sin\phi_1}{\sigma} & \dfrac{\cos\beta\sin\phi_1}{\sigma} & 0 \\ -\dfrac{\sin\beta\sin\phi_1}{\sigma} & 0 & \dfrac{-\cos\beta\cos\phi_1}{\sigma} & 0 \\ -\dfrac{\cos\beta\sin\phi_1}{\sigma} & \dfrac{\cos\beta\cos\phi_1}{\sigma} & 0 & 0 \\ 0 & 0 & 0 & 0 \end{bmatrix}$$

$$D_4 = \begin{bmatrix} 0 & -\cos\beta & \sin\beta & 0 \\ \cos\beta & 0 & 0 & 0 \\ -\sin\beta & 0 & 0 & 0 \\ 0 & 0 & 0 & 0 \end{bmatrix}$$

These can now be used in Eq. (12.26). Taking the elements from row 3, column 2, then row 1, column 3, and finally from row 2, column 1, we get the following three equations:

$$(\sin\phi_1)\dot\phi_2 + \left(\frac{\cos\beta\cos\phi_1}{\sigma}\right)\dot\phi_3 = 0$$

$$(-\cos\phi_1)\dot\phi_2 + \left(\frac{\cos\beta\sin\phi_1}{\sigma}\right)\dot\phi_3 + (\sin\beta)\dot\phi_4 = 0$$

$$\left(-\frac{\sin\beta\sin\phi_1}{\sigma}\right)\dot\phi_3 + (\cos\beta)\dot\phi_4 = -\dot\phi_1$$

and solving these simultaneously gives

$$\dot\phi_2 = \frac{-\sin\beta\cos\beta\cos\phi_1}{1 - \sin^2\beta\cos^2\phi_1}\dot\phi_1$$

$$\dot\phi_3 = \frac{\sin\beta\sin\phi_1}{\sqrt{1 - \sin^2\beta\cos^2\phi_1}}\dot\phi_1$$

$$\dot\phi_4 = \frac{-\cos\beta}{1 - \sin^2\beta\cos^2\phi_1}\dot\phi_1 \qquad\qquad Ans.$$

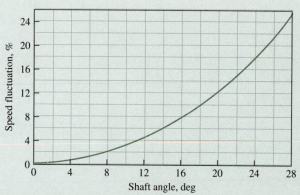

Figure 12.13 Relationship between shaft angle and speed fluctuation for Hooke universal shaft couplings.

Notice that when the input shaft has constant angular velocity $\dot{\phi}_1$, the angular velocity of the output shaft $\dot{\phi}_4$ is not constant unless the shafts are in line; if β is not equal to zero, the output angular velocity fluctuates as the shaft rotates. Because the shaft angle β is constant, but usually not zero, the maximum value of the output/input velocity ratio $\dot{\phi}_4/\dot{\phi}_1$ occurs when $\cos\phi_1 = 1$—that is, when $\phi_1 = 0°$, $180°$, $360°$, $540°$, and so on; the minimum value of this velocity ratio occurs when $\cos\phi_1 = 0$. If the difference between the maximum and minimum velocity ratio is expressed in percent and plotted versus shaft angle β, the graph shown in Fig. 12.13 results. This curve, showing percent speed fluctuation, is useful in evaluation of Hooke universal shaft coupling applications.

If we wanted to find the velocity of a moving point attached to link i, we would differentiate the equation for the position of the point, Eq. (12.14), with respect to time. Using the D_i operator matrices to do this, we are led to defining a set of velocity operator matrices

$$\omega_1 = 0 \qquad (12.27a)$$

$$\omega_{i+1} = \omega_i + D_i\dot{\phi}_i \qquad i = 1, 2, \ldots, n \qquad (12.27b)$$

and then the absolute velocity of a point is given by

$$\dot{R}_i = \omega_i T_{1,i} R_i \qquad (12.28)$$

Acceleration analysis follows the same approach as shown for velocities. Without presenting the details, the important equations are shown here. Differentiating Eq. (12.26) again with respect to time, we get a set of equations relating the pair-variable accelerations:

$$\sum D_i\ddot{\phi}_i = -\sum(\omega_i D_i - D_i\omega_i)\dot{\phi}_i \qquad (12.29)$$

These equations can be solved for the $\ddot{\phi}_i$ joint variable acceleration values once the position, velocity, and input acceleration values are known. The solution process is identical with that for Eq. (12.26); in fact, the matrix of coefficients of the unknowns is identical for both sets of equations.

This matrix of coefficients, called the *Jacobian,* is essential for the solution of any set of derivatives of the joint variables. As was pointed out in Section 3.16, if this matrix becomes singular, there is no unique solution for the velocities (or accelerations) of the joint variables. If this occurs, such a position of the mechanism is called a *singular* position; one example would be a dead-center position.

The derivatives of the velocity matrices of Eq. (12.27) give rise to definition of a set of acceleration operator matrices

$$\alpha_1 = 0 \tag{12.30a}$$

$$\alpha_{i+1} = \alpha_i + (\omega_i D_i - D_i \omega_i)\dot{\phi}_i \tag{12.30b}$$

and, from these, the absolute acceleration of a point on link i is given by

$$\ddot{R}_i = (\alpha_i + \omega_i \omega_i) T_{1,i} R_i \tag{12.31}$$

Much further detail and more power has been developed using this transformation matrix approach to the kinematic and dynamic analysis of rigid-body systems. However, these methods go far beyond the scope of this book. Further examples dealing with robotics are presented, however, in the next chapter.

12.9 GENERALIZED MECHANISM ANALYSIS COMPUTER PROGRAMS

It can probably be seen how the methods taken for the solution of each new problem are quite similar from one problem to the next. However, particularly in three-dimensional analysis, we also see that the number and complexity of the calculations can make solution by hand a very tedious task. These characteristics suggest that a general computer program might have a broad range of applications and that the development costs for such a program might be justified through repeated usage and increased accuracy, relief of human drudgery, and elimination of human errors. General computer programs for the simulation of rigid-body kinematic and dynamic systems have been under development for some years now, and some are available and are being used in industrial settings, particularly in the automotive and aircraft industries.

The first widely available program for mechanism analysis was named KAM (Kinematic Analysis Method) and was written and distributed by IBM. It included capabilities for position, velocity, acceleration, and force analysis of both planar and spatial mechanisms and was developed around the Chace vector-tetrahedron equation solutions discussed in Section 12.3. Released in 1964, this program was the first to recognize the need for a general program for mechanical systems exhibiting large geometric movements. Being first, however, it had limitations and has been superseded by more powerful programs, including those described next.

Powerful generalized programs have also been developed using finite element and finite difference methods; NASTRAN and ANSYS are two examples. In the realm of mechanical systems these programs have been developed primarily for stress analysis and have excellent capabilities for static- and dynamic-force analysis. These also allow the links of the simulated system to deflect under load and are capable of solving statically indeterminate force problems. They are very powerful programs with wide application in industry. Although they are sometimes used for mechanism analysis, they are limited by their inability to simulate the large geometric changes typical of kinematic systems.

The most widely used generalized programs for kinematic and dynamic simulation of three-dimensional rigid-body mechanical systems are ADAMS, DADS, and IMP. The ADAMS® program, standing for Automatic Dynamic Analysis of Mechanical Systems, grew from the research efforts of Chace, Orlandea, and others at the University of Michigan[11] and is available from Mechanical Dynamics, Inc. (MDI).[12] DADS, standing for Dynamic Analysis and Design System, was developed by Haug and others at the University of Iowa[13] and CAD Systems, Inc. (CADSI).[14] The Integrated Mechanisms Program (IMP) was developed by Uicker, Sheth, and others at the University of Wisconsin—Madison.[15] These and other similar programs are all applicable to single- or multiple-degree-of-freedom systems in both open- and closed-loop configurations. All will operate on mainframe computers or on workstations, and some will operate on microprocessors; all can display results with graphic animation. All are capable of solving position, velocity, acceleration, static force, and dynamic force analyses. All can formulate the dynamic equations of motion and predict the system response to a given set of initial conditions with prescribed motions or forces that may be functions of time. Some of these programs include collision detection, the ability to simulate impact, elasticity, or control system effects. Other commercial software in this area include the Pro/MECHANICA®[16] Motion Simulation Package and MSC Working Model®[17] systems.

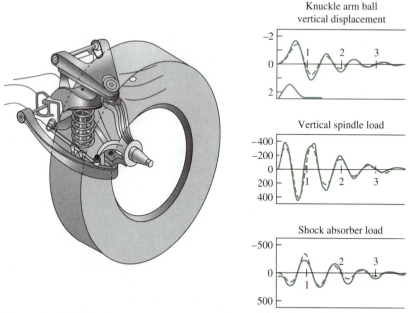

Figure 12.14 Example of a half-front automotive suspension simulated by both the ADAMS and IMP programs. The graphs show the comparison of experimental test data and numerical simulation results as the suspension encounters a 1-in hole. The units on the graphs are inches and pounds on the vertical axes and time in seconds on the horizontal axes. (JML Research, Inc., Madison, WI, and Mechanical Dynamics, Inc., Ann Arbor, MI.)

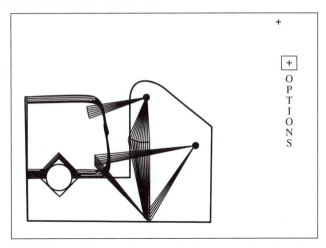

Figure 12.15 This pipe-clamp mechanism was designed in about 15 minutes using KINSYN III. KINSYN was developed at the Joint Computer Facility of the Massachusetts Institute of Technology under the direction of Dr. R. E. Kaufman, now Professor of Engineering at the George Washington University. (Courtesy of Prof. R. E. Kaufman.)

A typical application for any of these programs is the simulation of the automotive front suspension shown in Fig. 12.14.* Simulations of this type have been performed with several of these programs and they have been shown to compare well with experimental data.

Another type of generalized program available today is intended for kinematic synthesis. The earliest of such programs was KINSYN (KINematic SYNthesis) and this was followed by LINCAGES (*L*inkage *IN*teractive *C*omputer *A*nalysis and *G*raphically *En*hanced *S*ynthesis)[18] and others. These systems are directed toward the kinematic synthesis of planar linkages using methods analogous to those presented in Chapter 11. Users may input their motion requirements through a graphical user interface (GUI); the computer accepts the sketch and provides the required design information on the display screen. An example showing the use of KINSYN is shown in Fig. 12.15. A much more recent system of this type is the WATT Mechanism Design Tool from Heron Technologies,[19] a spinoff company from Twente University in Holland.

NOTES

1. L. Harrisberger, "A Number Synthesis Survey of Three-Dimensional Mechanisms," *J. Eng. Ind., ASME Trans.*, ser. B, vol. 87, no. 2, 1965.

2. For pictures of these see R. S. Hartenberg and J. Denavit, *Kinematic Synthesis of Linkages,* McGraw-Hill, New York, 1964, pp. 85–86.

3. L. Harrisberger and A. H. Soni, "A Survey of Three-Dimensional Mechanisms with One General Constraint," ASME Paper 66-MECH-44, October 1966. This paper contains 45 references on spatial mechanisms.

*Simulations by the ADAMS and IMP programs of the system shown in Fig. 12.14 were done for the Strain History Prediction Committee of the Society of Automotive Engineers. Vehicle data and experimental test results were provided by Chevrolet Division, General Motors Corp.

4.   An extremely detailed discussion of this entire topic forms one of the main themes of an excellent two volume set: Jack Phillips, *Freedom of Machinery, Volume 1, Introducing Screw Theory,* Cambridge University Press, 1984, and *Volume 2, Screw Theory Exemplified,* Cambridge University Press, 1990.

5.   Ibid, Section 20.16, The advantages of overconstraint, p. 151.

6.   M. A. Chace, "Vector Analysis of Linkages," *J. Eng. Ind., ASME Trans.,* ser. B, vol. 85, no. 3, 1963, pp. 289–297.

7.   A. T. Yang and F. Freudenstein, "Application of Dual-Number and Quaternion Algebra to the Analysis of Spatial Mechanisms," *J. Appl. Mech., ASME Trans.,* ser. E, vol. 86, 1964, pp. 300–308.

8.   J. J. Uicker, Jr., J. Denavit, and R. S. Hartenberg, "An Iterative Method for the Displacement Analysis of Spatial Linkages," *J. Appl. Mech., ASME Trans.,* ser. E, vol. 87, 1965, pp. 309–314.

9.   Ibid.

10.   J. Denavit and R. S. Hartenberg, "A Kinematic Notation for Lower-Pair Mechanisms Based on Matrices," *J. Appl. Mech., ASME Trans.,* ser. E, vol. 22, no. 2, June 1955, pp. 215–221.

11.   N. Orlandea, M. A. Chace, and D. A. Calahan, "A Sparsity-Oriented Approach to the Dynamic Analysis and Design of Mechanical Systems, Parts I and II," *J. Eng. Ind., ASME Trans.,* vol. 99, pp. 773–784, 1977.

12.   Mechanical Dynamics, Inc., 2300 Traverwood Drive, Ann Arbor, MI 48105.

13.   E. J. Haug, Computer-Aided Kinematics and Dynamics of Mechanical Systems, Allyn Bacon, Boston, MA, 1989.

14.   LMS International, Researchpark Z1, Interleuvenlaan 68, 3001 Leuven, Belgium; LMS CAE Division, 2651 Crosspark Road, Coralville, IA 52241.

15.   P. N. Sheth and J. J. Uicker, Jr., "IMP (Integrated Mechanisms Program), A Computer-Aided Design Analysis System for Mechanisms and Linkages," *J. Eng. Ind., ASME Trans.,* vol. 94, pp. 454–464, 1972.

16.   Parametric Technologies, Corp., 140 Kendrick Street, Needham, MA 02494.

17.   MSC Software Corp., 815 Colorado Boulevard, Los Angeles, CA 90041.

18.   LINCAGES is available through Dr. A. G. Erdman by sending e-mail to *agerdman@ me.umn.edu.*

19.   Heron Technologies b.v., P.O. Box 2, 7550 AA Hengelo, The Netherlands.

PROBLEMS

12.1   Use the Kutzbach criterion to determine the mobility of the *GGC* linkage shown in the figure. Identify any idle freedoms and state how they can be removed. What is the nature of the path described by point *B*?

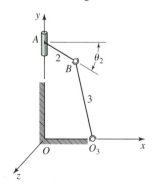

Figure P12.1 $R_{BA} = R_{O_3O} = 75$ mm, $R_{BO_3} = 150$ mm, $\theta_2 = 30°$.

12.2   For the *GGC* linkage shown express the position of each link in vector form.

12.3   With $V_A = -50\hat{j}$ mm/s, use vector analysis to find the angular velocities of links 2 and 3 and the velocity of point *B* at the position specified.

12.4   Solve Problem 12.3 using graphical techniques.

12.5   For the spherical *RRRR* shown in the figure, use vector algebra to make complete velocity and acceleration analyses at the position given.

12.6   Solve Problem 12.5 using graphical techniques.

12.7   Solve Problem 12.5 using transformation matrix techniques.

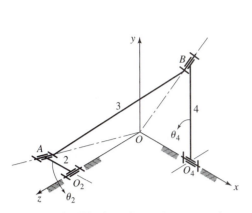

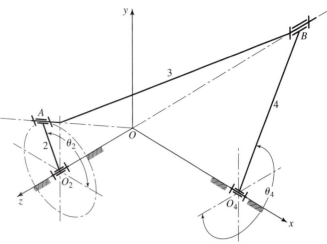

Figure P12.5 The figure is not drawn to scale; at the position shown, $R_{O_2O} = 7\hat{k}$ in, $R_{O_4O} = 2\hat{i}$ in, $R_{AO_2} = -3\hat{i}$ in, $R_{BO_4} = 9\hat{j}$ in, $R_{BA} = 5\hat{i} + 9\hat{j} - 7\hat{k}$ in, $\omega_2 = -60\hat{k}$ rad/s.

Figure P12.10 $R_{O_2O} = 150$ mm, $R_{O_4O} = 225$ mm, $R_{AO_2} = 37.5$ mm, $R_{BO_4} = 262$ mm, $R_{BA} = 412$ mm, $\theta_2 = 120°$, $\omega_2 = 30\hat{k}$ rad/s.

12.8 Solve Problem 12.5 except with $\theta_2 = 90°$.

12.9 Determine the advance-to-return time ratio for Problem 12.5. What is the total angle of oscillation of link 4?

12.10 For the spherical *RRRR* linkage shown, determine whether the crank is free to turn through a complete revolution. If so, find the angle of oscillation of link 4 and the advance-to-return time ratio.

12.11 Use vector algebra to make complete velocity and acceleration analyses of the linkage at the position specified.

12.12 Solve Problem 12.11 using graphical techniques.

12.13 Solve Problem 12.11 using transformation matrix techniques.

12.14 The figure shows the top, front, and auxiliary views of a spatial slider-crank *RGGP* linkage. In the construction of many such mechanisms provision is made to vary the angle β; thus the stroke of slider 4 becomes adjustable from zero, when $\beta = 0$, to twice the crank length, when $\beta = 90°$. With $\beta = 30°$, use vector algebra to make a complete velocity analysis of the linkage at the given position.

12.15 Solve Problem 12.14 using graphical techniques.

12.16 Solve Problem 12.14 using transformation matrix techniques.

12.17 Solve Problem 12.14 with $\beta = 60°$ using vector algebra.

12.18 Solve Problem 12.14 with $\beta = 60°$ using graphical techniques.

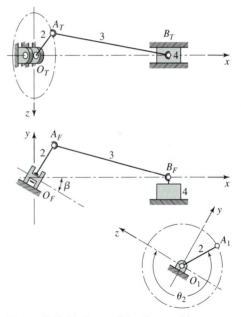

Figure P12.14 $R_{AO} = 2$ in, $R_{BA} = 6$ in, $\theta_2 = 240°$, $\omega_2 = 24\hat{i}$ rad/s.

12.19 Solve Problem 12.14 with $\beta = 60°$ using transformation matrix techniques.

12.20 The figure shows the top, front, and profile views of an *RGRC* crank and oscillating-slider linkage. Link 4, the oscillating slider, is rigidly attached to a round rod that rotates and slides in the two bearings.

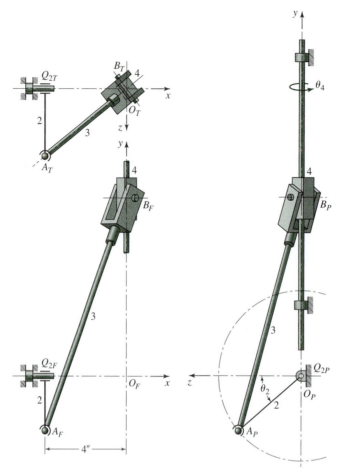

Figure P12.20 $R_{AO} = 4$ in, $R_{BA} = 12$ in, $\theta_2 = 40°$, $\omega_2 = -48\hat{i}$ rad/s.

(*a*) Use the Kutzbach criterion to find the mobility of this linkage. (*b*) With crank 2 as the driver, find the total angular and linear travel of link 4. (*c*) Write the loop-closure equation for this mechanism and use vector algebra to solve it for all unknown position data.

12.21 Use vector algebra to find V_B, ω_3, and ω_4 for Problem 12.20.

12.22 Solve Problem 12.21 using graphical techniques.

12.23 Solve Problem 12.21 using transformation matrix techniques.

13 | Robotics

13.1 INTRODUCTION

In the previous several chapters we have studied methods for analyzing the kinematics of machines. First we studied planar kinematics at great length, justifying this emphasis by pointing out that well over 90 percent of all machines in use today have planar motion. Then, in the most recent chapter, we showed how these methods extend to problems with spatial motion. Although the algebra became more lengthy with spatial problems, we found that there was no significant new block of theory required. We also concluded that as the mathematics became more tedious, there is a role for the computer to relieve the drudgery. Still, we found only a few problems with both practical application and spatial motion.

However, within the past decade or two, advancing technology has led to considerable attention on the development of robotic devices. Most of these are spatial mechanisms and require the attendant more extensive calculations. Fortunately, however, they also carry on-board computing capability and can deal with this added complexity. The one remaining requirement is that the engineers and designers of the robot itself have a clear understanding of their characteristics and have appropriate tools for their analysis. That is the purpose of this chapter.

The term *robot* comes from the Czech word *robota*, meaning work, and has been applied to a wide variety of computer-controlled electromechanical systems, from autonomous landrovers to underwater vehicles to teleoperated arms in industrial manipulators. The Robot Institute of America (RIA) defines a robot as *a reprogrammable multifunctional manipulator designed to move material, parts, tools, or specialized devices through variable programmed motions for the performance of a variety of tasks.* Many industrial manipulators bear a strong resemblance in their conceptual design to that of a human arm. However, this is not always true and is not essential; the key to the above definition is that a robot has flexibility through its programming, and its motion can be adapted to fit a variety of tasks.

13.2 TOPOLOGICAL ARRANGEMENTS OF ROBOTIC ARMS

Up to this point, we have studied machines with very few degrees of freedom. With other types of machines it is usually desirable to drive the entire machine from a single motor or a single source of power; thus they are designed to have mobility of $m = 1$. In keeping with the idea of flexibility of application, however, robots must have more. We know that, if a robot is to reach an arbitrary point in three-dimensional space, it must have mobility of at least $m = 3$ to adjust to the proper values of x, y, and z. In addition, if the robot is to be able to manipulate a tool or object it carries into an arbitrary orientation, once reaching the desired position, an additional 3 degrees of freedom will be required, giving a desired mobility of $m = 6$.

Along with the desire for six or more degrees of freedom, we also strive for simplicity, not only for reasons of good design and reliability, but also to minimize the computing burden in the control of the robot. Therefore, we often find that only three of the freedoms are designed into the robot arm itself and that another two or three may be included in the wrist. Because the tool, or *end effecter,* usually varies with the task to be performed, this portion of the total robot may be made interchangeable; the same basic robot arm might carry any of several different end effecters specially designed for particular tasks.

Based on the first three freedoms of the arm, one common arrangement of joints for a manipulator is an *RRR* linkage, also called an *articulated* configuration. Two examples are the Cincinnati Milacron T³ robot shown in Fig. 13.1 and the Intelledex articulated robot shown in Fig. 13.2.

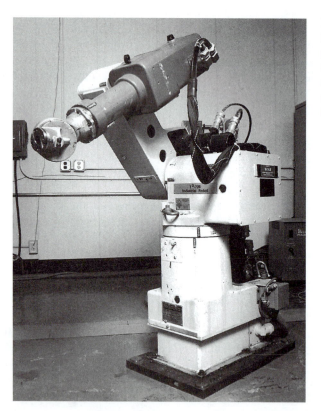

Figure 13.1 Cincinnati Milacron T³ articulated six-axis robot. This model T26 has a load capacity of 14 lb, horizontal reach of 40 in from the vertical centerline, and position repeatability of 0.004 in.

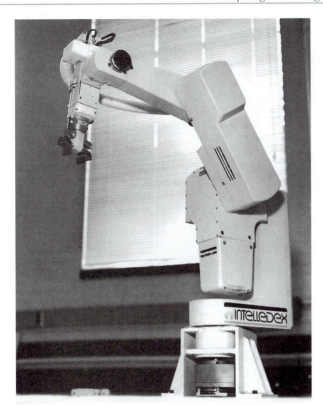

Figure 13.2 Intelledex articulated robot.

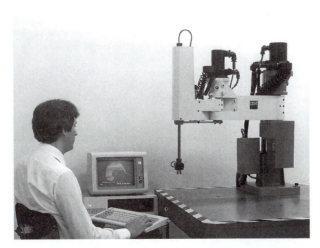

Figure 13.3 IBM model 7525 Selective Compliant Articulated Robot for Assembly (SCARA). (Courtesy of IBM Corp., Rochester, MN.)

The Selective Compliant Articulated Robot for Assembly, also called the SCARA robot, is a more recent but popular configuration based on the *RRP* linkage. An example is shown in Fig. 13.3. Although there are other *RRP* robot configurations, notice that the SCARA robot has all three joint variable axes parallel, thus restricting its freedom of movement but particularly suiting it to assembly operations.

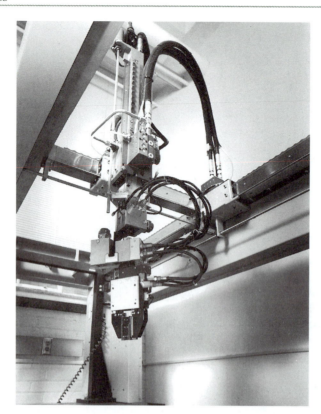

Figure 13.4 IBM model 7650 gantry robot.

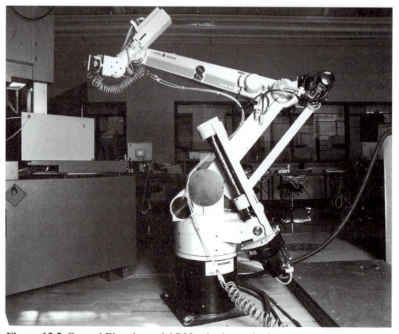

Figure 13.5 General Electric model P80 robotic manipulator.

A manipulator based on the *PPP* chain is pictured in Fig. 13.4. It has three mutually perpendicular prismatic joints and, therefore, is called a Cartesian configuration or a *gantry robot*. The kinematic analysis of and programming for this style is particularly simple because of the perpendicularity of the three joint axes. It has applications in table-top assembly and in transfer of cargo or material.

We notice that the arms depicted up to now are simple, series-connected kinematic chains. This is desirable because it reduces their kinematic complexity and eases their design and programming. However, it is not always true; there are robotic arms that include closed kinematic loops in their basic topology. One example is the robot of Fig. 13.5.

13.3 FORWARD KINEMATICS

The first kinematic analysis problem to be addressed for a robot is finding the position of the tool or end effector once we are given the geometry of each component and the positions of the several actuators controlling the degrees of freedom. This can easily be done by the methods presented in Chapter 12 on spatial mechanisms. First we should recognize three important characteristics of most robotics problems that are different from the spatial mechanisms of the previous chapter:

1. Knowledge of the position of a single point at the tip of the tool is often not enough. In order to have a complete knowledge of the tip of the tool, we must know the location and orientation of a coordinate system attached to the tool. This implies that vector methods are not sufficient and that the matrix methods of Section 12.7 are better suited.
2. Perhaps because of the previous observation, many robot manufacturers have found the Denavit–Hartenberg parameters (Section 12.6) for their particular robot designs. Thus, use of the transformation matrix approach is straightforward.
3. All joint variables of a serially connected chain are independent and are degrees of freedom. Thus, in the forward kinematics problem being discussed now, all joint variables are actuator variables and have given values; there are no loop-closure conditions and no unknown joint variable values to be found.

The conclusion implied by these three observations combined is that finding the position of the end effector for given positions of the actuators is a straightforward application of Eq. (12.16). The absolute position of any chosen point R_n in the tool coordinate system attached to link n is given by

$$R_1 = T_{1,n} R_n \tag{13.1}$$

where

$$T_{1,n} = T_{1,2} T_{2,3} \cdots T_{n-1,n} \tag{13.2}$$

and each T matrix is given by Eq. (12.12) once the Denavit–Hartenberg parameters (including the actuator positions) are known.

Of course, if the robot has $n = 6$ links, then symbolic multiplication of these matrices may become lengthy, unless some of the shape (constant) parameters are conveniently set to nice values. However, remembering that the robot will have computing capability on-board, numerical evaluation of Eq. (13.2) is no great challenge for a particular set of actuator values. Still, because of speed requirements for real-time computer control, it is desirable to simplify these expressions as much as possible before programming. Toward this goal, most robot manufacturers have chosen more simplified designs (having "nice" shape parameters) to simplify these expressions. Many have also worked out the final expressions for their particular robots and make these available in their technical documentation.

EXAMPLE 13.1

For the Microbot model TCM five-axis robot shown in Fig. 13.6, find the transformation matrix T_{16} relating the position of the tool coordinate system to the ground coordinate system when the joint actuators are set to the values $\phi_1 = 30°$, $\phi_2 = 60°$, $\phi_3 = -30°$, $\phi_4 = \phi_5 = 0°$. Also find the absolute position of the tool point which has coordinates $x_6 = y_6 = 0$, $z_6 = 2.5$ in.

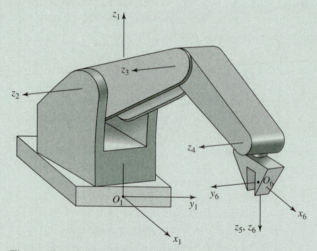

Figure 13.6 The Microbot model TCM five-axis robot.

SOLUTION

The coordinate system axes are shown on the figure. Notice that the $x_6 y_6 z_6$ coordinate system is not based on a joint axis, but was chosen arbitrarily to form a convenient tool coordinate system. Based on these axes and the data in the manufacturer's documentation, we find the Denavit–Hartenberg parameters to be as follows:

$a_{12} = 0$,	$\alpha_{12} = 90°$,	$\theta_{12} = \phi_1 = 30°$,	$s_{12} = 7.68$ in
$a_{23} = 7.00$ in,	$\alpha_{23} = 0°$,	$\theta_{23} = \phi_2 = 60°$,	$s_{23} = 0$
$a_{34} = 7.00$ in,	$\alpha_{34} = 0°$,	$\theta_{34} = \phi_3 = -30°$,	$s_{34} = 0$
$a_{45} = 0$,	$\alpha_{45} = 90°$,	$\theta_{45} = \phi_4 = 0°$,	$s_{45} = 0$
$a_{56} = 0$,	$\alpha_{56} = 0°$,	$\theta_{56} = \phi_5 = 0°$,	$s_{56} = 3.80$ in

Using Eqs. (12.12) and (12.15) we can now build the following matrices and matrix products:

$$T_{12} = \begin{bmatrix} 0.866 & 0 & 0.500 & 0 \\ 0.500 & 0 & -0.866 & 0 \\ 0 & 1 & 0 & 7.68 \\ 0 & 0 & 0 & 1 \end{bmatrix}$$

$$T_{23} = \begin{bmatrix} 0.500 & -0.866 & 0 & 3.50 \\ 0.866 & 0.500 & 0 & 6.06 \\ 0 & 0 & 1 & 0 \\ 0 & 0 & 0 & 1 \end{bmatrix}, \qquad T_{13} = \begin{bmatrix} 0.433 & -0.750 & 0.500 & 3.03 \\ 0.250 & -0.433 & -0.866 & 1.75 \\ 0.866 & 0.500 & 0 & 13.74 \\ 0 & 0 & 0 & 1 \end{bmatrix}$$

$$T_{34} = \begin{bmatrix} 0.866 & 0.500 & 0 & 6.06 \\ -0.500 & 0.866 & 0 & -3.50 \\ 0 & 0 & 1 & 0 \\ 0 & 0 & 0 & 1 \end{bmatrix}, \qquad T_{14} = \begin{bmatrix} 0.750 & -0.433 & 0.500 & 8.28 \\ 0.433 & -0.250 & -0.866 & 4.78 \\ 0.500 & 0.866 & 0 & 17.24 \\ 0 & 0 & 0 & 1 \end{bmatrix}$$

$$T_{45} = \begin{bmatrix} 1 & 0 & 0 & 0 \\ 0 & 0 & -1 & 0 \\ 0 & 1 & 0 & 0 \\ 0 & 0 & 0 & 1 \end{bmatrix}, \qquad T_{15} = \begin{bmatrix} 0.750 & 0.500 & 0.433 & 8.28 \\ 0.433 & -0.866 & 0.250 & 4.78 \\ 0.500 & 0 & -0.866 & 17.24 \\ 0 & 0 & 0 & 1 \end{bmatrix}$$

$$T_{56} = \begin{bmatrix} 1 & 0 & 0 & 0 \\ 0 & 1 & 0 & 0 \\ 0 & 0 & 1 & 3.80 \\ 0 & 0 & 0 & 1 \end{bmatrix}, \qquad T_{16} = \begin{bmatrix} 0.750 & 0.500 & 0.433 & 9.93 \\ 0.433 & -0.866 & 0.250 & 5.73 \\ 0.500 & 0 & -0.866 & 13.95 \\ 0 & 0 & 0 & 1 \end{bmatrix}$$

Ans.

The absolute position of the specified tool point is now found from Eq. (13.1):

$$R_1 = \begin{bmatrix} 0.750 & 0.500 & 0.433 & 9.93 \\ 0.433 & -0.866 & 0.250 & 5.73 \\ 0.500 & 0 & -0.866 & 13.95 \\ 0 & 0 & 0 & 1 \end{bmatrix} \begin{bmatrix} 0 \\ 0 \\ 2.5 \\ 1 \end{bmatrix} = \begin{bmatrix} 11.00 \text{ in} \\ 6.36 \text{ in} \\ 11.78 \text{ in} \\ 1 \end{bmatrix} \qquad \textit{Ans.}$$

The direct computation of velocities and accelerations of arbitrary points on the links of the robot can also be accomplished easily by the matrix methods of Section 12.8. In particular, Eqs. (12.27), (12.28), (12.30), and (12.31) are of direct use once the actuator velocities and accelerations are known. A continuation of the previous example should make this clear.

EXAMPLE 13.2

For the Microbot model TCM robot of Fig. 13.6 and in the position described in Example 13.1, find the instantaneous velocity and acceleration of the same tool point $x_6 = y_6 = 0$, $z_6 = 2.5$ in if the actuators are given constant velocities of $\dot{\phi}_1 = 0.20$ rad/s, $\dot{\phi}_4 = -0.35$ rad/s, and $\dot{\phi}_2 = \dot{\phi}_3 = \dot{\phi}_5 = 0$.

SOLUTION

Using Eq. (12.25) and the results from Example 13.1, we find

$$D_1 = \begin{bmatrix} 0 & -1 & 0 & 0 \\ 1 & 0 & 0 & 0 \\ 0 & 0 & 0 & 0 \\ 0 & 0 & 0 & 0 \end{bmatrix}, \quad D_4 = \begin{bmatrix} 0 & 0 & -0.866 & 14.93 \\ 0 & 0 & -0.500 & 8.62 \\ 0.866 & 0.500 & 0 & -9.56 \\ 0 & 0 & 0 & 0 \end{bmatrix}$$

and then, from Eq. (12.27), we obtain

$$\omega_1 = 0$$

$$\omega_2 = \begin{bmatrix} 0 & -0.200 & 0 & 0 \\ 0.200 & 0 & 0 & 0 \\ 0 & 0 & 0 & 0 \\ 0 & 0 & 0 & 0 \end{bmatrix}$$

$$\omega_3 = \omega_4 = \omega_2$$

$$\omega_5 = \begin{bmatrix} 0 & -0.200 & 0.303 & -5.225 \\ 0.200 & 0 & 0.175 & -3.017 \\ -0.303 & -0.175 & 0 & 3.346 \\ 0 & 0 & 0 & 0 \end{bmatrix}$$

$$\omega_6 = \omega_5$$

Now, using Eq. (12.28), we find the velocity of the tool point,

$$\dot{R}_6 = \omega_6 T_{16} R_6 = \begin{bmatrix} -2.93 \\ 1.24 \\ -1.10 \\ 0 \end{bmatrix} \text{ in/s} \qquad\qquad Ans.$$

which in vector notation is

$$\mathbf{V} = -2.93\hat{\mathbf{i}}_1 + 1.24\hat{\mathbf{j}}_1 - 1.10\hat{\mathbf{k}}_1 \text{ in/s} \qquad\qquad Ans.$$

The accelerations are done in similar fashion. From Eq. (12.30), we obtain

$$\alpha_1 = \alpha_2 = \alpha_3 = \alpha_4 = 0$$

$$\alpha_5 = \begin{bmatrix} 0 & 0 & -0.035 & 0.603 \\ 0 & 0 & 0.152 & -1.045 \\ 0.035 & -0.152 & 0 & 0 \\ 0 & 0 & 0 & 0 \end{bmatrix}$$

$$\alpha_6 = \alpha_5$$

Finally, using Eq. (12.31), we find the acceleration of the tool point.

$$\ddot{R}_6 = (\alpha_6 + \omega_6\omega_6)T_{16}R_6 = \begin{bmatrix} -0.390 \\ -0.033 \\ 0.088 \\ 0 \end{bmatrix} \text{in/s}^2 \qquad Ans.$$

which in vector notation is

$$\mathbf{A} = -0.390\hat{\mathbf{i}}_1 - 0.033\hat{\mathbf{j}}_1 + 0.088\hat{\mathbf{k}}_1 \text{ in/s}^2 \qquad Ans.$$

13.4 INVERSE POSITION ANALYSIS

The previous section shows the basic procedure for finding the positions, velocities, and accelerations of arbitrary points on the moving members of a robot once the positions, velocities, and accelerations of the joint actuators are known. The procedures are straightforward and simply applied if, as is usual, the robot is a simply connected chain and has no closed loops. In this simple case every joint variable is an independent degree of freedom, and no loop-closure equation needs to be solved. This avoids a major complication.

However, even though the robot itself might have no closed loops, the manner in which a problem or question is presented can sometimes lead to the same complications. Consider, for example, an open-loop robot that we desire to guide to follow a specified path; we will be given the desired path, but we will not know the actuator values (or, more precisely, the time functions) required to achieve this. This problem cannot be solved by the methods of the previous section. When the joint variables (or their derivatives) are the unknowns of the problem rather than given information, the problem is called an *inverse kinematics* problem. In general, this is a more complicated problem to solve, and it does arise repeatedly in robot applications.

When the inverse kinematics problem arises, we must of course find a set of equations which describe the given situation and which can be solved. In robotics this usually means that there will be a loop-closure equation, not necessarily defined by closed loops within the robot topology, but perhaps defined by the manner in which the problem is posed. For example, if we are told that the end effecter of the robot is to travel along a certain path with certain timing, then, in a sense, we are being told the values required for the transformation $T_{1,n}$ as known functions of time. Then Eq. (13.2) becomes a set of required loop-closure conditions which must be satisfied and which must be solved for the joint actuator values or functions of time. An example will make the procedure clear.

EXAMPLE 13.3

The Microbot robot of Fig. 13.6 and described in Example 13.1 is to guided along a path for which the origin of the end effecter O_6 follows the straight line given by

$$\mathbf{R}_{O_6}(t) = (4.0 + 1.6t)\hat{\mathbf{i}}_1 + (3.0 + 1.2t)\hat{\mathbf{j}}_1 + 2.0\hat{\mathbf{k}}_1 \text{ in}$$

with t varying from 0.0 to 5.0 s; the orientation of the end effector is to remain constant with $\hat{\imath}_6 = \hat{\mathbf{k}}_1$ (vertical) and $\hat{\mathbf{k}}_6$ radially outward from the base of the robot. Find expressions for how each of the joint actuators must be driven, as functions of time, to achieve this motion.

SOLUTION

From the problem requirements stated, we can express the required time history of the end effector coordinate system by the transformation matrix

$$T_{16}(t) = \begin{bmatrix} 0 & 0.600 & 0.800 & 4.0 + 1.6t \\ 0 & -0.800 & 0.600 & 3.0 + 1.2t \\ 1 & 0 & 0 & 2.0 \\ 0 & 0 & 0 & 1 \end{bmatrix} \tag{1}$$

We can also multiply out the matrix description for T_{16} from Eq. (13.2); this gives

$$T_{16} = \begin{bmatrix} \cos\phi_1 \cos\beta \cos\phi_5 & -\cos\phi_1 \cos\beta \sin\phi_5 & \cos\phi_1 \sin\beta & 7\cos\phi_1 \cos\phi_2 \\ \quad - \sin\phi_1 \sin\phi_5 & \quad + \sin\phi_1 \cos\phi_5 & & \quad + 7\cos\phi_1 \cos(\phi_2 + \phi_3) \\ & & & \quad + 3.8\cos\phi_1 \sin\beta \\[2mm] \sin\phi_1 \cos\beta \cos\phi_5 & -\sin\phi_1 \cos\beta \sin\phi_5 & \sin\phi_1 \sin\beta & 7\sin\phi_1 \cos\phi_2 \\ \quad - \cos\phi_1 \sin\phi_5 & \quad - \cos\phi_1 \cos\phi_5 & & \quad + 7\sin\phi_1 \sin(\phi_2 + \phi_3) \\ & & & \quad + 3.8\sin\phi_1 \sin\beta \\[2mm] \sin\beta \cos\phi_5 & -\sin\beta \sin\phi_5 & -\cos\beta & 7\sin\phi_2 \\ & & & \quad + 7\sin(\phi_2 + \phi_3) \\ & & & \quad - 3.8\cos\beta \\[2mm] 0 & 0 & 0 & 1 \end{bmatrix} \tag{2}$$

where in order to save space we have adopted the definition

$$\beta = \phi_2 + \phi_3 + \phi_4 \tag{3}$$

Now, in order to achieve the problem requirements, we must find expressions for the ϕ_i that will ensure that the elements of Eq. (2) are equal to those of Eq. (1) for all values of time; this is our equivalent loop-closure condition. Equating the ratios of the elements of the second row, third column with the elements of the first row, third column of Eqs. (2) and (1) gives

$$\tan\phi_1 = \frac{0.600}{0.800} = 0.750$$

$$\phi_1 = \tan^{-1} 0.750 = 36.87° \qquad\qquad Ans.$$

Similarly, from the ratios of the elements of the third row, second column with the elements of the third row, first column we find

$$\tan \phi_5 = 0$$

$$\phi_5 = 0° \qquad \qquad Ans.$$

Also, from the third row, third column elements, we obtain

$$\cos \beta = 0$$

$$\beta = \phi_2 + \phi_3 + \phi_4 = 90° \qquad (4)$$

Turning now to the elements of the fourth column, and simplifying according to already known results, we obtain two more independent equations:

$$7 \cos \phi_2 + 7 \cos(\phi_2 + \phi_3) = 1.2 + 2t \qquad (5)$$

$$7 \sin \phi_2 + 7 \sin(\phi_2 + \phi_3) = -5.68 \qquad (6)$$

Squaring and adding these, we get

$$98 + 98 \cos \phi_3 = 33.7024 + 4.8t + 4t^2$$

which gives a solution for ϕ_3

$$\phi_3 = \cos^{-1}(-0.65610 + 0.04898t + 0.040816t^2) \qquad Ans.$$

Because this form admits to multiple values, we will take the solution with $\phi_3 \le 0$ so that the robot elbow is kept above the path.

Knowing the solution for ϕ_3, we can now rewrite Eqs. (5) and (6) as follows:

$$\sin \phi_2 = \frac{-5.68}{14} - \frac{(1.2 + 2t)\sin \phi_3}{14(1 + \cos \phi_3)} \qquad (9)$$

$$\cos \phi_2 = \frac{(1.2 + 2t)}{14} - \frac{5.68 \sin \phi_3}{14(1 + \cos \phi_3)} \qquad (10)$$

Because ϕ_3 now has a known solution, either of these equations can be used to find ϕ_2; actually, both should be checked to ensure that the proper quadrant is found. These two equations are considered solutions even though they are not explicit, closed-form functions of time. Similarly, from Eq. (4), we can now find ϕ_4:

$$\phi_4 = 90° - \phi_2 - \phi_3 \qquad Ans. \quad (11)$$

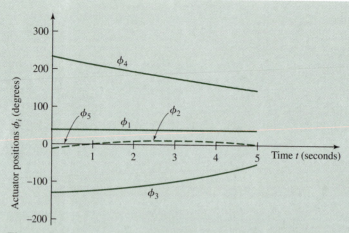

Figure 13.7 Example 13.3; graphs of inverse position solutions.

Graphs of the five joint variables versus time are shown in Fig. 13.7. Notice that through the path being followed, the joint variables follow very smooth but nonlinear curves. Thus it appears that the joint variable velocities and accelerations will not be zero, but should fall within very reasonable limits.

13.5 INVERSE VELOCITY AND ACCELERATION ANALYSIS

The above example presented the general approach for inverse position analysis. However, at least for control purposes, it may also be important to find the joint variable velocities and accelerations. One way to do this would be to directly differentiate the joint variable solution equations with respect to time. However, another, more general approach is to differentiate our loop-closure conditions.

As we saw in the above example, these were given by Eq. (13.2) as

$$T_{1,2}T_{2,3}\cdots T_{n-1,n} = T_{1,n}(t)$$

where, for the inverse kinematics problem, the right-hand side of this equation are known functions of time given by the problem specifications, the trajectory, or path to be followed. Now, using Eq. (12.24) and the D_i differentiation operator matrices of Eq. (12.25b), we can differentiate the left-hand side of the equation, and by straight differentiation with respect to time we can do the right-hand side. Differentiating both sides, we obtain the equation

$$\sum D_i \dot{\phi}_i T_{1,n} = \frac{dT_{1,n}(t)}{dt} \qquad (13.3)$$

If we now define a new (velocity) matrix Ω as

$$\Omega = \frac{dT_{1,n}(t)}{dt} T_{1,n}^{-1} \tag{13.4}$$

then, multiplying both sides of Eq. (13.3) by $T_{1,n}^{-1}$, we get

$$\sum D_i \dot{\phi}_i = \Omega \tag{13.5}$$

which, as we will see, can be solved for the joint actuator velocities $\dot{\phi}_i$.

Let us now look more carefully at the nature of Eq. (13.5). Each of the matrices has four rows and four columns. Yet, because they represent velocities in a single chain in three dimensions, there should only be a maximum of six independent equations, three for translations and three for rotations. Looking at the definitions of the Q_i differentiation operator matrices in Eq. (12.22) and reviewing the form of the D_i matrices seen in Examples 12.5 and 13.2, we can discover which of the elements of these matrices contain the six independent equations. Although the proof is left as an exercise, it can be shown that the D_i matrices always have a bottom row of zeroes and the upper-left three rows and three columns are always antisymmetric; that is, they are always of the form

$$D_i = \begin{bmatrix} 0 & -c & +b & d \\ +c & 0 & -a & e \\ -b & +a & 0 & f \\ 0 & 0 & 0 & 0 \end{bmatrix} \tag{13.6}$$

As expected, there are only six independent values $(a, b, c, d, e,$ and $f)$ in each of these matrices. It will now be convenient to rearrange these into a single column and to give this column a new symbol:

$$\{D_i\} = \begin{bmatrix} a \\ b \\ c \\ d \\ e \\ f \end{bmatrix} \tag{13.7}$$

where the braces indicate that the elements of the 4×4 matrix have been rearranged into a 6×1 column matrix. There is actually much more mathematical and physical significance to this 6×1 column vector than is explained here; they form the *Plücker coordinates* or *line coordinates* of the motion axis of the joint variable's freedom for motion.[1] The top three elements form the three-dimensional unit vector for the motion axis, while the bottom three uniquely locate that axis in the absolute coordinate system.

If we now reorganize all matrices in Eq. (13.5) into this column vector form, it becomes

$$\sum \{D_i\} \dot{\phi}_i = \{\Omega\} \tag{13.8}$$

which is a set of six simultaneous algebraic equations relating the joint variable velocities. If we write the equations out explicitly, they appear as

$$a_1\dot{\phi}_1 + a_2\dot{\phi}_2 + \cdots + a_n\dot{\phi}_n = a$$
$$b_1\dot{\phi}_1 + b_2\dot{\phi}_2 + \cdots + b_n\dot{\phi}_n = b$$
$$c_1\dot{\phi}_1 + c_2\dot{\phi}_2 + \cdots + c_n\dot{\phi}_n = c$$
$$d_1\dot{\phi}_1 + d_2\dot{\phi}_2 + \cdots + d_n\dot{\phi}_n = d$$
$$e_1\dot{\phi}_1 + e_2\dot{\phi}_2 + \cdots + e_n\dot{\phi}_n = e$$
$$f_1\dot{\phi}_1 + f_2\dot{\phi}_2 + \cdots + f_n\dot{\phi}_n = f$$

where a, b, and so on, are the elements of $\{\Omega\}$ taken in the order consistent with Eqs. (13.6) and (13.7). Regrouping the coefficients of this set of equations into matrix form, we get

$$J = \begin{bmatrix} a_1 & a_2 & \ldots & a_n \\ b_1 & b_2 & \ldots & b_n \\ c_1 & c_2 & \ldots & c_n \\ d_1 & d_2 & \ldots & d_n \\ e_1 & e_2 & \ldots & e_n \\ f_1 & f_2 & \ldots & f_n \end{bmatrix} \tag{13.9}$$

which is called the *Jacobian* of the system. Using this matrix, Eq. (13.8) now becomes

$$J\{\dot{\phi}\} = \{\Omega\} \tag{13.10}$$

where $\{\dot{\phi}\}$ is a column of the unknown joint actuator velocities $\dot{\phi}_i$ taken in order. This set of equations specifies the relations that must exist between the joint variable velocities in order to trace the specified path according to the given functions of time.

Assuming that the position analysis has already been completed, the Ω and D_i matrices are all known functions of time, and this set of equations can now be solved for the joint variable velocities as functions of time. If the robot has six independent joint variables, then the solution can be found by inverting the Jacobian matrix. In this case, the solution is

$$\{\dot{\phi}\} = J^{-1}\{\Omega\} \tag{13.11}$$

If the Jacobian is not a square matrix, then two cases are possible: (1) When there are less than six joint variables, other solution methods, such as Gaussian elimination, must be used. (2) When there are more than six joint variables, the problem has no unique solution; further equations, perhaps even arbitrary choices, must be supplied. If the matrix is singular, or if the equations are inconsistent, this shows that there is no finite unique solution for velocities; the problem specifications are not realistic for this robot to do the task specified. If none of these problems arises, the joint variable velocities become known functions of time.

Joint variable accelerations can be found by completely analogous methods. Taking the next time derivative of Eq. (13.5), we see that

$$\sum D_i\ddot{\phi}_i = \frac{d\Omega}{dt} - \sum(\omega_i D_i - D_i\omega_i)\dot{\phi}_i = A \tag{13.12}$$

where this equation defines the new matrix A and the ω_i matrices are defined as in Eq. (12.27). Extracting the same six independent equations, in the same order as before, we get

$$\sum\{D_i\}\ddot{\phi}_i = \{A\} \tag{13.13}$$

Now we notice that the coefficients of the unknown $\ddot{\phi}_i$ joint variable accelerations are identical to those of Eq. (13.8); thus we can use the same Jacobian matrix defined above. The solution to these equations now becomes

$$\{\ddot{\phi}\} = J^{-1}\{A\} \tag{13.14}$$

where $\{\ddot{\phi}\}$ is a column of the unknown joint variable accelerations $\ddot{\phi}_i$ taken in order. This set of equations specifies the relations which must exist between the joint variable accelerations required to trace the specified path according to the given functions of time.

As with direct kinematics problems, the velocity and acceleration of a tool point on one of the moving links can be found from Eqs. (12.27) through (12.31).

EXAMPLE 13.4

Continue with Example 13.3 and find the velocities required at the actuators as functions of time to achieve the motion described.

SOLUTION

Using Eq. (12.25) and the solutions of the position analysis from the above example, we can find the D_i matrices. These are

$$D_1 = \begin{bmatrix} 0 & -1 & 0 & 0 \\ 1 & 0 & 0 & 0 \\ 0 & 0 & 0 & 0 \\ 0 & 0 & 0 & 0 \end{bmatrix}, \quad D_2 = \begin{bmatrix} 0 & 0 & -0.8 & 6.144 \\ 0 & 0 & -0.6 & 4.608 \\ 0.8 & 0.6 & 0 & 0 \\ 0 & 0 & 0 & 0 \end{bmatrix}$$

$$D_3 = \begin{bmatrix} 0 & 0 & -0.8 & 5.6\sin\phi_2 + 6.144 \\ 0 & 0 & -0.6 & 4.2\sin\phi_2 + 4.608 \\ 0.8 & 0.6 & 0 & -7\cos\phi_2 \\ 0 & 0 & 0 & 0 \end{bmatrix}, \quad D_4 = \begin{bmatrix} 0 & 0 & -0.8 & 1.6 \\ 0 & 0 & -0.6 & 1.2 \\ 0.8 & 0.6 & 0 & -(2t+1.2) \\ 0 & 0 & 0 & 0 \end{bmatrix}$$

$$D_5 = \begin{bmatrix} 0 & 0 & 0.6 & -1.2 \\ 0 & 0 & -0.8 & 1.6 \\ -0.6 & 0.8 & 0 & 0 \\ 0 & 0 & 0 & 0 \end{bmatrix} \tag{12}$$

Using Eq. (13.4) and the trajectory specified in Example 13.3, we can also find the Ω matrix.

$$\Omega = \begin{bmatrix} 0 & 0 & 0 & 1.6 \\ 0 & 0 & 0 & 1.2 \\ 0 & 0 & 0 & 0 \\ 0 & 0 & 0 & 0 \end{bmatrix} \tag{13}$$

From these and Eq. (13.8) we can now construct the set of equations relating the joint variable velocities:

$$
\begin{bmatrix}
0 & 0.6 & 0.6 & 0.6 & 0.8 \\
0 & -0.8 & -0.8 & -0.8 & 0.6 \\
1 & 0 & 0 & 0 & 0 \\
0 & 6.144 & 5.6\sin\phi_2 + 6.144 & 1.6 & -1.2 \\
0 & 4.608 & 4.2\sin\phi_2 + 4.608 & 1.2 & 1.6 \\
0 & 0 & -7\cos\phi_2 & -(2t+1.2) & 0
\end{bmatrix}
\begin{bmatrix}
\dot{\phi}_1 \\
\dot{\phi}_2 \\
\dot{\phi}_3 \\
\dot{\phi}_4 \\
\dot{\phi}_5
\end{bmatrix}
=
\begin{bmatrix}
0 \\
0 \\
0 \\
1.6 \\
1.2 \\
0
\end{bmatrix}
\tag{14}
$$

Here we can see the Jacobian matrix for the robot following the specified trajectory. However, it is not square and cannot be inverted. This happened because the robot has only five degrees of freedom and, therefore, is not capable of arbitrary spatial motions; there are six equations and only five unknowns, the five actuator velocities.

Fortunately, however, the trajectory specified is within the capability of the robot. There is a set of solutions for the five unknown actuator velocities which fits all six equations; these solutions are:

$$\dot{\phi}_1 = 0 \qquad\qquad Ans.$$

$$\dot{\phi}_2 = \frac{(2t+1.2) - 7\cos\phi_2}{\Delta} \text{ rad/s} \qquad\qquad Ans.$$

$$\dot{\phi}_3 = \frac{-(2t+1.2)}{\Delta} \text{ rad/s} \qquad\qquad Ans.$$

$$\dot{\phi}_4 = \frac{7\cos\phi_2}{\Delta} \text{ rad/s} \qquad\qquad Ans.$$

$$\dot{\phi}_5 = 0 \qquad\qquad Ans.$$

where Δ is defined as

$$\Delta = -3.5(2t+1.2)\sin\phi_2 - 19.9\cos\phi_2 \tag{15}$$

As with the position solutions, we agree that the velocity expressions shown are solutions since the parameter ϕ_2 is already known from the position analysis.

13.6   ROBOT ACTUATOR FORCE ANALYSIS

Of course, the purpose of moving the robot along the planned trajectory is to perform some useful function, and this will almost certainly require the expenditure of work or power. The source of this work or power must come from the actuators at the joint variables. Therefore it would be very helpful to find a means of knowing what size forces and/or torques must be exerted by the actuators to perform a given task. Conversely, it would be useful to know how much force can be produced at the tool for a given force or torque capacity at the actuators.

If the force or torque capacity of the actuators is less than those demanded by the task being attempted, the robot is not capable of performing that task. It will probably not fail catastrophically, but it will deviate from the desired trajectory and perform a different motion than that desired. The purpose of this section is to find a means of evaluating the forces required of the robot actuators to perform a given trajectory with a given task loading, so that overloading of the actuators can be predicted and avoided.

The entire subject of force analysis in mechanical systems will be covered in depth in Chapters 14 through 16. Therefore, the treatment here will be more limited by simplifying assumptions. It will be suited specifically to the study of robots performing specified tasks with specified loads at slow speeds with no friction or other losses. If these assumptions do not apply, the more extensive treatments of the later chapters should be employed.

The interaction of the robot with the task being performed will produce a set of forces and torques at the end effecter or tool. Let us assume that these are known functions of time and are grouped into a given six-element column $\{F\}$ in the following order:

$$\{F\} = \begin{bmatrix} M^{x_1} \\ M^{y_1} \\ M^{z_1} \\ F^{x_1} \\ F^{y_1} \\ F^{z_1} \end{bmatrix} \tag{13.15}$$

where the first three elements are the components of the tool torques about the x_1, y_1, z_1 axes and the next three are the components of force along the same axes.

The source of the energy for these needs will be a set of unknown forces or torques τ_i, each one applied by an actuator at the ith joint variable location. We will arrange these into another column matrix τ in the order of numbering of the joint variables:

$$\tau = \begin{bmatrix} \tau_1 \\ \tau_2 \\ \dots \\ \tau_n \end{bmatrix} \tag{13.16}$$

Next we define a small (absolute) displacement of the tool against the loads by another six-element column matrix $\{\delta R\}$, arranged in the same order as the load vector $\{F\}$:

$$\{\delta R\} = \begin{bmatrix} \delta\theta^{x_1} \\ \delta\theta^{y_2} \\ \delta\theta^{z_1} \\ \delta R^{x_1} \\ \delta R^{y_1} \\ \delta R^{z_1} \end{bmatrix} \tag{13.17}$$

and yet another column matrix $\delta\phi$ of displacements of the joint variables $\delta\phi_i$ during this small displacement.

Now, if the actual robot were to undergo this small movement $\{\delta R\}$ against the loads $\{F\}$, the work would have to come from the torques τ of the actuators acting through their small displacements $\delta\phi$. Therefore, because work input must equal work output in the absence of other losses, we have

$$\{F\}^t\{\delta R\} = \tau^t\delta\phi \qquad (a)$$

where the superscript ^{t} signifies the transpose of the matrix. But we also know from Eq. (13.10) by integrating over a short time interval δt that the task displacement $\{\delta R\}$ is related to the joint actuator displacements $\{\delta\phi\}$ by

$$\{\delta R\} = J\delta\phi \qquad (13.18)$$

Putting this relationship into Eq. (a) and rearranging, we get

$$[\tau^t - \{F\}^t J]\delta\phi = 0 \qquad (b)$$

Because we assume that the small displacement $\delta\phi$ is arbitrary and not zero, we can solve for the actuator torques τ by setting the leading factor to zero. This shows that

$$\tau = J^t\{F\} \qquad (13.19)$$

which is the relationship we have been seeking. Under the simplifying conditions assumed above, we have found not only that the Jacobian matrix relates the actuator motions to the absolute tool motions, Eqs. (13.11), (13.14), and (13.18), but also that its transpose relates the actuator forces and tool forces, Eq. (13.19).

EXAMPLE 13.5

Continue with Example 13.4 and assume that the end effecter is working against a time-varying force loading of $10\hat{\mathbf{i}}_1 + 5t\hat{\mathbf{k}}_1$ lb and a constant torque loading of $25\hat{\mathbf{k}}_1$ in·lb. Find the torques required at the actuators as functions of time to achieve the motion described.

SOLUTION

From the data given, the tool force vector of Eq. (13.15) is found to be

$$\{F\} = \begin{bmatrix} 0 \\ 0 \\ 25 \\ 10 \\ 0 \\ 5t \end{bmatrix} \qquad (16)$$

Because the Jacobian matrix was already found in Example 13.4, Eq. (14), we can now use Eq. (13.19) to find the actuator torques directly. We find them to be

$$\tau_1 = 25.0 \text{ in·lb} \qquad\qquad\qquad \textit{Ans.}$$

$$\tau_2 = 61.4 \text{ in·lb} \qquad\qquad\qquad \textit{Ans.}$$

$$\tau_3 = -35t \cos \phi_2 + 56.0 \sin \phi_2 + 61.4 \text{ in·lb} \qquad\qquad\qquad \textit{Ans.}$$

$$\tau_4 = -10t^2 - 6t + 16.0 \text{ in·lb} \qquad\qquad\qquad \textit{Ans.}$$

$$\tau_5 = -12.0 \text{ in·lb} \qquad\qquad\qquad \textit{Ans.}$$

NOTE

[1.]   For further reading, an excellent reference is K. H. Hunt, *Kinematic Geometry of Mechanisms,* Oxford University Press, 1978, especially Chapter 11.

PROBLEMS

13.1   For the SCARA robot shown in the figure, find the transformation matrix T_{15} relating the position of the tool coordinate system to the ground coordinate system when the joint actuators are set to the values $\phi_1 = 30°$, $\phi_2 = -60°$, $\phi_3 = 2$ in, $\phi_4 = 0$. Also find the absolute position of the tool point which has coordinates $x_5 = y_5 = 0$, $z_5 = 1.5$ in.

13.2   Repeat Problem 13.1 using arbitrary (symbolic) values for the joint variables.

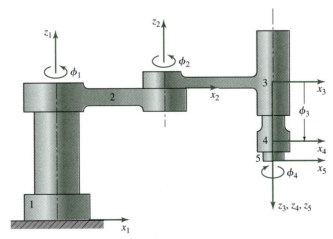

Figure P13.1, P13.2, P13.5, and P13.7 $a_{12} = a_{23} = 10$ in, $a_{34} = a_{45} = 0$, $\alpha_{12} = \alpha_{34} = \alpha_{45} = 0$, $\alpha_{23} = 180°$, $\theta_{12} = \phi_1$, $\theta_{23} = \phi_2$, $\theta_{34} = 0$, $\theta_{45} = \phi_4$, $s_{12} = 12$ in, $s_{23} = 0$, $s_{34} = \phi_3$, $s_{45} = 2$ in.

13.3 For the gantry robot shown in the figure, find the transformation matrix T_{15} relating the position of the tool coordinate system to the ground coordinate system when the joint actuators are set to the values $\phi_1 = 450$ mm, $\phi_2 = 180$ mm, $\phi_3 = 50$ mm, $\phi_4 = 0$. Also find the absolute position of the tool point which has coordinates $x_5 = y_5 = 0$, $z_5 = 45$ mm.

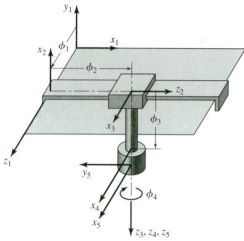

Figure P13.3, P13.4, P13.6, and P13.8
$a_{12} = a_{23} = a_{34} = a_{45} = 0$, $\alpha_{12} = 90°$, $\alpha_{23} = -90°$,
$\alpha_{34} = \alpha_{45} = 0$, $\theta_{12} = \theta_{23} = 90°$, $\theta_{34} = 0$, $\theta_{45} = \phi_4$,
$s_{12} = \phi_1$, $s_{23} = \phi_2$, $s_{34} = \phi_3$, $s_{45} = 50$ mm.

13.4 Repeat Problem 13.3 using arbitrary (symbolic) values for the joint variables.

13.5 For the SCARA robot of Problem 13.1 and in the position described, find the instantaneous velocity and acceleration of the same tool point, $x_5 = y_5 = 0$, $z_5 = 1.5$ in, if the actuators have (constant) velocities of $\dot{\phi}_1 = 0.20$ rad/s, $\dot{\phi}_2 = -0.35$ rad/s, and $\dot{\phi}_3 = \dot{\phi}_4 = 0$.

13.6 For the gantry robot of Problem 13.3 and in the position described, find the instantaneous velocity

and acceleration of the same tool point, $x_5 = y_5 = 0$, $z_5 = 45$ mm, if the actuators have (constant) velocities of $\dot{\phi}_1 = \dot{\phi}_2 = 0$, $\dot{\phi}_3 = 40$ mm/s, and $\dot{\phi}_4 = 20$ rad/s.

13.7 The SCARA robot of Problem 13.1 is to be guided along a path for which the origin of the end effecter O_5 follows the straight line given by

$$\mathbf{R}_{O_5}(t) = (1.6t + 4.0)\hat{\mathbf{i}}_1 + (1.2t + 3.0)\hat{\mathbf{j}}_1 + 2.0\hat{\mathbf{k}}_1 \text{ in}$$

with t varying from 0.0 to 5.0 s; the orientation of the end effecter is to remain constant with $\hat{\mathbf{k}}_5 = -\hat{\mathbf{k}}_1$ (vertically downward) and $\hat{\mathbf{i}}_5$ radially outward from the base of the robot. Find expressions for how each of the actuators must be driven, as functions of time, to achieve this motion.

13.8 The gantry robot of Problem 13.3 is to travel a path for which the origin of the end effecter O_5 follows the straight line given by

$$\mathbf{R}_{O_5}(t) = (120t + 300)\hat{\mathbf{i}}_1 - 150\hat{\mathbf{j}}_1 + (90t + 225)\hat{\mathbf{k}}_1 \text{ mm}$$

with t varying from 0.0 to 4.0 s; the orientation of the end effecter is to remain constant with $\hat{\mathbf{k}}_5 = -\hat{\mathbf{j}}_1$ (vertically downward) and $\hat{\mathbf{i}}_5 = \hat{\mathbf{i}}_1$. Find expressions for the positions of each of the actuators, as functions of time, for this motion.

13.9 The end effecter of the SCARA robot of Problem 13.1 is working against a force loading of $10\hat{\mathbf{i}}_1 + 5t\hat{\mathbf{k}}_1$ lb and a constant torque loading of $25\hat{\mathbf{k}}_1$ in·lb as it follows the trajectory described in Problem 13.7. Find the torques required at the actuators, as functions of time, to achieve the motion described.

13.10 The end effecter of the gantry robot of Problem 13.3 is working against a force loading of $20\hat{\mathbf{i}}_1 + 10t\hat{\mathbf{j}}_1$ N and a constant torque loading of $5\hat{\mathbf{j}}_1$ N·m as it follows the trajectory described in Problem 13.8. Find the torques required at the actuators, as functions of time, to achieve the motion described.

PART 3

Dynamics of Machines

14 Static Force Analysis

14.1 INTRODUCTION

We are now ready for a study of the dynamics of machines and systems. Such a study is usually simplified by starting with the statics of such systems. In our studies of kinematic analysis we limited ourselves to consideration of the geometry of the motions and of the relationships between displacement and time. The forces required to produce those motions or the motion that would result from the application of a given set of forces were not considered.

In the design of a machine, consideration of only those effects that are described by units of *length* and *time* is a tremendous simplification. It frees the mind of the complicating influence of many other factors that ultimately enter into the problem, and it permits our attention to be focused on the primary goal, that of designing a mechanism to obtain a desired motion. That was the problem of *kinematics,* covered in the previous chapters of this book.

The fundamental units in kinematic analysis are length and time; in dynamic analysis they are *length, time,* and *force.*

Forces are transmitted between machine members through mating surfaces—that is, from a gear to a shaft or from one gear through meshing teeth to another gear, from a connecting rod through a bearing to a lever, from a V-belt to a pulley, from a cam to a follower, or from a brake drum to a brake shoe. It is necessary to know the magnitudes of these forces for a variety of reasons. The distribution of these forces at the boundaries of mating surfaces must be reasonable, and their intensities must remain within the working limits of the materials composing the surfaces. For example, if the force operating on a sleeve bearing becomes too high, it will squeeze out the oil film and cause metal-to-metal contact, overheating, and rapid failure of the bearing. If the forces between gear teeth are too large, the oil film may be squeezed out from between them. This could result in flaking and spalling of the metal, noise, rough motion, and eventual failure. In our study of dynamics we are interested principally in determining the magnitudes, directions, and locations of the forces, but not in sizing the members on which they act.[1]

Some of the new terms used in this phase of our study are defined as follows.

Force   Our earliest ideas concerning forces arose because of our desire to push, lift, or pull various objects. So force is the action of one body acting on another. Our intuitive concept of force includes such ideas as *magnitude, direction,* and *place of application,* and these are called the *characteristics* of the force.

Matter   Matter is any material substance; if it is completely enclosed, it is called a body.

Mass   Newton defined mass as the *quantity of matter of a body as measured by its volume and density.* This is not a very satisfactory definition because density is the mass of a unit volume. We can excuse Newton by surmising that perhaps he did not mean it to be a definition. Nevertheless, he recognized the fact that all bodies possess some inherent property that is different than weight. Thus a moon rock has a certain constant amount of substance, even though its moon weight is different from its earth weight. The constant amount of substance, or quantity of matter, is called the *mass* of the rock.

Inertia   Inertia is the property of mass that causes it to resist any effort to change its motion.

Weight   Weight is the force that results from gravity acting upon a mass. The following quotation is pertinent:

> The great advantage of SI units is that there is one, and only one, unit for each physical quantity—the meter for length, the kilogram for mass, the newton for force, the second for time, etc. To be consistent with this unique feature, it follows that a given unit or word should not be used as an accepted technical name for two physical quantities. However, for generations the term "weight" has been used in both technical and nontechnical fields to mean either the force of gravity acting on a body or the mass of the body itself. The reason for the double use of the term for two physical quantities—force and mass—is attributed to the dual use of the pound units in our present customary gravitational system in which we often use weight to mean both force and mass.[2]

In this book we will always use the term *weight* to mean gravitational force.

Particle   A particle is a body whose dimensions are so small that they may be neglected. The dimensions are so small that a particle can be considered to be located at a single point; it is not a point, however, in the sense that a particle can consist of matter and can have mass, whereas a point cannot.

Rigid Body   All real bodies are either elastic or plastic and will deform, though perhaps only slightly, when acted upon by forces. When the deformation of such a body is small enough to be neglected, such a body is frequently assumed to be *rigid*—that is, incapable of deformation—in order to simplify the analysis. This assumption of rigidity was the key step that allowed the treatment of kinematics in all previous chapters of this book to be completed without consideration of the forces. Without this simplifying assumption of rigidity, forces and motions are interdependent and kinematic and dynamic analysis require simultaneous solution.

Deformable Body   The rigid-body assumption cannot be maintained when internal stresses and strains due to applied forces are to be analyzed. If stress is to be found, we must admit to the existence of strain; thus we must consider the body to be capable of deformation, even though small. If the deformations are small enough, in comparison to the gross dimensions and motion of the body, we can then still treat the body as rigid while treating the motion (kinematic) analysis, but must then consider it deformable when stresses are to be found. Using the additional assumption that the forces and stresses remain within the elastic range, such analysis is frequently called *elastic-body analysis*.

14.2 NEWTON'S LAWS

As stated in the *Principia,* Newton's three laws are:

[Law 1] Every body perseveres in its state of rest or uniform motion in a straight line, except in so far as it is compelled to change that state by impressed forces.

[Law 2] Change of motion is proportional to the moving force impressed, and takes place in the direction of the straight line in which such force is impressed.

[Law 3] Reaction is always equal and opposite to action; that is to say, the actions of two bodies upon each other are always equal and directly opposite.

For our purposes, it convenient to restate these laws:

Law 1 If all the forces acting on a particle are balanced, that is, sum to zero, then the particle will either remain at rest or will continue to move in a straight line at a uniform velocity.

Law 2 If the forces acting on a particle are not balanced, the particle will experience an acceleration proportional to the resultant force and in the direction of the resultant force.

Law 3 When two particles interact, a pair of reaction forces come into existence; these forces have the same magnitudes and opposite senses, and they act along the straight line common to the two particles.

Newton's first two laws can be summarized by the equation

$$\sum \mathbf{F}_{ij} = m_j \mathbf{A}_{Gj} \tag{14.1}$$

which is called the equation of motion for body j. In this equation, $\mathbf{A}_{Gj}$ is the absolute acceleration of the center of mass of the body j which has mass m_j, and this acceleration is produced by the sum of all forces acting on that body—that is, all values of the index i. Both $\mathbf{F}_{ij}$ and $\mathbf{A}_{Gj}$ are vector quantities.

14.3 SYSTEMS OF UNITS

An important use of Eq. (14.1) occurs in the standardization of systems of units. Let us employ the following symbols to designate units:

Length L
Time T
Mass M
Force F

These symbols are to stand for any unit we may choose to use for that respective quantity. Thus, possible choices for L are inches, feet, meters, kilometers, or miles. The symbols L, T, M, and F are not numbers, but they can be substituted in Eq. (14.1) as if they were. The equality sign then implies that the symbols on one side are equivalent to those on the other. Making the indicated substitutions then gives

$$F = MLT^{-2} \tag{14.2}$$

because the acceleration **A** has units of length divided by time squared. Equation (14.2) expresses an equivalence relationship among the four units force, mass, length, and time. We are free to choose units for three of these, and then the units used for the fourth depend on those used for the first three. It is for this reason that the first three units chosen are called *basic units,* while the fourth is called the *derived unit.*

When force, length, and time are chosen as basic units, then mass is the derived unit, and the system is called a *gravitational system of units.* When mass, length, and time are chosen as the basic units, then force is the derived unit, and the system is called an *absolute system of units.*

In most English-speaking countries the U.S. customary *foot-pound-second (fps) system* and the *inch-pound-second (ips) system* are the two gravitational systems of units most used by engineers.* In the fps system the (derived) unit of mass is

$$M = FT^2/L = \text{lbf} \cdot \text{s}^2/\text{ft} = \text{slug} \tag{14.3}$$

Thus force, length, and time are the three basic units in the fps gravitational system. The unit of force is the pound, more properly the *pound force.* We sometimes, but not always, abbreviate this unit as *lbf;* the abbreviation *lb* is also permissible because we shall be dealing with U.S. customary gravitational systems.**

Finally, we note in Eq. (14.3) that the derived unit of mass in the fps gravitational system is the lb·s²/ft, called a *slug;* there is no abbreviation for slug.

The derived unit of mass in the ips gravitational systems is

$$M = FT^2/L = \text{lbf} \cdot \text{s}^2/\text{in} \tag{14.4}$$

Note that this unit of mass has *not* been given a special name.

The International System of Units (SI) is an absolute system. The three basic units are the meter (m), the kilogram mass (kg), and the second (s). The unit of force is derived and

---

*Most engineers prefer to use gravitational systems; this helps to explain some of the resistance to the use of SI units, because the International System (SI) is an absolute system.

**The abbreviation lb for pound comes from the Latin word *Libra,* the balance, the seventh sign of the zodiac, which is represented as a pair of scales.

is called a *newton* to distinguish it from the kilogram, which, as indicated, is a unit of mass. The units of the newton (N) are

$$F = ML/T^2 = \text{kg·m/s}^2 = \text{N} \tag{14.5}$$

The weight of an object is the force exerted upon it by gravity. Designating the weight as **W** and the acceleration due to gravity as **g**, Eq. (14.1) becomes

$$\mathbf{W} = m\mathbf{g}$$

In the fps system, standard gravity is $g = 32.1740$ ft/s^2. For most cases this is rounded off to 32.2 ft/s^2. Thus the weight of a mass of one slug under standard gravity is

$$W = mg = (1 \text{ slug})(32.2 \text{ ft/s}^2) = 32.2 \text{ lb}$$

In the ips system, standard gravity is 386.088 in/s^2 or about 386 in/s^2. Thus, under standard gravity, a unit mass weighs

$$W = mg = (1 \text{ lbf·s}^2/\text{in})(386 \text{ in/s}^2) = 386 \text{ lb}$$

With SI units, standard gravity is 9.806 m/s^2 or about 9.80 m/s^2. Thus the weight of a one-kilogram mass under standard gravity is

$$W = mg = (1 \text{ kg})(9.80 \text{ m/s}^2) = 9.80 \text{ kg·m/s}^2 = 9.80 \text{ N}$$

It is convenient to remember that a large apple weighs about 1 N.

14.4 APPLIED AND CONSTRAINT FORCES

When two or more bodies are connected together to form a group or system, the pair of action and reaction forces between any two of the connected bodies are called *constraint forces*. These forces constrain the connected bodies to behave in a specific manner defined by the nature of the connection. Forces acting on this system of bodies from outside the system are called *applied* forces.

Electric, magnetic, and gravitational forces are examples of forces that may be applied without actual physical contact. However, a great many, if not most, of the forces with which we are concerned in mechanical equipment occur through direct physical or mechanical contact.

As we will see in the next section, the constraint forces of action and reaction at a mechanical contact occur in pairs, and thus have no *net* force effect on the system of bodies being considered. Such pairs of constraint forces, although they clearly exist and may be large, are usually not considered further when both the action and reaction forces act on bodies of the system being considered. However, when we try to consider a body or system of bodies to be isolated from its surroundings, only one of each pair of constraint forces acts on the system being considered at any point of contact on the separation boundary. The

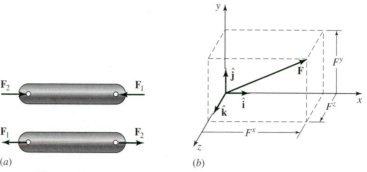

Figure 14.1 (*a*) Points of application of forces $\mathbf{F}_1$ and $\mathbf{F}_2$ to a rigid body may or may not be important. (*b*) The rectangular components of a force vector.

other constraint force, the reaction, is left acting on the surroundings. When we isolate the system being considered, these constraint forces at points of separation must be clearly identified, and they are essential to further study of the dynamics of the isolated system of bodies.

As indicated earlier, the *characteristics* of a force are its *magnitude,* its *direction,* and its *point of application.* The direction of a force includes the concept of a line along which the force is acting, and a *sense.* Thus a force may be directed either positively or negatively along its line of action.

Sometimes the position of the point of application along the line of action is not important, for example, when we are studying the equilibrium of a rigid body. Thus in Fig. 14.1*a* it does not matter whether we diagram the force pair $\mathbf{F}_1\mathbf{F}_2$ as if they compress the link, or if we diagram them as putting the link in tension, provided we are interested only in the equilibrium of the link. Of course, if we are also interested in the internal link stresses, the forces cannot be interchanged.

The notation to be used for force vectors is shown in Fig. 14.1*b*. Boldface letters are used for force vectors and lightface letters for their magnitudes. Thus the components of a force vector are

$$\mathbf{F} = F^x\hat{\mathbf{i}} + F^y\hat{\mathbf{j}} + F^z\hat{\mathbf{k}} \tag{a}$$

Note that the directions of the components in this book are indicated by superscripts, not subscripts.

Two equal and opposite forces along two parallel but *noncollinear* straight lines in a body cannot be combined to obtain a single resultant force on the body. Any two such forces acting on the body constitute a *couple.* The *arm of the couple* is the perpendicular distance between their lines of action, shown as h in Fig. 14.2*a*, and *the plane of the couple* is the plane containing the two lines of action.

The *moment* of a couple is another vector $\mathbf{M}$ directed normal to the plane of the couple; the sense of $\mathbf{M}$ is in accordance with the right-hand rule for rotation. The magnitude of the moment is the product of the arm h and the magnitude of one of the forces F. Thus

$$M = hF \tag{14.6}$$

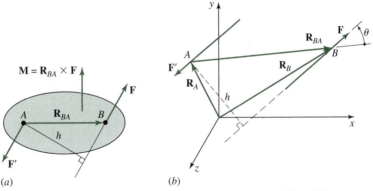

Figure 14.2 (*a*) $\mathbf{R}_{BA}$ is a position difference vector, while $\mathbf{F}$ and $\mathbf{F}'$ are force vectors; the free vector $\mathbf{M}$ is the moment of the couple formed by $\mathbf{F}$ and $\mathbf{F}'$. (*b*) A force couple formed by $\mathbf{F}$ and $\mathbf{F}'$.

In vector form the moment $\mathbf{M}$ is the cross product of the position difference vector $\mathbf{R}_{BA}$ and the force vector $\mathbf{F}$, and so it is defined by the equation

$$\mathbf{M} = \mathbf{R}_{BA} \times \mathbf{F} \tag{14.7}$$

Some interesting properties of couples can be determined by an examination of Fig. 14.2*b*. Here $\mathbf{F}$ and $\mathbf{F}'$ are two equal, opposite, and parallel forces. We can choose any point on each line of action and define the positions of these points by the vectors $\mathbf{R}_A$ and $\mathbf{R}_B$. Then the "relative position," or position difference, vector is

$$\mathbf{R}_{BA} = \mathbf{R}_B - \mathbf{R}_A \tag{b}$$

The moment of the couple is the sum of the moments of the two forces and is

$$\mathbf{M} = \mathbf{R}_A \times \mathbf{F}' + \mathbf{R}_B \times \mathbf{F} \tag{c}$$

But $\mathbf{F}' = -\mathbf{F}$, and so, using Eq. (*b*), Eq. (*c*) can be written as

$$\mathbf{M} = (\mathbf{R}_B - \mathbf{R}_A) \times \mathbf{F} = \mathbf{R}_{BA} \times \mathbf{F} \tag{d}$$

Equation (*d*) shows that:

1. The value of the moment of the couple is independent of the choice of the reference point about which the moments are taken, because the vector $\mathbf{R}_{BA}$ is the same for all positions of the origin.
2. Because $\mathbf{R}_A$ and $\mathbf{R}_B$ define any choices of points on the lines of action of the two forces, the vector $\mathbf{R}_{BA}$ is not restricted to perpendicularity with $\mathbf{F}$ and $\mathbf{F}'$. This is a very important characteristic of the vector product because it shows that the value of the moment is independent of how $\mathbf{R}_{BA}$ is chosen. The magnitude of the moment

can be obtained as follows: Resolve $\mathbf{R}_{BA}$ into two components $\mathbf{R}_{BA}^n$ and $\mathbf{R}_{BA}^t$, perpendicular and parallel to $\mathbf{F}$, respectively. Then

$$M = \left(\mathbf{R}_{BA}^n + \mathbf{R}_{BA}^t\right) \times \mathbf{F} \tag{e}$$

But $\mathbf{R}_{BA}^n$ is the perpendicular between the two lines of action and $\mathbf{R}_{BA}^t$ is parallel to $\mathbf{F}$. Therefore $\mathbf{R}_{BA}^t \times \mathbf{F} = 0$ and

$$M = \mathbf{R}_{BA}^n \times \mathbf{F}$$

is the moment of the couple. Because $R_{BA}^n = R_{BA}\sin\theta$, where θ is the angle between $\mathbf{R}_{BA}$ and $\mathbf{F}$, the magnitude of the moment is

$$M = (R_{BA}\sin\theta)F = hF \tag{f}$$

3. The moment vector $\mathbf{M}$ is independent of any particular origin or line of application and is thus a *free vector*.
4. The forces of a couple can be rotated together within their plane, keeping their magnitudes and the distance between their lines of action constant, or they can be translated to any parallel plane, without changing the magnitude, direction, or sense of the moment vector.
5. Two couples are equal if they have the same moment vectors, regardless of the forces or moment arms. This means that it is the vector product of the two that is significant, not the individual values.

14.5 FREE-BODY DIAGRAMS

The term "body" as used here may consist of an entire machine, several connected parts of a machine, a single part, or a portion of a machine part. A *free-body diagram* is a sketch or drawing of the body, isolated from the rest of the machine and its surroundings, upon which the forces and moments are shown in action. It is usually desirable to include on the diagram the known magnitudes and directions as well as other pertinent information.

The diagram so obtained is called "free" because the machine part, or parts, or portion of the body has been freed (or isolated) from the remaining machine elements, and their effects have been replaced by forces and moments. If the free-body diagram is of an entire machine part or system of parts, the forces shown on it are the external forces (applied forces) and moments exerted by adjacent or connecting parts that are not part of this free-body diagram. If the diagram is of a portion of a part, the forces and moments shown acting on the cut (separation) surface are the *internal* forces and moments—that is, the summation of the internal stresses exerted on the cut surface by the remainder of the part that has been cut away.

The construction and presentation of clear and neatly drawn free-body diagrams represent the heart of engineering communication. This is true because they represent a part of the thinking process, whether they are actually placed on paper or not, and because the construction of these diagrams is the *only* way the results of thinking can be communicated to others. You should acquire the habit of drawing free-body diagrams no matter how simple the problem may appear to be. They are a means of storing ideas while concentrating on the next step in the solution of a problem. Construction of or, as a very minimum, sketching of free-body diagrams speeds up the problem-solving process and dramatically decreases the chances of making mistakes.

The advantages of using free-body diagrams can be summarized as follows:

1. They make it easier for one to translate words, thoughts, and ideas into physical models.
2. They assist in seeing and understanding all facets of a problem.
3. They help in planning the approach to the problem.
4. They make mathematical relations easier to see or find.
5. Their use makes it easy to keep track of one's progress and help in making simplifying assumptions.
6. They are useful for storing the methods of solution for future reference.
7. They assist your memory and make it easier to present and explain your work to others.

In analyzing the forces in machines we shall almost always need to separate the machine into its individual components or subsystems and construct free-body diagrams showing the forces that act upon each. Many of these components will have been connected to each other by kinematic pairs (joints). Accordingly, Fig. 14.3 has been prepared to show the constraint forces acting between the elements of the lower pairs when friction forces are assumed to be zero. Note that these are not complete free-body diagrams in that they show only the forces at the mating surfaces and do not show the forces where the partial links have been severed from their remainders.

Upon careful examination of Fig. 14.3 we see that there is no component of the constraint force or moment transmitted along an axis where motion is possible—that is, along with the direction of the pair variable. This is the result of our assumption of no friction. If one of these pair elements were to try to transmit a force or torque to its mating element in the direction of the pair variable, in the absence of friction, the attempt would result in motion of the pair variable rather than in the transmission of force or torque. Similarly, in the case of higher pairs, the constraint forces are always normal to the contacting surfaces in the absence of friction.

The notation shown in Fig. 14.3 will be used consistently throughout the remainder of the book. $\mathbf{F}_{ij}$ denotes the force that link i exerts onto link j, while $\mathbf{F}_{ji}$ is the reaction to this force and is the force of link j acting on link i.

14.6 CONDITIONS FOR EQUILIBRIUM

A body or group of bodies is said to be in equilibrium if all the forces exerted on the system are in balance. In such a situation, Newton's laws as expressed in Eq. (14.1) show that no acceleration results. This may imply that no motion takes place, meaning that all velocities are zero. If so, the system is said to be in *static equilibrium*. On the other hand, no acceleration may imply that velocities do exist but remain constant; the system is then said

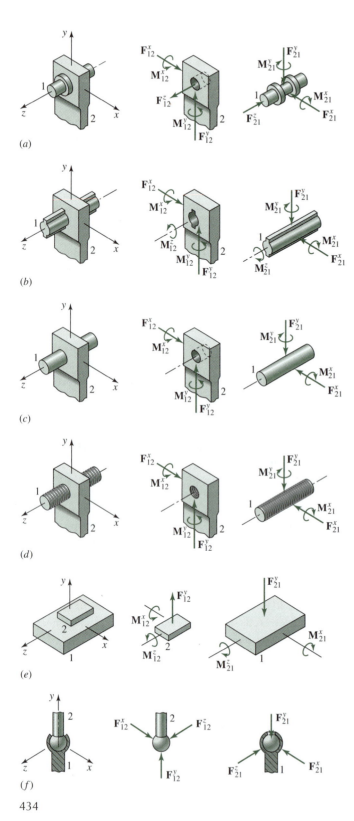

Figure 14.3 All lower pairs and their constraint forces: (*a*) revolute or turning pair with pair variable θ; (*b*) prismatic or sliding pair with pair variable z; (*c*) cylindric pair with pair variables θ and z; (*d*) screw or helical pair with pair variables θ or z; (*e*) planar or flat pair with pair variables x, z, and θ; (*f*) globular or spheric pair with pair variables θ, ϕ, and ψ.

to be in *dynamic equilibrium*. In either case, Eq. (14.1) shows that a system of bodies is in equilibrium if and only if:

1. *The vector sum of all forces acting upon it is zero.*
2. *The vector sum of the moments of all forces acting about any arbitrary axis is zero.*

Mathematically, these two statements are expressed as

$$\sum \mathbf{F} = 0 \quad \text{and} \quad \sum \mathbf{M} = 0 \tag{14.8}$$

Note how these statements are a result of Newton's first and third laws, it being understood that a body is composed of a collection of particles.

Many problems have forces acting in a single plane. When this is true, it is convenient to choose this as the xy plane. Then Eqs. (14.8) can be simplified to

$$\sum F^x = 0, \qquad \sum F^y = 0, \qquad \sum M = 0$$

where the z direction for the moment M components is implied by the fact that all forces act only in the xy plane.

14.7 TWO- AND THREE-FORCE MEMBERS

The free-body diagram of Fig. 14.4*a* shows an arbitrarily shaped body acted upon by two forces, $\mathbf{F}_A$ and $\mathbf{F}_B$, at points A and B, respectively. The forces have been shown with broken lines as a reminder that the magnitudes and directions of these forces are not yet known. Assuming that our sketched free-body diagram is complete and that no other forces or torques are active on the body, the first of Eqs. (14.8) gives

$$\sum \mathbf{F} = \mathbf{F}_A + \mathbf{F}_B = 0$$

This requires that $\mathbf{F}_A$ and $\mathbf{F}_B$ have *equal magnitudes* and *opposite directions*. Using the second of Eqs. (14.8) about either point A or point B shows that $\mathbf{F}_A$ and $\mathbf{F}_B$ must

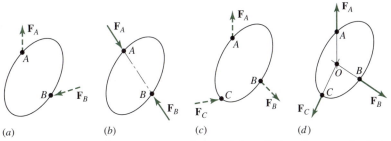

Figure 14.4 (*a*) Two-force member not in equilibrium; (*b*) two-force member is in equilibrium if $\mathbf{F}_A$ and $\mathbf{F}_B$ are equal, opposite, and share the same line of action; (*c*) three-force member is not in equilibrium; (*d*) three-force member is in equilibrium if $\mathbf{F}_A$, $\mathbf{F}_B$, and $\mathbf{F}_C$ are coplanar, if their lines of action intersect at a common point O, and if their vector sum is zero.

also have the *same line of action;* otherwise the moments could not sum to zero. Thus, for any two-force member with no applied torque, we have learned that the forces must be oppositely directed along the unique line of action defined by the points A and B, as shown in Fig. 14.4b. Once the magnitude of either becomes known, the magnitude of the other must be equal but of the opposite sense.

A similar study for a body with three active forces and no applied torques is shown in Figs. 14.4c and 14.4d. Here we also assume that the three forces, $\mathbf{F}_A$, $\mathbf{F}_B$, and $\mathbf{F}_C$, are all known to act in the plane defined by the three points, A, B, and C. Suppose further that we also know two of the lines of action, say, those of $\mathbf{F}_A$ and $\mathbf{F}_B$, and that they intersect at some point O. Because neither $\mathbf{F}_A$ nor $\mathbf{F}_B$ exerts any moment about point O, applying $\sum \mathbf{M} = 0$ to all three forces shows that the moment of $\mathbf{F}_C$ about point O must also be zero. Thus the lines of action of all three forces must intersect at the common point O; that is, the three forces acting on the body must be *concurrent.* This explains why, in two dimensions, a three-force member can be solved for only two force magnitudes, even though there are three scalar equations implied in Eqs. (14.8); the moment equation has already been satisfied once the lines of action become known. If one of the three force magnitudes is known, the other two can be found by using $\sum \mathbf{F} = 0$, as shown in the following example.

EXAMPLE 14.1

The four-bar linkage of Fig. 14.5a has crank 2 driven by an input torque $\mathbf{M}_{12}$; an external load $\mathbf{P} = 120\angle220°$ lb acts at point Q on link 4. For the particular position of the linkage shown, find all the constraint forces and their reactions necessary for this to be a position of equilibrium.

GRAPHIC SOLUTION

1. Draw the linkage and the given force or forces to scale, as shown in Fig. 14.5a. The size scale shown here is about 10 in/in, meaning that 1 in of the drawing represents 10 in of the linkage. The force scale shown is about 85 lb/in; that is, a vector shown 1 in long represents a force of 85 lb.

2. Begin drawings of the free-body diagrams of each of the moving links, as shown in Figs. 14.5b, 14.5e, and 14.5f. At this stage, all known information, such as the positions and the force $\mathbf{P}$, should be drawn to scale. Unknown information, such as all other applied forces on each link, may be sketched lightly on the free-body diagrams, but will be redrawn to scale once they become known.

3. Ensure that all free-body diagrams are complete—that is, that all pertinent information is shown. For example, we must now either find the weights and centers of gravity of all links and the direction of gravity or, as is done here, assume that the gravitational forces are small in comparison to the applied force $\mathbf{P}$ and that they can safely be ignored.

4. From the free-body diagram, we now observe that link 3 is a two-force member. Thus, according to the above discussion, we can now show the lines of action of $\mathbf{F}_{23}$ and $\mathbf{F}_{43}$ along the axis defined by points A and B in Fig. 14.5e. Because action and reaction forces must be equal and opposite, these also tell us the lines of action of $\mathbf{F}_{32}$ on link 2 in Fig. 14.5f and $\mathbf{F}_{34}$ on link 4 in Fig. 14.5b.

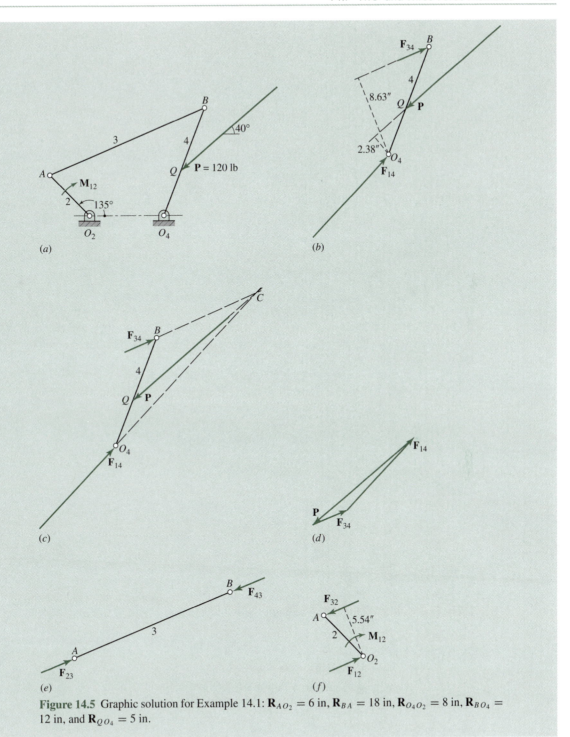

Figure 14.5 Graphic solution for Example 14.1: $\mathbf{R}_{AO_2} = 6$ in, $\mathbf{R}_{BA} = 18$ in, $\mathbf{R}_{O_4O_2} = 8$ in, $\mathbf{R}_{BO_4} = 12$ in, and $\mathbf{R}_{QO_4} = 5$ in.

5a. We proceed next to link 4, as shown in Fig 14.5*b*, where neither the magnitude nor the direction of the frame reaction $\mathbf{F}_{14}$ is known as yet. One method for continuation is to measure the moment arms $\mathbf{P}$ and $\mathbf{F}_{34}$ about O_4, which are found to be 2.38 in and 8.63 in, respectively. Then, taking counterclockwise moments as positive, Eqs. (14.8) gives

$$\sum M_{O_4} = (2.38 \text{ in})(120 \text{ lb}) - (8.63)F_{34} = 0$$

The solution is $F_{34} = 33.1$ lb, where the fact that the result is positive confirms the sense shown for $\mathbf{F}_{34}$ in the figure.

5b. As an alternative approach to step 5a, we could note that link 4 is a three-force member; therefore, when the lines of action of $\mathbf{P}$ and $\mathbf{F}_{34}$ are extended, they intersect at the point of concurrency C and define the line of action of $\mathbf{F}_{14}$, as shown in Fig. 14.5*c*.

6. No matter whether step 5a or step 5b is used, the force polygon shown in Fig. 14.5*d* can now be used to graphically solve the equation

$$\sum \mathbf{F} = \mathbf{P} + \mathbf{F}_{34} + \mathbf{F}_{14} = 0$$

giving $F_{34} = 33.1$ lb and $F_{14} = 89.0$ lb.

7. From the free-body diagram of link 3, Fig. 14.5*e*, and the nature of action and reaction forces, we note that $\mathbf{F}_{23} = -\mathbf{F}_{43} = \mathbf{F}_{34}$ and, therefore, $\mathbf{F}_{32} = -\mathbf{F}_{23}$ becomes known on the free-body diagram of link 2 in Fig. 14.5*f*.

8. From the summation of forces on the free-body diagram of link 2, we find $\mathbf{F}_{12} = -\mathbf{F}_{32}$. When the moment arm of $\mathbf{F}_{32}$ about point O_2 is measured, it is found to be 5.54 in. Therefore, taking counterclockwise moments as positive, we obtain

$$\sum M_{O_2} = (5.54 \text{ in})F_{32} - M_{12} = 0$$

which yields

$$M_{12} = (5.54 \text{ in})(33.1 \text{ lb}) = 183 \text{ in·lb} \qquad\qquad Ans.$$

where the positive sign confirms the clockwise sense shown in Fig. 14.5*f*.

ANALYTIC SOLUTION

1. First make a position analysis of the linkage at the position of interest, to determine the angular orientation of each link. The angles are shown in Fig. 14.6*a*. In addition, we find

$$\mathbf{R}_{AO_2} = 6.0\angle135° = -4.24\hat{\mathbf{i}} + 4.24\hat{\mathbf{j}} \text{ in}$$

$$\mathbf{R}_{BO_4} = 12.0\angle68.4° = 4.42\hat{\mathbf{i}} + 11.16\hat{\mathbf{j}} \text{ in}$$

$$\mathbf{R}_{QO_4} = 5.0\angle68.4° = 1.84\hat{\mathbf{i}} + 4.65\hat{\mathbf{j}} \text{ in}$$

$$\mathbf{F}_{34} = F_{34}\angle22.4° = 0.925F_{34}\hat{\mathbf{i}} + 0.381F_{34}\hat{\mathbf{j}} \text{ lb}$$

$$\mathbf{P} = 120\angle220° = -91.9\hat{\mathbf{i}} - 77.1\hat{\mathbf{j}} \text{ lb}$$

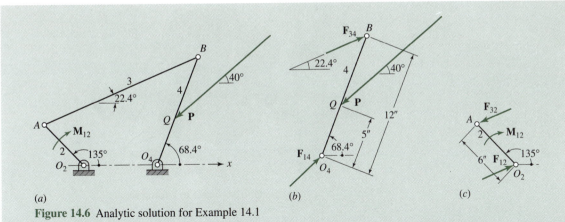

(a)   **(b)**   **(c)**

Figure 14.6 Analytic solution for Example 14.1

2. Referring to Fig. 14.6b, we sum moments about point O_4. Thus

$$\sum \mathbf{M}_{O_4} = \mathbf{R}_{QO_4} \times \mathbf{P} + \mathbf{R}_{BO_4} \times \mathbf{F}_{34} = 0$$

Performing the cross-product operations, we find the first term to be $\mathbf{R}_{QO_4} \times \mathbf{P} = 285.5\hat{\mathbf{k}}$ in·lb, while the second term is $\mathbf{R}_{BO_4} \times \mathbf{F}_{34} = -8.63 F_{34}\hat{\mathbf{k}}$ in·lb. Substituting these values into the above equation and solving gives

$$\mathbf{F}_{34} = 33.1\angle 22.4° = 30.6\hat{\mathbf{i}} + 12.6\hat{\mathbf{j}} \text{ lb} \qquad\qquad Ans.$$

3. The force $\mathbf{F}_{14}$ is found next from the equation

$$\sum \mathbf{F} = \mathbf{P} + \mathbf{F}_{34} + \mathbf{F}_{14} = 0$$

Solving gives

$$\mathbf{F}_{14} = 61.3\hat{\mathbf{i}} + 64.5\hat{\mathbf{j}} = 89.0\angle 46.5° \text{ lb.} \qquad\qquad Ans.$$

4. Next, from the free-body diagram of link 2 (see Fig. 14.6c), we write

$$\sum \mathbf{M}_{O_2} = -\mathbf{M}_{12} + \mathbf{R}_{AO_2} \times \mathbf{F}_{32} = 0$$

Solving $\mathbf{F}_{32} = -\mathbf{F}_{34} = -30.6\hat{\mathbf{i}} - 12.6\hat{\mathbf{j}} \text{ lb}$, we find

$$\mathbf{M}_{12} = -183\hat{\mathbf{k}} \text{ in·lb} \qquad\qquad Ans.$$

Notice that in both the graphic and the analytic procedures the free-body diagram of the frame, link 1, was never drawn and is not needed for the solution. If drawn, a force $\mathbf{F}_{21} = -\mathbf{F}_{12}$ would be shown at O_2, a force $\mathbf{F}_{41} = -\mathbf{F}_{14}$ at O_4, and a moment $\mathbf{M}_{21} = -\mathbf{M}_{12}$

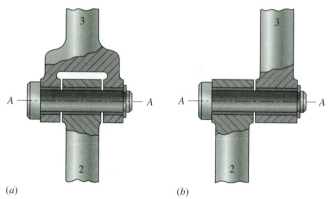

Figure 14.7 (*a*) A balanced revolute-pair connection. (*b*) An unbalanced revolute-pair connection produces a bending moment on the pin and on each link.

would be shown. In addition, a minimum of two more force components and a torque would be shown where link 1 is anchored to keep it stationary. Because this would introduce at least three additional unknowns and because Eqs. (14.9) would not allow the solution for more than three, drawing the free-body diagram of the frame will be of no benefit in the solution process. The only time that this is ever done is when these anchoring forces, called "shaking forces," and moments are sought; this is discussed further in Section 15.7 of the next chapter.

In the preceding example it has been assumed that the forces all act in the same plane. For the connecting link 3, for example, it was assumed that the line of action of the forces and the centerline of the link were coincident. A careful designer will sometimes go to extreme measures to approach these conditions as closely as possible. Note that if the pin connections are arranged as shown in Fig. 14.7*a*, such conditions are obtained theoretically. If, on the other hand, the connection is like the one shown in Fig. 14.7*b*, the pin itself as well as the link will have moments acting upon them. If the forces are not in the same plane, moments exist proportional to the distance between the force planes.

EXAMPLE 14.2

The slider-crank mechanism of Fig. 14.8*a* has an external load $P = 100$ lb acting horizontally at point Q on link 4. Determine the torque $\mathbf{M}_{12}$ that must be applied to link 2 to hold the mechanism in static equilibrium at the position shown in the figure. Assume that the gravitational forces are small in comparison to the applied force $\mathbf{P}$ and can be ignored.

GRAPHIC SOLUTION

1. Draw the mechanism and the given force to scale, as shown in Fig. 14.8*a*. The size scale is about 15 in/in and the force scale is about 140 lb/in.
2. Draw the free-body diagrams of each moving link, as shown in Figs. 14.8*b*, 14.8*c*, and 14.8*e*.

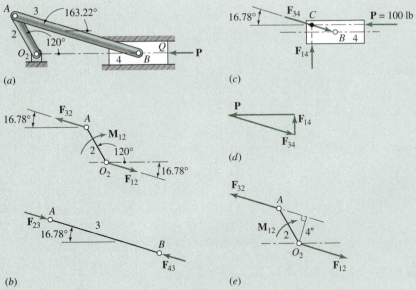

Figure 14.8 Graphic solution for Example 14.2: $R_{AO_2} = 6$ in, $R_{BA} = 18$ in, and $R_{QB} = \sqrt{17}$ in.

3. Observe that link 3 is a two-force member. Thus we can show the lines of action of F_{23} and F_{43} along the axis defined by points A and B in Fig. 14.8b. These also tell us the lines of action of F_{34} on link 4 in Fig. 14.8c and F_{32} on link 2 in Fig. 14.8e.

4. We proceed to link 4 where neither the magnitude nor the location of the frame reaction force F_{14} is known as yet. However, the direction of this force must be perpendicular to the ground because friction is neglected. Link 4 is a three-force member; therefore, when the lines of action of P and F_{34} are extended, they intersect at the point of concurrency C and define the line of action of F_{14}, as shown in Fig. 14.8c.

5. The force polygon shown in Fig. 14.8d is used to graphically solve the equation

$$\sum F = P + F_{34} + F_{14} = 0$$

giving

$$F_{34} = 105 \text{ lb} \quad \text{and} \quad F_{14} = 31 \text{ lb}$$

6. From the free-body diagram of link 3, Fig. 14.8b, and the nature of action and reaction forces, we note that $F_{23} = -F_{43} = F_{34}$ and, therefore, $F_{32} = -F_{23}$ becomes known on the free-body diagram of link 2 in Fig. 14.8e.

7. From the summation of forces on the free-body diagram of link 2, we find $F_{12} = -F_{32}$. When the moment arm of F_{32} about point O_2 is measured, it is found to be 4 in. Therefore, taking counterclockwise moments as positive, we obtain

$$\sum M_{O_2} = (4 \text{ in}) F_{32} - M_{12} = 0$$

which yields

$$M_{12} = (4 \text{ in})(105 \text{ lb}) = 420 \text{ in} \cdot \text{lb} \qquad \textit{Ans.}$$

where the positive sign confirms the clockwise sense shown in Fig. 14.8e.

ANALYTIC SOLUTION

1. From a position analysis of the mechanism at the position of interest, with the angles shown in Fig. 14.9a, we find

$$\mathbf{R}_{AO_2} = 6.0 \angle 120° = -3\hat{\mathbf{i}} + 5.196\hat{\mathbf{j}} \text{ in}$$

$$\mathbf{F}_{34} = F_{34} \angle 16.78° = 0.957\ 4 F_{34}\hat{\mathbf{i}} - 0.288\ 7 F_{34}\hat{\mathbf{j}} \text{ lb}$$

2. The forces $\mathbf{F}_{34}$ and $\mathbf{F}_{14}$ can be found from the equation

$$\sum \mathbf{F} = \mathbf{P} + \mathbf{F}_{34} + \mathbf{F}_{14} = \mathbf{0}$$

where $\mathbf{P} = -100\hat{\mathbf{i}}$ lb. This equation can be written as

$$-100\hat{\mathbf{i}} + 0.957\ 4 F_{34}\hat{\mathbf{i}} - 0.288\ 7 F_{34}\hat{\mathbf{j}} + F_{14}\hat{\mathbf{j}} = \mathbf{0}$$

Then equating the $\hat{\mathbf{i}}$ and $\hat{\mathbf{j}}$ components, respectively, gives

$$F_{34} = \frac{100}{0.957\ 4} = 104.45 \text{ lb}$$

$$F_{14} = 0.2887 F_{34} = 30.15 \text{ lb}$$

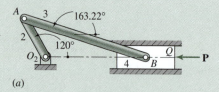

(a)

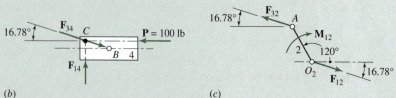

(b)   (c)

Figure 14.9 Analytic solution for Example 14.2.

The forces can be written as

$$\mathbf{F}_{34} = 100\hat{\mathbf{i}} - 30.15\hat{\mathbf{j}} = 104.45\angle 163.22° \text{ lb}$$

$$\mathbf{F}_{14} = 30.15\hat{\mathbf{j}} \text{ lb}$$

3. Next, from the free-body diagram of link 2 (see Fig. 14.9c), we write

$$\sum \mathbf{M}_{O_2} = -\mathbf{M}_{12} + \mathbf{R}_{AO_2} \times \mathbf{F}_{32} = 0$$

Solving $\mathbf{F}_{32} = -\mathbf{F}_{34} = -100\hat{\mathbf{i}} + 30.15\hat{\mathbf{j}}$ lb, we find

$$\mathbf{M}_{12} = -429.15\hat{\mathbf{k}} \text{ in·lb} \qquad\qquad \textit{Ans.}$$

14.8 FOUR-FORCE MEMBERS

When all free-body diagrams of a system have been constructed, we usually find that one or more represent either two- or three-force members without applied torques, and the techniques of the last section can be used for solution, at least for planar problems. Note that we started by seeking out such links and using them to establish known lines of action for other unknown forces. In doing this, we were implicitly using the summation-of-moments equation for such links. Then, by noting that action and reaction forces share the same line of action, we proceeded to establish more lines of action. Finally, upon reaching a link where one or more forces and all lines of action were known, we next applied the summation-of-forces equations to find the magnitudes and senses of the unknown forces. Proceeding from link to link in this way, we found the solutions for all unknowns.

In some problems, however, we find that this procedure reaches a point where the solution cannot proceed in this fashion, for example, when two- or three-force members cannot be found, and lines of action remain unknown. It is then wise to count the number of unknown quantities (magnitudes and directions) to be found for each free-body diagram. In planar problems, it is clear that Eqs. (14.8), the equilibrium conditions, cannot be solved for more than three unknowns on any single free-body diagram. Sometimes it is helpful to combine two or even three links and to draw a free-body diagram of the combined system. This approach can sometimes be used to eliminate unknown forces of the individual links, which combine with their reactions as internal stresses of the combined system.

In some problems this is still not sufficient. In such situations, it is always good practice to combine multiple forces on the same free-body diagram which are completely known by a single force representing the sum of the known forces. This sometimes simplifies the figures to reveal two- and three-force members that were otherwise not noticed.

In planar problems, the most general case of a system of forces which is solvable is one in which the three unknowns are the magnitudes of three forces. If all other forces are combined into a single force, we then have a four-force system. The following example shows how such a system can be treated.

EXAMPLE 14.3

A cam with a reciprocating roller follower are shown in Fig. 14.10a. The follower is held in contact with the cam by a spring pushing downward at C with a spring force of $F_C = 12$ N for this particular position. Also, an external load $F_E = 35$ N acts on the follower at E in the direction shown. For the position shown, determine the follower pin force at A and the bearing reactions at B and D. Assume no friction and a weightless follower.

SOLUTION

A free-body diagram of the follower is shown in Fig. 14.10b. Forces $\mathbf{F}_C$ and $\mathbf{F}_E$ are known, and their sum is obtained graphically in this figure. The free-body diagram is redrawn to show this in Fig. 14.10c. We show the resultant force $\mathbf{F}_E + \mathbf{F}_C$ with its point of application at E, where the original lines of action of $\mathbf{F}_E$ and $\mathbf{F}_C$ intersected, with line of action dictated by the direction of the vector sum. This placement of the point of application at E was not arbitrary, but required. Because the original forces $\mathbf{F}_E$ and $\mathbf{F}_C$ had no moments about point E, their resultant must not. The result of combining these two forces is that the free-body of Fig. 14.10c is now reduced to a four-force member with one known force, the resultant $\mathbf{F}_E + \mathbf{F}_C$, and three forces of unknown magnitudes.

In a similar manner, if the magnitude of $\mathbf{F}_D$ were known, it could be added to $\mathbf{F}_E + \mathbf{F}_C$ to produce a new resultant $\mathbf{F}_E + \mathbf{F}_C + \mathbf{F}_D$, which would act through point p.

Now consider the moment equation about point q. If we write $\sum \mathbf{M}_q = \mathbf{0}$, we see that the equation can be satisfied only if the resultant $\mathbf{F}_E + \mathbf{F}_C + \mathbf{F}_D$ has no moment about point q and thus has pq as its line of action. This is, therefore, the basis for the graphic solution shown in Fig. 14.10c. The direction of the line of action pq of the resultant $\mathbf{F}_E + \mathbf{F}_C + \mathbf{F}_D$ is used in the force polygon to find the force $\mathbf{F}_D$; the force polygon is then completed by finding $\mathbf{F}_A$ and $\mathbf{F}_B$ whose lines of action are known. The solutions are $F_A = 51.8$ N, $F_B = 32.8$ N, and $F_D = 5.05$ N when rounded to three significant figures.

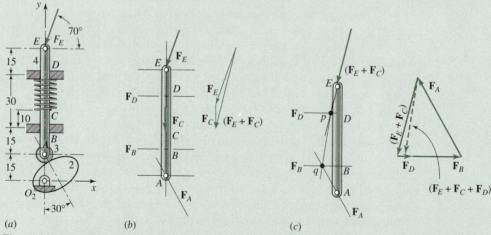

Figure 14.10 Graphic solution for Example 14.3: (a) A cam system with dimensions in millimeters. (b) Free-body diagram of link 4, a five-force member, and graphic summation of known forces F_C and F_E. (c) Free-body diagram of link 4, now reduced to a four-force member, and graphic solution.

Note that this approach defines a general concept, useful in either the graphic or the analytic approach. When there are three unknown force magnitudes on a single free-body diagram, choose a point such as q where the lines of action of two of the unknown forces meet and write the moment equation about that point, $\sum \mathbf{M}_q = \mathbf{0}$. This equation will have only one unknown remaining and can be solved directly. Only then should $\sum \mathbf{F} = \mathbf{0}$ be written, because the problem has then been reduced to two unknowns.

The following is a summary of the procedures for the graphic method of force analysis:

1. Identify each member in the mechanism; that is, identify two-force members, three-force members, four-force members, and so on. Draw complete free-body diagrams of each link. It is a good practice to combine all known forces on a free-body diagram into a single force; then commence with the lowest-force member. That is, draw two-force members first, then draw the three-force members, and, finally, draw any four-force members. If a member is acted upon by more than four forces, remember that either (a) it can be reduced to one of the above or (b) it has more than three unknowns and is not solvable.

2. Using the definitions of two-force, three-force, and four-force members, apply the following rules:
   (a) For a two-force member, the two forces must be equal, opposite, and collinear. Note that for a link with two forces and a torque, the forces are equal, opposite, and parallel.
   (b) For a three-force member, the three forces must intersect at a single point.
   (c) For a four-force member, the resultant of any two forces must be equal, opposite, and collinear with the resultant of the other two forces.

3. Draw force polygons (for three- and four-force members). Clearly state the scale of all force polygons. In order to be able to draw a force polygon, remember that you need three pieces of information (one of which must be a magnitude) for a three-force member. Five pieces of information (one of which must be a magnitude) are required for a four-force member. Note that, in general, it is less confusing if the force polygons are not superimposed on top of the mechanism diagram. Note that if a force polygon cannot be drawn for a particular member, then, in general, several links can be taken together and treated as a single free-body diagram.

14.9 FRICTION-FORCE MODELS

Over the years there has been much interest in the subjects of friction and wear, and many papers and books have been devoted to these subjects. It is not our purpose here to explore the mechanics of friction at all, but to present well-known simplifications which can be used in the analysis of the performance of mechanical devices. The results of any such analysis may not be theoretically perfect, but they do correspond closely to experimental measurements, so that reliable decisions can be made from them regarding a design and its operating performance.

Consider two bodies constrained to remain in contact with each other, with or without relative motion between them, such as the surfaces of block 3 and link 2 as shown in Fig. 14.11a. A force $\mathbf{F}_{43}$ may be exerted on block 3 by link 4, tending to cause the block 3

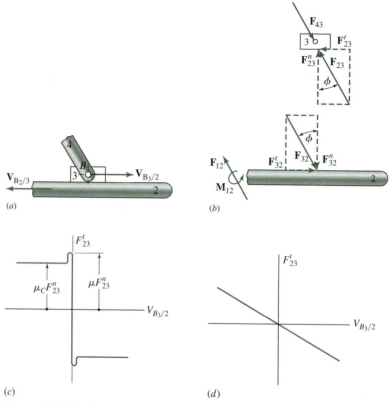

Figure 14.11 Mathematical representation of friction forces: (*a*) physical system; (*b*) free-body diagrams; (*c*) static and Coulomb friction models; and (*d*) viscous friction model.

to slide relative to link 2. Without the presence of friction between the interface surfaces of links 2 and 3, the block cannot transmit the component of $\mathbf{F}_{43}$ tangent to the surfaces. Instead of transmitting this force, the block would slide in the direction of this unbalanced force component; equilibrium would not be possible unless the line of action of $\mathbf{F}_{43}$ were normal to the surface.

With friction, however, a resisting force $\mathbf{F}_{23}^t$ is developed at the contact surface as shown in the free-body diagrams of Fig. 14.11*b*. This friction force $\mathbf{F}_{23}^t$ acts in addition to the usual constraint force $\mathbf{F}_{23}^n$ across the surface of the sliding joint; together, the force components $\mathbf{F}_{23}^n$ and $\mathbf{F}_{23}^t$ form the total constraint force $\mathbf{F}_{23}$ that balances the force $\mathbf{F}_{43}$ to keep the block 3 in equilibrium. Of course, the reaction force components $\mathbf{F}_{23}^n$ and $\mathbf{F}_{23}^t$ are also acting simultaneously on link 2, as shown in the other free-body diagram of Fig. 14.11*b*. The force component $\mathbf{F}_{23}^t$ and its reaction $\mathbf{F}_{32}^t$ are called *friction forces*.

Depending on the materials of links 2 and 3, there is a limit to the size of the force component $\mathbf{F}_{23}^t$, which can be developed by friction while still maintaining equilibrium. This limit is expressed by the relationship

$$F_{23}^t \leq \mu F_{23}^n \tag{14.9}$$

where μ, referred to as the *coefficient of static friction,* is a characteristic property of the materials in contact. Values of the coefficient μ have been determined experimentally for many materials and can be found in most engineering handbooks.[3]

If the force $\mathbf{F}_{43}$ is tipped too far, so that its tangential component and therefore the friction force component F_{23}^t would be too large to satisfy the inequality of Eq. (14.9), equilibrium is not possible and the block 3 will slide relative to link 2 with an apparent velocity $\mathbf{V}_{B_3/2}$. When sliding takes place, the friction force takes on the value

$$F_{23}^t = \mu_c F_{23}^n \tag{14.10}$$

where μ_c is the *coefficient of sliding friction.* This manner of approximating sliding friction is called *Coulomb friction,* and we shall often use this term to refer to the relationship of Eq. (14.10). The coefficient μ_c can also be found experimentally and is slightly less than μ, the static coefficient, for most materials.

To summarize what we have discussed so far, Fig. 14.11c shows a graph of the friction force F_{23}^t versus the apparent sliding velocity $\mathbf{V}_{B_3/2}$. Here it can be seen that when the sliding velocity is zero, the friction force F_{23}^t can have any magnitude between μF_{23}^n and $-\mu F_{23}^n$. When the velocity is not zero, the magnitude of the friction force F_{23}^t drops slightly to the value $\mu_c F_{23}^n$ and has a sense which opposes that of the sliding motion $\mathbf{V}_{B_3/2}$.

Looking again at the total force $\mathbf{F}_{23}$ in Fig. 14.11b, we see that it is tipped at an angle to the surface normal and is equal and opposite to $\mathbf{F}_{43}$ whenever the system is in equilibrium. Thus, the angle ϕ is given by

$$\tan \phi = \frac{F_{23}^t}{F_{23}^n}$$

$$\tan \phi \le \frac{\mu F_{23}^n}{F_{23}^n} = \mu$$

$$\phi \le \tan^{-1} \mu$$

When $\mathbf{F}_{43}$ is tipped so that the block 3 is just on the verge of sliding,

$$\phi = \tan^{-1} \mu \tag{14.11}$$

The limiting value of the angle ϕ, called the *friction angle,* defines the maximum angle through which the force $\mathbf{F}_{23}$ can tip from the surface normal before equilibrium will not be possible and sliding will take place. Notice that the limiting value of ϕ does not depend on the magnitude of the force $\mathbf{F}_{23}$ but only on the coefficient of friction of the materials involved.

We might also notice that, in the above discussion, we have treated only the effects of friction between surfaces with relative sliding motion. We might ask how this can be extended to treat relative rotational motion such as in a revolute joint. Because good low-friction rotational bearings are not expensive and because they can be easily lubricated to reduce friction, this is usually not a problem. Extensions of the above ideas are known for rotational bearings.[4] However, the additional difficulty does not often warrant their use, because the results are usually not affected by more than a very small percentage.

Even though the static friction or Coulomb friction models are often used and often do represent good approximations for friction forces in mechanical equipment, they are not the only models. Sometimes—for example, when representing a machine or its dynamic effects by its differential equation of motion—it is more convenient to analyze the machine's performance using another approximation for friction forces, called *viscous friction* or *viscous damping*. As shown in the graph of Fig. 14.11*d*, this model has a linear relationship between the magnitude of the friction force and the sliding velocity. This viscous friction model is especially useful when the dynamic analysis of a machine leads to the use of one or more differential equations. The nonlinear relationship of static and/or Coulomb friction, shown in Fig. 14.11*c*, leads to a nonlinear differential equation that is more difficult to treat.

Whether the friction effect is better represented by the viscous, Coulomb, or static friction models, it is important to recognize the sense of the friction force. As a mnemonic device, the rule is often stated that "friction opposes motion," as shown by the free-body diagram of links 3 in Fig. 14.11*b*, where the sense of $\mathbf{F}_{23}^t$ is opposite to that of $\mathbf{V}_{B_3/2}$. This rule of thumb is not wrong if it is applied carefully, but it can be dangerous and even misleading. It will be noted in Fig. 14.11*a* that there are two motions that might be thought of, namely, $\mathbf{V}_{B_3/2}$ and $\mathbf{V}_{B_2/3}$; there are also two friction forces, $\mathbf{F}_{23}^t$ and $\mathbf{F}_{32}^t$. Careful examination of Fig. 14.11*b* shows that $\mathbf{F}_{23}^t$ opposes the sense of $\mathbf{V}_{B_3/2}$ while $\mathbf{F}_{32}^t$ opposes the sense of $\mathbf{V}_{B_2/3}$. In machine systems, where both sides of a sliding joint are often in motion, it is very important to understand *which* friction force "opposes" *which* motion.

14.10 STATIC FORCE ANALYSIS WITH FRICTION

We will show the effect of including friction on our previous methods of static force analysis by the following examples.

EXAMPLE 14.4

Repeat the static force analysis of the slider-crank mechanism (Fig. 14.8*a*) that was analyzed in Example 14.2, assuming that there is a coefficient of static friction of $\mu = 0.25$ between the slider and the ground. Assume that the impending motion of the slider is to the left and that friction can be neglected in the revolute joints.

SOLUTION

As always, when we begin a force analysis with friction, it is necessary to solve the problem first without friction. The purpose is to find the sense of each of the normal force components, in this example $\mathbf{F}_{14}^n$. This was done in Example 14.2, where $\mathbf{F}_{14}^n$ was found to act vertically upward.

Because the impending motion of the slider is to the left—that is, $\mathbf{V}_{B_4/1}$ is to the left—the friction force $\mathbf{F}_{14}^t$ must act to the right. We can redraw the free-body diagram of link 4, Fig. 14.8*c*, and include the friction force as shown in Fig. 14.12*a*. Because of the static friction, the line of action of $\mathbf{F}_{14}$ is shown tipped through the angle ϕ, that is calculated from Eq. (14.11):

$$\phi = \tan^{-1} 0.25 = 14°$$

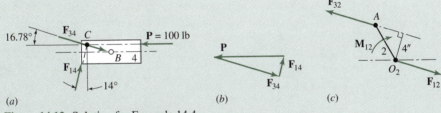

Figure 14.12 Solution for Example 14.4.

In deciding the direction of tip of the angle ϕ, it is necessary to know the sense of the normal force (upward) and the sense of the friction force (horizontal to the right) at the point of contact. This explains why the solution without friction should be completed first.

Once the new line of action of the force $\mathbf{F}_{14}$ is known, the solution can proceed exactly as in Example 14.2. The graphical solution, with friction effects, is shown in Fig. 14.12, where it is found that

$$F_{34} = 95 \text{ lb} \quad \text{and} \quad F_{14} = 30 \text{ lb} \qquad \qquad \textit{Ans.}$$

Note that the direction of $\mathbf{F}_{34}$ has not changed (there is no friction in the revolute joint B) and the magnitude is less than the no friction case. Therefore, less torque will be required to act on link 2 to maintain the mechanism in equilibrium. We find that

$$M_{12} = (4 \text{ in})(95 \text{ lb}) = 380 \text{ in·lb} \qquad \qquad \textit{Ans.}$$

EXAMPLE 14.5

Repeat the static force analysis of the cam-follower system analyzed in Example 14.3 (Fig. 14.10a), assuming that there is a coefficient of static friction of $\mu = 0.15$ between links 1 and 4 at both sliding bearings B and D. Friction in all other joints is considered negligible. Determine the minimum force necessary at A to hold the system in equilibrium.

SOLUTION

As always when we begin a force analysis with friction, it is necessary to solve the problem first without friction. The purpose is to find the sense of each of the normal force components, in our case $\mathbf{F}_B^n$ and $\mathbf{F}_D^n$. This was done in Example 14.3, where $\mathbf{F}_B^n$ and $\mathbf{F}_D^n$ were both found to act to the right in Fig. 14.10c.

The next step is to consider the problem statement carefully and to decide the direction of the impending motion. As stated, the problem asks for the minimum force at A necessary for equilibrium; that is, the problem statement implies that if $\mathbf{F}_A$ were any smaller, the system would move downward. Thus the impending motion is with velocities $\mathbf{V}_{D_4/1}$ and $\mathbf{V}_{B_4/1}$ downward. Thus the two friction forces at B and D must act upward on link 4.

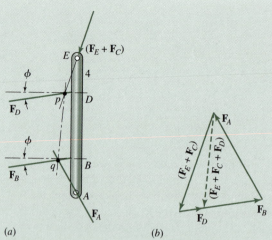

Figure 14.13 Graphic solution for Example 14.5: free-body diagram of link 4 with static friction.

Next we redraw the free-body diagram of link 4 (Fig. 14.10c) and include the friction forces as shown in Fig. 14.13a. Here, because of static friction, the lines of action of $\mathbf{F}_B$ and $\mathbf{F}_D$ are both shown tipped through the angle ϕ, which is calculated from Eq. (14.11):

$$\phi = \tan^{-1} 0.15 = 8.5°$$

In deciding the direction of tip of the angles ϕ, it is necessary to know the sense of each friction force (upward) and the sense of each normal force (toward the right) at B and D. This explains why the solution without friction must be done first.

Once the new lines of action of the forces $\mathbf{F}_B$ and $\mathbf{F}_D$ are known, the solution can proceed exactly as in Example 14.3. The graphic solution, with friction effects, is shown in Fig. 14.13b, where it is found that

$$F_B = 28.7 \text{ N}, \qquad F_D = 6.57 \text{ N}, \qquad F_A = 45.8 \text{ N} \qquad\qquad Ans.$$

Notice from this example that the normal components of the two forces at B and D are now $\mathbf{F}_B^n = 28.4$ N and $\mathbf{F}_D^n = 6.50$ N and are different from the values found without friction in Example 14.3. This should be a warning, whether one is working graphically or analytically, that it is incorrect simply to multiply the frictionless normal forces by the coefficient of friction to find the friction forces and then to add these to the frictionless solution. All forces may (and usually do) change magnitude when friction is included, and the problem must be completely reworked from the beginning with friction included. The effects of static or Coulomb friction *cannot* be added in afterward by superposition.

Notice also that if the problem statement had asked for the maximum force at A, the impending motion of link 4 would have been upward and the friction forces downward on link 4. This would have reversed the tilt of the two lines of action for both $\mathbf{F}_B$ and $\mathbf{F}_D$ and would have totally changed the final results. Now, in practice, if the actual value of the force $\mathbf{F}_A$ is between these minimum and maximum values, equilibrium will be maintained, and the values of other constraint forces will be between the two extreme values found by this type of analysis. Also, if the value of $\mathbf{F}_A$ is slowly increased from the minimum toward the maximum value, the follower will remain in equilibrium until the maximum value is reached, and then begin to move; this discontinuous action is sometimes referred to as "stiction."

14.11 SPUR- AND HELICAL-GEAR FORCE ANALYSIS

Figure 14.14a shows a pinion with center O_2 rotating clockwise at a speed of ω_2 and driving a gear with center at O_3 at a speed of ω_3. As brought out in Chapter 6, the reactions between the teeth occur along the pressure line AB, tipped by the pressure angle ϕ from the common tangent to the pitch circles. Free-body diagrams of the pinion and gear are shown in Fig. 14.14b. The action of the pinion on the gear is shown by the force $\mathbf{F}_{23}$ acting at the pitch point along the pressure line.[5] (See detailed explanation on page 463.) Because the gear is supported by its shaft, $\sum \mathbf{F} = \mathbf{0}$ shows that an equal and opposite force $\mathbf{F}_{13}$ must act at the centerline of the shaft. A similar analysis of the pinion shows that the same observations are true. In each case the forces are equal in magnitude, opposite in direction, parallel, and in the same plane. On either gear, therefore, they form a couple.

Note that the free-body diagram of the pinion shows the forces resolved into components. Here we employ the superscripts r and t to indicate the radial and tangential directions with respect to the pitch circle. It is expedient to use the same superscripts for the components of the force $\mathbf{F}_{12}$ that the shaft exerts on the pinion. The moment of the couple formed by $\mathbf{F}_{32}^t$ and $\mathbf{F}_{12}^t$ is in equilibrium with the torque $\mathbf{T}_{12}$ that must be applied

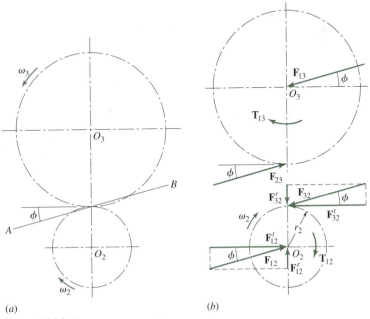

(a)   (b)

Figure 14.14 Forces on spur gears.

by the shaft to drive the gearset. When the pitch radius of the pinion is designated as R_2, $\sum \mathbf{M}_{O_2} = \mathbf{0}$ shows that the torque is

$$T_{12} = R_2 F_{32}^t \tag{14.12}$$

Note that the radial force component $\mathbf{F}_{32}^r$ serves no purpose as far as the transmission of power is concerned. For this reason, $\mathbf{F}_{32}^t$ is frequently called the *transmitted* force.

In applications involving gears, the power transmitted and the shaft speeds are often given. Remembering that power is the product of force times velocity, or torque times angular velocity, we can find the relation between power and the transmitted force. Using the symbol P to denote power, we obtain

$$P = T_{12}\omega_2 = R_2 F_{32}^t \omega_2 \tag{14.13a}$$

which can be solved for the transmitted force F_{32}^t as follows:

$$F_{32}^t = \frac{P}{R_2 \omega_2} \tag{14.13b}$$

In application of these formulae it is often necessary to remember that the U.S. customary units used for power are horsepower, abbreviated hp, where 1 hp = 33 000 ft·lb/min; and in SI units, power is measured in watts, abbreviated W, where 1 W = 1 N·m/s.

Once the transmitted force is known, the following relations for spur gears are evident from Fig. 14.14b:

$$F_{32}^r = F_{32}^t \tan \phi \quad \text{and} \quad F_{32} = \frac{F_{32}^t}{\cos \phi} \tag{14.14}$$

In the treatment of forces on helical gears it is convenient to determine the axial force, work with it independently, and treat the remaining force components the same as for straight spur gears. Figure 14.15 is a drawing of a helical gear with a portion of the face removed to show the tooth contact force and its components acting at the pitch point. The gear is imagined to be driven clockwise under load. The driving gear has been removed and its effect replaced by the force shown acting on the teeth.

The resultant force $\mathbf{F}$ is shown divided into three components $\mathbf{F}^a$, $\mathbf{F}^r$, and $\mathbf{F}^t$, which are the axial, radial, and tangential components, respectively. The tangential force component is the transmitted force and is the one that is effective in transmitting torque. When the transverse pressure angle is designated as ϕ_t and the helix angle as ψ, the following relations are evident from Fig. 14.15:

$$\mathbf{F} = \mathbf{F}^a + \mathbf{F}^r + \mathbf{F}^t \tag{14.15}$$

$$F^a = F^t \tan \psi \tag{14.16}$$

$$F^r = F^t \tan \phi_t \tag{14.17}$$

It is also expedient to make use of the resultant of $\mathbf{F}^r$ and $\mathbf{F}^t$. We shall designate this force as $\mathbf{F}^\phi$, defined by the equation

$$\mathbf{F}^\phi = \mathbf{F}^r + \mathbf{F}^t \tag{14.18}$$

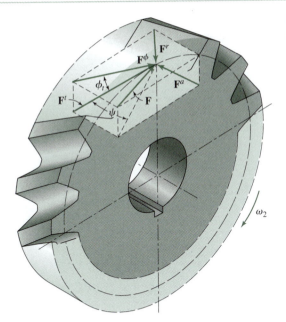

Figure 14.15 Force components on a helical gear at the tooth contact point.

EXAMPLE 14.6

A gear train is composed of three helical gears with shaft centers in line. The driver is a right-hand helical gear having a pitch radius of 2 in, a transverse pressure angle of 20°, and a helix angle of 30°. An idler gear in the train has the teeth cut left hand and has a pitch radius of 3.25 in. The idler transmits no power to its shaft. The driven gear in the train has the teeth cut right hand and has a pitch radius of 2.50 in. If the transmitted force is 600 lb, find the shaft forces acting on each gear. Gravitational forces can be neglected.

SOLUTION

First we consider only the axial forces as previously suggested. For each mesh, the axial component of the reaction, from Eq. (14.16), is

$$F^a = F^t \tan \psi = 600 \tan 30° = 346 \text{ lb}$$

Figure 14.16a is a top view of the three gears, looking down on the plane formed by the three axes of rotation. For each gear, rotation is considered to be about the z axis for this problem. In Fig. 14.16b, free-body diagrams of each of the three gears are drawn in perspective and the three coordinate axes are shown. As indicated, the idler exerts a force $\mathbf{F}_{32}^a$ on the driver. This is resisted by the axial shaft force $\mathbf{F}_{12}^a$. The forces $\mathbf{F}_{12}^a$ and $\mathbf{F}_{32}^a$ form a couple that is resisted by the moment $\mathbf{T}_{12}^y$. Note that this moment is negative (clockwise) about the $+y$ axis. Consequently it produces a bending moment in the shaft. The magnitude of this moment is

$$T_{12}^y = -R_2 F_{12}^a = -(2.00 \text{ in})(346 \text{ lb}) = -693 \text{ in·lb}$$

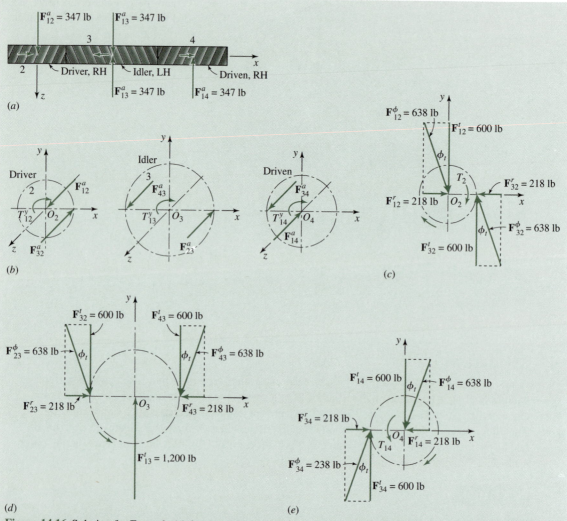

Figure 14.16 Solution for Example 14.6: (*a*) and (*b*) axial forces; (*c*) free-body diagram of driver gear 2; (*d*) free-body diagram of idler gear 3; (*e*) free-body diagram of driven gear 4.

Turning next to the idler, we see from Figs. 14.16*a* and 14.16*b* that the net axial force on the shaft of the idler is zero. The axial component of the force from the driver on the idler is $\mathbf{F}_{23}^a$, and that from the driven gear on the idler is $\mathbf{F}_{43}^a$. These two forces are equal and opposite and form a couple that is resisted by the bending moment in the shaft of the idler $\mathbf{T}_{13}^y$, of magnitude

$$T_{13}^y = -2R_3 F_{23}^a = -2(3.25\text{ in})(346\text{ lb}) = -2\,252\text{ in·lb}$$

The driven gear has the axial force component $\mathbf{F}_{34}^a$ acting along its pitch line due to the helix angle of the idler. This is resisted by the axial shaft reaction $\mathbf{F}_{14}^a$. As shown, these forces are equal and form a couple that is resisted by the moment $\mathbf{T}_{14}^y$. The magnitude of this moment is

$$T_{14}^y = -R_4 F_{34}^a = -(2.50 \text{ in})(346 \text{ lb}) = -866 \text{ in·lb}$$

It is emphasized again that the three resisting moments $\mathbf{T}_{12}^y$, $\mathbf{T}_{13}^y$, and $\mathbf{T}_{14}^y$ are due solely to the axial components of the reactions between the gear teeth. They produce static bearing reactions and have no effect on the amount of power transmitted.

Now that all of the reactions due to the axial components have been found, we turn our attention to the remaining force components and examine their effects as if they were operating independently of the axial forces. Free-body diagrams showing the force components in the plane of rotation for the driver, idler, and driven gears are shown, respectively, in Figs. 14.16c, 14.16d, and 14.16e. These force components can be obtained graphically as shown or by applying Eqs. (14.12) and (14.17). It is not necessary to combine the components to find the resultant forces, because the components are exactly those that are desired to proceed with machine design.

EXAMPLE 14.7

The planetary gear train shown in Fig. 14.17a has input shaft a which is driven by a torque of $\mathbf{T}_{a2} = -100\mathbf{k}$ in·lb. Note that input shaft a is connected directly to gear 2 and that the planetary arm 3 is connected directly to the output shaft b. Shafts a and b rotate about the same axis but are not connected. Gear 6 is fixed to the stationary frame 1 (not shown). All gears have a diametral pitch of 10 teeth per inch and a pressure angle of 20°. Assuming that the forces act in a single plane and that gravitational forces and centrifugal forces on the planet gears can be neglected, make a complete force analysis of the parts of the train and compute the magnitude and direction of the output torque delivered by shaft b.

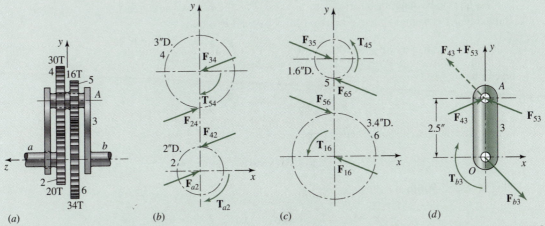

Figure 14.17 Solution for Example 14.7: (a) planetary gear train with tooth numbers; (b) free-body diagrams of gears 2 and 4; (c) free-body diagrams of gears 5 and 6; (d) free-body diagram of planet carrier arm 3.

SOLUTION

The pitch radii of the gears are $R_2 = 20/(2 \cdot 10) = 1.00$ in, $R_4 = 1.50$ in, $R_5 = 0.80$ in, and $R_6 = 1.70$ in. The distance between centers of meshing gear pairs is $(R_2 + R_4) = (1.00$ in $+ 1.50$ in$) = 2.50$ in. Because the torque that the input shaft exerts on gear 2 is $T_{a2} = 100$ in·lb, the transmitted force is $F_{42}^t = T_{a2}/R_2 = (100$ in·lb$)/(1$ in$) = 100$ lb. Therefore, $F_{42} = F_{42}^t/\cos\phi = 100/\cos 20° = 106$ lb. The free-body diagram of gear 2 is shown in Fig. 14.17b. In vector form the results are

$$\mathbf{F}_{a2} = -\mathbf{F}_{42} = 106\angle 20° \text{ lb}$$

Figure 14.17b also shows the free-body diagram of gear 4. The forces are

$$\mathbf{F}_{24} = -\mathbf{F}_{34} = 106\angle 20° \text{ lb}$$

where $\mathbf{F}_{34}$ is the force of planet arm 3 against gear 4. Gears 4 and 5 are connected to each other, but turn freely on the planet arm shaft. Thus $\mathbf{T}_{54}$ is the torque exerted by gear 5 onto gear 4. This torque is $T_{54} = (R_4)F_{24}^t = (1.50$ in$)(100$ lb$) = 150$ in·lb.

Turning next to the free-body diagram of gear 5 in Fig. 14.17c, we first find $F_{65}^t = T_{45}/R_5 = (150$ in·lb$)/(0.800$ in$) = 188$ lb. Therefore, $F_{65} = 188/\cos 20° = 200$ lb. In vector form the results for gear 5 are summarized as

$$\mathbf{F}_{65} = -\mathbf{F}_{35} = 200\angle 160° \text{ lb}, \quad \mathbf{T}_{45} = 150\hat{\mathbf{k}} \text{ in·lb}$$

For gear 6, shown in Fig. 14.17c, we have

$$\mathbf{F}_{15} = -\mathbf{F}_{56} = 200\angle 160° \text{ lb}$$

$$\mathbf{T}_{16} = R_6 F_{56}^t \hat{\mathbf{k}} = (1.70 \text{ in})(200\cos 20° \text{ lb})\hat{\mathbf{k}} = 319\hat{\mathbf{k}} \text{ in·lb}$$

Note that $\mathbf{F}_{16}$ and $\mathbf{T}_{16}$ are the force and torque, respectively, exerted by the frame on gear 6.

The free-body diagram of arm 3 is shown in Fig. 14.17d. As noted earlier, the forces are assumed to act in a single plane. The two forces $\mathbf{F}_{43}$ and $\mathbf{F}_{53}$ are

$$\mathbf{F}_{43} = -\mathbf{F}_{34} = 106\angle 20° \text{ lb}, \qquad \mathbf{F}_{53} = -\mathbf{F}_{35} = 200\angle 160° \text{ lb}$$

and can be summed to

$$\mathbf{F}_{43} + \mathbf{F}_{53} = 137\angle 130.2° \text{ lb}$$

We now find the output shaft reaction to be

$$\mathbf{F}_{b3} = -(\mathbf{F}_{43} + \mathbf{F}_{53}) = 137\angle -49.8° \text{ lb}$$

Using $\mathbf{R}_{AO} = 2.50\hat{\mathbf{j}}$ in and the equation

$$\sum \mathbf{M}_O = \mathbf{T}_{b3} + \mathbf{R}_{AO} \times (\mathbf{F}_{43} + \mathbf{F}_{53}) = 0$$

we find $\mathbf{T}_{b3} = -221\hat{\mathbf{k}}$ in·lb. Therefore, the output shaft torque is $\mathbf{T}_b = 221\hat{\mathbf{k}}$ in·lb.

14.12 STRAIGHT-BEVEL-GEAR FORCE ANALYSIS

In determining the tooth forces on bevel gears, it is customary to use the forces that would occur at the mid-thickness of the tooth on the pitch cone. The resultant tangential force probably occurs somewhere between the midpoint and the large end of the tooth, but there will be only a small error in making this approximation. The tangential or transmitted force is then given by

$$F^t = \frac{T}{R} \tag{14.19}$$

where R is the mid-radius of the pitch cone as shown in Fig. 14.18 and T is the shaft torque.

Figure 14.18 also shows all the components of the resultant force acting at the midpoint of the tooth. The following relationships can be derived by inspection of the figure:

$$\mathbf{F} = \mathbf{F}^a + \mathbf{F}^r + \mathbf{F}^t \tag{14.20}$$

$$F^r = F^t \tan\phi \cos\gamma \tag{14.21}$$

$$F^a = F^t \tan\phi \sin\gamma \tag{14.22}$$

Note, as in the case of helical gears, that the axial force $\mathbf{F}^a$ results in a couple on the shaft that produces a bending moment.

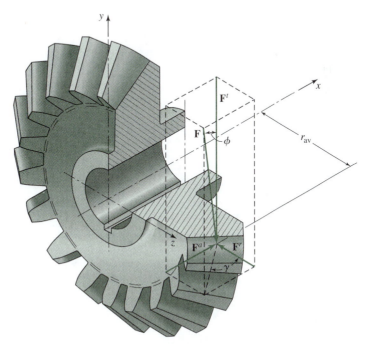

Figure 14.18 Force components on a straight bevel gear at the tooth contact point.

EXAMPLE 14.8

The bevel pinion shown in Fig. 14.19 rotates at 600 rev/min in the direction shown and transmits 5 hp to the gear. The mounting distances are shown, together with the locations of the bearings on each shaft. Bearings A and C are capable of taking both radial and axial loads, while bearings B and D are designed to receive only radial loads. The teeth of the gears have a 20° pressure angle. Find the components of the forces that the bearings exert on the shafts in the x, y, and z directions.

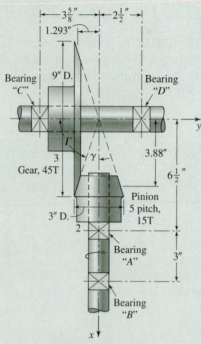

Figure 14.19 Bevel gearset and bearing locations for Example 14.8.

SOLUTION

The pitch angles for the pinion and gear are

$$\gamma = \tan^{-1}\left(\frac{3.0}{9.0}\right) = 18.4°$$

$$\Gamma = \tan^{-1}\left(\frac{9.0}{3.0}\right) = 71.6°$$

The radii to the midpoints of the teeth are shown on the drawing and are $R_2 = 1.293$ in and $R_3 = 3.875$ in for the pinion and gear, respectively.

Let us determine the forces acting upon the pinion first. Using Eq. (14.13b) we find the transmitted force to be

$$F_{32}^t = \frac{P}{R_2\omega_2} = \frac{(5 \text{ hp})(33\,000 \text{ ft·lb/min/hp})(12 \text{ in/ft})}{(1.293 \text{ in})(600 \text{ rev/min})(2\pi \text{ rad/rev})} = 406 \text{ lb}$$

This force acts in the negative z direction—that is, into the page in Fig. 14.19. The radial and axial components of $\mathbf{F}_{32}$ are obtained from Eqs. (14.21) and (14.22):

$$F_{32}^r = 406 \tan 20° \cos 18.4° = 140 \text{ lb}$$

$$F_{32}^a = 406 \tan 20° \sin 18.4° = 46.7 \text{ lb}$$

where $\mathbf{F}_{32}^r$ acts in the positive y direction and $\mathbf{F}_{32}^a$ in the positive x direction.

These three forces are components of the total force $\mathbf{F}_{32}$. Thus

$$\mathbf{F}_{32} = 46.7\hat{\mathbf{i}} + 140\hat{\mathbf{j}} - 406\hat{\mathbf{k}} \text{ lb}$$

The torque applied to the pinion shaft must be

$$\mathbf{T}_{12} = -\mathbf{R}_2 \times \mathbf{F}_{32}^t = -(-1.29\hat{\mathbf{j}} \text{ in}) \times (-406\hat{\mathbf{k}} \text{ lb}) = -525\hat{\mathbf{i}} \text{ in·lb}$$

A free-body diagram of the pinion and shaft is shown schematically in Fig. 14.20a. The bearing reactions $\mathbf{F}_A$ and $\mathbf{F}_B$ are to be determined; the dimensions, the torque $\mathbf{T}_{12}$, and the force $\mathbf{F}_{32}$ are the given elements of the problem. In order to find $\mathbf{F}_B$ we shall sum moments about A. This requires two position difference vectors, defined as

$$\mathbf{R}_{PA} = -2.625\hat{\mathbf{i}} - 1.293\hat{\mathbf{j}} \text{ in} \quad \text{and} \quad \mathbf{R}_{BA} = 3.0\hat{\mathbf{i}} \text{ in}$$

Summing moments about A gives

$$\sum \mathbf{M}_A = \mathbf{T}_{12} + \mathbf{R}_{BA} \times \mathbf{F}_B + \mathbf{R}_{PA} \times \mathbf{F}_{32} = 0 \tag{1}$$

The second and third terms of Eq. (1), respectively, are

$$\mathbf{R}_{BA} \times \mathbf{F}_B = 3.0\hat{\mathbf{i}} \times \left(F_B^y\hat{\mathbf{j}} + F_B^z\hat{\mathbf{k}}\right)$$
$$= -3.0F_B^z\hat{\mathbf{j}} + 3.0F_B^y\hat{\mathbf{k}} \tag{2}$$

$$\mathbf{R}_{PA} \times \mathbf{F}_{32} = (-2.625\hat{\mathbf{i}} - 1.293\hat{\mathbf{j}}) \times (46.7\hat{\mathbf{i}} + 140\hat{\mathbf{j}} - 406\hat{\mathbf{k}})$$
$$= 525\hat{\mathbf{i}} - 1\,066\hat{\mathbf{j}} - 307\hat{\mathbf{k}} \text{ in·lb} \tag{3}$$

Placing the value of $\mathbf{T}_{12}$ and Eqs. (2) and (3) into Eq. (1) and solving gives

$$\mathbf{F}_B = 102\hat{\mathbf{j}} - 355\hat{\mathbf{k}} \text{ lb} \qquad \textit{Ans.}$$

The magnitude of $\mathbf{F}_B$ is 370 lb.

Next, to find $\mathbf{F}_A$, we write

$$\sum \mathbf{F} = \mathbf{F}_{32} + \mathbf{F}_A + \mathbf{F}_B = 0$$

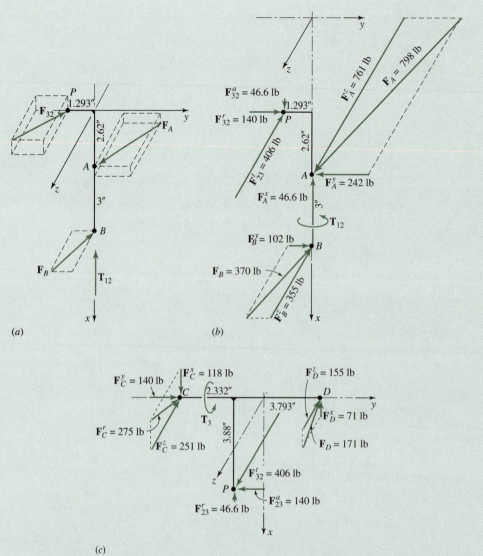

(a) (b)

(c)

Figure 14.20 Solution for Example 14.8: (a) free-body diagrams of piston and input shaft; (b) results and component values on pinion and input shaft; (c) free-body diagrams of gear and output shaft with results and component values.

When this equation is solved for $\mathbf{F}_A$, the result is

$$\mathbf{F}_A = -46.7\hat{\mathbf{i}} - 242\hat{\mathbf{j}} + 761\hat{\mathbf{k}} \text{ lb} \qquad\qquad Ans.$$

The magnitude of $\mathbf{F}_A$ is 800 lb. The results are shown in Fig. 14.20b.

A similar procedure is used for the gear shaft. The results are displayed in Fig. 14.20c.

14.13 THE METHOD OF VIRTUAL WORK

So far in this chapter we have learned to analyze problems involving the equilibrium of mechanical systems by the application of Newton's laws. Another fundamentally different approach to force-analysis problems is based on the principle of virtual work, first proposed by the Swiss mathematician J. Bernoulli in the eighteenth century.

The method is based on an energy balance of the system which requires that the net change in internal energy during a small displacement must be equal to the difference between the work input to the system and the work output including the work done against friction, if any. Thus, for a system of rigid bodies in equilibrium under a system of applied forces, if given an arbitrary small displacement from equilibrium, the net change in the internal energy, denoted here by dU, will be equal to the work dW done on the system:

$$dU = dW \tag{14.23}$$

Work and change in internal energy are positive when work is done on the system, giving it increased internal energy, and negative when work is lost from the system, such as when it is dissipated through friction. If the system has no friction or other dissipation losses, energy is conserved and the net change in internal energy during a small displacement from equilibrium is zero.

Of course, such a method requires that we know how to calculate the work done by each force during the small *virtual displacement* chosen. If some force **F** acts at a point of application Q which undergoes a small displacement $d\mathbf{R}_Q$, then the work done by this force on the system is given by

$$dU = \mathbf{F} \cdot d\mathbf{R}_Q \tag{14.24}$$

where dU is a scalar value, having units of work or energy, and will be positive for work done onto the system and negative for work output by the system.

The displacement considered is called a *virtual* displacement because it need not be one that truly happens on the physical machine. It need only be a small displacement that is hypothetically possible and consistent with the constraints imposed on the system. The small displacement relationships of the system can be found through the principles of kinematics covered in Part 1 of this book, and the input-to-output force relationships of the machine can therefore be found. This will become more clear through the following example.

EXAMPLE 14.9

Repeat the static force analysis of the four-bar linkage analyzed in Example 14.1. Find the input crank torque $\mathbf{M}_{12}$ required for equilibrium. Friction effects and the weights of the links may be neglected.

SOLUTION

From Example 14.1 we recall that

$$\mathbf{R}_{QO_4} = 1.84\hat{\mathbf{i}} + 4.65\hat{\mathbf{j}} \text{ in} \tag{1}$$

$$\mathbf{P} = 120\angle220° = -91.9\hat{\mathbf{i}} - 77.1\hat{\mathbf{j}} \text{ lb} \tag{2}$$

Figure 14.21 presents a scale diagram of the system at the position in question. If a small virtual displacement $d\theta_2$ is given to the input crank 2, this results in a small displacement $d\theta_4$ of the output crank 4, along with a small displacement $d\mathbf{R}_{QO_4}$ of point Q as shown.

Assuming that these virtual displacements are small and happen in a small time increment dt, the velocity difference equation, Eq. (3.3), can be multiplied by dt to yield the relationship between $d\mathbf{R}_{QO_4}$ and $d\theta_4$ as follows:

$$\begin{aligned} d\mathbf{R}_{QO_4} = d\theta_4\hat{\mathbf{k}} \times \mathbf{R}_{QO_4} &= (d\theta_4\hat{\mathbf{k}} \text{ rad}) \times (1.84\hat{\mathbf{i}} + 4.65\hat{\mathbf{j}} \text{ in}) \\ &= d\theta_4(-4.65\hat{\mathbf{i}} + 1.84\hat{\mathbf{j}}) \text{ in} \end{aligned} \tag{3}$$

Because the input torque $\mathbf{M}_{12}$ and the applied force $\mathbf{P}$ are the only forces doing work during the chosen displacement, Eq. (14.23) can be written for this problem as

$$dU = M_{12}\hat{\mathbf{k}} \cdot d\theta_2\hat{\mathbf{k}} + \mathbf{P} \cdot d\mathbf{R}_{QO_4} = 0$$

Substituting Eqs. (2) and (3) into this equation gives

$$M_{12}d\theta_2 + (-91.9\hat{\mathbf{i}} - 77.1\hat{\mathbf{j}}) \cdot (-4.65\hat{\mathbf{i}} + 1.84\hat{\mathbf{j}})d\theta_4 \text{ in·lb} = 0$$

Then dividing by $d\theta_2$, rearranging, and recognizing the first-order kinematic coefficient, we find

$$M_{12} = -285.5\frac{d\theta_4}{d\theta_2} = -285.5\theta_4' \text{ in·lb} \tag{4}$$

By dividing the numerator and denominator of this angular displacement ratio by dt, we recognize that this is equal to the angular-velocity ratio ω_4/ω_2, which can be found from the angular velocity ratio theorem of Eq. (3.25).[6] Using the location of the instant center P_{24} and

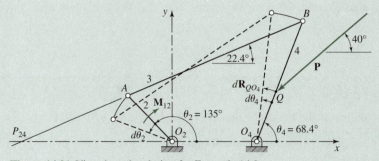

Figure 14.21 Virtual-work solution for Example 14.9.

measurements from Fig. 14.21, we find

$$\theta_4' = \frac{d\theta_4}{d\theta_2} = \frac{\omega_4}{\omega_2} = \frac{R_{P_{24}P_{12}}}{R_{P_{24}P_{14}}} = \frac{14.3 \text{ in}}{22.3 \text{ in}} = 0.641$$

Finally, using this in Eq. (4) gives the result

$$\mathbf{M}_{12} = (-285.5\hat{\mathbf{k}} \text{ in·lb})(0.641) = -183.1\hat{\mathbf{k}} \text{ in·lb} \qquad \qquad Ans.$$

The primary advantage of the method of virtual work for force analysis over the other methods shown above comes in problems where the input-to-output force relationships are sought. Notice that the constraint forces are not required in this solution technique because both their action and reaction forces are internal to the system, both move through identical displacements, and thus their virtual-work contributions cancel each other. This would not be true for friction forces where the displacements would be different for the action and reaction force components, with this work difference representing the energy dissipated. Otherwise, internal constraint forces need not be considered because they cause no net virtual work.

NOTES

[1.] The determination of the sizes of machine members is the subject of books usually titled machine design or mechanical design. See J. E. Shigley and C. R. Mischke, *Mechanical Engineering Design,* 6th ed., McGraw-Hill, New York, 2001.

[2.] From "S.I., The Weight/Mass Controversy," *Mech. Eng.,* vol. 101, no. 3, p. 42, March 1979.

[3.] See, for example, M. J. Neale (ed.), *Tribology Handbook,* Butterworths, London, 1975, p. C8.

[4.] See, for example, F. P. Beer and E. R. Johnston, *Vector Mechanics for Engineers,* 6th ed., McGraw-Hill, New York, 1997, pp. 426–432.

[5.] It is true that treating the force $\mathbf{F}_{23}$ in this manner ignores the possible effects of friction forces between the meshing gear teeth. Justification for this comes in four forms. (1) The actual point of contact is continually varying during the meshing cycle, but always remains near the pitch point, which is the instant center of velocity; thus the relative motion between the teeth is close to pure rolling motion with only a small amount of slip. (2) The fact that the friction forces are continually changing in both magnitude and direction throughout the meshing cycle, and that the total force is often shared by more than one tooth in contact, makes a more exact analysis impractical. (3) The machined surfaces of the teeth and the fact that they are usually well-lubricated produces a very small coefficient of friction. (4) Experimental data show that gear efficiencies are usually very high, often approaching 99 percent, showing that any errors produced by ignoring friction are quite small.

[6.] This technique of treating small displacement ratios as velocity ratios can often be extremely helpful because, as we remember from Section 3.8, velocity relationships lead to linear equations rather than nonlinear relations implied by position or displacement. Velocity polygons and instant-center methods are also helpful.

PROBLEMS*

14.1 The figure shows four mechanisms and the external forces ad torques exerted on or by the mechanisms. Sketch the free-body diagram of each part of each mechanism. Do not attempt to show the magnitudes of the forces, except roughly, but do sketch them in their proper locations and directions.

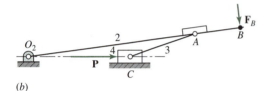

(a)

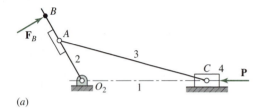

(b)

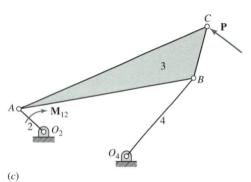

(c)

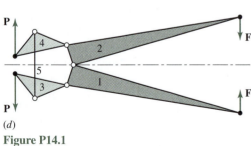

(d)

Figure P14.1

14.2 What moment M_{12} must be applied to the crank of the mechanism shown if $P = 0.9$ kN?

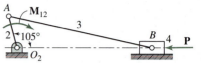

Figure P14.2, P14.3, and P14.25
$R_{AO_2} = 75$ mm; $R_{BA} = 350$ mm.

14.3 If $M_{12} = 100$ N·m for the mechanism shown, what force P is required to maintain static equilibrium?

14.4 Find the frame reactions and torque M_{12} necessary to maintain equilibrium of the four-bar linkage shown in the figure.

14.5 What torque must be applied to link 2 of the linkage shown to maintain static equilibrium?

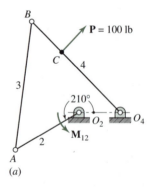

(a)

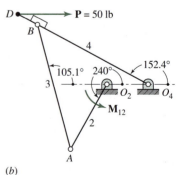

(b)

Figure P14.4, P14.5, and P14.26
$R_{AO_2} = 3.5$ in; $R_{BA} = R_{BO_4} =$
6 in; $R_{CO_4} = 4$ in; $R_{O_2O_4} = 2$ in.

---

*Unless otherwise stated, solve all problems without friction and without gravitational loads.

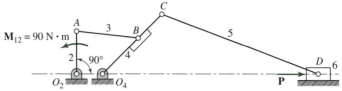

Figure P14.6 $R_{AO_2} = 100$ mm; $R_{BA} = 150$ mm; $R_{BO_4} = 125$ mm;
$R_{CO_4} = 200$ mm; $R_{CD} = 400$ mm; $R_{O_2O_4} = 60$ mm.

14.6 Sketch a complete free-body diagram of each link of the linkage shown. What force **P** is necessary for equilibrium?

14.7 Determine the torque $\mathbf{M}_{12}$ required to drive slider 6 of the figure against a load of $P = 100$ lb at a crank angle of $\theta = 30°$, or as specified by your instructor.

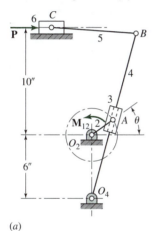

(a)

Figure P14.7, P14.15, and P14.27 (a) $R_{AO_2} = 2.5$ in; $R_{BO_4} = 16$ in; $R_{BC} = 8$ in.

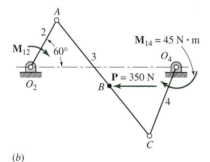

(b)

Figure P14.8 (b) $R_{AO_2} = 200$ mm; $R_{BA} = 400$ mm; $R_{CA} = R_{O_4O_2} = 700$ mm; $R_{CO_4} = 350$ mm.

14.8 Sketch complete free-body diagrams for the illustrated four-bar linkage. What torque $\mathbf{M}_{12}$ must be applied to 2 to maintain static equilibrium at the position shown?

14.9 Sketch free-body diagrams of each link and show all the forces acting. Find the magnitude and direction of the moment that must be applied to link 2 to drive the linkage against the forces shown.

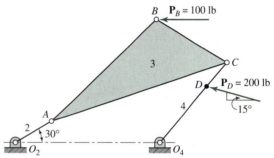

Figure P14.9 $R_{AO_2} = 4$ in; $R_{CA} = 14$ in; $R_{O_4O_2} = 14$ in; $R_{CO_4} = 10$ in; $R_{DO_4} = 7$ in; $R_{BA} = 14$ in; $R_{BC} = 8$ in.

14.10 The figure shows a four-bar linkage with external forces applied at points B and C. Draw a free-body diagram of each link and show all the forces acting on each. Find the torque that must be applied to link 2 to maintain equilibrium.

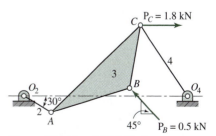

Figure P14.10 and P14.28 $R_{AO_2} = 75$ mm; $R_{CA} = 300$ mm; $R_{O_4O_2} = 400$ mm; $R_{CO_4} = R_{BA} = 200$ mm; $R_{BC} = 150$ mm.

14.11 Draw a free-body diagram of each of the members of the mechanism shown in the figure, and find the magnitude and the direction of all the forces and moments. Compute the magnitude and direction of the torque that must be applied to link 2 to maintain static equilibrium.

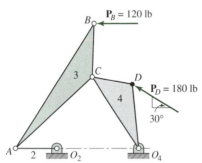

Figure P14.11 $R_{AO_2} = 4$ in;
$R_{CA} = 10$ in; $R_{O_4O_2} = R_{CO_4} = 8$ in;
$R_{DO_4} = 6$ in; $R_{DC} = 4$ in; $R_{BA} = 14$ in;
$R_{BC} = 5$ in.

14.12 Determine the magnitude and direction of the forces that must be applied to link 2 to maintain static equilibrium.

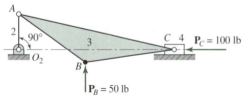

Figure P14.12 and P14.16 $R_{AO_2} = 3$ in; $R_{CA} = 14$ in; $R_{BA} = 7$ in; $R_{BC} = 8$ in.

14.13 The photograph shows the Figee floating crane with leminscate boom configuration. Also shown is a schematic diagram of the crane. The lifting capacity is 16 T (with 1 T = 1 metric ton = 1 000 kg) including the grab which is about 10 T. The maximum outreach is 30 m, which corresponds to $\theta_2 = 49°$. Minimum outreach is 10.5 m at $\theta_2 = 49°$. Other dimensions are given in the figure caption. For the maximum outreach position and a grab load of 10 T, find the bearing reactions at A, B, O_2, and O_4, and the moment $\mathbf{M}_{12}$ required. Notice that the photograph shows a counterweight on link 2; neglect this weight and also the weights of the members.

(a)

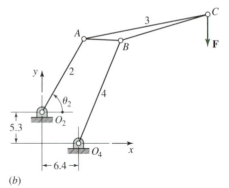

(b)

Figure P14.13 and P14.14 (a) Figee floating crane with lemniscate boom configuration. (Photograph by permission from B. V. Machinefabriek Figee, Haarlem, Holland). (b) Schematic diagram: $R_{AO_2} = 14.7$ m; $R_{BA} = 6.5$ m; $R_{BO_4} = 19.3$ m; $R_{CA} = 22.3$ m; $R_{CB} = 16$ m. (Dimensional details by permission from B. V. Machinefabriek Figee, Haarlem, Holland.)

14.14 Repeat Problem 14.12 for the minimum outreach position.

14.15 Repeat Problem 14.7 assuming coefficients of Coulomb friction $\mu_c = 0.20$ between links 1 and 6 and $\mu_c = 0.10$ between links 3 and 4. Determine the torque $\mathbf{M}_{12}$ necessary to drive the system, including friction, against the load $\mathbf{P}$.

14.16 Repeat Problem 14.12 assuming a coefficient of static friction $\mu = 0.15$ between links 1 and 4. Determine the torque $\mathbf{M}_{12}$ necessary to overcome friction.

14.17 In each case shown, pinion 2 is the driver, gear 3 is an idler, the gears have diametral pitch of 6 and 20° pressure angle. For each case, sketch the free-body diagram of gear 3 and show all forces acting. For (*a*) pinion 2 rotates at 600 rev/min and transmits 18 hp to the gearset. For (*b*) and (*c*), pinion 2 rotates at 900 rev/min and transmits 25 hp to the gearset.

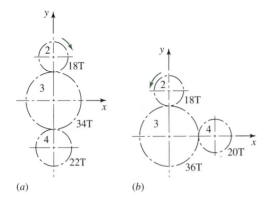

(*a*)   (*b*)

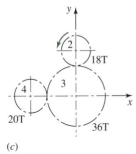

(*c*)

Figure P14.17

14.18 A 15-tooth spur pinion has a diametral pitch of 5 and 20° pressure angle, rotates at 600 rev/min, and drives a 60-tooth gear. The drive transmits 25 hp. Construct a free-body diagram of each gear showing upon it the tangential and radial components of the forces and their proper directions.

14.19 A 16-tooth pinion on shaft 2 rotates at 1 720 rev/min and transmits 5 hp to the double-reduction gear train. All gears have 20° pressure angle. The distances between centers of the bearings and gears for shaft 3 are shown in the figure. Find the magnitude and direction of the radial force that each bearing exerts against the shaft.

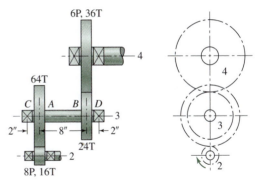

Figure P14.19

14.20 Solve Problem 14.17 if each pinion has right-hand helical teeth with a 30° helix angle and a 20° pressure angle. All gears in the train are helical, and, of course, the normal diametral pitch is 6 teeth per inch for each case.

14.21 Analyze the gear shaft of Example 14.6 and find the bearing reactions $\mathbf{F}_C$ and $\mathbf{F}_D$.

14.22 In each of the bevel gear drives shown in the figure, bearing A takes both thrust load and radial load,

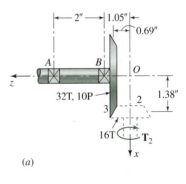

(*a*)

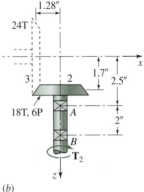

(*b*)

Figure P14.22

while bearing B takes only radial load. The teeth are cut with a $20°$ pressure angle. For (a) $\mathbf{T}_2 = -180\hat{\mathbf{i}}$ in·lb and for (b) $\mathbf{T}_2 = -240\hat{\mathbf{k}}$ in·lb. Compute the bearing loads for each case.

14.23 The figure shows a gear train composed of a pair of helical gears and a pair of straight bevel gears. Shaft 4 is the output of the train and delivers 6 hp to the load at a speed of 370 rev/min. All gears have pressure angles of $20°$. If bearing E is to take both thrust load and radial load, while bearing F is to take only radial load, determine the forces that each bearing exert against shaft 4.

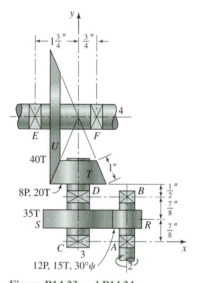

Figure P14.23 and P14.24

14.24 Using the data of Problem 14.23, find the forces exerted by bearings C and D onto shaft 3. Which of these bearings should take the thrust load if the shaft is to be loaded in compression?

14.25 Use the method of virtual work to solve the slider-crank mechanism of Problem 14.2.

14.26 Use the method of virtual work to solve the four-bar linkage of Problem 14.5.

14.27 Use the method of virtual work to analyze the crank-shaper linkage of Problem 14.7. Given that the load remains constant at $\mathbf{P} = 100\hat{\mathbf{i}}$ lb, find and plot a graph of the crank torque M_{12} for all positions in the cycle using increments of $30°$ for the input crank.

14.28 Use the method of virtual work to solve the four-bar linkage of Problem 14.10.

14.29 A car (link 2) which weighs 2 000 lb is slowly backing a 1 000-lb trailer (link 3) up a $30°$ inclined ramp as shown in the figure. The car wheels are of 13-in radius, and the trailer wheels have 10-in radius; the center of the hitch ball is also 13 in above the roadway. The centers of mass of the car and trailer are located at G_2 and G_3, respectively, and gravity acts vertically downward in the figure. The weights of the wheels and friction in the bearings are considered negligible. Assume that there are no brakes applied on the car or on the trailer, and that the car has front-wheel drive. Determine the loads on each of the wheels and the minimum coefficient of static friction between the driving wheels and the road to avoid slipping.

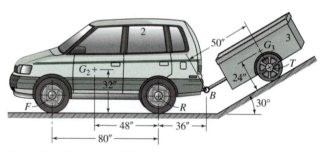

Figure P14.29 and P14.30

14.30 Repeat Problem 14.29, assuming that the car has rear-wheel drive rather than front-wheel drive.

14.31 The low-speed disk cam with oscillating flat-faced follower shown in the figure is driven at a constant shaft speed. The displacement curve for the cam has

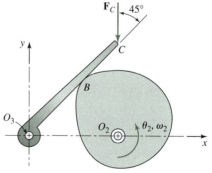

Figure P14.31, P14.32, P14.25, and P14.26 $R_{O_2O_3} = 50$ mm; $R_B = 42$ mm; $R_C = 150$ mm.

a full-rise cycloidal motion, defined by Eq. (5.19) with parameters $L = 30°$, $\beta = 30°$, and a prime circle radius $R_0 = 30$ mm; the instant pictured is at $\theta_2 = 112.5°$. A force of $F_C = 8$ N is applied at point C and remains at 45° from the face of the follower as shown. Use the virtual work approach to determine the moment $\mathbf{M}_{12}$ required on the crankshaft at the instant shown to produce this motion.

14.32 Repeat Problem 14.31 for the entire lift portion of the cycle, finding $\mathbf{M}_{12}$ has a function of θ_2.

14.33 A disk 3 of radius R is being slowly rolled under a pivoted bar 2 driven by an applied torque T as shown in the figure. Assume a coefficient of static friction of μ between the disk and ground and that all other joints are frictionless. A force $\mathbf{F}$ is acting vertically downward on the bar at a distance d from the pivot O_2. Assume that the weights of the links are negligible in comparison to F. Find an equation for the torque $\mathbf{T}$ required as a function of the distance $X = R_{CO_2}$, and find an equation for the final distance X that is reached when friction no longer allows further movement.

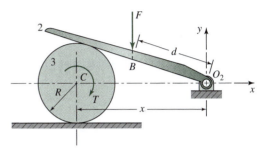

Figure P14.33

15 | Dynamic Force Analysis (Planar)

15.1 INTRODUCTION

In the previous chapter we studied the forces in machine systems in which all forces on the bodies were in balance, and therefore the systems were in either static or dynamic equilibrium. However, in real machines this is seldom, if ever, the case except when the machine is stopped. We learned in Chapter 4 that even though the input crank of a machine may be driven at constant speed, this does not mean that all points of the input crank have constant velocity vectors or even that other members of the machine will operate at constant speeds; there will be accelerations and therefore machines with moving parts having mass will not be in equilibrium.

Of course, techniques for static-force analysis are important, not only because stationary structures must be designed to withstand their imposed loads, but also because they introduce concepts and approaches that can be built upon and extended to nonequilibrium situations. That provides the purpose of this and the following chapter: to learn how much acceleration will result from a system of unbalanced forces and also to learn how these *dynamic* forces can be assessed for systems that are not in equilibrium.

15.2 CENTROID AND CENTER OF MASS

We recall from Section 14.2 that Newton's laws set forth the relationships between the net unbalanced force on a *particle,* its mass, and its acceleration. For that chapter, because we were only studying systems in equilibrium, we made use of the relationship for entire rigid bodies, arguing that they are made up of collections of particles and that the action and the reaction forces between the particles cancel each other. In this chapter we must be more careful: We must remember that each of these particles may have acceleration, and that the

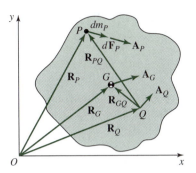

Figure 15.1 Particle of mass dm_P at location P on a rigid body.

accelerations of these particles may all be different from each other. Which of these many point accelerations are we to use? And why?

Referring to Fig. 15.1, we consider a particle with a mass dm_P at some arbitrary point P on the rigid body shown. For this single particle, Eq. (14.1) tells us that the net unbalanced force $d\mathbf{F}_P$ on that particle is proportional to its mass and its absolute acceleration $\mathbf{A}_P$:

$$d\mathbf{F}_P = \mathbf{A}_P \, dm_P \tag{a}$$

Our task now is to sum these effects—that is, to integrate over all particles of the body and to put the result in some usable form for rigid bodies other than single particles. As we did in the previous chapter, we can conclude that the action and the reaction forces between particles of the body balance each other, and therefore cancel in the process of the summation. The only net remaining forces are the constraint forces, those whose reactions are on some other body than this one. Thus, integrating Eq. (a) over all particles of mass in our rigid body (number j), we obtain

$$\sum \mathbf{F}_{ij} = \int \mathbf{A}_P \, dm_P \tag{b}$$

We find it difficult to perform this integration because each particle has a different acceleration. However, if we assume that we know (or can find) the acceleration of one particular point of the body—say, point Q—and also the angular velocity ω_j and angular acceleration α_j of the body, then we can express the various accelerations of all other particles of the body in terms of their acceleration differences from that of point Q. Thus, from Eqs. (4.8), (4.3), and (4.6), we obtain

$$\mathbf{A}_P = \mathbf{A}_Q + \omega_j \times (\omega_j \times \mathbf{R}_{PQ}) + \alpha_j \times \mathbf{R}_{PQ} \tag{c}$$

Substituting this into Eq. (b) gives

$$\sum \mathbf{F}_{ij} = \mathbf{A}_Q \int dm_P + \omega_j \times \left(\omega_j \times \int \mathbf{R}_{PQ} \, dm_P \right) + \alpha_j \times \int \mathbf{R}_{PQ} \, dm_P \tag{d}$$

where we have moved outside of the integrals all quantities that are the same for all particles of the body.

We can easily recognize the first integral of Eq. (d); the summation of all particle masses gives the total mass of the body m_j:

$$\int dm_P = m_j \tag{15.1}$$

The second and third terms of Eq. (d) both contain another integral, which does not look as easy. However, because of its frequent appearance in mechanics, a particular point G having a location given by the position vector $\mathbf{R}_G$ has been defined by the following integral equation:

$$\int \mathbf{R}_P \, dm_P = m_j \mathbf{R}_G \tag{15.2}$$

This point G is called the *centroid* or the *center of mass* of the body. More will be said about this special point in the material below.

Substituting Eqs. (15.1) and (15.2) into Eq. (d) gives

$$\sum \mathbf{F}_{ij} = \mathbf{A}_Q m_j + \boldsymbol{\omega}_j \times [\boldsymbol{\omega}_j \times (m_j \mathbf{R}_{GQ})] + \boldsymbol{\alpha}_j \times (m_j \mathbf{R}_{GQ})$$

$$= m_j [\mathbf{A}_Q + \boldsymbol{\omega}_j \times (\boldsymbol{\omega}_j \times \mathbf{R}_{GQ}) + \boldsymbol{\alpha}_j \times \mathbf{R}_{GQ}]$$

Finally, recognizing the sum of the three acceleration terms, we find

$$\sum \mathbf{F}_{ij} = m_j \mathbf{A}_G \tag{15.3}$$

This important equation is the integrated form of Newton's law for a particle, now extended to a rigid body. Notice that it is the same equation as was given in Eq. (14.1). However, careful derivation of the acceleration term was not done there because we were treating static problems where accelerations were to be set to zero.

We have now answered the question raised earlier in this section. Recognizing that each particle of a rigid body may have a different acceleration, which one should be used? Equation (15.3) shows clearly that the absolute acceleration of the center of mass of the body is the proper acceleration to be used in Newton's law. That particular point, and no other, is the proper point for which Newton's law for a rigid body has the same form as for a single particle.

In solving engineering problems, we frequently find that forces are distributed in some manner over a line, over an area, or over a volume. The resultant of these distributed forces is usually not too difficult to find. In order to have the same effect, this resultant must act at the centroid of the system. Thus, *the centroid of a system is a point at which a system of distributed forces may be considered concentrated with exactly the same effect.*

Instead of a system of forces, we may have a distributed mass, as in the above derivation. Then, by *center of mass* we mean *the point at which the mass may be considered concentrated so that the effect is the same.*

In Fig. 15.2a, a series of particles with masses are shown located at various positions along a line. The center of mass G is located at

$$\bar{x} = \frac{m_1 x_1 + m_2 x_2 + m_3 x_3}{m_1 + m_2 + m_3} = \frac{\sum m_i x_i}{\sum m_i} \tag{15.4}$$

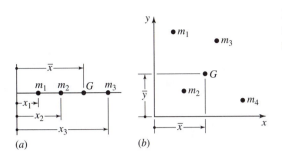

Figure 15.2 (*a*) Particles of mass distributed along a line. (*b*) Particles of mass distributed in a plane.

In Fig. 15.2*b*, the mass particles m_i are located at various positions $\mathbf{R}_i$ in a plane. The location of the center of mass G is now given by

$$\mathbf{R}_G = \frac{m_1\mathbf{R}_1 + m_2\mathbf{R}_2 + m_3\mathbf{R}_3 + m_4\mathbf{R}_4}{m_1 + m_2 + m_3 + m_4} = \frac{\sum m_i\mathbf{R}_i}{\sum m_i} \tag{15.5}$$

This procedure can also be extended to particles distributed in a volume by simply using Eq. (15.5) and treating position vectors $\mathbf{R}_i$ as three-dimensional rather than two-dimensional.

When the mass is distributed continuously along a line, or over a plane or volume, the concept of summation in Eq. (15.5) is replaced by integration over infinitesimal particles of mass. The result is

$$\mathbf{R}_G = \frac{1}{m_j} \int \mathbf{R}\, dm \tag{15.6}$$

where m_j is the total mass of the body j considered. It is from this definition that Eq. (15.2) was obtained.

When mass is evenly distributed over a plane area or a volume, the center of mass can often be found by symmetry. Figure 15.3 shows the locations of the centers of mass for a circular solid, a rectangular solid, and a triangular solid. Each is assumed to have constant thickness and a uniform density distribution.

When a body is of a more irregular shape, the center of mass can often still be found by considering it to be combination of simpler subshapes, as shown in the following example.

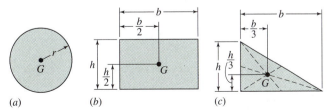

Figure 15.3 Center of mass location for (*a*) a right circular solid, (*b*) a rectangular solid, and (*c*) a triangular prism.

EXAMPLE 15.1

The volume shown in Fig. 15.4 consists of a rectangular solid, less a circular through hole, plus a triangular plate of a different thickness. The dimensions are as shown in the figure, and the entire part is made of cast iron. Find the center of mass of this composite shape.

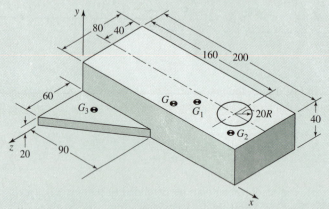

Figure 15.4 Composite shape for Example 15.1 with dimensions in millimeters.

SOLUTION

We can find the masses and centers of mass of each of the three subshapes with the help of the formulae shown in Fig. 15.3. For the rectangular solid subshape we get[1] (See note on page 511.)

$$m_1 = (80 \text{ mm})(200 \text{ mm})(40 \text{ mm})(\rho \text{ kg/mm}^3) = 640\,000\rho \text{ kg}$$

$$\mathbf{R}_{G_1} = 100\hat{\mathbf{i}} + 20\hat{\mathbf{j}} - 40\hat{\mathbf{k}} \text{ mm}$$

For the hole we treat the mass as negative, giving

$$m_2 = \pi(20 \text{ mm})^2(40 \text{ mm})(-\rho \text{ kg/mm}^3) = -50\,300\rho \text{ kg}$$

$$\mathbf{R}_{G_2} = 160\hat{\mathbf{i}} + 20\hat{\mathbf{j}} - 40\hat{\mathbf{k}} \text{ mm}$$

Finally, for the triangular subshape we have

$$m_3 = 0.5(60 \text{ mm})(90 \text{ mm})(20 \text{ mm})(\rho \text{ kg/mm}^3) = 54\,000\rho \text{ kg}$$

$$\mathbf{R}_{G_3} = 30\hat{\mathbf{i}} + 10\hat{\mathbf{j}} + 20\hat{\mathbf{k}} \text{ mm}$$

Now because we are completing a process of integration, we can combine the values of the subscripts; that is, from Eq. (15.5) we have

$$\mathbf{R}_G = \frac{m_1\mathbf{R}_1 + m_2\mathbf{R}_2 + m_3\mathbf{R}_3}{m_1 + m_2 + m_3} = \frac{\sum m_i \mathbf{R}_i}{\sum m_i} \tag{15.7}$$

This gives the following result:

$$\mathbf{R}_G = 89.4\hat{\mathbf{i}} + 19.1\hat{\mathbf{j}} - 35.0\hat{\mathbf{k}} \text{ mm} \qquad \qquad Ans.$$

15.3 MASS MOMENTS AND PRODUCTS OF INERTIA

Another problem that often arises when forces are distributed over an area is that of calculating their moment about a specified point or axis of rotation. Sometimes the force intensity varies according to its distance from the point or axis of rotation. Although we will save a more thorough derivation of these equations until the next chapter, we will point out here that such problems always give rise to integrals of the form $\int (\text{distance})^2 \, dm$.

In three-dimensional problems, three such integrals are defined as follows[2] (see note on page 511):

$$I^{xx} = \int (\hat{\mathbf{i}} \times \mathbf{R}) \cdot (\hat{\mathbf{i}} \times \mathbf{R}) \, dm = \int [(R^y)^2 + (R^z)^2] \, dm$$

$$I^{yy} = \int (\hat{\mathbf{j}} \times \mathbf{R}) \cdot (\hat{\mathbf{j}} \times \mathbf{R}) \, dm = \int [(R^z)^2 + (R^x)^2] \, dm \qquad (15.8)$$

$$I^{zz} = \int (\hat{\mathbf{j}} \times \mathbf{R}) \cdot (\hat{\mathbf{j}} \times \mathbf{R}) \, dm = \int [(R^x)^2 + (R^y)^2] \, dm$$

These three integrals are called the *mass moments of inertia* of the body. Another three similar integrals are

$$I^{xy} = I^{yx} = \int (\hat{\mathbf{i}} \times \mathbf{R}) \cdot (\hat{\mathbf{j}} \times \mathbf{R}) \, dm = \int (R^x R^y) \, dm$$

$$I^{yz} = I^{zy} = \int (\hat{\mathbf{j}} \times \mathbf{R}) \cdot (\hat{\mathbf{k}} \times \mathbf{R}) \, dm = \int (R^y R^z) \, dm \qquad (15.9)$$

$$I^{zx} = I^{xz} = \int (\hat{\mathbf{k}} \times \mathbf{R}) \cdot (\hat{\mathbf{i}} \times \mathbf{R}) \, dm = \int (R^z R^x) \, dm$$

and these three integrals are called the *mass products of inertia* of the body. Sometimes it is convenient to arrange these mass moments and products of inertia into a symmetric square array or matrix called the *inertia tensor* of the body:

$$\mathbf{I} = \begin{bmatrix} I^{xx} & -I^{xy} & -I^{xz} \\ -I^{yx} & I^{yy} & -I^{yz} \\ -I^{zx} & -I^{zy} & I^{zz} \end{bmatrix} \qquad (15.10)$$

A careful look at the above integrals will show that they represent the mass distribution of the body with respect to the coordinate system about which they are determined, but that they will change if evaluated in a different coordinate system. In order to keep their meaning direct and simple, we assume that the coordinate system chosen for each body is attached to that body in a convenient location and orientation. Therefore, for rigid bodies, the mass moments and products of inertia are constant properties of the body and its mass distribution and they do not change when the body moves; they do, however, depend on the coordinate system chosen.

An interesting property of these integrals is that it is always possible to choose the coordinate system so that its origin is located at the center of mass of the body and oriented such that all of the products of inertia become zero. Such a choice of the coordinate axes of the body is called its *principal axes,* and the corresponding values of Eqs. (15.8) are then called the *principal mass moments of inertia*. Appendix Table 5 shows a variety of simple geometric solids, the orientations of their principal axes, and formulae for their principal mass moments of inertia.

If we notice that mass moments of inertia have units of mass times distance squared, it seems natural to define a radius value for the body as follows:

$$I_G = k^2 m \quad \text{or} \quad k = \sqrt{\frac{I_G}{m}} \tag{15.11}$$

This distance k is called the *radius of gyration* of the body, and it is always calculated or measured from the center of mass of the part about one of the principal axes. For three-dimensional motions of parts there are three radii of gyration, k^x, k^y, and k^z, associated with the three principal axes, I^{xx}, I^{yy}, and I^{zz}.

It is often necessary to determine the moments and products of inertia of bodies which are composed of several simpler subshapes for which formulae are known, such as those shown in Appendix Table 5. The easiest method of finding these is to compute the mass moments about the principal axes of each subshape, and then to shift the origins of each to the mass center of the composite body, and then to sum the results. This requires that we develop methods of redefining mass moments and products of inertia when the axes are translated to a new position. The form of the *transfer*, or *parallel-axis formula* for mass moment of inertia, is written

$$I = I_G + md^2 \tag{15.12}$$

where I_G is one of the principal mass moments of inertia about some known principal axis and I is the mass moment of inertia about a parallel axis at distance d from that principal axis. Equation (15.12) must only be used for *translation* of inertia axes starting from a principal axis, however; the rotation of these axes results in the introduction of product of inertia terms. More will be said on general transformations of inertia axes in Chapter 16.

EXAMPLE 15.2

Figure 15.5 shows a connecting rod made of ductile iron with density of 0.260 lb/in³. Find the mass moment of inertia about the z axis.

SOLUTION

We solve by finding the mass moment of inertia of each of the cylinders at the ends of the rod, and of the central prismatic bar, all about their own mass centers. Then we use the parallel-axis formula to transfer each to the z axis.

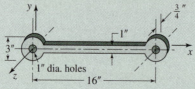

Figure 15.5 Connecting-rod shape for Example 15.2.

The mass of each hollow cylinder is

$$m_{cyl} = \rho\pi\left(r_o^2 - r_i^2\right)l$$

$$= \frac{(0.260 \text{ lb/in}^3)\pi[(1.5 \text{ in})^2 - (0.5 \text{ in})^2](0.75 \text{ in})}{386 \text{ in/s}^2}$$

$$= 0.003\ 17 \text{ lb·s}^2/\text{in}$$

The mass of the central prismatic bar is

$$m_{bar} = \rho whl = \frac{(0.260 \text{ lb/in}^3)(13 \text{ in})(1 \text{ in})(0.75 \text{ in})}{386 \text{ in/s}^2}$$

$$= 0.006\ 57 \text{ lb·s}^2/\text{in}$$

From Appendix Table 5 we find the mass moment of inertia of each element to be

$$I_{cyl} = \frac{m\left(r_o^2 + r_i^2\right)}{2}$$

$$= \frac{(0.003\ 17 \text{ lb·s}^2/\text{in})[(1.5 \text{ in})^2 + (0.5 \text{ in})^2]}{2}$$

$$= 0.003\ 96 \text{ in·lb·s}^2$$

$$I_{bar} = \frac{m(w^2 + h^2)}{12}$$

$$= \frac{(0.006\ 57 \text{ lb·s}^2/\text{in})[(1 \text{ in})^2 + (13 \text{ in})^2]}{12}$$

$$= 0.093\ 1 \text{ in·lb·s}^2$$

Then using Eq. (15.12) to transfer the axes, the mass moment of inertia about the z axis is

$$I^{zz} = I_{cyl} + \left(I_{bar} + m_{bar}d_{bar}^2\right) + \left(I_{cyl} + m_{cyl}d_{cyl}^2\right)$$

$$= 0.003\ 96 + [0.093\ 1 + 0.006\ 57(8)^2] + [0.003\ 96 + 0.003\ 17(16)^2]$$

$$= 1.33 \text{ in·lb·s}^2 \qquad\qquad\qquad\qquad\qquad\qquad\qquad\qquad\qquad \textit{Ans.}$$

It will be noted in this example that only one mass moment of inertia I^{zz} was asked for or found. This does not mean that I^{xx} and I^{yy} are zero, but rather that they will likely not be needed for further analysis. In problems with planar motion only, the main subject of this chapter, only I^{zz} is needed because I^{xx} and I^{yy} would be used only with rotations not in the xy plane. These other mass moments and products of inertia will be used in Chapter 16, where we treat problems with spatial motion, and would be solved for in identical fashion.

15.4 INERTIA FORCES AND D'ALEMBERT'S PRINCIPLE

Consider a moving rigid body of mass m acted upon by any system of forces, say, $\mathbf{F}_1$, $\mathbf{F}_2$, and $\mathbf{F}_3$, as shown in Fig. 15.6a. Designate the center of mass of the body as point G, and find the resultant of the system of forces from the equation

$$\sum \mathbf{F} = \mathbf{F}_1 + \mathbf{F}_2 + \mathbf{F}_3$$

In the general case the line of action of this resultant will *not* be through the mass center but will be displaced by some distance, shown here as distance h in the figure. In Eq. (15.3) above we showed that the effect of this unbalanced force system is to produce an acceleration of the center of mass of the body:

$$\sum \mathbf{F}_{ij} = m_j \mathbf{A}_G \tag{15.13}$$

In a very similar way, taking moments about the center of mass of the body, it has been proven that the unbalanced moment effect of this resultant force about the center of mass causes angular acceleration of the body that obeys the following equation:

$$\sum \mathbf{M}_{G_{ij}} = I_G \boldsymbol{\alpha}_j \tag{15.14}$$

The quantity $\sum \mathbf{F}$ is the resultant of all external forces acting upon the body, and $\sum \mathbf{M}_G$ is the sum of any applied external moments and the moments of all externally applied forces about point G. The mass moment of inertia is designated as I_G, signifying that it must also be taken with respect to the mass center G.

Equations (15.13) and (15.14) show that when an unbalanced system of forces act upon a rigid body, the body experiences a rectilinear acceleration $\mathbf{A}_G$ of its mass center in the same direction as the resultant force $\sum \mathbf{F}$. The body also experiences an angular acceleration $\boldsymbol{\alpha}_j$ in the same direction as the resultant moment $\sum \mathbf{M}_G$, due to the moments of the forces and the torques about the mass center. This situation is pictured in Fig. 15.6b.

If the forces and moments are known, Eqs. (15.13) and (15.14) may be used to determine the resulting acceleration pattern—that is, the resulting motion—of the body. During

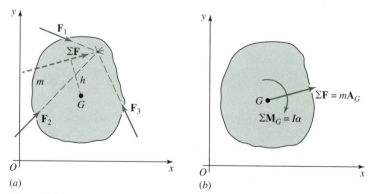

Figure 15.6 (*a*) An unbalanced set of forces on a rigid body. (*b*) The accelerations that result from the unbalanced forces.

engineering design, however, the desired motions of the machine members are often specified in advance by other machine requirements. The problem then is: Given the motion of the machine elements, what forces are required to produce these motions? The problem requires (1) a kinematic analysis in order to determine the translational and rotational accelerations of the various members and (2) definitions of the actual shapes, dimensions, and material specifications, in order to determine the centroids and mass moments of inertia of the members. In the examples to be discussed here, only the results of the kinematic analysis will be presented; methods of finding these were presented in Chapter 4. The selection of the materials, shapes, and many of the dimensions of machine members form the subject of machine design and is also not further discussed here.

Because, in the dynamic analysis of machines, the acceleration vectors are usually known, an alternative form of Eqs. (15.13) and (15.14) is often convenient in determining the forces required to produce these known accelerations. Thus, we can write

$$\sum \mathbf{F}_{ij} + (-m_j \mathbf{A}_G) = \mathbf{0} \tag{15.15}$$

$$\sum \mathbf{M}_{Gij} + (-I_G \boldsymbol{\alpha}_j) = \mathbf{0} \tag{15.16}$$

Both of these are vector equations applying to the planar motion of a rigid body. Equation (15.15) states that the vector sum of all external forces acting upon the body plus the fictitious force $-m_j \mathbf{A}_G$ sum to zero. This new fictitious force $-m_j \mathbf{A}_G$ is called an *inertia force*. It has the same line of action as the absolute acceleration $\mathbf{A}_G$, but is opposite in sense. Equation (15.16) states that the sum of all external moments and the moments of all external forces acting upon the body about an axis through G and perpendicular to the plane of motion plus the fictitious torque $-I_G \boldsymbol{\alpha}_j$ sum to zero. This new fictitious torque $-I_G \boldsymbol{\alpha}_j$ is called an *inertia torque*. The inertia torque is opposite in sense to the angular acceleration vector $\boldsymbol{\alpha}_j$. We recall that Newton's first law states that a body perseveres in its state of uniform motion except when compelled to change by impressed forces; in other words, bodies resist any change in motion. In a sense we can picture the fictitious inertia force and inertia torque vectors as resistances of the body to the change of motion required by the net unbalanced forces.

The equations above are known as *D'Alembert's principle,* because he[3] was the first to call attention to the fact that addition of the inertia force and inertia torque to the real system of forces and torques enables a solution from the equations of static equilibrium. We might note that the equations can also be written

$$\sum \mathbf{F} = \mathbf{0} \quad \text{and} \quad \sum \mathbf{M} = \mathbf{0} \tag{15.17}$$

where it is understood that both the external and the inertia forces and torques are to be included in the summations. Equations (15.17) are useful because they permit us to take the summation of moments about any axis perpendicular to the plane of motion.

D'Alembert's principle is summarized as follows: The vector sum of all external forces and inertia forces acting upon a system of rigid bodies is zero. The vector sum of all external moments and inertia torques acting upon a system of rigid bodies is also separately zero.

When a graphical solution by a force polygon is desired, Eqs. (15.17) can be combined. In Fig. 15.7a, a rigid link 3 is acted upon by the external forces $\mathbf{F}_{23}$ and $\mathbf{F}_{43}$. The resultant $\mathbf{F}_{23} + \mathbf{F}_{43}$ produces an acceleration $\mathbf{A}_G$ of the center of mass, and an angular

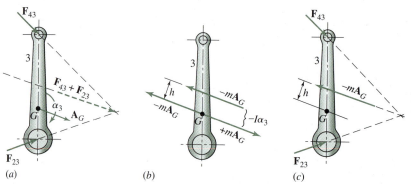

Figure 15.7 (*a*) Unbalanced forces and resulting accelerations. (*b*) Inertia force and inertia couple. (*c*) Inertia force offset from center of mass.

acceleration α_3 of member 3 because the line of action of the resultant does not pass through the center of mass. Representing the inertia torque $-I_G\alpha_3$ as a couple, as shown in Fig. 15.7*b*, we intentionally choose the two forces of this couple to be $\pm m_3\mathbf{A}_G$. In order for the moment of the couple to be of magnitude $-I_G\alpha_3$, the distance between the forces of the couple must be

$$h = \frac{I_G\alpha_3}{m_3 A_G} \tag{15.18}$$

Because of this particular choice for the couple, one force of the couple exactly cancels the inertia force itself and leaves only a single force, as shown in Fig. 15.7*c*. This force includes the combined effects of the inertia force and the inertia torque, yet appears as only a single inertia force offset by the distance h to give the effect of the inertia torque.

EXAMPLE 15.3

Determine the force $\mathbf{F}_A$ required for dynamic equilibrium at a constant velocity of $V_A = 12.6$ ft/s for the mechanism shown in Fig. 15.8*a*. Link 3 weighs 2.20 lb and $I_{G_3} = 0.047\,9$ in·lb·s². Assume no friction and that the linkage is in a horizontal plane so that gravity acts normal to the plane of motion.

SOLUTION

A kinematic analysis provides the acceleration information shown in the polygon of Fig. 15.8*b*. The angular acceleration of link 3 is

$$\alpha_3 = \frac{A_{BA}^t}{R_{BA}}$$

$$= \frac{(713 \text{ ft/s}^2)(12 \text{ in/ft})}{10 \text{ in}} = 856 \text{ rad/s}^2 \text{ cw}$$

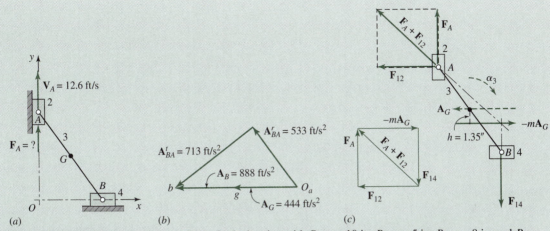

(a) (b) (c)

Figure 15.8 Solution for Example 15.3: (*a*) Scale drawing with $R_{BA} = 10$ in, $R_{GA} = 5$ in, $R_{AO} = 8$ in, and $R_{BO} = 6$ in. (*b*) Acceleration polygon. (*c*) Free-body diagram and force polygon.

The mass of link 3 is $m_3 = 2.20/386 = 0.005\ 70$ lb·s²/in. Therefore, Eq. (15.18) gives an offset distance of

$$h = \frac{(0.047\ 9\ \text{in·lb·s}^2)(856\ \text{rad/s}^2)}{(0.005\ 70\ \text{lb·s}^2/\text{in})(444\ \text{ft/s}^2)(12\ \text{in/ft})} = 1.35\ \text{in}$$

The free-body diagram of links 2, 3, and 4 and the resulting force polygon are shown in Fig. 15.8*c*. Notice that the inertia force $-m\mathbf{A}_G$ is offset from G by the distance h so as to produce the counterclockwise moment $-I_G\boldsymbol{\alpha}_3$ about G, and with the inertia force in the opposite sense to $\mathbf{A}_G$. The constraint reaction at B is $\mathbf{F}_{14}$ and, with no friction component, is vertically downward. The forces at A are the constraint reaction $\mathbf{F}_{12}$ and the actuating force $\mathbf{F}_A$. Recognizing this as a four-force system, as shown in Section 14.8, we find the concurrency point at the intersection of the offset inertia force and $\mathbf{F}_{14}$, the directions of both being known. The line of action of the total force $\mathbf{F}_{12} + \mathbf{F}_A$ at A must pass through the point of concurrency. This fact permits construction of the force polygon, where the unknown forces $\mathbf{F}_{12}$ and $\mathbf{F}_A$, having known directions, are found as components of $\mathbf{F}_{12} + \mathbf{F}_A$. The actuating force $\mathbf{F}_A$ is found by measurement to be

$$\mathbf{F}_A = 27\hat{\mathbf{j}}\ \text{lb} \qquad\qquad \textit{Ans.}$$

EXAMPLE 15.4

As an example of dynamic force analysis using SI units, we use the four-bar linkage of Fig. 15.9. The required data, based on a complete kinematic analysis, are shown in the figure and in the caption. At the crank angle shown, and assuming that gravity and friction effects are negligible, find all the constraint forces and the driving torque required to produce the velocity and acceleration conditions specified.

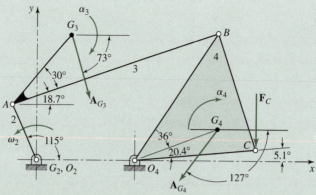

Figure 15.9 Example 15.4: $R_{AO_2} = 60$ mm, $R_{O_4O_2} = 100$ mm, $R_{BA} = 220$ mm, $R_{BO_4} = 150$ mm, $R_{CO_4} = R_{CB} = 120$ mm, $R_{G_3A} = 90$ mm, $R_{G_4O_4} = 90$ mm, $m_3 = 1.5$ kg, $m_4 = 5$ kg, $I_{G_2} = 0.025$ kg·m^2, $I_{G_3} = 0.012$ kg·m^2, $I_{G_4} = 0.054$ kg·m^2, $\alpha_2 = \mathbf{0}$, $\alpha_3 = -119\hat{\mathbf{k}}$ rad/s^2, $\alpha_4 = -625\hat{\mathbf{k}}$ rad/s^2, $\mathbf{A}_{G_3} = 162\angle-73.2°$ m/s^2, $\mathbf{A}_{G_4} = 104\angle233°$ m/s^2, $\mathbf{F}_C = -0.8\hat{\mathbf{j}}$ kN.

SOLUTION

We start with the following kinematic information:

$$\mathbf{R}_{AO_2} = 60\angle115° = -25.4\hat{\mathbf{i}} + 54.4\hat{\mathbf{j}} \text{ mm}$$

$$\mathbf{R}_{G_3A} = 90\angle48.7° = 59.4\hat{\mathbf{i}} + 67.6\hat{\mathbf{j}} \text{ mm}$$

$$\mathbf{R}_{BA} = 220\angle18.7° = 208.0\hat{\mathbf{i}} + 70.5\hat{\mathbf{j}} \text{ mm}$$

$$\mathbf{R}_{BO_4} = 150\angle56.4° = 83.0\hat{\mathbf{i}} + 125.0\hat{\mathbf{j}} \text{ mm}$$

$$\mathbf{R}_{G_4O_4} = 90\angle20.4° = 84.4\hat{\mathbf{i}} + 31.4\hat{\mathbf{j}} \text{ mm}$$

$$\mathbf{R}_{CO_4} = 120\angle5.1° = 120.0\hat{\mathbf{i}} + 10.7\hat{\mathbf{j}} \text{ mm}$$

$$\alpha_2 = \mathbf{0}, \qquad \alpha_3 = -119\hat{\mathbf{k}} \text{ rad/s}^2, \qquad \alpha_4 = -625\hat{\mathbf{k}} \text{ rad/s}^2$$

$$\mathbf{A}_{G_2} = \mathbf{0}, \qquad \mathbf{A}_{G_3} = 46.8\hat{\mathbf{i}} - 155\hat{\mathbf{j}} \text{ m/s}^2, \qquad \mathbf{A}_{G_4} = -62.6\hat{\mathbf{i}} - 83.1\hat{\mathbf{j}} \text{ m/s}^2$$

Now we calculate the inertia forces and inertia torques. Because this solution will be done analytically, we do not calculate offset distances, nor do we replace the inertia torques by couples.

$$-m_2\mathbf{A}_{G_2} = \mathbf{0}$$

$$-m_3\mathbf{A}_{G_3} = -(1.5 \text{ kg})(46.8\hat{\mathbf{i}} - 155\hat{\mathbf{j}} \text{ m/s}^2) = -70.2\hat{\mathbf{i}} + 233\hat{\mathbf{j}} \text{ N}$$

$$-m_4\mathbf{A}_{G_4} = -(5.0 \text{ kg})(-62.6\hat{\mathbf{i}} - 83.1\hat{\mathbf{j}} \text{ m/s}^2) = 313\hat{\mathbf{i}} + 415\hat{\mathbf{j}} \text{ N}$$

$$-I_{G_2}\alpha_2 = \mathbf{0}$$

$$-I_{G_3}\alpha_2 = -(0.012 \text{ kg·m}^2)(-119\hat{\mathbf{k}} \text{ rad/s}^2) = 1.43\hat{\mathbf{k}} \text{ N·m}$$

$$-I_{G_4}\alpha_4 = -(0.054 \text{ kg·m}^2)(-625\hat{\mathbf{k}} \text{ rad/s}^2) = 33.8\hat{\mathbf{k}} \text{ N·m}$$

In considering our next step, we should make at least rough sketches of the free-body diagrams of the separate links. In so doing we discover a problem. Each link has at least four unknown quantities, the magnitudes and directions of the reaction forces at each pivot. We know that because this analysis is planar, there are only three useful scalar equations for each body: the horizontal and vertical components of force and the in-plane moment equation. Worse yet, combining the bodies (say, links 3 and 4) into a single free-body diagram does not improve this. What can be done, or is this problem not solvable?

Fortunately, in looking at the two free-body diagrams for links 3 and 4, we notice that there are a total of six unknown force components between them. There are also six total scalar equations available: two components of the forces and one for torque on each of the two free-body diagrams. The problem is solvable, but these six equations must be solved simultaneously to find the six unknowns. Therefore, considering the free-body diagram of link 4 alone, we formulate the summation of moments about point O_4:

$$\sum \mathbf{M}_{O_4} = \mathbf{R}_{G_4 O_4} \times (-m_4 \mathbf{A}_{G_4}) + (-I_{G_4} \boldsymbol{\alpha}_4) + \mathbf{R}_{C O_4} \times \mathbf{F}_C + \mathbf{R}_{B O_4} \times \mathbf{F}_{34} = 0 \qquad (1)$$

Also, considering the free-body diagram of link 3 alone, we formulate the summation of moments about point A:

$$\sum \mathbf{M}_A = \mathbf{R}_{G_3 A} \times (-m_3 \mathbf{A}_{G_3}) + (-I_{G_3} \boldsymbol{\alpha}_3) + \mathbf{R}_{BA} \times \mathbf{F}_{43} = 0 \qquad (2)$$

When we remember that $\mathbf{F}_{43} = -\mathbf{F}_{34}$, the in-plane components of $\mathbf{F}_{34}$ are the only two unknowns and they are shared by these two equations.

The individual terms of the two equations are found to be

$$\mathbf{R}_{G_4 O_4} \times (-m_4 \mathbf{A}_{G_4}) = (84.4\hat{\mathbf{i}} + 31.4\hat{\mathbf{j}} \text{ mm}) \times (0.313\hat{\mathbf{i}} + 0.415\hat{\mathbf{j}} \text{ kN})$$

$$= 25.2\hat{\mathbf{k}} \text{ N·m}$$

$$\mathbf{R}_{C O_4} \times \mathbf{F}_C = (120\hat{\mathbf{i}} + 10.7\hat{\mathbf{j}} \text{ mm}) \times (-0.800\hat{\mathbf{j}} \text{ kN}) = -96.0\hat{\mathbf{k}} \text{ N·m}$$

$$\mathbf{R}_{B O_4} \times \mathbf{F}_{34} = (83\hat{\mathbf{i}} + 125\hat{\mathbf{j}} \text{ mm}) \times \left(F_{34}^x \hat{\mathbf{i}} + F_{34}^y \hat{\mathbf{j}} \text{ kN}\right)$$

$$= \left(-125 F_{34}^x + 83 F_{34}^y\right)\hat{\mathbf{k}} \text{ N·m}$$

$$\mathbf{R}_{G_3 A} \times (-m_3 \mathbf{A}_{G_3}) = (59.4\hat{\mathbf{i}} + 67.6\hat{\mathbf{j}} \text{ mm}) \times (-0.070\hat{\mathbf{i}} + 0.233\hat{\mathbf{j}} \text{ kN})$$

$$= 18.6\hat{\mathbf{k}} \text{ N·m}$$

$$\mathbf{R}_{BA} \times \mathbf{F}_{43} = (208\hat{\mathbf{i}} + 70.5\hat{\mathbf{j}} \text{ mm}) \times \left(-F_{34}^x \hat{\mathbf{i}} - F_{34}^y \hat{\mathbf{j}} \text{ kN}\right)$$

$$= \left(70.5 F_{34}^x - 208 F_{34}^y\right)\hat{\mathbf{k}} \text{ N·m}$$

Then, substituting these values into Eqs. (1) and (2) above, gives

$$\sum \mathbf{M}_{O_4} = 25.2\hat{\mathbf{k}} + 33.8\hat{\mathbf{k}} - 96.0\hat{\mathbf{k}} + \left(-125 F_{34}^x + 83 F_{34}^y\right)\hat{\mathbf{k}} = 0$$

$$\sum \mathbf{M}_A = 18.6\hat{\mathbf{k}} + 1.43\hat{\mathbf{k}} + \left(70.5 F_{34}^x - 208 F_{34}^y\right)\hat{\mathbf{k}} = 0$$

and rearranging terms gives two equations in two unknowns:

$$-125 F_{34}^x + 83 F_{34}^y = 37.0 \text{ N·m}$$

$$70.5 F_{34}^x - 208 F_{34}^y = -20.0 \text{ N·m}$$

Solving simultaneously, we find the two components and then the vector $\mathbf{F}_{34}$:

$$F_{34}^x = -0.300 \text{ kN} = -300 \text{ N} \quad \text{and} \quad F_{34}^y = -0.005 \ 39 \text{ kN} = -5.39 \text{ N}$$

$$F_{34} = -300\hat{\mathbf{i}} - 39\hat{\mathbf{j}} = 300\angle 181° \text{ N} \qquad \qquad Ans.$$

Next, summing forces on link 4 yields the equation

$$\sum \mathbf{F}_{i4} = \mathbf{F}_{14} + \mathbf{F}_{34} + \mathbf{F}_C + (-m_4 \mathbf{A}_{G_4}) = 0$$

and, upon solving, we find

$$\mathbf{F}_{14} = -\mathbf{F}_{34} - \mathbf{F}_C - (-m_4 \mathbf{A}_{G_4})$$
$$= -(-300\hat{\mathbf{i}} - 5.39\hat{\mathbf{j}} \text{ N}) - (-800\hat{\mathbf{j}} \text{ N}) - (313\hat{\mathbf{i}} + 415\hat{\mathbf{j}} \text{ N})$$
$$= -13\hat{\mathbf{i}} + 390\hat{\mathbf{j}} = 390\angle 91.9° \text{ N} \qquad \qquad Ans.$$

Similarly, summing forces on link 3 yields the equation

$$\sum \mathbf{F}_{13} = \mathbf{F}_{23} + \mathbf{F}_{43} + (-m_3 \mathbf{A}_{G_3}) = 0$$

and solving this, we get

$$\mathbf{F}_{23} = -(-\mathbf{F}_{34}) - (-m_3 \mathbf{A}_{G_3})$$
$$= -(300\hat{\mathbf{i}} + 5.39\hat{\mathbf{j}} \text{ N}) - (-70.2\hat{\mathbf{i}} + 233\hat{\mathbf{j}} \text{ N})$$
$$= -230\hat{\mathbf{i}} - 238\hat{\mathbf{j}} = 331\angle 226° \text{ N} \qquad \qquad Ans.$$

Now, for link 2, we have

$$\sum \mathbf{F}_{i2} = \mathbf{F}_{12} + \mathbf{F}_{32} + (-m_2 \mathbf{A}_{G_2}) = 0$$

or

$$\mathbf{F}_{12} = -\mathbf{F}_{32} = \mathbf{F}_{23} = -230\hat{\mathbf{i}} - 238\hat{\mathbf{j}} = 331\angle 226° \text{ N} \qquad \qquad Ans.$$

and, by summing moments on link 2 about O_2,

$$\sum \mathbf{M}_{O_2} = \mathbf{R}_{AO_2} \times \mathbf{F}_{32} + \mathbf{M}_{12} + (-I_{G_2} \boldsymbol{\alpha}_2) = 0$$

which can be solved for the driving crank torque:

$$\mathbf{M}_{12} = -\mathbf{R}_{AO_2} \times \mathbf{F}_{32} = -(-25.4\hat{\mathbf{i}} + 54.4\hat{\mathbf{j}} \text{ mm}) \times (230\hat{\mathbf{i}} + 238\hat{\mathbf{j}} \text{ N})$$

$$= 18.6\hat{\mathbf{k}} \text{ N·m} \qquad\qquad\qquad\qquad \textit{Ans.}$$

15.5 THE PRINCIPLE OF SUPERPOSITION

Linear systems are those in which effect is proportional to cause. This means that the response or output of a linear system is directly proportional to the drive or input to the system. An example of a linear system is a spring, where the deflection (output) is directly proportional to the force (input) exerted on the spring.

The *principle of superposition* may be used to solve problems involving linear systems by considering each of the inputs to the system separately. If the system is linear, the responses to each of these inputs can be summed or superposed on each other to determine the total response of the system. Thus the principle of superposition states that *for linear systems the individual responses to several disturbances or driving functions can be superposed on each other to obtain the total response of the system.*

The principle of superposition does not apply to nonlinear systems. Some examples of nonlinear systems, where superposition may not be used, are systems with static or Coulomb friction, systems with clearances or backlash, or systems with springs that change stiffness as they are deflected.

We have now reviewed all the principles necessary for making a complete dynamic-force analysis of a planar motion mechanism. The steps in using the principle of superposition for making such an analysis are summarized as follows:

1. Make a kinematic analysis of the mechanism. Locate the center of mass of each link and find the acceleration of each; also find the angular acceleration of each link.
2. Using the known value or values of the force or torque that must be delivered to the load, make a complete static-force analysis of the mechanism. The results of this step of the analysis include magnitudes and directions of input and constraint forces and torques acting upon each link. Observe particularly that the input and constraint forces and torques just found come from static-force analysis and do not yet include the effects of inertia forces or torques.
3. Employing the known values for the masses and moments of inertia of each link, along with the translational and angular accelerations found in step 1, calculate the inertia forces and inertia torques for each link or element of the mechanism. Taking these inertia loads as new applied forces and torques, but ignoring the applied loads used in step 2, make another complete force analysis of the mechanism. The results of this step of the analysis include new magnitudes and directions of input and constraint forces and torques acting upon each link which result from inertia effects.
4. Vectorially add the results of steps 2 and 3 to obtain the resultant forces and torques on each link.

EXAMPLE 15.5

Make a complete dynamic-force analysis of the four-bar linkage illustrated in Fig. 15.10. The known information is included in the figure caption.

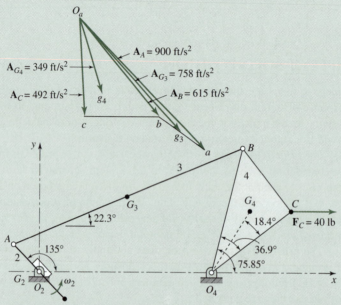

Figure 15.10 Example 15.5: $R_{AO_2} = 3$ in, $R_{O_4O_2} = 14$ in, $R_{BA} = 20$ in, $R_{BO_4} = 10$ in, $R_{CO_4} = 8$ in, $R_{CB} = 6$ in, $R_{G_3A} = 10$ in, $R_{G_4O_4} = 5.69$ in, $w_3 = 7.13$ lb, $w_4 = 3.42$ lb, $I_{G_2} = 0.25$ in·lb·s^2, $I_{G_3} = 0.625$ in·lb·s^2, $I_{G_4} = 0.037$ in·lb·s^2, $\omega_2 = 60$ rad/s, and $\alpha_2 = 0$.

SOLUTION

The first step is to make a kinematic analysis of the mechanism. This step is not all included here, but the resulting acceleration polygon is shown in Fig. 15.10. The numerical results are shown on the polygon in case you wish to verify them. By the methods of Chapter 4, the angular accelerations of link 3 and 4 are found to be

$$\alpha_3 = 148 \text{ rad/s}^2 \text{ ccw} \quad \text{and} \quad \alpha_4 = 604 \text{ rad/s}^2 \text{ cw}$$

A major portion of the analysis is concerned with links 3 and 4 because the center of mass of link 2 is located at O_2. Free-body diagrams of links 4 and 3 are shown separately in Figs. 15.11 and 15.12, respectively. Notice that these diagrams are arranged in equation form to emphasize the concept of superposition. Thus, in each illustration, the forces in (*a*) plus those in (*b*) and (*c*) produce the results shown in (*d*). The action and reaction forces in the two sets of illustrations are also correlated: For example, $\mathbf{F}'_{34}$ in Fig. 15.11*a* is equal to $-\mathbf{F}'_{43}$ in Fig. 15.12*a*, and so on. The following analysis is not difficult, but it is complex; it is important to read it slowly and examine the illustrations carefully, detail by detail.

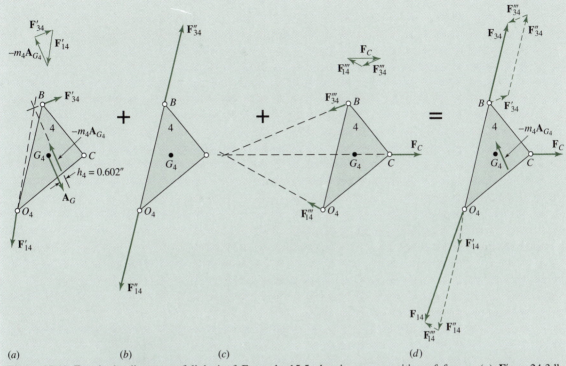

Figure 15.11 Free-body diagrams of link 4 of Example 15.5 showing superposition of forces: (*a*) $\mathbf{F}'_{34} = 24.3$ lb, $\mathbf{F}'_{14} = 44.3$ lb; (*b*) $\mathbf{F}''_{34} = -\mathbf{F}''_{14} = 94.8$ lb; (*c*) $\mathbf{F}'''_{34} = 25$ lb, $\mathbf{F}'''_{14} = 19.3$ lb; (*d*) $\mathbf{F}_{34} = 94.3$ lb, and $\mathbf{F}_{14} = 132$ lb.

Let us start with link 4 in Fig. 15.11*a*. Proceeding in accordance with our earlier investigations, we make the following calculations:

$$I_{G_4}\alpha_4 = (0.037 \text{ in·lb·s}^2)(604 \text{ rad/s}^2) = 22.3 \text{ in·lb cw}$$

$$m_4 A_{G_4} = \frac{3.42 \text{ lb}}{32.2 \text{ ft/s}^2}(349 \text{ ft/s}^2) = 37.1 \text{ lb}$$

$$h_4 = \frac{I_{G_4}\alpha_4}{m_4 A_{G_4}} = \frac{22.3 \text{ in·lb}}{37.1 \text{ lb}} = 0.602 \text{ in}$$

Now the inertia force $-m_4 A_{G_4} = 37.1$ lb is placed on the free-body diagram opposite in direction to $\mathbf{A}_{G_4}$ and offset from G_4 by the distance h_4. The direction of the offset is to the right of G_4 so that the inertia force $-m_4\mathbf{A}_{G_4}$ produces a counterclockwise inertia torque $-I_{G_4}\boldsymbol{\alpha}_4$ about G_4 opposed to the clockwise sense of $\boldsymbol{\alpha}_4$. The line of action of $\mathbf{F}'_{34}$ is taken along link 3, for part (*a*) of our superposition approach, because link 3 is a two-force member. The intersection of the inertia force $-m_4\mathbf{A}_{G_4}$ and $\mathbf{F}'_{34}$ gives the point of concurrency and establishes the line of action of $\mathbf{F}'_{14}$. The force polygon can now be constructed and the magnitudes of $\mathbf{F}'_{34}$ and $\mathbf{F}'_{14}$ can be found. These values are shown in the figure caption.

We proceed next to Fig. 15.12*a*. The forces $\mathbf{F}'_{43}$ and $\mathbf{F}'_{23}$ now become known from the preceding analysis.

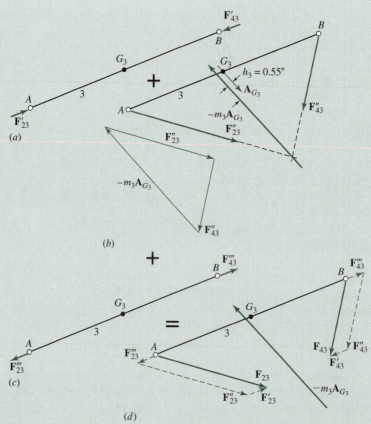

Figure 15.12 Free-body diagrams of link 3 of Example 15.5 showing super-position of forces: (a) $\mathbf{F}'_{23} = \mathbf{F}'_{43} = 24.3$ lb; (b) $\mathbf{F}''_{23} = 145$ lb, $\mathbf{F}''_{43} = 94.8$ lb; (c) $\mathbf{F}'''_{23} = \mathbf{F}'''_{43} = 25$ lb; (d) $\mathbf{F}_{23} = 145$ lb, $\mathbf{F}_{43} = 94.3$ lb.

Going next to Fig. 15.12b, we make the following calculations:

$$I_{G_3}\alpha_3 = (0.625 \text{ in·lb·s}^2)(148 \text{ rad/s}^2) = 92.5 \text{ in·lb ccw}$$

$$m_3 A_{G_3} = \frac{7.13 \text{ lb}}{32.3 \text{ ft/s}^2}(758 \text{ ft/s}^2) = 168 \text{ lb}$$

$$h_3 = \frac{I_{G_3}\alpha_3}{m_3 A_{G_3}} = \frac{92.5 \text{ in·lb}}{168 \text{ lb}} = 0.550 \text{ in}$$

Now we locate the inertia force $-m_3 A_{G_3} = 168$ lb on the free-body diagram opposite in direction to $\mathbf{A}_{G_3}$ and offset by a distance of $h_3 = 0.550$ in from G_3 so as to produce a clockwise torque about G_3, opposite to the counterclockwise sense of $\boldsymbol{\alpha}_3$. The line of action of $\mathbf{F}''_{43}$ is along link 4, for part (b) of our superposition approach, because link 4 is a two-force member. The intersection of the line of action of $\mathbf{F}''_{43}$ and the inertia force $-m_3 A_{G_3}$ gives the point of concurrency. Thus the

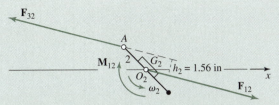

Figure 15.13 Free-body diagram of link 2 of Example 15.5: $F_{32} = F_{12} = 145$ lb, $M_{12} = 226$ in·lb.

line of action of $\mathbf{F}_{23}''$ becomes known and the force polygon can be constructed. The resulting magnitudes of $\mathbf{F}_{43}''$ and $\mathbf{F}_{23}''$ are included in the figure caption.

In Fig. 15.11*b* the forces $\mathbf{F}_{34}''$ and $\mathbf{F}_{14}''$ now become known from the preceding analysis.

Figures 15.11*c* and 15.12*c* show the results of the static-force analysis with $F_c = 40$ lb as the given loading. Recognizing that link 3 is again a two-force member, the force polygon in Fig. 15.11*c* determines the values of the forces acting on link 4. From these the magnitudes and directions of the forces acting on link 3 are found.

The next step is to perform the superposition of the results obtained; this is done by the vector additions shown in part (*d*) of each figure.

The analysis is completed by taking the resultant force $\mathbf{F}_{23}$ from Fig. 15.12*d* and applying its negative, $\mathbf{F}_{32}$, to link 2. This is shown in Fig. 15.13. The distance h_2 is found by measurement, and the external torque to be applied to link 2 is found to be

$$M_{12} = h_2 F_{32} = (1.56 \text{ in})(145 \text{ lb}) = 226 \text{ in·lb cw}$$

Note that this torque is opposite in sense to the direction of rotation of link 2; this will not be true for the entire cycle of operation, but it can occur at a particular crank angle. This torque must be in the reverse direction to continue rotation with constant input velocity.

15.6 PLANAR ROTATION ABOUT A FIXED CENTER

The previous sections have dealt with the general case of dynamic forces for a rigid body having a general planar motion. It is important to emphasize that the equations and methods of analysis investigated in these sections are general and apply to *all* problems with planar motion. It will be interesting now to study the application of these methods to a special case, that of a rigid body rotating about a fixed center.

Let us consider a rigid body constrained as shown in Fig 15.14*a* to rotate about some fixed center O not coincident with its center of mass G. A system of forces (not shown) is to be applied to the body, causing it to undergo an angular acceleration $\boldsymbol{\alpha}$. We also include the fact that the body is rotating with an angular velocity $\boldsymbol{\omega}$. This motion of the body implies that the mass center G has normal and tangential components of acceleration $\mathbf{A}_G^n$ and $\mathbf{A}_G^t$ whose magnitudes are $R_G \omega^2$ and $R_G \alpha$, respectively. Thus, if we resolve the resultant external force into normal and tangential components, these components must have magnitudes

$$F^n = m R_G \omega^2 \quad \text{and} \quad F^t = m R_G \alpha \qquad (a)$$

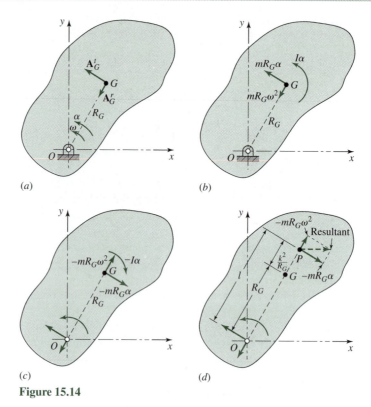

Figure 15.14

in accordance with Eq. (15.12). In addition, Eq. (15.14) states that an external torque must exist to create the angular acceleration and that the magnitude of this torque must be $T_G = I_G\alpha$. If we now sum the moments of these forces about O, we have

$$\sum \mathbf{M}_O = I_G\alpha + R_G(mR_G\alpha) = \left(I_G + mR_G^2\right)\alpha \tag{b}$$

But the quantity in parentheses in Eq. (b) is identical in form with Eq. (15.12) and transfers the moment of inertia to another axis not coincident with the center of mass. Therefore Eq. (b) can be written in the form

$$\sum \mathbf{M}_O = I_O\alpha$$

Equations (15.15) and (15.16) then become

$$\sum \mathbf{F} - m\mathbf{A}_G = \mathbf{0} \tag{15.19}$$

$$\sum \mathbf{M}_O - I_O\alpha = \mathbf{0} \tag{15.20}$$

by including the inertia force $-m\mathbf{A}_G$ and inertia torque $-I_O\alpha$ as shown in Fig. 15.14c. We observe particularly that the system of forces does *not* reduce to a single couple because of the existence of the inertia force component $-m\mathbf{R}_G\omega^2$, which has no moment arm about point O. Thus both Eqs. (15.19) and (15.20) are necessary.

A particular case arises when $\boldsymbol{\alpha} = \mathbf{0}$. Then the external moment $\sum \mathbf{M}_O$ is zero and the only inertia force is, from Fig. 15.14c, the centrifugal force $-m\mathbf{R}_G\omega^2$. A second special case exists under starting conditions, when $\omega = 0$ but α is not zero. Under these conditions the only inertia force is $-mR_G\alpha$, and the system reduces to a single couple.

When a rigid body has a motion of translation only, the resultant inertia force and the resultant external force share the same line of action, which passes through the center of mass of the body. When a rigid body has rotation and angular acceleration, the resultant inertia force and resultant external force have the same line of action, but this line does *not* pass through the center of mass but is offset from it. Let us now locate a point on this offset line of action of the resultant of the inertia forces.

The resultant of the inertia forces will pass through some point P of Fig. 15.14d on the line OG or its extension. This force can be resolved into two components, one of which will be the component $-m\mathbf{R}_G\omega^2$ acting along the line OG. The other component will be $-mR_G\alpha$ acting perpendicular to OG but not through point G. The distance, designated as l, to the unknown point P can be found by equating the moment of the component $-mR_G\alpha$ through P (Fig. 15.14d) to the sum of the inertia torque and the moment of the inertia forces which act through G (Fig. 15.14c). Thus, taking moments about O in each figure, we have

$$(-mR_Ga)l = -I_G\alpha + (-mR_G\alpha)R_G$$

or

$$l = \frac{I_G}{mR_G} + R_G$$

Remembering the definition of the radius of gyration and substituting the value of I_G from Eq. (15.11) gives

$$l = \frac{k^2}{R_G} + R_G \tag{15.21}$$

The point P located by Eq. (15.21) and shown in Fig. 15.14d is called the *center of percussion*. As shown, the resultant inertia force passes through P, and consequently the inertia force has zero moment about the center of percussion. If an external force is applied at P, perpendicular to OG, an angular acceleration α will result, but the bearing reaction at O will be zero except for the component caused by the centrifugal inertia force $-mR_G\omega^2$. It is the usual practice in shock testing machines to apply the force at the center of percussion in order to eliminate the tangential bearing reaction that would otherwise be caused by the externally applied shock load.

Equation (15.21) shows that the location of the center of percussion is independent of the values of ω and α.

If the axis of rotation is coincident with the center of mass, $R_G = 0$ and Eq. (15.21) shows that $l = \infty$. Under these conditions there is no resultant inertia force but a resultant inertia couple $-I_G\alpha$ instead.

15.7 SHAKING FORCES AND MOMENTS

Of special interest to the designer are the forces transmitted to the frame or foundation of a machine owing to the inertia of the moving links. When these forces vary in magnitude or direction, they tend to shake or vibrate the machine, and consequently such effects are called *shaking forces* and *shaking moments*.

If we consider some machine, say a four-bar linkage for an example, with links 2, 3, and 4 as the moving members and link 1 as the frame, then taking the entire group of moving parts as a system and drawing a free-body diagram, we can immediately write

$$\sum \mathbf{F} = \mathbf{F}_{12} + \mathbf{F}_{14} + (-m_2 \mathbf{A}_{G_2}) + (-m_3 \mathbf{A}_{G_3}) + (-m_4 \mathbf{A}_{G_4}) = \mathbf{0}$$

Using $\mathbf{F}_s$ as a symbol for the resulting shaking force, we define this as equal to the resultant of all the reaction forces on the ground link 1,

$$\mathbf{F}_s = \mathbf{F}_{21} + \mathbf{F}_{41}$$

Therefore, from the previous equation, we find

$$\mathbf{F}_s = (-m_2 \mathbf{A}_{G_2}) + (-m_3 \mathbf{A}_{G_3}) + (-m_4 \mathbf{A}_{G_4}) \tag{15.22}$$

Generalizing from this example, we can write for any machine that the shaking force is given by

$$\mathbf{F}_s = \sum (-m_j \mathbf{A}_{G_j}) \tag{15.23}$$

This makes sense because if we consider a free-body diagram of the entire machine *including the frame,* all other applied and constraint forces have equal and opposite reaction forces and these cancel within the free-body system; only the inertia forces, having no reactions, are ultimately external to the system and remain unbalanced.* These are not balanced by reaction forces and produce unbalanced shaking effects between the frame and whatever bench or other surface on which it is mounted. These are what requires the machine to be fastened down to prevent it from moving.

A similar derivation can be made for unbalanced moments. Using the symbol $\mathbf{M}_s$ for the shaking moment, we take the summation of moments about the coordinate system origin and find

$$\mathbf{M}_s = \sum [\mathbf{R}_{G_j} \times (-m_j \mathbf{A}_{G_j})] + \sum (-I_{G_j} \boldsymbol{\alpha}_j) \tag{15.24}$$

15.8 COMPLEX ALGEBRA APPROACH

This section presents the dynamic force analysis of a mechanism using the complex algebra approach. The general procedure that is adopted in this section is as follows:

1. Perform a complete kinematic analysis.
2. Draw free-body diagrams of each link in the mechanism.

---

*The same arguments can be made about gravitational, magnetic, electric, or other noncontact-force fields. However, because these are usually static forces, it is not customary to include them in the definition of shaking force.

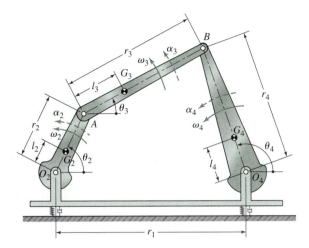

Figure 15.15 Four-bar linkage.

3. Write the dynamic force equations for each moving link.
4. Corresponding to each input angle, compute such items as:
   (a) the axial and the transverse forces exerted on each moving link;
   (b) the force components exerted on the ground bearings (henceforth referred to as the bearing loads);
   (c) the inertia torque; and
   (d) the shaking moment on the foundation.

To illustrate the complex algebra approach, we will consider the four-bar linkage shown in Fig. 15.15.

For a specified angular velocity and acceleration of the input link 2, the angular velocity and acceleration of the coupler link 3 and the output link 4 can be determined following the methods of Chapters 3 and 4. The acceleration of the mass centers of links 2, 3, and 4 will now be expressed in coordinates attached to and aligned along the respective links.

The acceleration of the mass center of the input link 2 can be written as

$$\mathbf{A}_{G_2} = l_2\left(-\omega_2^2 + j\alpha_2\right)e^{j\theta_2} = \left(A_{G_2}^{x_2} + jA_{G_2}^{y_2}\right)e^{j\theta_2} \tag{15.25a}$$

Therefore, the x- and y-components of the acceleration, respectively, are

$$A_{G_2}^{x_2} = -l_2\omega_2^2 \quad \text{and} \quad A_{G_2}^{y_2} = l_2\alpha_2 \tag{15.25b}$$

The acceleration of the mass center of the coupler link 3 can be written as

$$\mathbf{A}_{G_3} = r_2\left(-\omega_2^2 + j\alpha_2\right)e^{j\theta_2} + l_3\left(-\omega_3^2 + j\alpha_3\right)e^{j\theta_3}$$

or as

$$\mathbf{A}_{G_3} = \left[r_2\left(-\omega_2^2 + j\alpha_2\right)e^{j\eta_{23}} + l_3\left(-\omega_3^2 + j\alpha_3\right)\right]e^{j\theta_3} \tag{15.26}$$

where $\eta_{23} = \theta_2 - \theta_3$. The acceleration of the mass center of the coupler link can also be written as

$$\mathbf{A}_{G_3} = \left(A_{G_3}^{x_3} + jA_{G_3}^{y_3}\right)e^{j\theta_3} \tag{15.27}$$

Equating Eqs. (15.26) and (15.27), the x- and y-components of the acceleration are

$$A_{G_3}^{x_3} = r_2\left(-\omega_2^2 \cos \eta_{23} - \alpha_2 \sin \eta_{23}\right) - l_3\omega_3^2 \tag{15.28a}$$

and

$$A_{G_3}^{y_3} = r_2\left(-\omega_2^2 \sin \eta_{23} + \alpha_2 \cos \eta_{23}\right) - l_3\alpha_3 \tag{15.28b}$$

The acceleration of the mass center of the output link 4 can be written as

$$\mathbf{A}_{G_4} = l_4\left(-\omega_4^2 + j\alpha_4\right)e^{j\theta_4} = \left(A_{G_4}^{x_4} + jA_{G_4}^{y_4}\right)e^{j\theta_4} \tag{15.29a}$$

Therefore, the x- and y-components of the acceleration, respectively, are

$$A_{G_4}^{x_4} = -l_4\omega_4^2 \quad \text{and} \quad A_{G_4}^{y_4} = l_4\alpha_4 \tag{15.29b}$$

The free-body diagrams of the four links are presented in Figure 15.16. For the input link, a sum of the forces gives

$$\left[\left(F_{12}^{x_2} + F_{32}^{x_2}\right) + j\left(F_{12}^{y_2} + F_{32}^{y_2}\right)\right]e^{j\theta_2} = m_2\left(A_{G_2}^{x_2} + jA_{G_2}^{y_2}\right)e^{j\theta_2} \tag{15.30a}$$

and a sum of the moments about the bearing O_2 gives

$$r_2 F_{32}^{y_2} + T_2 = \left(I_{G_2} + m_2 l_2^2\right)\alpha_2 \tag{15.30b}$$

For the coupler link, a sum of the forces gives

$$\left[\left(F_{23}^{x_3} + F_{43}^{x_3}\right) + j\left(F_{23}^{y_3} + F_{43}^{y_3}\right)\right]e^{j\theta_3} = m_3\left(A_{G_3}^{x_3} + jA_{G_3}^{y_3}\right)e^{j\theta_3} \tag{15.31a}$$

and a sum of the moments about the mass center G_3 gives

$$F_{43}^{y_3}(r_3 - l_3) - F_{23}^{y_3}l_3 = I_{G_3}\alpha_3 \tag{15.31b}$$

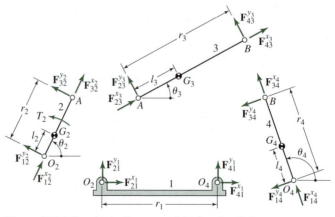

Figure 15.16 Free-body diagrams of the four-bar linkage.

For the output link, a sum of the forces gives

$$\left[\left(F_{14}^{x_4} + F_{34}^{x_4}\right) + j\left(F_{14}^{y_4} + F_{34}^{y_4}\right)\right]e^{j\theta_4} = m_4\left(A_{G_4}^{x_4} + jA_{G_4}^{y_4}\right)e^{j\theta_4} \tag{15.32a}$$

and a sum of the moments about the bearing O_4 gives

$$r_4 F_{34}^{y_4} = \left(I_{G_4} + m_4 l_4^2\right)\alpha_4 \tag{15.32b}$$

For the revolute joint A, a sum of the forces gives

$$\left(F_{32}^{x_2} + jF_{32}^{y_2}\right)e^{j\theta_2} + \left(F_{23}^{x_3} + jF_{23}^{y_3}\right)e^{j\theta_3} = 0 \tag{15.33a}$$

and for the revolute joint B, a sum of the forces gives

$$\left(F_{43}^{x_3} + jF_{43}^{y_3}\right)e^{j\theta_3} + \left(F_{34}^{x_4} + jF_{34}^{y_4}\right)e^{j\theta_4} = 0 \tag{15.33b}$$

For the bearing O_2, a sum of the forces gives

$$\left(F_{21}^{x_1} + jF_{21}^{y_1}\right) + \left(F_{12}^{x_2} + jF_{12}^{y_2}\right)e^{j\theta_2} = 0 \tag{15.33c}$$

and for the bearing O_4, a sum of the forces gives

$$\left(F_{41}^{x_1} + jF_{41}^{y_1}\right) + \left(F_{14}^{x_4} + jF_{14}^{y_4}\right)e^{j\theta_4} = 0 \tag{15.33d}$$

The procedure to obtain solutions to the dynamic force analysis is as follows. From Eq. (15.32b), the transverse force acting at point B on the output link 4 is

$$F_{34}^{y_4} = \left(I_{G_4} + m_4 l_4^2\right)\frac{\alpha_4}{r_4} \tag{15.34}$$

Equating the real and imaginary parts of Eq. (15.30a), the x and y components of the forces, respectively, are

$$F_{12}^{x_2} + F_{32}^{x_2} = m_2 A_{G_2}^{x_2} \tag{15.35a}$$

and

$$F_{12}^{y_2} + F_{32}^{y_2} = m_2 A_{G_2}^{y_2} \tag{15.35b}$$

Equating the real and imaginary parts of Eq. (15.31a), the x and y components of the forces, respectively, are

$$F_{23}^{x_3} + F_{43}^{x_3} = m_3 A_{G_3}^{x_3} \tag{15.36a}$$

and

$$F_{23}^{y_3} + F_{43}^{y_3} = m_3 A_{G_3}^{y_3} \tag{15.36b}$$

Equating the real and imaginary parts of Eq. (15.32a), the x and y components of the forces, respectively, are

$$F_{14}^{x_4} + F_{34}^{x_4} = m_4 A_{G_4}^{x_4} \tag{15.37a}$$

and

$$F_{14}^{y_4} + F_{34}^{y_4} = m_4 A_{G_4}^{y_4} \tag{15.37b}$$

Substituting Eq. (15.36b) into Eq. (15.31b), the transverse force acting at point B on the coupler link is

$$F_{43}^{y_3} = \frac{1}{r_3}\left(I_{G_3}\alpha_3 + m_3 l_3 A_{G_3}^{y_3}\right) \tag{15.38a}$$

Then from Eq. (15.36b), the transverse force acting at point A on the coupler link is

$$F_{23}^{y_3} = m_3 A_{G_3}^{y_3} - F_{43}^{y_3} \tag{15.38b}$$

Equation (15.33b) can be written as

$$\left(F_{43}^{x_3} + j F_{43}^{y_3}\right)e^{j\eta_{34}} + \left(F_{34}^{x_4} + j F_{34}^{y_4}\right) = 0 \tag{15.39a}$$

or as

$$\left(F_{43}^{x_3} + j F_{43}^{y_3}\right) + \left(F_{34}^{x_4} + j F_{34}^{y_4}\right)e^{-j\eta_{34}} = 0 \tag{15.39b}$$

where $\eta_{34} = \theta_3 - \theta_4$. Equating the imaginary parts of Eq. (15.39a) and equating the imaginary parts of Eq. (15.39b), respectively, gives

$$F_{43}^{x_3} \sin \eta_{34} + F_{43}^{y_3} \cos \eta_{34} + F_{34}^{y_4} = 0 \tag{15.40a}$$

and

$$F_{43}^{y_3} - \left(F_{34}^{x_4} \sin \eta_{34} - F_{34}^{y_4} \cos \eta_{34}\right) = 0 \tag{15.40b}$$

Then from these two equations we have

$$F_{43}^{x_3} = -\left(F_{43}^{y_3} \cot \eta_{34} + F_{34}^{y_4} \csc \eta_{34}\right) \tag{15.41a}$$

and

$$F_{34}^{x_4} = F_{43}^{y_3} \csc \eta_{34} + F_{34}^{y_4} \cot \eta_{34} \tag{15.41b}$$

Then from Eqs. (15.36a) and (15.37a), the axial forces acting on links 3 and 4, respectively, are

$$F_{23}^{x_3} = m_3 A_{G_3}^{x_3} - F_{43}^{x_3} \tag{15.42a}$$

and

$$F_{14}^{x_4} = m_4 A_{G_4}^{x_4} - F_{34}^{x_4} \tag{15.42b}$$

Rewriting Eq. (15.33a) as

$$F_{32}^{x_2} + j F_{32}^{y_2} = -\left(F_{23}^{x_3} + j F_{23}^{y_3}\right) e^{-j\eta_{23}} \tag{15.43}$$

and equating the real and imaginary parts, respectively, gives

$$F_{32}^{x_2} = -F_{23}^{x_3} \cos \eta_{23} - F_{23}^{y_3} \sin \eta_{23} \tag{15.44a}$$

and

$$F_{32}^{y_2} = F_{23}^{x_3} \sin \eta_{23} - F_{23}^{y_3} \cos \eta_{23} \tag{15.44b}$$

The input torque (commonly referred to as the inertia torque) can now be obtained by writing Eq. (15.30b) as

$$T_2 = -r_2 F_{32}^{y_2} + \left(I_{G_2} + m_2 l_2^2\right) \alpha_2 \tag{15.45}$$

If the four-bar linkage is in *steady-state operation*—that is, the input angular velocity is constant—then substituting $\alpha_2 = 0$ into Eq. (15.45) gives

$$T_2 = -r_2 F_{32}^{y_2} \tag{15.46}$$

From Eqs. (15.35) and (15.37) we have

$$F_{12}^{x_2} = m_2 A_{G_2}^{x_2} - F_{32}^{x_2} \quad \text{and} \quad F_{14}^{x_4} = m_4 A_{G_4}^{x_4} - F_{34}^{x_4} \tag{15.47a}$$

$$F_{12}^{y_2} = m_2 A_{G_2}^{y_2} - F_{32}^{y_2} \quad \text{and} \quad F_{14}^{y_4} = m_4 A_{G_4}^{y_4} - F_{34}^{y_4} \tag{15.47b}$$

If we write Eqs. (15.33c) and (15.33d), respectively, as

$$\left(F_{21}^{x_1} + j F_{21}^{y_1}\right) = -\left(F_{12}^{x_2} + j F_{12}^{y_2}\right) e^{j\theta_2}$$

and

$$\left(F_{41}^{x_1} + j F_{41}^{y_1}\right) = -\left(F_{14}^{x_4} + j F_{14}^{y_4}\right) e^{j\theta_4}$$

then the x- and y-components of the loads at the bearing O_2, respectively, are

$$F_{21}^{x_1} = -F_{12}^{x_2} \cos \theta_2 + F_{12}^{y_2} \sin \theta_2 \quad \text{and} \quad F_{21}^{y_1} = -F_{12}^{x_2} \sin \theta_2 - F_{12}^{y_2} \cos \theta_2 \tag{15.48a}$$

and the x- and y-components of the loads at the bearing O_4, respectively, are

$$F_{41}^{x_1} = -F_{14}^{x_4} \cos \theta_4 + F_{14}^{y_4} \sin \theta_4 \quad \text{and} \quad F_{41}^{y_1} = -F_{14}^{x_4} \sin \theta_4 - F_{14}^{y_4} \cos \theta_4 \tag{15.48b}$$

Finally, the dynamic shaking moment can be written as

$$M_S = r_1 F_{41}^{y_1}$$

Substituting Eq. (15.48b) into this equation gives

$$M_S = -r_1 \left(F_{14}^{x_4} \sin \theta_4 + F_{14}^{y_4} \cos \theta_4\right) \tag{15.49}$$

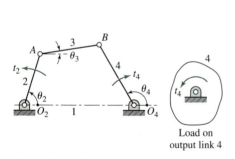

Figure 15.17 Output torque of the four-bar linkage.

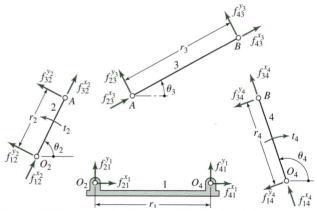

Figure 15.18 Free-body diagrams of the four-bar linkage.

Consider the special case where the four-bar linkage is in static equilibrium due to the output torque t_4 shown in Fig. 15.17—that is, the torque exerted on the load by the output axis through O_4. The free-body diagrams are as shown in Fig. 15.18.

From the dynamic force analysis, we can easily determine the following as functions of the input angle:

1. The axial and the transverse force components exerted on the three moving links
2. The bearing loads
3. The inertia torque
4. The shaking moment

From Eqs. (15.30), the sum of the forces and the sum of the moments about the bearing O_2 are

$$\left[\left(f_{12}^{x_2} + f_{32}^{x_2}\right) + j\left(f_{12}^{y_2} + f_{32}^{y_2}\right)\right]e^{j\theta_2} = 0 \qquad (15.50a)$$

and

$$r_2 f_{32}^{y_2} + t_2 = 0 \qquad (15.50b)$$

From Eqs. (15.31), the sum of the forces and the sum of the moments about the mass center G_3 are

$$\left[\left(f_{23}^{x_3} + f_{43}^{x_3}\right) + j\left(f_{23}^{y_3} + f_{43}^{y_3}\right)\right]e^{j\theta_3} = 0 \qquad (15.51a)$$

and

$$r_3 f_{43}^{y_3} = 0 \qquad (15.51b)$$

From Eq. (15.32), the sum of the forces and the sum of the moments about the bearing O_4 are

$$\left[\left(f_{14}^{x_4} + f_{34}^{x_4}\right) + j\left(f_{14}^{y_4} + f_{34}^{y_4}\right)\right]e^{j\theta_4} = 0 \qquad (15.52a)$$

and

$$r_4 f_{34}^{y_4} + t_4 = 0 \tag{15.52b}$$

For the revolute joint A, a sum of the forces gives

$$\left(f_{32}^{x_2} + j f_{32}^{y_2} \right) e^{j\theta_2} + \left(f_{23}^{x_3} + j f_{23}^{y_3} \right) e^{j\theta_3} = 0 \tag{15.53a}$$

and for the revolute joint B, a sum of the forces gives

$$\left(f_{43}^{x_3} + j f_{43}^{y_3} \right) e^{j\theta_3} + \left(f_{34}^{x_4} + j f_{34}^{y_4} \right) e^{j\theta_4} = 0 \tag{15.53b}$$

For the bearing O_2 a sum of the forces gives

$$\left(f_{21}^{x_1} + j f_{21}^{y_1} \right) + \left(f_{12}^{x_2} + j f_{12}^{y_2} \right) e^{i\theta_2} = 0 \tag{15.54a}$$

and for the bearing O_4 a sum of the forces gives

$$\left(f_{41}^{x_1} + j f_{41}^{y_1} \right) + \left(f_{14}^{x_4} + j f_{14}^{y_4} \right) e^{j\theta_4} = 0 \tag{15.54b}$$

From Eqs. (15.51) the axial and transverse forces at point A on the coupler link are

$$f_{23}^{x_3} = -f_{43}^{x_3} \quad \text{and} \quad f_{23}^{y_3} = -f_{43}^{y_3} = 0 \tag{15.55}$$

From Eq. (15.52), the transverse force acting at point B on the output link 4 is

$$f_{34}^{y_4} = -\frac{t_4}{r_4} \tag{15.56}$$

Substituting Eq. (15.55) into Eq. (15.53a), and rearranging, gives

$$f_{32}^{x_2} + j f_{32}^{y_2} = -f_{23}^{x_3} e^{-j\eta_{23}}$$

Equating the real and imaginary parts of this equation, respectively, gives

$$f_{32}^{x_2} = f_{43}^{x_3} \cos \eta_{23} \quad \text{and} \quad f_{32}^{y_2} = -f_{43}^{x_3} \sin \eta_{23} \tag{15.57}$$

Equation (15.53b) can be written as

$$f_{43}^{x_3} e^{j\eta_{34}} + \left(f_{34}^{x_4} + j f_{34}^{y_4} \right) = 0$$

Equating the real and imaginary parts of this equation, respectively, gives

$$f_{43}^{x_3} = -f_{23}^{x_3} = -\frac{t_4}{r_4} \frac{1}{\sin \eta_{34}} \quad \text{and} \quad f_{43}^{x_3} \cos \eta_{34} + f_{34}^{x_4} = 0 \tag{15.58}$$

Rearranging these two equations, the axial force on the output link can be written as

$$f_{34}^{x_4} = \frac{t_4}{r_4} \cot \eta_{34} \tag{15.59}$$

From Eqs. (15.50*a*) and (15.52*a*) we have

$$f_{12}^{x_2} = -f_{32}^{x_2} \quad \text{and} \quad f_{12}^{y_2} = -f_{32}^{y_2} \tag{15.60a}$$

$$f_{14}^{x_4} = -f_{34}^{x_4} \quad \text{and} \quad f_{14}^{y_4} = -f_{34}^{y_4} \tag{15.60b}$$

Adding Eqs. (15.53*a*) and (15.54*a*) and with the aid of Eqs. (15.60*a*), we obtain

$$f_{21}^{x_1} + j f_{21}^{y_1} = f_{43}^{x_3} e^{j\theta_3}$$

Equating the real and imaginary parts of this equation, respectively, gives

$$f_{21}^{x_1} = f_{43}^{x_3} \cos\theta_3 \quad \text{and} \quad f_{21}^{y_1} = f_{43}^{x_3} \sin\theta_3 \tag{15.61}$$

Adding Eqs. (15.53*b*) and (15.54*b*) and with the aid of Eqs. (15.60*b*), we obtain

$$f_{41}^{x_1} + j f_{41}^{y_1} = f_{43}^{x_3} e^{j\theta_3}$$

Equating the real and imaginary parts of this equation, respectively, gives

$$f_{41}^{x_1} = f_{43}^{x_3} \cos\theta_3 \quad \text{and} \quad f_{41}^{y_1} = f_{43}^{x_3} \sin\theta_3 \tag{15.62}$$

Substituting Eq. (15.58) into Eq. (15.57), the transverse force acting at point A on the input link can be written as

$$f_{32}^{y_2} = \frac{t_4 \sin\eta_{23}}{r_4 \sin\eta_{34}}$$

Then substituting this equation into (15.50*b*), and rearranging, the input torque can be written as

$$t_2 = -\frac{r_2 \sin\eta_{23}}{r_4 \sin\eta_{34}} t_4 \tag{15.63}$$

Recall from Chapters 1 and 3 (see Sections 1.10 and 3.20) that the mechanical advantage of a four-bar linkage is defined as the ratio of the output torque to the input torque. Therefore, rearranging Eq. (15.63), the mechanical advantage can be written as

$$MA = \frac{t_4}{t_2} = -\frac{r_4 \sin\eta_{34}}{r_2 \sin\eta_{23}}$$

Note that this result is consistent with Eq. (3.39).

Finally, the shaking moment can be written with the aid of Eq. (15.62) as

$$m_S = r_1 f_{41}^{y_1} = r_1 f_{43}^{x_3} \sin\theta_3$$

Substituting Eq. (15.58) into this equation, the shaking moment can be written as

$$m_S = \frac{r_1}{r_4} \frac{\sin\theta_3}{\sin\eta_{34}} t_4 \tag{15.64}$$

If friction is ignored in the mechanism, then the total load, which will be denoted here by an asterisk superscript, is equal to the sum of the static load and the dynamic load; this is the principle of superposition. Therefore, the load torque is

$$T_4^* = t_4 + T_4 \tag{15.65}$$

The axial and transverse loads at bearing O_2, respectively, are

$$\left(F_{12}^{x_2}\right)^* = -f_{32}^{x_2} + F_{12}^{x_2}, \qquad \left(F_{12}^{y_2}\right)^* = -f_{32}^{y_2} + F_{12}^{y_2} \tag{15.66}$$

For the input link, the axial and transverse loads at joint A, respectively, are

$$\left(F_{32}^{x_2}\right)^* = f_{32}^{x_2} + F_{32}^{x_2}, \qquad \left(F_{32}^{y_2}\right)^* = f_{32}^{y_2} + F_{32}^{y_2} \tag{15.67a}$$

For the coupler link, the axial and transverse loads at joint A, respectively, are

$$\left(F_{23}^{x_3}\right)^* = -f_{43}^{x_3} + F_{23}^{x_3}, \qquad \left(F_{23}^{y_3}\right)^* = F_{23}^{y_3} \tag{15.67b}$$

For the coupler link, the axial and transverse loads at joint B, respectively, are

$$\left(F_{43}^{x_3}\right)^* = -f_{43}^{x_3} + F_{43}^{x_3}, \qquad \left(F_{43}^{y_3}\right)^* = F_{43}^{y_3} \tag{15.68a}$$

For the output link, the axial and transverse loads at joint B, respectively, are

$$\left(F_{34}^{x_4}\right)^* = -f_{34}^{x_4} + F_{34}^{x_4}, \qquad \left(F_{34}^{y_4}\right)^* = f_{34}^{y_4} + F_{34}^{y_4} \tag{15.68b}$$

The axial and transverse loads at bearing O_4, respectively, are

$$\left(F_{14}^{x_4}\right)^* = -f_{34}^{x_4} + F_{14}^{x_4}, \qquad \left(F_{14}^{y_4}\right)^* = -f_{34}^{y_4} + F_{14}^{y_4} \tag{15.69}$$

The x- and y-components of the load at bearing O_2, respectively, are

$$\left(F_{21}^{x_1}\right)^* = f_{21}^{x_1} + F_{21}^{x_1} \quad \text{and} \quad \left(F_{21}^{y_1}\right)^* = f_{21}^{y_1} + F_{21}^{y_1} \tag{15.70a}$$

Therefore, the magnitude and the orientation, respectively, are

$$(F_{21})^* = \sqrt{\left(F_{21}^{x_1}\right)^{*2} + \left(F_{21}^{y_1}\right)^{*2}} \quad \text{and} \quad \psi_2 = \tan_2^{-1}\left(\frac{\left(F_{21}^{y_1}\right)^*}{\left(F_{21}^{x_1}\right)^*}\right) \tag{15.70b}$$

The x- and y-components of the load at bearing O_4, respectively, are

$$\left(F_{41}^{x_1}\right)^* = f_{41}^{x_1} + F_{41}^{x_1} \quad \text{and} \quad \left(F_{41}^{y_1}\right)^* = f_{41}^{y_1} + F_{41}^{y_1} \tag{15.71a}$$

Therefore, the magnitude and orientation, respectively, are

$$(F_{41})^* = \sqrt{\left(F_{41}^{x_1}\right)^{*2} + \left(F_{41}^{y_1}\right)^{*2}} \quad \text{and} \quad \psi_4 = \tan_2^{-1}\left(\frac{\left(F_{41}^{y_1}\right)^*}{\left(F_{41}^{x_1}\right)^*}\right) \tag{15.71b}$$

The total driving torque to the input link 2 is

$$T_2^* = T_2 + t_2$$

where the inertia input torque T_2 and the static input torque t_2 are given by Eqs. (15.45) and (15.63), respectively.

The total shaking moment is

$$M_S^* = M_S + m_S$$

where the dynamic shaking M_S and the static shaking moment m_S are given by Eqs. (15.51) and (15.64), respectively.

For stable operation we could vary the link lengths l_1, l_2, and l_3 such that the power $P = T_2^* \omega_2$, discussed in the following section, is maximized and the shaking moment is minimized. For a better suspension system we should try to minimize the vertical components of the loads at the two ground bearings, that is, $(F_{21}^{y_1})^*$ and $(F_{41}^{y_1})^*$.

15.9 EQUATION OF MOTION

This section presents a method to obtain the equation of motion of a mechanism. The method is based on energy considerations. We will first write the power equation for the mechanism and then express this equation in terms of kinematic coefficients, see Chapters 3 and 4.

The work-energy equation for a mechanism can be written as

$$W = \Delta T + \Delta U + W_f \tag{15.72}$$

where

W = the net work input to the mechanism = the work in less the work out
ΔT = the change in the kinetic energy of the moving links
ΔU = the change in the potential energy stored in the mechanism
W_f = the energy dissipated through damping and friction

Differentiating Eq. (15.72) with respect to time gives

$$P = \frac{dW}{dt} = \frac{dT}{dt} + \frac{dU}{dt} + \frac{dW_f}{dt} \tag{15.73}$$

where P is the *net power input* to the mechanism. Equation (15.73) is commonly referred to as the *power equation*. The net power is a scalar quantity and can be written as

$$P = Q \cdot \dot{\psi} \tag{15.74}$$

where Q is the generalized input force (that is, a force or a torque) and $\dot{\psi}$ is the generalized input velocity (that is, rectilinear velocity or angular velocity). If the product of these two signed quantities is positive—that is, if the signed quantities are acting in the same direction—then power is going into the system. If the product is negative—that is, if the two signed quantities are acting in opposite directions—then power is coming out of the system.

The kinetic energy of a link in the mechanism, say link j, can be written as

$$T_j = \tfrac{1}{2}m_j V_{G_j}^2 + \tfrac{1}{2}I_{G_j}\omega_j^2 \tag{15.75}$$

Recall from Chapter 3, Section 3.11, that the velocity of the mass center of link j can be written in terms of first-order kinematic coefficients as

$$\mathbf{V}_{G_j} = \left(x'_{G_j}\hat{\mathbf{i}} + y'_{G_j}\hat{\mathbf{j}}\right)\dot{\psi} \tag{15.76a}$$

and the angular velocity of link j can be written as

$$\omega_j = \theta'_j\dot{\psi} \tag{15.76b}$$

Substituting Eqs. (15.76) into Eq. (15.75), the kinetic energy of link j can be written as

$$T_j = \tfrac{1}{2}m_j\left(x_{G_j}'^2 + y_{G_j}'^2\right)\dot{\psi}^2 + \tfrac{1}{2}I_{G_j}\theta_j'^2\dot{\psi}^2 \tag{15.77}$$

Differentiating this equation with respect to time, the time rate of change in the kinetic energy of link j can be written as

$$\frac{dT_j}{dt} = A_j\dot{\psi}\ddot{\psi} + B_j\dot{\psi}^3 \tag{15.78}$$

where

$$A_j = m_j\left(x_{G_j}'^2 + y_{G_j}'^2\right) + I_{G_j}\theta_j'^2 \tag{15.79a}$$

and

$$B_j = m_j\left(x'_{G_j}x''_{G_j} + y'_{G_j}y''_{G_j}\right) + I_{G_j}\theta'_j\theta''_j \tag{15.79b}$$

Substituting Eq. (15.79a) into Eq. (15.77) we note that the kinetic energy can be written as

$$T_j = \tfrac{1}{2}A_j\dot{\psi}^2 \tag{15.80}$$

From Eq. (15.79), we also note the relationship between the coefficients, that is,

$$B_j = \frac{1}{2}\frac{dA_j}{d\psi} \tag{15.81}$$

For a mechanism with n links and with link 1 fixed, the time rate of change in kinetic energy can be written, from Eq. (15.78), as

$$\frac{dT}{dt} = \sum_{j=2}^{n} A_j\dot{\psi}\ddot{\psi} + \sum_{j=2}^{n} B_j\dot{\psi}^3 \tag{15.82}$$

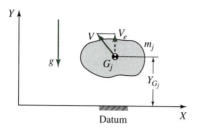

Figure 15.19 Potential energy due to elevation.

Next we consider the potential energy stored in link j of the mechanism:

1. *Due to gravity.* If we assume that the gravitational force is acting in the negative Y direction, then the work done against gravity is

$$U_{g_j} = m_j g Y_{G_j} \qquad (15.83)$$

where m_j is the mass of the link, g is the gravitational constant, and Y_{G_j} is the height of the mass center above an arbitrarily chosen datum, as shown in Fig. 15.19. Differentiating Eq. (15.83) with respect to time, the rate of change in the potential energy of link j can be written as

$$\frac{dU_{g_j}}{dt} = m_j g y'_{G_j} \dot{\psi} \qquad (15.84a)$$

where the first-order kinematic coefficient is

$$y'_{G_j} = \frac{dY_{G_j}}{d\psi} \qquad (15.84b)$$

Therefore, the time rate of change in the potential energy stored in the mechanism is

$$\frac{dU_g}{dt} = \sum_{j=2}^{n} m_j g y'_{G_j} \dot{\psi} \qquad (15.85)$$

2. *Due to a rectilinear spring.* If the mechanism contains a spring element then the work done in stretching the spring is

$$U_s = k(r_s - r_0)^2 \qquad (15.86)$$

where k is the spring rate, r_0 is the free length of the spring, and r_s is the stretched length of the spring as shown in Fig. 15.20. Differentiating Eq. (15.86) with respect to time, the time rate of change in the potential energy stored in the spring can be written as

$$\frac{dU_s}{dt} = k(r_s - r_0)r'_s \dot{\psi} \qquad (15.87a)$$

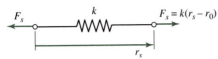

Figure 15.20 Vector across a linear spring.

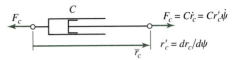

Figure 15.21 Vector across the damper.

where the first-order kinematic coefficient of the spring is

$$r'_s = \frac{dr_s}{d\psi} \tag{15.87b}$$

Next we consider the dissipative effects due to the following:

1. *A viscous damper.* The work done in overcoming the damping effect is

$$W_f = C\dot{r}_c \Delta r_c \tag{15.88}$$

where C is the damping coefficient and r_c is the length of the vector across the damper as shown in Fig. 15.21. Differentiating Eq. (15.88) with respect to time, the time rate of change of the dissipative effect can be written

$$\frac{dW_f}{dt} = Cr'^2_c \dot{\psi}^2 \tag{15.89a}$$

where the first-order kinematic coefficient of the damper is

$$r'_c = \frac{dr_c}{d\psi} \tag{15.89b}$$

2. *Coulomb friction.* The work done in overcoming the effect of Coulomb friction is

$$W_f = \mu N \Delta r_f \tag{15.90}$$

where μ is the coefficient of friction, N is the normal force between the contacting surfaces, and $\mathbf{r}_f$ is the vector along the link to the point of contact as shown in Fig. 15.22. Differentiating Eq. (15.90) with respect to time, the time rate of change of the dissipative effect can be written as

$$\frac{dW_f}{dt} = \mu N |r'_f \dot{\psi}| \tag{15.91a}$$

where the first-order kinematic coefficient is

$$r'_f = \frac{dr_f}{d\psi} \tag{15.91b}$$

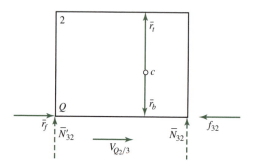

Figure 15.22 Vector to the point of contact.

Substituting Eqs. (15.74), (15.82), (15.85), (15.87a), (15.89a), and (15.91a) into Eq. (15.73), the net power can be written as

$$Q\dot\psi = \sum_{j=2}^{n} A_j \dot\psi \ddot\psi + \sum_{j=2}^{n} B_j \dot\psi^3 + \sum_{j=2}^{n} m_j g y'_{G_j} \dot\psi + k(r_s - r_0)r'_s \dot\psi + C r'^2_c \dot\psi^2 + \mu N |r'_f \dot\psi|$$

(15.92)

Note that each term in this equation contains the generalized input velocity. If we divide through by this common factor, then we obtain the *equation of motion* for the mechanism, that is,

$$Q = \sum_{j=2}^{n} A_j \ddot\psi + \sum_{j=2}^{n} B_j \dot\psi^2 + \sum_{j=2}^{n} m_j g y'_{G_j} + k(r_s - r_0)r'_s + C r'^2_c \dot\psi + \mu N |r'_f| \quad (15.93)$$

This result is also valid when the generalized input velocity is zero—that is, when the input link is instantaneously stopped or when the mechanism is starting from rest. In this case, Eq. (15.93) reduces to

$$Q = \sum_{j=2}^{n} A_j \ddot\psi + \sum_{j=2}^{n} m_j g y_{G_j} + k(r_s - r_0)r'_s + \mu N |r'_f| \quad (15.94)$$

and for this case there is no necessity to calculate $\sum_{j=2}^{n} B_j$ or the first-order kinematic coefficient of the viscous damper.

We will consider the two cases for the input link: (1) rotation, that is, $\dot\psi = \dot\theta_i$, and (2) translation, that is, $\dot\psi = \dot r_i$.

1. If the input link is only rotating, then Eq. (15.93) can be written as

$$M_i = \sum_{j=2}^{n} A_j \ddot\theta_i + \sum_{j=2}^{n} B_j \dot\theta_i^2 + \sum_{j=2}^{n} m_j g y'_{G_j} + k(r_s - r_0)r'_s + C r'^2_c \dot\theta_i + \mu N |r'_f| \quad (15.95)$$

where M_i is the torque acting on the input link. Note that the coefficient $\sum_{j=2}^{n} A_j$ must have the units of moment of inertia and is referred to as the *equivalent mass moment of inertia* of the mechanism and denoted as I_{EQ}. This moment of inertia, if attached to the input link, would have the same kinetic energy as the entire mechanism. From Eq. (15.80) the kinetic energy can be written as

$$T = \tfrac{1}{2} I_{EQ} \dot\theta^2$$

(15.96)

The equation of motion—that is, Eq. (15.95)—can, therefore, be written as

$$M_i = I_{EQ}\ddot{\theta}_i + \frac{1}{2}\frac{dI_{EQ}}{d\theta_i}\dot{\theta}_i^2 + \sum_{j=2}^{n} m_j g y'_{G_j} + k(r_s - r_0)r'_s + Cr_c'^2\dot{\theta}_i + \mu N|r'_f| \quad (15.97)$$

2. If the input link is only translating, then Eq. (15.94) can be written as

$$F_i = \sum_{j=2}^{n} A_j\ddot{r}_i + \sum_{j=2}^{n} B_j\dot{r}_i^2 + \sum_{j=2}^{n} m_j g y'_{G_j} + k(r_s - r_0)r'_s + Cr_c'^2\dot{r}_i + \mu N|r'_f| \quad (15.98)$$

where F_i is the force acting on the input link. For this case, the coefficient $\sum_{j=2}^{n} A_j$ has the units of mass and is referred to as the *equivalent mass* of the mechanism and denoted as m_{EQ}. This mass, if attached to the input link, would have the same kinetic energy as the entire mechanism. From Eq. (15.80) the kinetic energy can be written as

$$T = \tfrac{1}{2}m_{EQ}\dot{r}^2 \quad (15.99)$$

The equation of motion—that is, Eq. (15.98)— can therefore be written as

$$F_i = m_{EQ}\ddot{r}_i + \frac{1}{2}\frac{dm_{EQ}}{dr_i}\dot{r}_i^2 + \sum_{j=2}^{n} m_j g y'_{G_j} + k(r_s - r_0)r'_s + Cr_c'^2\dot{r}_i + \mu N|r'_f| \quad (15.100)$$

EXAMPLE 15.6

For the four-bar mechanism shown in Fig. 15.23, determine the torque M_2 that must be applied to the input link 2 to produce an angular velocity of $\omega_2 = 10$ rad/s ccw and an angular acceleration $\alpha_2 = 10$ rad/s² cw. In the position shown in the figure, $DA = BE = 80$ cm. The free length of the spring is $r_0 = 60$ cm, the spring rate is $k = 5$ N/cm, and the damping constant is $C = 10$ N·s/cm. The masses of the links are $m_2 = 2.5$ kg, $m_3 = 5$ kg, and $m_4 = 2.5$ kg, and the center of mass of each link is coincident with its geometric center. The mass moments of inertia of the links are $I_{G2} = I_{G4} = 2.5 \times 10^3$ kg·cm² and $I_{G3} = 4.5 \cdot 10^3$ kg·cm². Assume that gravity acts vertically downward and friction can be ignored.

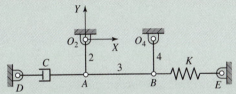

Figure 15.23 Example 15.6: The link dimensions are $r_1 = 80$ cm, $r_2 = 40$ cm, $r_3 = 80$ cm, $r_4 = 40$ cm.

SOLUTION

From the kinematic analysis of the four-bar linkage presented in Sections 3.11 and 4.11, the first- and second-order kinematic coefficients of links 3 and 4 are shown to be

$$\theta_3' = 0, \qquad \theta_4' = 1 \text{ rad/rad}, \qquad \theta_3'' = 0, \qquad \theta_4'' = 0 \tag{1}$$

Recall that the angular velocities and accelerations of links 3 and 4 can be written as

$$\omega_3 = \theta_3'\omega_2 \quad \text{and} \quad \omega_4 = \theta_4'\omega_2$$
$$\alpha_3 = \theta_3''\omega_2^2 + \theta_3'\alpha_2 \quad \text{and} \quad \alpha_4 = \theta_4''\omega_2^2 + \theta_4'\alpha_2 \tag{2}$$

Substituting Eqs. (1) and the input angular velocity and acceleration into Eqs. (2) gives

$$\omega_3 = 0, \qquad \omega_4 = 10 \text{ rad/s ccw}, \qquad \alpha_3 = 0, \qquad \alpha_4 = 10 \text{ rad/s}^2 \text{ cw} \tag{3}$$

The x and y components of the position vector of the mass center G_2 with respect to the ground pivot O_2 are

$$x_{G_2} = \tfrac{1}{2}r_2\cos\theta_2 \quad \text{and} \quad y_{G_2} = \tfrac{1}{2}r_2\sin\theta_2 \tag{4}$$

Differentiating these equations twice with respect to θ_2 and substituting $\theta_2 = 270°$, the first- and second-order kinematic coefficients of the mass center of link 2 are

$$x_{G_2}' = 20 \text{ cm}, \qquad y_{G_2}' = 0, \qquad x_{G_2}'' = 0, \qquad y_{G_2}'' = 20 \text{ cm} \tag{5}$$

The x and y components of the position vector of the mass center G_3 with respect to the ground pivot O_2 can be written as

$$x_{G_3} = r_2\cos\theta_2 + \tfrac{1}{2}r_3\cos\theta_3 \quad \text{and} \quad y_{G_3} = r_2\sin\theta_2 + \tfrac{1}{2}r_3\sin\theta_3 \tag{6}$$

Differentiating these equations twice with respect to θ_2 and substituting $\theta_2 = 270°$ and $\theta_3 = 0$, the first- and second-order kinematic coefficients of the mass center of link 3 are

$$x_{G_3}' = 40 \text{ cm}, \qquad y_{G_3}' = 0, \qquad x_{G_3}'' = 0, \qquad y_{G_3}'' = 40 \text{ cm} \tag{7}$$

The first- and second-order kinematic coefficients of the mass center of link 4 are the same as for the mass center of link 2, that is,

$$x_{G_4}' = 20 \text{ cm}, \qquad y_{G_4}' = 0, \qquad x_{G_4}'' = 0, \qquad y_{G_4}'' = 20 \text{ cm} \tag{8}$$

To obtain the first-order kinematic coefficient for the spring, the vector loop-closure equation, shown in Fig. 15.24, can be written as

$$\mathbf{r}_4 + \mathbf{r}_s - \mathbf{r}_6 = 0 \tag{9}$$

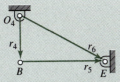

Figure 15.24 Vector loop for the spring.

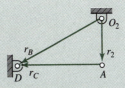

Figure 15.25 Vector loop for the viscous damper.

The two scalar equations are

$$r_4 \cos \theta_4 + r_s \cos \theta_s - r_6 \cos \theta_6 = 0$$
$$r_4 \sin \theta_4 + r_s \sin \theta_s - r_6 \sin \theta_6 = 0$$
(10)

Differentiating Eqs. (10) with respect to the input θ_2 gives

$$-r_4 \sin \theta_4 \theta_4' - r_s \sin \theta_s \theta_s' + r_s' \cos \theta_s = 0$$
(11a)

and

$$r_4 \cos \theta_4 \theta_4' + r_s \cos \theta_s \theta_s' + r_s' \sin \theta_s = 0$$
(11b)

Substituting $\theta_s = 0°$ and $\theta_4 = 270°$ into Eq. (11a), the first-order kinematic coefficient for the spring is

$$r_s' = -r_4 = -40 \text{ cm}$$
(12)

To obtain the first-order kinematic coefficient for the spring. The vector loop-closure equation, shown in Fig. 15.25, can be written as

$$\mathbf{r}_2 + \mathbf{r}_c - \mathbf{r}_8 = 0$$
(13)

The two scalar equations are

$$r_2 \cos \theta_2 + r_c \cos \theta_c - r_8 \cos \theta_8 = 0$$
$$r_2 \sin \theta_2 + r_c \sin \theta_c - r_8 \sin \theta_8 = 0$$
(14)

Differentiating Eqs. (13) with respect to the input θ_2 gives

$$-r_2 \sin \theta_2 - r_c \sin \theta_c \theta_c' + r_c' \cos \theta_c = 0$$
(15a)

and

$$r_2 \cos \theta_2 + r_c \cos \theta_c \theta_c' + r_c' \sin \theta_c = 0$$
(15b)

Substituting $\theta_2 = 270°$ and $\theta_c = 180°$ into Eqs. (15a), the first-order kinematic coefficient for the damper is

$$r'_c = r_2 = 40 \text{ cm} \tag{16}$$

The equivalent moment of inertia for the four-bar linkage is

$$I_{EQ} = \sum_{j=2}^{4} A_j$$

From Eq. (15.79a), the equivalent moment of inertia can be written as

$$I_{EQ} = m_2\left(x'^2_{G_2} + y'^2_{G_2}\right) + I_{G_2} + m_3\left(x'^2_{G_3} + y'^2_{G_3}\right) + I_{G_3}\theta'^2_3 + m_4\left(x'^2_{G_4} + y'^2_{G_4}\right) + I_{G_4}\theta'^2_4$$

Substituting the known data and the kinematic coefficients into this equation gives

$$I_{EQ} = 1\,000 + 2\,500 + 8\,000 + 1\,000 + 3\,500 = 15\,000 \text{ kg·cm}^2 \tag{17}$$

From Eq. (15.79b), we can write

$$\sum_{j=2}^{4} B_j = m_2\left(x'_{G_2}x''_{G_2} + y'_{G_2}y''_{G_2}\right) + m_3\left(x'_{G_3}x''_{G_3} + y'_{G_3}y''_{G_3}\right) + I_{G_3}\theta'_3\theta''_3$$
$$+ m_4\left(x'_{G_4}x''_{G_4} + y'_{G_4}y''_{G_4}\right) + I_{G_4}\theta'_4\theta''_4 \tag{18}$$

For the given position, the first-order kinematic coefficients are

$$y'_{G_2} = y'_{G_3} = y'_{G_4} = 0 \tag{19}$$

Substituting the known data and kinematic coefficients into Eq. (18) gives

$$\sum_{j=2}^{4} B_j = 0 \tag{20}$$

Then, substituting Eqs. (17), (19), and (20) into Eq. (15.97), the equation of motion can be written

$$M_i = 15\,000\ddot{\theta}_i + k(r_s - r_0)r'_s + Cr'^2_c\dot{\theta}_i$$

Finally, substituting the known data and Eqs. (12) and (16) into this equation, the input torque is

$$M_2 = -150\,000 + 5(80 - 60)(-40) + 10 \times 40^2 \times 10$$

$$M_2 = +6\,000 \text{ N·cm} = 60 \text{ N·m} \qquad\qquad Ans.$$

Because the answer is positive, the input torque acts in the same direction as the input angular velocity—that is, counterclockwise.

NOTES

1. Note that this example is done in SI units; in U.S. customary units, pounds, slugs, or other units might have been used. However, this is not critical here, because the density cancels in Eq. (15.7).

2. It should be carefully noted here that these integrals are not the same as those called *area moments of inertia* or *second moments of area,* which are integrals over dA, a differential area, rather than integrals over dm, a differential mass. These other integrals often arise in problems involving forces distributed over an area or volume, but are different. In two-dimensional problems of constant thickness, however, they are easily related because dA times the thickness times the mass density yields dm for the integral.

3. Jean leRond d'Alembert (1717–1783).

PROBLEMS*

15.1 The steel bell crank shown in the figure is used as an oscillating cam follower. Using 0.282 lb/in³ for the density of steel, find the mass moment of inertia of the lever about an axis through O.

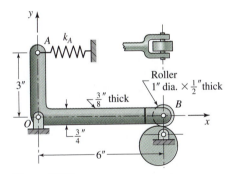

Figure P15.1

15.2 A 5- by 50- by 300-mm steel bar has two round steel disks, each 50 mm in diameter and 20 mm long, welded to one end as shown. A small hole is drilled 25 mm from the other end. The density of steel is 7.80 Mg/m³. Find the mass moment of inertia of this weldment about an axis through the hole.

15.3 Find the external torque which must be applied to link 2 of the four-bar linkage shown to drive it at the given velocity.

---

*Unless stated otherwise, solve all problems without friction and without gravitational loads.

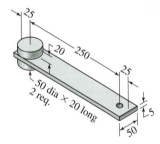

Figure P15.2 Dimensions are given in millimeters.

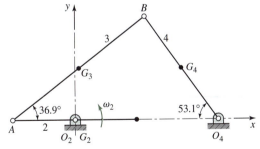

Figure P15.3 $R_{AO_2} = 3$ in, $R_{O_4O_2} = 7$ in, $R_{BA} = 8$ in, $R_{BO_4} = 6$ in, $R_{G_3A} = 4$ in, $R_{G_4O_4} = 3$ in, $w_3 = 0.708$ lb, $w_4 = 0.780$ lb, $I_{G_2} = 0.0258$ in·lb·s², $I_{G_3} = 0.0154$ in·lb·s², $I_{G_4} = 0.531$ in·lb·s², $\omega_2 = 180\hat{\mathbf{k}}$ rad/s, $\alpha_2 = \mathbf{0}$, $\alpha_3 = 4\,950\hat{\mathbf{k}}$ rad/s², $\alpha_4 = -8\,900\hat{\mathbf{k}}$ rad/s², $\mathbf{A}_{G_3} = 6\,320\hat{\mathbf{i}} + 750\hat{\mathbf{j}}$ ft/s², $\mathbf{A}_{G_4} = 2\,280\hat{\mathbf{i}} + 750\hat{\mathbf{j}}$ ft/s².

15.4 Crank 2 of the four-bar linkage shown in the figure is balanced. For the given angular velocity of link 2, find the forces acting at each joint and the external torque that must be applied to link 2.

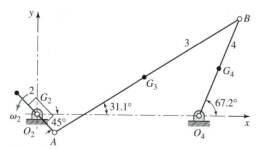

Figure P15.4 $R_{AO_2} = 2$ in, $R_{O_4O_2} = 13$ in, $R_{BA} = 17$ in, $R_{BO_4} = 8$ in, $R_{G_3A} = 8.5$ in, $R_{G_4O_4} = 4$ in, $w_3 = 2.65$ lb, $w_4 = 6.72$ lb, $I_{G_2} = 0.0239$ in·lb·s², $I_{G_3} = 0.06006$ in·lb·s², $I_{G_4} = 0.531$ in·lb·s², $\omega_2 = 200\hat{\mathbf{k}}$ rad/s, $\alpha_2 = 0$, $\alpha_3 = -6500\hat{\mathbf{k}}$ rad/s², $\alpha_4 = -240\hat{\mathbf{k}}$ rad/s², $\mathbf{A}_{G_3} = -3160\hat{\mathbf{i}} + 262\hat{\mathbf{j}}$ ft/s², $\mathbf{A}_{G_4} = -800\hat{\mathbf{i}} - 2110\hat{\mathbf{j}}$ ft/s².

15.5 For the angular velocity of crank 2 given in the figure, find the reactions at each joint and the external torque to be applied to the crank.

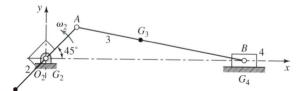

Figure P15.5 $R_{AO_2} = 3$ in, $R_{BA} = 12$ in, $R_{G_3A} = 4.5$ in, $w_3 = 3.40$ lb, $w_4 = 2.86$ lb, $I_{G_2} = 0.352$ in·lb·s², $I_{G_3} = 0.108$ in·lb·s², $\omega_2 = 210\hat{\mathbf{k}}$ rad/s, $\alpha_2 = 0$, $\alpha_3 = -7670\hat{\mathbf{k}}$ rad/s², $\mathbf{A}_{G_3} = -7820\hat{\mathbf{i}} - 4876\hat{\mathbf{j}}$ ft/s², $\mathbf{A}_{G_4} = -7850\hat{\mathbf{i}}$ ft/s².

15.6 The figure shows a slider-crank mechanism with an external force $\mathbf{F}_B$ applied to the piston. For the given crank velocity, find all the reaction forces in the joints and the crank torque.

15.7 The following data apply to the four-bar linkage shown in the figure: $R_{AO_2} = 0.3$ m, $R_{O_4O_2} = 0.9$ m, $R_{BA} = 1.5$ m, $R_{BO_4} = 0.8$ m, $R_{CA} = 0.85$ m, $\theta_C = 33°$, $R_{DO_4} = 0.4$ m, $\theta_D = 53°$, $R_{G_2O_2} = 0$, $R_{G_3A} = 0.65$ m, $\alpha = 16°$, $R_{G_4O_4} = 0.45$ m, $\beta = 17°$, $m_2 = 5.2$ kg, $m_3 = 65.8$ kg, $m_4 = 21.8$ kg,

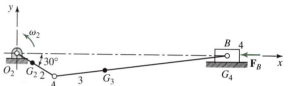

Figure P15.6 $R_{AO_2} = 3$ in, $R_{BA} = 12$ in, $R_{G_2O_2} = 1.25$ in, $R_{G_3A} = 3.5$ in, $w_2 = 0.95$ lb, $w_3 = 3.50$ lb, $w_4 = 2.50$ lb, $I_{G_2} = 0.00369$ in·lb·s², $I_{G_3} = 0.110$ in·lb·s², $\omega_2 = 160\hat{\mathbf{k}}$ rad/s, $\alpha_2 = 0$, $\alpha_3 = -3090\hat{\mathbf{k}}$ rad/s², $\mathbf{A}_{G_2} = 2640\angle150°$ ft/s², $\mathbf{A}_{G_3} = 6130\angle158.3°$ ft/s², $\mathbf{A}_{G_4} = 6280\angle180°$ ft/s², $\mathbf{F}_B = 800\angle180°$ lb.

$I_{G_2} = 2.3$ kg·m², $I_{G_3} = 4.2$ kg·m², $I_{G_4} = 0.51$ kg·m². A kinematic analysis at $\theta_2 = 53°$, $\omega_2 = 12\hat{\mathbf{k}}$ rad/s ccw, and $\alpha_2 = 0$, gave $\theta_3 = 0.7°$, $\theta_4 = 20.4°$, $\alpha_3 = 85.6$ rad/s² cw, $\alpha_4 = 172$ rad/s² cw, $\mathbf{A}_{G_3} = 96.4\angle259°$ m/s², and $\mathbf{A}_{G_4} = 97.8\angle270°$ m/s². Find all the pin reactions and the torque to be applied to link 2.

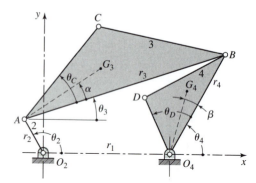

Figure P15.7

15.8 Solve Problem 15.7 with an additional external force $\mathbf{F}_D = 12\angle0°$ kN acting at point D.

15.9 Make a complete kinematic and dynamic analysis of the four-bar linkage of Problem 15.7 using the same data, but with $\theta_2 = 170°$, $\omega_2 = 12$ rad/s ccw, $\alpha_2 = 0$, and an external force $\mathbf{F}_D = 8.94\angle64.3°$ kN acting at point D.

15.10 Repeat Problem 15.9 using $\theta_2 = 200°$, $\omega_2 = 12$ rad/s ccw, $\alpha_2 = 0$, and an external force $\mathbf{F}_C = 8.49\angle45°$ kN acting at point C.

15.11 At $\theta_2 = 270°$, $\omega_2 = 18$ rad/s ccw, $\alpha_2 = 0$, a kinematic analysis of the linkage whose geometry is given in Problem 15.7 gives $\theta_3 = 46.6°$, $\theta_4 = 80.5°$,

$\alpha_3 = 178$ rad/s² cw, $\alpha_4 = 256$ rad/s² cw, $\mathbf{A}_{G_3} = 112\angle 22.7°$ m/s², $\mathbf{A}_{G_4} = 119\angle 352.5°$ m/s². An external force $\mathbf{F}_D = 8.94\angle 64.3°$ kN acts at point D. Make a complete dynamic analysis of the linkage.

15.12 The following data apply to the four-bar linkage shown in the figure: $R_{AO_2} = 120$ mm, $R_{O_4O_2} = 300$ mm, $R_{BA} = 320$ mm, $R_{BO_4} = 250$ mm, $R_{CA} = 360$ mm, $\theta_C = 15°$, $R_{DO_4} = 0$, $\theta_D = 0°$, $R_{G_2O_2} = 0$, $R_{G_3A} = 200$ mm, $\alpha = 8°$, $R_{G_4O_4} = 125$ mm, $\beta = 0°$, $m_2 = 0.5$ kg, $m_3 = 4$ kg, $m_4 = 1.5$ kg, $I_{G_2} = 0.005$ N·m·s², $I_{G_3} = 0.011$ N·m·s², $I_{G_4} = 0.0023$ N·m·s². A kinematic analysis at $\theta_2 = 90°$ and $\omega_2 = 32$ rad/s ccw with $\alpha_2 = 0$ gave $\theta_3 = 23.9°$, $\theta_4 = 91.7°$, $\alpha_3 = 221$ rad/s² ccw, $\alpha_4 = 122$ rad/s² ccw, $\mathbf{A}_{G_3} = 88.6\angle 255°$ m/s², and $\mathbf{A}_{G_4} = 32.6\angle 244°$ m/s². Using an external force $\mathbf{F}_C = 632\angle 342°$ N acting at point C, make a complete dynamic analysis of the linkage.

15.13 Repeat Problem 15.12 at $\theta_2 = 260°$. Analyze both the kinematics and the dynamics of the system at this position.

15.14 Repeat Problem 15.13 at $\theta_2 = 300°$.

15.15 Analyze the dynamics of the offset slider-crank mechanism shown in the figure using the following data: $a = 0.06$ m, $R_{AO_2} = 0.1$ m, $R_{AB} = 0.38$ m, $R_{CA} = 0.4$ m, $\theta_2 = 32°$, $R_{G_3A} = 0.26$ m, $\alpha = 22°$, $m_2 = 2.5$ kg, $m_3 = 7.4$ kg, $m_4 = 2.5$ kg, $I_{G_2} = 0.005$ N·m·s², $I_{G_3} = 0.0136$ N·m·s², $\theta_2 = 120°$, and $\omega_2 = 18$ rad/s cw with $\alpha = 0$, $\mathbf{F}_B = -2\,000\hat{\mathbf{i}}$ N, $\mathbf{F}_C = -1\,000\hat{\mathbf{i}}$ N. Assume a balanced crank and no friction forces.

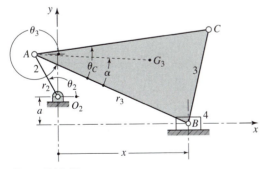

Figure P15.15

15.16 Analyze the system of Problem 15.15 for a complete rotation of the crank. Use $\mathbf{F}_C = \mathbf{0}$ and $\mathbf{F}_B = -1\,000\hat{\mathbf{i}}$ N when $\mathbf{V}_B$ is toward the right, but use $\mathbf{F}_B = \mathbf{0}$ when $\mathbf{V}_B$ is toward the left. Plot a graph of M_{12} versus θ_2.

15.17 A slider-crank mechanism similar to that of Problem 15.15 has $a = 0$, $R_{AO_2} = 0.1$ m, $R_{BA} = 0.45$ m, $R_{CA} = 0$, $\theta_C = 0$, $R_{G_3A} = 0.2$ m, $\alpha = 0$, $m_2 = 1.5$ kg, $m_3 = 3.5$ kg, $m_4 = 1.2$ kg, $I_{G_2} = 0.01$ N·m·s², $I_{G_3} = 0.060$ N·m·s², and $M_{12} = 60$ N·m. Corresponding to $\theta_2 = 120°$ and $\omega_2 = 24$ rad/s cw with $\alpha_2 = 0$, a kinematic analysis gave $\theta_3 = -9°$, $R_B = 0.374$ m, $\alpha_3 = 89.3$ rad/s² ccw, $\mathbf{A}_B = 40.6\hat{\mathbf{i}}$ m/s², and $\mathbf{A}_{G_3} = 40.6\hat{\mathbf{i}} - 22.6\hat{\mathbf{j}}$ m/s². Assume link 2 is balanced and find $\mathbf{F}_{14}$ and $\mathbf{F}_{23}$.

15.18 Repeat Prob. 15.17 for $\theta_2 = 240°$. The results of a kinematic analysis are $\theta_3 = 11.1°$, $R_B = 0.392$ m, $\alpha_3 = 112$ rad/s² cw, $\mathbf{A}_B = 35.2\hat{\mathbf{i}}$ m/s², and $\mathbf{A}_{G_3} = 31.6\hat{\mathbf{i}} + 27.7\hat{\mathbf{j}}$ m/s².

15.19 A slider-crank mechanism, as in Problem 15.15, has $a = 0.008$ m, $R_{AO_2} = 0.25$ m, $R_{BA} = 1.25$ m, $R_{CA} = 1.0$ m, $\theta_C = -38°$, $R_{G_3A} = 0.75$ m, $\alpha = -18°$, $m_2 = 10$ kg, $m_3 = 140$ kg, $m_4 = 50$ kg, $I_{G_2} = 2.0$ N·m·s², and $I_{G_3} = 8.42$ N·m·s² and has a balanced crank. Make a complete kinematic and dynamic analysis of this system at $\theta_2 = 120°$ with $\omega_2 = 6$ rad/s² ccw and $\alpha_2 = 0$, using $\mathbf{F}_B = 50\angle 180°$ kN and $\mathbf{F}_C = 80\angle -60°$ kN.

15.20 Cranks 2 and 4 of the cross-linkage shown in the figure are balanced. The dimensions of the linkage are $R_{AO_2} = 6$ in, $R_{O_4O_2} = 18$ in, $R_{BA} = 18$ in, $R_{BO_4} = 6$ in, $R_{CA} = 24$ in, and $R_{G_3A} = 12$ in. Also $w_3 = 4$ lb, $I_{G_2} = I_{G_4} = 0.063$ in·lb·s², and $I_{G_3} = 0.497$ in·lb·s². Corresponding to the position shown, and with $\omega_2 = 10$ rad/s ccw and $\alpha_2 = 0$, a kinematic analysis gave as results $\omega_3 = 1.43$ rad/s cw, $\omega_4 = 11.43$ rad/s cw, $\alpha_3 = \alpha_4 = 84.7$ rad/s² cw, and $\mathbf{A}_{G_3} = 47.6\hat{\mathbf{i}} + 70.3\hat{\mathbf{j}}$ ft/s². Find the driving torque and the pin reactions if $\mathbf{F}_C = -30\hat{\mathbf{j}}$ lb.

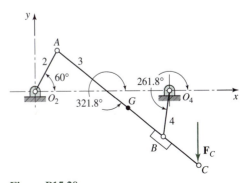

Figure P15.20

15.21 Find the driving torque and the pin reactions for the mechanism of Problem 15.20 under the same dynamic conditions, but with crank 4 as the driver.

15.22 A kinematic analysis of the mechanism of Problem 15.20 at $\theta_2 = 210°$ with $\omega_2 = 10$ rad/s ccw and $\alpha_2 = 0$ gave $\theta_3 = 14.7°$, $\theta_4 = 164.7°$, $\omega_3 = 4.73$ rad/s ccw, $\omega_4 = 5.27$ rad/s cw, $\alpha_3 = \alpha_4 = 10.39$ rad/s cw, and $\mathbf{A}_{G_3} = 26\angle20.85°$ ft/s². Compute the crank torque and the pin reactions for this phase of the motion using the same force $\mathbf{F}_C$ as in Problem 15.20.

15.23 The figure shows a linkage with an extended coupler having an external force of $\mathbf{F}_C$ acting during a portion of the cycle. The dimensions of the linkage are $R_{AO_2} = 16$ in, $R_{O_4O_2} = R_{BA} = 40$ in, $R_{BO_4} = 56$ in, $R_{G_3A} = 32$ in, and $R_{G_4O_4} = 20$ in. Also $w_3 = 222$ lb, $w_4 = 208$ lb, $I_{G_3} = 226$ in·lb·s², and $I_{G_4} = 264$ in·lb·s², and the crank is balanced. Make a kinematic and dynamic analysis for a complete rotation of the crank using $\omega_2 = 10$ rad/s ccw, $\mathbf{F}_C = -500\hat{\mathbf{i}} + 886\hat{\mathbf{j}}$ lb for $90° \le \theta_2 \le 300°$, and $\mathbf{F}_C = 0$ for other angles.

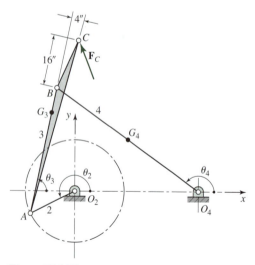

Figure P15.23

15.24 The figure shows a motor geared to a shaft on which a flywheel is mounted. The mass moments of inertia of the parts are as follows: flywheel, $I = 2.73$ in·lb·s²; flywheel shaft, $I = 0.015\,5$ in·lb·s²; gear, $I = 0.172$ in·lb·s²; pinion, $I = 0.003\,49$ in·lb·s²; motor, $I = 0.086\,4$ in·lb·s². If the motor has a starting torque of 75 in·lb, what is the angular acceleration of the flywheel shaft at the instant the motor is turned on ?

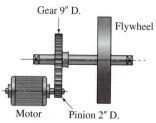

Gear 9" D.

Flywheel

Motor   Pinion 2" D.

Figure P15.24

15.25 The disk cam of Problem 14.31 is driven at a constant input shaft speed of $\omega_2 = 20$ rad/s ccw. Both the cam and the follower have been balanced so that the centers of mass of each are located at their respective fixed pivots. The mass of the cam is 0.075 kg with radius of gyration of 30 mm, and for the follower the mass is 0.030 kg with radius of gyration of 35 mm. Determine the moment $\mathbf{M}_{12}$ required on the camshaft at the instant shown in the figure to produce this motion.

15.26 Repeat Problem 15.25 with a shaftspeed of $\omega_2 = 40$ rad/s ccw.

15.27 A rotating drum is pivoted at O_2 and is decelerated by the double-shoe brake mechanism shown in the figure. The mass of the drum is 230 lb and its radius of gyration is 5.66 in. The brake is actuated by force $\mathbf{P} = -100\hat{\mathbf{j}}$ lb, and it is assumed that the contact between the two shoes and the drum act at points C and D, where the coefficients of coulomb friction are $\mu = 0.300$. Determine the angular acceleration of the drum and the reaction force at the fixed pivot $\mathbf{F}_{12}$.

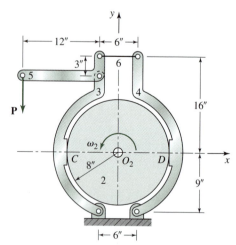

Figure P15.27

16 | Dynamic Force Analysis (Spatial)

16.1 INTRODUCTION

In the previous two chapters we studied methods for analyzing the forces in machines. First, in Chapter 14 we analyzed static or steady-state forces; then, in Chapter 15 we went on to analyze the time-varying dynamic forces caused by acceleration. A brief review will show that vector equations were used throughout and, therefore, most of the equations and techniques presented seem equally applicable in either two or three dimensions. The examples presented, however, were all limited to planar motion.

In this chapter, as implied by its title, we will extend our study to include spatial problems. Thus, the basic principles are not new, but the problems presented may seem more complex because of our difficulty in visualizing in three dimensions. In addition to this added complexity, we will see that our previous treatment of moments and angular motion were not presented in detail enough to deal with three-dimensional rotations.

We will also derive and study methods involving the use of translational and angular impulse and momentum. These additional techniques, while particularly valuable with spatial problems, also provide alternative methods of analysis for planar problems. These may improve our comprehension of the physical principles involved and, sometimes, may also greatly simplify the solution process for certain problems.

16.2 MEASURING MASS MOMENT OF INERTIA

Sometimes the shapes of machine parts are so complicated that it is extremely tedious and time-consuming to calculate the moment(s) of inertia. Consider, for example, the problem of finding the mass moment of inertia of an automobile body about a vertical axis through its center of mass. For such problems it is usually possible to determine the mass moment of inertia by observing the dynamic behavior of the body to a known rotational disturbance.

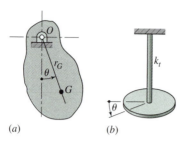

Figure 16.1 (a) Simple pendulum. (b) Torsional pendulum.

(a)   (b)

Many bodies, connecting rods and cranks for example, are shaped so that their masses can be assumed to lie in a single plane. Once such bodies have been weighed and their mass centers located, they can be suspended like a pendulum and caused to oscillate. The mass moment of inertia of such a body can then be computed from an observation of its period or frequency of oscillation. For best experimental results, the part should be suspended with the pivot located close to the mass center but not coincident with it. It is not usually necessary to drill a hole to suspend the body; for example, a spoked wheel or a gear can be suspended on a knife edge at its rim.

When the body of Fig. 16.1a is displaced through an angle θ, a gravity force mg acts at G. Summing moments about the pivot O gives

$$\sum M_O = -mg(r_G \sin\theta) - I_O\ddot{\theta} = 0 \qquad (a)$$

We intend that the pendulum be displaced only through small angles, so that $\sin\theta$ can be replaced by θ. Equation (a) can then be written as

$$\ddot{\theta} + \frac{mgr_G}{I_O}\theta = 0 \qquad (b)$$

This second-order, linear, differential equation has the well-known solution

$$\theta = c_1 \sin\sqrt{\frac{mgr_G}{I_O}}\,t + c_2 \cos\sqrt{\frac{mgr_G}{I_O}}\,t \qquad (c)$$

where c_1 and c_2 are the constants of integration.

We shall start the pendulum motion by displacing it through a small angle θ_0 and releasing it with no initial velocity from this position. Thus at time $t = 0$ we obtain $\theta = \theta_0$ and $\dot{\theta} = 0$. Substituting these conditions into Eq. (c) and its first time derivative enables us to evaluate the two constants. They are found to be $c_1 = 0$ and $c_2 = \theta_0$. Therefore

$$\theta = \theta_0 \cos\sqrt{\frac{mgr_G}{I_O}}\,t \qquad (16.1)$$

Because a cosine function repeats itself every $360°$ or 2π radians, the period of the motion is

$$\tau = 2\pi\sqrt{\frac{I_O}{mgr_G}} \qquad (d)$$

Therefore, the mass moment of inertia of the body about the pivot O is

$$I_O = mgr_G \left(\frac{\tau}{2\pi} \right)^2 \tag{16.2}$$

This equation shows that the body must be weighed to get mg, the distance r_G must be measured, and then the pendulum must be suspended and oscillated so that the period τ can be observed; Eq. (16.2) can then be solved to give I_O about O. If the moment of inertia about the mass center is desired, it can be obtained by using the parallel axis theorem, Eq. (15.12).

Figure 16.1*b* shows how the mass moment of inertia can be determined without actually weighing the body. The body is connected to a slender rod or wire at the mass center. A torsional stiffness k_t of the rod or the wire is defined as the torque necessary to twist the rod through a unit angle. If the body of Fig. 16.1*b* is turned through a small angle θ and released, the equation of motion becomes

$$\ddot{\theta} + \frac{k_t}{I_G} \theta = 0 \tag{e}$$

This is similar to Eq. (*b*) and, with the same starting conditions, has the solution

$$\theta = \theta_0 \cos \sqrt{\frac{k_t}{I_G}} t \tag{16.3}$$

The period of oscillation is then

$$\tau = 2\pi \sqrt{\frac{I_G}{k_t}} \tag{f}$$

and

$$I_G = k_t \left(\frac{\tau}{2\pi} \right)^2 \tag{16.4}$$

The torsional stiffness is often known or can be computed from a knowledge of the length and the diameter of the rod or the wire and its material. Then the oscillation of the body can be observed and Eq. (16.4) used to compute the mass moment of inertia I_G. Alternatively, when the torsional stiffness k_t is unknown, a body with known mass moment of inertia I_G can be mounted and Eq. (16.4) can be used to determine k_t.

A *trifilar pendulum,* also called a *three-string torsional pendulum,* illustrated in Fig. 16.2, can provide a very accurate method of measuring mass moment of inertia. Three strings of equal length support a lightweight platform and are equally spaced about its center. A round platform serves just as well as the triangular one shown. The part whose mass moment of inertia is to be determined is carefully placed on the platform so that the center of mass of the object coincides with the platform center. The platform is then made to oscillate, and the number of oscillations is counted over a specified period of time.[1]

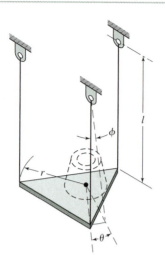

Figure 16.2 Trifilar pendulum.

The notation for the three-string torsional pendulum analysis is as follows:

m = mass of the part
m_p = mass of the platform
I_G = mass moment of inertia of the part
I_p = mass moment of inertia of the platform
r = platform radius
θ = platform angular displacement
l = string length
ϕ = string angular displacement
z = vertical axis through center of the platform

We begin by writing the summation of moments about the z axis. This gives

$$\sum M^z = -r(m + m_p)g \sin\phi - (I_G + I_p)\ddot{\theta} = 0 \qquad (g)$$

Since we are assuming small displacements, the sines of angles can be approximated by the angles themselves. Therefore

$$\phi = \frac{r}{l}\theta \qquad (h)$$

and Eq. (g) becomes

$$\ddot{\theta} + \frac{(m + m_p)gr^2}{(I_G + I_p)l}\theta = 0 \qquad (i)$$

This equation can be solved in the same manner as Eq. (b). The result is

$$I_G + I_p = \frac{(m + m_p)gr^2}{l}\left(\frac{\tau}{2\pi}\right)^2 \qquad (16.5)$$

This equation should be used first with an empty platform to find I_p. With m_p and I_p known, the equation can then be used to find I_G of the part being measured.

16.3 TRANSFORMATION OF INERTIA AXES

Up to this point, we have continually used the principal mass moments of inertia. This has an advantage because, in a coordinate system with origin at the center of mass and with axes aligned in the principal axis directions, all mass products of inertia are zero. This choice of axes is used to simplify the form of the equations from that required for other choices.

In Eq. (15.12), the parallel-axis formula, we showed how translation of axes could be performed. However, up to now, we have said little of how inertia properties could be rotated to a new coordinate system that is not parallel to the principal axes. This is the purpose of this section.

Let us assume that the principal axes are aligned along unit vector directions $\hat{\mathbf{i}}, \hat{\mathbf{j}}$, and $\hat{\mathbf{k}}$ and that the new rotated axes desired have directions denoted by $\hat{\mathbf{i}}', \hat{\mathbf{j}}'$, and $\hat{\mathbf{k}}'$. Then any point in the principal axis coordinate system is located by the position vector $\mathbf{R}$, and by $\mathbf{R}'$ in the rotated system. Because both are descriptions of the same point position, $\mathbf{R}' = \mathbf{R}$ and

$$R^{x'}\hat{\mathbf{i}}' + R^{y'}\hat{\mathbf{j}}' + R^{z'}\hat{\mathbf{k}}' = R^x\hat{\mathbf{i}} + R^y\hat{\mathbf{j}} + R^z\hat{\mathbf{k}}$$

Now, by taking dot products of this equation with $\hat{\mathbf{i}}', \hat{\mathbf{j}}'$, and $\hat{\mathbf{k}}'$, respectively, we find the transformation equations between the two sets of axes. They are

$$
\begin{aligned}
R^{x'} &= (\hat{\mathbf{i}}' \cdot \hat{\mathbf{i}})R^x + (\hat{\mathbf{i}}' \cdot \hat{\mathbf{j}})R^y + (\hat{\mathbf{i}}' \cdot \hat{\mathbf{k}})R^z \\
R^{y'} &= (\hat{\mathbf{j}}' \cdot \hat{\mathbf{i}})R^x + (\hat{\mathbf{j}}' \cdot \hat{\mathbf{j}})R^y + (\hat{\mathbf{j}}' \cdot \hat{\mathbf{k}})R^z \\
R^{z'} &= (\hat{\mathbf{k}}' \cdot \hat{\mathbf{i}})R^x + (\hat{\mathbf{k}}' \cdot \hat{\mathbf{j}})R^y + (\hat{\mathbf{k}}' \cdot \hat{\mathbf{k}})R^z
\end{aligned}
\qquad (a)
$$

But we may remember that the dot product of two unit vectors is equal to the cosine of the angle between them. For example, if the angle between $\hat{\mathbf{i}}'$, and $\hat{\mathbf{j}}$ is denoted by $\theta_{i'j}$, then

$$\hat{\mathbf{i}}' \cdot \hat{\mathbf{j}} = \cos \theta_{i'j}$$

Thus Eqs. (a) become

$$
\begin{aligned}
R^{x'} &= \cos\theta_{i'i}R^x + \cos\theta_{i'j}R^y + \cos\theta_{i'k}R^z \\
R^{y'} &= \cos\theta_{j'i}R^x + \cos\theta_{j'j}R^y + \cos\theta_{j'k}R^z \\
R^{z'} &= \cos\theta_{k'i}R^x + \cos\theta_{k'j}R^y + \cos\theta_{k'k}R^z
\end{aligned}
\qquad (b)
$$

By substituting these into the definitions of mass moments of inertia in Eqs. (15.8), expanding, and then performing the integration, we get

$$
\begin{aligned}
I^{x'x'} &= \cos^2\theta_{i'i}I^{xx} + \cos^2\theta_{i'j}I^{yy} + \cos^2\theta_{i'k}I^{zz} \\
&\quad + 2\cos\theta_{i'i}\cos\theta_{i'j}I^{xy} + 2\cos\theta_{i'i}\cos\theta_{i'k}I^{xz} + 2\cos\theta_{i'j}\cos\theta_{i'k}I^{yz}
\end{aligned}
\quad (16.6a)
$$

$$I^{y'y'} = \cos^2\theta_{j'i} I^{xx} + \cos^2\theta_{j'j} I^{yy} + \cos^2\theta_{j'k} I^{zz}$$
$$+ 2\cos\theta_{j'i}\cos\theta_{j'j} I^{xy} + 2\cos\theta_{j'i}\cos\theta_{j'k} I^{xz} + 2\cos\theta_{j'j}\cos\theta_{j'k} I^{yz} \quad (16.6b)$$

$$I^{z'z'} = \cos^2\theta_{k'i} I^{xx} + \cos^2\theta_{k'j} I^{yy} + \cos^2\theta_{k'k} I^{zz}$$
$$+ 2\cos\theta_{k'i}\cos\theta_{k'j} I^{xy} + 2\cos\theta_{k'i}\cos\theta_{k'k} I^{xz} + 2\cos\theta_{k'j}\cos\theta_{k'k} I^{yz} \quad (16.6c)$$

Similarly, for the mass products of inertia, starting from Eqs. (15.9), we get

$$I^{x'y'} = -\cos\theta_{i'i}\cos\theta_{j'i} I^{xx} - \cos\theta_{i'j}\cos\theta_{j'j} I^{yy} - \cos\theta_{i'k}\cos\theta_{j'k} I^{zz}$$
$$+ (\cos\theta_{i'i}\cos\theta_{j'j} + \cos\theta_{i'j}\cos\theta_{j'i}) I^{xy}$$
$$+ (\cos\theta_{i'j}\cos\theta_{j'k} + \cos\theta_{i'k}\cos\theta_{j'j}) I^{yz}$$
$$+ (\cos\theta_{i'k}\cos\theta_{j'i} + \cos\theta_{i'i}\cos\theta_{j'k}) I^{zx} \quad (16.7a)$$

$$I^{y'z'} = -\cos\theta_{j'i}\cos\theta_{k'i} I^{xx} - \cos\theta_{j'j}\cos\theta_{k'j} I^{yy} - \cos\theta_{j'k}\cos\theta_{k'k} I^{zz}$$
$$+ (\cos\theta_{j'i}\cos\theta_{k'j} + \cos\theta_{j'j}\cos\theta_{k'i}) I^{xy}$$
$$+ (\cos\theta_{j'j}\cos\theta_{k'k} + \cos\theta_{j'k}\cos\theta_{k'j}) I^{yz}$$
$$+ (\cos\theta_{j'k}\cos\theta_{k'i} + \cos\theta_{j'i}\cos\theta_{k'k}) I^{zx} \quad (16.7b)$$

$$I^{z'x'} = -\cos\theta_{k'i}\cos\theta_{i'i} I^{xx} - \cos\theta_{k'j}\cos\theta_{i'j} I^{yy} - \cos\theta_{k'k}\cos\theta_{i'k} I^{zz}$$
$$+ (\cos\theta_{k'i}\cos\theta_{i'j} + \cos\theta_{k'j}\cos\theta_{i'i}) I^{xy}$$
$$+ (\cos\theta_{k'j}\cos\theta_{i'k} + \cos\theta_{k'k}\cos\theta_{i'j}) I^{yz}$$
$$+ (\cos\theta_{k'k}\cos\theta_{i'i} + \cos\theta_{k'i}\cos\theta_{i'k}) I^{zx} \quad (16.7c)$$

Once we have rotated away from the principal axes, or if we had started with other than the principal axes, then the situation is different. The parallel-axis formulae of Eq. (15.12) are no longer sufficient for the translation of axes. Let us assume that the mass moments and products of inertia are known in one coordinate system and are desired in another coordinate system, which is translated from the first by the equations

$$R^{x''} = R^{x'} + d^{x'} \quad (16.8a)$$
$$R^{y''} = R^{y'} + d^{y'} \quad (16.8b)$$
$$R^{z''} = R^{z'} + d^{z'} \quad (16.8c)$$

Then, substituting in the original definition of I^{xx} of Eq. (15.8), for example, and integrating:

$$I^{x''x''} = \int [(R^{y''})^2 + (R^{z''})^2]\,dm$$

$$= \int [(R^{y'} + d^{y'})^2 + (R^{z'} + d^{z'})^2]\,dm$$

$$= \int [(R^{y'})^2 + (R^{z'})^2]\,dm + 2\left(\int R^{y'}\,dm\right)d^{y'}$$

$$+ 2\left(\int R^{z'}\,dm\right)d^{z'} + (d^{y'})^2\int dm + (d^{z'})^2\int dm$$

$$= I^{x'x'} + 2mR_G^{y'}d^{y'} + 2mR_G^{z'}d^{z'} + m(d^{y'})^2 + m(d^{z'})^2$$

Proceeding in this manner through each of Eqs. (15.8) and (15.9), we find

$$I^{x''x''} = I^{x'x'} + 2mR_G^{y'}d^{y'} + 2mR_G^{z'}d^{z'} + m(d^{y'})^2 + m(d^{z'})^2$$
$$I^{y''y''} = I^{y'y'} + 2mR_G^{z'}d^{z'} + 2mR_G^{x'}d^{x'} + m(d^{z'})^2 + m(d^{x'})^2$$
$$I^{z''z''} = I^{z'z'} + 2mR_G^{x'}d^{x'} + 2mR_G^{y'}d^{y'} + m(d^{x'})^2 + m(d^{y'})^2$$
$$I^{x''y''} = I^{x'y'} + mR_G^{x'}d^{y'} + mR_G^{y'}d^{x'} + md^{x'}d^{y'}$$
$$I^{y''z''} = I^{y'z'} + mR_G^{y'}d^{z'} + mR_G^{z'}d^{y'} + md^{y'}d^{z'}$$
$$I^{z''x''} = I^{z'x'} + mR_G^{z'}d^{x'} + mR_G^{x'}d^{z'} + md^{z'}d^{x'}$$

(16.9)

EXAMPLE 16.1

A circular steel disk is fastened at its center at an angle of 30° to a shaft as shown in Fig. 16.3. Using 0.282 lb/in³ as the density of steel, find the mass moments and products of inertia about the center of mass and aligned along the axes of the shaft.

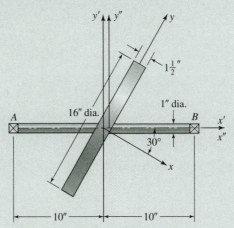

Figure 16.3 Shaft with swashplate for Example 16.1.

SOLUTION

First we calculate the mass of the disk and the principal mass moments of inertia of the disk using the equations of Appendix Table 4. These are

$$m = \frac{0.282 \text{ lb/in}^3}{386 \text{ in/s}^2} \pi (8 \text{ in})^2 (1.5 \text{ in}) = 0.220 \text{ lb·s}^2/\text{in}$$

$$I^{xx} = \frac{(0.220 \text{ lb·s}^2/\text{in})(8 \text{ in})^2}{2} = 7.04 \text{ in·lb·s}^2$$

$$I^{yy} = I^{zz} = \frac{(0.220 \text{ lb·s}^2/\text{in})(8 \text{ in})^2}{4} = 3.52 \text{ in·lb·s}^2$$

The products of inertia are zero since these are principal axes.

From the figure we now find the direction cosines between the coordinate systems. We find these to be

$$\cos\theta_{i'i} = 0.866, \qquad \cos\theta_{i'j} = 0.500, \qquad \cos\theta_{i'k} = 0$$
$$\cos\theta_{j'i} = -0.500, \qquad \cos\theta_{j'j} = 0.866, \qquad \cos\theta_{j'k} = 0$$
$$\cos\theta_{k'i} = 0, \qquad \cos\theta_{k'j} = 0, \qquad \cos\theta_{k'k} = 1.000$$

Substituting these results into Eqs. (16.6) and (16.7) gives the mass moments of inertia aligned with the axes of the shaft, that is,

$$I^{x'x'} = (0.866)^2(7.04) + (0.500)^2(3.52) = 6.16 \text{ in·lb·s}^2$$
$$I^{y'y'} = (-0.500)^2(7.04) + (0.866)^2(3.52) = 4.40 \text{ in·lb·s}^2$$
$$I^{z'z'} = I^{zz} = 3.52 \text{ in·lb·s}^2$$
$$I^{x'y'} = -(0.866)(-0.500)(7.04) - (0.500)(0.866)(3.52) = 1.52 \text{ in·lb·s}^2$$
$$I^{y'z'} = I^{z'x'} = 0$$

Next we find the mass of the shaft and the principal mass moments of inertia of the shaft using the equations of Appendix Table 4. These are

$$m = \frac{0.282 \text{ lb/in}^3}{386 \text{ in/s}^2}\pi(0.5 \text{ in})^2(24 \text{ in}) = 0.0138 \text{ lb·s}^2/\text{in}$$

$$I^{xx} = \frac{(0.0138 \text{ lb·s}^2/\text{in})(0.5 \text{ in})^2}{2} = 0.00086 \text{ in·lb·s}^2$$

$$I^{yy} = I^{zz} = \frac{(0.0138 \text{ lb·s}^2/\text{in})[3(0.5 \text{ in})^2 + (24 \text{ in})^2]}{12} = 0.662 \text{ in·lb·s}^2$$

The products of inertia are zero because these are principal axes. Translating the values to the same axes as the disk, Eq. (15.8) gives

$$I^{x'x'} = 0.00086 \text{ in·lb·s}^2$$
$$I^{y'y'} = I^{z'z'} = 0.662 + 0.0138(2)^2 = 0.717 \text{ in·lb·s}^2$$

The values of the disk and the shaft are now both in the same coordinate system and are combined to give

$$I^{x'x'} = 6.16 + 0.00086 = 6.16 \text{ in·lb·s}^2$$
$$I^{y'y'} = 4.40 + 0.717 = 5.12 \text{ in·lb·s}^2$$
$$I^{z'z'} = 3.52 + 0.717 = 4.24 \text{ in·lb·s}^2$$
$$I^{x'y'} = 1.52 + 0 = 1.52 \text{ in·lb·s}^2$$
$$I^{y'z'} = I^{z'x'} = 0$$

These are translated to the center of mass using Eqs. (16.9):

$$m = 0.220 + 0.0138 = 0.234 \text{ lb·s}^2/\text{in}$$

$$R_G^{x'} = \frac{(0.220)(0) + (0.0138)(2)}{0.235} = 0.118 \text{ in}$$

$$R_G^{y'} = R_G^{z'} = 0$$

$$d^{x'} = -0.127 \text{ in}$$

$$d^{y'} = d^{z'} = 0$$

$$I^{x''x''} = 6.16 \text{ in·lb·s}^2 \qquad\qquad Ans.$$

$$I^{y''y''} = 5.12 + 2(0.234)(0.118)(-0.118) + (0.234)(-0.118)^2 = 5.12 \text{ in·lb·s}^2 \qquad Ans.$$

$$I^{z''z''} = 4.24 + 2(0.234)(0.118)(-0.118) + (0.234)(-0.118)^2 = 4.24 \text{ in·lb·s}^2 \qquad Ans.$$

$$I^{x''y''} = 1.52 \text{ in·lb·s}^2 \qquad\qquad Ans.$$

$$I^{y''z''} = I^{z''x''} = 0 \qquad\qquad Ans.$$

16.4 EULER'S EQUATIONS OF MOTION

In Section 15.2 we presented a detailed derivation that showed the form of Newton's law for a rigid body by integrating over all particles of the body. This showed that the center of mass of the body was the single unique point of the body where Newton's law had the same form as that for a single particle. Repeating Eq. (15.13), this was

$$\sum \mathbf{F}_{ij} = m_j \mathbf{A}_{G_j} \tag{16.10}$$

Taking much less care with a derivation, we then gave the equivalent rotational form, Eq. (15.14), as

$$\sum \mathbf{M}_{ij} = I_{G_j} \boldsymbol{\alpha}_j$$

Although this equation is valid for all problems of Chapter 15, this is because those were problems with only planar effects. This equation is oversimplified and is *not* valid for spatial problems. Our task now is to more carefully derive an appropriate equation that *is* valid for problems which include three-dimensional effects.

We start, as before, with Newton's law written for a single particle at location P on a rigid body. This can be written as

$$d\mathbf{F} = \mathbf{A}_P \, dm$$

where $d\mathbf{F}$ is the net unbalanced force on the particle, $\mathbf{A}_P$ is the absolute acceleration of the particle; and dm is the mass of the particle. Next we find the net unbalanced moment that this particle contributes to the body by taking the moment of its net unbalanced force about the center of mass of the body, point G:

$$\mathbf{R}_{PG} \times d\mathbf{F} = \mathbf{R}_{PG} \times \mathbf{A}_P \, dm$$

$$= \mathbf{R}_{PG} \times [\mathbf{A}_G + \boldsymbol{\omega} \times (\boldsymbol{\omega} \times \mathbf{R}_{PG}) + \boldsymbol{\alpha} \times \mathbf{R}_{PG}] \, dm$$

On the left-hand side of this equation is the net unbalanced moment contribution of this single particle. If we rearrange the terms on the right-hand side, the equation becomes

$$d\mathbf{M} = \mathbf{R}_{PG} \times \mathbf{A}_G \, dm + \mathbf{R}_{PG} \times [\boldsymbol{\omega} \times (\boldsymbol{\omega} \times \mathbf{R}_{PG})] \, dm + \mathbf{R}_{PG} \times (\boldsymbol{\alpha} \times \mathbf{R}_{PG}) \, dm \qquad (a)$$

We now recall the following vector identity for the triple cross product of three arbitrary vectors, say $\mathbf{A}$, $\mathbf{B}$, and $\mathbf{C}$:

$$\mathbf{A} \times (\mathbf{B} \times \mathbf{C}) = (\mathbf{A} \cdot \mathbf{C})\mathbf{B} - (\mathbf{A} \cdot \mathbf{B})\mathbf{C}$$

Using this identity on the second term on the right-hand side of Eq. (a), we have

$$d\mathbf{M} = \mathbf{R}_{PG} \times \mathbf{A}_G \, dm + \mathbf{R}_{PG} \times (\boldsymbol{\omega} \cdot \mathbf{R}_{PG})\boldsymbol{\omega} \, dm$$

$$- (\boldsymbol{\omega} \cdot \boldsymbol{\omega})\mathbf{R}_{PG} \times \mathbf{R}_{PG} \, dm + \mathbf{R}_{PG} \times (\boldsymbol{\alpha} \times \mathbf{R}_{PG}) \, dm \qquad (b)$$

Now we wish to integrate this equation over all particles of the body. We will look at this term by term. The integral of the moment contributions of the individual particles gives the net externally applied unbalanced moment on the body:

$$\int d\mathbf{M} = \sum \mathbf{M}_{G_{ij}} \qquad (c)$$

Recognizing that $\mathbf{A}_G$ is common for all particles, we see that the first term on the right-hand side of Eq. (b) integrates to

$$\int \mathbf{R}_{PG} \times \mathbf{A}_G \, dm = \int \mathbf{R}_{PG} \, dm \times \mathbf{A}_G = \mathbf{0} \qquad (d)$$

where we have taken advantage of the definition of the center of mass and noted that $\int \mathbf{R}_{PG} \, dm = m\mathbf{R}_{GG} = \mathbf{0}$.

The second term on the right-hand side of Eq. (b) can be expanded according to its components along the coordinate axes and integrated. The integrals will be recognized as the mass moments and products of inertia of the body. Thus,

$$\int \mathbf{R}_{PG} \times (\boldsymbol{\omega} \cdot \mathbf{R}_{PG})\boldsymbol{\omega} \, dm$$

$$= \left[-\left(I_G^{yy} - I_G^{zz}\right)\omega^y \omega^z - I_G^{yz}\left(\omega^y \omega^y - \omega^z \omega^z\right) + \left(I_G^{xy}\omega^z - I_G^{xz}\omega^y\right)\omega^x \right]\hat{\mathbf{i}}$$

$$+ \left[-\left(I_G^{zz} - I_G^{xx}\right)\omega^z \omega^x - I_G^{zx}\left(\omega^z \omega^z - \omega^x \omega^x\right) + \left(I_G^{yz}\omega^x - I_G^{yx}\omega^z\right)\omega^y \right]\hat{\mathbf{j}}$$

$$+ \left[-\left(I_G^{xx} - I_G^{yy}\right)\omega^x \omega^y - I_G^{xy}\left(\omega^x \omega^x - \omega^y \omega^y\right) + \left(I_G^{zx}\omega^y - I_G^{zy}\omega^x\right)\omega^z \right]\hat{\mathbf{k}} \qquad (e)$$

Because $\mathbf{R}_{PG} \times \mathbf{R}_{PG} = \mathbf{0}$ for every particle, the third term on the right-hand side of Eq. (*b*) integrates to

$$(\boldsymbol{\omega} \cdot \boldsymbol{\omega}) \int \mathbf{R}_{PG} \times \mathbf{R}_{PG}\, dm = \mathbf{0} \tag{f}$$

The final term on the right-hand side of Eq. (*b*) is also expanded according to its components along the coordinate axes and integrated. Thus

$$\int \mathbf{R}_{PG} \times (\boldsymbol{\alpha} \times \mathbf{R}_{PG})\, dm = \left(I_G^{xx}\alpha^x - I_G^{xy}\alpha^y - I_G^{xz}\alpha^z\right)\hat{\mathbf{i}}$$
$$+ \left(-I_G^{yx}\alpha^x + I_G^{yy}\alpha^y - I_G^{yz}\alpha^z\right)\hat{\mathbf{j}}$$
$$+ \left(-I_G^{zx}\alpha^x + I_G^{zy}\alpha^y - I_G^{zz}\alpha^z\right)\hat{\mathbf{k}} \tag{g}$$

When we now substitute Eqs. (*c*) through (*g*) into Eq. (*b*) and equate components in the $\hat{\mathbf{i}}, \hat{\mathbf{j}}$, and $\hat{\mathbf{k}}$ directions, we obtain the following three equations:

$$\sum M_{G_{ij}}^x = I_G^{xx}\alpha^x - I_G^{xy}\alpha^y - I_G^{xz}\alpha^z$$
$$- \left(I_G^{yy} - I_G^{zz}\right)\omega^y\omega^z - I_G^{yz}(\omega^y\omega^y - \omega^z\omega^z) + \left(I_G^{xy}\omega^z - I_G^{xz}\omega^y\right)\omega^x \tag{16.11a}$$

$$\sum M_{G_{ij}}^y = -I_G^{yx}\alpha^x + I_G^{yy}\alpha^y - I_G^{yz}\alpha^z$$
$$- \left(I_G^{zz} - I_G^{xx}\right)\omega^z\omega^x - I_G^{zx}(\omega^z\omega^z - \omega^x\omega^x) + \left(I_G^{yz}\omega^x - I_G^{yx}\omega^z\right)\omega^y \tag{16.11b}$$

$$\sum M_{G_{ij}}^z = -I_G^{xx}\alpha^x - I_G^{zy}\alpha^y + I_G^{zz}\alpha^z$$
$$- \left(I_G^{xx} - I_G^{yy}\right)\omega^x\omega^y - I_G^{xy}(\omega^x\omega^x - \omega^y\omega^y) + \left(I_G^{zx}\omega^y - I_G^{zy}\omega^x\right)\omega^z \tag{16.11c}$$

This set of equations is the most general case of the moment equation. It allows for angular velocities and accelerations in three dimensions. Its only restriction is that the summation of moments and the mass moments and products of inertia must be taken about the mass center of the body.*

If we further require that the mass moments be measured around the principal axes of the body, then the mass products of inertia are zero. The above equations then reduce to

$$\sum M_{G_{ij}}^x = I_G^{xx}\alpha^x - \left(I_G^{yy} - I_G^{zz}\right)\omega^y\omega^z \tag{16.12a}$$

$$\sum M_{G_{ij}}^y = I_G^{yy}\alpha^y - \left(I_G^{zz} - I_G^{xx}\right)\omega^z\omega^x \tag{16.12b}$$

$$\sum M_{G_{ij}}^z = I_G^{zz}\alpha^z - \left(I_G^{xx} - I_G^{yy}\right)\omega^x\omega^y \tag{16.12c}$$

This important set of equations is called *Euler's equations of motion*. It should be emphasized that they govern any three-dimensional rotational motion of a rigid body, but that the principal axes of inertia must be used for the x, y, and z component directions.

---

*It should now be recognized that when we restricted ourselves to motions in the xy plane in Chapter 15, it was a special case of Eq. (16.11c) that we were using in Eq. (15.14) and what followed.

EXAMPLE 16.2

The shaft with the angularly mounted disk which was analyzed in Example 16.1 is shown again in Fig. 16.4. At the instant shown, the shaft is rotating at a speed of 1500 rev/min and this speed is being reduced at a rate of 100 rev/min/s. The shaft is supported by radial bearings at A and B. Calculate the bearing reactions.

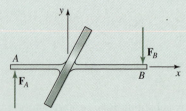

Figure 16.4 Shaft with disk for Example 16.2.

SOLUTION

From the choice of axes shown in the figure and the data specified in the problem statement, the angular velocity and acceleration, respectively, of the shaft are

$$\boldsymbol{\omega} = -(1500 \text{ rev/min})(2\pi \text{ rad/rev})/(60 \text{ s/min})\hat{\mathbf{i}} = -157\hat{\mathbf{i}} \text{ rad/s}$$

$$\boldsymbol{\alpha} = +(100 \text{ rev/min/s})(2\pi \text{ rad/rev})/(60 \text{ s/min})\hat{\mathbf{i}} = +10.5\hat{\mathbf{i}} \text{ rad/s}^2$$

The mass moments and products of inertia were found in Example 16.1.

Substituting these values into Eqs. (16.11) gives

$$\sum M^x = (6.16 \text{ in·lb·s}^2)(10.5 \text{ rad/s}^2) = 64.7 \text{ in·lb}$$

$$\sum M^y = -(1.52 \text{ in·lb·s}^2)(10.5 \text{ rad/s}^2) = -16 \text{ in·lb}$$

$$\sum M^z = -(1.52 \text{ in·lb·s}^2)(-157 \text{ rad/s})^2 = -37\,500 \text{ in·lb}$$

or

$$\sum \mathbf{M} = 64.7\hat{\mathbf{i}} - 16.0\hat{\mathbf{j}} - 37\,500\hat{\mathbf{k}} \text{ in·lb}$$

The $\sum M^x$ component of this result is due to the shaft torque causing the deceleration. The other two components are provided by forces at the bearings A and B. From these we find, by taking moments about the mass center, that

$$\sum M^y = (10 \text{ in})\left(F_A^z\right) - (14 \text{ in})\left(F_B^z\right) = -16.0 \text{ in·lb}$$

$$\sum M^z = -(10 \text{ in})\left(F_A^y\right) + (14 \text{ in})\left(F_B^y\right) = -37\,500 \text{ in·lb}$$

Remembering that $(F_A^y) + (F_B^y) = 0$ and $(F_A^z) + (F_B^z) = 0$, then we find that

$$F_A^z = -F_B^z = (-16.0 \text{ in·lb})/(24 \text{ in}) = -0.667 \text{ lb} \approx 0$$

$$F_A^y = -F_B^y = (37\,500 \text{ in·lb})/(24 \text{ in}) = 1\,560 \text{ lb}$$

Thus the two bearing reaction forces on the shaft are

$$\mathbf{F}_A = 1\,560\hat{\mathbf{j}} \text{ lb} \qquad\qquad Ans.$$

$$\mathbf{F}_B = -1\,560\hat{\mathbf{j}} \text{ lb} \qquad\qquad Ans.$$

Notice that these bearing reactions are caused almost entirely by the z component of the unbalanced moments. This, in turn, is caused by the angular velocity of the shaft, not by its angular acceleration; the bearing reaction loads would still be large even if the shaft were operating at constant speed.

Notice also that this example exhibits only planar motion, but if it had been analyzed by the methods of Chapter 15, we would have found *no* net bearing reaction forces! This should be sufficient warning of the danger of ignoring the third dimension in dynamic-force analysis, based solely on the justification that the *motion* is planar.

16.5 IMPULSE AND MOMENTUM

If we consider that both force and acceleration may be functions of time, we can multiply both sides of Eq. (16.10) by dt and integrate between t_1 and t_2. This results in

$$\sum \int_{t_1}^{t_2} \mathbf{F}_{ij}(t)\, dt = m_j \int_{t_1}^{t_2} \mathbf{A}_{Gj}(t)\, dt$$

$$= m_j \mathbf{V}_{Gj}(t_2) - m_j \mathbf{V}_{Gj}(t_1) \qquad (a)$$

where $\mathbf{V}_{Gj}(t_2)$ and $\mathbf{V}_{Gj}(t_1)$ are the velocities of the centers of mass at time t_2 and at time t_1, respectively.

The product of the mass and the velocity of the center of mass of a moving body is called its *momentum*. Momentum is a vector quantity and is given the symbol $\mathbf{L}$. Thus, for body j, the momentum is

$$\mathbf{L}_j(t) = m_j \mathbf{V}_{Gj}(t) \qquad (16.13)$$

Using this definition, Eq. (*a*) becomes

$$\sum \int_{t_1}^{t_2} \mathbf{F}_{ij}(t)\, dt = \mathbf{L}_j(t_2) - \mathbf{L}_j(t_1) = \Delta \mathbf{L}_j \qquad (16.14)$$

The integral of a force over an interval of time is called the *impulse* of the force. Therefore, Eq. (16.14) expresses the *principle of impulse and momentum,* which is that *the total impulse on a rigid body is equal to its change in momentum during the same time interval.*

Conversely, *any change in momentum of a rigid body over a time interval is due to the total impulse of the external forces on that body.*

If we now make the time interval between t_1 and t_2 infinitesimal, divide Eq. (16.14) by this interval, and take the limit, it becomes

$$\sum \mathbf{F}_{ij} = \frac{d\mathbf{L}_j}{dt} \tag{16.15}$$

Therefore, *the resultant external force acting upon a rigid body is equal to the time rate of change of its momentum.* If there is no net external force acting on a body, then there can be no change in its momentum. Thus, with no net external force,

$$\frac{d\mathbf{L}_j}{dt} = \mathbf{0} \quad \text{or} \quad \mathbf{L}_j = \text{constant} \tag{16.16}$$

is a statement of the law of *conservation of momentum.*

16.6 ANGULAR IMPULSE AND ANGULAR MOMENTUM

When a body translates, its motion is described in terms of its velocity and acceleration, and we are interested in the forces that produce this motion. When a body rotates, we are interested in the moments, and the motion is described by angular velocity and angular acceleration. Similarly, in dealing with rotational motion, we need to consider terms such as *angular impulse,* or *moment of impulse,* and *angular momentum,* also called *moment of momentum.*

Angular momentum is a vector quantity and is usually given the symbol $\mathbf{H}$. For a particle of mass dm at location P with momentum $d\mathbf{L}$, the moment of momentum $d\mathbf{H}$ of this single particle is defined, as implied by its name, as follows:

$$d\mathbf{H} = \mathbf{R}_P \times d\mathbf{L} = \mathbf{R}_P \times \mathbf{V}_P \, dm \tag{a}$$

Analogous to the development in Section 16.4, we can express this equation as

$$\begin{aligned} d\mathbf{H} &= \mathbf{R}_P \times (\mathbf{V}_G + \boldsymbol{\omega} \times \mathbf{R}_{PG}) \, dm \\ &= \mathbf{R}_P \times \mathbf{V}_G \, dm + \mathbf{R}_P \times (\boldsymbol{\omega} \times \mathbf{R}_{PG}) \, dm \\ &= \mathbf{R}_P \times \mathbf{V}_G \, dm + \mathbf{R}_G \times (\boldsymbol{\omega} \times \mathbf{R}_{PG}) \, dm + \mathbf{R}_{PG} \times (\boldsymbol{\omega} \times \mathbf{R}_{PG}) \, dm \end{aligned}$$

This equation can now be integrated over all particles of a rigid body:

$$\int d\mathbf{H} = \int \mathbf{R}_P \, dm \times \mathbf{V}_G + \mathbf{R}_G \times \left(\boldsymbol{\omega} \times \int \mathbf{R}_{PG} \, dm \right) + \int \mathbf{R}_{PG} \times (\boldsymbol{\omega} \times \mathbf{R}_{PG}) \, dm \tag{b}$$

The left-hand side of this equation integrates to give the total angular momentum of the body about the origin of the coordinates:

$$\int d\mathbf{H} = \mathbf{H}_O \tag{c}$$

Recognizing the definition of the center of mass, we see that the first two terms on the right-hand side of Eq. (*b*) become

$$\int \mathbf{R}_P\, dm \times \mathbf{V}_G = m\mathbf{R}_G \times \mathbf{V}_G = \mathbf{R}_G \times \mathbf{L} \tag{d}$$

$$\mathbf{R}_G \times \left(\boldsymbol{\omega} \times \int \mathbf{R}_{PG}\, dm\right) = \mathbf{R}_G \times (\boldsymbol{\omega} \times \mathbf{R}_{GG}) = \mathbf{0} \tag{e}$$

As in Section 16.4, the final term of Eq. (*b*) must be expanded according to its components along the coordinate axes and integrated. Thus,

$$\int \mathbf{R}_{PG} \times (\boldsymbol{\omega} \times \mathbf{R}_{PG})\, dm = \left(I_G^{xx}\omega^x - I_G^{xy}\omega^y - I_G^{xz}\omega^z\right)\hat{\mathbf{i}}$$
$$+ \left(-I_G^{yx}\omega^x + I_G^{yy}\omega^y - I_G^{yz}\omega^z\right)\hat{\mathbf{j}}$$
$$+ \left(-I_G^{zx}\omega^x - I_G^{zy}\omega^y + I_G^{zz}\omega^z\right)\hat{\mathbf{k}} \tag{f}$$

Upon substituting Eqs. (*c*) through (*f*) into Eq. (*b*), we obtain

$$\mathbf{H}_O = \mathbf{R}_G \times \mathbf{L}$$
$$+ \left(I_G^{xx}\omega^x - I_G^{xy}\omega^y - I_G^{xz}\omega^z\right)\hat{\mathbf{i}}$$
$$+ \left(-I_G^{yx}\omega^x + I_G^{yy}\omega^y - I_G^{yz}\omega^z\right)\hat{\mathbf{j}}$$
$$+ \left(-I_G^{zx}\omega^x - I_G^{zy}\omega^y + I_G^{zz}\omega^z\right)\hat{\mathbf{k}} \tag{g}$$

We can recognize that if we had chosen a coordinate system with origin at the center of mass of the body, then $\mathbf{R}_G = \mathbf{0}$, and the above equation gives the angular momentum as

$$\mathbf{H}_G = H_G^x\hat{\mathbf{i}} + H_G^y\hat{\mathbf{j}} + H_G^z\hat{\mathbf{k}} \tag{16.17}$$

where

$$H_G^x = I_G^{xx}\omega^x - I_G^{xy}\omega^y - I_G^{xz}\omega^z$$
$$H_G^y = -I_G^{yx}\omega^x + I_G^{yy}\omega^y - I_G^{yz}\omega^z \tag{16.18}$$
$$H_G^z = -I_G^{zx}\omega^x - I_G^{zy}\omega^y + I_G^{zz}\omega^z$$

However, when moments are not taken about the center of mass, then the moment of momentum about an arbitrary point O is

$$\mathbf{H}_O = \mathbf{H}_G + \mathbf{R}_{GO} \times \mathbf{L} \tag{16.19}$$

Now, taking a cross product with the position vector for each particle, as we did in Eq. (*a*) above, can also be done for the terms of Newton's laws for a particle. Using an inertial coordinate system,

$$\sum \mathbf{R}_P \times d\mathbf{F} = \mathbf{R}_P \times \mathbf{A}_P\, dm \tag{h}$$

If we consider that both force and acceleration may be functions of time, we can multiply both sides of this equation by dt and integrate between t_1 and t_2. The result is

$$\sum \int_{t_1}^{t_2} \mathbf{R}_P \times d\mathbf{F}(t)dt = \mathbf{R}_P \times \mathbf{V}_P(t_2)\, dm - \mathbf{R}_P \times \mathbf{V}_P(t_1)\, dm$$

After integrating over all particles of a rigid body, this yields

$$\sum \int_{t_1}^{t_2} \mathbf{M}_{ij}(t)\, dt = \mathbf{H}(t_2) - \mathbf{H}(t_1) = \Delta\mathbf{H} \tag{16.20}$$

The integral on the left-hand side of this equation is the net moment of all external impulses that occur on the body in the time interval t_1 to t_2 and is called the *angular impulse*. On the right is the change in the moment of momentum (angular momentum) that occurs over the same time interval as a result of the angular impulse.

If we now make the time interval between t_1 and t_2 an infinitesimal, divide Eq. (16.20) by this interval, and take the limit, it becomes

$$\sum \mathbf{M}_{ij} = \frac{d\mathbf{H}}{dt} \tag{16.21}$$

Therefore, *the resultant external moment acting upon a rigid body is equal to the time rate of change of its angular momentum.* If there is no net external moment acting on a body, then there can be no change in its angular momentum. Thus, with no net external moment,

$$\frac{d\mathbf{H}}{dt} = \mathbf{0} \quad \text{or} \quad \mathbf{H} = \text{constant} \tag{16.22}$$

is a statement of the *law of conservation of angular momentum.*

It would be wise now to review this section mentally, and to take careful note of how the coordinate axes may be selected for a particular application. This review will show that, whether stated or not, all results which were derived from Newton's law depend on the use of an inertial coordinate system. For example, the angular velocity used in Eq. (16.18) must be taken with respect to an absolute coordinate system so that the derivative of Eq. (16.21) will contain the required absolute angular acceleration terms. Yet this seems contradictory, because the integration performed in finding the mass moments and products of inertia can often be done only in a coordinate system attached to the body itself.

If the coordinate axes are chosen stationary, then the moments and products of inertia used in finding the angular momentum must be transformed to an inertial coordinate system, using the methods of Section 16.2, and will therefore become functions of time. This will complicate the use of Eqs. (16.20), (16.21), and (16.22). For this reason, it is usually preferable to choose the x, y, and z components to be directed along axes fixed to the moving body.

Equations (16.20), (16.21), and (16.22), being vector equations, can be expressed in any coordinate system, including body-fixed axes, and are still correct. This has the great advantage that the mass moments and products of inertia are constants for a rigid body. However, it must be kept in mind that the $\hat{\mathbf{i}}$, $\hat{\mathbf{j}}$, and $\hat{\mathbf{k}}$ unit vectors are moving. They are functions of time in Eq. (16.17), and they have derivatives when using Eqs. (16.21) or (16.22). Therefore, the components of Eqs. (16.21), for example, cannot be found directly by differentiating Eqs. (16.18), but will be

$$\sum M_G^x = I_G^{xx}\alpha^x - I_G^{xy}\alpha^y - I_G^{xz}\alpha^z + \omega^y H^z - \omega^z H^y$$

$$\sum M_G^y = -I_G^{yx}\alpha^x + I_G^{yy}\alpha^y - I_G^{yz}\alpha^z + \omega^z H^x - \omega^x H^z \qquad (16.23)$$

$$\sum M_G^z = -I_G^{zx}\alpha^x - I_G^{zy}\alpha^y + I_G^{zz}\alpha^z + \omega^x H^y - \omega^y H^x$$

Here, we must be careful to find the *components* of ω and α along the axes of the body, even though they are absolute angular velocity and acceleration terms. In addition, we must remember that these moment components are expressed along the body axes when results are interpreted.

EXAMPLE 16.3

Figure 16.5 illustrates a hypothetical problem typical of the situations occurring in the design or analysis of machines in which gyroscopic forces must be considered. A round plate designated as 2 rotates about its central axis with a constant angular velocity $\omega_2 = 5\hat{\mathbf{k}}$ rad/s. Mounted on this revolving round plate are two bearings A and B, which retain a shaft and mass 3 rotating at an angular velocity $\omega_{3/2} = 350\hat{\mathbf{i}}$ rad/s with respect to the rotating plate. The rotating body 3 has a mass of 4.5 kg and a radius of gyration of 50 mm about the spin axis. Its center of mass is located at G. Assume that the weight of the shaft is negligible, body 3 rotates at a constant angular velocity, and bearing B can only support a radial load. Calculate the bearing reactions at A and B.

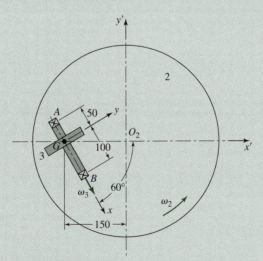

Figure 16.5 Offset flywheel on turntable for Example 16.3, with dimensions in millimeters.

SOLUTION

After considering the above discussion, we choose the coordinate system shown by axes x and y in the figure. Note that this coordinate system is fixed to body 2, not body 3, and it is not stationary. Yet because body 3 is assumed to be radially symmetric about this x axis, the mass properties will remain constant. Note also that the origin is chosen to be at the center of mass and that the axes are principal axes of inertia.

From the data given and a simple kinematic analysis, we have

$$I_{G_3}^{xx} = m_3\left(k_3^x\right)^2 = (4.5 \text{ kg})(0.050 \text{ m})^2 = 0.0113 \text{ kg·m}^2$$

$$\boldsymbol{\omega}_2 = 5\hat{\mathbf{k}} \text{ rad/s}$$

$$\boldsymbol{\omega}_3 = 350\hat{\mathbf{i}} + 5\hat{\mathbf{k}} \text{ rad/s}$$

$$\boldsymbol{\alpha}_2 = \boldsymbol{\alpha}_3 = \mathbf{0}$$

$$\mathbf{R}_{GO_2} = -(150\cos 60°)\hat{\mathbf{i}} - (150\sin 60°)\hat{\mathbf{j}} \text{ mm}$$

$$= -0.075\hat{\mathbf{i}} - 0.130\hat{\mathbf{j}} \text{ m}$$

$$\mathbf{V}_G = \boldsymbol{\omega}_2 \times \mathbf{R}_{GO_2} = 0.650\hat{\mathbf{i}} - 0.375\hat{\mathbf{j}} \text{ m/s}$$

$$\mathbf{R}_{AG} = -0.050\hat{\mathbf{i}} \text{ m}$$

$$\mathbf{R}_{BG} = 0.100\hat{\mathbf{i}} \text{ m}$$

and the time rates of change of the unit vectors are

$$\frac{d\hat{\mathbf{i}}}{dt} = \boldsymbol{\omega}_2 \times \hat{\mathbf{i}} = 5\hat{\mathbf{j}} \text{ rad/s}$$

$$\frac{d\hat{\mathbf{j}}}{dt} = \boldsymbol{\omega}_2 \times \hat{\mathbf{j}} = -5\hat{\mathbf{i}} \text{ rad/s}$$

$$\frac{d\hat{\mathbf{k}}}{dt} = \boldsymbol{\omega}_2 \times \hat{\mathbf{k}} = \mathbf{0}$$

From Eq. (16.13) we find the momentum, that is,

$$\mathbf{L}_3 = m_3\mathbf{V}_{G_3} = (4.5 \text{ kg})(0.650\hat{\mathbf{i}} - 0.375\hat{\mathbf{j}} \text{ m/s})$$

$$= 2.93\hat{\mathbf{i}} - 1.69\hat{\mathbf{j}} \text{ kg·m/s}$$

From Eq. (16.15) we find the dynamic force effects attributable to the motion of the center of mass. Remembering that $\hat{\mathbf{i}}$ and $\hat{\mathbf{j}}$ are moving and have derivatives, this equation gives

$$\sum \mathbf{F}_{i3} = \frac{d\mathbf{L}_3}{dt}$$

$$\mathbf{F}_{A_{23}} + \mathbf{F}_{B_{23}} = 2.93\frac{d\hat{\mathbf{i}}}{dt} - 1.69\frac{d\hat{\mathbf{j}}}{dt} = 8.45\hat{\mathbf{i}} + 14.65\hat{\mathbf{j}} \text{ N}$$

$$F^x_{A_{23}} + F^x_{B_{23}} = 8.45N \tag{1a}$$

$$F^y_{A_{23}} + F^y_{B_{23}} = 14.65N \tag{1b}$$

$$F^z_{A_{23}} + F^z_{B_{23}} = 0 \tag{1c}$$

From Eqs. (16.18) and (16.17) we find the angular momentum, which is,

$$H^x_{G_3} = (0.113 \text{ kg·m}^2)(350 \text{ rad/s}) = 3.96 \text{ kg·m}^2/\text{s}$$

$$H^y_{G_3} = 0$$

$$H^z_{G_3} = I^{zz}_G(5 \text{ rad/s}) = 5I^{zz}_G \text{ kg·m}^2/\text{s}$$

$$\mathbf{H}_{G_3} = 3.96\hat{\mathbf{i}} + 5I^{zz}_G\hat{\mathbf{k}} \text{ kg·m}^2/\text{s}$$

We can find the dynamic moment effects from Eq. (16.21), these are

$$\sum \mathbf{M}_{G_3} = \frac{d\mathbf{H}_{G_3}}{dt}$$

$$\mathbf{R}_{AG} \times \mathbf{F}_{A_{23}} + \mathbf{R}_{BG} \times \mathbf{F}_{B_{23}} = 3.96\frac{d\hat{\mathbf{i}}}{dt} + 5I^{zz}_G\frac{d\hat{\mathbf{k}}}{dt} \text{ N·m}$$

$$(-0.050\hat{\mathbf{i}}) \times \mathbf{F}_{A_{23}} + (0.100\hat{\mathbf{i}}) \times \mathbf{F}_{B_{23}} = 3.96(5\hat{\mathbf{j}}) \text{ N·m}$$

Equating the $\hat{\mathbf{j}}$ and $\hat{\mathbf{k}}$ components, respectively, gives

$$0.050F^z_{A_{23}} - 0.100F^z_{B_{23}} = 19.8 \text{ N·m} \tag{2a}$$

$$-0.050F^y_{A_{23}} + 0.100F^y_{B_{23}} = 0 \tag{2b}$$

Because bearing B can only support a radial load, $F^x_{B_{23}} = 0$. Then, solving Eqs. (1) and (2) simultaneously, we find

$$F^x_{A_{23}} = 8.45 \text{ N}$$

$$F^y_{A_{23}} = 9.76 \text{ N}$$

$$F^z_{A_{23}} = 132 \text{ N}$$

$$F^y_{B_{23}} = 4.88 \text{ N}$$

$$F^z_{B_{23}} = -132 \text{ N}$$

Therefore, the total bearing reactions at A and B are

$$\mathbf{F}_{A_{23}} = 8.45\hat{\mathbf{i}} + 9.76\hat{\mathbf{j}} + 132\hat{\mathbf{k}} \text{ N} \qquad \text{Ans.}$$

$$\mathbf{F}_{B_{23}} = 4.88\hat{\mathbf{j}} - 132\hat{\mathbf{k}} \text{ N} \qquad \text{Ans.}$$

We note that the effect of the gyroscopic couple is to lift the front bearing off the plate and to push the rear bearing against the plate. We can also note that there is no x component in $\sum \mathbf{M}_{23}$; this justifies our assumption that $\alpha_3 = 0$ even though there is no motor driving that shaft.

The general moment equation is given by Eq. (16.21) and can be written as

$$\sum \mathbf{M}_O = \dot{\mathbf{H}}_O$$

where O denotes either a fixed point or the center of mass of the system. Recall that, in the derivation, the time derivative of the angular momentum is with respect to an absolute coordinate system. In many problems it is helpful to express the time derivative of the angular momentum in terms of components measured relative to a moving coordinate system x, y, z which has an angular velocity $\mathbf{\Omega}$. The relation between the time derivative of the angular momentum in the moving x, y, z system and the absolute time derivative can be written as

$$\dot{\mathbf{H}}_O = \left(\frac{d\mathbf{H}_O}{dt} \right)_{xyz} + \mathbf{\Omega} \times \mathbf{H}_O \tag{16.24}$$

The first term on the right-hand side represents that part of $\dot{\mathbf{H}}_O$ due to the change in the magnitude of $\mathbf{H}_O$, and the cross product term represents that part due to the change in the direction of $\mathbf{H}_O$. An application of this equation to a dynamic problem will be demonstrated in the following example.

EXAMPLE 16.4

Figure 16.6 shows gear A rolling around the fixed horizontal gear B. Gear A is also free to rotate about the shaft OG with an angular speed ω_1. The shaft OG is attached to a vertical shaft by a clevis pin at point O and is initially at rest when it is given a constant angular acceleration $\dot{\mathbf{\Omega}}$. If the weight of gear A is 20 lb and the weight of shaft OG is negligible, determine the tangential force and the normal force between the two gears as a function of the angle θ.

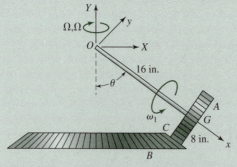

Figure 16.6 Meshing gears for Example 16.4, with dimensions in inches.

SOLUTION

The coordinate system shown by the x and y axes in the figure is fixed to shaft OG and it is not stationary. Because gear A is assumed to be symmetric about the x axis, the mass properties remain constant. Note also that the origin of the coordinate system is chosen at point O and that the axes are principal axes of inertia.

The mass moments of inertia of gear A about point O are

$$I_O^{xx} = \frac{1}{2}mR^2 = \frac{1}{2}\left(\frac{20}{386}\right)8^2 = 1.658 \text{ in·lb·s}^2 \tag{1}$$

$$I_O^{yy} = I_O^{zz} = I_G^{yy} + ml^2 = \frac{1}{4}mR^2 + ml^2 \tag{2}$$

$$= \frac{20}{386}\left(\frac{8^2}{4} + 16^2\right) = 14.093 \text{ in·lb·s}^2$$

and the products of inertia are

$$I_O^{xy} = I_O^{yz} = I_G^{xz} = 0 \tag{3}$$

The angular velocity and acceleration of gear A, respectively, are

$$\boldsymbol{\omega} = (\omega_1 + \Omega\cos\theta)\hat{\mathbf{i}} - \Omega\sin\theta\hat{\mathbf{j}} \text{ rad/s} \tag{4}$$

$$\boldsymbol{\alpha} = (\dot{\omega}_1 + \dot{\Omega}\cos\theta)\hat{\mathbf{i}} - \dot{\Omega}\sin\theta\hat{\mathbf{j}} + \omega_1\Omega\sin\theta\hat{\mathbf{k}} \text{ rad/s}^2 \tag{5}$$

The velocity of the point of contact between the two gears, point C, can be written as

$$\mathbf{V}_C = \mathbf{V}_O + \boldsymbol{\omega} \times \mathbf{R}_{CO}$$

where $\mathbf{R}_{CO} = 16\hat{\mathbf{i}} - 8\hat{\mathbf{j}}$. Because the velocity of points O and C are zero, this equation can be written, with the aid of Eq. (4), as

$$\boldsymbol{\omega} \times \mathbf{R}_{CO} = [-8(\omega_1 + \Omega\cos\theta) + 16\Omega\sin\theta]\hat{\mathbf{k}} = \mathbf{0}$$

Rearranging this equation, the angular velocity of gear A relative to the shaft OG can be written as

$$\omega_1 = \Omega(2\sin\theta - \cos\theta) \tag{6}$$

Differentiating this equation with respect to time gives

$$\dot{\omega}_1 = \dot{\Omega}(2\sin\theta - \cos\theta) \tag{7}$$

Substituting Eqs. (6) and (7) into Eqs. (4) and (5), respectively, gives

$$\boldsymbol{\omega} = \Omega\sin\theta(2\hat{\mathbf{i}} - \hat{\mathbf{j}}) \text{ rad/s} \tag{8}$$

$$\boldsymbol{\alpha} = \dot{\Omega}\sin\theta(2\hat{\mathbf{i}} - \hat{\mathbf{j}}) + \Omega^2\sin\theta(2\sin\theta - \cos\theta)\hat{\mathbf{k}} \text{ rad/s}^2 \tag{9}$$

From Eq. (16.24), the time rate of change of the angular momentum of gear A can be written as

$$\dot{\mathbf{H}}_O = \frac{d\mathbf{H}_O}{dt} + \boldsymbol{\omega} \times \mathbf{H}_O \tag{10}$$

From Eqs. (16.17) and (16.18), the angular momentum of gear A about the fixed point O can be written as

$$\mathbf{H}_O = I_O^{xx}\omega^x\hat{\mathbf{i}} + I_O^{yy}\omega^y\hat{\mathbf{j}} + I_O^{zz}\omega^z\hat{\mathbf{k}} \tag{11}$$

Substituting Eqs. (1), (2), and (8) into this equation gives

$$\begin{aligned}
\mathbf{H}_O &= 1.658(2\Omega\sin\theta)\hat{\mathbf{i}} + 14.093(-\Omega\sin\theta)\hat{\mathbf{j}} \\
&= \Omega\sin\theta(3.316\hat{\mathbf{i}} - 14.093\hat{\mathbf{j}}) \text{ in·lb·s}
\end{aligned} \tag{12}$$

Differentiating Eq. (11) with respect to time gives

$$\frac{d\mathbf{H}_O}{dt} = I_O^{xx}\alpha^x\hat{\mathbf{i}} + I_O^{yy}\alpha^y\hat{\mathbf{j}} + I_O^{zz}\alpha^z\hat{\mathbf{k}}$$

Substituting Eqs. (1), (2), and (9) into this equation gives

$$\frac{d\mathbf{H}_O}{dt} = 1.658(2\dot{\Omega}\sin\theta)\hat{\mathbf{i}} + 14.093(-\dot{\Omega}\sin\theta)\hat{\mathbf{j}} + 14.093[\Omega^2\sin\theta(2\sin\theta - \cos\theta)]\hat{\mathbf{k}}$$

which can be written as

$$\frac{d\mathbf{H}_O}{dt} = \sin\theta[3.316\dot{\Omega}\hat{\mathbf{i}} - 14.093\dot{\Omega}\hat{\mathbf{j}} + 14.093\Omega^2(2\sin\theta - \cos\theta)\hat{\mathbf{k}}] \text{ in·lb} \tag{13}$$

Substituting Eqs. (8), (12), and (13) into Eq. (10) gives

$$\dot{\mathbf{H}}_O = \sin\theta[3.316\dot{\Omega}\hat{\mathbf{i}} - 14.093\dot{\Omega}\hat{\mathbf{j}} + \Omega^2(3.316\sin\theta - 14.093\cos\theta)\hat{\mathbf{k}}] \text{ in·lb} \tag{14}$$

From the free-body diagram of the shaft and the gear shown in Fig. 16.7, the sum of the moments about the clevis pin O can be written as

$$\begin{aligned}
\sum\mathbf{M}_O &= M_O^y\hat{\mathbf{j}} + \mathbf{R}_{AO} \times 20(\cos\theta\hat{\mathbf{i}} - \sin\theta\hat{\mathbf{j}}) + \mathbf{R}_{CO} \times (N\hat{\mathbf{j}} + f\hat{\mathbf{k}}) \\
&= -8f\hat{\mathbf{i}} + (M_O^y - 16f)\hat{\mathbf{j}} + (-320\sin\theta + 16N)\hat{\mathbf{k}} \text{ in·lb}
\end{aligned} \tag{15}$$

where f is the tangential force between the two gears at the point of contact C.

Equating the $\hat{\mathbf{i}}$ components of Eqs. (14) and (15) gives

$$-8f = 3.316\sin\theta\dot{\Omega}$$

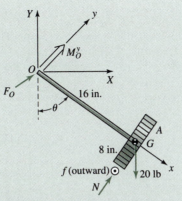

Figure 16.7 Free-body diagram of shaft OA and gear A.

Therefore, the tangential force between the two gears is

$$f = -0.415 \sin\theta \dot{\Omega} \text{ lb} \qquad \textit{Ans.}$$

Equating the $\hat{\mathbf{j}}$ components of Eqs. (14) and (15) gives

$$M_O^y - 16f = -14.093 \sin\theta \dot{\Omega}$$

Substituting the tangential force into this equation, and rearranging, the moment about point O can be written as

$$M_O^y = -20.725 \sin\theta \dot{\Omega} \text{ in·lb}$$

Finally, equating the $\hat{\mathbf{k}}$ components of Eqs. (14) and (15) gives

$$-320 \sin\theta + 16N = \Omega^2 \sin\theta(3.316 \sin\theta - 14.093 \cos\theta)$$

Rearranging this equation, the normal reaction force can be written as

$$N = 20 \sin\theta + \left(\frac{1}{4}\Omega\right)^2 \sin\theta(3.316 \sin\theta - 14.093 \cos\theta) \text{ lb} \qquad \textit{Ans.}$$

Note that the first term on the right-hand side of this equation represents the static gravitational effect, and the second term represents two opposing inertial effects. If gear A did not mesh with the fixed gear B, then the angular velocity ω_1 would be zero and the centripetal acceleration of point G would decrease the normal reaction force N. Due to ω_1, however, a gyroscopic moment is created that has the effect of forcing gear A downward. Note that if $\theta = \tan^{-1}\left(\frac{14.093}{3.316}\right) = 76.76°$, then the two dynamic effects counterbalance each other and the normal reaction force reduces to its static value of $N = 20 \sin\theta$ lb.

NOTES

1. Additional details can be found in F. E. Fisher and H. H. Alford, *Instrumentation for Mechanical Analysis*, The University of Michigan Summer Conferences, Ann Arbor, Michigan, 1977, p. 129. The analysis presented here is by permission of the authors.

PROBLEMS*

16.1 The figure shows a two-throw opposed-crank crankshaft mounted in bearings at A and G. Each crank has an eccentric weight of 6 lb, which may be considered as located at a radius of 2 in from the axis of rotation, and at the center of each throw (points C and E). It is proposed to locate weights at B and F in order to reduce the bearing reactions, caused by the rotating eccentric cranks, to zero. If these weights are to be mounted at 3 in from the axis of rotation, how much must they weigh?

and E. Calculate the magnitudes and angular locations of these weights.

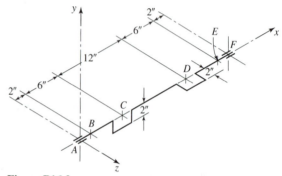

Figure P16.2

16.3 Solve Problem 16.2 with the angle between the two throws reduced from 90° to 0°.

16.4 The connecting rod shown in the figure weighs 7.90 lb and is pivoted on a knife edge and caused to oscillate as a pendulum. The rod is observed to

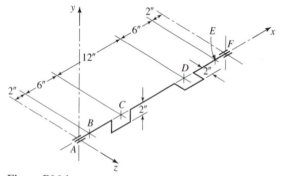

Figure P16.1

16.2 The figure illustrates a two-throw crankshaft, mounted in bearings at A and F, with the cranks spaced 90° apart. Each crank may be considered to have an eccentric weight of 6 lb at the center of the throw and 2 in from the axis of rotation. It is proposed to eliminate the rotating bearing reactions, which the crank would cause, by mounting additional correction weights on 3-in arms at points B

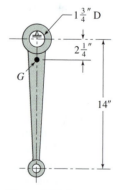

Figure P16.4

---

*Unless stated otherwise, solve all problems without friction and without gravitational loads.

complete 64.5 oscillations in 1 min. Determine the mass moment of inertia of the rod about its own center of mass.

16.5 A gear is suspended on a knife edge at the rim as shown in the figure and caused to oscillate as a pendulum. Its period of oscillation is observed to be 1.08 s. Assume that the center of mass and the center of the axis of rotation are coincident. If the weight of the gear is 41 lb, find the mass moment of inertia and the radius of gyration of the gear.

Figure P16.5

16.6 The figure shows a wheel whose mass moment of inertia I is to be determined. The wheel is mounted on a shaft in bearings with very low frictional resistance to rotation. At one end of the shaft and on the outboard side of the bearings is connected a rod with a weight W_b secured to its end. It is possible to measure the mass moment of inertia of the wheel by displacing the weight W_b from its equilibrium and permitting the assembly to oscillate. If the weight of the pendulum arm is neglected, show that the mass moment of inertia of the wheel can be obtained from the equation

$$I = W_b l \left(\frac{\tau^2}{4\pi^2} - \frac{l}{g} \right)$$

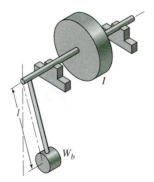

Figure P16.6

16.7 If the weight of the pendulum arm is not neglected in Problem 16.6, but is assumed to be uniformly distributed over the length l, show that the mass moment of inertia of the wheel can be obtained from the equation

$$I = l \left[\frac{\tau^2}{4\pi^2} \left(W_b + \frac{W_a}{2} \right) - \frac{l}{g} \left(W_b + \frac{W_a}{3} \right) \right]$$

where W_a is the weight of the arm.

16.8 Wheel 2 in the figure is a round disk which rotates about a vertical axis z through its center. The wheel carries a pin B at a distance R from the axis of rotation of the wheel, about which link 3 is free to rotate. Link 3 has its center of mass G located at a distance r from the vertical axis through B, and it has a weight W_3 and a mass moment of inertia I_G about its own mass center. The wheel rotates at an angular velocity ω_2 with link 3 fully extended. Develop an expression for the angular velocity ω_3 that link 3 would acquire if the wheel were suddenly stopped.

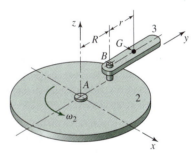

Figure P16.8

16.9 Repeat Problem 16.8, except assume that the wheel rotates with link 3 radially inward. Under these conditions, is there a value for the distance r for which the resulting angular velocity ω_3 is zero?

16.10 The figure illustrates a planetary gear-reduction unit which utilizes 7-pitch spur gears cut on the 20° full-depth system. The input to the reducer is driven with 25 hp at 600 rev/min. The mass moment of inertia of the resisting load is 5.83 lb·s²/in. Calculate the bearing reactions on the input, output, and planetary shafts. As a designer, what forces would you use in designing the mounting bolts? Why?

All parts are steel with density 0.282 lb/in³. The arm is rectangular and is 4 in wide by 14 in long with a 4-in-diameter central hub and two 3-in-diameter planetary hubs. The segment separating

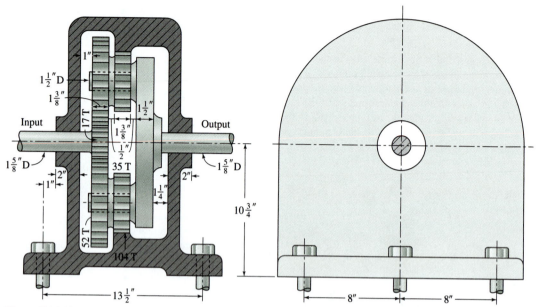

Figure P16.10 All parts are steel with density 0.282 lb/in³. The arm is rectangular and is 4 in wide by 14 in long with a 4-in-diameter central hub and two 3-in-diameter planetary hubs. The segment separating the planet gears is a 0.5-in by 4-in-diameter cylinder. The inertia of the gears can be obtained by treating them as cylinders equal in diameter to their respective pitch circles.

the planet gears is a 0.5-in- by 4-in-diameter cylinder. The inertia of the gears can be obtained by treating them as cylinders equal in diameter to their respective pitch circles.

16.11 It frequently happens in motor-driven machinery that the greatest torque is exerted when the motor is first turned on, because of the fact that some motors are capable of delivering more starting torque than running torque. Analyze the bearing reactions of Problem 16.10 again, but this time use a starting torque equal to 250% of the full-load torque. Assume a normal-load torque and a speed of zero. How does this starting condition affect the forces on the mounting bolts?

16.12 The gear-reduction unit of Problem 16.10 is running at 600 rev/min when the motor is suddenly turned off, without changing the resisting-load torque. Solve Problem 16.10 for this condition.

16.13 The differential gear train shown in the figure has gear 1 fixed and is driven by rotating shaft 5 at 500 rev/min cut in the direction shown. Gear 2 has fixed bearings constraining it to rotate about the positive y axis, which remains vertical; this is the output shaft. Gears 3 and 4 have bearings connecting

them to the ends of the carrier arm which is integral with shaft 5. The pitch diameters of gears 1 and 5 are both 8.0 in, while the pitch diameters of gears 3 and

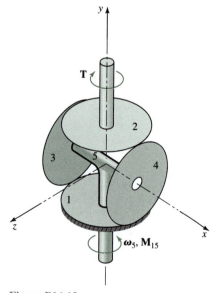

Figure P16.13

4 are both 6.0 in. All gears have the 20° pressure angles and are each 0.75 in thick, and all are made of steel with density 0.286 lb/in³. The mass of shaft 5 and all gravitational loads are negligible. The output shaft torque loading is $\mathbf{T} = -100\hat{\mathbf{j}}$ ft·lb as shown. Note that the coordinate axes shown rotate with the input shaft 5. Determine the driving torque required, and the forces and moments in each of the bearings. (*Hint:* It is reasonable to assume through symmetry that $F_{13}^t = F_{14}^t$. It is also necessary to recognize that only compressive loads, not tension, can be transmitted between gear teeth.)

16.14 The figure shows a flyball governor. Arms 2 and 3 are pivoted to block 6, which remains at the height shown but is free to rotate around the y axis. Block 7 also rotates about and is free to slide along the y axis. Links 4 and 5 are pivoted at both ends between the two arms and block 7. The two balls at the ends of links 2 and 3 weigh 3.5 lb each, and all other masses are negligible in comparison; gravity acts in the $-\hat{\mathbf{j}}$ direction. The spring between links 6 and 7 has a stiffness of 1.0 lb/in and would be unloaded if block 7 were at a height of $\mathbf{R}_D = 11\hat{\mathbf{j}}$ in. All moving links rotate about the y axis with angular velocities of $\omega\hat{\mathbf{j}}$. Make a graph of the height R_D versus the rotational speed ω in rev/min, assuming that changes in speed are slow.

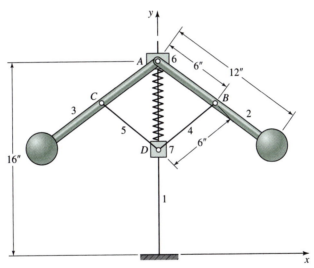

Figure P16.14

17

Vibration Analysis

The existence of vibrating elements in any mechanical system produces unwanted noise, high stresses, wear, poor reliability, and, frequently, premature failure of one or more of the parts. The moving parts of all machines are inherently vibration producers, and for this reason engineers must expect vibrations to exist in the devices they design. But there is a great deal they can do during the design of the system to anticipate a vibration problem and to minimize its undesirable effects.

Sometimes it is necessary to build a vibratory system into a machine—a vibratory conveyor, for example. Under these conditions the engineer must understand the mechanics of vibration in order to obtain an optimal design.

17.1 DIFFERENTIAL EQUATIONS OF MOTION

Any motion that exactly repeats itself after a certain interval of time is a periodic motion and is called a *vibration*. Vibrations may be either free or forced. A mechanical element is said to have a *free vibration* if the periodic motion continues after the cause of the original disturbance is removed, but if a vibratory motion persists because of the continuing existence of a disturbing force, then it is called a *forced vibration*. Any free vibration of a mechanical system will eventually cease because of loss of energy. In vibration analysis we often take account of these energy losses by using a single factor called the *damping factor*. Thus, a heavily damped system is one in which the vibration decays rapidly. The *period* of a vibration is the time for a single event or cycle; the *frequency* is the number of cycles or periods occurring in unit time. The *natural frequency* is the frequency of a free vibration. If the forcing frequency becomes equal to the natural frequency of a system, then *resonance* is said to occur.

We shall also use the terms *steady-state vibration,* to indicate that a motion is repeating itself exactly in each successive cycle, and *transient vibration,* to indicate a vibratory-type motion that is changing in character. If a periodic force operates on a mechanical system, the resulting motion will be transient in character when the force first begins to act, but after an interval of time the transient will decay, owing to damping, and the resulting motion is termed a *steady-state vibration*.

The word *response* is frequently used in discussing vibratory systems. The words *response, behavior,* and *performance* have roughly the same meaning when used in dynamic analysis. Thus we can apply an external force having a sine-wave relationship with time to a vibrating system in order to determine how the system "responds," or "behaves," when the frequency of the force is varied. A plot using the vibration amplitude along one axis and the forcing frequency along the other axis is then described as a *performance* or *response* curve for the system. Sometimes it is useful to apply arbitrary input disturbances or forces to a system. These may not resemble the force characteristics that a real system would receive in use at all; yet the response of the system to these arbitrary disturbances can provide much useful information about the system.

Vibration analysis is sometimes called *elastic-body analysis* or *deformable-body analysis,* because, as we shall see, a mechanical system must have elasticity in order to allow vibration. When a rotating shaft has a torsional vibration, this means that a mark on the circumference at one end of the shaft is successively ahead of and then behind a corresponding mark on the other end of the shaft. In order words, torsional vibration of a shaft is the alternate twisting and untwisting of the rotating material and requires elasticity for its existence. We shall begin our study of vibration by assuming that elastic parts have no mass and that heavy parts are absolutely rigid—that is, they have no elasticity. Of course these assumptions are never true, and so, in the course of our studies, we must also learn to correct for the effects of making these assumptions.

Figure 17.1 shows an idealized vibrating system having a mass m guided to move only in the x direction. The mass is connected to a fixed frame through the spring k and the dashpot c. The assumptions used are as follows:

1. The spring and the dashpot are massless.
2. The mass is absolutely rigid.
3. All the damping is concentrated in the dashpot.

It turns out that a great many mechanical systems can be analyzed quite accurately using these assumptions.

Figure 17.1

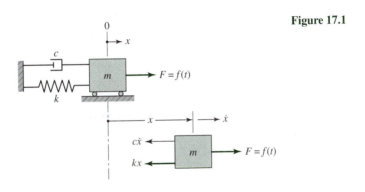

The elasticity of the system of Fig. 17.1 is completely represented by the spring. The *stiffness* or *scale* is designated as k and defined by the equation

$$k = \frac{F}{x} \tag{17.1}$$

where F is the force required to deflect the spring a distance x.

Similarly, the friction, or damping, is assumed to be entirely viscous damping (later we shall examine other kinds of friction) and is designated using a *coefficient of viscous damping c*.

Thus,

$$c = \frac{F}{\dot{x}} \tag{17.2}$$

where F is the force required to move the mass at a velocity of $\dot{x}$.

You should note that we are using the notation for a time derivative that is customary in vibration theory. Thus,

$$\dot{x} = \frac{dx}{dt} \quad \text{and} \quad \ddot{x} = \frac{d^2x}{dt^2}$$

This is called *Newton's notation,* because Newton was the first to use it. It is used only when the derivatives are with respect to time.

The vibrating system of Fig. 17.1 has one degree of freedom, because the position of the mass can be completely defined by a single coordinate. An external force $F = f(t)$ is shown acting upon the mass. Thus this system is classified as a forced, single-degree-of-freedom system with damping.

The equation of motion of the system is written by displacing the mass in the positive direction, giving it a positive velocity, and then summing all the forces, including the inertia force. As shown in the figure, there are three forces acting: a spring force kx acting in the negative direction, a damping force $c\dot{x}$ also acting in the negative direction, and an external force $f(t)$ acting in the positive direction. Summing these forces together with the inertia force gives

$$\sum F = -kx - c\dot{x} + f(t) + (-m\ddot{x}) = 0$$

or

$$m\ddot{x} + c\dot{x} + kx = f(t) \tag{17.3}$$

Equation (17.3) is an important equation in dynamic analysis, and we shall eventually solve it for many specialized conditions. It is a linear differential equation of the second order.

Consider next the idealized torsional vibrating system of Fig. 17.2. Here a disk having a mass moment of inertia I is mounted upon the end of a weightless shaft having a torsional spring constant k, defined by

$$k = \frac{T}{\theta} \tag{17.4}$$

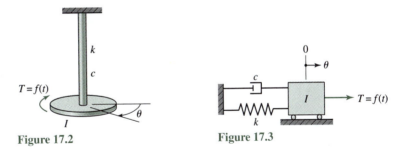

Figure 17.2   Figure 17.3

where T is the torque necessary to produce an angular deflection of the disk of θ. In a similar manner, the torsional damping factor is defined by

$$c = \frac{T}{\dot\theta} \qquad (17.5)$$

Note that we are using the same symbols for denoting these torsional parmeters as for rectilinear ones. The nature of the problem or of the differential equation will usually reveal whether the system is rectilinear or torsional, and so there will be no confusion in this usage.

Next, designating an external-torque forcing function by $T = f(t)$, we find that the differential equation for the torsional system is

$$\sum T = -k\theta - c\dot\theta + f(t) + (-I\ddot\theta) = 0$$

or

$$I\ddot\theta + c\dot\theta + k\theta = f(t) \qquad (17.6)$$

which is of the same form as Eq. (17.3). Thus, with appropriate substitutions, the solution of Eq. (17.6) will be the same as that of Eq. (17.3). Note, too, that this means we can *simulate* a torsional system, as in Fig. 17.3, merely by substituting torsional notation for the usual rectilinear notation.

It has been said that mathematics is a human invention and, therefore, that it can never perfectly describe nature.[1] But the linear differential equation frequently does such an excellent job of simulating the action of a mechanical system that one wonders whether differential equations are not an exception to this rule. Perhaps people just happened to discover them, as they have other curiosities of nature.

EXAMPLE 17.1

Figure 17.4a shows a vibrating system in which a time-dependent displacement $y = y(t)$ excites a spring-mass system through a viscous dashpot. Write the differential equation of this system.

SOLUTION

To write the differential equation, assume any arbitrary relationship between the coordinates and their first derivatives, say, $x > y$ and $\dot x > \dot y$. Also assume senses for x and $\dot x$, say, $x > 0$ and

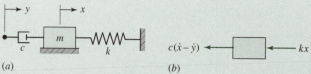

(a)   (b)

Figure 17.4 (a) The system. (b) Forces acting on the mass; note that these forces are acting in the negative direction because of the arbitrary assumptions that were made.

$\dot{x} > 0$. Regardless of the assumptions made, the result will be the same; try it. With these assumptions the forces acting on the mass are as shown in Fig. 17.4b. Summing the forces in the x direction and adding in the inertia forces gives

$$\sum F = -kx - c(\dot{x} - \dot{y}) + (-m\ddot{x}) = 0 \tag{1}$$

Such equations are usually rearranged with the forcing term on the right, that is

$$m\ddot{x} + c\dot{x} + kx = c\dot{y} \qquad\qquad Ans.$$

17.2 A VERTICAL MODEL

The system of Fig. 17.1 moves in the horizontal direction, and, as a result, gravity has no effect on its motion. Let us now turn the system to the vertical direction, and also incidentally, make the friction and the external force zero. This yields the model of Fig. 17.5. We choose the origin of the coordinate system corresponding to the equilibrium position of the mass. Thus, when the mass is given a positive displacement, the spring force has two components. One of these is $k\delta_{st} = W$, where δ_{st} is the distance the spring is deflected when the weight is suspended from it. The other component is kx. Summing the forces acting on the mass gives

$$\sum F = -k(x + \delta_{st}) + W + (-m\ddot{x}) = 0$$

Figure 17.5

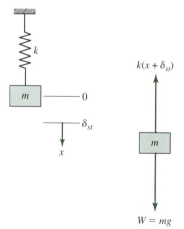

or

$$m\ddot{x} + kx = W - k\delta_{st} = 0 \qquad (17.7)$$

Note that Eq. (17.3) is identical to Eq. (17.7) if the damping and the external force are made zero.

17.3 SOLUTION OF THE DIFFERENTIAL EQUATION

If Eq. (17.7) is arranged in the form

$$\ddot{x} = -\frac{k}{m}x \qquad (a)$$

then we can see that the function used for x will have to be the negative of its second derivative. One such function is

$$x = A \sin bt \qquad (b)$$

Then

$$\dot{x} = Ab \cos bt \qquad (c)$$

$$\ddot{x} = -Ab^2 \sin bt \qquad (d)$$

Substituting Eqs. (b) and (d) into (a) produces

$$-Ab^2 \sin bt = -\frac{k}{m} A \sin bt \qquad (e)$$

Dividing both sides by $(-A \sin bt)$ leaves

$$b^2 = \frac{k}{m}$$

Thus Eq. (b) is a solution, provided that we set $b = \sqrt{k/m}$. As above, it can easily be shown that

$$x = B \cos bt \qquad (f)$$

is also a solution to Eq. (a). A general solution is obtained by adding Eqs. (b) and (f) to get

$$x = A \sin \sqrt{\frac{k}{m}}t + B \cos \sqrt{\frac{k}{m}}t \qquad (g)$$

Mathematically, the constants A and B are the constants of integration, because, theoretically, we could integrate Eq. (a) twice to get the solution. Unfortunately, most differential equations cannot be solved in this manner, and we must resort to little tricks to obtain the solution.

Physically, the constants A and B represent the manner in which the vibration started, or, to put it another way, the state of the motion at the instant $t = 0$.

We now define

$$\omega_n = \pm\sqrt{\frac{k}{m}} \tag{17.8}$$

where the $\pm$ sign means that the solution is valid for either positive or negative time. The solution can now be written in the form

$$x = A \sin \omega_n t + B \cos \omega_n t \tag{17.9}$$

As an illustration of the meaning of A and B, suppose we pull the mass downward a positive distance x_0 and release it with zero velocity at the instant $t = 0$. The starting conditions then are

$$t = 0, \qquad x = x_0, \qquad \dot{x} = 0$$

Substituting the first two of these into Eq. (17.9) gives

$$x_0 = A(0) + B(1)$$

or

$$B = x_0$$

Next, taking the first time derivative of Eq. (17.9),

$$\dot{x} = A\omega_n \cos \omega_n t - B\omega_n \sin \omega_n t \tag{h}$$

and substituting the first and third conditions in this equation yields

$$0 = A\omega_n(1) - B\omega_n(0)$$

or

$$A = 0$$

Then, substituting A and B back into Eq. (17.9) produces

$$x = x_0 \cos \omega_n t \tag{17.10}$$

If we start the motion with

$$t = 0, \qquad x = 0, \qquad \dot{x} = v_0$$

we get

$$x = \frac{v_0}{\omega_n} \sin \omega_n t \tag{17.11}$$

But if we start the motion with

$$t = 0, \qquad x = x_0, \qquad \dot{x} = v_0$$

we get

$$\dot{x} = \frac{v_0}{\omega_n} \sin \omega_n t + x_0 \cos \omega_n t \qquad (17.12)$$

which is, therefore, the most general form of the solution. You should substitute Eq. (17.12) together with its second derivative back into the differential equation and so demonstrate its validity.

The three solutions [Eqs. (7.10), (17.11), and (17.12)] are represented graphically in Fig. 17.6 using *phasors* to generate the trigonometric functions. Phasors are not vectors in the classic sense, because they can also be manipulated in ways that are not defined for vectors. They are complex numbers, however, and they can be added and subtracted just as vectors can.

The ordinate of the graph of Fig. 17.6 is the displacement x, and the abscissa can be considered as the time axis or as the angular displacement $\omega_n t$ of the phasors for a given time after the motion has started. The phasors x_0 and v_0/ω_n are shown in their initial positions, and as time passes, these rotate counterclockwise with an angular velocity of ω_n and generate the displacement curves shown. The figure shows that the phasor x_0 starts from a maximum positive displacement and the phasor v_0/ω_n starts from a zero displacement. These, therefore, are very special, and the most general form of start is that given by Eq. (17.12), in which motion begins at some intermediate point.

The quantity

$$\omega_n = \sqrt{\frac{k}{m}} \qquad (17.13)$$

is called the *natural circular frequency* of the *undamped free vibration,* and its units are radians per second. Note that this is *not* quite the same as the *natural frequency* defined earlier in this chapter, which has the units of cycles per second. Nevertheless, we shall often

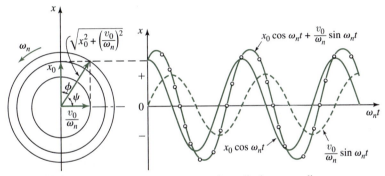

Figure 17.6 The use of phasors to generate time–displacement diagrams.

describe ω_n as the natural frequency too, omitting the word "circular" for convenience, because its circular character follows from the units used. For most systems ω_n is a constant because the mass and spring constant do not vary. Because one cycle of motion is completed in an angle of 2π rad, the period of a vibration is given by the equation

$$\tau = \frac{2\pi}{\omega_n} = 2\pi\sqrt{\frac{m}{k}} \tag{17.14}$$

where τ is in seconds. The frequency is the reciprocal of the period and is

$$f = \frac{\omega_n}{2\pi} = \frac{1}{2\pi}\sqrt{\frac{k}{m}} \tag{17.15}$$

where f is in hertz (cycles/s or s^{-1}) and is abbreviated as Hz.

Study of Fig. 17.6 suggests that one should also be able to express the motion by the equation[2] (see note on page 593)

$$x = X_0 \cos(\omega_n t - \phi) \tag{17.16}$$

where X_0 and ϕ are the constants of integration and their values depend upon the initial conditions. These constants can be obtained directly from the trigonometry of Fig. 17.6 and are

$$X_0 = \sqrt{x_0^2 + \left(\frac{v_0}{\omega_n}\right)^2} \quad \text{and} \quad \phi = \tan^{-1}\frac{v_0}{\omega_n x_0} \tag{17.17}$$

Equation (17.16) can now be written in the form

$$x = \sqrt{x_0^2 + \left(\frac{v_0}{\omega_n}\right)^2}\cos(\omega_n t - \phi) \tag{17.18}$$

This is a particularly convenient form of the equation because the coefficient is the *amplitude* of the vibration. The amplitude is the maximum displacement of the mass. The angle ϕ is called a *phase angle,* and it denotes the angular lag of the motion with respect to the cosine function.

The velocity and acceleration can be obtained by successively differentiating Eq. (17.16), that is,

$$\dot{x} = -X_0\omega_n \sin(\omega_n t - \phi) \quad \text{and} \quad \ddot{x} = -X_0\omega_n^2 \cos(\omega_n t - \phi)$$

where the velocity amplitude is $X_0\omega_n$ and the amplitude of the acceleration is $X_0\omega_n^2$. The phasor displacement, velocity, and acceleration are plotted in Fig. 17.7 to show their phase relationships. All three phasors maintain fixed phase angles as they rotate at constant angular velocity ω_n. It is seen that the velocity leads the displacement by 90° and that the acceleration is 180° out of phase with the displacement.

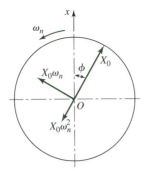

Figure 17.7 Phase relationship of displacement, velocity, and acceleration.

17.4 STEP INPUT FORCING

Let us consider the vibrating system of Fig. 17.1 again, this time adding to the system a constant force F applied to the mass and acting in the positive x direction. As before, we consider the damping to be zero. For this condition Eq. (17.3) is written

$$\ddot{x} + \frac{k}{m}x = \frac{F}{m} \tag{17.19}$$

The solution to this equation is

$$x = A \cos \omega_n t + B \sin \omega_n t + \frac{F}{k} \tag{17.20}$$

where A and B are the constants of integration and where

$$\omega_n = \sqrt{\frac{k}{m}} \tag{a}$$

as before. It is noted that the dimensions of the quantity F/k are in units of length. The physical interpretation of this is that a force F applied to a spring of index k produces an elongation (or deformation) of the spring of magnitude F/k.

It will be interesting to start the motion when the system is at rest. Thus we apply a force F at the instant $t = 0$ and observe the behavior of the system. Because the system is motionless and in equilibrium the instant the force is applied, the starting conditions are $x = 0$, $\dot{x} = 0$, when $t = 0$. Substituting these conditions into Eq. (17.20) produces the following values for the constants of integration:

$$A = -\frac{F}{k} \quad \text{and} \quad B = 0$$

The equation of motion is obtained by substituting these back into Eq. (17.20). This gives

$$x = \frac{F}{k}(1 - \cos \omega_n t) \tag{17.21}$$

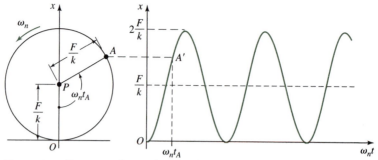

Figure 17.8 Response of an undamped vibrating system to a constant force.

This equation is plotted in Fig. 17.8 to show how the system behaves. The figure shows that the application of a constant force F produces a vibration of amplitude F/k about a position of equilibrium displaced a distance F/k from the origin. This is evident from Eq. (17.21), because it contains a positive constant term

$$x_1 = \frac{F}{k}$$

and a negative trigonometric term

$$x_2 = -\frac{F}{k}\cos\omega_n t$$

For $t = 0$ these equations become

$$x_1 = \frac{F}{k} \quad \text{and} \quad x_2 = -\frac{F}{k}$$

On the phase diagram (Fig. 17.8) the motion is represented by a phasor PA of length F/k rotating counterclockwise at ω_n rad/s. This phasor starts from the position PO when $t = 0$, rotates about P as a center, and generates the circle of radius F/k.

The velocity and acceleration are obtained by successive differentiation of Eq. (17.21) and are

$$\dot{x} = \frac{F}{k}\omega_n \sin\omega_n t \tag{17.22}$$

$$\ddot{x} = \frac{F}{k}\omega_n^2 \cos\omega_n t \tag{17.23}$$

These can be represented as phasors too, as shown in Fig. 17.9. Their starting positions are found by calculating their values for $\omega_n t = 0$. Velocity–time and acceleration–time graphs can also be plotted for these phasors employing the same methods used to obtain the displacement–time plot of Fig. 17.8.

Figure 17.10 shows the results when the force is just as suddenly removed. If, for example, the force is removed when the phasor occupies the position PB, the resulting motion has an amplitude about the zero axis of $2F/k$. The diagram shows an amplitude of

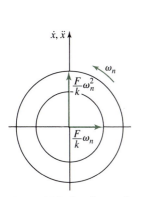

Figure 17.9 Starting positions of the velocity and acceleration phasors.

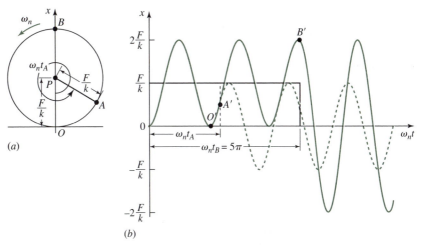

(a)

(b)

Figure 17.10 (a) Phase diagram. (b) Displacement diagram.

F/k about the zero axis if the force is removed at position A' on the displacement diagram. Finally, note that if the force is removed at point O' on the displacement diagram, the resulting motion is zero.

17.5 PHASE-PLANE REPRESENTATION

The phase-plane method is a graphical means of solving transient vibration problems which is quite easy to understand and to use. The method eliminates the necessity for solving differential equations, some of which are very difficult, and even enables solutions to be obtained when the functions involved are not expressed in algebraic form. Engineers must concern themselves as much with transient disturbances and motions of machine parts as with steady-state motions. The phase-plane method presents the physics of the problem with so much clarity that it will serve as an excellent vehicle for the study of mechanical transients.

Before introducing the details of the phase-plane method, it will be of value to show how the displacement-time and the velocity-time relations are generated by a single rotating phasor. We have already observed that a free undamped vibrating system has an equation of motion which can be expressed in the form

$$x = X_0 \cos(\omega_n t - \phi) \tag{17.24}$$

and that its velocity is

$$\dot{x} = -X_0 \omega_n \sin(\omega_n t - \phi) \tag{17.25}$$

The displacement, as given by Eq. (17.24), can be represented by the projection on a vertical axis of a phasor of length X_0 rotating at ω_n rad/s in the counterclockwise direction (Fig. 17.11a). The angle $\omega_n t - \phi$, in this example, is measured from the vertical axis. Similarly, the velocity can be represented on the same vertical axis as the projection of another phasor of length $X_0 \omega_n$ rotating at the same angular velocity but leading X_0 by a phase

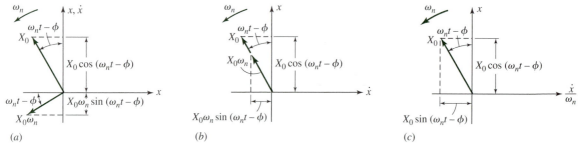

Figure 17.11

angle of 90°, as shown in Fig. 17.11a. Therefore, the angular location of the velocity phasor is measured from the horizontal axis. If we take the coordinate system containing the velocity phasor and rotate it backward (clockwise) through an angle of 90°, then the velocity and displacement phasors will be coincident and their angular locations can be measured from the same vertical axis. This step has been taken in Fig. 17.11b, where it is seen that the displacement is still measured along the same vertical axis. But, having rotated the coordinate system in which the velocity is measured, we see that the velocity is obtained by projecting the $X_0\omega_n$ phasor to the horizontal axis. Thus we can now measure velocities on a separate axis from displacements. Note, too, that the direction of positive velocities is to the right.

Our final step is taken by noting that the velocity phasor differs in length from the displacement phasor by the constant factor ω_n. Thus, instead of plotting velocities, if we plot the quantity $\dot{x}/\omega_n$, then we will have a quantity which is proportional to velocity. This step has been taken in Fig. 17.11c, where the horizontal axis is designated as the $\dot{x}/\omega_n$ axis. With this change the projection of the phasor X_0 on the x axis gives the displacement, and its projection on the $\dot{x}/\omega_n$ axis gives a quantity that is directly proportional to the velocity.

The displacement–time and the velocity–time graphs of a free undamped vibration have been plotted on Fig. 17.12 to show how the vector diagram is related to them. A point A on the phase diagram corresponds with A' on the displacement diagram and with A'' on the velocity diagram. Note that quantities obtained from the velocity plot must be multiplied by ω_n in order to obtain the actual velocity.

It is possible to arrive at these same conclusions in a different manner. If Eq. (17.24) is squared, we obtain

$$x^2 = X_0^2 \cos^2(\omega_n t - \phi) \tag{a}$$

Next, dividing Eq. (17.25) by ω_n and squaring the result gives

$$\left(\frac{\dot{x}}{\omega_n}\right)^2 = X_0^2 \sin^2(\omega_n t - \phi) \tag{b}$$

Then adding Eqs. (a) and (b) gives

$$x^2 + \left(\frac{\dot{x}}{\omega_n}\right)^2 = X_0^2 \tag{17.26}$$

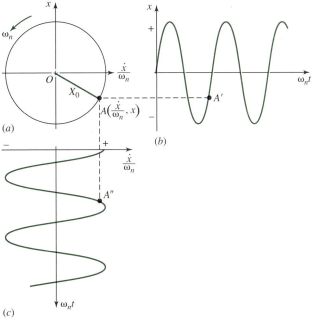

Figure 17.12 (*a*) Phase diagram. (*b*) Displacement diagram. (*c*) Velocity diagram.   **Figure 17.13**

This is the equation of a circle having the amplitude X_0 as its radius and with its center at the origin of a coordinate system having the axes x and $\dot{x}/\omega_n$. Thus Eq. (17.26) describes the circle of Fig. 17.12*a*, where OA is the amplitude X_0 of the motion. The coordinate system x, $\dot{x}/\omega_n$ defines the position of points in a region which is called the *phase plane*.

The phase plane is widely used in the solution of nonlinear differential equations. Such equations occur frequently in the study of vibrations and feedback-control systems. When the phase plane is used to solve nonlinear differential equations, it is customary to arrange the axes as shown in Fig. 17.13, with ω_n considered positive in the clockwise direction instead of the counterclockwise direction as we are using it here. This arrangement of the axes seems more logical, because $\dot{x}/\omega_n$ is a function of x. However, for the analysis of transient disturbances to mechanical systems and for analyzing cam mechanisms, the arrangement of the axes as in Fig. 17.12 is more appropriate, and this is the one we shall employ throughout this chapter. The reason for this is that we shall employ the phase-plane method, not as an end in itself, as is done when it is used for the solution of nonlinear problems, but as a tool to obtain the transient response.

17.6 PHASE-PLANE ANALYSIS

As a first example of the use of the method we shall consider the vibrating system of Fig. 17.14. This is a spring–mass system having a spring attached to a frame that can be positioned at will. To begin the motion we might consider the system as initially at rest and then, at $t = 0$, suddenly move the frame a distance x_1 to the right. The effect of this sudden motion is to compress the spring an amount x_1 and to shift the equilibrium position of the

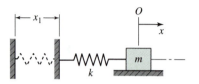

Figure 17.14

mass a distance x_1 to the right. For $t < 0$ the mass is in equilibrium at the position shown in the figure. For $t > 0$ the equilibrium position of the mass is a distance x_1 to the right of the position shown in the figure. If the motion of the frame occurs instantaneously, the initial conditions are

$$x = -x_1 \quad \text{and} \quad \dot{x} = 0 \quad \text{at} \quad t = 0 \qquad (a)$$

where the motion is now measured from the shifted equilibrium position. If these initial conditions are used to evaluate the constants of integration of Eq. (17.16), we obtain

$$X_0 = -x_1 \quad \text{and} \quad \phi = 0$$

so that

$$x = -x_1 \cos \omega_n t \qquad (b)$$

Next, note that the application of a constant force to an undamped spring–mass system gave as the equation of motion [Eq. (17.21)]

$$x = \frac{F}{k} - \frac{F}{k} \cos \omega_n t \qquad (c)$$

By rearranging, we obtain

$$x - \frac{F}{k} = -\frac{F}{k} \cos \omega_n t \qquad (d)$$

If we let $F/k = x_1$, then Eq. (d) becomes

$$x - x_1 = -x_1 \cos \omega_n t \qquad (e)$$

But if the origin of x is shifted a distance x_1, then Eq. (e) is the same as Eq. (b). Thus shifting of the frame of a vibrating system through a distance x_1 is equivalent to adding a constant force $F = kx_1$ acting upon the mass. This problem is illustrated in Fig. 17.15, where the old origin is O and the new origin is O_1. These are separated, then, by the distance x_1. At the instant $t = 0$, the origin O is shifted to O_1, initiating the motion. A phasor $O_1 A$, equal in magnitude to the distance x_1, begins its rotation from the position $O_1 O$. Rotating at ω_n rad/s, the projection of this phasor on the x axis describes the displacement of the mass. This phasor then continues to rotate until something else happens to the system. Arriving at point A it has traversed an angle $\omega_n t_A$, and the corresponding point on the displacement-time diagram is A'.

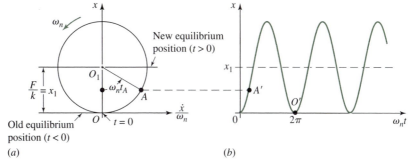

Figure 17.15 (*a*) Phase-plane diagram. (*b*) Displacement-time diagram.

The phase-plane method permits us to move the origin in any manner we choose and at any instant in time that we may choose. In the example above, the origin was shifted a distance x_1, and we have seen that this is equivalent to the sudden application of a force $F = kx_1$ applied to the mass in the positive direction. Suppose that we permit the phasor to make one complete revolution of $360°$. It then generates one complete cycle of displacement and returns to point O. If, at this instant, we shift the frame back to its original position, we might reasonably ask: What are the displacement and velocity at the instant of making this shift? The phase-plane diagram shows that both the displacement and velocity are zero at the end of a cycle. The displacement and velocity are exactly the quantities we require in order to determine the motion of the system in the next era. Because these are both zero, there is no motion after returning the frame to its original position. It will be recalled that these are exactly the results that we obtained in analyzing square-ware forcing functions when the force endured for an integral number of cycles. Thus the phasor generates the displacement diagram through the angle $\omega_n t = 2\pi$ in this case, and at this point we return the frame to its original position. The initial conditions are now $x = 0, \dot{x} = 0$; consequently, the mass stops its motion completely.

Let us now create a vibration by shifting the frame from $x = 0$ to $x = x_1$, waiting a period of time Δt, then shifting the frame back to $x = 0$. This constitutes a square-wave forcing function, or a step disturbance to the system, and the phase-plane and displacement diagrams for such a motion are illustrated in Fig. 17.16. The motion can be described in three eras. During the first, from t_0 to t_1, everything is at rest. At $t = t_1$ the frame suddenly moves to the right a distance $x = x_1$, compressing the spring. The duration of the second era is from t_1 to $t_2(\Delta t)$, and at time t_2 the frame suddenly returns to its original position. The third era represents time when $t \geq t_2$. The position of the frame during these three eras is shown in the displacement diagram. The displacement diagram also describes the motion of the mass, and the phase-plane diagram explains why it moves as it does. Between t_0 and t_1 the mass is at rest and nothing happens. At time t_1 the frame shifts from O to O_1 on the phase plane diagram, which is the distance x_1. If we designate the original frame position as the origin of the $x, \dot{x}/\omega_n$ system, then the conditions at $t = t_1$ are $x = +x_1, \dot{x} = 0$. However, the motion of the mass about O is equal to its motion about O_1 plus the distance from O to O_1. Therefore

$$x = -x_1 \cos \omega_n t' + x_1 = x_1(1 - \cos \omega_n t') \qquad (f)$$

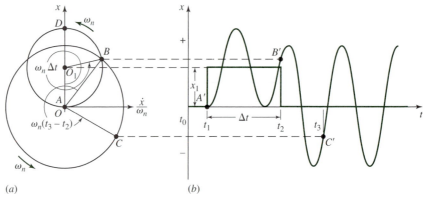

(a) (b)

Figure 17.16 A step disturbance.

Also

$$\dot{x} = x_1 \omega_n \sin \omega_n t' \tag{g}$$

where t' is understood to begin at time t_1. As shown in Fig. 17.16b, the mass vibrates about the position $x = x_1$ during this era. The vibration begins at point A on the phase-plane diagram and continues as a free vibration for the time Δt. During this period the line $O_1 A$ rotates counterclockwise with angular velocity ω_n rad/s until at time t_2 it occupies the position $O_1 B$. At this instant the third era begins when the frame suddenly returns through the distance x_1 to its original position. Thus the starting conditions for the third era are the same as the ending conditions for the second and are

$$x_2 = x_1(1 - \cos \omega_n \Delta t) \quad \text{and} \quad \dot{x}_2 = x_1 \omega_n \sin \omega_n \Delta t \tag{h}$$

Substituting these conditions into Eq. (17.12) and rearranging gives the equation of motion for the third era as

$$x = x_1(1 - \cos \omega_n \Delta t) \cos \omega_n t'' + x_1 \sin \omega_n \Delta t \sin \omega_n t'' \tag{i}$$

where t'' is the time measured from the start of the third era. Equation (i) can be transformed into an equivalent expression containing only a single trigonometric term and a phase angle as in Eq. (17.16), but we shall not do so here. The third era, we have seen, begins at point B on the phase-plane diagram (B' on the displacement diagram); the motion of point B as it moves about a circle with center at O in the counterclockwise direction describes the motion of the mass. Thus at instant t_3 the line OB will have rotated through the angle $\omega_n(t_3 - t_2)$ and be located at C. The corresponding point on the displacement diagram is C'.

The extension of the phase-plane method to any number of steps, taken in either or both the positive or negative x direction, should now be apparent. In each case the starting conditions for the next era are taken equal to the ending conditions for the previous era. The equations of motion should be written for each era with time counted from the start of that era. You can now understand that if a third era of Fig. 17.16 begins at point D on the phase-plane diagram, then the resulting motion will have an amplitude twice as large as that in the second era.

17.7 TRANSIENT DISTURBANCES

Any action that destroys the static equilibrium of a vibrating system may be called a *disturbance* to that system. A *transient disturbance* is any action that endures for only a relatively short period of time. The analyses in the several preceding sections have dealt with transient disturbances having a stepwise relationship to time. Because all machine parts have elasticity and inertia, forces do not come into existence instantaneously in real life.* Consequently, we can usually expect to encounter forcing functions that vary smoothly with time. Although the step forcing function is not true to nature, it is our purpose in this section to demonstrate how the step function is used with the phase-plane method to obtain very good approximations of the vibration of systems excited by "natural" disturbances.

The procedure is to plot the disturbance as a function of time, to divide this into steps, and then to use the steps successively to make a phase-plane plot. The resulting displacement and velocity diagrams can then be obtained by graphically projecting points from the phase-plane diagram as previously explained. It turns out that very accurate results can frequently be obtained using only a small number of steps. Of course, as in any graphical solution, better results are obtained when a large number of steps are employed and when the work is plotted to a large scale. It is difficult to set up general rules for selecting the size of the steps to be used. For slowly vibrating systems and for relatively smooth forcing functions the step width can be quite large, but even a slow system will require narrow steps if the forcing function has numerous sharp peaks and valleys—that is, if it has a great deal of frequency content. For smooth forcing functions and slowly vibrating systems a step width such that the phasor sweeps out an angle of 180° is probably about the largest that one should use. It is a good idea to check the step width during the construction of the phase-plane diagram. Too great a width will cause a discontinuity in the slope of two curves at the point of adjacency of the two curves. If this occurs, then the step can immediately be narrowed and the procedure resumed.

Figure 17.17 shows how to find the heights of the steps. The first step for the forcing function of this figure has been given a width of Δt_1 and a height of h_1. This height is

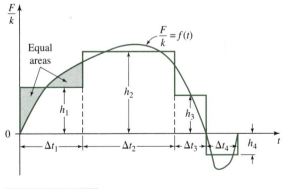

Figure 17.17 Finding the heights of the steps.

---

*If a force could truly be applied instantaneously, then, according to Newton, an acceleration would appear instantaneously. It would still take at least one instant to integrate this to become a velocity, and another instant to integrate into a change in position. Finally, according to Hooke's law, our elastic body would display a force.

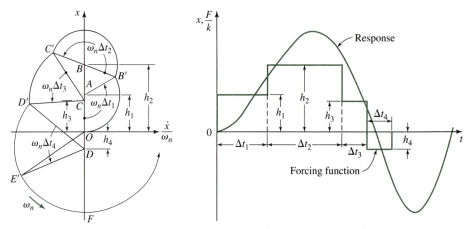

Figure 17.18 Construction of the phase-plane and displacement diagrams for a four-step forcing function.

obtained by constructing a horizontal line across the force curve such that the areas of the two shaded triangles are equal. For a step width of Δt_1 the phasor sweeps out an angle of $\omega_n \Delta t_1$. Thus the angular rotation of the phasor is obtained simply by multiplying the step width in seconds by the natural circular frequency of the system.

Figure 17.18 shows a four-step forcing function that we may assume has been deduced from a "natural" function. The width and height of the steps have been plotted to scale and are Δt_1 and h_1 for the first step, Δt_2 and h_2 for the second step, and so on. Notice that the fourth step has a negative h_4. The motion begins at $t = 0$ when a phasor of length h_1 starts rotating about A as a center from the initial position AO. This phasor rotates through an angle $\omega_n \Delta t_1$ and generates the portion of the response curve contained in the interval Δt_1. At the end of this period of time the second step begins with the center of rotation of the phasor shifting from A to B. The length of the phasor for the duration of the second step is the distance BB'. This phasor then rotates through the angle $\omega_n \Delta t_2$ about a center at B until it arrives at the position BC'. At this instant the third step begins. The center of rotation shifts to C, and the phasor CC' rotates through the angle $\omega_n \Delta t_3$. At the end of the fourth step the phasor has arrived at E'. The center of rotation now shifts to the origin O, and the motion continues as a free vibration of amplitude OE' until something else (not shown) happens to the system.

EXAMPLE 17.2

Measurements on a mechanical vibrating system show that the mass has a weight of 16.90 lb and that the springs can be combined to give an equivalent spring index of 30 lb/in. This system is observed to vibrate quite freely, and so damping can be neglected. A transient force resembling the first half-cycle of a sine wave operates on the system. If the maximum value of the force is 10 lb, determine the response when the force is applied for 0.120 s.

SOLUTION

The natural frequency is

$$\omega_n = \sqrt{\frac{k}{m}} = \sqrt{\frac{(30 \text{ lb/in})(386 \text{ in/s}^2)}{16.90 \text{ lb}}} = 26.2 \text{ rad/s}$$

The period and frequency are

$$\tau = \frac{2\pi}{\omega_n} = \frac{2\pi \text{ rad/cycle}}{26.2 \text{ rad/s}} = 0.240 \text{ s/cycle} \quad \text{and} \quad f = \frac{1}{\tau} = \frac{1}{0.240 \text{ s/cycle}} = 4.17 \text{ Hz}$$

We shall first determine the response by replacing the entire forcing function by a single step disturbance. The height of the step should be the time average of the forcing function for the duration of the step. The average ordinate of a half-cycle of the sine wave is

$$h_1 = \frac{1}{\pi} \int_0^{\pi} \left[\frac{F_{max}}{k} \sin bt \right] d(bt) = 0.637 \frac{F_{max}}{k} \tag{1}$$

or

$$h_1 = (0.637) \frac{10 \text{ lb}}{30 \text{ lb/in}} = 0.212 \text{ in}$$

This is all we need to solve the problem. The graphical solution is shown in Fig. 17.19, together with the force and the step, all plotted to the same time scale. As shown, the amplitude of a vibration resulting from a single step is $X_0 = 0.424$ in.

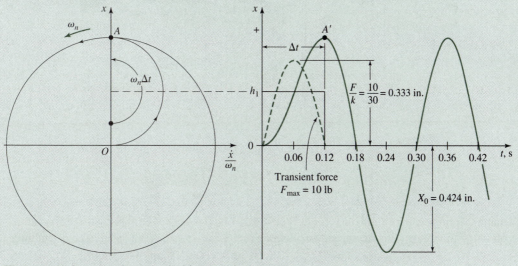

Figure 17.19 Response determined with only a single step.

In order to check the accuracy of a single step we shall solve the example again using three steps. These steps need not be equal, but it will be convenient in this example to make them so. For the three equal steps Eq. (1) is written as

$$h_1 = h_3 = \frac{1}{\pi/3} \int_0^{\pi/3} \left[\frac{F_{max}}{k} \sin bt \right] d(bt) = \frac{3}{2\pi} \frac{F_{max}}{k}$$

$$= \frac{3}{2\pi} \frac{10\ lb}{30\ lb/in} = 0.159\ in$$

$$h_2 = \frac{1}{\pi/3} \int_{\pi/3}^{2\pi/3} \left[\frac{F_{max}}{k} \sin bt \right] d(bt) = \frac{3}{\pi} \frac{F_{max}}{k}$$

$$= \frac{3}{\pi} \frac{10\ lb}{30\ lb/in} = 0.318\ in$$

The construction and results are shown in Fig. 17.20. Note that the amplitude is slightly larger than that for a single step. In this case it is very doubtful if the extra labor of a five-step solution would be justified.

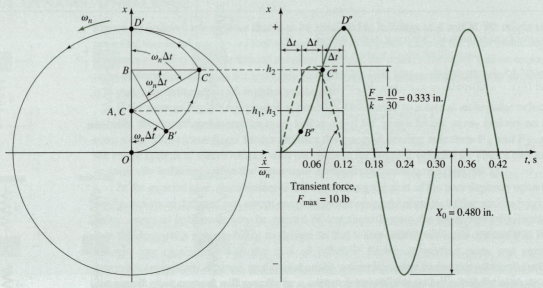

Figure 17.20 Response determined in three steps. The first step has a height of h_1 and is taken from O to B' with a radius AO; the second has a height h_2 and is from B' to C' with a radius BB'; the third has a height h_3 and is from C' to D' with a radius CC'. The vibration is a free motion after point D'.

17.8 FREE VIBRATION WITH VISCOUS DAMPING

Lumping all the friction in a system and representing it as pure viscous damping is the method most widely used in the simulation of vibration in mechanical systems. This is primarily because it results in a linear differential equation that is convenient to solve, not necessarily because this truly represents the real system. The differential equation for an unforced, viscously damped system was derived previously, and is

$$m\ddot{x} + c\dot{x} + kx = 0 \tag{a}$$

Dividing by m gives

$$\ddot{x} + \frac{c}{m}\dot{x} + \frac{k}{m}x = 0 \tag{17.27}$$

Inspection of this equation shows that x and its derivatives must be alike if the relation is to be satisfied. The exponential function satisfies this requirement; so it is not unreasonable to guess that the solution might be in the form

$$x = Ae^{st} \tag{b}$$

where A and s are constants still to be determined. The first and second derivatives of Eq. (*b*) are

$$\dot{x} = Ase^{st} \quad \text{and} \quad \ddot{x} = As^2e^{st}$$

Substituting these into Eq. (17.27) and canceling gives

$$s^2 + \frac{c}{m}s + \frac{k}{m} = 0 \tag{c}$$

Thus Eq. (*b*) is a solution of Eq. (17.27), provided that the quantity s is selected to satisfy Eq. (*c*). The roots of Eq. (*c*) are

$$s = -\frac{c}{2m} \pm \sqrt{\left(\frac{c}{2m}\right)^2 - \frac{k}{m}} \tag{d}$$

and so Eq. (*b*) can be written

$$x = Ae^{s_1t} + Be^{s_2t} \tag{e}$$

where s_1 and s_2 are the two roots of Eq. (*d*), and A and B are the constants of integration.

The value of the damping that makes the radical of Eq. (*d*) zero has special significance; we shall call it the *critical-damping coefficient* and designate it by the symbol c_c. With critical damping the radical is zero, and so

$$\frac{c_c}{2m} = \sqrt{\frac{k}{m}} = \omega_n \quad \text{or} \quad c_c = 2m\omega_n \tag{17.28}$$

It is also convenient to define a *damping ratio* ζ, which is the ratio of the actual to the critical damping. Thus

$$\zeta = \frac{c}{c_c} = \frac{c}{2m\omega_n} \tag{17.29}$$

After some algebraic manipulation, Eq. (*d*) can be written

$$s = \left(-\zeta \pm \sqrt{\zeta^2 - 1}\right)\omega_n \tag{f}$$

In the case in which the actual damping is larger than the critical, $\zeta > 1$ and the radical is real. This is analogous to movement of the mass in a very thick viscous fluid, such as molasses, and there is no vibration. Although this case does have application in certain mechanical systems, because of space limitations we shall pursue it no further here. It is not difficult to learn that if the mass is displaced and released in a system with more than critical damping, the mass will return slowly to its position of equilibrium without overshooting.

When the damping is less than critical, $\zeta < 1$ and the radical of Eq. (*f*) is imaginary. It is then written as

$$s = \left(-\zeta \pm j\sqrt{1 - \zeta^2}\right)\omega_n \tag{g}$$

where we have used the operator notation $j = \sqrt{-1}$. Substituting these roots into Eq. (*e*) gives

$$x = e^{-\zeta\omega_n t}\left(Ae^{j\sqrt{1-\zeta^2}\omega_n t} + Be^{-j\sqrt{1-\zeta^2}\omega_n t}\right) \tag{h}$$

Since A and B are complex conjugates, this equation can be transformed to

$$x = e^{-\zeta\omega_n t}[(A + B)\cos\omega_d t + j(A - B)\sin\omega_d t] \tag{i}$$

where ω_d is the natural frequency of the damped vibration, and

$$\omega_d = \omega_n\sqrt{1 - \zeta^2} \tag{17.30}$$

It is expedient to transform Eq. (*i*) into a form having only a single trigonometric term. Making this transformation gives

$$x = X_0 e^{-\zeta\omega_n t}\cos(\omega_d t - \phi) \tag{17.31}$$

where X_0 and ϕ are the new constants of integration. It is apparent that Eq. (17.31) reduces to Eq. (17.16) if the damping is made zero. The constants of integration can also be found in exactly the same manner.

Equation (17.31) is the product of a trigonometric function and a decreasing exponential function. The resulting motion is therefore oscillatory with an exponentially decreasing amplitude, as shown in Fig. 17.21. The frequency does *not* depend upon amplitude, and is less than that of an undamped system, as is indicated by the factor $\sqrt{1 - \zeta^2}$.

Figure 17.21

17.9 DAMPING OBTAINED BY EXPERIMENT

The classical method of obtaining the damping coefficient is by an experiment in which the system is disturbed in some manner—say, by hitting it with a sledge hammer—and then the decaying response is recorded by means of a strain gauge, rotary potentiometer, solar cell, or other transducer. If the friction is mostly viscous, the result should resemble Fig. 17.21. The rate of decay is measured, and, by means of the analysis to follow, the viscous-damping coefficient is easily calculated.

The record also provides a qualitative guide to the predominating friction. In many mechanical systems, records reveal that the first portion of the decay is curved as in Fig. 17.21, but the smaller part decays at a linear rate instead. A linear rate of decay identifies Coulomb, or sliding, friction. Still another type of decay sometimes found is curved at the beginning but then flattens out for small amplitudes and requires a much greater time to decay to zero. This is a kind of damping that is proportional to the square of the velocity.

Most mechanical systems have several kinds of friction present, and the investigator is usually interested only in the predominant kind. For this reason he or she should analyze that portion of the decay record in which the amplitudes are closest to those actually experienced in operating the system.

Perhaps the best method of obtaining an average damping coefficient is to use quite a number of cycles of decay, if they can be obtained, rather than a single cycle.

If we take any response curve, such as that of Fig. 7.21, and measure the amplitude of the nth and also of the $(n + N)$th cycle, then these measurements are taken when the cosine term of Eq. (17.31) is approximately unity; so

$$x_n = X_0 e^{-\zeta \omega_n t_n} \quad \text{and} \quad x_{n+N} = X_0 e^{-\zeta \omega_n (t_n + N\tau)}$$

where τ is the period of vibration and N is the number of cycles of motion between the amplitude measurements. The logarithmic decrement δ_N is defined as the natural logarithm of the ratio of these two amplitudes and is

$$\delta_N = \ln \frac{x_n}{x_{n+N}} = \ln \frac{X_0 e^{-\zeta \omega_n t_n}}{X_0 e^{-\zeta \omega_n (t_n + N\tau)}} = \ln e^{\zeta \omega_n N\tau} = \zeta \omega_n N\tau \qquad (17.32)$$

The period of the vibration is

$$\tau = \frac{2\pi}{\omega_d} = \frac{2\pi}{\omega_n \sqrt{1 - \zeta^2}} \qquad (17.33)$$

and then the decrement can be written in the form

$$\delta = \frac{\delta_N}{N} = \frac{2\pi\zeta}{\sqrt{1-\zeta^2}} \tag{17.34}$$

Measurements of many damping ratios indicate that a value of under 20 percent can be expected for most machine systems, with a value of 10 percent or less being the most probable figure. For this range of values the radical in Eq. (17.34) can be taken as unity, giving

$$\delta = 2\pi\zeta \tag{17.35}$$

as an approximate formula.

EXAMPLE 17.3

Let the vibrating system of Example 17.2 have a dashpot attached which exerts a force of 0.25 lb on the mass when the mass has a velocity of 1 in/s. Find the critical damping constant, the logarithmic decrement, and the ratio of two consecutive maxima.

SOLUTION

The data from Example 17.2 are repeated here for convenience:

$$W = 16.90 \text{ lb}, \qquad k = 30 \text{ lb/in}, \qquad \omega_n = 26.2 \text{ rad/s}$$

The critical damping constant is

$$c_c = 2m\omega_n = (2)\left(\frac{16.90}{386}\right)(26.2) = 2.29 \text{ lb·s/in} \qquad Ans.$$

The actual damping constant is $c = 0.25$ lb·s/in. Therefore the damping ratio is

$$\zeta = \frac{c}{c_c} = \frac{0.25}{2.29} = 0.109$$

The logarithmic decrement is obtained from Eq. (17.34):

$$\delta = \frac{2\pi\zeta}{\sqrt{1-\zeta^2}} = \frac{2\pi(0.109)}{\sqrt{1-(0.109)^2}} = 0.689 \qquad Ans.$$

The ratio of two consecutive maxima is

$$\frac{x_{n+1}}{x_n} = e^{-\delta} = e^{-0.689} = 0.502 \qquad Ans.$$

Therefore, for this amount of damping each amplitude is approximately 50 percent of that of the previous cycle.

17.10 PHASE-PLANE REPRESENTATION OF DAMPED VIBRATION

We have seen that when undamped mechanical systems are subjected to transient forces, the resulting motion (mathematically, at least) endures forever without decreasing in amplitude. We know, however, that energy losses always exist, though sometimes only in minute amounts, and that these losses will eventually cause a vibration to stop. In the case of vibrating systems known to be acted upon by transient forces, it is often desirable to introduce additional damping as one means of decreasing the number of cycles of vibration. For this reason the phase-plane solution of a damped vibration is particularly important.

We have seen that the phase-plane diagram of an undamped free vibration is generated by a phasor of constant length rotating at a constant angular velocity. In the case of a damped free vibration the phase-plane diagram is generated by a phasor whose length is decreasing exponentially. Thus the trajectory for such a motion is a spiral instead of a circle.

It turns out that a simple spiral will give incorrect values for the velocity if plotted on the same x, $\dot{x}/\omega_n$ axes as used for undamped free vibration. The reason for this is that the phase relationship between the velocity and displacement is not $90°$ as it is for an undamped system. Instead, for a damped system, the phase angle depends upon the amount of damping present.

An easy way to obtain a phase-plane diagram for damped motion is to tilt or rotate the $\dot{x}/\omega_n$ axis through an angle σ, as shown in Fig. 17.22. The angle σ is called the *damping phase angle,* and the following analysis shows how this angle comes about.

Taking the time derivative of Eq. (17.31) gives

$$\dot{x} = -X_0\zeta\omega_n e^{-\zeta\omega_n t}\cos\left(\sqrt{1-\zeta^2}\omega_n t - \phi\right) - X_0\sqrt{1-\zeta^2}\omega_n e^{-\zeta\omega_n t}\sin\left(\sqrt{1-\zeta^2}\omega_n t - \phi\right)$$

or

$$\frac{\dot{x}}{\omega_n} = -X_0 e^{-\zeta\omega_n t}\left[\zeta\cos\left(\sqrt{1-\zeta^2}\omega_n t - \phi\right) + \sqrt{1-\zeta^2}\sin\left(\sqrt{1-\zeta^2}\omega_n t - \phi\right)\right] \quad (a)$$

Designating $\cos\sigma = \sqrt{1-\zeta^2}$ and $\sin\sigma = \zeta$ and noting the trigonometric identity

$$\sin(\alpha + \beta) = \sin\alpha\cos\beta + \cos\alpha\sin\beta$$

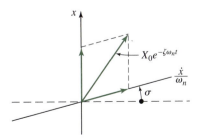

Figure 17.22 The angle σ is the damping phase angle and is applied only to the velocity term.

we see that Eq. (a) can be written as

$$\frac{\dot{x}}{\omega_n} = -X_0 e^{-\zeta \omega_n t} \sin\left(\sqrt{1 - \zeta^2}\omega_n t - \phi + \sigma\right) \tag{17.36}$$

If the damping is zero, note that $\sqrt{1 - \zeta^2} = 1$, $e^{-\zeta \omega_n t} = 1$, and $\sigma = 0$, and Eqs. (17.31) and (17.36) reduce to the same set of equations that we employed to develop the phase-plane analysis of undamped motion. As shown in Fig. 17.22, the $\dot{x}/\omega_n$ axis should be rotated counterclockwise through the angle σ in plotting the phase-plane diagram. The velocity diagram is then obtained by projecting perpendicular to the $\dot{x}/\omega_n$ axis.

In analyzing a damped system that is acted upon by transient forces by the phase-plane method, it is very helpful to construct transparent spiral templates as shown in Fig. 17.23. These are made from a sheet of clear plastic that is placed over a drawing of the spiral so that the spiral and its center can be scribed on the plastic using the point of a pair of dividers. The scribed line can then be trimmed with a pair of manicurist's scissors or a sharp razor blade. Because only one template is needed for each damping ratio, they can be stored for use in future problems. Do not forget to label each template with the value of the damping ratio for which it was constructed.

The spirals can be plotted directly from a table calculated using Eq. (17.31), but this is very tedious. A much more rapid method, which is accurate enough for graphical purposes, is to approximate each quadrant of the spiral with a circle arc. To do this it is necessary only to calculate the change in the length of the phasor in a quarter of a turn. From Eq. (17.31), the length of the phasor is $x = X_0 c^{-\zeta \omega_n t}$ and its angular velocity is $\omega_n\sqrt{1 - \zeta^2}$. Therefore, for 90° rotation,

$$\omega_n\sqrt{1 - \zeta^2}\,t = \frac{\pi}{2} \quad \text{or} \quad \omega_n t = \frac{\pi}{2\sqrt{1 - \zeta^2}} \tag{b}$$

Thus the length, after rotation through 90°, is

$$x_{90°} = X_0 e^{-\zeta \pi / 2\sqrt{1 - \zeta^2}} \tag{c}$$

To demonstrate the construction of the spirals, let us employ a damping ratio $\zeta = 0.15$ and begin with a phasor 2.50 in long. Then, after 90° of rotation, the length, from Eq. (c), is

$$x_{90°} = (2.50)e^{-0.15\pi / 2}\sqrt{1 - (0.15)^2} = 1.97 \text{ in}$$

Figure 17.23 A spiral template for $\zeta = 0.15$.

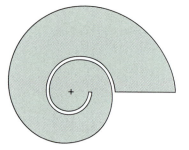

Figure 17.24

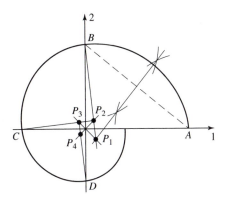

The construction is shown in Fig. 17.24 and is explained as follows: Construct the axes 1 and 2 at right angles to each other and lay off 2.50 in to A and 1.97 in to B on axes 1 and 2, respectively. Draw two lines through the origin at 45° to the axes. The perpendicular bisector of AB intersects one of these lines at P_1; using P_1 as a center, strike an arc from A to B. A line P_1B crosses another 45° line defining the center P_2 of arc BC. This process is continued in the same fashion with the point P_3 defining the center of the arc CD, and so on.

The template is used by placing its center at the same point from which a circle arc would be constructed for undamped motion. It is then turned until the template spiral coincides with the point from which the spiral diagram is to be started.

EXAMPLE 17.4

A potential vibrating system has an 80-lb mass mounted upon springs with an equivalent spring constant of 360 lb/in. A dashpot is included in the system and is estimated to produce 15 percent of critical damping. A transient force, which operates on the system, is assumed to act in three steps as follows: 720 lb for 0.0508 s, −270 lb for 0.0635 s, and 180 lb for 0.0381 s. The negative sign on the second step force simply means that its direction is opposite to the direction of the first and third. Determine the response of the system assuming no motion when the force function begins to act.

SOLUTION

The undamped natural frequency is

$$\omega_n = \sqrt{\frac{k}{m}} = \sqrt{\frac{(360)(386)}{80}} = 41.7 \text{ rad/s}$$

and so the damped frequency is

$$\omega_d = \omega_n\sqrt{1 - \zeta^2} = 41.7\sqrt{1 - (0.15)^2} = 41.2 \text{ rad/s}$$

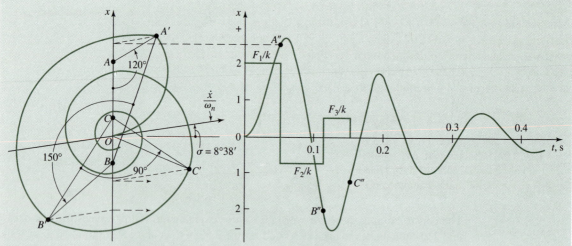

Figure 17.25

The period of the motion is $2\pi/\omega_n\sqrt{1-\zeta^2}$, which gives 0.152 s. The F/k values are

$$\frac{F_1}{k} = \frac{720}{360} = 2 \text{ in}, \qquad \frac{F_2}{k} = \frac{270}{360} = 0.75 \text{ in}, \qquad \frac{F_3}{k} = \frac{180}{360} = 0.50 \text{ in}$$

These three steps are plotted in Fig. 17.25 to scale. The angular duration of each step for the phase-plane diagram is obtained by multiplying the damped natural frequency by the time of each step and converting to degrees. This gives

Step 1
$$\omega_n\sqrt{1-\zeta^2}\Delta t_1 \frac{180}{\pi} = (41.2)(0.0508)\frac{180}{\pi} = 120°$$

Step 2
$$\omega_n\sqrt{1-\zeta^2}\Delta t_2 \frac{180}{\pi} = (41.2)(0.0635)\frac{180}{\pi} = 150°$$

Step 3
$$\omega_n\sqrt{1-\zeta^2}\Delta t_3 \frac{180}{\pi} = (41.2)(0.0381)\frac{180}{\pi} = 90°$$

The $\dot{x}/\omega_n$ axis must be turned $\sigma = \sin^{-1}(0.15) = 8°38'$ in the counterclockwise direction, as shown in the phase-plane diagram.

The construction of Fig. 17.25 is explained as follows: Project F_1/k to A on the x axis. A phasor AO, with center at A, begins rotating at the instant $t = 0$. This phasor rotates through an angle of 120° to AA' while exponentially decreasing its length. At A' the step function changes the origin of the phasor to B, and it rotates 150° about this point to B', also exponentially decreasing. At B' the origin again shifts, this time to C, and the phasor rotates 90° to C'. Now the origin shifts to O, and the phasor continues its rotation with O as the origin until the motion becomes too small to follow. Note that various points on the spiral diagram are projected parallel to the $\dot{x}/\omega_n$ axis to the x axis and then horizontally to the displacement diagram. In a similar manner the velocity diagram, though not shown, would be projected perpendicular to the $\dot{x}/\omega_n$ axis. Points on the spiral would project parallel to the x axis until they intersect the $\dot{x}/\omega_n$ axis. The displacement diagram shows the resulting response. The maximum amplitude occurs on the first positive half-cycle and is 2.72 in.

17.11 RESPONSE TO PERIODIC FORCING

Differential equations of the form

$$m\ddot{x} + kx = ky(\omega t) \tag{17.37}$$

occur often in mechanical systems. We define $\omega t = n\pi$, where n is an integer; we shall then be concerned in this section only with small values of n, say, less than 8. An example of Eq. (17.37) is shown in Fig. 17.26a, in which a roller riding in the groove of a face cam generates the motion from point O (Fig. 17.26b) of

$$y(\omega t) = y_0(1 - \cos \omega t) \tag{17.38}$$

This, then, is a motion applied to the far end of a spring k connected to a sliding mass with friction assumed to be zero. Note, in Fig. 17.26b, that we choose Eq. (17.38) to represent the driving function because we have assumed that the roller is in its extreme left position when the cam begins its rotation.

Alternatively, we could have chosen point A in Fig. 17.26b to start the motion. For this choice, the driving function would have been

$$y_A(\omega t) = y_0 \sin \omega t \tag{a}$$

One of the reasons for introducing this problem is to illustrate an interesting relationship: Figure 17.26a represents a single-degree-of-freedom vibrating system; but, when Eq. (17.37) (or a similar equation) is solved, the solution often looks more like that of a two-degree-of-freedom system. Begineers are sometimes quite surprised when they first view this solution and are unable to explain what is happening.

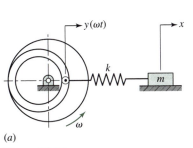

(a)

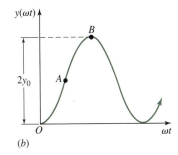

(b)

Figure 17.26

Substituting Eq. (17.38) into Eq. (17.37) gives the equation to be solved as

$$m\ddot{x} + kx = ky_0(1 - \cos \omega t) \tag{17.39}$$

The complementary solution is

$$x' = A \cos \omega_n t + B \sin \omega_n t \tag{b}$$

and the particular solution is

$$x'' = C + D \cos \omega t \tag{c}$$

The second derivative of this equation is

$$\ddot{x}'' = -D\omega^2 \cos \omega t \tag{d}$$

By substituting Eqs. (c) and (d) into Eq. (17.39), we find that $C = y_0$ and

$$D = -\frac{ky_0}{k - m\omega^2} = -\frac{y_0}{1 - \left(\omega^2/\omega_n^2\right)}$$

Therefore, the complete solution is

$$x = A \cos \omega_n t + B \sin \omega_n t + y_0\left[1 - \frac{\cos \omega t}{1 - \left(\omega^2/\omega_n^2\right)}\right] \tag{17.40}$$

The velocity is

$$\dot{x} = -A\omega_n \sin \omega_n t + B\omega_n \cos \omega_n t + \frac{\omega y_0 \sin \omega t}{1 - \left(\omega^2/\omega_n^2\right)} \tag{17.41}$$

Because we are starting from rest, where $x(0) = \dot{x}(0) = 0$, Eqs. (17.40) and (17.41) give

$$A = \frac{\left(\omega^2/\omega_n^2\right)y_0 \cos \omega_n t}{1 - \left(\omega^2/\omega_n^2\right)} \quad \text{and} \quad B = 0$$

Equation (17.40) then becomes

$$x = \frac{\left(\omega^2/\omega_n^2\right)y_0 \cos \omega_n t}{1 - \left(\omega^2/\omega_n^2\right)} - \frac{y_0 \cos \omega t}{1 - \left(\omega^2/\omega_n^2\right)} + y_0 \tag{17.42}$$

The first term on the right-hand side of Eq. (17.42) is called the *starting transient*. Notice that this is a vibration at the natural frequency ω_n, not at the forcing frequency. The usual physical system will contain a certain amount of friction, which, as we shall see in the sections to follow, will cause this term to die out after a certain period of time. The second and third terms on the right represent the *steady-state* solution and contain another component

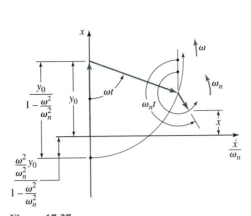

Figure 17.27

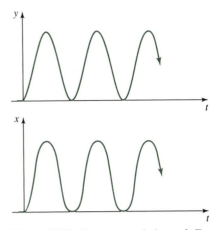

Figure 17.28 Computer solution of Eq. (17.39) for $\omega_n = 3\omega$; $S_x = S_y$.

of the vibration at the forcing frequency ω. Thus the total motion is composed of a constant plus two vibrations of differing amplitudes and frequencies. If Fig. 17.26a were a cam-and-follower system, spring k would be quite stiff; consequently, ω_n would be large compared with ω. This would cause the amplitude of the starting transient to be relatively small, and a phase-diagram plot would resemble Fig. 17.27. Here we can see that the motion is formed by the sum of a fixed phasor of length y_0 and two phasors of different lengths and rotational velocities.

A computer solution of the particular case in which $\omega_n = 3\omega$ is shown in Fig. 17.28. This is also of interest because of the nearly perfect dwells that occur at the extremes of the motion. One can easily prove the existence of these dwells by constructing a plot similar to Fig. 17.27 or by substituting $\omega_n = 3\omega$ in Eq. (17.42) and investigating the solution in the region for which ωt is zero or π.

Examination of Eq. (17.42) shows that, for $\omega/\omega_n = 0$, the solution becomes

$$x = y_0(1 - \cos \omega t)$$

which is the rigid-body solution. If we replace the spring with a rigid member, then k and, therefore, ω_n become very large, and so the mass exactly follows the cam motion.

Resonance is defined by the case in which $\omega = \omega_n$. Rearranging Eq. (17.42) to

$$x = y_0 \frac{\left(\omega^2/\omega_n^2\right)\cos \omega_n t - \cos \omega t + 1 - \left(\omega^2/\omega_n^2\right)}{1 - \left(\omega^2/\omega_n^2\right)} \tag{e}$$

we see that $x = 0/0$ for $\omega/\omega_n = 1$. L'Hôpital's rule enables us to find the solution. Thus

$$\lim_{\omega \to \omega_n} (x) = y_0 \left(1 - \cos \omega_n t + \frac{\omega_n t}{2} \sin \omega_n t \right) \tag{f}$$

A plot of this equation, obtained by solving the differential equation on the computer, is shown in Fig. 17.29. Notice that the sine term very quickly predominates in the response.

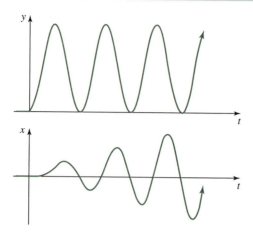

Figure 17.29 Response at resonance. In this figure the amplitude scale S_x has been greatly increased to retain the response on the screen.

17.12 HARMONIC FORCING

We have seen that the action of any transient forcing function on a damped system is to create a vibration, but the vibration decays when the applied force is removed. A machine element that is connected to or is a part of any rotating or moving machinery is often subject to forces that vary periodically with time. Because all metal machine parts have both mass and elasticity, the opportunity for vibration exists. Many machines do operate at fairly constant speeds and constant output, and it is not difficult to see that vibratory forces may exist which have a fairly constant amplitude over a period of time. Of course, these varying forces do change in magnitude when the machine speed or output changes, but there is a rather broad class of vibration problems that can be analyzed and corrected using the assumption of a periodically varying force of constant amplitude. Sometimes these forces exhibit a time characteristic that is very similar to that of a sine wave. At other times they are quite complex and have to be analyzed as a Fourier series. Such a series is the sum of a number of sine and cosine waves, and the resultant motion is the sum of the responses to the individual terms. In this text we shall study only the motion resulting from the application of a single sinusoidal force. In Section 17.11 we examined an undamped system subjected to a periodic force and discovered that the solution contained components of the motion at two frequencies. One of these components contained the forcing frequency, whereas the other contained the natural frequency. Actual systems always have damping present, and this causes the component at the (damped) natural frequency to become insignificant after a certain period of time; the motion that remains contains only the driving frequency and is termed the *steady-state motion*.

To illustrate steady-state motion, we shall solve the equation

$$m\ddot{x} + c\dot{x} + kx = F_0 \cos \omega t \tag{17.43}$$

The solution is in two parts, as before. The first, or complementary, part is the solution to the homogeneous equation

$$m\ddot{x} + c\dot{x} + kx = 0 \tag{17.44}$$

This we solved in Section 17.8, and we obtained [see Eq. (i)]

$$x' = e^{-\zeta \omega_n t}(C_1 \cos \omega_d t + C_2 \sin \omega_d t) \tag{17.45}$$

where C_1 and C_2 are the constants of integration. The particular part is obtained by assuming a solution in the form

$$x'' = A \cos \omega t + B \sin \omega t \tag{17.46}$$

where ω is the frequency of the driving function, and where A and B are constants yet to be determined, but are not constants of integration. Taking successive derivatives of Eq. (17.46) produces

$$\dot{x}'' = -A\omega \sin \omega t + B \cos \omega t \tag{a}$$

$$\ddot{x}'' = -A\omega^2 \cos \omega t + B\omega^2 \sin \omega t \tag{b}$$

and substituting Eqs. (17.46), (a), and (b) back into the differential equation (17.43) yields

$$(kA + c\omega B - m\omega^2 A) \cos \omega t + (kB + c\omega A - m\omega^2 B) \sin \omega t = F_0 \cos \omega t$$

Next, separating the terms, we obtain

$$(k - m\omega^2)A + c\omega B = F_0$$

$$-c\omega A + (k - m\omega^2)B = 0 \tag{c}$$

Solving Eqs. (c) for A and B and substituting the results back into Eq. (17.46), we obtain

$$x'' = \frac{F_0}{(k - m\omega^2)^2 + c^2\omega^2}[(k - m\omega^2) \cos \omega t + c\omega \sin \omega t] \tag{d}$$

Now, when we create the phase diagram (Fig. 17.30) and plot the amplitude of the two trigonometric terms, we find the resultant is

$$R = \sqrt{(k - m\omega^2)^2 + c^2\omega^2} \tag{e}$$

Figure 17.30

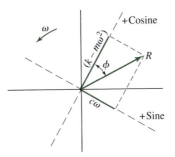

lagging the direction of the positive cosine by a phase angle of

$$\phi = \tan^{-1}\left(\frac{c\omega}{k - m\omega^2}\right) \tag{17.47}$$

Substituting Eq. (e) into Eq. (d) and canceling gives

$$x'' = \frac{F_0 \cos(\omega t - \phi)}{\sqrt{(k - m\omega^2)^2 + c^2\omega^2}} \tag{17.48}$$

and so the complete solution is

$$x = x' + x'' = e^{-\zeta \omega_n t}(C_1 \cos \omega_d t + C_2 \sin \omega_d t) + \frac{F_0 \cos(\omega t - \phi)}{\sqrt{(k - m\omega^2)^2 + c^2\omega^2}} \tag{17.49}$$

The first part of Eq. (17.49) is called the *transient* part because the exponential term will cause it to decay in a short period of time. When this happens the second term will be all that remains; and because this has neither a beginning nor an end, it is called the *steady-state solution*.

When we are interested only in the steady-state term, we shall write the solution in the form of Eq. (17.48), omitting the prime marks on x.

Now write Eq. (17.48) in the form

$$x = X \cos(\omega t - \phi) \tag{17.50}$$

and find the successive derivatives to be

$$\dot{x} = -\omega X \sin(\omega t - \phi) \tag{f}$$

$$\ddot{x} = -\omega^2 X \cos(\omega t - \phi) \tag{g}$$

These are then substituted into Eq. (17.43), and the forcing term is brought to the left-hand side of the equation. This gives

$$-m\omega^2 X \cos(\omega t - \phi) - c\omega X \sin(\omega t - \phi) + kX \cos(\omega t - \phi) - F_0 \cos \omega t = 0 \tag{h}$$

The amplitudes of the trigonometric terms are identified as follows:

$$m\omega^2 X = \text{inertia force}$$
$$c\omega X = \text{damping force}$$
$$kX = \text{spring force}$$
$$F_0 = \text{exciting force}$$

Our next step is to plot these phasors on the phase diagram. We are interested only in the steady-state solution; it has no beginning and no end, and so, although the phasors have a definite phase relation with each other, they can exist anywhere on the diagram at the instant we examine them.

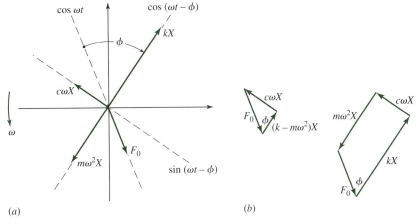

(a)   (b)

Figure 17.31

We begin Fig. 17.31a by selecting one of the directions, say, the direction of $\cos(\omega t - \phi)$. This automatically defines the direction of $\sin(\omega t - \phi)$. The direction of $\cos \omega t$ will be ahead of $\cos(\omega t - \phi)$ by the angle ϕ. Having defined this set of directions, we lay off each phasor in the proper direction, being sure not to forget the signs.

Because Eq. (h) states that the sum of these four phasors is zero, we can also construct a phasor or vector polygon of them. This has been done in Fig. 17.31b; from the trigonometry, we see

$$[(k - m\omega^2)^2 + c^2\omega^2]X^2 = F_0^2$$

or

$$X = \frac{F_0}{\sqrt{(k - m\omega^2)^2 + c^2\omega^2}} \tag{i}$$

and

$$\phi = \tan^{-1}\left(\frac{c\omega}{k - m\omega^2}\right) \tag{17.51}$$

Consequently, Eq. (17.50) becomes

$$x = \frac{F_0 \cos(\omega t - \phi)}{\sqrt{(k - m\omega^2)^2 + c^2\omega^2}} \tag{17.52}$$

These equations can be simplified by introducing the expressions

$$\omega_n = \sqrt{\frac{k}{m}}, \qquad \zeta = \frac{c}{c_c}, \qquad c_c = 2m\omega_n$$

They can then be written in the form

$$\frac{X}{F_0/k} = \frac{1}{\sqrt{\left(1 - \omega^2/\omega_n^2\right)^2 + (2\zeta\omega/\omega_n)^2}} \tag{17.53}$$

and

$$\phi = \tan^{-1}\left(\frac{2\zeta\omega/\omega_n}{1 - \omega^2/\omega_n^2}\right) \tag{17.54}$$

which is a dimensionless form that is more convenient for analysis. It is interesting to note that the quantity F_0/k is the deflection that a spring of rate k would experience if acted upon by the force F_0.

If various values are selected for the frequency and damping ratios, curves can be plotted showing the effects of varying these quantities upon the motion. This has been done in Figs. 17.32 and 17.33. Notice that the amplitude is very large near resonance (that is, where

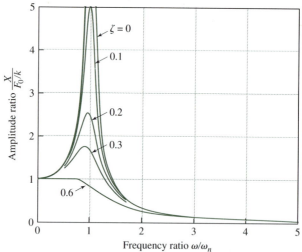

Figure 17.32 Relative displacement of a damped forced system as a function of the damping and frequency ratios.

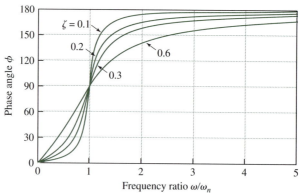

Figure 17.33 Relationship of the phase angle to the damping and frequency ratios.

ω/ω_n is near unity) for small damping ratios. Notice, too, that the peak amplitudes occur slightly before the resonance point is reached. Figure 17.33 also indicates that the phase angle changes very rapidly near the resonance condition.

17.13 FORCING CAUSED BY UNBALANCE

Quite frequently the exciting force in a machine (see Chapter 19) is due to the rapid rotation of a small unbalanced mass. In this case, the differential equation is written as

$$m\ddot{x} + c\dot{x} + kx = m_u e\omega^2 \cos \omega t \tag{17.55}$$

where m = vibrating mass
m_u = unbalanced mass
e = eccentricity

The solution to Eq. (17.55) is the same as that for Eq. (17.52) with F_0 replaced by $m_u e\omega^2$. Thus the motion of the main mass is

$$x = \frac{m_u e\omega^2 \cos(\omega t - \phi)}{\sqrt{(k - m\omega^2)^2 + c^2\omega^2}} \tag{17.56}$$

If we make the same transformation we used to obtain Eq. (17.53), we get

$$\frac{mX}{m_u e} = \frac{\omega^2/\omega_n^2}{\sqrt{\left[1 - \left(\omega^2/\omega_n^2\right)\right]^2 + (2\zeta\omega/\omega_n)^2}} \tag{17.57}$$

This is plotted in Fig. 17.34. This graph shows that for high speeds where the frequency ratio is greater than unity, we can reduce the amplitude only by reducing the mass and eccentricity of the rotating unbalance.

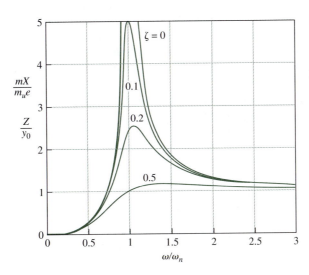

Figure 17.34 A plot of magnification factor versus frequency ratio for two systems. In one system the main mass is excited by a rotating unbalanced mass m_u. The chart gives the nondimensional absolute response of the main mass. In the other system the mass is excited by moving the far end of the spring and dashpot, which is connected to the mass, with harmonic motion. The chart gives the nondimensional response Z/y_0 relative to the far end of the spring–dashpot combination.

17.14 RELATIVE MOTION

In many cases—in vibration-measuring instruments, for example—it is useful to know the response of the system relative to a moving system. Thus, in Fig. 17.35, we might wish to learn the relative response z instead of the absolute response x.

The differential equation for this system is

$$m\ddot{x} + c\dot{x} + kx = ky + c\dot{y} \tag{17.58}$$

Note that because $z = x - y$ and $\ddot{z} = \ddot{x} - \ddot{y}$, in terms of z and y, Eq. (17.58) becomes

$$m\ddot{z} + c\dot{z} + kz = -m\ddot{y}$$

or

$$m\ddot{z} + c\dot{z} + kz = -m\omega^2 y_0 \cos \omega t \tag{17.59}$$

So the solution is the same as that of Eq. (17.55), and we can write

$$\frac{Z}{y_0} = \frac{\omega^2/\omega_n^2}{\sqrt{\left[1 - \left(\omega^2/\omega_n^2\right)\right]^2 + (2\zeta\omega/\omega_n)^2}} \tag{17.60}$$

The response chart of Fig. 17.34 also applies to the relative response of such a system.

17.15 ISOLATION

The investigations so far enable the engineer to design mechanical equipment so as to minimize vibration and other dynamic problems; but because machines are inherently vibration generators, in many cases it is impractical to eliminate all such motion. In such cases, a more practical and economical vibration approach is that of reducing as much as possible the annoyances the motion causes. This may take any of several directions, depending upon the nature of the problem. For example, everything economically feasible may have been done toward elimination of vibrations in a machine, but the residuals may still be strong and cause objectionable noise by transmitting these vibrations to the base structure. Again, an item of equipment, such as a computer monitor, which is not of itself a vibration generator, may receive objectionable vibrations from another source. Both of these problems can be solved by isolating the equipment from the support. In analyzing these problems we may be interested, therefore, in the *isolation of forces* or in the *isolation of motions,* as shown in Fig. 17.36.

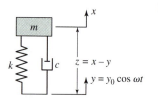

Figure 17.35

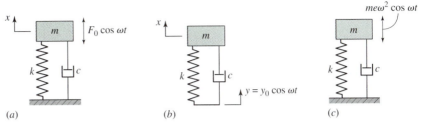

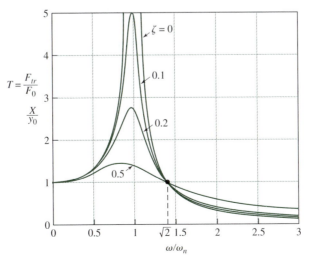

Figure 17.36 In (*a*) the exciting-force amplitude is constant, but in (*c*) it depends upon ω because it is due to rotating unbalance. In both (*a*) and (*c*) the support is to be isolated from the vibrating mass.

In Figs. 17.36*a* and 17.36*c* the forces acting against the support must be transmitted through the spring and dashpot because these are the only connections. The spring and damping forces are phasors that are always at right angles to each other, and so the force transmitted through these connections is

$$F_u = \sqrt{(kX^2) + (c\omega X)^2} = kX\sqrt{1 + \left(\frac{2\zeta\omega}{\omega_n}\right)^2} \qquad (17.61)$$

The *transmissibility T* is a nondimensional ratio that defines the percentage of the exciting force transmitted to the frame. For the system of Fig. 17.36*a*, X is given by Eq. (17.53), and so the transmissibility is

$$T = \frac{F_u}{F_0} = \frac{\sqrt{1 + (2\zeta\omega/\omega_n)^2}}{\sqrt{\left[1 - (\omega^2/\omega_n^2)\right]^2 + (2\zeta\omega/\omega_n)^2}} \qquad (17.62)$$

Figure 17.37 shows that the isolator must be designed so that ω/ω_n is large and so that the damping is small.

Figure 17.37 This is a plot of force transmissibility versus frequency ratio for a system in which a steady-state periodic forcing function is applied directly to the mass. The transmissibility is the percentage of the exciting force that is transmitted to the frame. This is also a plot of the amplitude–magnification factor for a system in which the far end of a spring and dashpot are driven by a sinusoidal displacement of amplitude y_0. The graph gives the nondimensional absolute response of the mass.

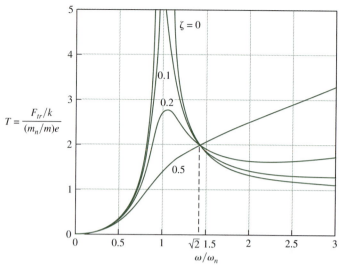

Figure 17.38 A plot of acceleration transmissibility versus frequency ratio. In a system in which the exciting force is produced by a rotating unbalanced mass, this plot gives the percentage of this force transmitted to the frame of the machine.

By using Eqs. (17.57) and (17.61), we find that the transmissibility for the system of Fig. 17.36c is

$$T = \frac{F_{tr}/k}{em_n/m} = \frac{(\omega^2/\omega_n^2)\sqrt{1 + (2\zeta\omega/\omega_n)^2}}{\sqrt{[1 - (\omega^2/\omega_n^2)]^2 + (2\zeta\omega/\omega_n)^2}} \tag{17.63}$$

The plot of Fig. 17.38 is discouraging. For large values of ω/ω_n the damping actually serves to increase the transmissibility. The only practical solution, in general, is to balance the equipment as perfectly as possible in order to minimize the product em_n.

We shall choose the complex-operator method for the solution of the system of Fig. 17.36b. Note that this system is the same as that of Section 17.14, except that now we are interested in the absolute motion of the mass. We begin by defining the forcing function as

$$y = y_0 e^{j\omega t} \tag{a}$$

Then

$$\dot{y} = j\omega y_0 e^{j\omega t} \tag{b}$$

Also, assuming a solution in the form

$$x = X e^{j\omega t} \tag{c}$$

gives for the velocity and acceleration

$$\dot{x} = j\omega X e^{j\omega t} \quad \text{and} \quad \ddot{x} = -\omega^2 X e^{j\omega t} \tag{d}$$

Substituting all these into Eq. (17.58) and canceling the exponential produces

$$-m\omega^2 X + jc\omega X + kX = ky_0 + jc\omega y_0 \tag{e}$$

and so

$$\frac{X}{y_0} = \frac{k + jc\omega}{(k - m\omega^2) + jc\omega} \tag{f}$$

The product of two complex numbers is a new quantity whose magnitude is the product of the magnitudes of the original numbers and whose angle is the sum of the angles of the original numbers. Complex numbers can be divided too; in this case, divide their magnitudes and subtract their angles.

In this case we are not interested in the phase and so Eq. (f) becomes

$$\frac{X}{y_0} = \frac{\sqrt{k^2 + c^2\omega^2}}{\sqrt{(k - m\omega^2)^2 + c^2\omega^2}} \tag{g}$$

Dividing both numerator and denominator by k, we finally obtain

$$\frac{X}{y_0} = \frac{\sqrt{1 + (2\zeta\omega/\omega_n)^2}}{\sqrt{\left[1 - (\omega^2/\omega_n^2)\right]^2 + (2\zeta\omega/\omega_n)^2}} \tag{17.64}$$

which is plotted in Fig. 17.37.

17.16 RAYLEIGH'S METHOD

The static deflection y_0 of a cantilever due to a concentrated weight W on the end is

$$y_0 = \frac{Wl^3}{3EI} \tag{a}$$

where l is the length of the cantilever, E is its modulus of elasticity, and I is its second moment of area. Because spring constant is defined as the force required to produce unit deflection, Eq. (a) can be arranged to

$$k = \frac{W}{y_0} = \frac{3EI}{l^3} \tag{b}$$

Substituting this value for k in the equation for natural frequency produces

$$\omega_n = \sqrt{\frac{k}{m}} = \sqrt{\frac{3EI/l^3}{W/g}} = \sqrt{\frac{3EIg}{Wl^3}} \tag{c}$$

so that

$$f = \frac{1}{2\pi}\sqrt{\frac{3EIg}{Wl^3}} \tag{d}$$

Next, note that for any single-degree-of-freedom system, the quantity W/k represents the static deflection of a spring due to the weight W acting upon it. Rearranging the natural frequency equation by substituting $m = W/g$ gives

$$\omega_n = \sqrt{\frac{k}{m}} = \sqrt{\frac{g}{W/k}} = \sqrt{\frac{g}{y_0}} \tag{17.65}$$

where y_0 is the static deflection due to the weight W acting upon the spring. If we substitute into Eq. (17.65) the value of y_0 from Eq. (*a*), we obtain

$$\omega_n = \sqrt{\frac{g}{Wl^3/3EI}} = \sqrt{\frac{3EIg}{Wl^3}} \tag{e}$$

which agrees with Eq. (*c*). Equation (17.65) is, incidentally, very useful for determining natural frequencies of mechanical systems, because static deflections can usually be measured quite easily.

In our investigation of the mechanics of vibration we have been concerned with systems whose motions can be described using a single coordinate. In the cases of vibrating beams and rotating shafts, there may be many masses involved or the mass may be distributed. A coupled set of differential equations must be written, with one equation for each mass or element of mass of the system, and these equations must be solved simultaneously if we are to obtain the equations of motion of any multimass system. While a number of optional approaches are available, here we shall present an energy method, which is due to Lord Rayleigh,[3] because of its importance in the study of vibration.

It is probable that Eq. (17.65) first suggested to Rayleigh the idea of employing the static deflection to find the natural frequency of a system. If we consider a freely vibrating system without damping, then, during motion, no energy is added to the system, nor is any taken away. Yet when the mass has velocity, kinetic energy exists, and when the spring is compressed or extended, potential energy exists. Because no energy is added or taken away, the maximum kinetic energy of a system must be the same as the maximum potential energy. This is the basis of Rayleigh's method; a mathematical statement of it is

$$u_{K,\max} = u_{P,\max} \tag{17.66}$$

In order to see how it works, let us apply the method to the simple cantilever. We begin by assuming that the motion is harmonic and of the form

$$y = y_0 \sin \omega_n t \tag{f}$$

The potential energy is a maximum when the spring is fully extended or compressed and occurs when $\sin \omega_n t = 1$. It is

$$u_{P,\max} = \tfrac{1}{2} W y_0 \tag{g}$$

The velocity of the weight is given by

$$\dot{y} = y_0 \omega_n \cos \omega_n t \tag{h}$$

The kinetic energy reaches a maximum when the velocity is a maximum, that is, when $\cos \omega_n t = 1$. The kinetic energy is

$$u_{K,\max} = \frac{1}{2} m v_{\max}^2 = \frac{1}{2} \frac{W}{g} (y_0 \omega_n)^2 \qquad (i)$$

Applying Eq. (17.66), we obtain

$$\frac{1}{2} \frac{W}{g} (y_0 \omega_n)^2 = \frac{1}{2} W y_0 \quad \text{or} \quad \omega_n = \sqrt{\frac{g}{y_0}} \qquad (j)$$

which is identical with Eq. (17.65).

It is true in multimass systems that the dynamic deflection curves are not the same as the static deflection curves. The importance of the method, however, is that any *reasonable* deflection curve can be used in the process. The static deflection curve is a reasonable one, and hence it gives a good approximation. Rayleigh also shows that the correct curve will always give the lowest value for the natural frequency (a lower bound), though we shall not demonstrate this fact here. It is sufficient to know that if many deflection curves are assumed, the one giving the lowest natural frequency is the best.

Rayleigh's method is applied to a multimass system composed of weights W_1, W_2, W_3, and so on, by assuming, as before, a deflection of each mass according to the equation

$$y_i = y_{0i} \sin \omega_n t \qquad (k)$$

The maximum deflections are therefore y_{0_1}, y_{0_2}, y_{0_3}, and so on, and the maximum velocities are $y_{0_1} \omega_n$, $y_{0_2} \omega_n$, $y_{0_3} \omega_n$, and so on. The maximum potential energy for the system is

$$u_{P,\max} = \tfrac{1}{2} W_1 y_{01} + \tfrac{1}{2} W_2 y_{02} + \tfrac{1}{2} W_2 y_{03} + \cdots$$

$$= \frac{1}{2} \sum W_i y_{0i} \qquad (l)$$

The maximum kinetic energy for the system is

$$u_{K,\max} = \frac{1}{2g} W_1 y_{01}^2 \omega_n^2 + \frac{1}{2g} W_2 y_{02}^2 \omega_n^2 + \frac{1}{2g} W_3 y_{03}^2 \omega_n^2 + \cdots$$

$$= \frac{\omega_n^2}{2g} \sum W_i y_{0i}^2 \qquad (m)$$

Applying Rayleigh's principle and solving for the frequency yields

$$\omega_n = \sqrt{\frac{g \sum W_i y_{0i}}{\sum W_i y_{0i}^2}} \quad \text{or} \quad f = \frac{1}{2\pi} \sqrt{\frac{g \sum W_i y_{0i}}{\sum W_i y_{0i}^2}} \qquad (17.67)$$

Equation (17.67), commonly called the Rayleigh-Ritz equation, can be applied to a beam consisting of several masses or to a rotating shaft having mounted upon it masses in the form of gears, pulleys, flywheels, and the like. If the speed of the rotating shaft should become equal to the natural frequency given by Eq. (17.67), then violent vibrations will occur. For this reason it is common practice to designate this frequency as the *critical speed*.

17.17 FIRST AND SECOND CRITICAL SPEEDS OF A SHAFT

This section presents an approach to obtain exact solutions to the first and second critical speeds of a rotating shaft using the method of *influence coefficients*.

The deflection at location i of the shaft due to a load of unit magnitude at location j is denoted as a_{ij} and is referred to as an *influence coefficient*. Note the reciprocal relationship between the definition of an influence coefficient and the definition of spring stiffness, that is,

$$a_{ij} = \frac{1}{k_{ij}} \tag{17.68}$$

So for a shaft of negligible mass supporting two disks (for example, gears, or flywheels), the influence coefficients are as follows:

$a_{11} =$ the deflection at location 1 due to a load of unit magnitude at location 1
$a_{12} =$ the deflection at location 1 due to a load of unit magnitude at location 2
$a_{21} =$ the deflection at location 2 due to a load of unit magnitude at location 1
$a_{22} =$ the deflection at location 2 due to a load of unit magnitude at location 2

Maxwell's Reciprocity Theorem   This theorem states that the deflection at location i due to a unit load at location j is equal to the deflection at location j due to a unit load at location i.

$$a_{ij} = a_{ji} \tag{17.69}$$

The total deflection at location 1 due to both disks can be written, from superposition, as

$$x_1 = a_{11}W_1 + a_{12}W_2 \tag{17.70}$$

Similarly, the total deflection at location 2 due to both disks can be written as

$$x_2 = a_{21}W_1 + a_{22}W_2 \tag{17.71}$$

Therefore, the total deflections at locations 1 and 2 can be written from Eqs. (17.70) and (17.71) as

$$\begin{aligned} x_1 &= a_{11}(m_1 x_1 \omega^2) + a_{12}(m_2 x_2 \omega^2) \\ x_2 &= a_{21}(m_1 x_1 \omega^2) + a_{22}(m_2 x_2 \omega^2) \end{aligned} \tag{17.72}$$

Rearranging these two equations and writing them in matrix form gives

$$\begin{bmatrix} a_{11}m_1\omega^2 - 1 & a_{12}m_2\omega^2 \\ a_{21}m_1\omega^2 & a_{22}m_2\omega^2 - 1 \end{bmatrix} \begin{bmatrix} x_1 \\ x_2 \end{bmatrix} = \begin{bmatrix} 0 \\ 0 \end{bmatrix} \tag{17.73}$$

A nontrivial solution requires that the determinant of the (2×2) coefficient matrix must be zero, that is

$$(a_{11}m_1\omega^2 - 1)(a_{22}m_2\omega^2 - 1) - (a_{12}m_2\omega^2)(a_{21}m_1\omega^2) = 0 \qquad (17.74a)$$

which can be written as

$$m_1m_2(a_{11}a_{22})\omega^4 - (a_{11}m_1 + a_{22}m_2\omega^2) + 1 - m_1m_2(a_{12}a_{21})\omega^4 = 0 \qquad (17.74b)$$

Rearranging this equation and dividing by ω^4, we have

$$1/\omega^4 - (a_{11}m_1 + a_{22}m_2)1/\omega^2 + m_1m_2(a_{11}a_{22} - a_{12}a_{21}) = 0 \qquad (17.75)$$

Equation (17.75) is commonly referred to as the *frequency equation*. If we substitute $X = 1/\omega^2$ then we have a quadratic equation in X, that is

$$X^2 - (a_{11}m_1 + a_{22}m_2)X + m_1m_2(a_{11}a_{22} - a_{12}a_{21}) = 0 \qquad (17.76)$$

Then the two solutions for X can be written as

$$X_1, X_2 = \frac{(a_{11}m_1 + a_{22}m_2) \pm \sqrt{(a_{11}m_1 + a_{22}m_2)^2 - 4m_1m_2(a_{11}a_{22} - a_{12}a_{21})}}{2} \qquad (17.77)$$

This equation is commonly referred to as the *exact equation* for the first two critical speeds of the system.

This procedure can easily be extended to a rotating shaft supporting n disks. In this case, Eq. (17.73) would be an $(n \times n)$ matrix and Eq. (17.76) would be an nth-order polynomial in ω^2. The n critical speeds of the system could then be determined from this polynomial.

Dunkerley's Method   The Dunkerley equation is an approximation for the first critical speed of a rotating shaft supporting mass disks. Note that the coefficient of X in Eq. (17.76) is the sum of the roots of the polynomial, therefore; for two mass disks

$$\frac{1}{\omega_1^2} + \frac{1}{\omega_2^2} = a_{11}m_1 + a_{22}m_2 \qquad (17.78)$$

To obtain an approximation to the first critical speed assume that

$$\frac{1}{\omega_1^2} > \frac{1}{\omega_2^2} \qquad (17.79)$$

Therefore, Eq. (17.78) can be written as

$$\frac{1}{\omega_1^2} = a_{11}m_1 + a_{22}m_2 \qquad (17.80)$$

From Eq. (17.80), the Dunkerley approximation for the first critical speed can be written as

$$\omega_1 = \frac{1}{\sqrt{a_{11}m_1 + a_{22}m_2}} \qquad (17.81)$$

It is interesting to note that, in general, Rayleigh's method will overestimate the first critical speed and Dunkerley's method will underestimate the first critical speed.

EXAMPLE 17.5

A steel shaft that is 4 ft long and has 0.5 in diameter is simply supported in two bearings as shown in the Fig. 17.39. Two 80-lb gears are attached to the shaft. One gear is 12 in to the right of the left bearing, and the other is 12 in to the left of the right bearing. The mass of the shaft is negligible compared to the mass of the gears. The influence coefficients are known to be $a_{11} = a_{22} = 5.504 \times 10^{-5}$ in·lb^{-1} and $a_{12} = a_{21} = 4.278 \times 10^{-5}$ in·lb^{-1}. Determine the first critical speed of the system using (i) the Rayleigh equation, (ii) the exact equation, and (iii) the Dunkerley equation.

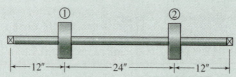

Figure 17.39 Example 17.5

SOLUTION

(i) Rayleigh's equation, see Eq. (17.67), can be written as

$$\omega_1^2 = \frac{g(W_1 x_1 + W_2 x_2)}{W_1 x_1^2 + W_2 x_2^2} \qquad (1)$$

Because the two gears have the same weight $W_1 = W_2$, and Eq. (1) can be written

$$\omega_1^2 = \frac{g(x_1 + x_2)}{x_1^2 + x_2^2} \qquad (2)$$

The total deflections at locations 1 and 2 from Eqs. (17.70) and (17.71) are

$$x_1 = a_{11}W_1 + a_{12}W_2 \qquad (3)$$
$$x_2 = a_{21}W_1 + a_{22}W_2 \qquad (4)$$

Note from symmetry that the deflection is $x_1 = x_2 = x$; therefore, Eq. (2) can be written

$$\omega_1^2 = \frac{g}{x} \qquad (5)$$

Substituting the given data into Eq. (3) or Eq. (4) gives

$$x = x_1 = x_2 = 80(a_{11} + a_{12}) \text{ in} \tag{6a}$$

Therefore, the deflection is

$$x = 80(5.504 + 4.278) \times 10^{-5} = 782.56 \times 10^{-5} \text{ in} \tag{6b}$$

Substituting Eq. (6b) into Eq. (5) gives

$$\omega_1^2 = \frac{386 \times 10^5}{782.56} = 49\,325 \text{ rad}^2/\text{s}^2$$

Therefore, the first critical speed from Rayleigh's method is

$$\omega_1 = 222 \text{ rad/s} \qquad\qquad Ans.$$

(ii) From Eq. (17.77), the exact equation for the first two critical speeds of the system is

$$X_1, X_2 = \frac{(a_{11}m_1 + a_{22}m_2) \pm \sqrt{(a_{11}m_1 + a_{22}m_2)^2 - 4m_1m_2(a_{11}a_{22} - a_{12}a_{21})}}{2} \tag{7}$$

Substituting the given data into this equation gives

$$X_1, X_2 = \frac{2.282 \times 10^{-5} \pm \sqrt{(2.282^2 - 2.060) \times 10^{-10}}}{2} \text{ s}^2 \tag{8a}$$

which can be written as

$$\frac{1}{\omega_1^2}, \frac{1}{\omega_2^2} = (1.141 \pm 0.887) \times 10^{-5} \text{ s}^2 \tag{8b}$$

or as

$$\omega_1^2 = 49\,310 \text{ rad}^2/\text{s}^2 \quad \text{and} \quad \omega_2^2 = 393\,701 \text{ rad}^2/\text{s}^2 \tag{8c}$$

Therefore, the first and second critical speeds from the exact equation are

$$\omega_1 = 222 \text{ rad/s} \quad \text{and} \quad \omega_2 = 627 \text{ rad/s} \qquad\qquad Ans.$$

Note that for this example the answer for the first critical speed from the Rayleigh equation is the same as the answer from the exact equation.

(iii) The Dunkerley equation for the rotating shaft supporting the two mass discs is

$$\frac{1}{\omega_1^2} = a_{11}m_1 + a_{22}m_2 \tag{9}$$

Substituting the given data into this equation gives

$$\frac{1}{\omega_1^2} = 2 \left(\frac{80}{386} \right) 5.500 \times 10^{-5} = 2.280 \times 10^{-5} \text{ s}^2 \tag{10}$$

Therefore, the first critical speed for the system is

$$\omega_1 = 209 \text{ rad/s} \qquad \qquad Ans.$$

Note that this answer is less than the first critical speed obtained from the exact equation.

EXAMPLE 17.6

The rotating shaft shown in the Fig. 17.40 supports two flywheels each of mass m. The first critical speed of the system has been found experimentally to be 300 rad/s and the first critical speed of the shaft alone is known to be 600 rad/s. If one of the flywheels is removed from the shaft then determine the first critical speed of the system.

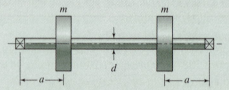

Figure 17.40 Example 17.6

SOLUTION

The first natural frequency of the shaft carrying the two flywheels can be written as

$$\frac{1}{\omega_1^2} = \frac{1}{\omega_{11}^2} + \frac{1}{\omega_{22}^2} + \frac{1}{\omega_{ss}^2} \tag{1a}$$

or as

$$\frac{1}{\omega_1^2} = ma_{11} + ma_{22} + \frac{1}{\omega_{ss}^2} \tag{1b}$$

where ω_{ss} is the critical speed of the shaft alone.

From symmetry we note that the influence coefficient $a_{11} = a_{22}$; therefore Eq. (1b) can be written as

$$2ma_{11} = \frac{1}{\omega_1^2} - \frac{1}{\omega_{ss}^2} \tag{2}$$

Substituting the given critical speeds into Eq. (2) gives

$$2ma_{11} = \frac{1}{(300)^2} - \frac{1}{(600)^2} \, \text{s}^2 \qquad (3)$$

For the massless shaft with only one flywheel, the first natural frequency can be written from Eq. (1a) as

$$\frac{1}{\omega_1^2} = \frac{1}{\omega_{11}^2} + \frac{1}{\omega_{ss}^2} \qquad (4)$$

or as

$$\frac{1}{\omega_1^2} = ma_{11} + \frac{1}{\omega_{ss}^2} \qquad (5)$$

Substituting Eq. (3) into Eq. (5) gives

$$\frac{1}{\omega_1^2} = \frac{1}{2} \left[\frac{1}{300^2} - \frac{1}{600^2} \right] + \frac{1}{\omega_{ss}^2} \qquad (6)$$

Finally, substituting the critical speed of the shaft $\omega_{ss} = 600$ rad/s into Eq. (6) gives

$$\frac{1}{\omega_1^2} = \frac{1}{2} \left[\frac{1}{300^2} + \frac{1}{600^2} \right] = 0.06944 \times 10^{-4} \, \text{s}^2 \qquad (7)$$

Therefore, the first critical speed of the system is

$$\omega_1 = 379.5 \text{ rad/s} \qquad \qquad \textit{Ans.}$$

In the investigation of rectilinear vibrations to follow, we shall show that there are as many natural frequencies in a multimass vibrating system as there are degrees of freedom. Similarly, a three-mass torsional system will have three degrees of freedom and, consequently, three natural frequencies. Equation (17.67) gives only the first or lowest of these frequencies.

It is probable that the critical speed can be determined by Eq. (17.67), for most cases, within about 5 percent. This is because of the assumption that the static and dynamic deflection curves are identical. Bearings, couplings, belts, and so on, all have an effect upon the spring constant and the damping. Shaft vibration usually occurs over an appreciable range of shaft speed, and for this reason Eq. (17.67) is accurate enough for many engineering purposes.

As a general rule, shafts have many discontinuities—such as shoulders, keyways, holes, and grooves—to locate and secure various gears, pulleys, and other shaft-mounted masses. These diametral changes usually require deflection analysis using numerical integration. For such problems, Simpson's rule is easy to apply using a computer, or calculator, or by hand calculations.[4]

17.18 TORSIONAL SYSTEMS

Figure 17.41 shows a shaft supported on bearings at A and B with two masses connected at the ends. The masses represent any rotating machine parts—an engine and its flywheel, for example. We wish to study the possibilities of free vibration of the system when it rotates at constant angular velocity. In order to investigate the motion of each mass, it is necessary to picture a reference system fixed to the shaft and rotating with the shaft at the same angular velocity. Then we can measure the angular displacement of either mass by finding the instantaneous angular location of a mark on the mass relative to one of the rotating axes. Thus we define θ_1 and θ_2 as the angular displacements of mass 1 and mass 2, respectively, with respect to the rotating axes.

Now, assuming no damping, Eq. (17.6) is written for each mass:

$$I_1\ddot{\theta}_1 + k_t(\theta_1 - \theta_2) = 0$$
$$I_2\ddot{\theta}_2 + k_t(\theta_2 - \theta_1) = 0 \tag{a}$$

where the angle $(\theta_1 - \theta_2)$ represents the total twist of the shaft. A solution is assumed in the form

$$\theta_i = \gamma_i \sin \omega_n t \tag{b}$$

Then, the acceleration is

$$\ddot{\theta}_i = -\gamma_i \omega_n^2 \sin \omega_n t \tag{c}$$

If these equations are substituted into Eqs. (a), we obtain

$$\left(k_t - I_1\omega_n^2\right)\gamma_1 - k_t\gamma_2 = 0$$
$$-k_t\gamma_1 + \left(k_t - I_2\omega_n^2\right)\gamma_2 = 0 \tag{d}$$

which can be written in matrix form as

$$\begin{bmatrix} k_t - I_1\omega_n^2 & -k_t \\ -k_t & k_t - I_2\omega_n^2 \end{bmatrix} \begin{bmatrix} \gamma_1 \\ \gamma_2 \end{bmatrix} = \begin{bmatrix} 0 \\ 0 \end{bmatrix} \tag{e}$$

For a nontrivial solution, the determinant of the coefficient matrix must be zero; that is,

$$\left(k_t - I_1\omega_n^2\right)\left(k_t - I_2\omega_n^2\right) - k_t^2 = 0 \tag{f}$$

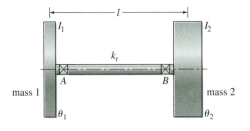

Figure 17.41

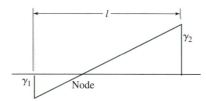

Figure 17.42

so that

$$\omega_n^2 \left(\omega_n^2 - k_t \frac{I_1 + I_2}{I_1 I_2} \right) = 0 \tag{g}$$

and

$$\omega_{n1} = 0 \quad \text{and} \quad \omega_{n2} = \sqrt{k_t \frac{I_1 + I_2}{I_1 I_2}} \tag{17.82}$$

Substituting $\omega_{n1} = 0$ into Eqs. (d), we find that $\gamma_1 = \gamma_2$. Therefore the masses rotate together without relative displacement and there is no vibration.

Substituting the value of ω_{n2} into Eqs. (d), it is found that

$$\gamma_1 = -\frac{I_2}{I_1}\gamma_2 \tag{h}$$

indicating that the motion of the second mass is opposite to that of the first and proportional to I_1/I_2. Figure 17.42 is a graphical representation of the vibration. Here l is the distance between the two masses and γ_1 and γ_2 are the instantaneous angular displacements plotted to scale. If a line is drawn connecting the ends of the angular displacements, this line crosses the axis at a *node,* which is a location on the shaft having zero angular displacement. It is convenient to designate the configuration of the vibrating system as a *mode* of vibration. This system has two modes of vibration, corresponding to the two frequencies, although we have seen that one of them is degenerate. A system of n masses would have n modes corresponding to n different frequencies.

For multimass torsional systems, the *Holzer tabulation method* can be used to find all the natural frequencies. It is easy to use and avoids solving the simultaneous differential equations.[5]

NOTES

[1.]   P. W. Bridgman, *The Logic of Modern Physics,* Macmillan, New York, 1946, p. 62.

[2.]   The motion can also be expressed in the form

$$x = X_0 \sin(\omega_n t + \psi)$$

where X_0 and ψ are the constants of integration. This is probably not a very good way of expressing it, however, because, in the study of forced vibration, it might imply that the output can lead the input, which, of course, is not possible.

3.   John William Strutt (Baron Rayleigh), *Theory of Sound,* republished by Dover, New York, 1945. This book is in two volumes and is the classic treatise on the theory of vibrations. It was originally published in 1877–1878.

4.   See J. E. Shigley and C. R. Mischke, *Mechanical Engineering Design,* 5th ed., McGraw-Hill, New York, 1989, pp. 101–105, for methods and examples of such calculations.

5.   A detailed description of this method with a worked-out numerical example can be found in T. S. Sankar and R. B. Bhat, "Viration and Control of Vibration," in J. Shigley and C. R. Mischke (eds.), *Standard Handbook of Machine Design,* McGraw-Hill, New York, 1986, pp. 36.22–38.27.

PROBLEMS

17.1   Derive the differential equation of motion for each of the systems shown in the figure and write the formula for the natural frequency ω_n for each system.

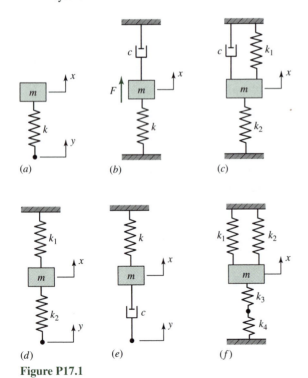

(a)   (b)   (c)

(d)   (e)   (f)

Figure P17.1

17.2   Evaluate the constants of integration of the solution to the differential equation for an undamped free system, using the following sets of starting conditions:

(a) $x = x_0$, $\dot{x} = 0$
(b) $x = 0$, $\dot{x} = v_0$
(c) $x = x_0$, $\ddot{x} = a_0$
(d) $x = x_0$, $\dddot{x} = b_0$

For each case, transform the solution to a form containing a single trigonometric term.

17.3   A system like Fig. 17.5 has $m = 1$ kg and an equation of motion $x = 20\cos(8\pi t - \pi/4)$ mm. Determine:

(a) The spring constant k
(b) The static deflection δ_{st}
(c) The period
(d) The frequency in hertz
(e) The velocity and acceleration at the instant $t = 0.20$ s
(f) The spring force at $t = 0.20$ s

Plot a phase diagram to scale showing the displacement, velocity, acceleration, and spring-force phasors at the instant $t = 0.20$ s.

17.4   The weight W_1 in the figure drops through the distance h and collides with W_2 with inelastic impact (a coefficient of restitution of zero). Derive the differential equation of motion of the system, and determine the amplitude of the resulting motion of W_2.

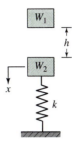

Figure P17.4

17.5 The vibrating system shown in the figure has $k_1 = k_3 = 875$ N/m, $k_2 = 1\,750$ N/m, and $W = 40$ N. What is the natural frequency in hertz?

Figure P17.5

17.6 The figure shows a weight $W = 15$ lb connected to a pivoted rod which is assumed to be weightless and very rigid. A spring having a rate of $k = 60$ lb/in is connected to the center of the rod and holds the system in static equilibrium at the position shown. Assuming that the rod can vibrate with a small amplitude, determine the period of the motion.

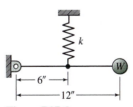

Figure P17.6

17.7 The figure shows an upside-down pendulum of length l retained by two springs connected a distance a from the pivot. The springs have been positioned such that the pendulum is in static equilibrium when it is in the vertical position.
(a) For small amplitudes, find the natural frequency of this system.

(b) Find the ratio l/a at which the system becomes unstable.

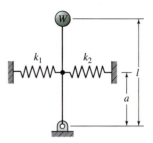

Figure P17.7

17.8 (a) Write the differential equation for the system shown in the figure and find the natural frequency.
(b) Find the response x if y is a step input of height y_0.
(c) Find the relative response $z = x - y$ to the step input of part (b).

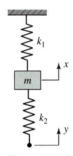

Figure P17.8

17.9 An undamped vibrating system consists of a spring whose scale is 35 kN/m and a mass of 1.2 kg. A step force $F = 50$ N is exerted on the mass for 0.040 s.
(a) Write the equations of motion of the system for the era in which the force acts and for the era after.
(b) What are the amplitudes in each era?
(c) Sketch a displacement time plot of the motion.

17.10 The figure shows a round shaft whose torsional spring constant is k_t lb·in/rad connecting two wheels having the mass moments of inertia I_1 and I_2. Show that the system is capable of vibrating torsionally with a frequency of

$$\omega_n = \left[\frac{k_t(I_1 + I_2)}{I_1 I_2}\right]^{1/2}$$

Figure P17.10

17.11 A motor is connected to a flywheel by a 5/8-in-diameter steel shaft 36 in long, as shown. Using the methods of this chapter, it can be demonstrated that the torsional spring constant of the shaft is 4 700 lb·in/rad. The mass moments of inertia of the motor and flywheel are 24.0 and 56.0 lb·s²·in, respectively. The motor is turned on for 2 s, and during this period it exerts a constant torque of 200 lb·in on the shaft.
   (a) What speed in revolutions per minute does the shaft attain?
   (b) What is the natural circular frequency of vibration of the system?
   (c) Assuming no damping, what is the amplitude of the vibration of the system indegrees during the first era? During the second era?

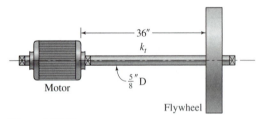

Figure P17.11

17.12 The weight of the mass of a vibrating system is 10 lb, and it has a natural frequency of 1 Hz. Using the phase-plane method, plot the response of the system to the force function shown in the figure. What is the final amplitude of the motion?

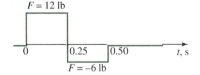

Figure P17.12

17.13 An undamped vibrating system has a spring scale of 200 lb/in and a weight of 50 lb. Find the response and the final amplitude of vibration of the system if it is acted upon by the forcing function shown in the figure. Use the phase-plane method.

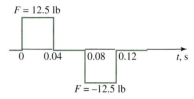

Figure P17.13

17.14 A vibrating system has a spring $k = 400$ lb/in and a weight of $W = 80$ lb. Plot the response of this system to the forcing function shown in the figure:
   (a) Using three steps
   (b) Using six steps

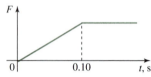

Figure P17.14 $F_{max} = 200$ lb

17.15 (a) What is the value of the coefficient of critical damping for a spring–mass–damper system in which $k = 56$ kN/m and $m = 40$ kg?
   (b) If the actual damping is 20 percent of critical, what is the natural circular frequency of the system?
   (c) What is the period of the damped system?
   (d) What is the value of the logarithmic decrement?

17.16 A vibrating system has a spring $k = 3.5$ kN/m and a mass $m = 15$ kg. When disturbed, it was observed that the amplitude decayed to one-fourth of its original value in 4.80 s. Find the damping coefficient and the damping factor.

17.17 A vibrating system has $k = 300$ lb/in, $W = 90$ lb, and damping equal to 20 percent of critical.
   (a) What is the damped natural frequency ω_d of the system?
   (b) What are the period and the logarithmic decrement?

17.18 Solve Problem 17.14 using damping equal to 15 percent of critical.

17.19 A damped vibrating system has an undamped natural frequency of 10 Hz and a weight of 800 lb. The damping ratio is 0.15. Using the phase-plane method, determine the response of the system to the forcing function shown in the figure.

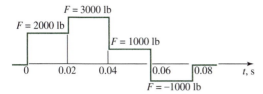

Figure P17.19

17.20 A vibrating system has a spring rate of 3 000 lb/in, a damping factor of 55 lb·s/in, and a weight of 800 lb. It is excited by a harmonically varying force $F_0 = 100$ lb at a frequency of 435 cycles per minute.

(*a*) Calculate the amplitude of the forced vibration and the phase angle between the vibration and the force.

(*b*) Plot several cycles of the displacement-time and force-time diagrams.

17.21 A spring-mounted mass has $k = 525$ kN/m, $c = 9640$ N·s/m, and $m = 360$ kg. This system is excited by a force having an amplitude of 450 N at a frequency of 4.80 Hz. Find the amplitude and phase angle of the resulting vibration and plot several cycles of the force-time and displacement-time diagrams.

17.22 When a 6 000-lb press is mounted upon structural-steel floor beams, it causes them to deflect 0.75 in. If the press has a reciprocating unbalance of 420 lb and it operates at a speed of 80 rev/min, how much of the force will be transmitted from the floor beams to other parts of the building? Assume no damping. Can this mounting be improved?

17.23 Four vibration mounts are used to support a 450-kg machine that has a rotating unbalance of 0.35 kg·m and runs at 300 rev/min. The vibration mounts have damping equal to 30 percent of critical. What must the spring constant of the mounting be if 20 percent of the exciting force is transmitted to the foundation? What is the resulting amplitude of motion of the machine?

18 | Dynamics of Reciprocating Engines

The purpose of this chapter is to apply fundamentals—kinematic and dynamic analysis—in a complete investigation of a particular group of machines. The reciprocating engine has been selected for this purpose because it has reached a high state of development and is of more general interest than other machines. For our purposes, however, any machine or group of machines involving interesting dynamic situations would serve just as well. The primary objective is to demonstrate methods of applying fundamentals to the analysis of machines.

18.1 ENGINE TYPES

The description and characteristics of all the engines that have been conceived and constructed would fill many books. Here our purpose is to outline very briefly a few of the engine types that are currently in general use. The exposition is not intended to be complete. Furthermore, because you are expected to be mechanically inclined and generally familiar with internal combustion engines, the primary purpose of this section is merely to record things that you know and to furnish a nomenclature for the balance of the chapter.

In this chapter we classify engines according to their intended use, the combustion cycle used, and the number and arrangement of the cylinders. Thus, we refer to aircraft engines, automotive engines, marine engines, and stationary engines, for example, all so named because of the purpose for which they were designed. Similarly, one might have in mind an engine designed on the basis of the *Otto cycle,* in which the fuel and air are mixed before compression and in which combustion takes place with no excess air, or the *diesel engine,* in which the fuel is injected near the end of compression and combustion takes place with much excess air. The Otto-cycle engine uses quite volatile fuels, and ignition is by spark, but the diesel-cycle engine operates on fuels of low volatility and ignition occurs because of compression.

The diesel- and Otto-cycle engines may be either *two-stroke-cycle* or *four-stroke-cycle,* depending upon the number of piston strokes required for the complete combustion cycle. Many outboard engines use the two-stroke-cycle (or simply two-stroke) process, in which the piston uncovers exhaust ports in the cylinder wall near the end of the expansion stroke and permits the exhaust gases to flow out. Soon after the exhaust ports are opened, the inlet ports open too and permit entry of a precompressed fuel–air mixture that also assists in expelling the remaining exhaust gases. The ports are then closed by the piston moving upward, and the fuel mixture is again compressed. Then the cycle begins again. Note that the two-cycle engine has an expansion and a compression cycle and that they occur during one revolution of the crank.

The four-cycle engine has four piston strokes in a single combustion cycle corresponding to two revolutions of the crank. The events corresponding to four strokes are (1) expansion, or power, stroke, (2) exhaust, (3) suction, or intake, stroke, and (4) compression.

Multicylinder engines are broadly classified according to how the cylinders are arranged with respect to each other and the crankshaft. Thus, an *in-line* engine is one in which the piston axes form a single plane coincident with the crankshaft and in which the pistons are all on the same side of the crankshaft. Figure 18.1 is a schematic drawing of a three-cylinder in-line engine with the cranks spaced 120°; a firing-order diagram for four-cycle operation is included for interest.

Figure 18.2 shows a cutaway view of a five-cylinder in-line engine for a modern passenger car. We can see here that the plane formed by the centerlines of the cylinders is not mounted vertically in order to keep the height of the engine smaller to fit the low-profile styling requirements. This figure also shows a good view of the relative location of the camshaft and the overhead valves.

A V-type engine uses two banks of one or more in-line cylinders each, all connected to a single crankshaft. Figure 18.3 illustrates several common crank arrangements. The pistons in the right and left banks of (*a*) and (*b*) are in the same plane, but those in (*c*) are in different planes.

If the V angle is increased to 180°, the result is called an *opposed-piston engine.* An opposed engine may have the two piston axes coincident or offset, and the rods may connect to the same crank or to separate cranks 180° apart.

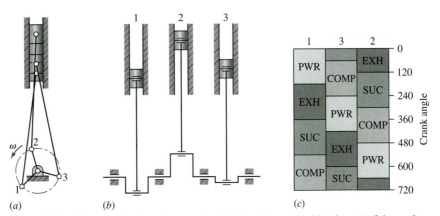

Figure 18.1 A three-cylinder in-line engine: (*a*) front view; (*b*) side view; (*c*) firing order.

Figure 18.2 A five-cylinder in-line engine from the 1990 Audi Coupe Quattro. (Audi of America, Inc., Troy, MI.)

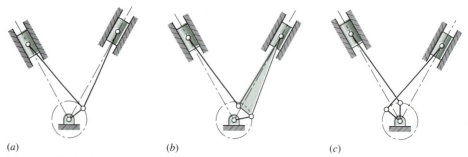

(a)   (b)   (c)

Figure 18.3 Crank arrangements of V engines: (a) single crank per pair of cylinders—connecting rods interlock with each other and are of *fork-and-blade* design; (b) single crank per pair of cylinders—the *master connecting rod* carries a bearing for the *articulated rod;* (c) separate crank throws connect to staggered rods and pistons.

A *radial* engine is one having the pistons arranged in a circle about the crank center. Radial engines use a master connecting rod for one cylinder, and the remaining pistons are connected to the master rod by articulated rods somewhat the same as for the V engine of Fig. 18.3b.

Figures 18.4 to 18.6 illustrate, respectively, the piston-connecting-rod assembly, the crankshaft, and the block of a V6 truck engine. These are included as typical of modern engine design to show the form of important parts of an engine. The following specifications also give a general idea of the performance and design characteristics of typical engines, together with the sizes of parts used in them.

Figure 18.4 A piston-connecting-rod assembly for a 351-in^3 V6 truck engine. (GMC Truck and Coach Division, General Motors Corporation, Pontiac, MI.)

Figure 18.5 A cast crankshaft for a 305-in^3 V6 truck engine. (GMC Truck and Coach Division, General Motors Corporation, Pontiac, MI.)

Figure 18.6 Block for a 305-in^3 V6 truck engine. The same casting is used for a 351-in^3 engine by boring for larger pistons. (GMC Truck and Coach Division, General Motors Corporation, Pontiac, MI.)

GMC Truck and Coach Division, General Motors Corporation, Pontiac, Michigan. One of the GMC V6 truck engines is illustrated in Fig. 18.7. These engines are manufactured in four displacements, and they include one model, a V12 (702 in³), which is described as a "twin six" because many of the V6 parts are interchangeable with it. Data included here are for the 401-in³ engine. Typical performance curves are exhibited in Figs. 18.8, 18.9, and 18.10. The specifications are as follows: 60° vee design; bore = 4.875 in; stroke = 3.56 in;

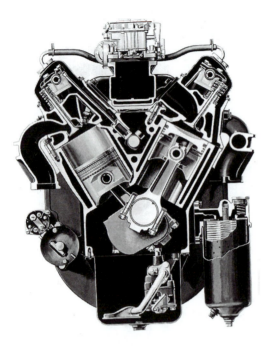

Figure 18.7 Cross-sectional view of a 401-in³ V6 truck engine. (GMC Truck and Coach Division, General Motors Corporation, Pontiac, MI.)

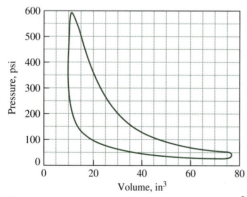

Figure 18.8 Typical indicator diagram for a 401-in³ V6 truck engine; conditions unknown. (GMC Truck and Coach Division, General Motors Corporation, Pontiac, MI.)

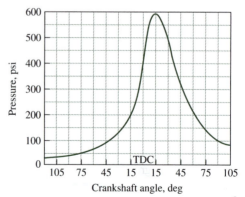

Figure 18.9 A pressure–time curve for the 401-in³ V6 truck engine; these data were taken from a running engine. (GMC Truck and Coach Division, General Motors Corporation, Pontiac, MI.)

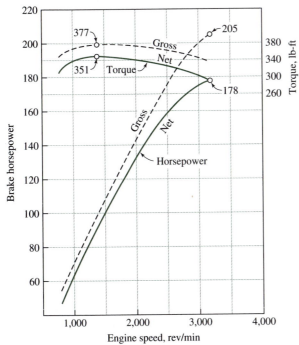

Figure 18.10 Horsepower and torque characteristics of the 401-in³ V6 truck engine. The solid curve is the net output as installed; the dashed curve is the maximum output without accessories. Notice that the maximum torque occurs at a very low engine speed. (GMC Truck and Coach Division, General Motors Corporation, Pontiac, MI.)

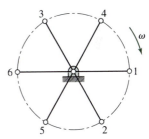

Figure 18.11 Front view of a V6 truck engine showing crank arrangement and direction of rotation.

connecting rod length = 7.19 in; compression ratio = 7.50 : 1; cylinders numbered 1, 3, 5 from front to rear on the left bank, and 2, 4, 6 on the right bank; firing order 1, 6, 5, 4, 3, 2; crank arrangement is shown in Fig. 18.11.

18.2 INDICATOR DIAGRAMS

Experimentally, an instrument called an *engine indicator* is used to measure the variation in pressure within a cylinder. The instrument constructs a graph, during operation of the engine, which is known as an *indicator diagram*. Known constraints of the indicator make it possible to study the diagram and determine the relationship between the gas pressure and the crank angle for the particular set of running conditions in existence at the time the diagram was taken.

When an engine is in the design stage, it is necessary to estimate a diagram from theoretical considerations. From such an approximation a pilot model of the proposed engine can be designed and built, and the actual indicator diagram can be taken and compared with the theoretically devised one. This provides much useful information for the design of the production model.

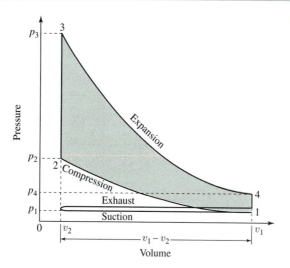

Figure 18.12 An ideal indicator diagram for a four-cycle engine.

An indicator diagram for the ideal air-standard cycle is shown in Fig. 18.12 for a four-stroke-cycle engine. During compression the cylinder volume changes from v_1 to v_2 and the cylinder pressure changes from p_1 to p_2. The relationship, at any point of the stroke, is given by the polytropic gas law as

$$p_x v_x^k = p_1 v_1^k = \text{constant} \tag{18.1}$$

In an actual indicator card the corners at points 2 and 3 are rounded and the line joining these points is curved. This is explained by the fact that combustion is not instantaneous and ignition occurs before the end of the compression stroke. An actual card is also rounded at points 4 and 1 because the values do not operate simultaneously.

The polytropic exponent k in Eq. (18.1) is often taken to be about 1.30 for both compression and expansion, although differences probably do exist.

The relationship between the horsepower developed and the dimensions of the engine is given by

$$\text{bhp} = \frac{p_b l a n}{(33\,000 \text{ ft·lb/hp·s})(12 \text{ in/ft})} \tag{18.2}$$

where bhp = brake horsepower per cylinder
  p_b = brake mean effective pressure
  l = length of stroke, in
  a = piston area, in²
  n = number of working strokes per minute

The amount of horsepower that can be obtained from 1 in³ of piston displacement varies considerably, depending upon the engine type. For typical automotive engines it ranges from 0.55 up to 1.00 hp/in³, with an average of perhaps 0.70 at present. On the other hand, many marine diesel engines have ratios varying from 0.10 to 0.20 hp/in³. About the best that can be done in designing a new engine is to use standard references to discover what others have done with the same types of engines and then to choose a value that seems to be reasonably attainable.

For many engines the ratio of bore to stroke varies from about 0.75 to 1.00. The tendency in automotive-engine design seems to be toward shorter-stroke engines in order to reduce engine height.

Decisions on bore–stroke ratio and horsepower per unit displacement volume are helpful in solving Eq. (18.2) to obtain suitable dimensions when the horsepower, speed, and number of cylinders have been decided upon.

The ratio of the brake mean effective pressure p_b to the indicated mean effective pressure p_i, obtained experimentally from an indicator card, is the mechanical efficiency e_m:

$$e_m = \frac{p_b}{p_i} \tag{18.3}$$

Differences between a theoretical and an experimentally determined indicator diagram can be accounted for by applying a correction called a *card factor*. The card factor is defined by

$$f_c = \frac{p_i}{p_i'} \tag{18.4}$$

where p_i' is the theoretical indicated mean effective pressure and the card factor is usually about 0.90 to 0.95.

If the compression ratio (Fig. 18.12) is defined as

$$r = \frac{v_1}{v_2} \tag{18.5}$$

the work done during compression is

$$U_c = \int_{v_2}^{v_1} p \, dv = p_1 v_1^k \int_{v_2}^{v_1} \frac{dv}{v^k} = \frac{p_1 v_1}{k-1}(r^{k-1} - 1) \tag{a}$$

The displacement volume can be written as

$$v_1 - v_2 = v_1 - \frac{v_1}{r} = \frac{v_1(r-1)}{r} \tag{b}$$

Substituting v_1 from Eq. (b) into Eq. (a) yields

$$U_c = \frac{p_1(v_1 - v_2)}{k-1} \frac{r^k - r}{r-1} \tag{c}$$

The work done during expansion is the area under the curve between points 3 and 4 of Fig. 18.12. This is found in the same manner; the result is

$$U_e = \frac{p_4(v_1 - v_2)}{k-1} \frac{r^k - r}{r-1} \tag{d}$$

The net amount of work accomplished in a cycle is the difference in the amounts given by Eqs. (c) and (d), and it must be equal to the product of the indicated mean and effective pressure and the displacement volume. Thus

$$U = U_e - U_c = p_i'(v_1 - v_2)$$

$$= \frac{p_4(v_1 - v_2)}{k - 1} \frac{r^k - r}{r - 1} - \frac{p_1(v_1 - v_2)}{k - 1} \frac{r^k - r}{r - 1} \tag{18.6}$$

If the exponent is the same for expansion as for compression, Eq. (18.6) can be solved to give

$$p_4 = p_1'(k - 1) \frac{r - 1}{r^k - r} + p_1 \tag{e}$$

Substituting p_i' from Eq. (18.4) produces

$$p_4 = (k - 1) \frac{r - 1}{r^k - r} \frac{p_i}{f_c} + p_1 \tag{18.7}$$

Equations (18.1) and (18.7) can be used to create the theoretical indicator diagram. The corners are then rounded off so that the pressure at point 3 is made about 75 percent of that given by Eq. (18.1). As a check, the area of the diagram can be measured and divided by the displacement volume; the result should equal the indicated mean effective pressure.

18.3 DYNAMIC ANALYSIS—GENERAL

The balance of this chapter is devoted to an analysis of the dynamics of a single-cylinder engine. To simplify this work, it is assumed that the engine is running at a constant crankshaft speed and that gravitational forces and friction forces can be neglected in comparison to dynamic-force effects. The gas forces and the inertia forces are found separately; then these forces are combined, using the principle of superposition (see Section 15.5), to obtain the total bearing forces and crankshaft torque.

The subject of engine balancing is treated in Chapter 19.

18.4 GAS FORCES

In this section we assume that the moving parts are massless so that gravity and inertia forces and torques are zero, and also that there is no friction. These assumptions allow us to trace the force effects of the gas pressure from the piston to the crankshaft without the complicating effects of other forces. These other effects will be added later, using superposition.

In Chapter 14, both graphical and analytical methods for finding the forces in any mechanism were presented. Any of those approaches could be used to solve this gas-force problem. The advantage of the analytic methods is that they can be programmed for automatic computation throughout a cycle. But the graphical technique must be repeated for each crank position until a complete cycle of operation (720° for a four-cycle engine) is completed. Because we prefer not to duplicate the studies of Chapter 14, however, we present here an algebraic approach.

Figure 18.13

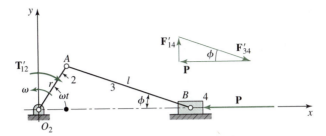

In Fig. 18.13 we designate the crank angle as ωt, taken positive in the counterclockwise direction, and the connecting-rod angle as ϕ, taken positive when the crank pivot A is in the first quadrant ($R^y_{AO_2} > 0$) as shown. A relation between these two angles is seen from the figure:

$$l \sin \phi = r \sin \omega t \quad \text{or} \quad \sin \phi = \frac{r}{l} \sin \omega t \qquad (a)$$

where r and l designate the lengths of the crank and the connecting rod, respectively. Designating the piston position by the coordinate $x = R_{BO_2}$, we find

$$x = r \cos \omega t + l \cos \phi = r \cos \omega t + l \sqrt{1 - \left(\frac{r}{l} \sin \omega t\right)^2} \qquad (18.8)$$

For most engines the ratio r/l is about $1/4$, and so the maximum value of the second term under the radical is about $1/16$, or perhaps less. If we expand the radical using the binomial theorem and neglect all but the first two terms, we obtain

$$\sqrt{1 - \left(\frac{r}{l} \sin \omega t\right)^2} = 1 - \frac{r^2}{2l^2} \sin^2 \omega t \qquad (b)$$

Because

$$\sin^2 \omega t = \frac{1 - \cos 2\omega t}{2} \qquad (c)$$

Eq. (18.8) can be written as

$$x = l - \frac{r^2}{4l} + r\left(\cos \omega t + \frac{r}{4l} \cos 2\omega t\right) \qquad (18.9)$$

Differentiating this equation successively to obtain the velocity and acceleration gives

$$\dot{x} = -r\omega\left(\sin \omega t + \frac{r}{2l} \sin 2\omega t\right) \qquad (18.10)$$

$$\ddot{x} = -r\alpha\left(\sin \omega t + \frac{r}{2l} \sin 2\omega t\right) - r\omega^2\left(\cos \omega t + \frac{r}{l} \cos 2\omega t\right) \qquad (18.11)$$

Referring again to Fig. 18.13, we designate a gas-force vector **P** as defined or obtained using the methods of Section 18.2. Reactions due to this force are designated using a single prime. Thus $\mathbf{F}'_{14}$ is the force of the cylinder wall acting against the piston and $\mathbf{F}'_{34}$ is the force of the connecting rod acting against the piston at the piston pin. The force polygon in Fig. 18.13 shows the relation among **P**, $\mathbf{F}'_{14}$, and $\mathbf{F}'_{34}$. Thus, we have

$$\mathbf{F}'_{14} = P \tan \phi \hat{\mathbf{j}} \tag{18.12}$$

The quantity $\tan \phi$ appears frequently in expressions throughout this chapter. It is therefore convenient to develop an expression in terms of the crank angle ωt. Thus

$$\tan \phi = \frac{(r/l) \sin \omega t}{\cos \phi} = \frac{(r/l) \sin \omega t}{\sqrt{1 - [(r/l) \sin \omega t]^2}} \tag{d}$$

Now, using the binomial expansion again, we find that

$$\frac{1}{\sqrt{1 - [(r/l) \sin \omega t]^2}} = 1 + \frac{r^2}{2l^2} \sin^2 \omega t \tag{e}$$

where only the first two terms have been retained. Then substituting Eq. (e) into Eq. (d) we have

$$\tan \phi = \frac{r}{l} \sin \omega t \left(1 + \frac{r^2}{2l^2} \sin^2 \omega t \right) \tag{18.13}$$

The trigonometry of Fig. 18.13 shows that the wrist-pin (piston-pin) bearing force has a magnitude of

$$F'_{34} = \frac{P}{\cos \phi} = \frac{P}{\sqrt{1 - [(r/l) \sin \omega t]^2}} = P \left(1 + \frac{r^2}{2l^2} \sin^2 \omega t \right) \tag{f}$$

or, in vector notation,

$$\mathbf{F}'_{34} = P\hat{\mathbf{i}} - F'_{14}\hat{\mathbf{j}} = P\hat{\mathbf{i}} - P \tan \phi \hat{\mathbf{j}} \tag{18.14}$$

By taking moments about the crank center we find the torque $\mathbf{T}'_{21}$ delivered by the crank to the shaft is the product of the force $\mathbf{F}'_{14}$ and the piston coordinate x. Using Eqs. (18.9), (18.12), and (18.13) yields

$$\mathbf{T}'_{21} = F'_{14} x \hat{\mathbf{k}}$$

$$= P \left(\frac{r}{l} \sin \omega t \right) \left(1 + \frac{r^2}{2l^2} \sin^2 \omega t \right) \left[l - \frac{r^2}{4l} + r \left(\cos \omega t + \frac{r}{4l} \cos 2\omega t \right) \right] \hat{\mathbf{k}} \tag{g}$$

When the terms of Eq. (g) are multiplied, we can neglect those containing second or higher powers of r/l with only a very small error. Equation (g) then becomes

$$\mathbf{T}'_{21} = Pr \sin \omega t \left(1 + \frac{r}{l} \cos \omega t \right) \hat{\mathbf{k}} \tag{18.15}$$

This is the torque delivered to the crankshaft by the gas force; the counterclockwise direction is positive.

18.5 EQUIVALENT MASSES

Problems 15.5 and 15.6, at the end of Chapter 15, focused on the slider-crank mechanism, which is an example of an engine mechanism. The dynamics of this mechanism were analyzed using the methods presented in that chapter. In this chapter we are concerned with the same problem. However, here we will show certain simplifications that are customarily used to reduce the complexity of the algebraic solution process. These simplifications are approximations and do introduce certain errors into the analysis. In this chapter we will show the simplifications and comment on the errors which they introduce.

In analyzing the inertia forces due to the connecting rod of an engine, it is often convenient to picture a portion of the mass as concentrated at the crankpin A and the remaining portion at the wrist pin B (Fig. 18.14). The reason for this is that the crankpin moves on a circle and the wrist pin on a straight line. Both of these motions are quite easy to analyze. However, the center of gravity G of the connecting rod is somewhere between the crankpin and the wrist pin, and its motion is more complicated and consequently more difficult to determine in algebraic form.

The mass of the connecting rod m_3 is assumed to be concentrated at the center of gravity G_3. We divide this mass into two parts; one, m_{3P}, is concentrated at the center of percussion P for oscillation of the rod about point B. This disposition of the mass of the rod is dynamically equivalent to the original rod if the total mass is the same, if the position of the center of gravity G_3 is unchanged, and if the moment of inertia is the same. Writing these three conditions, respectively, in equation form produces

$$m_3 = m_{3B} + m_{3P} \qquad (a)$$

$$m_{3B}l_B = m_{3P}l_P \qquad (b)$$

$$I_G = m_{3B}l_B^2 + m_{3P}l_P^2 \qquad (c)$$

Solving Eqs. (a) and (b) simultaneously gives the portion of mass to be concentrated at each point:

$$m_{3B} = m_3\frac{l_P}{l_B + l_P} \quad \text{and} \quad m_{3P} = m_3\frac{l_B}{l_B + l_P} \qquad (18.16)$$

Figure 18.14

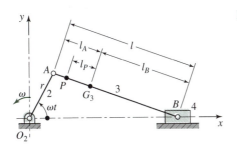

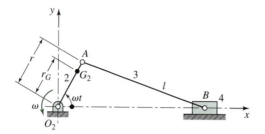

Figure 18.15

Substituting Eq. (18.16) into Eq. (c) gives

$$I_G = m_3 \frac{l_P}{l_B + l_P} l_B^2 + m_3 \frac{l_B}{l_B + l_P} l_P^2 = m_3 l_P l_B \qquad (d)$$

or

$$l_P l_B = \frac{I_G}{m_3} \qquad (18.17)$$

Equation (18.17) shows that the two distances l_P and l_B are dependent on each other. Thus, if l_B is specified in advance, l_P is fixed in length by Eq. (18.17).

In the usual connecting rod, the center of percussion P is close to the crankpin A and it is assumed that they are coincident. Thus, if we let $l_A = l_P$, Eqs. (18.16) reduce to

$$m_{3B} = \frac{m_3 l_A}{l} \quad \text{and} \quad m_{3A} = \frac{m_3 l_B}{l} \qquad (18.18)$$

We note again that the equivalent masses, obtained by Eqs. (18.18), are not exact because of the assumption made, but are close enough for ordinary connecting rods. The approximation, for example, is not valid for the master connecting rod of a radial engine, because the crankpin end has bearings for all of the other connecting rods as well as its own bearing.

For estimating and checking purposes, about two-thirds of the mass should be concentrated at A and the remaining third at B.

Figure 18.15 illustrates an engine linkage in which the mass of the crank m_2 is not balanced, as evidenced by the fact that the center of gravity G_2 is displaced outward along the crank a distance r_G from the axis of rotation. In the inertia-force analysis, simplification is obtained by locating an equivalent mass m_{2A} at the crankpin. Thus, for equivalence,

$$m_2 r_G = m_{2A} r \quad \text{or} \quad m_{2A} = m_2 \frac{r_G}{r} \qquad (18.19)$$

18.6 INERTIA FORCES

Using the methods of the preceding section, we begin by locating equivalent masses at the crankpins and at the wrist pin. Thus,

$$m_A = m_{2A} + m_{3A} \qquad (18.20)$$

$$m_B = m_{3B} + m_4 \qquad (18.21)$$

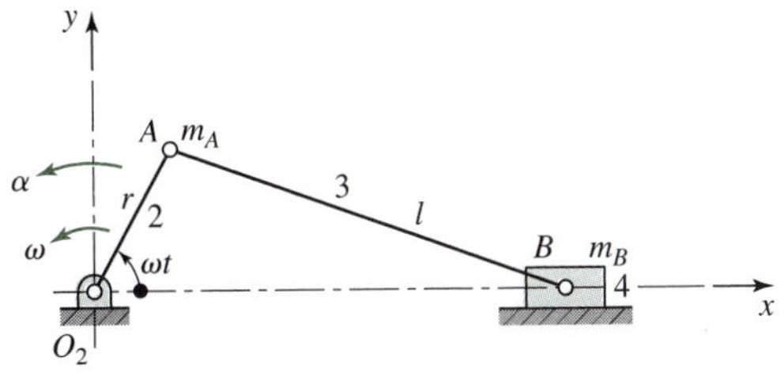

Figure 18.16

Equation (18.20) states that the mass m_A located at the crankpin is made up of the equivalent masses m_{2A} of the crank and m_{3A} of part of the connecting rod. Of course, if the crank is balanced, all its mass is assumed to be located at the axis of rotation and m_{2A} is then zero. Equation (18.21) indicates that the reciprocating mass m_B located at the wrist pin is composed of the equivalent mass m_{3B} of the other part of the connecting rod and the mass m_4 of the piston assembly.

Figure 18.16 shows the slider-crank mechanism with masses m_A and m_B located at points A and B, respectively. If we designate the angular velocity of the crank as ω and the angular acceleration as α, the position vector of the crankpin relative to the origin O_2 is

$$\mathbf{R}_A = r \cos \omega t \hat{\mathbf{i}} + r \sin \omega t \hat{\mathbf{j}} \tag{a}$$

Differentiating this equation twice to obtain the acceleration of point A gives

$$\mathbf{A}_A = (-r\alpha \sin \omega t - r\omega^2 \cos \omega t)\hat{\mathbf{i}} + (r\alpha \cos \omega t - r\omega^2 \sin \omega t)\hat{\mathbf{j}} \tag{18.22}$$

The inertia force of the rotating parts is then

$$-m_A\mathbf{A}_A = m_A r(\alpha \sin \omega t + \omega^2 \cos \omega t)\hat{\mathbf{i}} + m_A r(-\alpha \cos \omega t + \omega^2 \sin \omega t)\hat{\mathbf{j}} \tag{18.23}$$

Because the analysis is usually made at constant angular velocity ($\alpha = 0$), Eq. (18.23) reduces to

$$-m_A\mathbf{A}_A = m_A r\omega^2(\cos \omega t \hat{\mathbf{i}} + \sin \omega t \hat{\mathbf{j}}) \tag{18.24}$$

The acceleration of the piston has already been determined [Eq. (18.11)] and is repeated here for convenience in a slightly different form:

$$\mathbf{A}_B = \left[-r\alpha \left(\sin \omega t + \frac{r}{2l} \sin 2\omega t \right) - r\omega^2 \left(\cos \omega t + \frac{r}{l} \cos 2\omega t \right) \right] \hat{\mathbf{i}} \tag{18.25}$$

The inertia force of the reciprocating parts is therefore

$$-m_B\mathbf{A}_B = \left[m_B r\alpha \left(\sin \omega t + \frac{r}{2l} \sin 2\omega t \right) + m_B r\omega^2 \left(\cos \omega t + \frac{r}{l} \cos 2\omega t \right) \right] \hat{\mathbf{i}} \tag{18.26}$$

or, for constant angular velocity,

$$-m_B\mathbf{A}_B = m_B r\omega^2 \left(\cos \omega t + \frac{r}{l} \cos 2\omega t \right) \hat{\mathbf{i}} \tag{18.27}$$

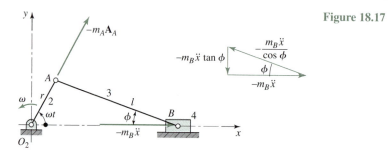

Figure 18.17

Adding Eqs. (18.24) and (18.27) gives the total inertia force for all the moving parts. The components in the x and y directions are

$$F^x = (m_A + m_B)r\omega^2 \cos \omega t + \left(m_B \frac{r}{l}\right)r\omega^2 \cos 2\omega t \qquad (18.28)$$

$$F^y = m_A r\omega^2 \sin \omega t \qquad (18.29)$$

It is customary to refer to the portion of the force occurring at the frequency ω rad/s as the *primary inertia force* and the portion occurring at 2ω rad/s as the *secondary inertia force*. We note that the vertical component has only a primary part and that it therefore varies directly with the crankshaft speed. On the other hand, the horizontal component, which is in the direction of the cylinder axis, has a primary part varying directly with the crankshaft speed and a secondary part varying at twice the crankshaft speed.

We proceed now to a determination of the inertia torque. As shown in Fig. 18.17, the inertia force due to the mass at A has no moment arm about O_2 and therefore produces no torque. Consequently, we need consider only the inertia force given by Eq. (18.27) due to the reciprocating part of the mass.

From the force polygon of Fig. 18.17, the inertia torque exerted by the engine on the crankshaft is

$$\mathbf{T}''_{21} = -(-m_B \ddot{x} \tan \phi)x\hat{\mathbf{k}} \qquad (b)$$

Expressions for x, $\ddot{x}$, and $\tan \phi$ appear in Section 18.4. Making appropriate substitutions for these yields the following equation for the torque:

$$\mathbf{T}''_{21} = -m_B r\omega^2 \left(\cos \omega t + \frac{r}{l} \cos 2\omega t\right)\left[l - \frac{r^2}{4l} + r\left(\cos \omega t + \frac{r}{4l} \cos 2\omega t\right)\right]$$
$$\cdot \frac{r}{l} \sin \omega t \left(1 + \frac{r^2}{2l^2} \sin^2 \omega t\right)\hat{\mathbf{k}} \qquad (c)$$

Terms that are proportional to the second and higher powers of r/l can be neglected in performing the indicated multiplication. Equation (c) can then be written as

$$\mathbf{T}''_{21} = -m_B r^2 \omega^2 \sin \omega t \left(\frac{r}{2l} + \cos \omega t + \frac{3r}{2l} \cos 2\omega t\right)\hat{\mathbf{k}} \qquad (d)$$

Then, using the identities

$$2 \sin \omega t \cos 2\omega t = \sin 3\omega t - \sin \omega t \tag{e}$$

and

$$2 \sin \omega t \cos \omega t = \sin 2\omega t \tag{f}$$

results in an equation having only sine terms, and Eq. (d) finally becomes

$$\mathbf{T}''_{21} = \frac{m_B}{2} r^2 \omega^2 \left(\frac{r}{2l} \sin \omega t - \sin 2\omega t - \frac{3r}{2l} \sin 3\omega t \right) \hat{\mathbf{k}} \tag{18.30}$$

This is the inertia torque exerted by the engine on the shaft in the positive direction. A clockwise or negative inertia torque of the same magnitude is, of course, exerted on the frame of the engine.

The assumed distribution of the connecting-rod mass results in a moment of inertia that is greater than the true value. Consequently, the torque given by Eq. (18.30) is not the exact value. In addition, terms proportional to the second- and higher-order powers of r/l were dropped in simplifying Eq. (c). These two errors are about the same magnitude and are quite small for ordinary connecting rods having r/l ratios near $1/4$.

18.7 BEARING LOADS IN A SINGLE-CYLINDER ENGINE

The designer of a reciprocating engine must know the values of the forces acting upon the bearings and how these forces vary in a cycle of operation. This is necessary to proportion and select the bearings properly, and it is also needed for the design of other engine parts. This selection is an investigation of the force exerted by the piston against the cylinder wall and the forces acting against the piston pin and against the crankpin. Main bearing forces will be investigated in a later section because they depend upon the action of the cylinders of the engine.

The resultant bearing loads are made up of the following components:

1. The gas-force components, designated by a single prime
2. Inertia force due to the mass of the piston assembly, designated by a double prime
3. Inertia force of that part of the connecting rod assigned to the piston-pin end, designated by a triple prime
4. Connecting-rod inertia force at the crankpin end, designated by a quadruple prime

Equations for the gas-force components have been determined in Section 18.4, and references will be made to them in finding the total bearing loads.

Figure 18.18 is a graphical analysis of the forces in the engine mechanism with zero gas force and subjected to an inertia force resulting only from the mass of the piston assembly. Figure 18.18a shows the position of the mechanism selected for analysis, and the inertia force $-m_4 \mathbf{A}_B$ is shown acting upon the piston. In Fig. 18.18b the free-body diagram of the piston forces is shown together with the force polygon from which they were obtained. Figures 18.18c through 18.18e illustrate, respectively, the free-body diagrams of forces acting upon the connecting rod, crank, and frame.

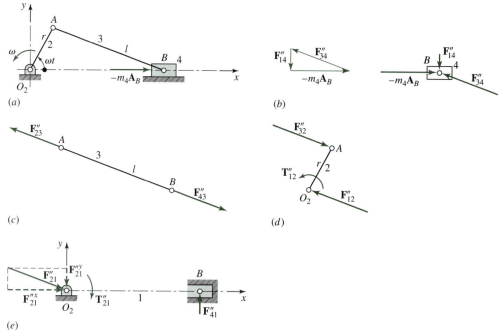

Figure 18.18 Graphical analysis of the forces in the engine mechanism when only the inertia force due to the mass of the piston assembly is considered.

In Fig. 18.18e, notice that the torque $\mathbf{T}''_{21}$ balances the force couple formed by the forces $\mathbf{F}''_{41}$ and $\mathbf{F}''^y_{21}$. But the force $\mathbf{F}''^x_{21}$ at the crank center remains unopposed by any other force. This is a very important observation that we shall reserve for discussion in a separate section.

The following forces are of interest to us:

1. The force $\mathbf{F}''_{41}$ of the piston against the cylinder wall
2. The force $\mathbf{F}''_{34}$ of the connecting rod against the piston pin
3. The force $\mathbf{F}''_{32}$ of the connecting rod against the crankpin
4. The force $\mathbf{F}''_{12}$ of the shaft against the crank

By methods similar to those used earlier in this chapter, the analytical expressions are found to be

$$\mathbf{F}''_{41} = -m_4\ddot{x}\tan\phi\hat{\mathbf{j}} \tag{18.31}$$

$$\mathbf{F}''_{34} = m_4\ddot{x}\hat{\mathbf{i}} - m_4\ddot{x}\tan\phi\hat{\mathbf{j}} \tag{18.32}$$

$$\mathbf{F}''_{32} = -\mathbf{F}''_{34} \tag{18.33}$$

$$\mathbf{F}''_{12} = -\mathbf{F}''_{32} = \mathbf{F}''_{34} \tag{18.34}$$

where $\ddot{x}$ is the acceleration of the piston as given by Eq. (18.11) and m_4 is the mass of the piston assembly. The quantity $\tan\phi$ can be evaluated in terms of the crank angle with the use of Eq. (18.13).

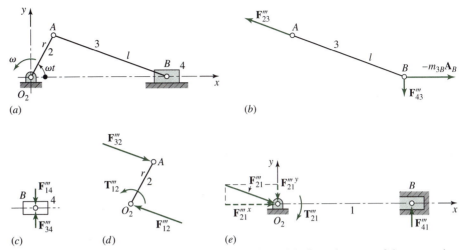

Figure 18.19 Graphical analysis of the forces resulting solely from the mass of the connecting rod, assumed to be concentrated at the wrist-pin end.

In Fig. 18.19 we neglect all forces except those that result because of that part of the mass of the connecting rod which is assumed to be located at the piston-pin center. Thus Fig. 18.19b is a free-body diagram of the connecting rod showing the inertia force $-m_{3B}\mathbf{A}_B$ acting at the piston-pin end.

We now note that it is incorrect to add m_{3B} and m_4 together and then to compute a resultant inertia force in finding the bearing loads, although such a procedure would seem to be simpler. The reason for this is that m_4 is the mass of the piston assembly and the corresponding inertia force acts on the piston side of the wrist pin. But m_{3B} is part of the connecting-rod mass, and hence its inertia force acts on the connecting-rod side of the wrist pin. Thus, adding the two will yield correct results for the crankpin load and the force of the piston against the cylinder wall but will give *incorrect* results for the piston-pin load.

The forces on the piston pin, the crank, and the frame are illustrated in Figs. 18.19c, 18.19d, and 18.19e, respectively. The equations for these forces for a crank having uniform angular velocity are found to be

$$\mathbf{F}'''_{41} = -m_{3B}\ddot{x}\tan\phi\hat{\mathbf{j}} \tag{18.35}$$

$$\mathbf{F}'''_{34} = \mathbf{F}'''_{41} \tag{18.36}$$

$$\mathbf{F}'''_{32} = -m_{3B}\ddot{x}\hat{\mathbf{i}} + m_{3B}\ddot{x}\tan\phi\hat{\mathbf{j}} \tag{18.37}$$

$$\mathbf{F}'''_{12} = -\mathbf{F}'''_{32} \tag{18.38}$$

Figure 18.20 illustrates the forces that result because of that part of the connecting-rod which is concentrated at the crankpin end. While a counterweight attached to the crank balances the reaction at O_2, it cannot make F''''_{32} zero. Thus the crankpin force exists no matter whether the rotating mass of the connecting rod is balanced or not. This force is

$$\mathbf{F}''''_{32} = m_{3A}r\omega^2(\cos\omega t\hat{\mathbf{i}} + \sin\omega t\hat{\mathbf{j}}) \tag{18.39}$$

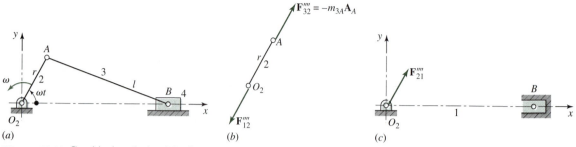

Figure 18.20 Graphical analysis of the forces resulting solely from the mass of the connecting rod, assumed to be concentrated at the crankpin end.

The last step is to sum these expressions to obtain the resultant bearing loads. The total force of the piston against the cylinder wall, for example, is found by summing Eqs. (18.12), (18.31), and (18.35), with due regard for subscripts and signs. When simplified, the answer is

$$\mathbf{F}_{41} = \mathbf{F}'_{41} + \mathbf{F}''_{41} + \mathbf{F}'''_{41} = -[(m_{3B} + m_4)\ddot{x} + P]\tan\phi\hat{\mathbf{j}} \qquad (18.40)$$

The forces on the piston pin, the crankpin, and the crankshaft are found in a similar manner and are

$$\mathbf{F}_{34} = (m_4\ddot{x} + P)\hat{\mathbf{i}} - [(m_{3B} + m_4)\ddot{x} + P]\tan\phi\hat{\mathbf{j}} \qquad (18.41)$$

$$\mathbf{F}_{32} = [m_{3A}r\omega^2\cos\omega t - (m_{3B} + m_4)\ddot{x} - P]\hat{\mathbf{i}} + \{m_{3A}r\omega^2\sin\omega t$$
$$+ [(m_{3B} + m_4)\ddot{x} + P]\tan\phi\}\hat{\mathbf{j}} \qquad (18.42)$$

$$\mathbf{F}_{21} = \mathbf{F}_{32} \qquad (18.43)$$

18.8 CRANKSHAFT TORQUE

The torque delivered by the crankshaft to the load is called the *crankshaft torque,* and it is the negative of the moment of the couple formed by the forces $\mathbf{F}_{41}$ and $\mathbf{F}^y_{21}$. Therefore, it is obtained from the equation

$$\mathbf{T}_{21} = -F_{41}x\hat{\mathbf{k}} = [(m_{3B} + m_4)\ddot{x} + P]x\tan\phi\hat{\mathbf{k}} \qquad (18.44)$$

18.9 ENGINE SHAKING FORCES

The inertia force due to the reciprocating masses is shown acting in the positive direction in Fig. 18.21a. In Fig. 18.21b the forces acting upon the engine block due to these inertia forces are shown. The resultant forces are $\mathbf{F}_{21}$, the force exerted by the crankshaft on the main bearings, and a positive couple formed by the forces $\mathbf{F}_{41}$ and $\mathbf{F}^y_{21}$. The force $\mathbf{F}^x_{21} = -m_B\mathbf{A}_B$ is frequently termed a *shaking force,* and the couple $T = xF_{41}$ is called a *shaking couple.* As indicated by Eqs. (18.27) and (18.30), the magnitude and the direction of this force and couple change with ωt; consequently, the shaking force induces linear vibration of the block in the x direction, and the shaking couple induces a torsional vibration of the block about the crank center.

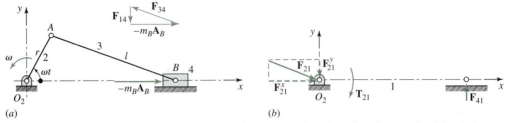

Figure 18.21 Dynamic forces due to the reciprocating masses; the primes have been omitted for clarity.

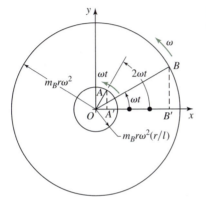

Figure 18.22 Circle diagram for finding inertia forces. Total inertia force is $OA' + OB'$.

A graphical representation of the inertia force is possible if Eq. (18.27) is rearranged as

$$F = m_B r\omega^2 \cos\omega t + m_B r\omega^2 \frac{r}{l} \cos 2\omega t \qquad (18.45)$$

where $F = F_{21}^x$ for simplicity of notation. The first term of Eq. (18.45) is represented by the x projection of a vector $m_B r\omega^2$ in length rotating at ω rad/s. This is the primary part of the inertia force. The second term is similarly represented by the x projection of a vector $m_B r\omega^2(r/l)$ in length rotating at 2ω rad/s; this is the secondary part. Such a diagram is shown in Fig. 18.22 for $r/l = 1/4$. The total inertia or shaking force is the algebraic sum of the horizontal projections of the two vectors.

18.10 COMPUTATION HINTS

This section contains suggestions for using computers and programmable calculators in solving the dynamics of engine mechanisms. Many of the ideas, however, will be useful for readers using non-programmable machines as well as for checking purposes.

Indicator Diagrams   It would be very convenient if a subprogram for computing the gas forces could be devised and the results used directly in a main program to compute all the resultant bearing forces and crankshaft torques. Unfortunately, the theoretical indicator diagram must be manipulated by hand in order to obtain a reasonable approximation to the experimental data. This manipulation can be done graphically or with a computer having a graphical display. The procedure is illustrated by the following example.

EXAMPLE 18.1

Determine the pressure-versus-piston displacement relation for a six-cylinder engine having a displacement of 140 in^3, a compression ratio of 8, and a brake horsepower of 57 at 2 400 rev/min. Use a mechanical efficiency of 75 percent, a card factor of 0.85, a suction pressure of 14.7 lb/in^2, and a polytropic exponent of 1.30.

SOLUTION

Rearranging Eq. (18.2), we find the brake mean effective pressure as follows:

$$p_b = \frac{(33\,000)(12)(\text{bhp})}{lan} = \frac{(33\,000)(12)(57/6)}{(140/6)(2\,400/2)} = 135 \text{ lb/in}^2$$

Then, from Eq. (18.3), the indicated mean effective pressure is

$$p_i = \frac{p_b}{e_m} = \frac{135}{0.75} = 180 \text{ lb/in}^2$$

We must now determine p_4 on the theoretical diagram of Fig. 18.12. Employing Eq. (18.7), we find

$$p_4 = (k-1)\frac{r-1}{r^k - r}\frac{p_i}{f_c} + p_1$$

$$= (1.3 - 1)\frac{8-1}{8^{1.3} - 8}\frac{180}{0.85} + 14.7 = 78.2 \text{ lb/in}^2$$

The volume difference $v_1 - v_2$ in Fig. 18.12 is the volume swept out by the piston. Therefore

$$v_1 - v_2 = la = \frac{140}{6} = 23.3 \text{ in}^3$$

Then, from Eq. (*b*) of Section 18.2, we have

$$v_1 = \frac{r(v_1 - v_2)}{r - 1} = \frac{8(23.3)}{8 - 1} = 26.6 \text{ in}^3$$

Therefore

$$v_2 = 26.6 - 23.3 = 3.3 \text{ in}^3$$

Then the percentage clearance C is

$$C = \frac{3.3(100)}{23.3} = 14.2\%$$

Expressing volumes as percentages of the displacement volume enables us to write Eq. (18.1) in the form

$$p_x(X + C)^k = p_1(100 + C)^k$$

where X is the percentage of piston travel measured from the head end of the stroke. Thus the formula

$$p_x = p_1 \left(\frac{100 + C}{X + C} \right)^k = 14.7 \left(\frac{100 + 14.2}{X + 14.2} \right)^{1.3} \qquad (1)$$

is used to compute the pressure during the compression stroke for any piston position between $X = 0$ and $X = 100$ percent. For the expansion stroke Eq. (18.1) becomes

$$p_x = p_4 \left(\frac{100 + C}{X + C} \right)^k = 78.2 \left(\frac{100 + 14.2}{X + 14.2} \right)^{1.3} \qquad (2)$$

Equations (1) and (2) are easy to program for machine computation. The results can be displayed and recorded, or printed, for graphical use. Alternatively, the results can be displayed on a CRT for hand calculation.

Figure 18.23 shows the plotted results of the computation using $\Delta X = 5$ percent increments. Note particularly how the results have been rounded to obtain a smooth indicator diagram. This rounding will, of course, produce results that will not be exactly duplicated in subsequent trials. The greatest differences will occur in the vicinity of point B.

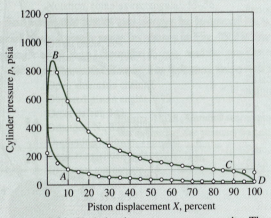

Figure 18.23 Circled points are computer results. The diagram was rounded by hand from A to B and from C to D. Point B is about 75 percent of maximum computed pressure at the beginning of the expansion stroke.

Force Analysis   In a computer analysis the values of the pressure will be read from a diagram like Fig. 18.23. Because most analysts will want to tabulate this data, a table should be constructed with the first column containing values of the crank angle ωt. For a four-cycle engine, values of this angle should be entered from $0°$ to $720°$.

Values of x corresponding to each ωt are obtained from Eq. (18.9). Then the corresponding piston displacement X in percent is obtained from the equation

$$X = \frac{r - l - x}{2r}(100) \qquad (18.46)$$

Some care must be taken in tabulating X and the corresponding pressures. Then the gas forces corresponding to each value of ωt are computed using the piston area.

The balance of the analysis is perfectly straightforward; use Eqs. (18.11), (18.13), and (18.40) through (18.44) in that order.

PROBLEMS

18.1 A one-cylinder, four-cycle engine has a compression of 7.6 and develops 3 bhp at 3 000 rev/min. The crank length is 0.875 in with a 2.375-in bore. Develop and plot a rounded indicator diagram using a card factor of 0.90, a mechanical efficiency of 72 percent, a suction pressure of 14.7 lb/in², and a polytropic exponent of 1.30.

18.2 Construct a rounded indicator diagram for a four-cylinder, four-cycle gasoline engine having a 3.375-in bore, a 3.5-in stroke, and a compression ratio of 6.25. The operating conditions to be used are 30 hp at 1 900 rev/min. Use a mechanical efficiency of 72 percent, a card factor of 0.90, and a polytropic exponent of 1.30.

18.3 Construct an indicator diagram for a V6 four-cycle gasoline engine having a 100-mm bore, a 90-mm stroke, and a compression ratio of 8.40. The engine develops 150 kW at 4 400 rev/min. Use a mechanical efficiency of 72 percent, a card factor of 0.88, and a polytropic exponent of 1.30.

18.4 A single-cylinder, two-cycle gasoline engine develops 30 kW at 4 500 rev/min. The engine has an 80-mm bore, a stroke of 70 mm, and a compression ratio of 7.0. Develop a rounded indicator diagram for this engine using a card factor of 0.990, a mechanical efficiency of 65 percent, a suction pressure of 100 kPa, and a polytropic exponent of 1.30.

18.5 The engine of Problem 18.1 has a connecting rod 3.125 in long and a weight of 0.124 lb, with the mass center 0.40 in from the crankpin end. Piston weight is 0.393 lb. Find the bearing reactions and the crankshaft torque during the expansion stroke corresponding to a piston displacement of $X = 30$ percent ($\omega t = 60°$). To find p_e, see the answer to Problem 18.3 in the answers to selected problems.

18.6 Repeat Problem 18.5, but do the computations for the compression cycle ($\omega t = 660°$).

18.7 Make a complete force analysis of the engine of Problem 18.5. Plot a graph of the crankshaft torque versus crank angle for 720° of crank rotation.

18.8 The engine of Problem 18.3 uses a connecting rod 300 mm long. The masses are $m_{3A} = 0.80$ kg, $m_{3B} = 0.38$ kg, and $m_4 = 1.64$ kg. Find all the bearing reactions and the crankshaft torque for one cylinder of the engine during the expansion stroke at a piston displacement of $X = 30$ percent ($\omega t = 63.2°$). The pressure should be obtained from the indicator diagram, Fig. AP18.3 in the answers to selected problems.

18.9 Repeat Problem 18.8, but do the computations for the same position in the compression cycle ($\omega t = 656.8°$).

18.10 Additional data for the engine of Problem 18.4 are $l_3 = 110$ mm, $R_{G3A} = 15$ mm, $m_4 = 0.24$ kg, and $m_3 = 0.13$ kg. Make a complete force analysis of the engine and plot a graph of the crankshaft torque versus crank angle for 360° of crank rotation.

18.11 The four-cycle engine of Problem 18.1 has a stroke of 2.60 in and a connecting rod length of 7.20 in. The weight of the rod is 0.850 lb, and the center of mass center is 1.66 in from the crankpin. The piston assembly weighs 1.27 lb. Make a complete force analysis for one cylinder of this engine for 720° of crank rotation. Use 16 lb/in² for the exhaust pressure and 10 lb/in² for the suction pressure. Plot a graph to show the variation of the crankshaft torque with the crank angle. Use Fig. 18.23 for the pressures.

19 | Balancing

Balancing is the technique of correcting or eliminating unwanted inertia forces and moments in rotating machinery. In previous chapters we have seen that shaking forces on the frame can vary significantly during a cycle of operation. Such forces can cause vibrations that at times may reach dangerous amplitudes. Even if they are not dangerous, vibrations increase the component stresses and subject bearings to repeated loads that may cause parts to fail prematurely by fatigue. Thus, in the design of machinery it is not sufficient merely to avoid operation near the critical speeds; we must eliminate, or at least reduce, the dynamic forces that produce these vibrations in the first place.

Production tolerances used in the manufacture of machinery are adjusted as closely as possible without increasing the cost of manufacture prohibitively. In general, it is more economical to produce parts that are not quite true and then to subject them to a balancing procedure than it is to produce such perfect parts that no correction is needed. Because of this, each part produced is an individual case in that no two parts can normally be expected to require the same corrective measures. Thus determining the unbalance and the application of corrections is the principal problem in the study of balancing.

19.1 STATIC UNBALANCE

The arrangement shown in Fig. 19.1a consists of a disk-and-shaft combination resting on rigid rails so that the shaft, which is assumed to be perfectly straight, can roll without friction. A reference system xyz is attached to the disk and moves with it. Simple experiments to determine whether the disk is statically unbalanced can be conducted as follows. Roll the disk gently by hand and permit it to coast until it comes to rest. Then mark with chalk the lowest point of the periphery of the disk. Repeat four or five times. If the chalk marks are scattered at different places around the periphery, the disk is in static balance. If

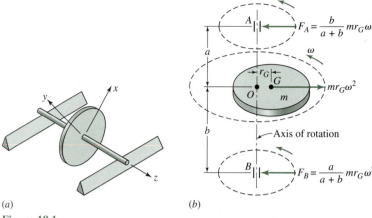

(a)   (b)

Figure 19.1

all the chalk marks are coincident, the disk is statically unbalanced, which means that the axis of the shaft and center of mass of the disk are not coincident. The position of the chalk marks with respect to the xy system indicates the angular location of unbalance but *not* the amount.

It is unlikely that any of the marks will be located 180° from the remaining ones even though it is theoretically possible to obtain static equilibrium with the unbalance above the shaft axis.

If static unbalance is found to exist, it can be corrected by drilling out material at the chalk mark or by adding mass to the periphery 180° from the mark. Because the amount of unbalance is unknown, these corrections must be made by trial and error.

19.2 EQUATIONS OF MOTION

If an unbalanced disk and shaft is mounted in bearings and caused to rotate, the centrifugal force $mr_G\omega^2$ exists as shown in Fig. 19.1b. This force acting upon the shaft produces the rotating bearing reactions shown in the figure.

In order to determine the equation of motion of the system, we specify m as the total mass and m_u as the unbalanced mass. Also, let k be the shaft stiffness, a number that describes the magnitude of a force necessary to bend the shaft a unit distance when applied at O. Thus, k has units of pounds force per inch or newtons per meter. Let c be the coefficient of viscous damping as defined in Eq. (17.2). Selecting any x coordinate normal to the shaft axis, we can now write

$$\sum F_o = -kx - c\dot{x} - m\ddot{x} + m_u r_G \omega^2 \cos \omega t = 0 \qquad (a)$$

The solution to this differential equation was studied in Section 17.12. From Eq. (17.52) it is

$$x = \frac{m_u r_G \omega^2 \cos(\omega t - \phi)}{\sqrt{(k - m\omega^2) + c^2\omega^2}} \qquad (b)$$

where ϕ is the angle between the force $m_u r_G \omega^2$ and the amplitude X of the shaft vibration; thus ϕ is the *phase angle*. Its value is

$$\phi = \tan^{-1}\left(\frac{c\omega}{k - m\omega^2}\right) \qquad (c)$$

Certain simplifications can be made with Eq. (b) to clarify its meaning.

First consider the term $k - m\omega^2$ in the denominator of Eq. (b). If this term were zero, the amplitude of x would be very large because it would be limited only by the damping constant c, which is usually very small. The value of ω that makes the term $k - m\omega^2$ zero is called the *natural angular velocity*, the *critical speed*, and also the *natural circular frequency*. It is designated as ω_n and is seen to be

$$\omega_n = \sqrt{\frac{k}{m}} \qquad (19.1)$$

In the study of free or unforced vibrations it is found that a certain value of the viscous damping factor c will result in no vibration at all. This special value is called the *critical coefficient of viscous damping* and is given by the equation

$$c_c = 2m\omega_n \qquad (19.2)$$

The *damping ratio* ζ is the ratio of the actual to the critical and is

$$\zeta = \frac{c}{c_c} = \frac{c}{2m\omega_n} \qquad (19.3)$$

For most machine systems in which damping is not deliberately introduced, ζ will be in the approximate range $0.015 \le \zeta \le 0.120$.

Next, note that Eq. (b) can be expressed in the form

$$x = X \cos(\omega t - \phi) \qquad (d)$$

If we now divide the numerator and denominator of the amplitude X of Eq. (b) by k, designate the eccentricity as $e = r_G$, and introduce Eqs. (19.1) and (19.3), we obtain the ratio

$$\frac{mX}{m_u e} = \frac{(\omega/\omega_n)^2}{\sqrt{\left(1 - \omega^2/\omega_n^2\right)^2 + (2\zeta\omega/\omega_n)^2}} \qquad (19.4)$$

This is the equation for the amplitude ratio of the vibration of a rotating disk-and-shaft combination. If we neglect damping, let $m = m_u$, and replace e with r_G again, we obtain

$$X = r_G \frac{(\omega/\omega_n)^2}{1 - (\omega/\omega_n)^2} \qquad (19.5)$$

where r_G is the eccentricity and X is the amplitude of the vibration corresponding to any frequency ratio ω/ω_n. Now if, in Fig. 19.1b, we designate O as the center of the shaft at the

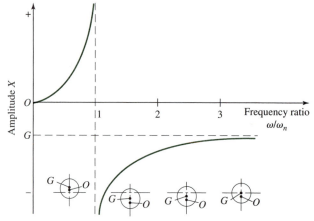

Figure 19.2 The small figures below the graph indicate the relative position of three points for various frequency ratios. The mass center of the disk is at G, the center of the shaft is at O, and the axis of rotation is at the intersection of the centerlines. Thus this figure shows both amplitude and phase relationships.

disk and G as the mass center of the disk, we can draw some interesting conclusions by plotting Eq. (19.5). This is done in Fig. 19.2, where the amplitude is plotted on the vertical axis and the frequency ratio along the abscissa. The natural frequency is ω_n, which corresponds to the critical speed, while ω is the actual speed of the shaft. When rotation is just beginning, ω is much less than ω_n and the graph shows that the amplitude of the vibration is very small. As the shaft speed increases, the amplitude also increases and becomes infinite at the critical speed. As the shaft goes through the critical, the amplitude changes over to a negative value and decreases as the shaft speed increases. The graph shows that the amplitude never returns to zero no matter how much the shaft speed is increased but reaches a limiting value of $-r_G$. Note that in this range the disk is rotating about its own center of gravity, which is then coincident with the bearing centerline.

The preceding discussion demonstrates that statically unbalanced rotating systems produce undesirable vibrations and rotating bearing reactions. Using static balancing equipment, the eccentricity r_G can be reduced, but it is impossible to make it zero. Therefore, no matter how small r_G is made, trouble can always be expected whenever $\omega = \omega_n$. When the operating frequency is higher than the natural frequency, the machine should be designed to pass through the natural frequency as rapidly as possible in order to prevent dangerous vibrations from building.

19.3 STATIC BALANCING MACHINES

The purpose of a balancing machine is first to indicate whether a part is in balance. If it is out of balance, the machine should measure the unbalance by indicating its *magnitude* and *location*.

Static balancing machines are used only for parts whose axial dimensions are small, such as gears, fans, and impellers, and the machines are often called *single-plane balancers* because the mass must lie practically in a single plane. In the sections to follow we discuss balancing in several planes, but it is important to note here that if several wheels are to be mounted upon a shaft that is to rotate, the parts should be individually statically balanced before mounting. While it is possible to balance the assembly in two planes after the parts are mounted, additional bending moments inevitably come into existence when this is done.

Static balancing is essentially a weighting process in which the part is acted upon by either a gravity force or a centrifugal force. We have seen that the disk and shaft of the preceding section could be balanced by placing it on two parallel rails, rocking it, and permitting it to seek equilibrium. In this case the location of the unbalance is found through the aid of the force of gravity. Another method of balancing the disk would be to rotate it at a predetermined speed. Then the bearing reactions could be measured and their magnitudes used to indicate the amount of unbalance. Because the part is rotating while the measurements are taken, a stroboscope is used to indicate the location of the required connection.

When machine parts are manufactured in large quantities, a balancer is required which will measure both the amount and location of the unbalance and give the correction directly and quickly. Time can also be saved if it is not necessary to rotate the part. Such a balancing machine is shown in Fig. 19.3. This machine is essentially a pendulum which can tilt in any direction, as illustrated by the schematic drawing of Fig. 19.4a.

Figure 19.3 A helicopter-rotor assembly balancer. (Micro-Poise Engineering and Sales Company, Detroit, MI.)

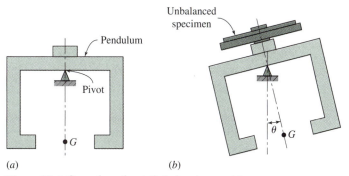

Figure 19.4 Operation of a static balancing machine.

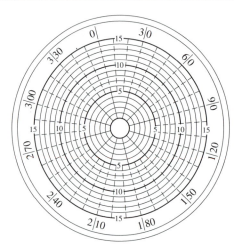

Figure 19.5 Drawing of the universal level used in the Micro-Poise balancer. The numbers on the periphery are degrees; the radial distances are calibrated in units proportional to ounce-inches. The position of the bubble indicates both the direction and the magnitude of the unbalance. (Micro-Poise Engineering and Sales Company, Detroit, MI.)

When an unbalanced specimen is mounted on the platform of the machine, the pendulum tilts. The direction of the tilt gives the location of the unbalance, while the angle θ (Fig. 19.4b) indicates the magnitude of the unbalance. Some damping is employed to eliminate oscillations of the pendulum. Figure 19.5 shows a universal level that is mounted on the platform of the balancer. A bubble, shown at the center, moves and shows both the location and the magnitude of the correction.

19.4 DYNAMIC UNBALANCE

Figure 19.6 shows a long rotor that is to be mounted in bearings at A and B. We might suppose that two equal masses m_1 and m_2 are placed at opposite ends of the rotor and at equal distances r_1 and r_2 from the axis of rotation. Because the masses are equal and on opposite sides of the rotational axis, the rotor can be placed on rails as described earlier to show that it is statically balanced in all angular positions.

If the rotor of Fig. 19.6 is placed in bearings and caused to rotate at an angular velocity ω rad/s, the centrifugal forces $m_1 r_1 \omega^2$ and $m_2 r_2 \omega^2$ act, respectively, at m_1 and m_2 on the rotor ends. These centrifugal forces produce the unequal bearing reactions $\mathbf{F}_A$ and $\mathbf{F}_B$, and the entire system of forces rotates with the rotor at the angular velocity ω. Thus a part may be statically balanced and at the same time dynamically unbalanced (Fig. 19.7).

In the general case, distribution of the mass along the axis of the part depends upon the configuration of the part, but errors occur in machining and also in casting and in forging. Other errors or unbalance may be caused by improper boring, by keys, and by assembly. It is the designer's responsibility to design so that a line joining all mass centers will be a straight line coinciding with the axis of rotation. However, perfect parts and perfect assembly are seldom attained, and consequently a line from one end of the part to the other, joining all mass centers, will usually be a space curve which may occasionally cross or coincide with the axis of rotation. An unbalanced part, therefore, will usually be out of balance both statically and dynamically. This is the most general kind of unbalance, and if the part is supported by two bearings, one can then expect the magnitudes as well as the directions of these rotating bearing reactions to be different.

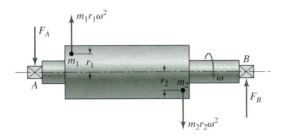

Figure 19.6 If $m_1 = m_2$ and $r_1 = r_2$, the rotor is statically balanced but dynamically unbalanced.

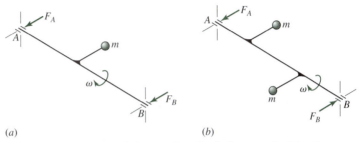

(a)   (b)

Figure 19.7 (*a*) Static unbalance; when the shaft rotates, both bearing reactions are in the same plane and in the same direction. (*b*) Dynamic unbalance; when the shaft rotates, the unbalance creates a couple tending to turn the shaft end over end. The shaft is in equilibrium because of the opposite couple formed by the bearing reactions. Note that the bearing reactions are still in the same plane but in opposite directions.

19.5 ANALYSIS OF UNBALANCE

In this section we show how to analyze any unbalanced rotating system and determine the proper corrections using graphical methods, vector methods, and computer or calculator programming.

Graphic Analysis   The two equations

$$\sum \mathbf{F} = 0 \quad \text{and} \quad \sum \mathbf{M} = 0 \qquad (a)$$

are used to determine the amount and location of the corrections. We begin by noting that the centrifugal force is proportional to the product $m\mathbf{r}$ of a rotating eccentric mass. Thus vector quantities, proportional to the centrifugal force of each of the three masses $m_1\mathbf{R}_1$, $m_2\mathbf{R}_2$, and $m_3\mathbf{R}_3$ of Fig. 19.8*a* will act in radial directions as shown. The first of Eqs. (*a*) is applied by constructing a force polygon (Fig. 19.8*b*). Because this polygon requires another vector $m_c\mathbf{R}_c$ for closure, the magnitude of the correction is $m_c\mathbf{R}_c$ and its direction is parallel to $\mathbf{R}_c$. The three masses of Fig. 19.8 are assumed to rotate in a single plane and so this is a case of static unbalance.

When the rotating masses are in different planes, both Eqs. (*a*) must be used. Figure 19.9*a* is an end view of a shaft having mounted upon it the three masses m_1, m_2, and m_3 at the radial distances R_1, R_2, and R_3, respectively. Figure 19.9*b* is a side view of the

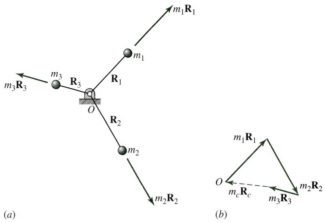

Figure 19.8 (*a*) A three-mass system rotating in a single plane. (*b*) Centrifugal force polygon gives $m_c\mathbf{R}_c$ as the required correction.

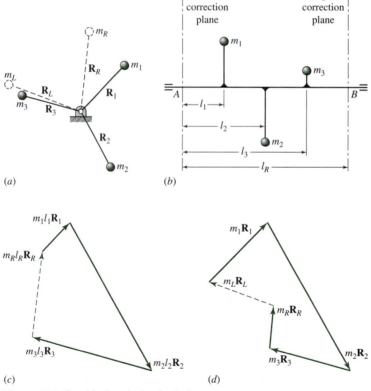

Figure 19.9 Graphical analysis of unbalance.

same shaft showing left and right correction planes and the distances to the three masses. We want to find the magnitude and angular location of the corrections for each plane.

The first step in the solution is to take a summation of the moments of the centrifugal forces, including the corrections, about some point. We choose to take this summation about A in the left correction plane in order to eliminate the moment of the left correction mass. Thus applying the second of Eqs. (a) gives

$$\sum \mathbf{M}_A = m_1 l_1 \mathbf{R}_1 + m_2 l_2 \mathbf{R}_2 + m_3 l_3 \mathbf{R}_3 + m_R l_R \mathbf{R}_R = \mathbf{0} \qquad (b)$$

This is a vector equation in which the directions of the vectors are parallel, respectively, to the vectors $\mathbf{R}_N$ in Fig. 19.9a. Consequently, the moment polygon in Fig. 19.9c can be constructed. The closing vector $m_R l_R \mathbf{R}_R$ gives the magnitude and direction of the correction required for the right-hand plane. The quantities m_R and $\mathbf{R}_R$ can now be found because the magnitude of $\mathbf{R}_R$ is ordinarily given in the problem. Therefore the equation

$$\sum \mathbf{F} = m_1 \mathbf{R}_1 + m_2 \mathbf{R}_2 + m_3 \mathbf{R}_3 + m_R \mathbf{R}_R + m_L \mathbf{R}_L = \mathbf{0} \qquad (c)$$

can be written. The magnitude of $\mathbf{R}_L$ being given, this equation is solved for the left-hand correction $m_L \mathbf{R}_L$ by constructing the force polygon of Fig. 19.9d.

Though Fig. 19.9c has been called a moment polygon, it is worth noting that the vectors making up this polygon consist of the moment magnitude and the position vector directions. A true moment polygon would be obtained by rotating the polygon 90° cw, because each moment vector is equal to $ml\hat{\mathbf{k}} \times \mathbf{R}\omega^2$.

Vector Analysis   The following two examples illustrate the vector algebra approach.

EXAMPLE 19.1

Figure 19.10 represents a rotating system that has been idealized for illustrative purposes. A weightless shaft that is supported in bearings at A and B rotates with an angular velocity $\omega = 100\hat{\mathbf{i}}$ rad/s. When U.S. customary units are used, the unbalances are given in ounces. The three weights w_1, w_2, and w_3 are connected to the shaft and rotate with it, causing an unbalance. Determine the bearing reactions at A and B for the angular orientation shown.

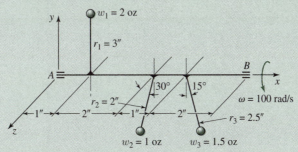

Figure 19.10 Example 19.1.

SOLUTION

We begin by calculating the centrifugal force due to each rotating weight:

$$m_1 r_1 \omega^2 = \frac{(2 \text{ oz})(3 \text{ in})(100 \text{ rad/s})^2}{(386 \text{ in/s}^2)(16 \text{ oz/lb})} = 9.72 \text{ lb}$$

$$m_2 r_2 \omega^2 = \frac{(1 \text{ oz})(2 \text{ in})(100 \text{ rad/s})^2}{(386 \text{ in/s}^2)(16 \text{ oz/lb})} = 3.24 \text{ lb}$$

$$m_3 r_3 \omega^2 = \frac{(1.5 \text{ oz})(2.5 \text{ in})(100 \text{ rad/s})^2}{(386 \text{ in/s}^2)(16 \text{ oz/lb})} = 6.07 \text{ lb}$$

These three forces are parallel to the yz plane, and we can now write them in vector form by inspection:

$$\mathbf{F}_1 = m_1 r_1 \omega^2 \angle \theta_1 = 9.72 \angle 0° = 9.72 \hat{\mathbf{j}} \text{ lb}$$

$$\mathbf{F}_2 = m_2 r_2 \omega^2 \angle \theta_2 = 3.24 \angle 120° = -1.62 \hat{\mathbf{j}} + 2.81 \hat{\mathbf{k}} \text{ lb}$$

$$\mathbf{F}_3 = m_3 r_3 \omega^2 \angle \theta_3 = 6.07 \angle 195° = -5.86 \hat{\mathbf{j}} - 1.57 \hat{\mathbf{k}} \text{ lb}$$

where θ in this example, is measured counterclockwise from y when viewed from the positive end of x. The moments of these forces taken about the bearing at A must be balanced by the moment of the bearing reaction at B. Therefore,

$$\sum \mathbf{M}_A = 1\hat{\mathbf{i}} \times 9.72\hat{\mathbf{j}} + 3\hat{\mathbf{i}} \times (-1.62\hat{\mathbf{j}} + 2.81\hat{\mathbf{k}}) + 4\hat{\mathbf{i}} \times (-5.86\hat{\mathbf{j}} - 1.57\hat{\mathbf{k}}) + 6\hat{\mathbf{i}} \times \mathbf{F}_B = 0$$

Solving this gives the bearing reaction at B as

$$\mathbf{F}_B = -3.10\hat{\mathbf{j}} - 0.36\hat{\mathbf{k}} \text{ lb} = 3.12 \angle 186.6° \text{ lb} \qquad \textit{Ans.}$$

To find the reaction at A we repeat the analysis. Taking moments about B gives

$$\sum \mathbf{M}_B = -2\hat{\mathbf{i}} \times (-5.86\hat{\mathbf{j}} - 1.57\hat{\mathbf{k}}) + (-3\hat{\mathbf{i}}) \times (-1.62\hat{\mathbf{j}} + 2.81\hat{\mathbf{k}})$$

$$+ (-5\hat{\mathbf{i}}) \times 9.72\hat{\mathbf{j}} + (-6\hat{\mathbf{i}}) \times \mathbf{F}_A = 0$$

Solving again gives

$$\mathbf{F}_A = -5.34\hat{\mathbf{j}} - 0.88\hat{\mathbf{k}} \text{ lb} = 5.41 \angle 189.4° \text{ lb} \qquad \textit{Ans.}$$

Note that these are rotating reactions and that the static or stationary components due to the gravity force are not included.

EXAMPLE 19.2

(*a*) What are the bearing reactions for the system shown in Figure 19.11 if the speed is 750 rev/min? (*b*) Determine the location and the magnitude of a balancing mass if it is to be placed at a radius of 0.25 m.

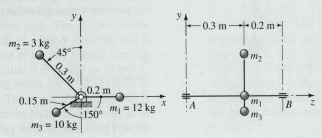

Figure 19.11 Example 19.2.

SOLUTION

(*a*) The angular velocity of this system is $\omega = 2\pi(750)/60 = 78.5$ rad/s. The centrifugal forces due to the masses are

$$\mathbf{F}_1 = m_1 r_1 \omega^2 = (12 \text{ kg})(0.2 \text{ m})(78.5 \text{ rad/s})^2(10)^{-3} = 14.80 \text{ kN}$$

$$\mathbf{F}_2 = m_2 r_2 \omega^2 = (3 \text{ kg})(0.3 \text{ m})(78.5 \text{ rad/s})^2(10)^{-3} = 5.55 \text{ kN}$$

$$\mathbf{F}_3 = m_3 r_3 \omega^2 = (10 \text{ kg})(0.15 \text{ m})(78.5 \text{ rad/s})^2(10)^{-3} = 9.25 \text{ kN}$$

In vector form these forces are

$$\mathbf{F}_1 = 14.80\angle 0° = 14.80\hat{\mathbf{i}} \text{ kN}, \qquad \mathbf{F}_2 = 5.55\angle 135° = -3.93\hat{\mathbf{i}} + 3.93\hat{\mathbf{j}} \text{ kN},$$

$$\mathbf{F}_3 = 9.25\angle -150° = -8.01\hat{\mathbf{i}} - 4.63\hat{\mathbf{j}} \text{ kN}$$

To find the bearing reaction at *B* we take moments about the bearing *A*. This equation is written

$$\sum \mathbf{M}_A - 0.3\hat{\mathbf{k}} \times [(14.8\hat{\mathbf{i}}) + (-3.93\hat{\mathbf{i}} + 3.93\hat{\mathbf{j}}) + (-8.00\hat{\mathbf{i}} - 4.63\hat{\mathbf{j}})] + 0.5\hat{\mathbf{k}} \times \mathbf{F}_B = 0$$

Taking the cross products and rearranging gives

$$0.5\hat{\mathbf{k}} \times \mathbf{F}_B = -0.21\hat{\mathbf{i}} - 0.86\hat{\mathbf{j}}$$

When this equation is solved for $\mathbf{F}_B$, we obtain

$$\mathbf{F}_B = 1.72\hat{\mathbf{i}} + 0.42\hat{\mathbf{j}} = 1.77\angle 13.7° \text{ kN} \qquad\qquad\qquad Ans.$$

We can find the reaction at *A* by summing forces. Thus

$$\mathbf{F}_A = -\mathbf{F}_1 - \mathbf{F}_2 - \mathbf{F}_3 - \mathbf{F}_B$$

$$= -14.8\hat{\mathbf{i}} - (-3.93\hat{\mathbf{i}} + 3.93\hat{\mathbf{j}}) - (-8.01\hat{\mathbf{i}} - 4.63\hat{\mathbf{j}}) - (1.72\hat{\mathbf{i}} + 0.42\hat{\mathbf{j}})$$

and

$$\mathbf{F}_A = -4.59\hat{\mathbf{i}} + 0.28\hat{\mathbf{j}} = 4.59\angle 176.5° \text{ kN} \qquad \textit{Ans.}$$

(*b*) Let $\mathbf{F}_C$ be the correcting force. Then for zero bearing reactions

$$\sum \mathbf{F} = \mathbf{F}_1 + \mathbf{F}_2 + \mathbf{F}_3 + \mathbf{F}_C = \mathbf{0}$$

Thus

$$\mathbf{F}_C = -14.8\hat{\mathbf{i}} - (-3.92\hat{\mathbf{i}} + 3.92\hat{\mathbf{j}}) - (-8.00\hat{\mathbf{i}} - 4.62\hat{\mathbf{j}})$$
$$= -2.87\hat{\mathbf{i}} + 0.70\hat{\mathbf{j}} = 2.95\angle 166.3° \text{ kN}$$

and so

$$m_C = \frac{F_C}{r_C \omega^2} = \frac{2.95(10)^3}{0.25(78.5)^2} = 1.91 \text{ kg} \qquad \textit{Ans.}$$

Scalar Equations   For a rotating system with n discrete mass particles and 2 correcting planes ($j = 1$ and 2) we can write the following four scalar equations:

$$\sum_{i=1}^{n} m_i r_i \cos \phi_i + \sum_{j=1}^{2} m_{cj} r_{cj} \cos \phi_{cj} = 0 \qquad (19.6a)$$

$$\sum_{i=1}^{n} m_i r_i \sin \phi_i + \sum_{j=1}^{2} m_{cj} r_{cj} \sin \phi_{cj} = 0 \qquad (19.6b)$$

$$\sum_{i=1}^{n} Z_i m_i r_i \cos \phi_i + \sum_{j=1}^{2} Z_{cj} m_{cj} r_{cj} \cos \phi_{cj} = 0 \qquad (19.7a)$$

$$\sum_{i=1}^{n} Z_i m_i r_i \sin \phi_i + \sum_{j=1}^{2} Z_{cj} m_{cj} r_{cj} \sin \phi_{cj} = 0 \qquad (19.7b)$$

Note that the equations do not depend on the angular velocity of the rotating system; that is, if the system is balanced at one speed, then it is balanced for all speeds.

For a distributed mass system and 2 correcting planes we can write the following four scalar equations:

$$F_{A_{21x}} + F_{B_{21x}} + \omega^2 \sum_{j=1}^{2} m_{cj} r_{cj} \cos \phi_{cj} = 0 \qquad (19.8a)$$

$$F_{A_{21y}} + F_{B_{21y}} + \omega^2 \sum_{j=1}^{2} m_{cj} r_{cj} \sin \phi_{cj} = 0 \qquad (19.8b)$$

$$Z_A F_{A_{21x}} + Z_B F_{B_{21x}} + \omega^2 \sum_{j=1}^{2} Z_j m_{cj} r_{cj} \cos \phi_{cj} = 0 \tag{19.9a}$$

$$Z_A F_{A_{21y}} + Z_B F_{B_{21y}} + \omega^2 \sum_{j=1}^{2} Z_j m_{cj} r_{cj} \sin \phi_{cj} = 0 \tag{19.9b}$$

A Direct Method of Dynamic Balancing   To avoid solving four equations in four un-knowns, that is Eqs. (19.6) and (19.7) or Eqs. (19.8) and (19.9), we can take moments about one correction plane and solve for two unknowns then take moments about the other correction plane and solve for the remaining two unknowns. Recall that if a rotating system is dynamically balanced, then it is automatically statically balanced. However, if the system is statically balanced there is no guarantee that it is dynamically balanced (see Section 19.4).

EXAMPLE 19.3

The distributed mass system shown in Fig. 19.12 has been tested for unbalance by rotating the rotor at an angular velocity of 100 rad/s. The bearing reactions on the frame at the two bearings A and B are as shown in the figure. Determine the location and the magnitude of mass corrections to be removed in the specified planes 1 and 2 in order to achieve dynamic balance. The correct-ing masses are to be located at the radii $r_1 = r_2 = r = 4$ in.

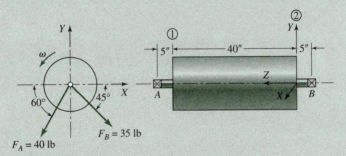

Figure 19.12 Example 19.3.

SOLUTION

The x and y components of the forces acting on the frame at the bearings A and B are

$$F_{A_{21x}} = 40 \cos 240° = -20 \text{ lb} \tag{1a}$$

$$F_{A_{21y}} = 40 \sin 240° = -20\sqrt{3} \text{ lb} \tag{1b}$$

$$F_{B_{21x}} = 35 \cos(-45°) = 17.5\sqrt{2} \text{ lb} \tag{1c}$$

$$F_{B_{21y}} = 35 \sin(-45°) = -17.5\sqrt{2} \text{ lb} \tag{1d}$$

Equations (19.8) and (19.9) can be written as

$$m_1 r\omega^2 \cos\phi_1 + m_2 r\omega^2 \cos\phi_2 = -F_{A_{21x}} - F_{B_{21x}} \tag{2a}$$

$$m_1 r\omega^2 \sin\phi_1 + m_2 r\omega^2 \sin\phi_2 = -F_{A_{21y}} - F_{B_{21y}} \tag{2b}$$

$$Z_1 m_1 r_1 \omega^2 \cos\phi_1 + Z_2 m_2 r_2 \omega^2 \cos\phi_2 = -Z_A F_{A_{21x}} - Z_B F_{B_{21x}} \tag{3a}$$

$$Z_1 m_1 r_1 \omega^2 \sin\phi_1 + Z_2 m_2 r_2 \omega^2 \sin\phi_2 = -Z_A F_{A_{21y}} - Z_B F_{B_{21y}} \tag{3b}$$

To simplify the analysis take moments about one of the two balancing planes. Here we take moments about plane 2 and note that the distances are

$$Z_1 = +40 \text{ in}, \qquad Z_2 = 0, \qquad Z_A = +45 \text{ in}, \qquad Z_B = -5 \text{ in} \tag{4}$$

Substituting Eqs. (4) and the known data into Eqs. (2) and (3) gives

$$(4 \text{ in})(100 \text{ rad/s})^2(m_1 \cos\phi_1 + m_2 \cos\phi_2) = -4.749 \text{ lb} \tag{5a}$$

$$(4 \text{ in})(100 \text{ rad/s})^2(m_1 \sin\phi_1 + m_2 \sin\phi_2) = 59.390 \text{ lb} \tag{5b}$$

$$(40 \text{ in})(4 \text{ in})(100 \text{ rad/s})^2 m_1 \cos\phi_1 = 1\,023.744 \text{ in·lb} \tag{6a}$$

$$(40 \text{ in})(4 \text{ in})(100 \text{ rad/s})^2 m_1 \sin\phi_1 = 1\,435.102 \text{ in·lb} \tag{6b}$$

Dividing Eq. (6b) by Eq. (6a) gives

$$\phi_1 = \tan^{-1}\left(\frac{1\,435.102}{1\,023.744}\right) = 54.5° \tag{7a}$$

Substituting Eq. (7a) into Eq. (6a) or (6b) gives

$$m_1 = \frac{(1\,435.102 \text{ lb})(386 \text{ in/s}^2)}{(1\,600\,000 \text{ in}^2/\text{s}^2)\sin 54.5°} = 0.425 \text{ lb} \tag{7b}$$

Substituting Eqs. (7) into Eqs. (5), rearranging and dividing Eq. (5b) by Eq. (5a) gives

$$\phi_2 = \tan^{-1}\left(\frac{23.512}{-30.342}\right) = 142.2° \tag{8a}$$

Substituting Eq. (8a) into Eq. (5a) or (5b) gives

$$m_2 = \frac{(-30.342 \text{ lb})(386 \text{ in/s}^2)}{(40\,000 \text{ in/s}^2)\cos 142.23°} = 0.370 \text{ lb} \qquad \textit{Ans.} \tag{8b}$$

Therefore, to balance the rotating system we must do the following:

At Plane 1:   Remove the mass $m_1 = 0.425$ lb at the angle $-125.5°$ *Ans.*

At Plane 2:   Remove the mass $m_2 = 0.370$ lb at the angle $-37.8°$ *Ans.*

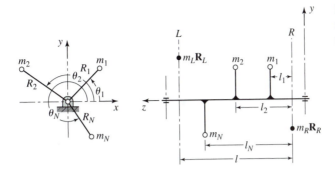

Figure 19.13 Notation for numerical solution; corrections are not shown in end view.

Computer Solution   For a computer analysis it is convenient to choose the xy plane as the plane of rotation with z as the axis of rotation, as shown in Fig. 19.13. In this manner the unbalance vectors $m_i\mathbf{R}_i$ and the two correction vectors, $m_L\mathbf{R}_L$ in the left plane and $m_R\mathbf{R}_R$ in the right plane, can be expressed in the two-dimensional polar notation $m\mathbf{R} = mR\angle\theta$. This makes it easy to use the polar-rectangular conversion feature and its inverse, found on most programmable calculators.

Note that Fig. 19.13 has $m_1, m_2, \ldots, m_N$ unbalances. By solving Eqs. (b) and (c) (in the section entitled "Graphic Analysis," above) for the corrections, we get

$$m_L\mathbf{R}_L = -\sum_{i=1}^{N} \frac{m_i l_i}{l} \mathbf{R}_i \tag{19.10}$$

$$m_R\mathbf{R}_R = -m_L\mathbf{R}_L - \sum_{i=1}^{N} m_i \mathbf{R}_i \tag{19.11}$$

These two equations can easily be programmed for computer solution. If a programmable calculator is used, it is suggested that the summation key be employed with each term of the summation entered using a user-defined key.

19.6 DYNAMIC BALANCING

The units in which each unbalance is measured have customarily been the ounce·inch (oz·in), the gram·centimeter (g·cm), and the bastard unit of gram·inch (g·in). If correct practice is followed in the use of SI units, the most appropriate unite of unbalance is the milligram·meter (mg·m) because prefixes in multiples of 1 000 are preferred in SI; thus, the prefix centi- is not recommended. Furthermore, not more than one prefix should be used in a compound unit; preferably, the first-named quantity should be prefixed. Thus neither the gram·centimeter nor the kilogram·millimeter, both acceptable in size, should be used. In this book we use the ounce·inch (oz·in) and the milligram·meter (mg·m) for units of unbalance.

We have seen that static balancing is sufficient for rotating disks, wheels, gears, and the like, when the mass can be assumed to exist in a single rotating plane. In the case of longer machine elements, such as turbine rotors or motor armatures, the unbalanced centrifugal forces result in couples whose effect tends to cause the rotor to turn end over end.

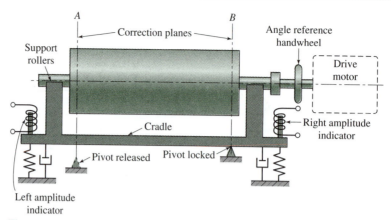

Figure 19.14 Schematic drawing of a pivoted-cradle balancing machine.

The purpose of balancing is to measure the unbalanced couple and to add a new couple in the opposite direction of the same magnitude. The new couple is introduced by the addition of masses in two preselected correction planes or by subtracting masses (drilling out) of these two planes. A rotor to be balanced usually has both static and dynamic unbalance, and consequently the correction masses, their radial locations, or both are not the same for the two correction masses; also their radial locations are not the same for the two correction planes. This also means that the angular separation of the correction masses on the two planes is usually not 180°. Thus, to balance a rotor, one must measure the magnitude and angular location of the correction mass for each of the two correction planes.

Three methods of measuring the corrections for two planes are in general use, the *pivoted-cradle,* the *nodal-point,* and the *mechanical-compensation methods.*

Figure 19.14 shows a specimen to be balanced mounted on half-bearings or rollers attached to a cradle. The right end of the specimen is connected to a drive motor through a universal joint. The cradle can be rocked about either of two points that are adjusted to coincide with the correction planes on the specimen to be balanced. In the figure the left pivot is shown in the released position and the cradle and specimen are free to rock or oscillate about the right pivot, which is shown in the locked position. Springs and dashpots are secured at each end of the cradle to provide a single-degree-of-freedom vibrating system. Often they are made adjustable so that the natural frequency can be tuned to the motor speed. Also shown are amplitude indicators at each end of the cradle. These transducers are differential transformers, or they may consist of a permanent magnet mounted on the cradle which moves relative to a stationary coil to generate a voltage proportional to the unbalance.

With the pivots located in the two correction planes, one can lock either pivot and take readings of the amount and orientation angle of the correction. The readings obtained are completely independent of the measurements taken in the other correction plane because an unbalance in the plane of the locked pivot has no moment about that pivot. With the right-hand pivot locked, an unbalance correctable in the left correction plane causes vibration whose amplitude is measured by the left amplitude indicator. When this correction is made (or measured), the right-hand pivot is released, the left pivot is locked, and another set of measurements is made for the right-hand correction plane using the right-hand amplitude indicator.

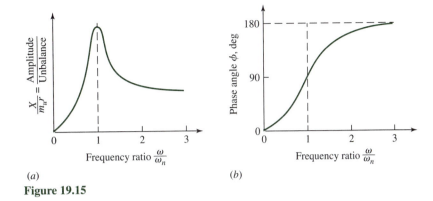

Figure 19.15

The relation between the amount of unbalance and the measured amplitude is given by Eq. (19.4). Rearranging and substituting r for e gives

$$X = \frac{m_u r (\omega/\omega_n)^2}{m\sqrt{\left(1 - \omega^2/\omega_n^2\right)^2 + (2\zeta\omega/\omega_n)^2}} \tag{19.12}$$

where $m_u r$ = unbalance
$\quad\quad m$ = mass of cradle and specimen
$\quad\quad X$ = amplitude

This equation shows that the amplitude of the motion X is directly proportional to the unbalance $m_u r$. Figure 19.15a shows a plot of this equation for a particular damping ratio ζ. The figure shows that the machine is most sensitive near resonance ($\omega = \omega_n$), because in this region the greatest amplitude is recorded for a given unbalance. Damping is deliberately introduced in balancing machines to filter noise and other vibrations that might affect the results. Damping also helps to maintain calibration against effects of temperature and other environmental conditions.

Not shown in Fig. 19.14 is a sine-wave signal generator that is attached to the drive shaft. If the resulting sine-wave signal generator is compared on a dual-beam oscilloscope with the wave generated by one of the amplitude indicators, a phase difference is found. The angular phase difference is the angular orientation of the unbalance. In a balancing machine an electronic phasemeter measures the phase angle and gives the result on another meter calibrated in degrees. To locate the correction on the specimen (Fig. 19.14), the angular reference handwheel is turned by hand until the indicated angle is in line with a reference pointer. This places the heavy side of the specimen in a preselected position and permits the correction to be made.

By manipulating Eq. (c) of Section 19.2, we obtain the equation for the phase angle in parameter form. Thus

$$\phi = \tan^{-1} \frac{2\zeta\omega/\omega_n}{1 - \omega^2/\omega_n^2} \tag{19.13}$$

A plot of this equation for a single damping ratio and for various frequency ratios is shown in Fig. 19.15b. This curve shows that, at resonance, when the speed ω of the shaft and the natural frequency ω_n of the system are the same, the displacement lags the unbalance by the angle $\phi = 90°$. If the top of the specimen is turning away from the operator, the unbalance will be horizontal and directly in front of the operator when the displacement is maximum downward. The figure also shows that the angular location approaches $180°$ as the shaft speed ω is increased above resonance.

19.7 BALANCING MACHINES

A pivoted-cradle balancing machine for high-speed production is illustrated in Fig. 19.16. The shaft-mounted signal generator can be seen at the extreme left.

Nodal-Point Balancing   Plane separation using a point of zero or minimum vibration is called the *nodal-point method of balancing*. To see how this method is used, examine Fig. 19.17. Here the specimen to be balanced is shown mounted on bearings that are fastened to a nodal bar. We assume that the specimen is already balanced in the left-hand correction plane and that an unbalance still exists in the right-hand plane, as shown. Because of this unbalance a vibration of the entire assembly takes place, causing the nodal bar to oscillate about some point O, occupying first position CC and then DD. Point O is easily located by sliding a dial indicator along the nodal bar; a point of zero motion or minimum motion is then readily found. This is the null or nodal point. Its location is the center of oscillation for a center of percussion in the right-hand correction plane.

We assumed, at the beginning of this discussion, that no unbalance existed in the left-hand correction plane. However, if unbalance is present, its magnitude is given by the dial

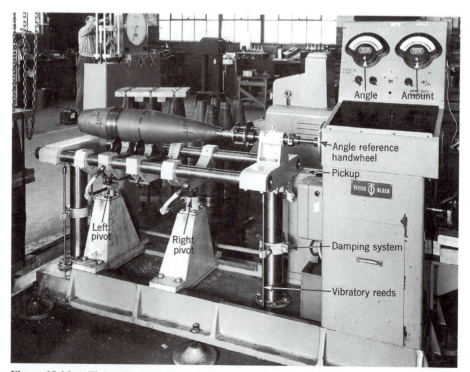

Figure 19.16 A Tinius Olsen static-dynamic pivoted-cradle balancing machine with specimen mounted for balancing. (Tinius Olsen Testing Machine Company, Willow Grove, PA.)

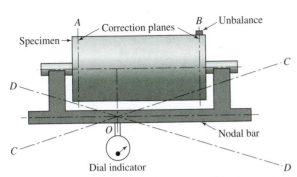

Figure 19.17 Plane separation by the nodal-point method. The nodal bar experiences the same vibration as the specimen.

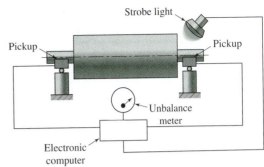

Figure 19.18 Diagram of the electrical circuit in a Micro Dynamic Balancer. (Micro-Balancing, Inc., Garden City Park, NY.)

indicator located at the nodal point just found. Thus, by locating the dial indicator at this nodal point, we measure the unbalance in the left-hand plane without any interference from that in the right-hand plane. In a similar manner, another nodal point can be found which will measure only the unbalance in the right-hand correction plane without any interference from that in the left-hand plane.

In commercial balancing machines employing the nodal-point principle, the plane separation is accomplished in electrical networks. Typical of these is the Micro Dynamic Balancer; a schematic is shown in Fig. 19.18. On this machine a switching knob selects either correction plane and displays the unbalance on a voltmeter, which is calibrated in appropriate unbalance units.

The computer of Fig. 19.18 contains a filter that eliminates bearing noise and other frequencies not related to the unbalance. A multiplying network is used to give the sensitivity desired and to cause the meter to read in preselected balancing units. The strobe light is driven by an oscillator that is synchronized to the rotor speed.

The rotor is driven at a speed that is much greater than the natural frequency of the system, and because the damping is quite small, Fig. 19.15b shows that the phase angle is approximately 180°. Marked on the right-hand end of the rotor are degrees or numbers that are readable and stationary under the strobe light during rotation of the rotor. Thus it is only necessary to observe the particular station number of the degree marking under the strobe light to locate the heavy spot. When the switch is shifted to the other correction plane, the meter again reads the amount and the strobe light illuminates the station. Sometimes as few as five station numbers distributed uniformly around the periphery are adequate for balancing.

The direction of the vibration is horizontal, and the phase angle is nearly 180°. Thus, rotation such that the top of the rotor moves away from the operator will cause the heavy spot to be in a horizontal plane and on the near side of the axis when illuminated by the strobe lamp. A pointer is usually placed here to indicate its location. If, during production balancing, it is found that the phase angle is less than 180°, the pointer can be shifted slightly to indicate the proper position to observe.

Mechanical Compensation   An unbalanced rotating rotor located in a balancing machine develops a vibration. One can introduce in the balancing machine counterforces in each correction plane which exactly balance the forces causing the vibration. The result

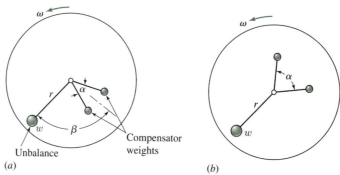

Figure 19.19 Correction plane viewed along the axis of rotation to show the unbalance and the compensator weights: (*a*) position of compensator weights increases the vibration; (*b*) compensated.

of introducing these forces is a smooth running motor. Upon stopping, the location and amount of the counterforce are measured to give the exact correction required. This is called *mechanical compensation.*

When mechanical compensation is used, the speed of the rotor during balancing is not important because the equipment is in calibration for all speeds. The rotor may be driven by a belt, from a universal joint, or it may be self-driven if, for example, it is a gasoline engine. The electronic equipment is simple, no built-in damping is necessary, and the machine is easy to operate because the unbalance in both correction planes is measured simultaneously and the magnitude and location are read directly.

We can understand how mechanical compensation is applied by examining Fig. 19.19*a*. Looking at the end of the rotor, we see one of the correction planes with the unbalance to be corrected represented by wr. Two compensator weights are also shown in the figure. All three of these weights are to rotate with the same angular velocity ω, but the position of the compensator weights relative to one another and their position relative to the unbalanced weight can be varied by two controls. One of these controls changes the angle α—that is, the angle between the compensator weights. The other control changes the angular position of the compensator weights relative to the unbalance—that is, the angle β. The knob that changes the angle β is the *location control;* and when the rotor is compensated (balanced) in this plane, a pointer on the knob indicates the exact angular location of the unbalance. The knob that changes the angle α is the *amount control,* and it also gives a direct reading when the rotor unbalance is compensated. The magnitude of the vibration is measured electrically and displayed on a voltmeter. Thus compensation is secured when the controls are manipulated to make the voltmeter read zero.

19.8 FIELD BALANCING WITH A PROGRAMMABLE CALCULATOR[1]

It is possible to balance a machine in the field by balancing a single plane at a time. But cross effects and correction-plane interference often require balancing each end of a rotor two or three times to obtain satisfactory results. Some machines may require as much as an hour to bring them up to full speed, resulting in even more delays in the balancing procedure.

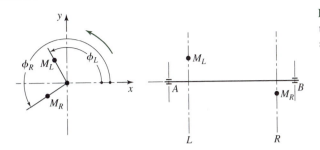

Figure 19.20 Notation for two-plane field balancing. The xy system is the rotating reference.

Field balancing is necessary for very large rotors for which balancing machines are impractical. And even though high-speed rotors are balanced in the shop during manufacture, it is frequently necessary to rebalance them in the field because of slight deformations brought on by shipping, by creep, or by high operating temperatures.

Both Rathbone and Thearle[2] have developed methods of two-plane field balancing which can be expressed in complex-number notation and solved using a programmable calculator. The time saved by using a programmable calculator is several hours when compared with graphical methods or analysis with complex numbers using an ordinary scientific calculator.

In the analysis that follows, boldface letters are used to represent complex numbers:

$$\mathbf{R} = R\angle\theta = Re^{j\theta} = x + jy$$

In Fig. 19.20, unknown unbalances $\mathbf{M}_L$ and $\mathbf{M}_R$ are assumed to exist in the left- and right-hand corrections planes, respectively. The magnitudes of these unbalances are M_L and M_R, and they are located at angles ϕ_L and ϕ_R from the rotating reference. When these unbalances have been found, their negatives are located in the left and right planes to achieve balance.

The rotating unbalances $\mathbf{M}_L$ and $\mathbf{M}_R$ produce disturbances at bearings A and B. Using commercial field-balancing equipment, it is possible to measure the amplitudes and the angular locations of these disturbances. The notation $\mathbf{X} = X\angle\phi$, with appropriate subscripts, will be used to designate these quantities.

In field balancing, three runs or tests are made, as follows:

- *First run:* Measure vector $\mathbf{X}_A = X_A\angle\phi_A$, at bearing A and vector $\mathbf{X}_B = X_B\angle\phi_B$, at bearing B due only to the original unbalances $\mathbf{M}_L = M_L\angle\phi_L$ and $\mathbf{M}_R = M_R\angle\phi_R$.
- *Second run:* Add trial mass $\mathbf{m}_L = m_L\angle\theta_L$ to the left correction plane and measure the amplitudes $\mathbf{X}_{AL} = X_{AL}\angle\phi_{AL}$, and $\mathbf{X}_{BL} = X_{BL}\angle\phi_{BL}$, at the left and right bearings (A and B) respectively.
- *Third run:* Remove trial mass $\mathbf{m}_L = m_L\angle\theta_L$. Add trial mass $\mathbf{m}_R = m_R\angle\theta_R$ to the right-hand correction plane and again measure the bearing amplitudes. These results are designated $\mathbf{X}_{AR} = X_{AR}\angle\phi_{AR}$ for bearing A, and $\mathbf{X}_{BR} = X_{BR}\angle\phi_{BR}$ for bearing B.

Note in the above runs that the term "trial mass" means the same as a trial unbalance and a unit distance from the axis of rotation is used.

To develop the equations for the unbalance to be found we first define the *complex stiffness,* by which we mean the amplitude that would result at either bearing due to a unit unbalance located at the intersection of the rotating reference mark and one of the correction planes. Thus we need to find the complex stiffnesses $\mathbf{A}_L$ and $\mathbf{B}_L$ due to a unit unbalance located at the intersection of the rotating reference mark and plane L. And we require the complex stiffnesses $\mathbf{A}_R$ and $\mathbf{B}_R$ due to a unit unbalance located at the intersection of the rotating reference mark and plane R.

If these stiffnesses were known, we could write the following sets of complex equations:

$$\mathbf{X}_{AL} = \mathbf{X}_A + \mathbf{A}_L \mathbf{m}_L, \qquad \mathbf{X}_{BL} = \mathbf{X}_B + \mathbf{B}_L \mathbf{m}_L \qquad (a)$$

$$\mathbf{X}_{AR} = \mathbf{X}_A + \mathbf{A}_R \mathbf{m}_R, \qquad \mathbf{X}_{BR} = \mathbf{X}_B + \mathbf{B}_R \mathbf{m}_R \qquad (b)$$

After the three runs are made, the stiffnesses will be the only unknowns in these equations. Therefore

$$\mathbf{A}_L = \frac{\mathbf{X}_{AL} - \mathbf{X}_A}{\mathbf{m}_L}, \qquad \mathbf{B}_L = \frac{\mathbf{X}_{BL} - \mathbf{X}_B}{\mathbf{m}_L}$$

$$\mathbf{A}_R = \frac{\mathbf{X}_{AR} - \mathbf{X}_A}{\mathbf{m}_R}, \qquad \mathbf{B}_R = \frac{\mathbf{X}_{BR} - \mathbf{X}_B}{\mathbf{m}_R} \qquad (19.14)$$

Then, from the definition of stiffness, we have from the first run

$$\mathbf{X}_A = \mathbf{A}_L \mathbf{M}_L + \mathbf{A}_R \mathbf{M}_R, \qquad \mathbf{X}_B = \mathbf{B}_L \mathbf{M}_L + \mathbf{B}_R \mathbf{M}_R \qquad (c)$$

Solving this pair of equations simultaneously gives

$$\mathbf{M}_L = \frac{\mathbf{X}_A \mathbf{B}_R - \mathbf{A}_R \mathbf{X}_B}{\mathbf{A}_L \mathbf{B}_R - \mathbf{A}_R \mathbf{B}_L}, \qquad \mathbf{M}_R = \frac{\mathbf{A}_L \mathbf{X}_B - \mathbf{X}_A \mathbf{B}_L}{\mathbf{A}_L \mathbf{B}_R - \mathbf{A}_R \mathbf{B}_L} \qquad (19.15)$$

These equations can be programmed either in complex polar form or in complex rectangular form. The suggestions that follow were formed assuming a complex rectangular form for the solution.

Because the original data are formulated in polar coordinates, a subprogram should be written to transform the data into rectangular coordinates before storage.

The equations reveal that complex subtraction, division, and multiplication are used often. These operations can be set up as subprograms to be called from the main program. If $\mathbf{A} = a + jb$ and $\mathbf{B} = c + jd$, the formula for complex subtraction is

$$\mathbf{A} - \mathbf{B} = (a - c) + j(b - d) \qquad (19.16)$$

For complex multiplication we have

$$\mathbf{AB} = (ac - bd) + j(bc + ad) \qquad (19.17)$$

and for complex division the formula is

$$\frac{\mathbf{A}}{\mathbf{B}} = \frac{(ac + bd) + j(bc - ad)}{c^2 + d^2} \tag{19.18}$$

With these subprograms it is an easy matter to program Eqs. (19.14) and (19.15).

As a check on your programming, you may wish to use the following data: $\mathbf{X}_A = 8.6\angle 63°$, $\mathbf{X}_B = 6.5\angle 206°$, $\mathbf{m}_L = 10\angle 270°$, $\mathbf{m}_R = 12\angle 180°$, $\mathbf{X}_{AL} = 5.9\angle 123°$, $\mathbf{X}_{BL} = 4.5\angle 228°$, $\mathbf{X}_{AR} = 6.2\angle 36°$, $\mathbf{X}_{BR} = 10.4\angle 162°$. The correct results are $\mathbf{M}_L = 10.76\angle 146.6°$ and $\mathbf{M}_R = 6.20\angle 245.4°$.

According to Fagerstrom, the vibration angles used can be expressed in two different systems. The first is the rotating-protractor–stationary-mark system (RPSM). This is the system used in the preceding analysis and is the one a theoretician would prefer. In actual practice, however, it is usually easier to have the protractor stationary and use a rotating mark like a key or keyway. This is called the rotating-mark–stationary-protractor system (RMSP). The only difference between the two systems is in the sign of the vibration angle, but there is no sign change on the trial or correction masses.

19.9 BALANCING A SINGLE-CYLINDER ENGINE

The rotating masses in a single-cylinder engine can be balanced using the methods already discussed in this chapter. The reciprocating masses, however, cannot be completely balanced at all, and so our studies in this section are really concerned with minimizing the unbalance.

Though the reciprocating masses cannot be totally balanced using a simple counterweight, it is possible to modify the shaking forces (see Section 18.9) by unbalancing the rotating masses. As an example of this let us add a counterweight opposite the crankpin whose mass exceeds the rotating mass by one-half of the reciprocating mass (from one-half to two-thirds of the reciprocating mass is usually added to the counterweight to alter the balance characteristics in single-cylinder engines). We designate the mass of the counterweight by m_C, substitute this mass in Eq. (18.24), and use a negative sign because the counterweight is opposite the crankpin; then the inertia force due to this counterweight is

$$\mathbf{F}_C = -m_C r\omega^2 \cos \omega t \,\hat{\mathbf{i}} - m_C r\omega^2 \sin \omega t \,\hat{\mathbf{j}} \tag{a}$$

Note that the balancing mass and the crankpin both have the same radius. Designating m_A and m_B as the masses of the rotating and the reciprocating parts, respectively, as in Chapter 18, we have

$$m_C = m_A + \tfrac{1}{2}m_B \tag{b}$$

according to the supposition above. Equation (a) can now be written

$$\mathbf{F}_C = -\left(m_A + \tfrac{1}{2}m_B\right)r\omega^2 \cos \omega t \,\hat{\mathbf{i}} - \left(m_A + \tfrac{1}{2}m_B\right)r\omega^2 \sin \omega t \,\hat{\mathbf{j}} \tag{c}$$

The inertia force due to the rotating and reciprocating masses is, from Eqs. (18.28) and (18.29)

$$\mathbf{F}_{A,B} = F^x\hat{\mathbf{i}} + F^y\hat{\mathbf{j}}$$

$$= \left[(m_A + m_B)r\omega^2 \cos\omega t + m_B r\omega^2\frac{r}{l}\cos 2\omega t \right]\hat{\mathbf{i}} + m_A r\omega^2 \sin\omega t\hat{\mathbf{j}} \qquad (d)$$

Adding Eqs. (c) and (d) gives the resultant inertia force as

$$\mathbf{F} = \left(\frac{1}{2}m_B r\omega^2 \cos\omega t + m_B r\omega^2\frac{r}{l}\cos 2\omega t \right)\hat{\mathbf{i}} - \frac{1}{2}m_B r\omega^2 \sin\omega t\hat{\mathbf{j}} \qquad (19.19)$$

The vector

$$\tfrac{1}{2}m_B r\omega^2(\cos\omega t\hat{\mathbf{i}} - \sin\omega t\hat{\mathbf{j}})$$

is called the *primary component* of Eq. (19.19). This component has a magnitude $\frac{1}{2}m_B r\omega^2$ and can be represented as a backward (clockwise)-rotating vector with angular velocity ω. The remaining component in Eq. (19.19) is called the *secondary component;* it is the x projection of a vector of length $m_B r\omega^2(r/l)$ rotating forward (counterclockwise) with an angular velocity of 2ω.

The maximum inertia force occurs when $\omega t = 0$ and from Eq. (19.19) is seen to be

$$F_{\max} = m_B r\omega^2 \left(\frac{r}{l} + \frac{l}{2} \right) \qquad (e)$$

because $\cos\omega t = \cos 2\omega t = 1$ when $\omega t = 0$. Before the extra counterweight is added, the maximum inertia force is

$$F_{\max} = m_B r\omega^2 \left(\frac{r}{l} + 1 \right) \qquad (f)$$

Thus in this instance the effect of the additional counterweight is to reduce the maximum shaking force by 50 percent of the primary component and to add vertical inertia forces where formerly none existed. Equation (19.19) is plotted as a polar diagram in Fig. 19.21 for an r/l value of $\frac{1}{4}$. Here, the vector OA rotates counterclockwise at angular velocity of 2ω. The horizontal projection of this vector OA' is the secondary component. The vector OB, the primary component, rotates clockwise at angular velocity ω. The total shaking force F is shown for the 30° position and is the sum of the vectors OB and $BB' = OA'$.

Imaginary-Mass Approach

Stevensen has refined and extended a method of engine balancing which is here called the *imaginary-mass approach.* It is possible that the method is known in some circles as the *virtual-rotor approach*[3] because it uses what might be called a virtual rotor that counter-rotates to accommodate part of the piston effect in a reciprocating engine.

Before going into details, it is necessary to explain a change in the method of viewing the crank circle of an engine. In developing the imaginary-mass approach in this section and the one to follow, we use the coordinate system of Fig. 19.22a. This might appear to be

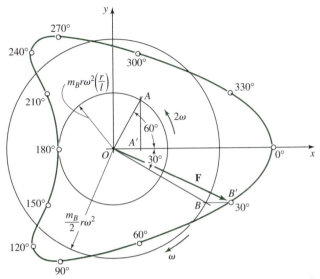

Figure 19.21 Polar diagram of inertia forces in a single-cylinder engine for $r/l = 1/4$. The counterweight includes one-half of the reciprocating mass.

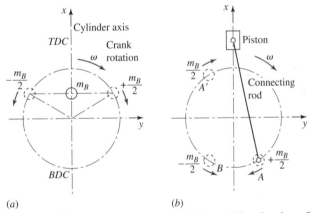

(a)   (b)

Figure 19.22 Note that the axes are right-handed but the view of the crank circle is from the negative z axis.

a left-handed system because the y axis is located clockwise from the x axis and because positive rotation is shown as clockwise. We adopt this notation because it has been used for so long by the automotive industry.* If you like, you can think of this system as a right-handed three-dimensional system viewed from the negative z axis.

The imaginary-mass approach uses two fictitious masses, each equal to half the equivalent reciprocating mass at the particular harmonic studied. The purpose of these fictitious masses is to replace the effects of the reciprocating mass. These imaginary masses rotate

---

*Readers who are antique buffs will understand this convention because it is the direction in which an antique engine is hand-cranked.

about the crank center in opposite directions and with equal velocities. They are arranged so that they come together at both the top dead center (TDC) and bottom dead center (BDC), as shown in Fig. 19.22a. The mass $+\frac{1}{2}m_B$ rotates *with* the crank motion; the other mass $-\frac{1}{2}m_B$ rotates opposite to the crank motion. The mass rotating with the crank is designated on the figure by a plus sign, and the mass rotating in the opposite direction is designated by a minus sign. The center of mass of these two rotating masses always lies on the cylinder axis. The imaginary-mass approach was conceived because the piston motion and the resulting inertia force can always be represented by a Fourier series. Such a series has an infinite number of terms, each term representing a simple harmonic motion having a known frequency and amplitude. It turns out that the higher-frequency amplitudes are so small that they can be neglected and hence only a small number of the lower-frequency amplitudes are needed. Also, the odd harmonics (third, fifth, etc.) are not present because of the symmetry of the piston motion.

Each harmonic (the first, second, fourth, and so on) is represented by a pair of imaginary masses. The angular velocities of these masses are $\pm\omega$ for the first harmonic, $\pm2\omega$ for the second, $\pm4\omega$ for the fourth, and so on. It is rarely necessary to consider the sixth or higher harmonics.

Stevensen gives the following rule for placing the imaginary masses:

> For any given position of the cranks the positions of the imaginary masses are found, first, by determining the angles of travel of each crank from its top dead center and, second, by moving their imaginary masses, one clockwise and the other counterclockwise, by angles equal to the crank angle times the number of the harmonic.

All of these angles must be measured from the same dead-center crank position.

Let us apply this approach to the single-cylinder engine, considering only the first harmonic. In Fig. 19.22b the mass $\pm\frac{1}{2}m_B$ at A rotates at the velocity ω with the crank while mass $-\frac{1}{2}m_B$ at B rotates at the velocity $-\omega$ opposite to crank rotation. The imaginary mass at A can be balanced by adding an equal mass at A' to rotate with the crankshaft. However, the mass at B can be balanced neither by the addition nor by the subtraction of mass from any part of the crankshaft because it is rotating in the opposite direction. When half the mass of the reciprocating parts is balanced in this manner—that is, by adding the mass at A'—the unbalanced part of the first harmonic, due to the mass at B, causes the engine to vibrate in the plane of rotation equally in all directions like a true unbalanced rotating mass.

It is interesting to know that in one-cylinder motorcycle engines a fore-and-aft unbalance is less objectionable than an up-and-down unbalance. For this reason such engines are overbalanced by using a counterweight whose mass is more than half the reciprocating mass.

It is impossible to balance the second and higher harmonics with masses rotating at crankshaft speeds, because the frequency of the unbalance is higher than that of crankshaft rotation. Balancing of second harmonics has been accomplished by using shafts geared to run at twice the engine crankshaft speed, as in the case of the 1976 Plymouth Arrow engine, but at the cost of complication. It is not usually done.

For ready reference we now summarize the inertia forces in the single-cylinder engine, with balancing masses, in Table 19.1. The equations shown have been obtained from Eqs. (18.24) and (18.27), rewritten so that the effect of the second harmonic is presented as a mass equal to $\frac{1}{4}m_B r/l$ reciprocating at a speed of 2ω. Note that the subscript C is used to

TABLE 19.1 One-Cylinder Engine Inertia Forces

Type	Equivalent Mass	Radius	With Cylinder Axis (x)	Across Cylinder Axis (y)
Centrifugal	m_A	r	$m_A r \omega^2 \cos \omega t$	$m_A r \omega^2 \sin \omega t$
	m_{AC}	r_C	$m_{AC} r_C \omega^2 \cos(\omega t + \pi)$	$m_{AC} r_C \omega^2 \sin(\omega t + \pi)$
Reciprocating	m_B	r	$m_B r \omega^2 \cos \omega t$	0
First harmonic	m_{BC}	r_C	$m_{BC} r_C \omega^2 \cos(\omega t + \pi)$	$m_{BC} r_C \omega^2 \sin(\omega t + \pi)$
Second harmonic	$\dfrac{m_B r}{4l}$	r	$\dfrac{m_B r}{4l}(r)(2\omega)^2 \cos 2\omega t$	0

designate the counterweights (balancing masses) and their radii. Because the centrifugal balancing masses will be selected and placed to counterbalance the centrifugal forces, the only unbalance that results along the cylinder axis will be the sum of the last three entries. Similarly, the only unbalance across the cylinder axis will be the value in the fourth row. The maximum values of the unbalance in these two directions can be predetermined in any desired ratio to each other, as indicated previously, and a solution can be obtained for m_{BC} at the radius r_C.

If this approach is used to include the effect of the fourth harmonic, there will result an additional mass of $m_B r^3 / 16 l^3$ reciprocating at the speed of 2ω and a mass of $-m_B r^3 / 64 l^3$ reciprocating at a speed of 4ω, illustrating the decreasing significance of the higher harmonics.

19.10 BALANCING MULTICYLINDER ENGINES

To obtain a basic understanding of the balancing problem in multi-cylinder engines, let us consider a two-cylinder in-line engine having cranks 180° apart and rotating parts already balanced by counterweights. Such an engine is shown in Fig. 19.23. Applying the imaginary-mass approach for the first harmonics results in the diagram of Fig. 19.23a. This figure shows that masses +1 and +2, rotating clockwise, balance each other, as do masses −1 and −2, rotating counterclockwise. Thus, the first harmonic forces are inherently balanced for this crank arrangement. Figure 19.23b shows, however, that these forces are not in the same plane. For this reason, unbalanced couples will be set up which tend to rotate the engine about the y axis. The values of these couples can be determined using the force expressions in Table 19.1 together with the coupling distance because the equations can be applied to each cylinder separately. It is possible to balance the couple due to the real rotating masses as well as the imaginary half-masses that rotate with the engine; however, the couple due to the half-mass of the first harmonic that is counterrotating cannot be balanced.

Figure 19.24a shows the location of the imaginary masses for the second harmonic using Stevensen's rule. This diagram shows that the second-harmonic forces are not balanced. Because the greatest unbalance occurs at the dead-center positions, the diagrams are usually drawn for this extreme position, with crank 1 at TDC as in Fig. 19.24b. This unbalance causes a vibration in the xz plane having the frequency 2ω.

The diagram for the fourth harmonics, not shown, is the same as in Fig. 19.24b but, of course, the speed is 4ω.

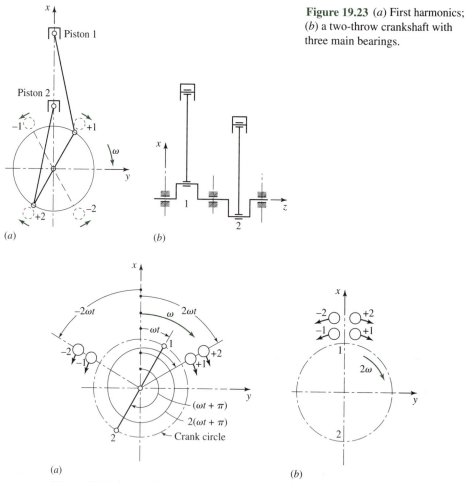

Figure 19.23 (*a*) First harmonics; (*b*) a two-throw crankshaft with three main bearings.

(*a*)   (*b*)

Figure 19.24 Second-harmonic positions of the imaginary masses: (*a*) positions for same crank angle in Fig. 19.23*a*; (*b*) extreme, or dead-center, positions.

Four-Cylinder Engine   A four-cylinder in-line engine with cranks spaced 180° apart is shown in Fig. 19.25*c*. This engine can be treated as two two-cylinder engines back to back. Thus the first harmonic forces still balance and, in addition, from Figs. 19.25*a* and 19.25*c*, the first-harmonic couples also balance. These couples will tend, however, to deflect the center bearing of a three-bearing crankshaft up and down and to bend the center of a two-bearing shaft in the same manner.

Figure 19.25*b* shows that when cranks 1 and 4 are at top dead center, all the masses representing the second harmonic traveling in both directions accumulate at top dead center, giving an unbalanced force. The center of mass of all the masses is always on the *x* axis, and so the unbalanced second harmonics cause a vertical vibration with a frequency of twice the engine speed. This characteristic is typical of all four-cylinder engines with this crank arrangement. Because the masses and forces all act in the same direction, there is no coupling action.

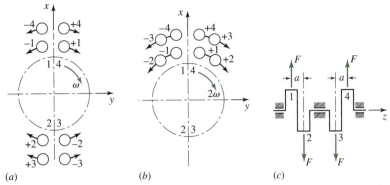

Figure 19.25 Four-cycle engine: (*a*) first-harmonic positions; (*b*) second-harmonic positions; (*c*) crankshaft with first-harmonic couples added.

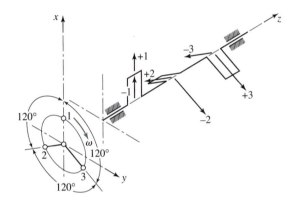

Figure 19.26 Crank arrangement of three-cylinder engine; first-harmonic forces shown.

A diagram of the fourth harmonics is identical to that of Fig. 19.25*b*, and the effects are the same, but they do have a higher frequency and exert less force.

Three-Cylinder Engine   A three-cylinder in-line engine with cranks spaced 120° apart is shown in Fig. 19.26. Note that the cylinders are numbered according to the order in which they arrive at top dead center. Figure 19.27 shows that the first, second, and fourth harmonic forces are completely balanced, and only the sixth harmonic forces are completely unbalanced. These unbalanced forces tend to create a vibration in the plane of the centerlines of the cylinders, but the magnitude of the forces is very small and can be neglected as far as vibration is concerned.

An analysis of the couples of the first harmonic forces shows that when crank 1 is at top dead center (Fig. 19.26), there is a vertical component of the forces on cranks 2 and 3 equal in magnitude to half of the force on crank 1. The resultant of these two downward components is equivalent to a force downward, equal in magnitude to the force on crank 1 and located halfway between cranks 2 and 3. Thus a couple is set up with an arm equal to the distance between the center of crank 1 and the centerline between cranks 2 and 3. At the same time, the horizontal components of the +2 and −2 forces cancel each other, as do the horizontal components of the +3 and −3 forces (Fig. 19.27). Therefore, no horizontal

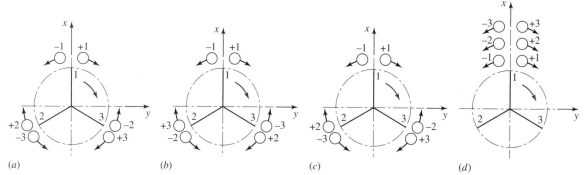

Figure 19.27 Positions of the imaginary masses for the three-cylinder engine (*a*) first harmonics; (*b*) second harmonics; (*c*) fourth harmonics; (*d*) sixth harmonics.

couple exists. Similar couples are found for both the second and fourth harmonics. Thus a three-cylinder engine, although inherently balanced for forces in the first, second, and fourth harmonics, still is not free from vibration due to the presence of couples at these harmonics.

Six-Cylinder Engine   A six-cylinder in-line engine is conceived as a combination of two three-cylinder engines back to back with parallel cylinders. It has the same inherent balance of the first, second, and fourth harmonics. And, by virtue of symmetry, the couples of each three-cylinder engine acts in opposite directions and balance the other. These couples, although perfectly balanced, tend to bend the crankshaft and crankcase and necessitate the use of rigid construction for high-speed operation. As before, the sixth harmonics are completely unbalanced and tend to create a vibration in the vertical plane with a frequency of 6ω. The magnitude of these forces, however, is very small and practically negligible as a source of vibration.

Other Engines   Taking into consideration the cylinder arrangement and crank spacing permits a great many configurations. For any combination, the balancing situation can be investigated to any harmonic desired by the methods outlined in this section. Particular attention must be paid when analyzing to that part of Stevensen's rule which calls for determining the angle of travel from the top dead center of the cylinder under consideration and moving the imaginary masses through the appropriate angles from that same top dead center. This is especially important when radial and opposed-piston engines are investigated.

As practice problems you may wish to use these methods to confirm the following facts:

1. In a three-cylinder radial engine with one crank and three connecting rods having the same crankpin, the negative masses are inherently balanced for the first harmonic forces while the positive masses are always located at the crankpin. These two findings are inherently true for all radial engines. Also, because the radial engine has its cylinders in a single plane, unbalanced couples do not occur. The three-cylinder engine has unbalanced forces in the second and higher harmonics.

2. A two-cylinder opposed-piston engine with a crank spacing of 180° is balanced for forces in the first, second, and fourth harmonics but unbalanced for couples.
3. A four-cylinder in-line engine with cranks at 90° is balanced for forces in the first harmonic but unbalanced for couples. In the second harmonic it is balanced for both forces and couples.
4. An eight-cylinder in-line engine with the cranks at 90° is inherently balanced for both forces and couples in the first and second harmonics but unbalanced in the fourth harmonic.
5. An eight-cylinder V-engine with cranks at 90° is inherently balanced for forces in the first and second harmonics and for couples in the second. The unbalanced couples in the first harmonic can be balanced by counterweights that introduce an equal and opposite couple. Such an engine is unbalanced for forces in the fourth harmonic.

19.11 ANALYTICAL TECHNIQUE FOR BALANCING MULTICYLINDER RECIPROCATING ENGINES

First Harmonic Forces   The X and Y components of the resultant of the first harmonic forces for any multicylinder reciprocating engine can be written in the form

$$F_{PX} = A\cos\theta + B\sin\theta$$
$$F_{PY} = C\cos\theta + D\sin\theta$$

(19.20)

The first harmonic forces can always be balanced by a pair of rotating masses, or in some special cases a single mass, as shown in Fig. 19.28. The mass m_1 creates a force F_1 at a location angle of $(\theta + \gamma_1)$ from the X axis, and the mass m_2 creates a force F_2 at a location angle of $-(\theta + \gamma_2)$ from the X axis. For balance, these two forces plus the resultant of the

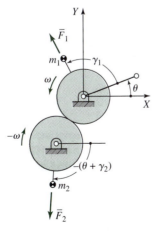

Figure 19.28 General arrangement of counter-rotating masses for balancing the primary forces.

first harmonic forces must be equal to zero; that is,

$$A \cos \theta + B \sin \theta + F_1 \cos(\theta + \gamma_1) + F_2 \cos[-(\theta + \gamma_2)] = 0$$
$$C \cos \theta + D \sin \theta + F_1 \sin(\theta + \gamma_1) + F_2 \sin[-(\theta + \gamma_2)] = 0$$

(19.21)

Expanding these two equations, in terms of functions of θ, γ_1, and γ_2, and rearranging gives

$$(F_1 \cos \gamma_1 + F_2 \cos \gamma_2)\cos \theta - (F_1 \sin \gamma_1 + F_2 \sin \gamma_2)\sin \theta = -A \cos \theta - B \sin \theta$$
$$(F_1 \sin \gamma_1 - F_2 \sin \gamma_2)\cos \theta + (F_1 \cos \gamma_1 - F_2 \cos \gamma_2)\sin \theta = -C \cos \theta - D \sin \theta$$

(19.22)

To satisfy Eqs. (19.22), for all values of the crank angle θ, the necessary conditions are

$$F_1 \cos \gamma_1 + F_2 \cos \gamma_2 = -A$$

(19.23a)

$$F_1 \sin \gamma_1 + F_2 \sin \gamma_2 = B$$

(19.23b)

$$F_1 \sin \gamma_1 - F_2 \sin \gamma_2 = -C$$

(19.23c)

$$F_1 \cos \gamma_1 - F_2 \cos \gamma_2 = -D$$

(19.23d)

Solving Eqs. (19.23), the forces are

$$F_1 = \tfrac{1}{2}\sqrt{(A + D)^2 + (B - C)^2}$$

(19.24a)

$$F_2 = \tfrac{1}{2}\sqrt{(A - D)^2 + (B + C)^2}$$

(19.24b)

and the location angles are

$$\tan \gamma_1 = (B - C)/-(D + A)$$

(19.25a)

$$\tan \gamma_2 = (B + C)/(D - A)$$

(19.25b)

Second Harmonic Forces   The X and Y components of the resultant of the second harmonic forces can be written in the form

$$F_{SX} = A' \cos 2\theta + B' \sin 2\theta$$
$$F_{SY} = C' \cos 2\theta + D' \sin 2\theta$$

(19.26)

The second harmonic forces can always be balanced by a pair of gears rotating at twice the crankshaft speed, as shown in Fig. 19.29. The mass m'_1 creates a force F'_1 at an orientation angle of $(2\theta + \lambda_1)$ from the X axis, and the mass m'_2 creates a force F'_2 at an orientation angle of $-(2\theta + \lambda_2)$ from the X axis. For balance, these two forces plus the resultant of the second harmonic forces must be equal to zero.

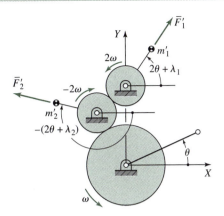

Figure 19.29 General arrangement of counter-rotating masses for balancing the secondary forces.

Note that the analysis for the second harmonic forces is exactly the same as for the first harmonic forces, with A, B, C, and D, replaced by A', B', C', and D', θ replaced by 2θ, and the orientation angles γ_1 and γ_2 replaced by the orientation angles λ_1 and λ_2, respectively.

EXAMPLE 19.4

Consider the three-cylinder engine shown in Fig. 19.30.

SOLUTION

The resultant first harmonic force vector is

$$\mathbf{F}_P = 1.5P\cos\theta\,\hat{\mathbf{i}} - 1.5P\sin\theta\,\hat{\mathbf{j}} \tag{1}$$

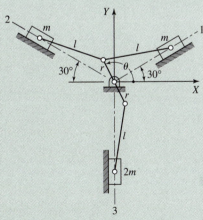

Figure 19.30 Example 19.4: Three-cylinder arrangement.

where

$$P = mr\omega^2$$

Comparing Eq. (1) with Eqs. (19.20), the coefficients are

$$A = 1.5P, \qquad B = 0, \qquad C = 0, \qquad D = -1.5P \tag{2}$$

Therefore, the forces from Eqs. (19.24), are

$$F_1 = \tfrac{1}{2}\sqrt{(A+D)^2 + (B-C)^2} = 0 \tag{3}$$

$$F_2 = \tfrac{1}{2}\sqrt{(A-D)^2 + (B+C)^2} = 1.5P \tag{4}$$

Because the first force is zero, there is only one correcting mass. The orientation angle for this mass, from Eq. (19.24b), is

$$\tan \gamma_2 = (B+C)/(D-A) = 0/-3P = 0 \tag{5}$$

Therefore, the answer is $\gamma_2 = 180°$, since the denominator (the cosine of the angle) is negative.

The conclusion is that we need one mass rotating opposite to the crankshaft (at crankshaft speed) and oriented at an angle $-(\theta + 180°)$. Figure 19.31 shows the location of the correcting mass for $\theta = 0°$. If the correcting mass m_2 is to be placed at a radius r_2, then the inertia force is

$$F_2 = m_2 r_2 \omega^2 = 1.5 mr\omega^2 \tag{6}$$

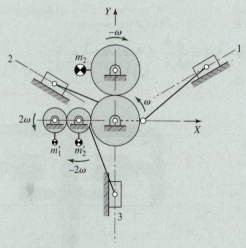

Figure 19.31 Example 19.4: Balancing arrangement (shown at $\theta = 0°$).

Therefore, the product is

$$m_2 r_2 = 1.5mr \qquad\qquad Ans. \quad (7)$$

The second harmonic force vector is

$$\mathbf{F}_s = 1.5Q \sin 2\theta \hat{\mathbf{i}} + 2.5Q \cos 2\theta \hat{\mathbf{j}} \qquad\qquad (8)$$

where

$$Q = \frac{mr^2\omega^2}{l}$$

Comparing Eq. (8) with Eqs. (19.22), the coefficients are

$$A' = 0, \qquad B' = 1.5Q, \qquad C' = 2.5Q, \qquad D' = 0 \qquad\qquad (9)$$

Therefore, the inertia forces, from Eqs. (19.24), are

$$F_1' = \tfrac{1}{2}\sqrt{(A' + D')^2 + (B' - C')^2} = 0.5Q \qquad\qquad (10)$$

$$F_2' = \tfrac{1}{2}\sqrt{(A' - D')^2 + (B' + C')^2} = 2.0Q \qquad\qquad (11)$$

The forces can be expressed as

$$F_1' = m_1' r_1' (2\omega)^2 = 0.5m\frac{r^2}{l}\omega^2 \qquad\qquad (12)$$

$$F_2' = m_2' r_2' (2\omega)^2 = 2m\frac{r^2}{l}\omega^2 \qquad\qquad (13)$$

Therefore, the correcting mass m_1' can be placed at a radius r_1' such that

$$m_1' r_1' = 0.125mr^2/l \qquad\qquad Ans. \quad (14)$$

and the correcting mass m_2' can be placed at a radius r_2' such that

$$m_2' r_2' = 0.5mr^2/l \qquad\qquad Ans. \quad (15)$$

From Eq. (19.25a), the first orientation angle is

$$\tan \lambda_1 = (B' - C')/-(D' + A') = -1.0Q/0 = \infty \qquad\qquad (16)$$

Therefore, the answer is $\lambda_1 = 270°$, because the numerator (the sine of the angle) is negative. From Eq. (19.25b), the second location angle is

$$\tan \lambda_2 = (B' + C')/(D' - A') = +4.0Q/0 = \infty \tag{17}$$

The answer is $\lambda_2 = 90°$, because the numerator (or the sine of the angle) is positive.

The conclusion is that we need the arrangement as shown in Fig. 19.31. The mass m_1' is located on a shaft turning in the same direction as the crankshaft and at twice the crankshaft speed. The location angle is $2\theta + 270°$. The mass m_2' is located on a shaft turning opposite to the crankshaft and at twice the crankshaft speed. The location angle is $-(2\theta + 90°)$.

19.12 BALANCING LINKAGES[4]

The two problems that arise in balancing linkages are balancing the shaking force and balancing the shaking moment.

In force balancing a linkage we must concern ourselves with the position of the total center of mass. If a way can be found to cause this total center of mass to remain stationary, the vector sum of all the frame forces will always be zero. Lowen and Berkof[5] have catalogued five methods of force balancing:

1. The method of static balancing, in which concentrated link masses are replaced by systems of masses that are statically equivalent
2. The method of principal vectors, in which an analytical expression is obtained for the center of mass and then manipulated to learn how its trajectory can be influenced
3. The method of linearly independent vectors, in which the center of mass of a mechanism is made stationary, causing the coefficients of the time-dependent terms of the equation describing the trajectory of the total center of mass to vanish
4. The use of cam-driven masses to keep the total center of mass stationary
5. The addition of an axially symmetric duplicated mechanism by which the new combined total center of mass is made stationary

Lowen and Berkof state that very few studies have been reported on the problem of balancing the shaking moment. This problem is discussed further in Section 19.13.

Here we present only the Berkof–Lowen method,[6] which employs the method of linearly independent vectors. The method is developed completely for the four-bar linkage, but only the final results are given for a typical six-bar linkage. Here is the procedure. First, find the equation that describes the trajectory of the total center of mass of the linkage. This equation will contain certain terms whose coefficients are time-dependent. Then the total center of mass is made stationary by changing the position of the individual link masses so that the coefficients of all the time-dependent terms vanish. In order to accomplish this, it is necessary to write the equation in such a form that the time-dependent unit vectors contained in the equation are linearly independent.

In Fig. 19.32 a general four-bar linkage is shown having link masses m_2 located at G_2, m_3 located at G_3, and m_4 located at G_4. The coordinates a_i, ϕ_i describe the positions of

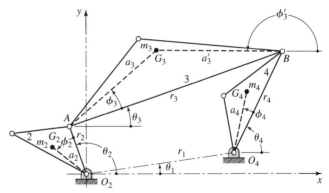

Figure 19.32 Four-bar linkage showing irregular positions of the link masses.

these points within each link. We begin by defining the position of the total center of mass of the linkage by the vector $\mathbf{r}_s$:

$$\mathbf{r}_s = \frac{1}{M}(m_2\mathbf{r}_{s_2} + m_3\mathbf{r}_{s_3} + m_4\mathbf{r}_{s_4}) \qquad (a)$$

where $\mathbf{r}_{s_2}$, $\mathbf{r}_{s_3}$, and $\mathbf{r}_{s_4}$ are the vectors that describe the positions of m_2, m_3, and m_4, respectively, in the xy coordinate system. Thus, from Fig. 19.27,

$$\mathbf{r}_{s_2} = a_2 e^{j(\theta_2+\phi_2)}$$
$$\mathbf{r}_{s_3} = r_2 e^{j\theta_2} + a_3' e^{j(\theta_3+\phi_3)} \qquad (b)$$
$$\mathbf{r}_{s_4} = r_1 e^{j\theta_1} + a_4 e^{j(\theta_4+\phi_4)}$$

The total mass M of the four-bar linkage is

$$M = m_2 + m_3 + m_4 \qquad (c)$$

Substituting Eq. (b) into (a) gives

$$M\mathbf{r}_s = (m_2 a_2 e^{j\phi_2} + m_3 r_2)e^{j\theta_2} + (m_3 a_3 e^{j\phi_3})e^{j\theta_3} + (m_4 a_4 e^{j\phi_4})e^{j\theta_4} + m_4 r_1 e^{j\theta_1} \qquad (d)$$

where we have used the identity $e^{j(\alpha+\beta)} = e^{j\alpha}e^{j\beta}$. For a four-bar linkage the vector loop-closure equation can be written as

$$r_2 e^{j\theta_2} + r_3 e^{j\theta_3} - r_4 e^{j\theta_4} - r_1 e^{j\theta_1} = 0 \qquad (e)$$

Thus, the time-dependent terms $e^{j\theta_2}$, $e^{j\theta_3}$, and $e^{j\theta_4}$ in Eq. (d) are not linearly independent. To make them so, we solve Eq. (e) for one of the unit vectors, say $e^{j\theta_3}$, and substitute the result into Eq. (d). Thus

$$e^{j\theta_3} = \frac{1}{r_3}(r_1 e^{j\theta_1} - r_2 e^{j\theta_2} + r_4 e^{j\theta_4}) \qquad (f)$$

Equation (d) now becomes

$$M\mathbf{r}_s = \left(m_2 a_2 e^{j\phi_2} + m_3 r_2 - m_3 a_3 \frac{r_2}{r_3} e^{j\phi_3} \right) e^{j\theta_2} + \left(m_4 a_4 e^{j\phi_4} + m_3 a_3 \frac{r_4}{r_3} e^{j\phi_3} \right) e^{j\theta_4}$$

$$+ \left(m_4 r_1 + m_3 a_3 \frac{r_1}{r_3} e^{j\phi_3} \right) e^{j\theta_1} \tag{g}$$

Equation (g) shows that the center of mass can be made stationary at the position

$$\mathbf{r}_s = \frac{r_1}{r_3 M} (m_4 r_3 + m_3 a_3 e^{j\phi_3}) e^{j\theta_1} \tag{19.27}$$

if the following coefficients of the time-dependent terms vanish:

$$m_2 a_2 e^{j\phi_2} + m_3 r_2 - m_3 a_3 \frac{r_2}{r_3} e^{j\phi_3} = 0 \tag{h}$$

$$m_4 a_4 e^{j\phi_4} + m_3 a_3 \frac{r_4}{r_3} e^{j\phi_3} = 0 \tag{i}$$

But Eq. (h) can be simplified by locating G_3 from point B instead of point A (Fig. 19.32). Thus

$$a_3 e^{j\phi_3} = r_3 + a_3' e^{j\phi_3'}$$

With this substitution, Eq. (h) becomes

$$m_2 a_2 e^{j\phi_2} - m_3 a_3' \frac{r_2}{r_3} e^{j\phi_3'} = 0 \tag{j}$$

Equations (i) and (j) must be satisfied to obtain total force balance. These equations yield the two sets of conditions:

$$m_2 a_2 = m_3 a_3' \frac{r_2}{r_3} \quad \text{and} \quad \phi_2 = \phi_3'$$

$$\tag{19.28}$$

$$m_4 a_4 = m_3 a_3 \frac{r_4}{r_3} \quad \text{and} \quad \phi_4 = \phi_3 + \pi$$

A study of these conditions shows that the mass and its location can be specified in advance for any single link; and then full balance can be obtained by rearranging the mass of the other two links.

The usual problem in balancing a four-bar linkage is that the link lengths r_i are specified in advance because of the function performed. For this situation, counterweights can be added to the input and output links in order to redistribute their masses, while the geometry of the third moving link is undisturbed.

When adding counterweights, the following relations must be satisfied:

$$m_i a_i \angle \phi_i = m_i^o a_i^o \angle \phi_i^o = m_i^* a_i^* \angle \phi_i^* \tag{19.29}$$

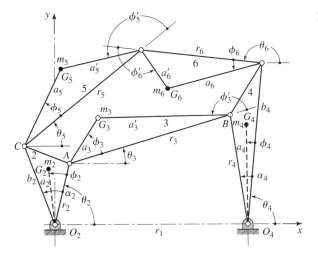

Figure 19.33 Notation for a six-bar linkage.

where m_i^o, a_i^o, ϕ_i^o are the parameters of the unbalanced linkage, m_i^*, a_i^*, ϕ_i^* are the parameters of the counterweights, and m_i, a_i, ϕ_i are the parameters that result from Eqs. (19.28). A second condition that must generally be satisfied is

$$m_i = m_i^o + m_i^* \tag{19.30}$$

If the solution to a balancing problem can remain as the mass distance product $m_i^* a_i^*$, Eq. (19.30) need not be used and Eq. (19.29) can be solved to yield

$$m_i^* a_i^* = \sqrt{(m_i a_i)^2 + \left(m_i^o a_i^o\right)^2 - 2(m_i a_i)\left(m_i^o a_i^o\right)\cos\left(\phi_i - \phi_i^o\right)} \tag{19.31}$$

$$\phi_i^* = \tan^{-1}\left(\frac{m_i a_i \sin\phi_i - m_i^o a_i^o \sin\phi_i^o}{m_i a_i \cos\phi_i - m_i^o a_i^o \cos\phi_i^o}\right) \tag{19.32}$$

Figure 19.33 shows a typical six-bar linkage and the notation. For this, the Berkof–Lowen conditions for total balance are:

$$m_2\frac{a_2}{r_2}e^{j\phi_2} = m_5\frac{a_5'}{r_5}\frac{b_2}{r_2}e^{j(\phi_1+\alpha_2)} + m_3\frac{a_3'}{r_3}e^{j\phi_1} \tag{19.33a}$$

$$m_4\frac{a_4}{r_4}e^{j\phi_4} = m_6\frac{a_6'}{r_6}\frac{b_4}{r_4}e^{j\phi_6'} - m_3\frac{a_3}{r_3}e^{j(\phi_3+\alpha_4)} \tag{19.33b}$$

$$m_5\frac{a_5}{r_5}e^{j\phi_5} = -m_6\frac{a_6}{r_6}e^{j\phi_6} \tag{19.33c}$$

Similar relations can be devised for other six-bar linkages. For total balance, Eqs. (19.33) show that a certain mass-geometry relation between links 5 and 6 must be satisfied, after which the masses of any two links and their locations can be specified. Balance is then achieved by a redistribution of the masses of the remaining three movable links.

It is important to note that the addition of counterweights to balance the shaking forces will probably increase the internal bearing forces as well as the shaking moment. Thus only a partial balance may represent the best compromise between these three effects.

EXAMPLE 19.5

Table 19.2 is a tabulation of the dimensions, masses, and the locations of the mass centers of a four-bar mechanism having link 2 as the input and link 4 as the output. Complete force balancing is desired by adding counterweights to the input and the output links. Find the mass-distance value and the angular orientation of each counterweight.

TABLE 19.2 Parameters of an Unbalanced Four-Bar Linkage

Link i	1	2	3	4
r_i, mm	140	50	150	75
a_i^0, mm	—	25	80	40
ϕ_i^0	—	0°	15°	0°
$a_i'^0$, mm	—	—	75.6	—
$\phi_i'^0$	—	—	164.1°	—
m_i^0, kg	—	0.046	0.125	0.054

SOLUTION

From Eqs. (19.28) we first find

$$m_2 a_2 = m_3^0 a_3'^0 \frac{r_2}{r_3} = (0.125)(75.6)\frac{50}{150} = 3.150 \text{ g·m}$$

$$\phi_2 = \phi_3' = 164.1°$$

$$m_4 a_4 = m_4^0 a_3^0 \frac{r_4}{r_3} = (0.125)(80)\frac{75}{150} = 5.000 \text{ g·m}$$

$$\phi_4 = \phi_4^0 + 180° = 15° + 180° = 195°$$

Note that $m_2 a_2$ and $m_4 a_4$ are the mass–distance values *after* the counterweights have been added. Also, note that the link 3 parameters will not be altered. We next compute

$$m_2^0 a_2^0 = (0.046)(25) = 1.150 \text{ g·m} \quad \text{and} \quad m_4^0 a_4^0 = (0.054)(40) = 2.160 \text{ g·m}$$

Using Eq. (19.31), we compute the mass–distance value for the link 2 counterweight as

$$m_2^* a_2^* = \sqrt{(m_2 a_2)^2 + \left(m_2^0 a_2^0\right)^2 - 2(m_2 a_2)\left(m_2^0 a_2^0\right)\cos\left(\phi_2 - \phi_2^0\right)}$$

$$= \sqrt{(3.150)^2 + (1.150)^2 - 2(3.150)(1.150)\cos(164.1° - 0°)} = 4.268 \text{ g·m}$$

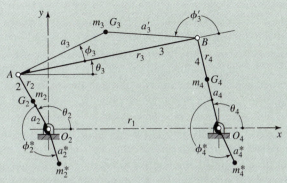

Figure 19.34 Four-bar crank-and-rocker linkage showing counterweights added to input and output links to achieve complete force balance.

From Eq. (19.21), we find the orientation of this counterweight to be

$$\phi_2^* = \tan^{-1}\left(\frac{m_2 a_2 \sin\phi_2 - m_2^o a_2^o \sin\phi_2^o}{m_2 a_2 \cos\phi_2 - m_2^o a_2^o \cos\phi_2^o}\right)$$

$$= \tan^{-1}\left(\frac{3.150\sin 164.1° - 1.150\sin 0°}{3.150\cos 164.1° - 1.150\cos 0°}\right) = 168.3°$$

Using the same procedure for link 4 yields

$$m_4^* a_4^* = 7.11 \text{ g·m} \qquad \text{at } \phi_4^* = 190.5°$$

Figure 19.34 is a scale drawing of the complete linkage with the two counterweights added.

19.13 BALANCING OF MACHINES[7]

In the previous section we learned how to force balance a simple linkage by using two or more counterweights, depending upon the number of links composing the linkage. Unfortunately, this does not balance the shaking moment and, in fact, may even make this worse because of the addition of the counterweights. If a machine is imagined to be composed of several mechanisms, one might consider balancing the machine by balancing each mechanism separately. But this may not result in the best balance for the machine, because the addition of a large number of counterweights may cause the inertia torque to be completely unacceptable. Furthermore, unbalance of one mechanism may counteract the unbalance in another, eliminating the need for some of the counterweights in the first place.

Stevensen shows that any single harmonic of unbalanced forces, moments of forces, and torques in a machine can be balanced by the addition of six counterweights. They are arranged on three shafts, two per shaft, driven at the constant speed of the harmonic, and have axes parallel, respectively, to each of the three mutually perpendicular axes through the center of mass of the machine. The method is too complex to be included in this book, but it is worthwhile to look at the overall approach.

Using the methods of this book together with current computing facilities, the linear and angular accelerations of each of the moving mass centers of a machine are computed for points throughout a cycle of motion. The masses and mass moments of inertia of the machine must also be computed or determined experimentally. Then the inertia forces, the inertia torques, and the moments of the forces are computed with reference to the three mutually perpendicular coordinate axes through the center of mass of the machine. When these are summed for each point in the cycle, six functions of time are the result: three for the forces and three for the moments. With the aid of a digital computer it is then possible to use numerical harmonic analysis to define the component harmonics of the unbalanced forces parallel to the three axes, and of the unbalanced moments about the three axes.

To balance a single harmonic, each component of the unbalance of the machine is represented in the form $A \cos \omega t + B \sin \omega t$ with appropriate subscripts. Six equations of equilibrium are then written which include the unbalances as well as the effects of the six unknown counterweights. These equations are arranged so that each of the $\sin \omega t$ and $\cos \omega t$ terms is multiplied by the parenthetical coefficients. Balance is then achieved by setting the parenthetical terms in each equation equal to zero, much in the same manner as in the preceding section. This results in 12 equations, all linear, in 12 unknowns. With the locations for the balancing counterweights on the three shafts specified, the 12 equations can be solved for the six mr products and the six phase angles needed for the six balancing weights. Stevensen goes on to show that when less than the necessary three shafts are available, it becomes necessary to optimize some effect of the unbalance, such as the motion of a point on the machine.

NOTES

[1.]   The authors are grateful to W. B. Fagerstrom, of E. I. DuPont de Nemours, Wilmington, DE, for contributing some of the ideas of this section.

[2.]   T. C. Rathbone, "Turbine Vibration and Balancing," *Trans. ASME,* p. 267, 1929; E. L. Thearle, "Dynamic Balancing in the Field," *Trans. ASME,* p. 745, 1934.

[3.]   The presentation here is from Prof. E. N. Stevensen, University of Hartford, class notes, with his permission. While some changes have been made to conform to the notation of this book, the material is all Stevensen's. He refers to Maleev and Lichty [V. L. Maleev, *Internal Combustion Engines,* McGraw-Hill, New York, 1993; and L. C. Lichty, *Internal Combustion Engines,* 5th ed., McGraw-Hill, 1939] and states that the method first came to his attention in both the Maleev and Lichty books.

[4.]   Those who wish to investigate this topic in detail should begin with the following reference, in which an entire issue is devoted to the subject of linkage balancing: G. G. Lowen and R. S. Berkof, "Survey of Investigations into the Balancing of Linkages," *J. Mech.,* vol. 3, no. 4, p. 221, 1968. This issue contains 11 translations on the subject from the German and Russian literature.

[5.]   *Ibid.*

[6.]   R. S. Berkof and G. G. Lowen, "A New Method for Completely Force Balancing Simple Linkages," *J. Eng. Ind., Trans. ASME,* ser. B., vol. 91, no. 1, pp. 21–26, February 1969.

[7.]   The material for this section is from E. N. Stevensen, Jr., "Balancing of Machines," *J. Eng. Ind., Trans. ASME,* ser. B, vol. 95, pp. 650–656, May 1973. It is included with the advice and consent of Professor Stevensen.

PROBLEMS

19.1 Determine the bearing reactions at A and B for the system shown in the figure if the speed is 300 rev/min. Determine the magnitude and the angular orientation of the balancing mass if it is located at a radius of 50 mm.

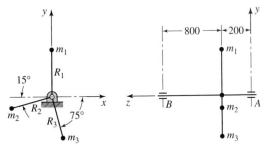

Figure P19.1 $R_1 = 25$ mm, $R_2 = 35$ mm, $R_3 = 40$ mm, $m_1 = 2$ kg, $m_2 = 1.5$ kg, $m_3 = 3$ kg.

19.2 The figure shows three weights connected to a shaft that rotates in bearings at A and B. Determine the magnitude of the bearing reactions if the shaft speed is 300 rev/min. A counterweight is to be located at a radius of 10 in. Find the value of the weight and its angular orientation.

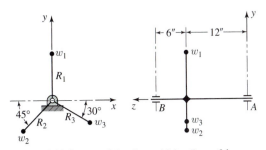

Figure P19.2 $R_1 = 8$ in, $R_2 = 12$ in, $R_3 = 6$ in, $w_1 = 2$ oz, $w_2 = 1.5$ oz, $w_3 = 3$ oz.

19.3 The figure shows two weights connected to a rotating shaft and mounted outboard of bearings A and B. If the shaft rotates at 120 rev/min, what are the magnitudes of the bearing reactions at A and B? Suppose the system is to be balanced by reducing a weight at a radius of 5 in. Determine the amount and the angular orientation of the weight to be removed.

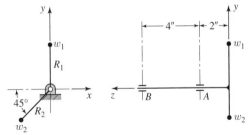

Figure P19.3 $R_1 = 4$ in, $R_2 = 6$ in, $w_1 = 4$ lb, $w_2 = 3$ lb.

19.4 For a speed of 220 rev/min, calculate the magnitudes and relative angular orientations of the bearing reactions at A and B for the two-mass system shown.

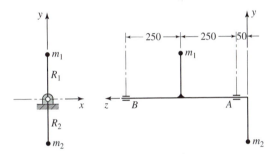

Figure P19.4 $R_1 = 60$ mm, $R_2 = 40$ mm, $m_1 = 2$ kg, $m_2 = 1.5$ kg.

19.5 The rotating system shown in the figure has $R_1 = R_2 = 60$ mm, $a = c = 300$ mm, $b = 600$ mm, $m_1 = 1$ kg, and $m_2 = 3$ kg. Find the bearing reactions at A and B and their angular orientations measured from a rotating reference mark if the shaft speed is 100 rev/min.

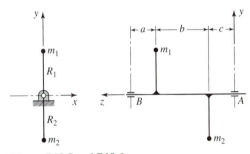

Figure P19.5 and P19.6

19.6 The rotating shaft shown in the figure supports two masses m_1 and m_2 whose weights are 4 lb and 5 lb, respectively. The dimensions are $R_1 = 4$ in, $R_2 = 3$ in, $a = 2$ in, $b = 8$ in, and $c = 3$ in. Find the magnitudes of the rotating-bearing reactions at A and B and their angular orientations measured from a rotating reference mark if the shaft speed is 360 rev/min.

19.7 The shaft shown in the figure is to be balanced by placing masses in the correction planes L and R. The weights of the three masses m_1, m_2, and m_3 are 4, 3, and 4 oz, respectively. The dimensions are $R_1 = 5$ in, $R_2 = 4$ in, $R_3 = 5$ in, $a = 1$ in, $b = e = 8$ in, $c = 10$ in, and $d = 9$ in. Calculate the magnitudes of the corrections in ounce-inches and their angular orientations.

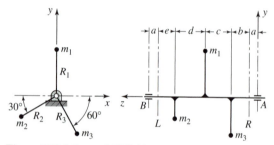

Figure P19.7 through P19.10

19.8 The shaft of Problem 19.7 is to be balanced by removing weight from the two correction planes.

Determine the corrections to be subtracted in ounce-inches and their angular orientations.

19.9 The shaft shown in the figure is to be balanced by placing masses in the two correction planes L and R. The three masses are $m_1 = 6$ g, $m_2 = 7$ g, and $m_3 = 5$ g. The dimensions are $R_1 = 125$ mm, $R_2 = 150$ mm, $R_3 = 100$ mm, $a = 25$ mm, $b = 300$ mm, $c = 600$ mm, $d = 150$ mm, and $e = 75$ mm. Calculate the magnitudes and angular locations of the corrections.

19.10 Repeat Problem 19.9 if masses are to be added in the two correction planes.

19.11 Solve the two-plane balancing problem as stated in Section 19.8.

19.12 A rotor to be balanced in the field yielded an amplitude of 5 at an angle of 142° at the left-hand bearing and an amplitude of 3 at an angle of −22° at the right-hand bearing due to unbalance. To correct this, a trial mass of 12 was added to the left-hand correction plane at an angle of 210° from the rotating reference. A second run then gave left-hand and right-hand responses of 8∠160° and 4∠260°, respectively. The first trial mass was then removed and a second mass of 6 added to the right-hand correction plane at an angle of −70°. The responses to this were 2∠74° and 4.5∠−80° for the left- and right-hand bearings, respectively. Determine the original unbalances.

20 | Cam Dynamics

20.1 RIGID- AND ELASTIC-BODY CAM SYSTEMS

Figure 20.1*a* is a cross-sectional view showing the overhead valve arrangement in an automobile engine. In analyzing the dynamics of this or any other cam system, we would expect to determine the contact force at the cam surface, the spring force, and the cam-shaft torque, all for a complete rotation of the cam. In one method of analysis the entire cam-follower train, consisting of the push rod, the rocker arm, and the valve stem together with the cam shaft, are considered rigid. If the members are in fact fairly rigid and if the speed is moderate, such an analysis will usually produce quite satisfactory results. In any event, such rigid-body analysis should always be made first.

Sometimes the speeds are so high or the members are so elastic (perhaps because of extreme length) that an elastic-body analysis must be used. This fact is usually discovered when troubles are encountered with the cam system. Such troubles are usually evidenced by noise, chatter, unusual wear, poor product quality, or perhaps fatigue failure of some of the parts. In other cases, laboratory investigation of the performance of a prototype cam system may reveal substantial differences between the theoretical and the observed performance.

Figure 20.1*b* is a mathematical model of an elastic-body cam system. Here m_3 is the mass of the cam and a portion of the cam shaft. The motion machined into the cam is the coordinate y, a function of the cam-shaft angle θ. The bending stiffness of the cam shaft is designated as k_4. The follower retaining spring is k_1. The masses m_1 and m_2 and the stiffnesses k_2 and k_3 are lumped characteristics of the follower train. The dashpots c_i ($i = 1, 2, 3,$ and 4) are inserted to represent friction, which, in the analysis, may indicate either viscous or sliding friction or any combination of the two. The system of Fig. 20.1*b* is a rather sophisticated one requiring the solution of three simultaneous differential equations. We will deal with simpler systems in this chapter.

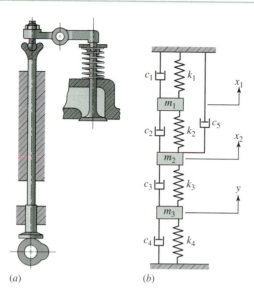

Figure 20.1 An overhead valve arrangement for an automotive engine.

(a)   (b)

20.2 ANALYSIS OF AN ECCENTRIC CAM

An eccentric plate cam is a circular disk with the cam-shaft hole drilled off-center. The distance e between the center of the disk and the center of the shaft is called the eccentricity. Figure 20.2a shows a simple reciprocating follower eccentric-cam system. It consists of a plate cam, a flat-face follower mass, and a retaining spring of stiffness k. The coordinate y designates the motion of the follower as long as cam contact is made. We arbitrarily select $y = 0$ at the bottom of the stroke. Then the kinematic quantities of interest are

$$y = e - e \cos \omega t, \qquad \dot{y} = e\omega \sin \omega t, \qquad \ddot{y} = e\omega^2 \cos \omega t \qquad (20.1)$$

where ωt is the same as the cam angle θ.

To make a rigid-body analysis, we assume no friction and construct a free-body diagram of the follower (Fig. 20.2b). In this figure, F_{23} is the cam contact force and F_S is the spring force. In general, F_{23} and F_S do not have the same line of action; and so a pair of frame forces, $F_{13,A}$ and $F_{13,B}$, act at bearings A and B.

Before writing the equation of motion, let us investigate the spring force in more detail. By *spring stiffness* k, also called the *spring rate,* we mean the amount of force necessary to deform the spring a unit length. Thus the units of k will usually be expressed in newtons per meter or in pounds per inch. The purpose of the spring is to keep or retain the follower in contact with the cam. Thus, the spring should exert some force even at the bottom of the stroke where it is extended the most. This force, called the *preload P,* is the force exerted by the spring when $y = 0$. Thus, $P = k\delta$, where δ is the distance the spring must be compressed in order to assemble it.

Summing forces on the follower mass in the y direction gives

$$\sum F^y = F_{23} - k(y + \delta) = m\ddot{y} \qquad (a)$$

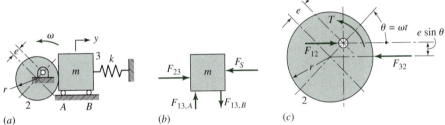

Figure 20.2 (*a*) Eccentric plate cam and flat-face follower; (*b*) free-body diagram of the follower; (*c*) free-body diagram of the cam.

Note particularly that F_{23} can have only positive values. Solving Eq. (*a*) for the contact force and substituting the first and third equations of Eqs. (20.1) gives

$$F_{23} = (ke + P) + (m\omega^2 - k)e \cos \omega t \tag{20.2}$$

Figure 20.2*c* is a free-body diagram of the cam. The torque T, applied by the shaft to the cam, is

$$T = F_{23}e \sin \omega t = [(ke + P) + (m\omega^2 - k)e \cos \omega t]e \sin \omega t$$

$$= e(ke + P) \sin \omega t + \tfrac{1}{2}e^2(m\omega^2 - k) \sin 2\omega t \tag{20.3}$$

Equation (20.2) and Fig. 20.3*a* show that the contact force F_{23} consists of a constant term $ke + P$ with a cosine wave superimposed on it. The maximum occurs at $\theta = 0°$ and the minimum occurs at $\theta = 180°$. The cosine or variable component has an amplitude that depends upon the square of the cam-shaft velocity. Thus, as the velocity increases, this term increases at a greater rate. At a certain speed the contact force could become zero at $\theta = 180°$. When this happens, there is usually some impact between the cam and the follower, resulting in clicking, rattling, or very noisy operation. In effect the sluggishness or inertia of the follower prevents it from following the cam. The result is often called *jump* or *float*. The noise occurs when contact is reestablished. Of course, the purpose of the retaining spring is to prevent this. Because the contact force consists of a cosine wave superimposed on a constant term, all we need to do to prevent jump is to move or elevate the cosine wave away from the zero position. To do this we can increase the term $ke + P$, by increasing the preload P or the spring rate k or both.

Having learned that jump begins at $\cos \omega t = -1$ with $F_{23} = 0$, we can solve Eq. (20.2) for the jump speed. The result is

$$\omega = \sqrt{\frac{2ke + P}{me}} \tag{20.4}$$

Using the same procedure, we find that jump will not occur if

$$P > e(m\omega^2 - 2k) \tag{20.5}$$

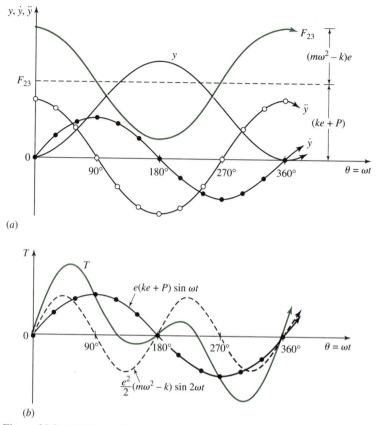

(a)

(b)

Figure 20.3 (a) Plot of displacement, velocity, acceleration, and contact force for an eccentric cam system; (b) graph of torque components and total cam-shaft torque.

Figure 20.3b is a plot of Eq. (20.3). Note that the torque consists of a double-frequency component, whose amplitude is a function of the cam velocity squared, superimposed on a single-frequency component, whose amplitude is independent of velocity. In this example, the area of the torque-displacement diagram in the positive T direction is the same as in the negative T direction. This means that the energy required to drive the follower in the forward direction is recovered when the follower returns. A flywheel or inertia on the cam shaft can be used to handle this fluctuating energy requirement. Of course, if an external load is connected in some manner to the follower system, the energy required to drive this load will lift the torque curve in the positive direction and increase the area in the positive loop of the T curve.

EXAMPLE 20.1

A cam-and-follower mechanism similar to Fig. 20.2a has the cam machined so that it will move the follower to the right through a distance of 40 mm with parabolic motion in 120° of cam rotation, dwell for 30°, then return with parabolic motion to the starting position in the remaining cam angle. The spring rate is 5 kN/m, and the mechanism is assembled with a 35 N preload. The follower mass is 18 kg. Assume no friction. (*a*) Without computing numerical values, sketch approximate graphs of the displacement motion, the acceleration, and the cam-contact force, all versus cam angle for the entire cycle of events from $\theta = 0°$ to $\theta = 360°$ of cam rotation. On this graph, show where jump or liftoff is most likely to begin. (*b*) At what speed would jump begin? Show all calculations.

SOLUTION

(*a*) Solving Eq. (*a*) of Section 20.2 for the contact force gives

$$F = ky + P + m\ddot{y} \qquad (1)$$

which is composed of a constant term P, a term ky that varies like the displacement, and a term $m\ddot{y}$ that varies like the acceleration. Figure 20.4 shows the displacement diagram, the acceleration $\ddot{y}$, and the cam contact force F. Note that jump will first occur at $\theta = \omega t = 60°$ because this is the closest approach of F to the zero position.

(*b*) Liftoff will occur at the half-point of rise where $\theta = \beta/2 = 60°$ (that is, $\beta = 120°$) when the acceleration becomes negative. The acceleration at this position is

$$\ddot{y} = -\frac{4L\omega^2}{\beta^2} = -\frac{4(0.040 \text{ m})\omega^2}{(120\pi/180 \text{ rad})^2} = -0.0365\omega^2 \text{ m/s}^2$$

Substituting this value, $P = 35$ N, and $ky = (5 \text{ kN/m})(20 \text{ mm}) = 100$ N into Eq. (1) with $F = 0$ gives

$$0 = 100 \text{ N} + 35 \text{ N} + (18 \text{ kg})(-0.0365\omega^2 \text{ m/s}^2)$$

Then rearranging this equation gives

$$\omega = \sqrt{\frac{100 \text{ N} + 35 \text{ N}}{(18 \text{ kg})(0.0365 \text{ m/rad}^2)}} = 14.3 \text{ rad/s}$$

$$= \frac{(14.3 \text{ rad/s})(60 \text{ s/min})}{2\pi \text{ rad/rev}} = 137 \text{ rev/min} \qquad \qquad Ans.$$

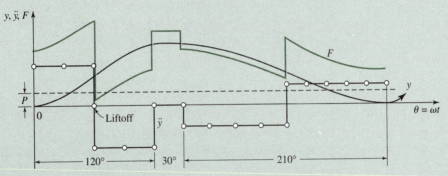

Figure 20.4 Example 20.1.

20.3 EFFECT OF SLIDING FRICTION

Let F_μ be the force of sliding (Coulomb) friction as defined by Eq. (14.10). Because the friction force is always opposite in direction to the velocity, let us define a sign function as follows:

$$\text{sign } \dot{y} = \begin{cases} +1 & \text{for } \dot{y} \geq 0 \\ -1 & \text{for } \dot{y} < 0 \end{cases} \tag{20.6}$$

With this notation, Eq. (a) of Section 20.2 can be written

$$\sum F^y = F_{23} - F_\mu \text{ sign } \dot{y} - k(y + \delta) - m\ddot{y} = 0$$

or

$$F_{23} = F_\mu \text{ sign } \dot{y} + k(y + \delta) + m\ddot{y} \tag{20.7}$$

This equation is plotted for simple harmonic motion with no dwells in Fig. 20.5. By studying both parts of this diagram we can note that F_μ is positive when $\dot{y}$ is positive and we can see how F_{23} is obtained by summing the four component curves graphically.

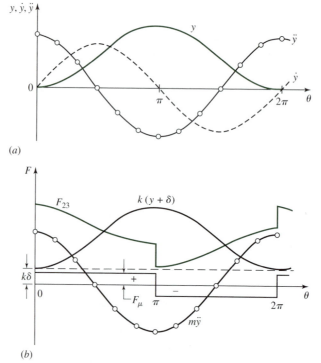

Figure 20.5 Effect of sliding friction on a cam system with harmonic motion: (a) graph of displacement, velocity, and acceleration for one motion cycle; (b) graph of force components F_μ, $k\delta$, $k(y + \delta)$, $m\ddot{y}$, and the resultant contact force F_{23}.

20.4 ANALYSIS OF DISK CAM WITH RECIPROCATING ROLLER FOLLOWER

In Chapter 14 we analyzed a cam system incorporating a reciprocating roller follower. In this section we present an analytical approach to a similar problem in which sliding friction is also included. The geometry of such a system is shown in Fig. 20.6a. In the analysis to follow, the effect of follower weight on bearings B and C is neglected.

Figure 20.6b is a free-body diagram of the follower and roller. If y is any motion machined into the cam and $\theta = \omega t$ is the cam angle, at $y = 0$ the follower is at the bottom of its stroke and so $O_2 A = R + r$. Therefore

$$a = R + r + y \tag{20.8}$$

In Fig. 20.6b the roller contact force forms the angle ϕ, the pressure angle, with the x axis. Because the direction of $\mathbf{F}_{23}$ is the same as the normal to the contacting surfaces, the intersection of this line with the x axis is the common instant center of the cam and follower. This means that the velocity of this point is the same no matter whether it is considered as a point on the follower or a point on the cam. Therefore

$$\dot{y} = a\omega \tan \phi$$

and so

$$\tan \phi = \frac{\dot{y}}{a\omega} \tag{20.9}$$

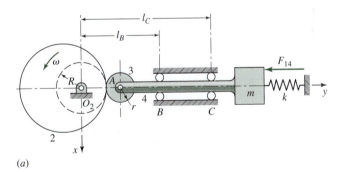

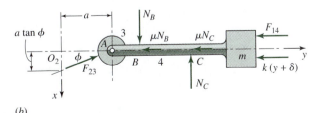

(b)

Figure 20.6 (a) Plate cam driving a reciprocating roller-follower system. (b) Free-body diagram of the follower system.

In the analysis to follow, the two bearing reactions are N_B and N_C, the coefficient of sliding friction is μ, and δ is the precompression of the retaining spring. Summing forces in the x and y directions gives

$$\sum F_{i,4}^x = -F_{23}^x + N_B - N_C = 0 \tag{a}$$

$$\sum F_{i,4}^y = F_{23}^y - \mu \text{ sign } \dot{y}(N_B + N_C) - F_{14} - k(y + \delta) - m\ddot{y} = 0 \tag{b}$$

A third equation is obtained by taking moments about A:

$$\sum M_A = -N_B(l_B - a) + N_C(l_C - a) = 0 \tag{c}$$

With the help of Eq. (20.9), these three equations can be solved for the unknowns F_{23}, N_B, and N_C.

First, solve Eq. (c) for N_C. This gives

$$N_C = N_B \frac{l_B - a}{l_C - a} \tag{d}$$

Now substitute Eq. (d) into Eq. (a) and solve for N_B. The result is

$$N_B = \frac{l_C - a}{l_C - l_B} F_{23}^x \tag{e}$$

But because $F_{23}^x = F_{23}^y \tan \phi$, we obtain

$$N_B = \frac{F_{23}^y(l_C - a) \tan \phi}{l_C - l_B} \tag{20.10}$$

Next, substitute Eqs. (d) and (20.10) into the friction term of Eq. (b):

$$\mu \text{ sign } \dot{y}(N_B + N_C) = \mu F_{23}^y \tan \phi \text{ sign } \dot{y} \left[\frac{l_C + l_B - 2a}{l_C - l_B} \right] \tag{f}$$

Substituting this result back into Eq. (b) and solving for F_{23}^y gives

$$F_{23}^y = \frac{F_{14} + k(y + \delta) + m\ddot{y}}{1 - \mu \tan \phi \text{ sign } \dot{y}[(l_C + l_B - 2a)/(l_C - l_B)]} \tag{20.11}$$

For computer or calculator solution, a simple solution to the sign function is

$$\text{sign } \dot{y} = \frac{\dot{y}}{|\dot{y}|} \tag{20.12}$$

Finally, the cam-shaft torque is

$$T = -a F_{23}^y \tan \phi \tag{20.13}$$

The equations of this section require the kinematic relations for the appropriate rise and return motions, developed in Chapter 5.

20.5 ANALYSIS OF ELASTIC CAM SYSTEMS

Figure 20.7 illustrates the effect of follower elasticity upon the displacement and velocity characteristics of a follower system driven by a cycloidal cam. To see what has happened, you should compare these diagrams with the theoretical ones in Chapter 5. Though the effect of elasticity is most pronounced for the velocity characteristic, it is usually the modification of the displacement characteristic, especially at the top of rise, that causes the most trouble in practical situations. These troubles are usually evidenced by poor or unreliable product quality when the systems are used in manufacturing or assembly lines, and result in noise, unusual wear, and fatigue failure.

A complete analysis of elastic cam systems requires a good background in vibration studies. To avoid the necessity for this background while still developing a basic understanding, we will use an extremely simplified cam system using a linear-motion cam. It must be observed, however, that such a cam system should never be used for high-speed applications.

In Fig. 20.8a, k_1 is the stiffness of the retaining spring, m is the lumped mass of the follower, and k_2 represents the stiffness of the follower. Because the follower is usually a rod or a lever, k_2, is many times greater than k_1.

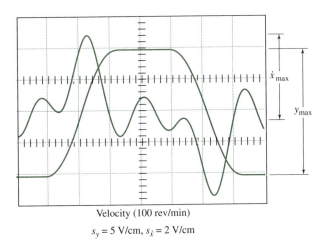

Velocity (100 rev/min)

$s_y = 5$ V/cm, $s_{\dot{x}} = 2$ V/cm

Figure 20.7 Photograph of the oscilloscope traces of the displacement and velocity characteristics of a dwell-rise–dwell-return cam-and-follower system machined for cycloidal motion. The zero axis of the displacement diagram has been translated downward to obtain a larger diagram in the space available.

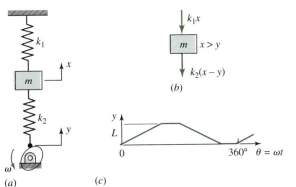

Figure 20.8 (*a*) Undamped model of a cam-and-follower system. (*b*) Free-body diagram of the follower mass. (*c*) Displacement diagram.

Spring k_1 is assembled with a preload. The coordinate x of the follower motion is chosen at the equilibrium position of the mass after spring k_1 is assembled. Thus k_1 and k_2 exert equal and opposite preload forces on the mass. Assuming no friction, the free-body diagram of the mass is as shown in Fig. 20.8b. To determine the directions of the forces the coordinate x, representing the motion of the follower, has been assumed to be larger than the coordinate y, representing the motion machined into the cam. However, the same result is obtained if y is assumed to be larger than x.

Using Fig. 20.8b, we find the equation of motion to be

$$\sum F = -k_1 x - k_2(x - y) - m\ddot{x} = 0 \tag{a}$$

or

$$\ddot{x} + \frac{k_1 + k_2}{m}x = \frac{k_2}{m}y \tag{20.14}$$

This is the differential equation for the motion of the follower. It can be solved as shown in Chapter 17 when the function y is specified. This equation must be solved piecewise for each cam event; that is, the ending conditions for one event, or era of motion, must be used as the beginning or starting conditions for the next era.

Let us analyze the first era of motion using uniform motion as illustrated in Fig. 20.8c. First we use the notation

$$\omega_n = \sqrt{\frac{k_1 + k_2}{m}} \tag{20.15}$$

Do not confuse ω_n with the angular velocity of the cam ω. The quantity ω_n here is called the *undamped natural frequency*. The units of ω_n are reciprocal seconds and are usually written as radians per second. This is implied by the circular nature of the quantity.

Equation (20.14) can now be written as

$$\ddot{x} + \omega_n^2 x = \frac{k_2 y}{m} \tag{20.16}$$

The solution to this equation is

$$x = A \cos \omega_n t + B \sin \omega_n t + \frac{k_2 y}{m \omega_n^2} \tag{b}$$

where

$$y = \frac{L}{\beta}\theta = \frac{L\omega t}{\beta} \tag{20.17}$$

Of course, Eq. (20.17) is valid only during the rise era. You can verify Eq. (b) as the solution by substituting it and its second derivative into Eq. (20.16).

The first derivative of Eq. (b) is

$$\dot{x} = -A\omega_n \sin \omega_n t + B\omega_n \cos \omega_n t + \frac{k_2 \dot{y}}{m \omega_n^2} \tag{c}$$

Using $t = 0$ at the beginning of rise with $x = \dot{x} = 0$, we find from Eqs. (b) and (c) that

$$A = 0 \qquad B = -\frac{k_2 \dot{y}}{m \omega_n^3}$$

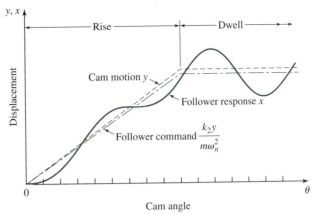

Figure 20.9 Displacement diagram of a uniform-motion cam mechanism showing the follower response.

Thus Eq. (*b*) becomes

$$x = \frac{k_2}{m\omega_n^2} \left(y - \frac{\dot{y}}{\omega_n} \sin \omega_n t \right) \tag{20.18}$$

This equation is plotted in Fig. 20.9. Note that the motion consists of a negative sine term superimposed upon a ramp representing the uniform rise. Because of the additional compression of k_2 during rise, the ramp term $k_2 y / m\omega_n^2$, called the follower command in the figure, becomes less than the rise motion y.

At the end of rise, Eqs. (20.16) through (20.18) are no longer valid, and a second era of motion begins. The follower response for this era is shown in Fig. 20.9, but we will not solve for it.

Equation (20.18) shows that the vibration amplitude $\dot{y}/\omega_n$ can be reduced by making ω_n large, and Eq. (20.15) shows that this can be done by increasing k_2, which means that a very rigid follower should be used.

20.6 UNBALANCE, SPRING SURGE, AND WINDUP

As shown in Fig. 20.10*a*, a disk cam produces unbalance because its mass is not symmetrical with the axis of rotation. This means that two sets of vibratory forces exist, one due to the eccentric cam mass and the other due to the reaction of the follower against the cam. By keeping these effects in mind during design, the engineer can do much to guard against difficulties during operation.

Figures 20.10*b* and 20.10*c* show that face and cylindrical cams have good balance characteristics. For this reason, these are good choices when high-speed operation is involved.

Spring Surge   It is shown in texts on spring design that helical springs may themselves vibrate when subjected to rapidly varying forces. For example, poorly designed automotive valve springs operating near the critical frequency range permit the valve to open for short intervals during the period the valve is supposed to be closed. Such conditions result in very poor operation of the engine and rapid fatigue failure of the springs themselves. This

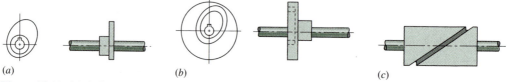

Figure 20.10 (*a*) A disk cam is inherently unbalanced. (*b*) A face cam is usually well-balanced. (*c*) A cylindrical cam has good balance.

vibration of the retaining spring, called *spring surge,* has been photographed with high-speed motion-picture cameras, and the results exhibited in slow motion. When serious vibrations exist, a clear wave motion can be seen traveling up and down the valve spring.

Windup   Figure 20.3*b* is a plot of cam-shaft torque showing that the shaft exerts torque on the cam during a portion of the cycle and that the cam exerts torque on the shaft during another portion of the cycle. This varying torque requirement may cause the shaft to twist, or wind up, as the torque increases during follower rise. Also, during this period, the angular cam velocity is slowed and so is the follower velocity. Near the end of rise the energy stored in the shaft by the windup is released, causing both the follower velocity and acceleration to rise above normal values. The resulting kick may produce follower jump or impact.

This effect is most pronounced when heavy loads are being moved by the follower, when the follower moves at a high speed, and when the shaft is flexible.

In most cases a flywheel must be employed in cam systems, to provide for the varying torque requirements. Cam-shaft windup can be prevented to a large extent by mounting the flywheel as close as possible to the cam. Mounting it a long distance from the cam may actually worsen matters.

PROBLEMS

20.1   In part (*a*) of the figure, the mass *m* is constrained to move only in the vertical direction. The eccentric cam has an eccentricity of 2 in, a speed of 20 rad/s, and the weight of the mass is 8 lb. Neglecting friction, find the angle $\theta = \omega t$ at the instant the cam jumps.

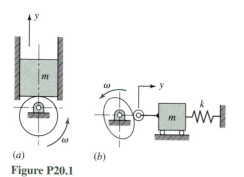

(*a*)   (*b*)
Figure P20.1

20.2   In part (*a*) of the figure, the mass *m* is driven up and down by the eccentric cam and it has a weight of 10 lb. The cam eccentricity is 1 in. Assume no friction.
  (*a*) Derive the equation for the contact force.
  (*b*) Find the cam velocity ω corresponding to the beginning of the cam jump.

20.3   In part (*a*) of the figure, the slider has a mass of 2.5 kg. The cam is a simple eccentric and causes the slider to rise 25 mm with no friction. At what cam speed in revolutions per minute will the slider first lose contact with the cam? Sketch a graph of the contact force at this speed for 360° of cam rotation.

20.4   The cam-and-follower system shown in part (*b*) of the figure has $k = 1$ kN/m, $m = 0.90$ kg, $y = 15 - 15 \cos \omega t$ mm, and $\omega = 60$ rad/s. The retaining spring is assembled with a preload 2.5 N.

(a) Compute the maximum and minimum values of the contact force.

(b) If the follower is found to jump off the cam, compute the angle ωt corresponding to the very beginning of jump.

20.5 Part (b) of the figure shows the mathematical model of a cam-and-follower system. The motion machined into the cam is to move the mass to the right through a distance of 2 in with parabolic motion in 150° of cam rotation, dwell for 30°, return to the starting position with simple harmonic motion, and dwell for the remaining 30° of cam angle. There is no friction or damping. The spring rate is 40 lb/in, and the spring preload is 6 lb, corresponding to the $y = 0$ position. The weight of the mass is 36 lb.

(a) Sketch a displacement diagram showing the follower motion for the entire 150° of cam rotation. Without computing numerical values, superimpose graphs of the acceleration and cam contact force onto the same axes. Show where jump is most likely to begin.

(b) At what speed in revolutions per minute would jump begin?

20.6 A cam-and-follower mechanism is shown in abstract form in part (b) of the figure. The cam is cut so that it causes the mass to move to the right a distance of 25 mm with harmonic motion in 150° of cam rotation, dwell for 30°, then return to the starting position in the remaining 180° of cam rotation, also with harmonic motion. The spring is assembled with a 22-N preload and it has a rate of 4.4 kN/m. The follower mass is 17.5 g. Compute the cam speed in revolutions per minute at which jump would begin.

20.7 The figure shows a lever OAB driven by a cam cut to give the roller a rise of 1 in with parabolic motion and a parabolic return with no dwells. The lever and roller are to be assumed as weightless, and there is no friction. Calculate the jump speed if $l = 5$ in and the mass weighs 5 lb.

20.8 A cam-and-follower system similar to the one of Fig. 20.6 uses a plate cam driven at a speed of

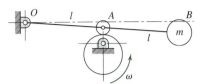

Figure P20.7

600 rev/min and employs simple harmonic rise and parabolic return motions. The events are rise in 150°, dwell for 30°, and return in 180°. The retaining spring has a rate $k = 14$ kN/m with a precompression of 12.5 mm. The follower has a mass of 1.6 kg. The external load is related to the follower motion y by the relation $F = 0.325 - 10.75y$, where y is in meters and F is in kilonewtons. Dimensions corresponding to Fig. 20.6 are $R = 20$ mm, $r = 5$ mm, $l_B = 60$ mm, and $l_C = 90$ mm. Using a rise of $L = 20$ mm and assuming no friction, plot the displacement, cam-shaft torque, and radial component of the cam force for one complete revolution of the cam.

20.9 Repeat Problem 20.8 with the speed of 900 rev/min, $F = 0.110 + 10.75y$ kN, where y is in meters, and the coefficient of sliding friction is $\mu = 0.025$.

20.10 A plate cam drives a reciprocating roller follower through the distance $L = 1.25$ in with parabolic motion in 120°, then dwells for the remaining cam angle. The external load on the follower is $F_{14} = 36$ lb during rise and zero during the dwells and the return. In the notation of Fig. 20.6, $R = 3$ in, $r = 1$ in, $l_B = 6$ in, $l_C = 8$ in, and $k = 150$ lb/in. The spring is assembled with a preload of 37.5 lb when the follower is at the bottom of its stroke. The weight of the follower is 1.8 lb, and the cam velocity is 140 rad/s. Assuming no friction, plot the displacement, the torque exerted on the cam by the shaft, and the radial component of the contact force exerted by the roller against the cam surface for one complete cycle of motion.

20.11 Repeat Problem 20.10 if friction exists with $\mu = 0.04$ and the cycloidal return takes place in 180°.

21 | Flywheels

A flywheel is an energy storage device. It absorbs mechanical energy by increasing its angular velocity and delivers energy by decreasing its angular velocity. Commonly, the flywheel is used to smooth the flow of energy between a power source and its load. If the load happens to be a punch press, the actual punching operation requires energy for only a fraction of its motion cycle. If the power source happens to be a two-cylinder four-cycle engine, the engine delivers energy during only about half of its motion cycle. More recent applications under investigation involve using a flywheel to absorb braking energy and deliver accelerating energy for an automobile and to act as energy-smoothing devices for electric utilities as well as solar and wind-power generating facilities. Electric railways have long used regenerative braking by feeding braking energy back into power lines, but newer and stronger materials now make the flywheel more feasible for such purposes.

21.1 DYNAMIC THEORY

Figure 21.1 is a mathematical representation of a flywheel. The flywheel, whose motion is measured by the angular coordinate θ, has a mass moment of inertia I. An input torque T_i, corresponding to a coordinate θ_i, will cause the flywheel speed to increase. A load or output torque T_o, with corresponding coordinate θ_o, will absorb energy from the flywheel and cause it to slow down. If the work into the system is considered positive and work output is negative, the equation of motion of the flywheel is

$$\sum M = T_i(\theta_i, \dot{\theta}_i) - T_o(\theta_o, \dot{\theta}_o) - I\ddot{\theta} = 0$$

or

$$I\alpha = T_i(\theta_i, \omega_i) - T_o(\theta_o, \omega_o) \qquad (a)$$

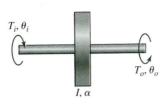

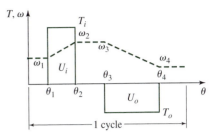

Figure 21.1 Mathematical representation of a flywheel.

Figure 21.2

Note that both T_i and T_o may depend for their values on the angular displacements θ_i and θ_o, as well as on their angular velocities ω_i and ω_o. Typically, the torque characteristic depends upon only one of these. Thus the torque delivered by an induction motor depends on the speed of the motor. In fact, electric-motor manufacturers publish charts detailing the torque-speed characteristics of their various motors.

When the input and output torque functions are given, Eq. (a) can be solved for the motion of the flywheel using well-known techniques for solving linear and nonlinear differential equations. The resulting equations can easily be solved using the methods of Chapter 17. In this chapter we assume rigid shafting, giving $\theta_i = \theta = \theta_o$. Thus Eq. (a) becomes

$$I\alpha = T_i(\theta, \omega) - T_o(\theta, \omega) \qquad (b)$$

When the two torque functions are known and the starting values of the displacement θ and velocity ω are given, Eq. (b) can be solved for θ, ω, and α as functions of time. However, we are not really interested in the instantaneous values of the kinematic quantities. Primarily, we want to know the overall performance of the flywheel. What should its moment of inertia be? How do we match the power source to the load to get an optimum size of motor or engine? Finally, what are the resulting performance characteristics of the system?

To gain insight into the problem, a hypothetical situation is diagrammed in Fig. 21.2. An input power source subjects a flywheel to a constant torque T_i while the output shaft rotates from θ_1 to θ_2. This is a positive torque and is plotted upward. Equation (b) indicates that a positive acceleration α will be the result, and so the shaft angular velocity increases from ω_1 to ω_2. As shown, the shaft now rotates from θ_2 to θ_3 with zero torque and hence, from Eq. (b), with zero angular acceleration. Therefore, $\omega_3 = \omega_2$. From θ_3 to θ_4 a load, or output torque, of constant magnitude is applied, causing the shaft to slow down from ω_3 to ω_4. Note that the output torque is plotted in the negative direction in accordance with Eq. (b).

The work input to the flywheel is the area of the rectangle between θ_1 and θ_2, or

$$U_i = T_i(\theta_2 - \theta_1) \qquad (c)$$

The work output of the flywheel is the area of the rectangle from θ_3 to θ_4, or

$$U_o = T_o(\theta_4 - \theta_3) \qquad (d)$$

If U_o is greater than U_i, the load uses more energy than has been delivered to the flywheel and so ω_4 will be less than ω_1. If $U_o = U_i$, then ω_4 will equal ω_1 because the gain and loss are equal; we are assuming no friction losses. And finally ω_4 will be greater than ω_1 if U_i is greater than U_o.

We can also write these relations in terms of kinetic energy. At $\theta = \theta_1$, the flywheel has an angular velocity of ω_1 rad/s, and so its kinetic energy is

$$U_1 = \tfrac{1}{2}I\omega_1^2 \qquad\qquad (e)$$

At $\theta = \theta_2$, the angular velocity is ω_2, and so

$$U_2 = \tfrac{1}{2}I\omega_2^2 \qquad\qquad (f)$$

Thus the change in kinetic energy is

$$U_2 - U_1 = \tfrac{1}{2}I\left(\omega_2^2 - \omega_1^2\right) \qquad\qquad (21.1)$$

21.2 INTEGRATION TECHNIQUE

Many of the torque–displacement functions encountered in practical engineering situations are so complicated that they must be integrated by approximate methods. Figure 21.3, for example, is a plot of the engine torque for one cycle of motion of a single-cylinder engine. Because a part of the torque curve is negative, the flywheel must return part of the energy back to the engine. Approximate integration of this curve for a cycle of 4π rad yields a mean torque T_m available to drive a load.

The simplest integration routine is Simpson's rule; this approximation can be handled on any computer and is short enough for the crudest programmable calculators. In fact, this routine is usually found as a part of the library for most calculators and personal computers. The equation used is

$$\int_{x_0}^{x_n} f(x)\,dx = \frac{h}{3}(f_0 + 4f_1 + 2f_2 + 4f_3 + 2f_4 + \cdots + 2f_{n-2} + 4f_{n-1} + f_n) \quad (21.2)$$

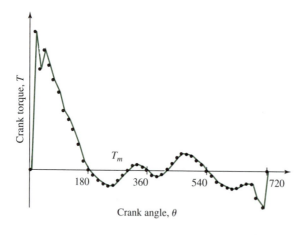

Figure 21.3 Relation between torque and crank angle for a one-cylinder four-cycle internal combustion engine.

where

$$h = (x_n - x_0)/n \qquad \text{with } x_n > x_0$$

and n is the number of subintervals used, 2, 4, 6, ..., which must be even. If memory is limited, Eq. (21.2) can be solved in two or more steps, say from 0 to $n/2$ and then from $n/2$ to n.

For flywheel applications it is convenient to define a *coefficient of speed fluctuation* as

$$C_s = \frac{\omega_2 - \omega_1}{\omega} \tag{21.3}$$

where ω is the nominal or average angular velocity, given by

$$\omega = \frac{\omega_2 + \omega_1}{2} \tag{21.4}$$

Equation (21.1) can be factored to give

$$U_2 - U_1 = \tfrac{1}{2}I(\omega_2 - \omega_1)(\omega_2 + \omega_1)$$

and, because $(\omega_2 - \omega_1) = C_s\omega$ and $(\omega_2 + \omega_1) = 2\omega$, we have

$$U_2 - U_1 = C_s I \omega^2 \tag{21.5}$$

Equation (21.5) can be used to obtain an appropriate flywheel inertia corresponding to the energy change $U_2 - U_1$.

EXAMPLE 21.1

Table 21.1 lists values of the torque plotted in Fig. 21.3. The nominal speed of the engine is to be 250 rad/s.

(a) Integrate the torque–displacement function for one cycle and find the energy that can be delivered to the load during the cycle.

(b) Determine the mean torque T_m (see Fig. 21.3).

(c) The greatest energy fluctuation will occur approximately between $\theta = 15°$ and $\theta = 150°$ on the $T_i - T_o$ diagram; see Fig. 21.3 and note that $T_o = -T_m$. Using a coefficient of speed fluctuation of $C_s = 0.1$, find a suitable value for the flywheel inertia.

(d) Find ω_2 and ω_1.

SOLUTION

(a) Using $n = 48$ and $h = 4\pi/48$, we enter the data of Table 21.1 into our calculator and compute the integral by Simpson's rule as defined in Eq. (21.2); for the amount of energy that can be delivered to the load we get $U = 3\,490$ in·lb.   *Ans.*

(b) $T_m = 3490/4\pi = 278$ in·lb   *Ans.*

TABLE 21.1 Example 21.1: Torque Data for Fig. 21.3

θ_i deg	T_i in·lb	θ_i deg	T_i in·lb	θ_i deg	T_i in·lb	θ_i deg	T_i in·lb	θ_i deg	T_i in·lb
0	0	150	532	300	−8	450	242	600	−355
15	2 800	165	184	315	89	465	310	615	−371
30	2 090	180	0	330	125	480	323	630	−362
45	2 430	195	−107	345	85	495	280	645	−312
60	2 160	210	−206	360	0	510	206	660	−272
75	1 840	225	−280	375	−85	525	107	675	−274
90	1 590	240	−323	390	−125	540	0	690	−548
105	1 210	255	−310	405	−89	555	−107	705	−760
120	1 066	270	−242	420	8	570	−206		
135	803	285	−126	435	126	585	−292		

(c) The largest positive loop in the torque-displacement diagram occurs between $\theta = 0°$ and $\theta = 180°$. We select this loop as yielding the largest speed change. Subtracting 278 in·lb from the values in Table 21.1 for this loop gives, respectively, −278, 2 522, 1 812, 2 152, 1 882, 1 562, 1 312, 932, 788, 535, 254, −94, and −278 in·lb. Entering the Simpson's rule again, using $n = 12$ and $h = \pi/12$, gives $U_2 - U_1 = 3\,663$ in·lb. We now solve Eq. (21.5) for I. This gives

$$I = \frac{U_2 - U_1}{C_s \omega^2} = \frac{3\,663 \text{ in·lb}}{0.1(250 \text{ rad/s})^2} = 0.586 \text{ in·lb·s}^2 \qquad \textit{Ans.}$$

(d) Equations (21.3) and (21.4) can now be solved simultaneously for ω_2 and ω_1. Substituting appropriate values into these two equations yields

$$\omega_2 = \tfrac{1}{2}(2 + C_s)\omega = \tfrac{1}{2}(2 + 0.1)250 = 262.5 \text{ rad/s} \qquad \textit{Ans.}$$

$$\omega_1 = 2\omega - \omega_2 = 2(250) - 262.5 = 237.5 \text{ rad/s} \qquad \textit{Ans.}$$

These two speeds occur at $\theta = 180°$ and $\theta = 0$, respectively.

21.3 MULTICYLINDER ENGINE TORQUE SUMMATION

After the torque–displacement relation has been defined for a single cylinder, it is easy to assume these for multicylinder engines. If, for example, we wish to find the torque function for a three-cylinder engine, then we examine the firing order shown in Fig. 18.1 and write

$$T_{total} = T_\theta + T_{\theta+240} + T_{\theta+480} \qquad (a)$$

because the torque events are spaced 240° apart. Here θ is the crank angle of the first cylinder. Applying this same procedure for a four-cylinder engine, along with using the torque values in Table 21.1, gives the results in Table 21.2. Figure 21.4 shows how this torque varies during each 180° of crank rotation.

TABLE 21.2 Torque Data for a Four-Cylinder Four-Cycle Internal Combustion Engine

θ_i deg	T_θ in·lb	$T_{\theta+180}$, in·lb	$T_{\theta+360}$, in·lb	$T_{\theta+540}$, in·lb	T_{total}, in·lb
0	0	0		0	0
15	2 800	−107	−85	−107	2 501
30	2 090	−206	−125	−206	1 553
45	2 430	−280	−89	−292	1 769
60	2 160	−323	8	−355	1 490
75	1 840	−310	126	−371	1 285
90	1 590	−242	242	−362	1 228
105	1 210	−126	310	−312	1 082
120	1 066	−8	323	−272	1 109
135	803	89	280	−274	898
150	532	125	206	−548	315
165	184	85	107	−760	−384

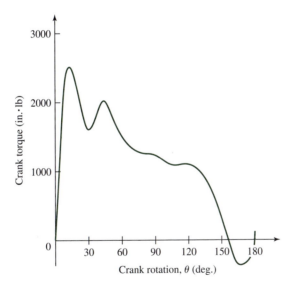

Figure 21.4 Relation between torque and crank angle for a four-cylinder four-cycle internal combustion engine.

PROBLEMS

21.1 Table P21.1 lists the output torque for a one-cylinder engine running at 4 600 rev/min.
(a) Find the mean output torque.
(b) Determine the mass moment of inertia of an appropriate flywheel using $C_s = 0.025$.

21.2 Using the data of Table 21.2, determine the moment of inertia for a flywheel for a two-cylinder 90° V engine having a single crank. Use $C_s = 0.0125$ and a nominal speed of 4 600 rev/min. If a cylindrical or disk-type flywheel is to be used, what should be the thickness if it is made of steel and has an outside diameter of 400 mm? Use $\rho = 7.8 \, \text{Mg/m}^3$ as the density of steel.

21.3 Using the data of Table 21.1, find the mean output torque and the flywheel inertia required for a three-cylinder in-line engine corresponding to a nominal speed of 2 400 rev/min. Use $C_s = 0.03$.

TABLE P21.1 Torque Data for Problem 21.1

θ_i deg	T_i N·m	θ_i deg	T_i N·m	θ_i deg	T_i N·m	θ_i deg	T_i N·m
0	0	180	0	360	0	540	0
10	17	190	−344	370	−145	550	−344
20	812	200	−540	380	−150	560	−540
30	963	210	−576	390	7	570	−577
40	1 016	220	−570	400	164	580	−572
50	937	230	−638	410	235	590	−643
60	774	240	−785	420	203	600	−793
70	641	250	−879	430	490	610	−893
80	697	260	−814	440	424	620	−836
90	849	270	−571	450	571	630	−605
100	1 031	280	−324	460	814	640	−379
110	1 027	290	−190	470	879	650	−264
120	902	300	−203	480	785	660	−300
130	712	310	−235	490	638	670	−368
140	607	320	−164	500	570	680	−334
150	594	330	−7	510	576	690	−198
160	544	340	150	520	540	700	−56
170	345	350	145	530	344	710	−2

TABLE P21.4 Torque Data for Problem 21.4

θ_i deg	T_i in·lb	θ_i deg	T_i in·lb	θ_i deg	T_i in·lb	θ_i deg	T_i in·lb
0	857	90	7 888	180	1 801	270	857
10	857	100	8 317	190	1 629	280	857
20	857	110	8 488	200	1 458	290	857
30	857	120	8 574	210	1 372	300	857
40	857	130	8 403	220	1 115	310	857
50	1 287	140	7 717	230	1 029	320	857
60	2 572	150	3 515	240	943	330	857
70	5 144	160	2 144	250	857	340	857
80	6 859	170	1 972	260	857	350	857

21.4 The load torque required by a 200-ton punch press is displayed in Table P21.4 for one revolution of the flywheel. The flywheel is to have a nominal angular velocity of 2 400 rev/min and to be designed for a coefficient of speed fluctuation of 0.075.

(a) Determine the mean motor torque required at the flywheel shaft and the motor horsepower needed, assuming a constant torque–speed characteristic for the motor.

(b) Find the moment of inertia needed for the flywheel.

21.5 Find T_m for the four-cylinder engine whose torque displacement is that of Fig. 21.4.

22 Governors

In Chapter 21 we learned that flywheels are used to regulate speed over short intervals of time such as a single revolution or the duration of an engine cycle. *Governors,* too, are devices used to regulate speed. In contrast to the flywheel, however, governors are used to regulate speed over a much longer interval of time; in fact, they are intended to maintain a balance between the energy supplied to a moving system and the external load or resistance applied to that system.

22.1 CLASSIFICATION

When the speed of a machine must, during its lifetime, always be kept at the same level, or approximately so, then a shaft-mounted mechanical device may be an appropriate speed regulator. Such governors may be classified as

- Centrifugal governors
- Inertia governors

As the name indicates, centrifugal force plays the important role in centrifugal governors. In inertia governors it is more the angular acceleration, or change in speed, that dominates the regulating action.

The availability today of a wide variety of low-priced solid-state electronic devices and transducers makes it possible to regulate mechanical systems to a finer degree and at less cost than with the older all-mechanical governors. The electronic governor also has the advantage that the speed to be regulated can be changed quite easily and at will.

22.2 CENTRIFUGAL GOVERNORS

Figure 22.1 shows a simple spring-controlled centrifugal shaft governor. Masses attached to the bell-crank levers are driven outward by centrifugal force against springs. The motion of the shaft-mounted sleeve is dependent on the motions of the masses and the ratio of the bell-crank lengths.

Selecting the nomenclature of Fig. 22.2, we let

k = spring rate
r = instantaneous location of mass center
P = spring force
W = sleeve weight corresponding to a vertical shaft
a = position of mass at zero spring force

Then the spring force at any position r is

$$P = k(r - a) \qquad (a)$$

Taking moments about pivot O gives

$$\sum M_O = -Pl_2 \cos\alpha - \frac{W}{2}l_1 \cos\alpha - (-m\ddot{r})l_2 \cos\alpha = 0 \qquad (b)$$

Now, with $\ddot{r} = r\omega^2$ and $m = W/g$, summing forces gives

$$\frac{W}{g}\omega^2 r l_2 = Pl_2 + \frac{W}{2}l_1 \qquad (c)$$

The controlling force is

$$F = \frac{W}{g}\omega^2 r = P + \frac{W}{2}\frac{l_1}{l_2} = k(r - a) + \frac{W}{2}\frac{l_1}{l_2} \qquad (22.1)$$

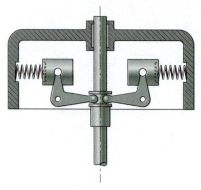

Figure 22.1 A centrifugal shaft governor.

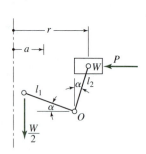

Figure 22.2

This corresponds to an angular velocity of

$$\omega^2 = \frac{kg}{Wr}\left[r - \left(a - \frac{W\,l_1}{2k\,l_2}\right)\right]$$ (22.2)

22.3 INERTIA GOVERNORS

In a centrifugal governor an increase in speed causes the rotating masses to move radially outward as we have seen. Thus it is the radial (normal) component of the acceleration that is primarily responsible for creating the controlling force. In an inertia shaft governor, as shown in Fig. 22.3, the mass pivot is located at point A very close to the shaft center at O. Thus the radial component of acceleration is smaller and much less effective. But a sudden change of speed, producing an angular acceleration, will cause the masses to lag and produce tension in the spring. This spring tension produces the transverse acceleration force. Consequently, it is the angular acceleration that determines the position of the masses.

Compared to the centrifugal governor, the inertia type is more sensitive because it acts at the very beginning of a speed change.

22.4 MECHANICAL CONTROL SYSTEMS

Many mechanical control systems are represented as in Fig. 22.4. Here, θ_i and θ_o represent any set of input and output functions. In the case of a governor, θ_i represents the desired speed and θ_o the actual output speed. For control systems in general the input and output functions could just as easily represent force, torque, or displacement, either rectilinear or angular.

The system shown in Fig. 22.4 is called a *closed-loop* or *feedback control system* because the output θ_o is fed back to the detector at the input so as to measure the error ε, which is the difference between the input and the output. The purpose of the controller is to cause this error to become as close to zero as possible. The mechanical characteristics of the system, that is, the mechanical clearances, friction, inertias, and stiffnesses, sometimes

Figure 22.3 An inertia governor.

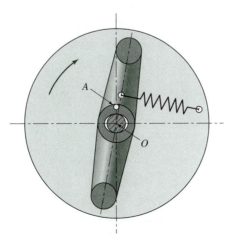

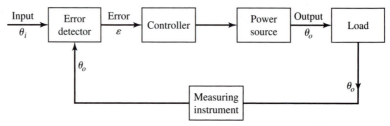

Figure 22.4 Block diagram of a closed-loop control system.

cause the output to differ somewhat from the input, and so it is the designer's responsibility to examine these mechanical effects in an effort to minimize the error for all operating conditions.

The differential equation for a control system is always written as an expression of the dynamic equilibrium of the elements of the system. Thus, for the system of Fig. 22.4, we express mathematically that the inertia torque is equal to the sum of all other torques acting upon the system. If the load has both inertia and viscous damping, then the expression is

$$I\ddot{\theta}_o = -c\dot{\theta}_o + k'f(\varepsilon) \tag{a}$$

The function $f(\varepsilon)$ depends upon the characteristics of the controller and the power source. A very simple system would result if the components had characteristics such that the relation between $f(\varepsilon)$ and ε is linear. For this condition we can make the substitution

$$k'f(\varepsilon) = k\varepsilon \tag{b}$$

Equation (a) can now be rearranged to give

$$I\ddot{\theta}_o + c\dot{\theta}_o = k\varepsilon \tag{c}$$

But because, by definition, we have

$$\varepsilon = \theta_i - \theta_o \tag{d}$$

then Eq. (c) becomes

$$I\ddot{\theta}_o + c\dot{\theta}_o + k\theta_o = k\theta_i \tag{22.3}$$

Equation (22.3) can be simplified by the following substitutions:

$$\omega_n = \sqrt{\frac{k}{I}} \tag{22.4}$$

$$2\zeta\omega_n = \frac{c}{I} \tag{22.5}$$

$$\zeta = \frac{c}{2\sqrt{kI}} \tag{22.6}$$

where I = mass moment of inertia

c = torsional damping coefficient

k = torsional stiffness

ω_n = natural frequency

ζ = damping ratio, c/c_{cr}

After review of this notation for linear systems in Chapter 17, Eq. (22.3) can now be written in the form

$$\ddot{\theta}_o + 2\zeta\omega_n\dot{\theta}_o + \omega_n^2\theta_o = \omega_n^2\theta_i \tag{22.7}$$

22.5 STANDARD INPUT FUNCTIONS

There is a great deal of useful information that can be gained from a mathematical analysis as well as a laboratory analysis of feedback control systems when standard input functions are used to study their performance. Standard input functions result in differential equations that are much easier to analyze mathematically than they would be if the actual operating conditions were used as input. Furthermore, the use of standard inputs makes it possible to compare the performance of various proposed control systems.

One of the most useful of the standard input functions is the *unit-step function* shown in Fig. 22.5a. This function is not continuous, and consequently we cannot define initial conditions at $t = 0$. It is customary to specify the conditions at $t = 0+$ and $t = 0-$, where the signs indicate the conditions for values of time slightly greater than or slightly less than zero. Thus for the unit-step function we have

$$\theta_i = 0, \quad \dot{\theta}_i = 0 \qquad \text{when } t = 0-$$

$$\theta_i = 1, \quad \dot{\theta}_i = 0 \qquad \text{when } t = 0+$$

The performance resulting from the application of these input conditions is called the *unit-step response*.

Another standard input function is the *unit-step velocity function,* as shown in Fig. 22.5b. This function is sometimes called the *unit-slope ramp function* because of the

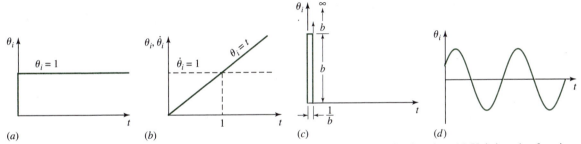

Figure 22.5 Standard input functions. (*a*) Unit-step function. (*b*) Unit-step velocity function. (*c*) Unit-impulse function. (*d*) Harmonic input function.

form of its displacement diagram. This function is defined as follows:

$$\theta_i = 0, \quad \dot{\theta}_i = 0 \qquad \text{when } t = 0-$$

$$\theta_i = t, \quad \dot{\theta}_i = 1 \qquad \text{when } t = 0+$$

Also shown in Figs. 22.5c and 22.5d are the unit-impulse function and the harmonic input function, both of which are useful in studying a system's response.

It sometimes happens in control systems that the input function is relatively constant but the output is subjected to suddenly varying forces or torques. For example, an automatically controlled machine tool may run into a deeper cut, causing a greater load torque to be exerted on the machine, thus tending to slow it down. The differential equation for this condition is written as

$$I\ddot{\theta}_o + c\dot{\theta}_o + k\theta_o = k\theta_i - T \tag{a}$$

When this equation is simplified by substituting the values from Eqs. (22.4) to (22.6) and rearranged, it becomes

$$\ddot{\theta}_o + 2\zeta\omega_n\dot{\theta}_o + \omega_n^2\theta_o = \omega_n^2\theta_i - \frac{T}{I} \tag{22.8}$$

When the solution to Eq. (22.7) is obtained for a given function θ_i, the solution to Eq. (22.8) can be obtained by superposition.

22.6 SOLUTION OF LINEAR DIFFERENTIAL EQUATIONS

In the analysis of feedback control systems nth-order linear differential equations are very frequently encountered. The general form of these equations is

$$a_n\frac{d^n\theta}{dt^n} + a_{n-1}\frac{d^{n-1}\theta}{dt^{n-1}} + \cdots + a_1\frac{d\theta}{dt} + a_0\theta = f(t) \tag{22.9}$$

The function $f(t)$ is the driving or forcing function, and so t and θ are the independent and dependent variables, respectively. The coefficients $a_0, a_1, \ldots, a_n$ are constants and are independent of t and θ.

The solution to equations of the form of Eq. (22.9) is composed of two parts. The first part is called the *complimentary solution,* and it is the solution to the homogeneous equation

$$a_n\frac{d^n\theta}{dt^n} + a_{n-1}\frac{d^{n-1}\theta}{dt^{n-1}} + \cdots + a_1\frac{d\theta}{dt} + a_0\theta = 0 \tag{22.10}$$

The complementary solution is also called the *transient solution* in the literature of automatic controls because, for a damped stable system, this part of the solution quickly dies away. If the control system should happen to be unstable, then we shall find that the controlled quantity quickly increases without limit. Note that the transient solution is obtained by making the forcing function zero.

The other part of the solution is called the *particular solution* by mathematicians and called the *steady-state solution* by control engineers. It is any particular solution of Eq. (22.9).

The complete solution is the sum of the transient and the steady-state solutions. In the transient part it will contain n arbitrary constants that are evaluated from the initial conditions. The amplitude of the transient solution depends upon both the forcing function and the initial conditions, but all other characteristics of the transient solution are completely independent of the forcing function. We shall find, for linear systems, that neither the forcing function nor the amplitude has any effect on the system stability, this being dependent only on the transient portion of the solution.

The Transient Solution   The following steps are used to obtain the transient solution:

1. Set the forcing function equal to zero and arrange the equation in the form of Eq. (22.10).
2. Assume a solution of the form

$$\theta = Ae^{st} \tag{22.11}$$

3. Substitute Eq. (22.11) and its derivatives into the differential equation, and simplify to obtain the characteristic equation.
4. Solve the characteristic equation for its roots.
5. Obtain the transient solution by substituting the roots back into the assumed solution.

As an example of this procedure let us solve Eq. (22.7) for the transient solution. For step 1 we write

$$\ddot{\theta}_o + 2\zeta\omega_n\dot{\theta}_o + \omega_n^2\theta_o = 0 \tag{a}$$

and for step 2 we have

$$\theta_o = Ae^{st}, \qquad \dot{\theta}_o = Ase^{st}, \qquad \ddot{\theta}_o = As^2e^{st} \tag{b}$$

For step 3 we substitute Eqs (b) into Eq. (a). This gives

$$As^2e^{st} + 2\zeta\omega_n Ase^{st} + \omega_n^2 Ae^{st} = 0 \tag{c}$$

The characteristic equation is then obtained by dividing out common terms. Thus

$$s^2 + 2\zeta\omega_n s + \omega_n^2 = 0 \tag{d}$$

is the characteristic equation. Step 4 is to solve this for its roots. This yields

$$\begin{aligned} s_1 &= -\zeta\omega_n + \omega_n\sqrt{\zeta^2 - 1} \\ s_2 &= -\zeta\omega_n - \omega_n\sqrt{\zeta^2 - 1} \end{aligned} \tag{e}$$

We shall not consider the situation in which $\zeta > 1$. Consequently, for $\zeta < 1$, the radical in Eq. (e) has an imaginary value, and we prefer to write the roots in the form

$$
\begin{aligned}
s_1 &= -\zeta\omega_n + j\omega_n\sqrt{1-\zeta^2} \\
s_2 &= -\zeta\omega_n - j\omega_n\sqrt{1-\zeta^2}
\end{aligned}
\tag{f}
$$

where $j = \sqrt{-1}$. Finally, for step 5, we substitute the roots back into the assumed solution. Then, after factoring, the transient solution becomes

$$
\theta_{o,t} = e^{-\zeta\omega_n t}\left(Ae^{j\omega_n\sqrt{1-\zeta^2}t} + Be^{-j\omega_n\sqrt{1-\zeta^2}t}\right)
\tag{g}
$$

where A and B are the constants of integration and are two in number, the same as the order of the highest derivative in the differential equation. These cannot be evaluated until the complete solution is found. At that time they will be found from the initial conditions. We can, however, transform Eq. (g) into trigonometric form by using DeMoivre's theorem. The result of this transformation is

$$
\theta_{o,t} = e^{-\zeta\omega_n t}\left(A\cos\omega_n\sqrt{1-\zeta^2}t + B\sin\omega_n\sqrt{1-\zeta^2}t\right)
\tag{22.12}
$$

The coefficient $e^{-\zeta\omega_n t}$ of Eq. (22.12) indicates its transient nature, because this term becomes approximately zero for large values of t.

It can sometimes happen that two or more roots of the characteristic equation are identical. For the case of two identical roots $s_1 = s_2$ the transient solution is

$$
\theta_{o,t} = Ae^{s_1 t} + Bte^{s_1 t}
\tag{h}
$$

For other special cases the reader should refer to a text on the solution of linear differential equations.

The Steady-State Solution

Here we shall solve Eq. (22.7) for various input conditions. The equation is repeated for convenience:

$$
\ddot{\theta}_o + 2\zeta\omega_n\dot{\theta}_o + \omega_n^2\theta_o = \omega_n^2\theta_i
\tag{22.7}
$$

For the unit-step function $\theta_i = 1$, the equation to be solved for the particular solution is

$$
\ddot{\theta}_o + 2\zeta\omega_n\dot{\theta}_o + \omega_n^2\theta_o = \omega_n^2
\tag{i}
$$

The solution is

$$
\theta_{o,s} = 1
\tag{22.13}
$$

which can be verified by substituting Eq. (22.13) and its derivatives into Eq. (i).

For the unit-step velocity function, $\theta_i = t$, the differential equation becomes

$$
\ddot{\theta}_o + 2\zeta\omega_n\dot{\theta}_o + \omega_n^2\theta_o = \omega_n^2 t
\tag{j}
$$

We assume a solution of the form

$$
\theta_{o,s} = At + B
$$

The successive derivatives are

$$\dot{\theta}_{o,s} = A, \qquad \ddot{\theta}_{o,s} = 0$$

Substituting the assumed solution and its derivatives into Eq. (j) gives

$$2\zeta\omega_n A + \omega_n^2(At + B) = \omega_n^2 t$$

We now arrange the equation into the form

$$\left(2\zeta\omega_n A + \omega_n^2 B\right) + \left(\omega_n^2 A\right)t = \omega_n^2 t$$

which we solve for A and B. Thus

$$A = 1, \qquad B = -\frac{2\zeta}{\omega_n}$$

Therefore the steady-state solution for a unit step velocity function is

$$\theta_{o,s} = t - \frac{2\zeta}{\omega_n} \tag{22.14}$$

The Complete Solution   The complete solution is the sum of the transient and the steady-state solution terms. Thus

$$\theta_o = \theta_{o,t} + \theta_{o,s}$$

For the unit-step input this is formed by summing Eqs. (22.12) and (22.13). The result is

$$\theta_o = e^{-\zeta\omega_n t}\left(C_1 \cos\omega_n\sqrt{1-\zeta^2}t + C_2 \sin\omega_n\sqrt{1-\zeta^2}t\right) + 1 \tag{22.15}$$

The conditions for $t = 0+$ can now be applied in order to evaluate the constants C_1 and C_2. Imposing the condition that $\theta_o = 0$ at the beginning of the step gives

$$0 = 1(C_1 \cos 0 + C_2 \sin 0) + 1$$

or

$$C_1 = -1$$

The second condition to be imposed is that $\dot{\theta}_o = 0$ at $t = 0$. To apply this condition it is necessary to take the derivative of Eq. (22.15). This is

$$\dot{\theta}_o = -\zeta\omega_n e^{-\zeta\omega_n t}\left(C_1 \cos\omega_n\sqrt{1-\zeta^2}t + C_2 \sin\omega_n\sqrt{1-\zeta^2}t\right)$$

$$+ e^{-\zeta\omega_n t}\left(-C_1\omega_n\sqrt{1-\zeta^2}\sin\omega_n\sqrt{1-\zeta^2}t + C_2\omega_n\sqrt{1-\zeta^2}\cos\omega_n\sqrt{1-\zeta^2}t\right)$$

Substituting $\dot{\theta}_o = 0$ and $t = 0$ gives

$$0 = -\zeta\omega_n(C_1\cos 0 + C_2\sin 0) + 1\left(-C_1\omega_n\sqrt{1-\zeta^2}\sin 0 + C_2\omega_n\sqrt{1-\zeta^2}\cos 0\right)$$

Substituting the value of C_1 and solving yields

$$C_2 = -\frac{\zeta}{\sqrt{1-\zeta^2}}$$

The values of C_1 and C_2 are now replaced in Eq. (22.15), giving the complete solution

$$\theta_o = 1 - e^{-\zeta\omega_n t}\left(\cos\omega_n\sqrt{1-\zeta^2}t + \sin\omega_n\sqrt{1-\zeta^2}t\right) \tag{22.16}$$

If transformed to a single trigonometric term and a phase angle as demonstrated in Chapter 17, the solution is

$$\theta_o = 1 - \frac{e^{-\zeta\omega_n t}}{\sqrt{1-\zeta^2}}\cos\left(\omega_n\sqrt{1-\zeta^2}t - \phi\right) \tag{22.17}$$

where

$$\phi = \tan^{-1}\left(\frac{\zeta}{\sqrt{1-\zeta^2}}\right)$$

The complete solution for the unit-step velocity function input is obtained in a similar manner. The initial conditions to be applied in order to evaluate the constants of integration are

$$\theta_o = 0 \quad \text{and} \quad \dot{\theta}_o = 0 \qquad \text{when } t = 0-$$

The equation to be solved for the integration constants is obtained by adding Eqs. (22.12) and (22.14); it is

$$\theta_o = e^{-\zeta\omega_n t}\left(C_1\cos\omega_n\sqrt{1-\zeta^2}t + C_2\sin\omega_n\sqrt{1-\zeta^2}t\right) + t - \frac{2\zeta}{\omega_n} \tag{22.18}$$

The two constants are found to be

$$C_1 = \frac{2\zeta}{\omega_n}, \qquad C_2 = -\frac{1-2\zeta^2}{\omega_n\sqrt{1-\zeta^2}}$$

and the complete solution is

$$\theta_o = t - \frac{2\zeta}{\omega_n} + e^{-\zeta\omega_n t}\left(\frac{2\zeta}{\omega_n}\cos\omega_n\sqrt{1-\zeta^2}t - \frac{1-2\zeta^2}{\omega_n\sqrt{1-\zeta^2}}\sin\omega_n\sqrt{1-\zeta^2}t\right) \tag{22.19}$$

or

$$\theta_o = t - \frac{2\zeta}{\omega_n} + \frac{e^{-\zeta\omega_n t}}{\omega_n\sqrt{1 - \zeta^2}}\cos\left(\omega_n\sqrt{1 - \zeta^2}t - \phi\right) \tag{22.20}$$

where

$$\phi = \tan^{-1}\left(\frac{2\zeta^2 - 1}{2\zeta\sqrt{1 - \zeta^2}}\right)$$

22.7 ANALYSIS OF PROPORTIONAL-ERROR FEEDBACK SYSTEMS

As the name implies, a proportional-error feedback control system is one that operates by applying a correction that is directly proportional to any error which might exist. Such a system must be analyzed in order to determine the following:

1. The response time, or the time required to reach steady state operation after the application of a disturbance to the system. Because a system with viscous damping never reaches a steady-state condition, this is usually defined as the time required for the transient to decay to 5 percent of its initial value.
2. The natural frequency.
3. The steady-state error, that is, the difference between the input and output during steady-state operation.
4. The maximum overshoot, or the maximum deviation between output and input during transient conditions.

Output response of the system to a unit step input is given by Eq. (22.17) and has been plotted in Fig. 22.6 for various values of the damping ratio. The abscissa is measured in dimensionless time $\omega_n t/2\pi$, and so the curves can be applied to any physical system. Thus, if the undamped natural frequency ω_n is known for a given system, then the response time can be calculated simply by multiplying dimensionless time by the quantity by $2\pi/\omega_n$. The response for zero damping is included for its academic interest; an automatic control system with no damping at all would give completely unsatisfactory performance.

The response for critical damping $\zeta = 1$ shows no overshoot, and the steady-state condition is reached at about $\omega_n t/2\pi = 1.0$.

It is convenient to define an *overshoot ratio* as the ratio of overshoot with damping to the overshoot that would exist if no damping were present. Figure 22.7 shows a plot of this ratio versus the damping ratio. For $\zeta = 0.40$, the overshoot ratio is approximately 0.25; this quantity can also be read from Fig. 22.6. Notice that the steady-state condition is not reached for $\zeta = 0.40$ until $\omega_n t/2\pi = 2.0$. The response time, therefore, is twice that for critical damping.

Because the response time is proportional to $\omega_n t/2\pi$, we can make this time short simply by making the undamped natural frequency ω_n large.

In practice, ζ is usually between 0.4 and 0.7 in physical systems. If a certain deviation from steady-state is permitted, say 5 percent, then a system with $\zeta = 0.7$ will reach steady state before one having $\zeta = 1.0$ because the early portion of its response curve is steeper.

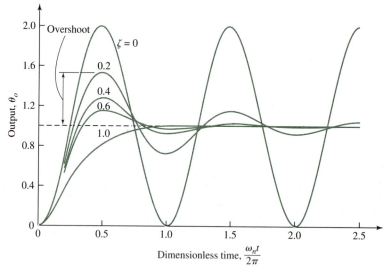

Figure 22.6 Response to a unit step input.

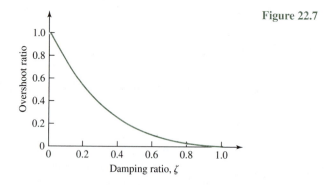

Figure 22.7

Figure 22.6 shows that the steady-state value of the output is unity. Consequently, there is no steady-state error for step-input functions when applied to proportional-error feedback systems.

Local Disturbance   Another look at the behavior of the system can be obtained by considering that the input signal is constant or zero and that the load is subjected to a disturbance. Selecting the zero input condition, for simplicity, Eq. (22.8) then becomes

$$\ddot{\theta}_o + 2\zeta\omega_n\dot{\theta}_o + \omega_n^2\theta_o = -\frac{T}{I} \tag{22.21}$$

where T represents a constant torque suddenly applied to the input. The steady-state component of the solution is

$$\theta_{o,s} = -\frac{T}{I\omega_n^2} \tag{a}$$

so that the complete solution must be

$$\theta_o = e^{-\zeta \omega_n t} \left(C_1 \cos \omega_n \sqrt{1 - \zeta^2} t + C_2 \sin \omega_n \sqrt{1 - \zeta^2} t \right) - \frac{T}{I \omega_n^2} \qquad (b)$$

The initial conditions are $\theta_o = 0$ and $\dot{\theta}_o = 0$ at $t = 0$. Using these we can solve Eq. (b) for the two constants

$$C_1 = \frac{T}{I \omega_n^2}, \qquad C_2 = \frac{\zeta T}{I \omega_n^2 \sqrt{1 - \zeta^2}}$$

When these are substituted into Eq. (b) and it is transformed to a single trigonometric term the result is

$$\theta_o = \frac{T}{I \omega_n^2} \left[\frac{e^{-\zeta \omega_n t}}{\sqrt{1 - \zeta^2}} \cos \left(\omega_n \sqrt{1 - \zeta^2} t - \phi \right) - 1 \right] \qquad (22.22)$$

where

$$\phi = \tan^{-1} \left(\frac{\zeta}{\sqrt{1 - \zeta^2}} \right)$$

This equation is plotted in Fig. 22.8 for three values of the damping ratio to show what happens. As shown, the disturbance decays at a rate which is dependent on the amount of damping. It finally reaches a steady-state condition which is not zero, but of an amount

$$\varepsilon_s = \frac{T}{I \omega_n^2} \qquad (22.23)$$

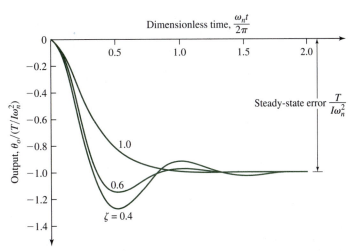

Figure 22.8 Response of proportional-error feedback control to a load disturbance.

This is the steady-state error. If we substitute $\omega_n^2 = k/I$, the equation becomes

$$\varepsilon_s = \frac{T}{k} \tag{22.24}$$

so the only method of reducing the magnitude of this error is to increase the gain k of the system.

We have seen that the control system is defined by the parameters I, c, and k. Of these three, the gain is usually the easiest to change. Some variation in the damping is usually possible, but the employment of dashpots or friction dampers is not often a good solution. Variation in the inertia I is the most difficult change to make because this is fixed by the design of the driven element and, furthermore, improvement always requires a decrease in the inertia.

Unit-Step Velocity Input   For the unit-step velocity input we have $\theta_i = t$, and from Eq. (22.14) we obtain

$$\theta_{o,s} = t - \frac{2\zeta}{\omega_n}$$

after the transient decays. Therefore the steady-state error is

$$\varepsilon_s = \frac{2\zeta}{\omega_n} = \frac{c}{k} \tag{22.25}$$

which is obtained by substitution of the value of ζ from Eq. (22.6). This error occurs only as long as the input function signals for a constant velocity. Again we see that the magnitude can be reduced by increasing the gain k.

The widely used automotive cruise-control system is an excellent example of an electromechanical governor. A transducer is attached to the speedometer cable, and the electrical output of this transducer is the signal θ_o fed to the error detector of Fig. 22.4. In some cases, magnets are mounted on the driveshaft of the car to activate the transducer. In the cruise-control system, the error detector is an electronic regulator, usually mounted under the dashboard. The regulator is turned on by an engagement switch under or near the steering wheel. A power unit is connected to the carburetor throttle linkage; the power unit is controlled by the regulator and gets its power from a vacuum port on the engine. Such systems have one or two brake-release switches as well as the engagement switch. The accelerator pedal can also be used to override the system.

23 | Gyroscopes

23.1 INTRODUCTION

A gyroscope may be defined as a rigid body capable of three-dimensional rotation with high angular velocity about any axis that passes through a fixed point called the center, which may or may not be its center of mass. A child's toy top fits this definition and is a form of gyroscope; its fixed point is the point of contact of the top with the floor or table on which it spins.

The usual form of a gyroscope is a mechanical device in which the essential part is a rotor having a heavy rim spinning at high speed about a fixed point on the rotor axis. The rotor is then mounted so as to turn freely about its center of mass by means of a double gimbals called a Cardan suspension; an example is pictured in Fig. 23.1.

The gyroscope has fascinated students of mechanics and applied mathematics for many years. In fact, once the rotor is set spinning, a gyroscope appears to act like a device possessing intelligence. If we attempt to move some of its parts, it seems not only to resist this motion but even to evade it. We shall see that it apparently fails to conform to the laws of static equilibrium and of gravitation.

The early history of the gyroscope is rather obscure. Probably the earliest gyroscope of the type now in use was constructed by Bohenburger in Germany in 1810. In 1852 Leon Foucault, of Paris, constructed a very refined version to show the rotation of the earth; it was Foucault who named it the gyroscope from the Greek words *gyros,* circle or ring, and *skopien,* to view. The mathematical foundations of gyroscopic theory were laid by Leonhard Euler in 1765 in his work on the dynamics of rigid bodies. Gyroscopes were not put to practical and industrial use until the beginning of the twentieth century in the United States, at which time the gyroscopic compass, the ship stabilizer, and the monorail car were all invented. The subsequent uses of the gyroscope as a turn-and-bank indicator, artificial horizon, and automatic pilot in aircraft and missles are well known.

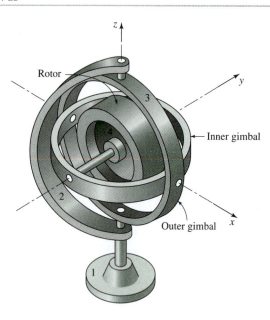

Figure 23.1 A laboratory gyroscope.

We may also become concerned with gyroscopic effects in the design of machines, although not always intentionally. Such effects are present in the riding of a motorcycle or bicycle; they are always present, owing to the rotating masses, when an airplane or automobile is making a turn. Sometimes these gyroscopic effects are desirable, but more often they are undesirable and designers must account for them in their selection of bearings and rotating parts. It is certainly true that, as machine speeds increase to higher and higher values and as factors of safety decrease, we must stop neglecting gyroscopic forces in our design calculations because their values will become more significant.

23.2 THE MOTION OF A GYROSCOPE

Although we have noted above that a gyroscope seems to have intelligence and to avoid compliance with the fundamental laws of mechanics, this is not truly the case. In fact, we have thoroughly covered the basic theory involved in Chapter 16, where we studied dynamic forces with spatial motion. Still, because gyroscopic forces are of increasing importance in higher-speed machines, we will look at them again and try to explain the apparent paradoxes they seem to raise.

To provide a vehicle for the explanation of the simpler motions of a gyroscope, we will consider a series of experiments to be performed on the one pictured in Fig. 23.1. In the following discussion we assume that the rotor is already spinning rapidly and that any pivot friction is negligible.

As a first experiment we transport the entire gyroscope about the table or the room. We find that even though we travel along a curved path, the orientation of the rotor's axis of rotation does not change as we move. This is a consequence of the law of conservation of angular momentum. If the axis of rotation is to change its orientation, the angular momentum vector must also change its orientation. But this requires an externally applied moment that, with the three sets of frictionless bearings, have not been supplied for this experiment. Therefore the orientation of the rotation axis does not change.

For a second experiment, while the rotor is still spinning, we lift the inner gimbal out of its bearings and move it about. We again find that it can be translated anywhere but that we meet with definite resistance if we attempt to rotate the axis of spin. In other words, the rotor persists in maintaining its plane of rotation.

For our third experiment, we replace the inner gimbal back into its bearings with the axis of *spin* horizontal as shown in Fig. 23.1. If we now apply a steady downward force to the inner gimbal at one end of the spin axis, say by pushing on it with a pencil, we find that the end of the spin axis does not move downward as we might expect. Not only do we meet with resistance to the force of the pencil, but the outer gimbal begins to rotate about the vertical axis, causing the rotor to skew around in the horizontal plane, and it continues this rotation until we remove the force of the pencil. This skewing motion of the spin axis is called *precession*.

Although it may seem strange and unexpected, this precession motion is in strict obedience to the laws of dynamic equilibrium as expressed by the Euler equations of motion in Eqs. (16.11). Yet, it is easier to understand and explain this phenomena if we again think in terms of the angular momentum vector. The downward force applied by the pencil to the inner gimbal produces a net external moment **M** on the rotor shaft through a pair of equal and opposite forces at the bearings and results in a time rate of change of the angular momentum vector **H**. As we showed in Eq. (16-21),

$$\frac{d\mathbf{H}}{dt} = \sum \mathbf{M} \tag{23.1}$$

Because this applied external moment cannot change the rate of spin of the rotor about its own axis, it changes the angular momentum by making the rotor rotate about the vertical axis as well, thus causing the precession.

As our next set of experiments we repeat the previous experiment watching carefully the directions involved. We first cause the rotor to spin in the positive direction—that is, with its angular velocity vector in the positive y direction. If we next apply a positive torque (moment vector in the positive x direction) by using downward force of our pencil on the negative y end of the rotor axis, the precession (rotation) of the outer gimbal is found to be in the negative z direction. Further experiments show that either a negative spin velocity or a negative moment caused by the pencil result in a positive direction for the precession.

For a final experiment we apply a moment to the outer gimbal in an attempt to cause the rotor to rotate about the z axis. Such an attempt meets with definite resistance and causes the inner gimbal and the spin axis to rotate. Note again in this case that the angular momentum vector is changing as the result of the application of external torque. If the spin axis starts in the vertical position (aligned along z), however, then the gyroscope is in stable equilibrium and the outer and inner gimabls can be turned together quite freely.

23.3 STEADY OR REGULAR PRECESSION

Let us now suppose that a heavy rotor is spinning about its spin axis with a constant angular velocity of ω_s about a spin axis which is tipped at a constant angle θ from the vertical. At the same time we assume that the spin axis has a constant angular velocity of ω_p as shown in Fig. 23.2. The state of motion just described is called *steady* or *regular* precession.

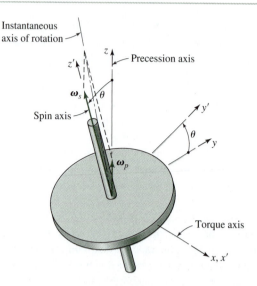

Figure 23.2 Angular velocity components of a gyroscope rotor.

We should take careful note of several things in this figure. First, we have shown the rotor at a position for which the inner gimbal is rotated to quite a different angle than in Fig. 23.1; yet the xyz coordinate system is still oriented with x along the axis of rotation between the inner and outer gimbals, and z is still along the true axis of precession. Second, we have designated a new coordinate system $x'y'z'$ with the x' axis coincident with x and the z' axis aligned along the axis of the rotor, the spin axis. Third, we should note that neither of these two coordinate systems is stationary and, similarly, that $x'y'z'$ is not attached to the rotor and does not experience the angular velocity $\boldsymbol{\omega}_s$. The $x'y'z'$ coordinate system remains fixed to the inner gimbal while xyz remains fixed to the outer gimbal.

In spite of this we can see that, for a disk shaped rotor mounted as described, the $x'y'z'$ axes are the principal axes of inertia of the rotor. We will designate the corresponding principal mass moments of inertia $I^{z'z'} = I^s$ and $I^{x'x'} = I^{y'y'} = I$.

If we now designate the rotor as body 4, with the inner and outer gimbals being links 3 and 2, respectively, then we can recognize that the constant angular spin velocity $\boldsymbol{\omega}_s$ is not truly an absolute angular velocity at all; it is the "relative" or *apparent* angular velocity of the rotor with respect to the inner gimbal.

$$\boldsymbol{\omega}_{4/3} = \boldsymbol{\omega}_s = \omega_s \hat{\mathbf{k}}' \qquad (a)$$

Similarly, the angular velocity of precession is really the angular velocity of the outer gimbal.

$$\boldsymbol{\omega}_2 = \boldsymbol{\omega}_p = \omega_p \hat{\mathbf{k}} = (\omega_p \sin\theta)\hat{\mathbf{j}}' + (\omega_p \cos\theta)\hat{\mathbf{k}}' \qquad (b)$$

Because we are interested in steady precession here, the angle θ has been assumed constant; thus there is no rotation between the inner and outer gimbals.

$$\boldsymbol{\omega}_{3/2} = \dot{\theta}\hat{\mathbf{i}}' = 0 \qquad (c)$$

We can now use Eqs. (*a*) through (*c*) to find the absolute angular velocity of the rotor.[1]

$$\boldsymbol{\omega}_4 = \boldsymbol{\omega}_2 + \boldsymbol{\omega}_{3/2} + \boldsymbol{\omega}_{4/3} = \boldsymbol{\omega}_p + \boldsymbol{\omega}_s$$

$$= (\omega_p \sin\theta)\hat{\mathbf{j}}' + (\omega_s + \omega_p \cos\theta)\hat{\mathbf{k}}' \tag{23.2}$$

If we assume that the masses of the gimbals are negligible in comparison with the rotor, then the angular momentum of the system is

$$\mathbf{H} = I(\omega_p \sin\theta)\hat{\mathbf{j}}' + I^s(\omega_s + \omega_p \cos\theta)\hat{\mathbf{k}}' \tag{23.3}$$

In order to find the time rate of change of this expression for angular momentum we need to recognize that the $\hat{\mathbf{j}}'$ and $\hat{\mathbf{k}}'$ are not constant; they rotate with the angular velocity of the $x'y'z'$ coordinate system. From Eqs. (*b*) and (*c*), this is

$$\boldsymbol{\omega}_3 = \boldsymbol{\omega}_2 + \boldsymbol{\omega}_{3/2} = \boldsymbol{\omega}_p$$

$$= (\omega_p \sin\theta)\hat{\mathbf{j}}' + (\omega_p \cos\theta)\hat{\mathbf{k}}' \tag{d}$$

Thus the time rate of change of $\hat{\mathbf{j}}'$ and $\hat{\mathbf{k}}'$ are

$$\frac{d\hat{\mathbf{j}}'}{dt} = \boldsymbol{\omega}_p \times \hat{\mathbf{j}}' = -(\omega_p \cos\theta)\hat{\mathbf{i}}' \tag{e}$$

$$\frac{d\hat{\mathbf{k}}'}{dt} = \boldsymbol{\omega}_p \times \hat{\mathbf{k}}' = (\omega_p \sin\theta)\hat{\mathbf{i}}' \tag{f}$$

Finally, differentiating Eq. (23.3) and using Eqs. (*e*) and (*f*), we find the net external moment that must be applied to the rotor to sustain the steady precession motion.

$$\sum\mathbf{M} = \frac{d\mathbf{H}}{dt} = \boldsymbol{\omega}_p \times \mathbf{H}$$

$$= -I(\omega_p \sin\theta)(\omega_p \cos\theta)\hat{\mathbf{i}}' + I^s(\omega_s + \omega_p \cos\theta)(\omega_p \sin\theta)\hat{\mathbf{i}}' \tag{23.4}$$

$$\sum\mathbf{M} = [I^s\omega_s + (I^s - I)\omega_p \cos\theta]\omega_p \sin\theta\,\hat{\mathbf{i}}' \tag{23.5}$$

$$\sum\mathbf{M} = \left[I^s + (I^s - I)\frac{\omega_p}{\omega_s}\cos\theta\right](\boldsymbol{\omega}_p \times \boldsymbol{\omega}_s) \tag{23.6}$$

We note that when the angular velocity of spin ω_s is much larger than that of precession ω_p, which is usually the case, then the second term in the square brackets is negligible with respect to the first. In this case

$$\sum\mathbf{M} = I^s\omega_p\omega_s \sin\theta\,\hat{\mathbf{i}}' = I^s(\boldsymbol{\omega}_p \times \boldsymbol{\omega}_s) \tag{23.7}$$

Figure 23.3 shows the same rotor in the same orientation as Fig. 23.2, but this time the angular momentum vector **H** rather than the angular velocities are shown. We have already

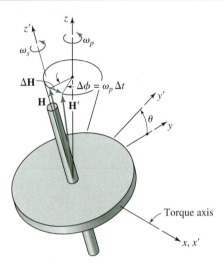

Figure 23.3 Angular momentum vector of a gyroscope rotor during steady precession.

noted that the angular momentum vector includes the effects of both the spin and the precession angular velocities. We can also see that it continually precesses, sweeping out a cone with apex angle θ about the precession axis.

In Fig. 23.3 we see the angular momentum vector **H** at some instant t and also its changed orientation **H**′ after a short time interval Δt, and we note that it has not changed magnitude, only orientation. Spanning the tips of these two vectors we see the vector change in angular momentum Δ**H** over this short time interval. We note that in the limit as Δt approaches zero, the direction of Δ**H** vector approaches the direction of the positive x and x' axes. This is totally consistent with Eq. (23.5), which showed that Δ**H**$/\Delta t$ was a vector in the $\hat{\mathbf{i}}'$ direction.

In order to maintain a steady precession, Eq. (23.6) shows that an external moment must be continually applied to the rotor; if this moment is not applied, regular precession will not continue. Notice that the axis of the applied moment must be along the x' axis, which is continually changing during the precession. Notice also that the sense of the moment is the same as that which would seem required to increase the angle θ. Thus we see that there really is no paradox at all; the externally applied moment $\sum$**M** about the x axis causes a change in angular momentum in exactly the same direction as the moment is applied.

23.4 FORCED PRECESSION

We have already observed that as speeds are increased in modern machinery, the engineer must be mindful of the increased importance of gyroscopic torques. Common mechanical equipment, which, in the past, were not thought of as exhibiting gyroscopic effects, do, in fact, often experience such torques, and these will become more significant as speeds increase. The purpose of this section is to show a few examples that, although they may not look like the standard gyroscope, do present gyroscopic torques that should be considered during the design of the equipment.

Much was presented in the earlier sections on the phenomena of precession. This type of motion is almost certainly accompanied by gyroscopic torques. Yet, lest we think that precession is only an unintended wobbling motion of a toy top, we will look at examples where this precession is recognized and even designed into the operation of some mechanical devices.

Let us first consider a vehicle such as a train, automobile, or racing car, moving at high speed on a straight road. The gyroscopic effect of the spinning wheels is to keep the vehicle moving straight ahead and to resist changes in direction. But when an external moment is applied which forces the wheel to change its direction, gyroscopic reaction forces immediately come into play.

To study this gyroscopic reaction in the case of vehicles, let us consider a pair of wheels connected by an axle, rounding a curve as shown in Fig. 23.4. This wheel ensemble may be considered a gyroscope. Such rounding of the curve is a *forced* precession of the wheel-axle assembly around a vertical axis through the center of curvature of the track.

EXAMPLE 23.1

A pair of wheels of radius r and combined mass moment of inertia I^s about their axis of rotation are connected by a straight axle, and this assembly is rounding a curve of constant radius R with the center of the axle travelling at a velocity $\mathbf{V}$ as shown in Fig. 23.4. For simplicity it is assumed that the roadbed is not banked. Find the gyroscopic torque exerted on the wheel-axle assembly.

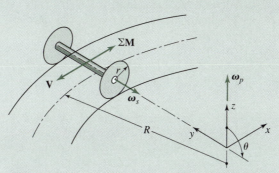

Figure 23.4 Example 23.1. Automobile axle and tires following a curved path.

SOLUTION

First, from the radius of the turn and the velocity, we can find the angular velocity of precession of the assembly. It is

$$\omega_p = V/R \qquad (1)$$

in the direction shown in the figure. Similarly, from the radius of the wheels and the velocity, we can find the average angular velocity of spin for the two wheels. It is

$$\omega_s = V/r \qquad (2)$$

and is also shown in the figure.

From the figure we see that the angle of the precession cone, the angle between $\boldsymbol{\omega}_p$ and $\boldsymbol{\omega}_s$, is $\theta = 90°$. Using this in Eq. (23.6), we find

$$\sum \mathbf{M} = I^s \boldsymbol{\omega}_p \times \boldsymbol{\omega}_s \tag{3}$$

and, using Eqs. (1) and (2), this becomes

$$\sum \mathbf{M} = \frac{I^s V^2}{Rr}\hat{\mathbf{i}} \qquad\qquad Ans. \tag{4}$$

This is the additional external moment applied on the wheel-axle assembly caused by the gyroscopic action of the wheels while rounding the turn (forced precession) over and above the static and other steady-state dynamic loads. We note that the direction of this additional moment is such as to increase the tendency of the vehicle to roll over during the turn. This moment is applied by the tires by increasing the upward force on the outside tire and decreasing the upward force on the inside tire.

In order to make a quantitative comparison of the gyroscopic and centrifugal moments on the vehicle, let us consider the case of a racing car with weight of $W = 1\,800$ lb including the driver, and with its center of mass at $h = 20$ in above the road; we estimate that its wheels have $r = 18$-in radius, $k = 15$-in radius of gyration, and $w = 100$-lb weight each. From these data we find the centrifugal moment $\mathbf{T}$ tending to cause roll-over of the vehicle. Taking moments about the contact point of the outer tire, we get

$$\mathbf{T} = \frac{hWV^2}{gR}\hat{\mathbf{i}} = \frac{(20\text{ in})(1\,800\text{ lb})V^2}{gR}\hat{\mathbf{i}} = 36\,000\frac{V^2}{gR}\hat{\mathbf{i}}\text{ in}\cdot\text{lb} \tag{5}$$

Because the vehicle has two axles and each has two wheels, we obtain

$$I^s = 2mk^2 = \frac{2(100\text{ lb})(15\text{ in})^2}{g} = \frac{45\,000\text{ in}\cdot\text{lb}}{g} \tag{6}$$

and, using Eq. (4) for two axles, we find

$$\sum \mathbf{M} = 2\frac{I^s V^2}{Rr}\hat{\mathbf{i}} = 2\frac{(45\,000\text{ in}\cdot\text{lb})V^2}{gR(18\text{ in})}\hat{\mathbf{i}} = (5\,000\text{ lb})\frac{V^2}{gR}\hat{\mathbf{i}} = 0.138\,9\mathbf{T} \tag{7}$$

Thus we see that the gyroscopic effects of the tires add almost 14 percent to the tendency of the vehicle to roll over in a turn, and we see that this is independent of both the radius of the turn and the speed of the vehicle.

We can note from this example that the problem itself looks nothing like a gyroscope; yet it has gyroscopic effects and these may be skipped in an oversimplified analysis. We notice also that the precession motion is not just an unexpected result of the motion characteristics of the system. It was knowingly caused by the driver who chose to turn the car; it was a forced precession.

Another gyroscopic effect in automobiles is that due to the flywheel. Because the rotation of the flywheel of a rear-wheel-drive vehicle is along the longitudinal axis, and because the flywheel rotates counterclockwise as viewed from the rear, the spin vector ω_s points toward the rear of the vehicle. When the vehicle makes a turn, the axis of the flywheel is forced to precess about a vertical axis as in the previous example. This forced precession brings into existence a gyroscopic applied moment about a horizontal axis through the center of the turn. The effect of this gyroscopic moment is to produce a bending moment on the driveshaft, tending to bend it in a vertical plane. The size of this gyroscopic moment causing bending in the driveshaft is usually of minor importance compared to the torsion loading because of the relatively small mass of the flywheel.

Lest we should think that gyroscopic forces are always a disadvantage with which we must cope, we shall now look at a problem in which the gyroscopic effect is put to good use.

EXAMPLE 23.2

The edge mill shown schematically in Fig. 23.5 is a gyroscopic grinder used for crushing ore, seeds, grain, and so on. It consists of a large steel pan in which one or more heavy conical rollers, called *mullers,* roll without slipping on the bottom of the pan and at the same time revolve about a vertical shaft passing through the central axis of the pan. The mullers rotate about either horizontal or inclined axles that are attached to the vertical shaft, which is rotated under power. The radius of the muller is $r = 18$ in and the length of its axle is $l = 30$ in; its weight is $W = 2\,200$ lb and its mass moment of inertia can be approximated by that for a cylindrical disk. Assuming that the vertical shaft is driven at a constant angular velocity of $\omega_p = 40$ rev/min and that the muller rolls without slipping, find the optimum inclination angle θ for the muller axle, which maximizes the crushing force between the muller and the pan.

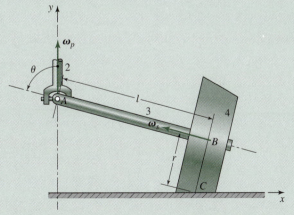

Figure 23.5 Example 23.2. Schematic diagram of an edge mill.

SOLUTION

First we must use the assumption of rolling without slipping to find the angular velocity of the muller. From the figure we see that the angular velocity of the muller axis, link 3, is

$$\boldsymbol{\omega}_p = \boldsymbol{\omega}_3 = 40\hat{\mathbf{j}} \text{ rev/min} = 4.19\hat{\mathbf{j}} \text{ rad/s} \tag{1}$$

while the angular velocity of spin of the muller about its axle is

$$\boldsymbol{\omega}_s = \boldsymbol{\omega}_{4/3} = \omega_s(-\sin\theta\hat{\mathbf{i}} + \cos\theta\hat{\mathbf{j}}) \tag{2}$$

But, because the muller undergoes the precession in addition to the spin about its axle, the absolute angular velocity of the muller is

$$\boldsymbol{\omega}_4 = \boldsymbol{\omega}_3 + \boldsymbol{\omega}_{4/3} = (-\omega_s \sin\theta)\hat{\mathbf{i}} + (4.19 + \omega_s \cos\theta)\hat{\mathbf{j}} \tag{3}$$

Because point A is fixed, we can find the velocity of point B of link 3 as follows:

$$\begin{aligned}
\mathbf{V}_B &= \boldsymbol{\omega}_3 \times \mathbf{R}_{BA} \\
&= (4.19\hat{\mathbf{j}} \text{ rad/s}) \times (30\cos\theta\hat{\mathbf{i}} - 30\sin\theta\hat{\mathbf{j}} \text{ in}) \\
&= -125.7\hat{\mathbf{k}} \text{ in/s}
\end{aligned} \tag{4}$$

Also, because there is assumed to be no slip at point C, we find the velocity of point B of link 4 to be

$$\begin{aligned}
\mathbf{V}_B &= \boldsymbol{\omega}_4 \times \mathbf{R}_{BC} \\
&= [(-\omega_s \sin\theta)\hat{\mathbf{i}} + (4.19 + \omega_s \cos\theta)\hat{\mathbf{j}}] \times (18\cos\theta\hat{\mathbf{i}} + 18\sin\theta\hat{\mathbf{j}} \text{ in}) \\
&= -18(\omega_s + 4.19\cos\theta)\hat{\mathbf{k}}
\end{aligned} \tag{5}$$

Equating Eqs. (4) and (5), we now find a solution for the angular velocity of spin of the muller about its axle:

$$\omega_s = 6.98\sin\theta - 4.19\cos\theta \text{ rad/s} \tag{6}$$

Hoping to use Eq. (23.6), we find the mass moments of inertia. Using the formula for a round disk from Appendix Table 5, we find

$$I^s = \frac{mr^2}{2} = \frac{(2\,200 \text{ lb})(18 \text{ in})^2}{(386 \text{ in/s}^2)2} = 923.3 \text{ in} \cdot \text{lb} \cdot \text{s}^2$$

$$I = \frac{mr^2}{4} + ml^2 = 5\,590 \text{ in} \cdot \text{lb} \cdot \text{s}^2$$

Next we find a value for the cross product $(\boldsymbol{\omega}_p \times \boldsymbol{\omega}_s)$. From Eqs. (1) and (2) this is

$$(\boldsymbol{\omega}_p \times \boldsymbol{\omega}_s) = (4.19\hat{\mathbf{j}} \text{ rad/s}) \times \omega_s(-\sin\theta\hat{\mathbf{i}} + \cos\theta\hat{\mathbf{j}})$$

$$= 4.19\omega_s \sin\theta\hat{\mathbf{k}}$$

and, using Eq. (6), it becomes

$$(\boldsymbol{\omega}_p \times \boldsymbol{\omega}_s) = (6.98\sin\theta - 4.19\cos\theta)4.19\sin\theta\hat{\mathbf{k}} \text{ rad/s}^2 \qquad (7)$$

Now, using Eq. (23.6), we can find the net externally applied moment on the muller. We get

$$\sum \mathbf{M} = \left[923.3 + \frac{-4\,667(4.19)\cos\theta}{6.98\sin\theta - 4.19\cos\theta} \right](6.98\sin\theta - 4.19\cos\theta)4.19\sin\theta\hat{\mathbf{k}}$$

and, after rearrangement, this becomes

$$\sum \mathbf{M} = (27\,001\sin^2\theta - 98\,082\sin\theta\cos\theta)\hat{\mathbf{k}} \text{ in·lb} \qquad (8)$$

This moment must be externally applied to the muller in order to sustain the forced precession. It must come from the net effect of the weight of the muller and the crushing force between the muller and the pan. Formulating these effects for the other side of the equation, we get

$$\sum \mathbf{M} = \mathbf{R}_{BA} \times \mathbf{W}_4 + \mathbf{R}_{CA} \times \mathbf{F}_c$$

$$= (30\sin\theta\hat{\mathbf{i}} - 30\cos\theta\hat{\mathbf{j}} \text{ in}) \times (-2\,200\hat{\mathbf{j}} \text{ lb})$$

$$+ [(30\sin\theta - 18\cos\theta)\hat{\mathbf{i}} + (18\sin\theta - 30\cos\theta)\hat{\mathbf{j}}] \times (F_c\hat{\mathbf{j}})$$

$$= (-66\,000\sin\theta + 30F_c\sin\theta - 18F_c\cos\theta)\hat{\mathbf{k}} \text{ in·lb} \qquad (9)$$

Now, setting Eqs. (8) and (9) equal to each other, we can solve for the crushing force F_c. This becomes

$$F_c = \frac{27\,001\sin^2\theta - 98\,082\sin\theta\cos\theta + 66\,000\sin\theta}{30\sin\theta - 18\cos\theta} \qquad (10)$$

We notice that it is a function of the angle θ of inclination of the muller axis; we now hope to choose this angle θ to maximize the crushing force. To do this we differentiate the expression for F_c and set it equal to zero. This gives

$$(54\,002\sin\theta\cos\theta - 98\,082\cos^2\theta + 98\,082\sin^2\theta + 66\,000\cos\theta)(30\sin\theta - 18\cos\theta)$$

$$- (30\cos\theta + 18\sin\theta)(27\,001\sin^2\theta - 98\,082\sin\theta\cos\theta + 66\,000\sin\theta) = 0$$

which reduces to

$$2\,456\,442\sin^3\theta + 810\,030\sin^2\theta\cos\theta$$

$$- 972\,036\sin\theta\cos^2\theta + 1\,765\,476\cos^3\theta - 1\,188\,000 = 0$$

This equation may be solved numerically and has multiple roots. When this is done, the root that maximizes F_c of Eq. (10) is found to be

$$\theta = 115.9° \qquad \qquad Ans.$$

This is the optimum angle of inclination of the muller axle. At this angle, Eq. (10) gives the crushing force of the mill as

$$F_c = 3437 \text{ lb} \qquad \qquad Ans.$$

We should note that this crushing force is more than one and one-half times the weight of the muller; the additional force is attributable to choosing the inclination angle θ for which both the gyroscopic and the centrifugal force effects contribute as much as possible to the crushing action of the mill. There are other designs for crushing machines in which the pan rotates under a muller that has a fixed axis. In such a design there is no gyroscopic action and the crushing action is due solely to the weight of the muller.

An airplane propeller spinning at high speed is another example of a mechanical system exhibiting gyroscopic torque effects. In the case of a single propeller airplane, a turn in compass heading, for example, is a forced precession of the propeller's spin axis about a vertical axis. If the propeller is rotating clockwise as viewed from the rear, this forced precession will induce a gyroscopic moment, causing the nose of the plane to move upward or downward as the heading is changed to the left or right, respectively. Turning the nose upward rather suddenly will cause the plane to turn to the right, while a sudden turn downward will cause a turn to the left.

Notice that it is not the propeller forces that cause this effect, but the spinning mass and its gyroscopic effect during a change in direction of the plane (forced precession). The very substantial spinning mass of the radial engine (see Section 1.8, Fig. 1.22b) for early aircraft markedly exaggerated this gyroscopic effect and was, at least in part, responsible for the disappearance of use of rotary engines on airplanes. Because the effect comes from the precession of the spinning mass and not the propeller, does the same danger not exist from the spinning mass of the rotor of a turbojet engine? When an airplane is equipped with two propellers rotating at equal speeds in opposite directions (or with counter-rotating turbines), however, the gyroscopic torques of the two can annul each other and leave no appreciable net effect on the plane as a whole.

NOTE

[1.] We can notice here that the true instantaneous axis of rotation of the rotor is not the spin axis, but accounts for the precession rotation also; this axis is shown in Fig. 23.2. It may be our tendency to picture the spin axis as the true axis of rotation of the rotor that leads us to the intuitive feeling that a gyroscope does not follow the laws of mechanics.

PROBLEMS

23.1 A pendulum mill is shown schematically in the figure. In such a mill, grinding is done by a conical muller that is free to spin about a pendulous axle that, in turn, is connected to a powered vertical shaft by a Hooke universal joint. The muller presses against the inner wall of a heavy steel pan, and it rolls around the inside of the pan without slipping. The weight of the muller is $W = 980$ lb; its principal mass moments of inertia are $I^s = 121$ in·lb·s^2 and $I = 88$ in·lb·s^2. The length of the muller axle is $l = R_{GA} = 40$ in and the radius of the muller at its center of mass is $R_{GB} = 10$ in. Assuming that the vertical shaft is to be inclined at $\theta = 30°$ and will be driven at a constant angular velocity of $\omega_p = 240$ rev/min, find the crushing force between the muller and the pan. Also determine the minimum angular velocity ω_p required to ensure contact between the muller and the pan.

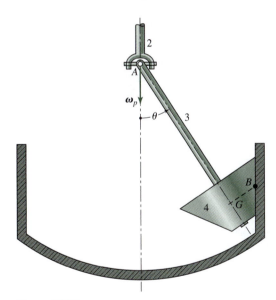

Figure P23.1

23.2 Use the gyroscopic formulae of this chapter to solve again the problem presented in Example 16.3 of Chapter 16.

23.3 The oscillating fan shown in Fig. P23.3 precesses sinusoidally according to the equation $\theta_p = \beta \sin 1.5t$, where $\beta = 30°$; the fan blade spins at $\omega_s = 1800\hat{\mathbf{i}}$ rev/min. The weight of the fan and motor armature is 5.25 lb, and other masses can be assumed negligible; gravity acts in the $-\hat{\mathbf{j}}$ direction. The principal mass moments of inertia are $I^s = 0.065$ in·lb·s^2 and $I = 0.025$ in·lb·s^2; the center of mass is located at $R_{GC} = 4$ in to the front of the precession axis. Determine the maximum moment M^z that must be accounted for in the clamped tilting pivot at C.

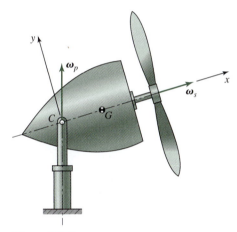

Figure P23.3

23.4 The propeller of an outboard motorboat is spinning at high speed and is caused to precess by steering to the right or left. Do the gyroscopic effects tend to raise or lower the rear of the boat? What is the effect and is it of noticeable size?

23.5 A large and very high-speed turbine is to operate at an angular velocity of $\omega = 18\,000$ rev/min and will have a rotor with a principal mass moment of inertia of $I^s = 225$ in·lb·s^2. It has been suggested that because this turbine will be installed at the north pole with its axis horizontal, perhaps the rotation of the earth will cause significant gyroscopic loads on it bearings. Estimate the size of these additional loads.

Appendix A: Tables

TABLE 1 Standard SI Prefixes[a,b]

Name	Symbol	Factor
exa	E	$1\,000\,000\,000\,000\,000\,000 = 10^{18}$
peta	P	$1\,000\,000\,000\,000\,000 = 10^{15}$
tera	T	$1\,000\,000\,000\,000 = 10^{12}$
giga	G	$1\,000\,000\,000 = 10^{9}$
mega	M	$1\,000\,000 = 10^{6}$
kilo	k	$1\,000 = 10^{3}$
hecto[c]	h	$100 = 10^{2}$
deka[c]	da	$10 = 10^{1}$
deci[c]	d	$0.1 = 10^{-1}$
centi[c]	c	$0.01 = 10^{-2}$
milli	m	$0.001 = 10^{-3}$
micro	μ	$0.000\,001 = 10^{-6}$
nano	n	$0.000\,000\,001 = 10^{-9}$
pico	p	$0.000\,000\,000\,001 = 10^{-12}$
femto	f	$0.000\,000\,000\,000\,001 = 10^{-15}$
atto	a	$0.000\,000\,000\,000\,000\,001 = 10^{-18}$

[a]If possible, use multiple and submultiple prefixes in steps of 1000. For example, specify lengths in millimeters, meters, or kilometers, say. In a combination unit, use prefixes only in the numerator; for example, use meganewton per square meter (MN/m^2), but not newton per square millimeter (N/mm^2).

[b]Spaces are used in SI instead of commas to group numbers to avoid confusion with the practice in some European countries to use commas for decimal points.

[c]Not recommended but sometimes encountered.

TABLE 2 Conversion from U.S. Customary Units to SI Units

| To Convert from | To | Multiply by | |
		Accurate[a]	Common
Foot (ft)	Meter (m)	0.304 800*	0.305
Horsepower (hp)	Watt (W)	745.699 9	746
Inch (in)	Meter (m)	0.025 400*	0.025 4
Mile, U.S. statute (mi)	Meter (m)	1 609.344*	1 610
Pound force (lb)	Newton (N)	4.448 222	4.45
Pound mass (lb)	Kilogram (kg)	0.453 592 4	0.454
Pound·foot (lb·ft)	Newton·meter (N·m)	1.355 818	1.36
	Joule (J)	1.355 818	1.36
Pound·foot/second (lb·ft/s)	Watt (W)	1.355 818	1.36
Pound·inch (lb·in)	Newton·meter (N·m)	0.112 818 2	0.113
	Joule (J)	0.112 818 2	0.113
Pound·inch/second (lb·in/s)	Watt (W)	0.112 818 2	0.113
Pound/ft^2 (lb/ft^2)	Pascal (Pa)	47.880 26	47.9
Pound/in^2 (lb/in^2), (psi)	Pascal (Pa)	6 894.757	6 890
Revolutions/min (rev/min)	Radian/second (rad/s)	0.104 719 8	0.105
Ton, short (2 000 lb)	Kilogram (kg)	907.184 7	907

[a]An asterisk indicates that the conversion is exact.

TABLE 3 Conversion from SI Units to U.S. Customary Units

| To Convert from | To | Multiply by | |
		Accurate	Common
Joule (J)	Pound·foot (lb·ft)	0.737 562 0	0.738
	Pound·inch (lb·in)	8.850 744	8.85
Kilogram (kg)	Pound mass (lb)	2.204 622	2.20
	Ton, short (2000 lb)	0.001 102 311	0.001 10
Meter (m)	Foot (ft)	3.280 840	3.28
	Inch (in)	39.370 08	39.4
	Mile (mi)	621.371 2	621
Newton (N)	Pound (lb)	0.224 808 9	0.225
Newton·meter (N·m)	Pound·foot (lb·ft)	0.737 562 0	0.738
	Pound·inch (lb·in)	8.850 744	8.85
Newton·meter/second (N·m/s)	Horsepower (hp)	0.001 341 022	0.001 34
Pascal (Pa)	Pound/foot^2 (lb/ft^2)	0.020 885 43	0.020 9
	Pound/inch^2 (lb/in^2, psi)	0.000 145 0370	0.000 145
Radian/second (rad/s)	Revolutions/minute (rev/min)	9.549 297	9.55
Watt (W)	Horsepower (hp)	0.001 341 022	0.001 34
	Pound·foot/second (lb·ft/s)	0.737 562 0	0.738
	Pound·inch/second (lb·in/s)	8.850 744	8.85

TABLE 4 Properties of Areas

A = area
I = area moment of inertia
J = polar area moment of inertia
k = radius of gyration
$\bar{y}$ = centroidal distance

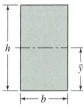

Rectangle

$A = bh$
$I = bh^3/12$
$k = 0.289h$
$\bar{y} = h/2$

Triangle

$A = bh/2$
$I = bh^3/36$
$k = 0.236h$
$\bar{y} = h/3$

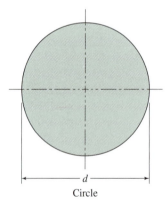

Circle

$A = \pi d^2/4$
$I = \pi d^4/64$
$J = \pi d^4/32$
$k = d/4$
$\bar{y} = d/2$

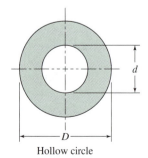

Hollow circle

$A = \pi(D^2 - d^2)/4$
$I = \pi(D^4 - d^4)/64$
$J = \pi(D^4 - d^4)/32$
$k = \sqrt{D^2 + d^2}/4$
$\bar{y} = D/2$

TABLE 5 Mass Moments of Inertia

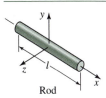
$$i_y = i_z = ml^2/12$$

Rod

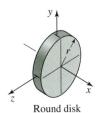
$$i_x = mr^2/2$$
$$i_y = i_z = mr^2/4$$

Round disk

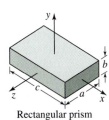
$$i_x = m(a^2 + b^2)/12$$
$$i_y = m(a^2 + c^2)/12$$
$$i_z = m(b^2 + c^2)/12$$

Rectangular prism

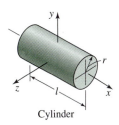
$$i_x = mr^2/2$$
$$i_y = i_z = m(3r^2 + l^2)/12$$

Cylinder

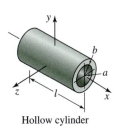
$$i_x = m(a^2 + b^2)/2$$
$$i_y = i_z = m(3a^2 + 3b^2 + l^2)/12$$

Hollow cylinder

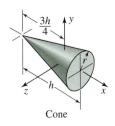
$$i_x = 3mr^2/10$$
$$i_y = i_z = m(12r^2 + 3h^2)/80$$

Cone

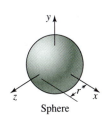
$$i_x = i_y = i_z = 2mr^2/5$$

Sphere

TABLE 6 Involute Function

ϕ (deg)	Inv (ϕ)	Inv ($\phi + 0.1°$)	Inv ($\phi + 0.2°$)	Inv ($\phi + 0.3°$)	Inv ($\phi + 0.4°$)
0.0	0.000000	0.000000	0.000000	0.000000	0.000000
0.5	0.000000	0.000000	0.000000	0.000000	0.000001
1.0	0.000002	0.000002	0.000003	0.000004	0.000005
1.5	0.000006	0.000007	0.000009	0.000010	0.000012
2.0	0.000014	0.000016	0.000019	0.000022	0.000025
2.5	0.000028	0.000031	0.000035	0.000039	0.000043
3.0	0.000048	0.000053	0.000058	0.000064	0.000070
3.5	0.000076	0.000083	0.000090	0.000097	0.000105
4.0	0.000114	0.000122	0.000132	0.000141	0.000151
4.5	0.000162	0.000173	0.000184	0.000197	0.000209
5.0	0.000222	0.000236	0.000250	0.000265	0.000280
5.5	0.000296	0.000312	0.000329	0.000347	0.000366
6.0	0.000384	0.000404	0.000424	0.000445	0.000467
6.5	0.000489	0.000512	0.000536	0.000560	0.000586
7.0	0.000612	0.000638	0.000666	0.000694	0.000723
7.5	0.000753	0.000783	0.000815	0.000847	0.000880
8.0	0.000914	0.000949	0.000985	0.001022	0.001059
8.5	0.001098	0.001137	0.001178	0.001219	0.001283
9.0	0.001305	0.001349	0.001394	0.001440	0.001488
9.5	0.001536	0.001586	0.001636	0.001688	0.001740
10.0	0.001794	0.001849	0.001905	0.001962	0.002020
10.5	0.002079	0.002140	0.002202	0.002265	0.002329
11.0	0.002394	0.002461	0.002528	0.002598	0.002668
11.5	0.002739	0.002812	0.002894	0.002962	0.003039
12.0	0.003117	0.003197	0.003277	0.003360	0.003443
12.5	0.003529	0.003615	0.003712	0.003792	0.003882
13.0	0.003975	0.004069	0.004164	0.004261	0.004359
13.5	0.004459	0.004561	0.004664	0.004768	0.004874
14.0	0.004982	0.005091	0.005202	0.005315	0.005429
14.5	0.005545	0.005662	0.005782	0.005903	0.006025
15.0	0.006150	0.006276	0.006404	0.006534	0.006665
15.5	0.006799	0.006934	0.007071	0.007209	0.007350
16.0	0.007493	0.007637	0.007784	0.007932	0.008082
16.5	0.008234	0.008388	0.008544	0.008702	0.008863
17.0	0.009025	0.009189	0.009355	0.009523	0.009694
17.5	0.009866	0.010041	0.010217	0.010396	0.010577
18.0	0.010760	0.010946	0.011133	0.011323	0.011515
18.5	0.011709	0.011906	0.012105	0.012306	0.012509
19.0	0.012715	0.012923	0.013134	0.013346	0.013562
19.5	0.013779	0.013999	0.014222	0.014447	0.014674
20.0	0.014904	0.015137	0.015372	0.015609	0.015850
20.5	0.016092	0.016337	0.016585	0.016836	0.017089
21.0	0.017345	0.017603	0.017865	0.018129	0.018395
21.5	0.018665	0.018937	0.019212	0.019490	0.019770
22.0	0.020054	0.020340	0.020630	0.020921	0.021216
22.5	0.021514	0.021815	0.022119	0.022426	0.022736

TABLE 6 (*continued*)

φ (deg)	Inv (φ)	Inv (φ + 0.1°)	Inv (φ + 0.2°)	Inv (φ + 0.3°)	Inv (φ + 0.4°)
23.0	0.023049	0.023365	0.023684	0.024006	0.024332
23.5	0.024660	0.024992	0.025326	0.025664	0.026005
24.0	0.026350	0.026697	0.027048	0.027402	0.027760
24.5	0.028121	0.028485	0.028852	0.029223	0.029598
25.0	0.029975	0.030357	0.030741	0.031130	0.031521
25.5	0.031917	0.032315	0.032718	0.033124	0.033534
26.0	0.033947	0.034364	0.034785	0.035209	0.035637
26.5	0.036069	0.036505	0.036945	0.037388	0.037835
27.0	0.038287	0.038696	0.039201	0.039664	0.040131
27.5	0.040602	0.041076	0.041556	0.042039	0.042526
28.0	0.043017	0.043513	0.044012	0.044516	0.045024
28.5	0.045537	0.046054	0.046575	0.047100	0.047630
29.0	0.048164	0.048702	0.049245	0.049792	0.050344
29.5	0.050901	0.051462	0.052027	0.052597	0.053172
30.0	0.053751	0.054336	0.054924	0.055519	0.056116
30.5	0.056720	0.057267	0.057940	0.058558	0.059181
31.0	0.059809	0.060441	0.061779	0.061721	0.062369
31.5	0.063022	0.063680	0.064343	0.065012	0.065685
32.0	0.066364	0.067048	0.067738	0.068432	0.069133
32.5	0.069838	0.070549	0.071266	0.071988	0.072716
33.0	0.073449	0.074188	0.074932	0.075683	0.076439
33.5	0.077200	0.077968	0.078741	0.079520	0.080305
34.0	0.081097	0.081974	0.082697	0.083506	0.084321
34.5	0.085142	0.085970	0.086804	0.087644	0.088490
35.0	0.089342	0.090201	0.091066	0.091938	0.092816
35.5	0.093701	0.094592	0.095490	0.096395	0.097306
36.0	0.098224	0.099149	0.100080	0.101019	0.101964
36.5	0.102916	0.103875	0.104841	0.105814	0.106795
37.0	0.107782	0.108777	0.109779	0.110788	0.111805
37.5	0.112828	0.113860	0.114899	0.115945	0.116999
38.0	0.118060	0.119130	0.120207	0.121291	0.122384
38.5	0.123484	0.124592	0.125709	0.126833	0.127965
39.0	0.129106	0.130254	0.131411	0.132576	0.133749
39.5	0.134931	0.136122	0.137320	0.138528	0.139743
40.0	0.140968	0.142201	0.143443	0.144694	0.145954
40.5	0.147222	0.148500	0.149787	0.151082	0.152387
41.0	0.153702	0.155025	0.156358	0.157700	0.159052
41.5	0.160414	0.161785	0.163165	0.164556	0.165956
42.0	0.167366	0.168786	0.170216	0.171656	0.173106
42.5	0.174566	0.176037	0.177518	0.179009	0.180511
43.0	0.182023	0.183546	0.185080	0.186625	0.188180
43.5	0.189746	0.191324	0.192912	0.194511	0.196122
44.0	0.197744	0.199377	0.201022	0.202678	0.204346
44.5	0.206026	0.207717	0.209420	0.211135	0.212863
45.0	0.214602				

Appendix B

Answers to Selected Problems

1.3 $\gamma_{max} = 53°$; $\gamma_{min} = 98°$; at $\theta = 40°$, $\gamma = 59°$; at $\theta = 229°$, $\gamma = 91°$

1.5 (a) $m = 1$; (b) $m = 1$; (c) $m = 0$; (d) $m = 1$

1.7 $Q = 1.10$

2.1 Spiral

2.3 $\mathbf{R}_{QP} = -7\hat{\mathbf{i}} - 14\hat{\mathbf{j}}$

2.5 $\mathbf{R}_A = -4.5a\hat{\mathbf{i}}$

2.7 Clockwise; $\mathbf{R} = 4\angle 0°$; $t = 20$; $\mathbf{R} = 404\angle 0°$; $\Delta\mathbf{R} = 400\angle 0°$

2.9 $\mathbf{R}_{P_3} = -2.12\hat{\mathbf{i}}_1 + 3.88\hat{\mathbf{j}}_1$; $\mathbf{R}_{P_3/2} = 3\hat{\mathbf{i}}_2$

2.11 $\Delta\mathbf{R}_Q = 1.90\hat{\mathbf{i}} + 1.10\hat{\mathbf{j}} = 2.20\angle 30°$

2.13 $R_C^x = 2.50\cos\theta_2 + \sqrt{(41.75 - 5\sin\theta_2 + 6.25\cos^2\theta_2)}$

2.17 $\theta_2 = 90°$; $270°$

2.19 (a) First, $\theta_2 = 29°$, $\theta_4 = 58°$; second, $\theta_2 = 248°$, $\theta_4 = 136°$; (b) $78°$; (c) $29°$, $68°$

3.1 $\dot{\mathbf{R}} = 314\angle 162°$ in/s

3.3 $\mathbf{V}_{BA} = \mathbf{V}_{B_3/2} = 82.6$ mi/h N 25° E

3.5 (a) d = 1 400 mm; (b) $\mathbf{V}_{AB} = 60\hat{\mathbf{j}}$ m/s; $\omega_2 = 200$ rad/s cw

3.7 (a) Straight line at N 48° E; (b) no change.

3.9 $\omega_3 = 1.43$ rad/s ccw; $\omega_4 = 15.4$ rad/s ccw

3.11 $\mathbf{V}_C = 23.7\angle 284°$ ft/s; $\omega_3 = 0.33$ rad/s ccw

3.13 $\mathbf{V}_C = 0.402\angle 151°$ m/s; $\mathbf{V}_D = 0.290\angle 249°$ m/s

3.15 $\mathbf{V}_B = 4.79\angle 96°$ m/s; $\omega_3 = \omega_4 = 22$ rad/s ccw

3.17 $\omega_6 = 4$ rad/s ccw; $\mathbf{V}_B = 0.963\angle 180°$ ft/s; $\mathbf{V}_C = 2.02\angle 208°$ ft/s; $\mathbf{V}_D = 2.01\angle 206°$ ft/s

3.19 $\omega_3 = 3.23$ rad/s ccw; $\mathbf{V}_B = 16.9\angle -56°$ ft/s

3.21 $\mathbf{V}_C = 9.03\angle 138°$ m/s

3.23 $\mathbf{V}_B = 35.5\angle 240°$ ft/s; $\mathbf{V}_C = 40.6\angle 267°$ ft/s; $\mathbf{V}_D = 31.6\angle -60°$ ft/s

3.25 $\mathbf{V}_B = 1.04\angle -23°$ ft/s

3.27 $\omega_3 = \omega_4 = 14.4$ rad/s ccw; $\omega_5 = \omega_6 = 9.67$ rad/s cw; $\mathbf{V}_E = 77.4\angle -100°$ in/s

3.29 $\omega_3 = 1.57$ rad/s cw

3.31 $\mathbf{V}_E = 10.0\angle 221°$ in/s; $\mathbf{V}_G = 11.9\angle -57°$ in/s; $\omega_3 = \omega_4 = 3.33$ rad/s ccw; $\omega_5 = 25.6$ rad/s cw; $\omega_6 = 3.77$ rad/s cw

3.33 $\omega_3 = 30.0$ rad/s cw

4.1 $\ddot{\mathbf{R}} = -4\hat{\mathbf{i}}$ in/s^2

4.3 $\hat{\tau} = 0.300\hat{\mathbf{i}} - 0.954\hat{\mathbf{j}}$; $A^n = 43.7$ mm/s^2; $A^t = 12.6$ mm/s^2; $\rho = -405$ mm

4.5 $\mathbf{A}_A = -7\,200\hat{\mathbf{i}} + 2\,400\hat{\mathbf{j}}$ m/s^2

4.7 $V_B = 12.0\angle 270°$ ft/s; $V_C = 8.37\angle 12°$ ft/s; $A_B = 395\angle 165°$ ft/s²; $A_C = 210\angle 240°$ ft/s²

4.9 $\omega_2 = 38.6$ rad/s cw; $\alpha_2 = 5571$ rad/s² cw

4.11 $\alpha_3 = 563$ rad/s² ccw; $\alpha_4 = 124$ rad/s² ccw

4.13 $A_C = 3\,104\angle 113°$ ft/s²; $\alpha_3 = 1\,740$ rad/s² ccw; $\alpha_4 = 3\,050$ rad/s² ccw

4.15 $A_C = 2\,610\angle -69°$ ft/s²; $\alpha_4 = 1\,490$ rad/s² ccw

4.17 $A_B = 16.7\angle 0°$ ft/s²; $\alpha_3 = 17.5$ rad/s² ccw; $\alpha_6 = 10.8$ rad/s² cw

4.19 $\alpha_2 = 4\,180$ rad/s² ccw

4.21 $A_C = 450\angle 256°$ m/s²; $\alpha_3 = 74.1$ rad/s² cw

4.23 $A_B = 2\,440\angle 240°$ ft/s²; $A_D = 4\,030\angle 120°$ ft/s²

4.25 $\theta_3 = 171°$; $\theta_4 = 195°$; $\omega_3 = 70.5$ rad/s ccw; $\omega_4 = 47.6$ rad/s ccw; $\alpha_3 = 3\,200$ rad/s² ccw; $\alpha_4 = 3\,330$ rad/s² ccw

4.27 $\theta_3 = 28.3°$; $\theta_4 = 55.9°$; $\omega_3 = 0.633$ rad/s cw; $\omega_4 = 2.16$ rad/s cw; $\alpha_3 = 7.82$ rad/s² ccw; $\alpha_4 = 6.70$ rad/s² ccw

4.29 $\theta_3 = 38.4°$; $\theta_4 = 156°$; $\omega_3 = 6.85$ rad/s cw; $\omega_4 = 1.24$ rad/s cw; $\alpha_3 = 62.5$ rad/s² ccw; $\alpha_4 = 96.5$ rad/s² cw

4.31 $V_B = 184\angle -19°$ m/s; $A_B = 2\,700\angle -172°$ m/s²; $\omega_4 = 6.57$ rad/s cw; $\alpha_4 = 86.4$ rad/s² ccw

4.33 $A_{A_4} = 290\angle 180°$ ft/s²

4.35 $A_B = 412\angle -26°$ m/s²; $\alpha_4 = 45.8$ rad/s² ccw

4.37 $A_{P_4} = 125\angle -67°$ m/s²

4.39 $A_{C_4} = 600\angle 91°$ in/s²; $\alpha_3 = 6.00$ rad/s² ccw

4.41 $A_G = 351\angle -64°$ in/s²; $\alpha_5 = 0$ rad/s² ccw; $\alpha_6 = 110$ rad/s² cw

5.3 Face = 195 mm from pivot

5.5 $y'(\beta/2) = \pi L/2\beta$; $y'''(\beta/2) = -\pi^3 L/2\beta^3$; $y''(0) = \pi^2 L/2\beta^2$; $y''(\beta) = -\pi^2 L/2\beta^2$

5.7 AB: dwell, $L_1 = 0$, $\beta_1 = 60°$; BC: full-rise eighth-order polynomial motion, Eq. (5.20), $L_2 = 2.5$ in, $\beta_2 = 62.10°$; CD: half-harmonic return motion, Eq. (5.26), $L_3 = 0.081$ in, $\beta_3 = 7.66°$; DE: uniform motion, $L_4 = 1.0$ in, $\beta_4 = 60°$; EA: half-cycloidal return motion, Eq. (5.31), $L_5 = 1.419$ in, $\beta_5 = 170.25°$

5.9 $t_{AB} = 0.025$ s; $\dot{y}_{max} = 171$ in/s; $\dot{y}_{min} = -40$ in/s; $\ddot{y}_{max} = 19\,700$ in/s²; $\ddot{y}_{min} = -19\,700$ in/s²

5.11 $\dot{y}_{max} = 41.9$ rad/s; $\ddot{y}_{max} = 7\,890$ rad/s²

5.13 Face width = 2.200 in; $\rho_{min} = 3.000$ in

5.15 $R_0 > 9.21$ in; face width > 5.35 in

5.17 $\phi_{max} = 12°$; $R_r < 14.0$ in

5.19 $R_0 > 56$ mm; $\ddot{y}_{max} = 37$ m/s²

5.21 $R_0 > 71$ mm; $\ddot{y}_{max} = 39.5$ m/s²

5.23 $u = (R_0 + R_C + y)\sin\theta + y'\cos\theta$; $v = (R_0 + R_C + y)\cos\theta - y'\sin\theta$;
$$R = \sqrt{(R_0 + R_C + y)^2 + (y')^2};\ \psi = \frac{\pi}{2} - \theta - \tan^{-1}\left(\frac{y'}{R_0 + R_C + y}\right)$$

6.1 $P = 16$ teeth/in

6.3 $m = 2$ mm/tooth

6.5 $P = 0.8976$ teeth/in; $D = 44.563$ in

6.7 $m = 12.732$ mm/tooth; $D = 458.4$ mm

6.9 $D = 9.191$ in

6.11 $N_2 = 17$ teeth; $N_3 = 51$ teeth

6.13 $a = 0.250$ in; $d = 0.3125$ in; $c = 0.0625$ in; $p_c = 0.785$ in/tooth;
$p_b = 0.738$ in/tooth; $t = 0.393$ in; $r_2 = 2.819$ in; $r_3 = 4.229$ in;
$CP = 0.625$ in; $PD = 0.591$ in; $m_c = 1.647$ teeth

6.15 $\alpha_2 = 18.58°$; $\alpha_3 = 6.32°$; $\beta_2 = 16.04°$; $\beta_3 = 5.45°$; $m_c = 1.635$ teeth

6.17 $CP = 18.930$ mm; $PD = 16.243$ mm; $m_c = 1.544$ teeth

6.19   (a) $CP = 1.462$ in; $PD = 1.196$ in; $m_c = 1.80$ teeth
   (b) $CP = 1.096$ in; $PD = 1.196$ in; $m_c = 1.55$ teeth; no change in pressure
      angle

6.25 $t_b = 17.142$ mm; $t_a = 6.737$ mm; $\varphi_a = 32.78°$

6.27 $t_b = 1.146$ in

6.29 $t_b = 0.1594$ in; $\varphi_a = 35.33°$; $t_a = 0.0415$ in

6.31   (a) 0.1682 in; (b) 9.8270 in

6.33 $m_c = 1.664$

6.35 $m_c = 1.771$

6.37 $\phi = 26.24°$

6.39 $a_3 = 1.344$ in

7.1 $p_t = 0.524$ in/tooth, $p_n = 0.370$ in/tooth, $P_n = 8.49$ teeth/in, $D_2 = 2.5$ in,
$D_3 = 4.0$ in, $N_{e2} = 42.43$ teeth, $N_{e3} = 67.88$ teeth

7.3 $\psi_b = 33.34°$, $m_x = 1.672$ teeth

7.5 $N_3 = 72$ teeth, $D_2 = 4.0$ in, $D_3 = 18.0$ in, $P_n = 4.345$ teeth/in,
$p_n = 0.723$ in/tooth, $F = 3.700$ in

7.7 $m_n = 1.79$ teeth, $m_{total} = 2.87$ teeth

7.9 $\psi_3 = 15°$ RH, $N_3 = 39$ teeth, $D_2 = 1.932$ in, $D_3 = 3.365$ in

8.1 $\gamma_2 = 18.43°$, $\gamma_3 = 71.57°$

8.3 $\gamma_2 = 27.00°$, $\gamma_3 = 93.00°$

8.5 $D_2 = 2.125$ in, $D_3 = 3.500$ in, $\gamma_2 = 34.83°$, $\gamma_3 = 70.17°$, $a_2 = 0.1681$ in,
$a_3 = 0.0819$ in, $d_2 = 0.1054$ in, $d_3 = 0.1916$ in, $F = 0.563$ in, $N_{e2} = 20.71$,
$N_{e3} = 82.54$

9.1 $R_2 = 0.556$ in, $R_3 = 1.194$ in

9.3 $l = 3.75$ in, $\lambda = \psi = 34.298°$, $R_3 = 7.958$ in

10.1 $\omega_8 = 68.18$ rev/min cw. $\theta'_{8/2} = +5/88$

10.3 $\omega_9 = 11.84$ rev/min cw

10.5   One solution: $N_3 = 30$ T, $N_4 = 25$ T, $N_5 = 30$ T, $N_6 = 20$ T, $N_7 = 25$ T,
$N_8 = 35$ T, $N_{10} = 35$ T

10.7 $\omega_7 = 222.1$ rev/min ccw

10.9 $\omega_2 = 644.8$ rev/min cw

10.11 $\omega_3 = (-5/22)\omega_2$ or opposite in direction; replace gears 4 and 5 with a single
gear

10.13   (a) $N_5 = 84$ T, $R_A = 156$ mm; (b) $\omega_A = 8.08$ rev/min ccw

10.15 (*a*) $\omega_R = 651.26$ rev/min, $\omega_L = 693.28$ rev/min; (*b*) $\omega_A = 672.27$ rev/min

10.17 $\omega_A = -(9/34)\omega_2$; Lévai type-F

10.19 59.1%

11.1 For six points, 0.170, 1.464, 3.706, 6.294, 8.536, and 9 830.

11.3 Typical solution: $r_2 = 7.4$ in, $r_3 = 20.9$ in, $e = 8$ in

11.5 Typical solution: $r_1 = 7.63$ ft, $r_2 = 3.22$ ft, $r_3 = 8.48$ ft

11.7 Typical solution: O_2 at $x = -1\,790$ mm, $y = 320$ mm, $r_2 = 360$ mm, $r_3 = 1\,990$ mm

11.9 $r_1 = 12$ in, $r_2 = 9$ in, $r_3 = 6$ in, $r_4 = 9$ in; seat locks in open or toggle position, which is a 3–4–5 triangle

11.13 and 11.23 $r_2/r_1 = -3.352$, $r_3/r_1 = 0.845$, $r_4/r_1 = 3.485$

11.15 and 11.25 $r_2/r_1 = -2.660$, $r_3/r_1 = 7.430$, $r_4/r_1 = 8.685$

11.17 and 11.27 $r_2/r_1 = -0.385$, $r_3/r_1 = 1.030$, $r_4/r_1 = 0.384$

11.19 and 11.29 $r_2/r_1 = 2.523$, $r_3/r_1 = 3.329$, $r_4/r_1 = -0.556$

11.21 and 11.31 $r_2/r_1 = -1.606$, $r_3/r_1 = 0.925$, $r_4/r_1 = 1.107$

12.1 $m = 2$ including one idle freedom. Path of B is the intersection of cylinder of radius of BA about y axis and sphere of radius BO_3 about O_3.

12.3 $\omega_2 = -2.58\hat{\mathbf{j}}$ rad/s; $\omega_3 = 1.16\hat{\mathbf{i}} - 0.09\hat{\mathbf{j}} + 0.64\hat{\mathbf{k}}$ rad/s;
$\mathbf{V}_B = -0.096\hat{\mathbf{i}} - 0.050\hat{\mathbf{j}} + 0.168\hat{\mathbf{k}}$ m/s

12.5 and 12.7 $\omega_3 = \omega_4 = -25.7\hat{\mathbf{i}}$ rad/s; $\mathbf{V}_A = 180\hat{\mathbf{j}}$ in/s; $\mathbf{V}_B = -231\hat{\mathbf{k}}$ in/s;
$\alpha_3 = 1\,543\hat{\mathbf{j}}$ rad/s^2; $\alpha_4 = -343\hat{\mathbf{i}}$ rad/s^2; $\mathbf{A}_A = 10\,800\hat{\mathbf{i}}$ in/s^2;
$\mathbf{A}_B = -5\,950\hat{\mathbf{j}} - 3\,087\hat{\mathbf{k}}$ in/s^2.

12.9 $\Delta\theta_4 = 48°$; $Q = 1.0$

12.11 and 12.13 $\omega_3 = 4.08\hat{\mathbf{i}} - 7.07\hat{\mathbf{j}} + 3.34\hat{\mathbf{k}}$ rad/s; $\omega_4 = 10.80\hat{\mathbf{i}}$ rad/s;
$\mathbf{V}_A = -1.15\hat{\mathbf{i}} - 0.65\hat{\mathbf{j}}$ m/s; $\mathbf{V}_B = 1.21\hat{\mathbf{j}} + 2.56\hat{\mathbf{k}}$ m/s;
$\alpha_3 = 273\hat{\mathbf{i}} + 115\hat{\mathbf{j}} - 148\hat{\mathbf{k}}$ rad/s^2; $\alpha_4 = 130\hat{\mathbf{i}}$ rad/s^2;
$\mathbf{A}_A = 23.3\hat{\mathbf{i}} - 41.4\hat{\mathbf{j}}$ m/s^2; $\mathbf{A}_B = -13.04\hat{\mathbf{j}} + 43.91\hat{\mathbf{k}}$ m/s^2

12.15 $\omega_2 = 20.8\hat{\mathbf{i}} - 12.0\hat{\mathbf{j}}$ rad/s; $\omega_3 = 2.31\hat{\mathbf{i}} + 6.66\hat{\mathbf{j}} - 3.23\hat{\mathbf{k}}$ rad/s;
$\mathbf{V}_A = 12.0\hat{\mathbf{i}} + 20.8\hat{\mathbf{j}} + 41.6\hat{\mathbf{k}}$ in/s; $\mathbf{V}_B = 13.8\hat{\mathbf{i}}$ in/s

12.17 and 12.19 $\omega_2 = 12.0\hat{\mathbf{i}} - 20.8\hat{\mathbf{j}}$ rad/s; $\omega_3 = 1.33\hat{\mathbf{i}} + 6.91\hat{\mathbf{j}} - 1.82\hat{\mathbf{k}}$ rad/s;
$\mathbf{V}_A = 20.8\hat{\mathbf{i}} + 12.0\hat{\mathbf{j}} + 41.6\hat{\mathbf{k}}$ in/s; $\mathbf{V}_B = 26.1\hat{\mathbf{i}}$ in/s.

12.21 and 12.23 $\mathbf{V}_B = 112$ in/s; $\omega_3 = -4.19\hat{\mathbf{i}} + 19.4\hat{\mathbf{j}} - 5.47\hat{\mathbf{k}}$ rad/s;
$\omega_4 = 19.4\hat{\mathbf{j}}$ rad/s.

13.1 $T_{15} = \begin{bmatrix} 0.866 & -0.500 & 0 & 17.32 \\ -0.500 & -0.866 & 0 & 0 \\ 0 & 0 & 1 & 8.00 \\ 0 & 0 & 0 & 1 \end{bmatrix}$; $\mathbf{R} = 17.32\hat{\mathbf{i}}_1 + 6.50\hat{\mathbf{k}}_1$ in

13.3 $T_{15} = \begin{bmatrix} 0 & -1 & 0 & 180 \\ 0 & 0 & -1 & -100 \\ 1 & 0 & 0 & 450 \\ 0 & 0 & 0 & 1 \end{bmatrix}$; $\mathbf{R} = 180\hat{\mathbf{i}}_1 - 145\hat{\mathbf{j}}_1 + 450\hat{\mathbf{k}}_1$ mm

13.5 $\mathbf{V} = -1.750\hat{\mathbf{i}}_1 + 0.433\hat{\mathbf{j}}_1$ in/s; $\mathbf{A} = -0.541\hat{\mathbf{i}}_1 - 0.088\hat{\mathbf{j}}_1$ in/s^2

13.7 $\phi_2 = \cos^{-1}(0.020t^2 + 0.050t - 0.875)$ with $\phi_2 \leq 0$

$$\phi_1 = \tan^{-1}\frac{3(1 + \cos\phi_2) - 4\sin\phi_2}{4(1 + \cos\phi_2) + 3\sin\phi_2}; \phi_3 = 8 \text{ in}; \phi_4 = \phi_1 + \phi_2 - 36.87°$$

13.9 $\tau_1 = 25$ in·lb; $\tau_2 = 100\sin\phi_1 - 50t\cos\phi_1 + 25$ in·lb; $\tau_3 = -5$ lb;
$\tau_4 = 100\sin\phi_1 - 100\sin(\phi_1 + \phi_2)$ in·lb

14.3 $P = 1\,460$ N.
14.5 $\mathbf{M}_{12} = 90.9\hat{\mathbf{k}}$ in·lb
14.7 $\mathbf{M}_{12} = 377\hat{\mathbf{k}}$ in·lb
14.9 $\mathbf{M}_{12} = -276\hat{\mathbf{k}}$ in·lb
14.11 $\mathbf{M}_{12} = -761\hat{\mathbf{k}}$ in·lb, $\mathbf{F}_{12} = \mathbf{F}_{23} = 228\angle56.6°$ lb, $\mathbf{F}_{14} = 318\angle-61.7°$ lb, $\mathbf{F}_{34} = 190\angle88.4°$ lb.
14.13 $\mathbf{F}_{12} = \mathbf{F}_{23} = 306\angle230.4°$ kN; $\mathbf{F}_{14} = \mathbf{F}_{43} = 387\angle59.7°$ kN; $\mathbf{M}_{12} = -110\hat{\mathbf{k}}$ kN·m
14.15 $\mathbf{M}_{12} = 420\hat{\mathbf{k}}$ in·lb
14.17 (a) $\mathbf{F}_{13} = 2\,520\angle0°$ lb; (b) $\mathbf{F}_{13} = 1\,049\angle225°$ lb; (c) $\mathbf{F}_{13} = 2\,250\angle135°$ lb
14.19 $\mathbf{F}_C = 216\angle189°$ lb; $\mathbf{F}_D = 350\angle163.4°$ lb
14-21 $\mathbf{F}_C = 118\hat{\mathbf{i}} + 140\hat{\mathbf{j}} - 251\hat{\mathbf{k}}$ lb; $\mathbf{F}_D = -71\hat{\mathbf{i}} - 155\hat{\mathbf{k}}$ lb
14.23 $\mathbf{F}_E = 163\hat{\mathbf{i}} - 192\hat{\mathbf{j}} + 355\hat{\mathbf{k}}$ lb; $\mathbf{F}_F = 110\hat{\mathbf{j}} + 145\hat{\mathbf{k}}$ lb
14.25 $\mathbf{M}_{12} = -61.5\hat{\mathbf{k}}$ N·m
14.29 $\mathbf{F}_F = 378\hat{\mathbf{i}} + 1\,106\hat{\mathbf{j}}$ lb; $\mathbf{F}_R = 1\,239\hat{\mathbf{j}}$ lb; $\mathbf{F}_T = -378\hat{\mathbf{i}} + 655\hat{\mathbf{j}}$ lb; $\mu = 0.342$
14.31 $\mathbf{M}_{12} = 0.170\hat{\mathbf{k}}$ N·m
14.33 $T = F_B d(R/X)^2$; $X \geq R\sqrt{1 + 1/\mu^2}$

15.1 $I_O = 0.030\,9$ in·lb·s^2
15.3 $\mathbf{M}_{12} = -190\hat{\mathbf{k}}$ in·lb
15.5 $\mathbf{F}_{23} = \mathbf{F}_{12} = 1\,535\angle188°$ lb, $\mathbf{F}_{14} = 300\angle-90°$ lb, $\mathbf{F}_{34} = 755\angle156.6°$ lb, $\mathbf{M}_{12} = 2.870\hat{\mathbf{k}}$ in·lb
15.7 $\mathbf{F}_{23} = \mathbf{F}_{12} = 9.98\angle-20°$ kN, $\mathbf{F}_{14} = 11.7\angle205°$ kN, $\mathbf{F}_{34} = 11.0\angle14.8°$ kN, $\mathbf{M}_{12} = -2.950\hat{\mathbf{k}}$ N·m
15.9 $\mathbf{F}_{23} = \mathbf{F}_{12} = 2.59\angle230°$ kN, $\mathbf{F}_{14} = 6.98\angle-84.9°$ kN, $\mathbf{F}_{34} = 4.37\angle196°$ kN, $\mathbf{M}_{12} = 674\hat{\mathbf{k}}$ N·m
15.11 $\mathbf{F}_{23} = \mathbf{F}_{12} = 20.9\angle45.4°$ kN, $\mathbf{F}_{14} = 7.57\angle282.7°$ kN, $\mathbf{F}_{34} = 14.4\angle56.7°$ kN, $\mathbf{M}_{12} = 4.400\hat{\mathbf{k}}$ N·m
15.13 $\boldsymbol{\alpha}_3 = -200\hat{\mathbf{k}}$ rad/s^2, $\mathbf{F}_{14} = 689\angle47.2°$ N, $\mathbf{M}_{12} = 11.1\hat{\mathbf{k}}$ N·m
15.15 $\mathbf{F}_{23} = 3.190\hat{\mathbf{i}} - 0.705\hat{\mathbf{j}}$ kN, $\mathbf{F}_{14} = 0.646\hat{\mathbf{j}}$ kN, $\mathbf{M}_{12} = -241\hat{\mathbf{k}}$ N·m
15.17 $\mathbf{F}_{23} = -0.372\hat{\mathbf{i}} - 0.476\hat{\mathbf{j}}$ kN, $\mathbf{F}_{14} = 0.546\hat{\mathbf{i}} + 0.397\hat{\mathbf{j}}$ kN
15.19 $\mathbf{F}_{23} = 8.42\hat{\mathbf{i}} + 47\hat{\mathbf{j}}$ kN, $\mathbf{F}_{14} = 22.1\hat{\mathbf{j}}$ kN, $\mathbf{M}_{12} = 9.750\hat{\mathbf{k}}$ kN·m
15.21 $\mathbf{F}_{23} = 6.80\angle240°$ lb, $\mathbf{F}_{43} = 45.6\angle78.2°$ lb, $\mathbf{M}_{14} = 22.4\hat{\mathbf{k}}$ in·lb
15.23 At $\theta_2 = 0°$, $\theta_3 = 120°$, $\theta_4 = 141.8°$, $\omega_3 = 6.67$ rad/s cw, $\omega_4 = 6.67$ rad/s cw, $\alpha_3 = 141$ rad/s^2 cw, $\alpha_4 = 64.1$ rad/s^2 cw, $\mathbf{F}_{21} = 6\,734\angle-56°$ lb, $\mathbf{F}_{21} = 7\,883\angle142.8°$ lb, $\mathbf{M}_{12} = 7\,468\hat{\mathbf{k}}$ ft·lb
15.25 $\mathbf{M}_{12} = 0.187\hat{\mathbf{k}}$ N·m
15.27 $\boldsymbol{\alpha}_2 = -262\hat{\mathbf{k}}$ rad/s^2, $\mathbf{F}_{12} = 132\hat{\mathbf{i}} + 40\hat{\mathbf{j}}$ lb

16.1 $w_B = w_F = 1.33$ lb
16.3 $w_B = w_E = 2.67$ lb, $\theta_B = \theta_E = 180°$

16.5 $I_O = 9.69$ in·lb·s^2, $k_G = 5.22$ in

16.9 $\omega_3 = 0$ for $r = 3R/4$.

16.11 $\alpha_{in} = 366$ rad/s^2; unbalanced moment on housing of 4 190 in·lb must be absorbed by mounting bolts during startup.

16.13 $\mathbf{F}_{12} = -87.4\hat{\mathbf{j}}$ lb, $\mathbf{M}_{12} = -1200\hat{\mathbf{j}}$ in·lb, $\mathbf{F}_{35} = -238\hat{\mathbf{i}} - 300\hat{\mathbf{k}}$ lb,
$\mathbf{M}_{35} = 258\hat{\mathbf{k}}$ in·lb, $\mathbf{F}_{45} = 238\hat{\mathbf{i}} + 300\hat{\mathbf{k}}$ lb, $\mathbf{M}_{45} = -258\hat{\mathbf{k}}$ in·lb, $\mathbf{F}_{15} = \mathbf{0}$,
$\mathbf{M}_{15} = 2\,400\hat{\mathbf{j}}$ in·lb

17.3 (a) 632 N/m; (b) 15.5 mm; (c) 0.25 s; (d) 4 Hz; (e) $\dot{x} = 0.448$ m/s;
$\ddot{x} = -5.74$ m/s^2; (f) −5.74 N

17.6 0.32 s

17.8 (a) $\omega_n = \sqrt{(k_1 + k_2)/m}$; (b) $x = [k_2 y_0/(k_1 + k_2)](1 - \cos\omega_n t)$;
(c) $z = [k_1 y_0/(k_1 + k_2)](1 - \cos\omega_n t)$

17.11 (a) 47.7 rev/min; (b) 16.9 rad/s; (c) 1.68°, 3.18°

17.15 (a) $c_c = 2\,990$ N·s/m; (b) $\omega_n = 37.4$ rad/s; (c) $\omega_d = 36.6$ rad/s; (d) $\delta = 1.29$

17.20 (a) $X = 0.035\,3$ in, $\phi = 119°$

17.23 $k = 35.4$ kN/m, $X = 0.831$ mm

18.1 At $X = 30\%$, $p_e = 251$ lb/in^2, $p_c = 50$ lb/in^2

18.3 See figure.

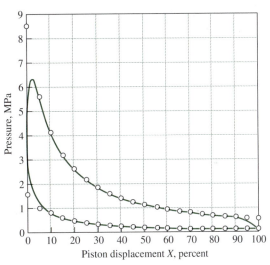

Figure AP18.3

18.5 $F_{41} = -230$ lb, $F_{34} = 935$ lb, $\mathbf{F}_{32} = 941\angle164°$ lb, $T_{21} = 800$ in · lb

18.9 $F_{41} = -0.52$ kN, $\mathbf{F}_{34} = 3.19\angle189°$ kN,
$\mathbf{F}_{32} = 10.2\angle-38.4°$ kN, $T_{21} = 191$ N·m.

18.11 At $\omega t = 60°$, $X = 28.4\%$, $P = 2\,600$ lb, $x = -33\,600$ in/s^2, $F_{41} = -392$ lb,
$\mathbf{F}_{34} = 2\,520\angle-8.9°$ lb, $\mathbf{F}_{32} = 2\,460\angle168°$ lb, $T_{21} = 3\,040$ in·lb

19.1 $\mathbf{F}_A = 64.7\angle76.1°$ N, $\mathbf{F}_B = 16.2\angle76.1°$ N, $m_c = 1.64$ kg

19.3 $\mathbf{F}_A = 8.06\angle-14.4°$ lb, $\mathbf{F}_B = 2.68\angle165.5°$ lb, $w_C = 2.63$ lb at $\theta_C = -14.4°$

19.5 $\mathbf{F}_A = 13.15 \angle 90°$ kN, $\mathbf{F}_B = 0$

19.7 $m_L \mathbf{R}_L = 5.98 \angle -16.5°$ in·oz, $m_R \mathbf{R}_R = 7.33 \angle 136.8°$ in·oz

19.9   Remove $m_L \mathbf{R}_L = 782.1 \angle 180.4°$ mg·m and $m_R \mathbf{R}_R = 236.8 \angle 301.2°$ mg·m

19.11 See Section 19.8 for answers.

20.1 $119°$

20.3   267 rev/min

20.5   Jump begins at $\theta = 75°$; $\omega = 242$ rev/min

20.7   21.8 rad/s

20.9   At $\theta = 120°$, $F_{32}^y = -572$ N, $T = 4.04$ N·m; at $\theta = 225°$, $F_{32}^y = -608$ N, $T = -3.87$ N·m

20.11 At $\theta = 60°$, $F_{32}^y = -278$ lb, $T = 332$ in·lb; at $\theta = 255°$, $F_{32}^y = -139$ lb, $T = -104$ in·lb

21.1 $T_m = 65.6$ N·m, $I = 0.336$ kg·m·s^2

21.3 $T_m = 833$ in·lb, $I = 1.013$ in·lb·s^2

21.5 $T_m = 1111$ in·lb.

23.1 $F = 42\,300$ lb; $\omega_p \geq 30.6$ rev/min

23.3 $M^z = 30.6$ in·lb

23.5 $\Sigma M = 30.8$ in·lb (negligble)

Index